Material										
Magnesium, extrusion, AZ80X	0.066	35	26	b	21	19	6.5	2.4	12	14.4
Magnesium, sand cast, AZ63-HT	0.066	14	14	b	19	14	6.5	2.4	12	14.4
Monel, wrought, hot-rolled	0.319	50	b	b		40	26	9.5	35	7.8
Red brass, cold-rolled	0.316	60	b				15	5.6	4	9.8
Red brass, annealed	0.316	15	b	b			15	5.6	50	9.8
Bronze, cold-rolled	0.320	75					15	6.5	3	9.4
Bronze, annealed	0.320	20	b	b			15	6.5	50	9.4
Titanium alloy, annealed	0.167	135	b	b			14	5.3	13	9.4
Invar, annealed	0.292	42	b	b			21	8.1	41	0.6
Nonmetallic materials										
Douglas fir, green[e]	0.022	4.8	3.4	3.9	0.9		1.6			
Douglas fir, air dry[e]	0.020	8.1	6.4	7.4	1.1		1.9			
Red oak, green[e]	0.037	4.4	2.6	3.5	1.2		1.4			1.9
Red oak, air dry[e]	0.025	8.4	4.6	6.9	1.8		1.8			
Concrete, medium strength	0.087	1.2	3.0				3.0			6.0
Concrete, fairly high strength	0.087	2.0	5.0				4.5			6.0

[a] Elastic strength may be represented by proportional limit, yield point, or yield strength at a specified offset (usually 0.2 percent for ductile metals).

[b] For ductile metals (those with an appreciable ultimate elongation), it is customary to assume the properties in compression have the same values as those in tension.

[c] Rotating beam.

[d] Elongation in 8 in.

[e] All timber properties are parallel to the grain.

Mechanics of Materials
Sixth Edition

WILLIAM F. RILEY

LEROY D. STURGES

Associate Professor
Aerospace Engineering and Engineering Mechanics
Iowa State University

DON H. MORRIS

Professor Emeritus
Engineering Science and Mechanics
Virginia Polytechnic Institute and State University

JOHN WILEY & SONS, INC.

ACQUISITIONS EDITOR	Joseph Hayton
SENIOR EDITORIAL ASSISTANT	Maureen Clendenny
SENIOR PRODUCTION EDITOR	Sandra Dumas
MARKETING MANAGER	Phyllis Cerys
DESIGNER	Hope Miller
SENIOR ILLUSTRATION EDITOR	Sigmund Malinowski
MEDIA EDITOR	Stefanie Liebman
PRODUCTION SERVICES	Suzanne Ingrao/Ingrao Associates

Cover photo by Bill Gover provided courtesy of The Angle Ring Company;
©The Angle Ring Company Limited, West Midlands, UK, http://www.anglering.com.

This book was set in Times New Roman by GTS Companies (TechBooks) and printed and bound by R. R. Donnelley/Willard. The cover was printed by Phoenix Color Corporation.

This book is printed on acid free paper.

To order books or for customer service, please call 1-800-CALL WILEY (225-5945).

ISBN 978-0-471-70511-6

Printed in the United States of America

10 9 8 7 6 5

Preface

INTRODUCTION

The primary objectives of a course in mechanics of materials are: (1) to develop a working knowledge of the relations between the loads applied to a nonrigid body made of a given material and the resulting deformations of the body; (2) to develop a thorough understanding of the relations between the loads applied to a nonrigid body and the stresses produced in the body: (3) to develop a clear insight into the relations between stress and strain for a wide variety of conditions and materials; and (4) to develop adequate procedures for finding the required dimensions of a member of a specified material to carry a given load subject to stated specifications of stress and deflection. These objectives involve the concepts and skills that form the foundation of all structural and machine design.

The principles and methods used to meet the general objectives are drawn largely from prerequisite courses in mechanics, physics, and mathematics together with the basic concepts of the theory of elasticity and the properties of engineering materials. This book is designed to emphasize the required fundamental principles, with numerous applications to demonstrate and develop logical, orderly methods of procedure. Instead of deriving numerous formulas for all types of problems, we have stressed the use of free-body diagrams and the equations of equilibrium, together with the geometry of the deformed body, and the observed relations between stress and strain, for the analysis of the force system acting on a body.

This book is designed for a first course in mechanics of deformable bodies. Because of the extensive subdivision into different topics, the book will provide flexibility in the choice of assignments to cover courses of different length and content. The developments of structural applications include the inelastic as well as the elastic range of stress; however, the material is organized so that the book will be found satisfactory for elastic coverage only.

NEW TO THIS EDITION

Content changes

- A sign convention for internal forces is established in Chapter 1 and followed consistently throughout the text.

- Discussion of the stress element in Chapter 2 is expanded.
- Section 4-6 of the Fifth Edition has been moved and combined with section 5-11 on Design
- Updated coverage of combined loadings provided in multiple chapters—Chapters 5, 6, and 7—to offer continuous reinforcement of this difficult topic
- New and revised example problems and homework problems throughout.

New visualization tool

MecMovies, by Tim Philpot of University of Missouri-Rolla is integrated at appropriate places in the text. Winner of the Premier Award for Excellence in Engineering Education Software, *MecMovies* offers a series of interactive tutorials, quizzes, problems, and games to support lectures and aid student self-study. Icons in the margin refer to appropriate sections of MecMovies, and MecMovie Problems and Activities are provided with most end of section problem sets. Available by accessing the companion site www.wiley.com/college/riley (student companion site) and using the registration code that accompanies new copies of the text.

ORGANIZATION OF THE TEXT

Since most mechanics of materials problems begin with a statics problem (finding the forces in structural members or the forces in pins connecting structural members), we have included a review of statics in Chapter 1 of this book. The coverage is perhaps more complete and comprehensive than would be necessary for review so that the book could be used for both statics and mechanics of materials if desired.

After the review of statics in Chapter 1, Chapters 2 and 3 consist of a thorough discussion of material stress and strain including principal stresses and principal strains. We choose to present principal stresses and principal strains at this early position to make it easier to talk about maximum stresses in the axial, torsional, and flexural applications that follow. It also allows us to talk about the maximum stresses in combined loading situations in Chapters 5, 6, and 7, rather than waiting until the end of the book. The ideas of principal values also allow for continuous reinforcement throughout the book of the state of stress and strain at a point.

Material properties and the relationship between stress and strain are presented in Chapter 4. In Chapters 5, 6, and 7, we consider the stresses and strains in axial, torsional, and flexural loading applications. In addition to calculating the stresses in members subjected to axial loading and the stresses in pressure vessels subjected to internal pressure, in Chapter 5 we also calculate the stresses in pressure vessels subjected to axial and pressure loading. In addition to calculating the stresses in circular shafts subjected to torsional loading, in Chapter 6, we also calculate the stresses in circular shafts subjected to axial and torsional loads and in pressure vessels subjected to torsional loads. In Chapter 7, we first calculate the normal and shear stresses in beams subjected to flexural loading. We conclude Chapter 7 with the calculation of stresses in beams and circular shafts subjected to a combination of axial, torsional, and flexural loads.

In Chapter 8, we calculate the deflection of beams due to various loading situations and also cover the calculation of support reactions for and stresses in statically indeterminate beams. In Chapter 9, we consider the tendency of columns to buckle. Finally, in Chapter 10 we discuss theories of failure and the use of energy methods.

Every chapter opens with a brief Introduction and ends with a Summary of important concepts covered in the chapter followed by a set of Review Problems. All principles are illustrated by one or more Example Problems and several Homework Problems. The Homework Problems are graded in difficulty and are separated into groups of Introductory, Intermediate, and Challenging problems. Several sections of Homework Problems also have a set of Computer Problems. While the computation could be accomplished by the student writing a FORTRAN program, the computation could just as easily be carried out using MathCAD, Mathematica, or a spreadsheet program. The important concept of the Computer problems is that they require students to analyze how the solution depends on some parameter of the problem.

Most chapters conclude with a section on Design, which includes Example Problems and a set of Homework Problems. The emphasis in these problems is that there are often more than just one criteria to be satisfied in a design specification. An acceptable design must satisfy all specified criteria. In addition, standard lumber, pipes, beams, etc. come in specific sizes. The student must choose an appropriate structural member from these standard materials. Since each different choice of a beam or a piece of lumber has a different specific weight and affects the overall problem differently, students are also introduced to the idea that design is an iterative process.

FREE-BODY DIAGRAMS

We strongly feel that a proper free-body diagram is just as important in mechanics of materials as it is in statics. It is our approach that, whenever an equation of equilibrium is written, it must be accompanied by a complete, proper free-body diagram. Furthermore, since the primary purpose of a free-body diagram is to show the forces acting on a body, the free-body diagram should not be used for any other purpose. We encourage students to draw separate diagrams to show deformation and compatibility relationships.

PROBLEMS-SOLVING PROCEDURES

Students are urged to develop the ability to reduce problems to a series of simpler component problems that can be easily analyzed and combined to give the solution of the initial problem. Along with an effective methodology for problem decomposition and solution, the ability to present results in a clear, logical, and neat manner is emphasized throughout the text.

HOMEWORK PROBLEMS

The illustrative examples and problems have been selected with special attention devoted to problems that require an understanding of the principles of mechanics of materials without demanding excessive time for computational work. A large number of homework problems are included so that problem assignments may be varied from term to term. The problems in each set represent a considerable range of difficulty and are grouped according to this range of difficulty. Mastery, in general, is not achieved by solving a large number of simple but similar problems. While the solution of simple problems is necessary to build a student's problem-solving skills and confidence, we believe that a student gains mastery of a subject through application of basic theory to the solution of problems that appear somewhat difficult.

SIGNIFICANT FIGURES

Results should always be reported as accurately as possible. However, results should not be reported to 10 significant figures merely because the calculator displays that many digits. One of the tasks in all engineering work is to determine the accuracy of the given data and the expected accuracy of the final answer. Results should always reflect the accuracy of the given data.

In a textbook, however, it is not possible for students to examine or question the accuracy of the given data. It is also impractical, in an introductory course, to give error bounds on every number. Therefore, since an accuracy greater than about 0.2% is seldom possible for practical engineering problems, all given data in Example Problems and Homework Problems, regardless of the number of figures shown, will be assumed sufficiently accurate to justify rounding off the final answer to approximately this degree of accuracy (three to four significant figures).

SI VERSUS USCS UNITS

U.S. customary units and SI units are used in approximately equal proportions in the text for both Example Problems and Homework Problems. To help the instructor who wants to assign problems of one type or the other, odd-numbered Homework Problems are in U.S. customary units and even-numbered Homework Problems are in SI units.

ANSWERS PROVIDED

Answers to about half of the Homework Problems are provided on the student companion site: www.wiley.com/college/Riley. Since the convenient designation of problems for which answers are provided is of great value to those who make up assignment sheets, the problems for which answers are provided are indicated by means of an asterisk (*) after the problem number.

INSTRUCTOR RESOURCES

In addition to a fully worked solutions manual, all figures from the text are available in electronic format for instructors who adopt this book for use in their classes. All resources will be available for download from the book's website: www.wiley.com/college/riley.

ACKNOWLEDGMENTS

We are grateful for comments and suggestions received from colleagues and from users of the earlier editions of this book. Special thanks go to the following people who provided input and comments:

Candace M. Ammerman, Colorado School of Mines, James N. Craddock, Southern Illinois University, Leonard De Rooy, Calvin College, Xin-Lin Gao, Michigan Technological University, John B. Ligon, Michigan Technological University, Charles E. Bakis, Pennsylvania State University, Shashi S. Marikunte, Southern Illinois University, Timothy A. Philpot, University of Missouri-Rolla, Ray Ruichong Zhang, Colorado School of Mines, Jiang Zhe, University of Akron.

Final judgments concerning organization of material and emphasis of topics, however, were made by the authors. We will be pleased to receive comments from readers and will attempt to acknowledge all such communications. Comments can be sent by email to sturges@ iastate.edu or to dhmorris @ vt.edu.

William F. Riley
Leroy D. Sturges
Don H. Morris

Contents

Chapter 7
Flexural Loading: Stresses in Beams 349

Chapter 8
Flexural Loading: Beam Deflections 487

Chapter 9
Columns 578

Chapter 10
Energy Methods and Theories of Failure 614

Appendices

*Answers**

Index 705

*Available online at the Wiley website www.wiley.com

Chapter 1
Introduction and Review of Statics

1-1 INTRODUCTION

The primary objective of a course in mechanics of materials is the development of relationships between the loads applied to a nonrigid body and the internal forces and deformations induced in the body. Ever since the time of Galileo Galilei (1564–1642), scientists and engineers have studied the problem of the load-carrying capacity of structural members and machine components, and have developed mathematical and experimental methods of analysis for determining the internal forces and the deformations induced by the applied loads. The experiences and observations of these scientists and engineers of the last three centuries are the heritage of the engineer of today. The fundamental knowledge gained over the last three centuries, together with the theories and analysis techniques developed, permit the modern engineer to design, with complete competence and assurance, structures and machines of unprecedented size and complexity.

The subject matter of this book forms the basis for the solution of three general types of problems:

1. Given a certain function to perform (transporting traffic over a river by means of a bridge, conveying scientific instruments to Mars in a space vehicle, converting water power into electric power), of what materials should the machine or structure be constructed, and what should be the sizes and proportions of the various elements? This is the designer's task, and obviously there is no single solution to any given problem.
2. Given the completed design, is it adequate? That is, does it perform the function economically and without excessive deformation? This is the checker's problem.
3. Given a completed structure or machine, what is its actual load-carrying capacity? The structure may have been designed for some purpose other than the one for which it is now to be used. Is it adequate for the proposed use? For example, a building may have been designed as an office building but is later found to be desirable for use as a warehouse. In such a case, what maximum loading may the floor safely support? This is the rating problem.

Because the complete scope of these problems is obviously too comprehensive for mastery in a single course, this book is restricted to a study of individual members and very simple structures or machines. The design courses that follow will consider the entire structure or machine, and will provide essential background for the complete analysis of the three problems.

1

The principles and methods used to meet the objective stated at the beginning of this chapter depend to a great extent on prerequisite courses in mathematics and mechanics, supplemented by additional concepts from the theory of elasticity and the properties of engineering materials. The equations of equilibrium from statics are used extensively, with one major change in the free-body diagrams; namely, most free bodies are isolated by cutting through a member instead of simply removing a pin or some other connection. The internal force on the cut section is related to the stresses (force per unit area) generated by the cohesive forces holding the member together. The size and shape of the member must be adjusted to keep the stress below the limiting value for the type of material from which the member is constructed.

In some instances, the specified maximum deformation, not the specified maximum stress, will govern the maximum load that a member may carry. In other instances, it may be found that the equations of equilibrium (or motion) are not sufficient to determine all of the unknown loads or reactions acting on a body. In such cases it is necessary to consider the geometry (the change in size or shape) of the body after the loads are applied. The deformation per unit length in any direction or dimension is called *strain*.

Some knowledge of the physical and mechanical properties of materials is required in order to create a design, to properly evaluate a given design, or even to write the correct relation between an applied load and the resulting deformation of a loaded member. Essential information will be introduced as required, and more complete information can be obtained from textbooks and handbooks on properties of materials.

1-2 CLASSIFICATION OF FORCES

Force is one of the most important of the basic concepts in the study of mechanics of materials (or the mechanics of deformable bodies). *Force* is the action of one body on another; forces always exist in equal magnitude, opposite direction pairs. Forces may result from direct physical contact between two bodies or from two bodies that are not in direct contact. For example, consider a person standing on a sidewalk. The person exerts a force on the sidewalk through direct physical contact between the soles of his or her shoes and the sidewalk; the sidewalk in turn exerts an equal magnitude, opposite direction force on the soles of the person's shoes. If the person were to jump, the contact force would vanish but there would still be a gravitational attraction (force between two bodies not in direct contact) between the person and the earth. The gravitational attraction force exerted on the person by the earth is called the *weight* of the person; an equal magnitude, opposite direction, attraction force is exerted on the earth by the person. Another type of force that exists without direct physical contact is an electromagnetic force.

Contact forces are called *surface* forces, since they exist at surfaces of contact between two bodies. If the area of contact is small compared to the size of the body, the force is called a *concentrated* force; this type of force is assumed to act at a point. For example, the force applied by a car wheel to the pavement on a bridge (see Fig. 1-1) is often modeled as a concentrated force. Also, a contact force may be distributed over a narrow region in a uniform or nonuniform manner. This situation would exist where floor decking contacts a floor joist, as shown in Fig. 1-2a. Here, the floor decking exerts a uniformly distributed load (force) on the joist, as shown in Fig. 1-2b. The intensity of the distributed load is *w* and has dimensions of force per unit length.

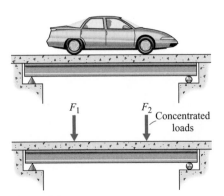

F_1 F_2 Concentrated loads

Figure 1-1

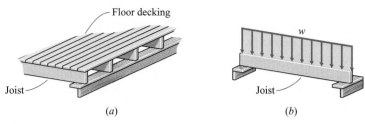

(a)   (b)

Figure 1-2

Other common types of forces are external, internal, applied, and reaction. To illustrate, consider the beam loaded and supported, as shown in Fig. 1-3*a*. A free-body diagram of the beam is shown in Fig. 1-3*b*. All forces acting on the free-body diagram are *external* forces; that is, they represent the interaction between the beam (the object shown in the free-body diagram) and the external world (everything else that has been discarded). Force F is a concentrated force, whereas w is a uniformly distributed load with dimensions of force/length. The forces F and w are called *applied* forces or loads. They are the forces that the beam is designed to carry. Forces A_x, A_y, and B are necessary to prevent movement of the beam. Such supporting forces are called *reactions*. Force distributions at supports are complicated, and reactions are usually modeled as concentrated forces.

Once again, all the forces shown in Figure 1-3 are external forces. At every section along the beam, there also exists a system of equal magnitude, opposite direction pairs of internal forces between the atoms on either side of the section. The study of mechanics of materials or mechanics of deformable bodies depends on the calculation of these internal forces at various sections of a structure or machine element and how these forces are distributed over the sections. The determination of internal forces is discussed in Section 1-5.

In our previous discussion of loads (forces), we saw that the loads might be concentrated forces or distributed forces. Furthermore, we assumed that the forces did not vary with time, that is, they were *static* loads. In many situations, loads may be a function of time. For example, a sustained load is a load that is constant over a long period of time, such as the weight of a structure (called *dead* load). This type of load is treated in the same manner as a static load; however, for some materials and conditions of temperature and stress, the resistance to failure may

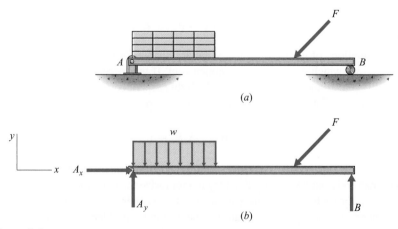

Figure 1-3

be different under short-time loading and sustained loading. An impact load is a rapidly applied load which transfers a large amount of energy in a short period of time. Vibration normally results from an impact load, and equilibrium is not established until the vibration is eliminated, usually by natural damping forces. A repeated load is a load that is applied and removed many thousands of times. The helical springs that close the valves on an automobile engine are subjected to repeated loading.

1-3 EQUILIBRIUM OF A RIGID BODY

A *rigid* body (a body that does not deform under the action of applied loads) is in equilibrium when the resultant of the system of forces acting on the body is zero. This condition is satisfied if

$$\Sigma \mathbf{F} = \mathbf{0} \tag{1-1}$$

$$\Sigma \mathbf{M}_O = \mathbf{0} \tag{1-2}$$

Equation 1-1 states that the vector sum of all external forces acting on the body is zero, whereas Eq. 1-2 states that the vector sum of the moments of the external forces about any point O (on or off the body) is zero. Equations 1-1 and 1-2 are the necessary and the sufficient conditions for equilibrium of a rigid body. The two vector equations of equilibrium may be written as six scalar equations. Selecting a right-handed, *xyz*-rectangular coordinate system, the equations of equilibrium may be written

$$\Sigma F_x = 0 \qquad \Sigma F_y = 0 \qquad \Sigma F_z = 0 \tag{1-3}$$

$$\Sigma M_x = 0 \qquad \Sigma M_y = 0 \qquad \Sigma M_z = 0 \tag{1-4}$$

Equation 1-3 states that the sum of all external forces acting on the body in the x-, y-, and z-directions is zero. Equation 1-4 states that the sum of the moments of all of the external forces acting on a body about the x-, y-, and z-axes is zero. Many problems encountered in mechanics of materials are two-dimensional in nature. Selecting the x- and y-axes in the plane of the forces and the z-axis perpendicular to the plane, the equations of equilibrium reduce to

$$\Sigma F_x = 0 \qquad \Sigma F_y = 0 \qquad \Sigma M_z = 0 \tag{1-5}$$

These equations of equilibrium would be applicable for the force system shown in Fig. 1-3b, which is coplanar and noncurrent. If the force system acting on the body is coplanar and concurrent, as in the light suspended from the ceiling by two wires shown in Fig. 1-4, Eqs. 1-3 and 1-4 reduce to

$$\Sigma F_x = 0 \qquad \Sigma F_y = 0 \tag{1-6}$$

Other types of force systems exist for problems encountered in the study of mechanics of materials. Students should not try to memorize the equations of equilibrium that apply for each of the various force systems. Rather, Eqs. 1-3 and 1-4 should be reduced to equations appropriate for the particular problem at hand. This will be illustrated in the example problems presented at the end of this section.

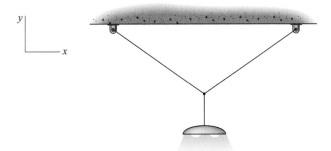

Figure 1-4

Note from previous discussions that the equations of equilibrium are applied to a system of forces. The system of forces may act on a single body or on a system of connected bodies. A free-body diagram is a carefully prepared drawing that shows a "body of interest" separated from all other interacting bodies and that shows all external forces, both known and unknown, that are applied to the body. The word "free" in the name "free-body diagram" emphasizes the idea that all bodies exerting forces on the body of interest are removed or withdrawn and are replaced by the forces that they exert. At each position on the free-body diagram where other bodies have been removed, the equal magnitude, opposite direction pairs of forces have been broken, and the forces which act on the free-body diagram must be shown. These forces may be either surface forces or body forces, or both. An important body force is the gravitational attraction of the earth, that is, the weight of the body.

The following examples illustrate the use of free-body diagrams together with the equations of equilibrium to determine unknown forces acting on rigid bodies. *The importance of drawing a correct free-body diagram cannot be overemphasized.* The free-body diagram clearly establishes which body or portion of the body is being studied. A correct free-body diagram clearly identifies all forces (both known and unknown) that must be included in the equations of equilibrium. The methods commonly used to find the unknown forces which act on rigid bodies must be thoroughly mastered, since these methods, as well as an extension of these methods to deformable bodies, are used throughout this book.

Example Problem 1-1   The rigid structure shown in Fig. 1-5a is subjected to a 5000-lb force P. The connections at joints A, B, and C are frictionless pins. Determine the forces at A and B on member AB.

SOLUTION
We first draw a free-body diagram of member AB, as shown in Fig. 1-5b. Member AB is "freed" from interacting bodies: the bracket and pin at A and member BC. At the places where AB is separated from interacting bodies, we show external forces acting on AB. Since member BC is a straight two-force member, the force T must lie along the member. The force at pin A has an unknown magnitude and direction. We show this force as two components, A_x and A_y, where the directions have been assumed. The weight of member AB is assumed to be small compared to the applied force P and is not shown on the free-body diagram. The free-body diagram contains three unknowns, A_x, A_y, and T. These unknowns, together with

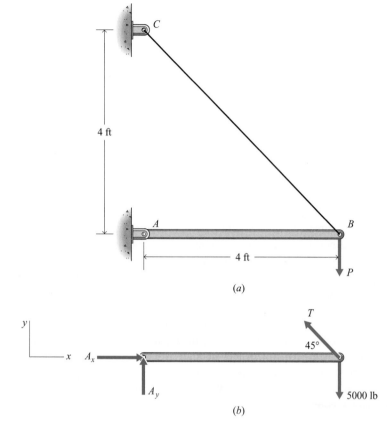

4 ft

A

4 ft

B

P

(a)

y

x

A_x

A_y

T

45°

5000 lb

(b)

Figure 1-5

the 5000-lb applied load, constitute a coplanar, nonconcurrent force system. The equations of equilibrium 1-3 and 1-4 for this system of forces reduce to

$$\Sigma F_x = 0 \qquad \Sigma F_y = 0 \qquad \Sigma M_A = 0 \qquad (a)$$

where $\Sigma M_z = 0$ has been replaced with $\Sigma M_A = 0$ (point A is the intersection of the z-axis and the plane of the two-dimensional structure). Point A was selected for the moment equation; any other point on or off body AB could have been selected. There is no particular order in which we write the equations of equilibrium; mathematical convenience usually dictates the order. In this example, we use the order given in Eq. (a).

▶ Point A was chosen for the moment equation since two of the unknown forces (A_x and A_y) intersect at point A. Therefore, the moment equation relative to point A contains only one unknown (T) and can be solved immediately for the value of T. Two of the unknown forces also intersect at points B (forces A_x and T) and C (forces A_y and T), and these would also be good points to use for the moment equation.

$$+ \rightarrow \Sigma F_x = 0: \qquad A_x - T \, \cos 45° = 0 \qquad (b)$$

$$+ \uparrow \Sigma F_y = 0: \qquad A_y + T \sin 45° - 5000 = 0 \qquad (c)$$

$$+ \downarrow \Sigma M_A = 0: \qquad (T \sin 45°)(4) - 5000(4) = 0 \qquad (d)$$

Equation (d) is solved for T, which is then substituted into Eqs. (b) and (c) to find the components of the pin forces at A. The results are

$$A_x = 5000 \, \text{lb} \qquad A_y = 0.04795 \, \text{lb} \qquad T = 7071 \, \text{lb} \qquad (e)$$

Before proceeding further, we examine the results. Why were the forces written with the number of significant figures shown in Eq. (*e*)? For example, consider the solution of Eq. (*d*):

$$T = \frac{5000}{\sin 45} = 7071.067812 \text{ lb} \qquad (f)$$

Although results should always be reported as accurately as possible, the numbers to the right of the decimal point in Eq. (*f*) have meaning only if the original data (dimensions and applied load) are known to the same relative accuracy as the solution for the force *T*. One of the tasks in all engineering work is to determine the accuracy of the given data and the expected accuracy of the final answer. Results should always reflect the accuracy of the given data.

    It is not possible, however, for students to examine or question the accuracy of the given data in a textbook. It is also impractical in an introductory course to give error bounds on every number. Therefore, since an accuracy greater than about 0.2 percent is seldom possible for practical engineering problems, all given data in Example Problems and Homework Problems, regardless of the number of figures shown, will be assumed sufficiently accurate to justify rounding off the final answer to approximately this degree of accuracy (three to four significant figures). One commonly used rounding scheme uses the leading digit to determine how many significant figures to keep in the final answer. If the first nonzero digit of the result is a 1, then the answer is reported with four significant figures; otherwise the answer is reported with three significant figures.

    Of course, all intermediate steps in the solution must maintain more significant figures than are used to represent the final results so as to reduce the effect of roundoff errors on the final results. Using the value of *T* from Eq. (*f*) in Eq. (*c*) would yield $A_y = 0$ instead of the value shown in Eq. (*e*). The point of this discussion is: Don't report final results with more accuracy than is justified by the data and don't round off numbers too much too soon. For this example problem, then, the answer is

$$A_x = 5000 \text{ lb} \qquad A_y = 0 \text{ lb} \qquad T = 7070 \text{ lb} \qquad \textbf{Ans.}$$

Example Problem 1-2 A 900-kg mass is supported by a roller that can move along a beam, as shown in Fig. 1-6a. The beam is supported by a pin at *A* and a roller at *B*.

(a) Neglect the mass of the beam and determine the reactions at *A* and *B*.

(b) If the mass of the beam is 8.5 kg/m, determine the reactions at *A* and *B*.

SOLUTION
The beam can be modeled as a rigid member with frictionless pin and roller supports at *A* and *B*. A free-body diagram for the beam is constructed by "freeing" the beam from its supports at *A* and *B* and from the roller that supports the 900-kg mass, as shown in Fig. 1-6b.

(a) As in Example Problem 1-1, the components A_x and A_y of the pin reaction at *A* are shown. Rollers exert forces on the beam that are perpendicular to the

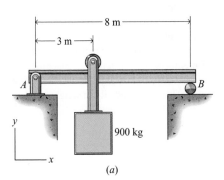

Figure 1-6(a)

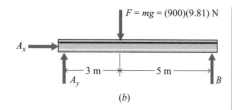

(b)

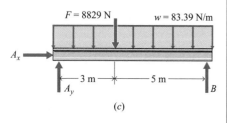

(c)

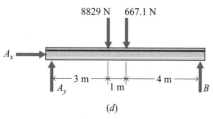

(d)

Figure 1-6(b–d)

▶ The maximum difference in the results from parts (a) and (b) is less than 10 percent. In many problems in engineering, we neglect the weight of members as being small when compared to the applied loads. As you gain experience in solving problems, you will be able to judge when you can safely neglect and when you must include the weights of members of a structure or machine. Of course, if you have any doubts, the safest approach is to include these weights.

beam. The masses of the bar and roller connecting the 900-kg mass to the beam are neglected. The force F in Fig. 1-6b is the weight of the 900-kg mass, $F = mg = 8829$ N. There are three unknown forces (A_x, A_y, and B) shown on the free-body diagram (Fig. 1-6b) for the beam. The three equations of equilibrium available to solve for the unknowns are

$$\Sigma F_x = 0 \qquad \Sigma F_y = 0 \qquad \Sigma M_A = 0$$

$$+ \rightarrow \Sigma F_x = 0: \qquad A_x = 0$$

$$+ \uparrow \Sigma F_y = 0: \qquad A_y + B - 8829 = 0$$

$$+ \downarrow \Sigma M_A = 0: \qquad B(8) - 8829(3) = 0$$

Solving for A_y and B gives

$$A_y = 5518 \text{ N} \qquad B = 3311 \text{ N}$$

Thus, the reactions at supports A and B are

$$A_x = 0 \text{ N} \qquad A_y = 5520 \text{ N} \qquad B = 3310 \text{ N} \qquad \textbf{Ans.}$$

(b) For a beam mass of 8.5 kg/m, the uniformly distributed force on the beam resulting from its weight is $w = mg = (8.5)(9.81) = 83.39$ N/m. A free-body diagram for this beam is shown in Fig. 1-6c. In the equilibrium equations, the distributed force is statically equivalent to a single force whose magnitude is equal to the area under the load diagram (the area of a rectangle, 8 m × 83.39 N/m = 667.1 N) and which acts through the centroid of the load diagram (which is 4 m to the right of A). The free-body diagram of Fig. 1-6d and the equations of equilibrium give

$$+ \rightarrow \Sigma F_x = 0: \qquad A_x = 0$$

$$+ \uparrow \Sigma F_y = 0: \qquad A_y + B - 8829 - 8(83.39) = 0$$

$$+ \downarrow \Sigma M_A = 0: \qquad B(8) - 8829(3) - 8(83.39)(4) = 0$$

Solving gives

$$A_y = 5852 \text{ N} \qquad B = 3644 \text{ N}$$

Thus, the reactions at A and B are

$$A_x = 0 \text{ N} \qquad A_y = 5850 \text{ N} \qquad B = 3640 \text{ N} \qquad \textbf{Ans.}$$

Example Problem 1-3 The truss shown in Fig. 1-7a supports one side of a bridge; an identical truss supports the other side. Floor beams carry vehicle loads to the truss joints. A 3400-lb car is stopped on the bridge. Assume that the weight of the car is evenly distributed among the wheels and that the center of gravity of the car is 16 ft from support A. Calculate the support reactions and the forces in members BD, DE, and CE of the truss.

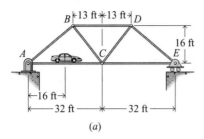

(a)

Figure 1-7(a)

SOLUTION

We model an actual truss by making four assumptions:

1. Truss members are connected only at their ends.

2. Truss members are connected by frictionless pins.

3. The truss is loaded only at the joints.

4. The weights of the members may be neglected.

These assumptions are idealizations of actual structures, but real trusses behave according to the idealizations to a high degree of approximation. As a result of the assumptions, each member of a truss is a two-force member. Since truss members are also usually straight, the force is along the member, and a member is subjected to either tension or compression.

According to assumption 3, we must proportion the weight of the car between the joints of the truss. Half of the car's weight, $3400/2 = 1700$ lb, is carried by the truss shown and the other half is carried by the truss on the other side of the bridge. Since the weight is evenly distributed to each wheel and the center of gravity of the car is midway between A and C, 850 lb will be applied to joint A and 850 lb will be applied to joint C. A free-body diagram of the entire truss is shown in Fig. 1-7b. The equations of equilibrium yield

$$+ \rightarrow \Sigma F_x = 0: \qquad A_x = 0$$

$$+ \uparrow \Sigma F_y = 0: \qquad A_y + E - 850 - 850 = 0$$

$$+ \downarrow \Sigma M_A = 0: \qquad E(64) - 850(32) = 0$$

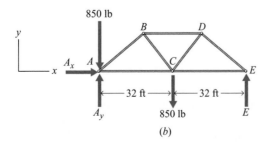

(b)

Figure 1-7(b)

Thus, the support reactions are

$$A_x = 0 \text{ lb} \qquad A_y = 1275 \text{ lb} \qquad E = 425 \text{ lb} \qquad \textbf{Ans.}$$

The forces in the various members of the truss can be found using either the method of joints or the method of sections. We choose the method of joints to calculate the forces in members CE and DE. A free-body diagram of pin E is shown in Fig. 1-7c. The force system is concurrent, and there are two equations of equilibrium

$$+\rightarrow \Sigma F_x = 0: \qquad -T_{CE} - T_{DE} \cos \theta = 0$$
$$+\uparrow \Sigma F_y = 0: \qquad E + T_{DE} \sin \theta = 425 + T_{DE} \sin \theta = 0$$

► By Newton's third law (action-reaction), a force that points away from (pulls on) a joint also points away from (pulls on) a member. That is, the corresponding member is in tension. If all member forces are shown in tension (pointing away from the joints) on free-body diagrams, then a positive sign for a force will indicate that the corresponding member is in tension (as assumed on the free-body diagram), while a negative sign for a force will indicate that the corresponding member is in compression.

(c)

Figure 1-7(c)

Since $\theta = \tan^{-1}(16/19) = 40.10°$, the forces in members CE and DE are

$$T_{DE} = -659.8 \text{ lb} \qquad T_{CE} = 504.7 \text{ lb}$$

The minus sign for T_{DE} indicates that the direction of the force on Fig. 1-7c should be reversed. Figure 1-7d shows the correct directions for T_{CE} and T_{DE}, along with the directions of the forces in members CE and DE. Clearly, member CE is in tension (T) and member DE is in compression (C). The final results are

$$T_{CE} = 505 \text{ lb } (T) \qquad T_{DE} = 660 \text{ lb } (C) \qquad \textbf{Ans.}$$

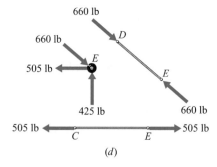

(d)

Figure 1-7(d)

The force in member BD could be found using the method of joints (joint D) since T_{DE} is now known. However, as a review of the method of sections, we select this method to calculate T_{BD}. We "section" through members BD, BC, and AC and draw the free-body diagram shown in Fig. 1-7e. The force system is

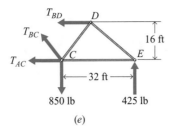

Figure 1-7(e)

coplanar and noncurrent, and there are three equations of equilibrium. Summing moments about point C eliminates all unknowns except the desired force in BD.

$$+\downarrow \Sigma M_C = 0: \qquad T_{BD}(16) + 425(32) = 0$$

$$T_{BD} = -850 \text{ lb } = 850 \text{ lb } (C) \qquad \qquad \textbf{Ans.}$$

Example Problem 1-4   A bag of potatoes is sitting on the chair of Fig. 1-8a. The force exerted by the potatoes on the frame at one side of the chair is equivalent to horizontal and vertical forces of 24 N and 84 N, respectively, at E and a force of 28 N perpendicular to member BH at G (as shown in the free-body diagram of Fig. 1-8b). Find the forces acting on member BH.

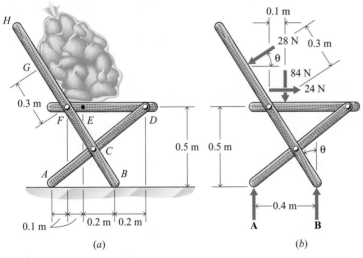

Figure 1-8

SOLUTION
The equations of equilibrium for the entire chair (Fig. 1-8b) are

$$+ \rightarrow \Sigma F_x = 0: \qquad\qquad\qquad\qquad\qquad 24 - 28 \cos\theta = 0$$

$$+ \uparrow \Sigma F_y = 0: \qquad\qquad\qquad\qquad A + B - 84 - 28 \sin\theta = 0$$

$$+\downarrow \Sigma M_B = 0: \qquad 0.2(84) - 0.5(24) - 0.4A + \left(0.3 + \frac{0.5}{\cos\theta}\right)(28) = 0$$

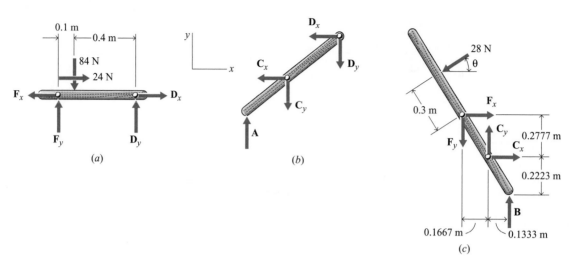

Figure 1-9

where $\theta = \tan^{-1}(\frac{3}{5}) = 30.96°$. The first equation is satisfied identically. The remaining two equations give

$$A = 73.82 \text{ N} \qquad B = 24.58 \text{ N}$$

Next the chair is disassembled and free-body diagrams are drawn for each part (Fig. 1-9). For member DF, the equilibrium equations can be written

$$+ \rightarrow \Sigma F_x = 0: \qquad D_x - F_x + 24 = 0$$

$$+ \uparrow \Sigma F_y = 0: \qquad F_y + D_y - 84 = 0$$

$$+ \downarrow \Sigma M_D = 0: \qquad 0.4(84) - 0.5F_y = 0$$

which gives

$$F_y = 67.2 \text{ N} \qquad D_y = 16.80 \text{ N} \qquad D_x = F_x - 24 \text{ N}$$

Now the equations of equilibrium for member BH are

$$+ \rightarrow \Sigma F_x = 0: \qquad F_x + C_x - 28 \cos \theta = 0$$

$$+ \uparrow \Sigma F_y = 0: \qquad 24.58 + C_y - 67.2 - 28 \sin \theta = 0$$

$$+ \downarrow \Sigma M_C = 0: \qquad \left(0.3 + \frac{0.1667}{\sin \theta}\right)(28) + 0.1333(24.58)$$

$$+ 0.1667(67.2) - 0.2777F_x = 0$$

which have only three unknowns remaining and can be solved to get

$$F_x = 115.1 \text{ N} \qquad C_x = -91.0 \text{ N} \qquad C_y = 57.0 \text{ N}$$

Then the forces acting on member *BH* are

$$\mathbf{B} = 24.6 \ \mathbf{j} \ \text{N} \qquad \qquad \textbf{Ans.}$$

$$\mathbf{C} = -91.0 \ \mathbf{i} + 57.0 \ \mathbf{j} \ \text{N} \qquad \textbf{Ans.}$$

$$\mathbf{F} = 115.1 \ \mathbf{i} - 67.2 \ \mathbf{j} \ \text{N} \qquad \textbf{Ans.}$$

plus the applied force of 28 N perpendicular to the bar at *G*. These forces are shown on the "report diagram" of Fig. 1-10.

The "report diagram," though not necessary, may be used to check the results. For example,

$$+ \rightarrow \Sigma F_x = 0: \qquad -28 \cos 30.96° + 115.1 - 91.0 = 0.08925 \cong 0$$

$$+ \uparrow \Sigma F_y = 0: \quad -28 \sin 30.96° - 67.2 + 57.0 + 24.6 = -4.307 \times 10^{-3} \cong 0$$

$$+ \downarrow \Sigma M_F = 0: \quad 28(0.3) + 57.0(0.1667) - 91.0(0.2777) + 24.6(0.3)$$

$$= 0.0112 \cong 0$$

The force and moment equations of equilibrium do not exactly equate to zero due to roundoff error.

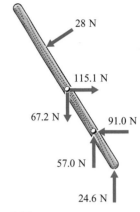

28 N

115.1 N

67.2 N

91.0 N

57.0 N

24.6 N

Figure 1-10

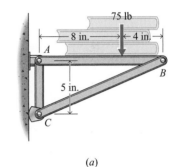

Example Problem 1-5 The weight of books on a shelf bracket is equivalent to a vertical force of 75 lb as shown on Fig. 1-11*a*. All members are made of 195-T6 cast aluminum, and all pins have ¼-in. diameters. Determine all forces acting on all three members of this frame.

SOLUTION
First draw the free-body diagram of the entire shelf bracket as in Fig. 1-11*b*. Here the "body of interest" is the frame *ABC*. The pins at *A*, *B*, and *C* remain attached to the frame, and thus the forces that would result from removal of the pins are not shown on the free-body diagram. The bracket at *A* has been removed from the frame, and the forces that the bracket exerts at *A* are shown as A_x and A_y (directions assumed). Similarly, the rocker at *C* has been removed, and the force of the rocker on the frame is shown as **C**. The equations of equilibrium are

$$+ \downarrow \Sigma M_A = 0: \qquad 5C - 8(75) = 0$$

$$+ \rightarrow \Sigma F_x = 0: \qquad A_x + C = 0$$

$$+ \uparrow \Sigma F_y = 0: \qquad A_y - 75 = 0$$

which are solved to get the support reactions

$$A_x = -120.0 \ \text{lb} \qquad A_y = 75.0 \ \text{lb} \qquad C = 120.0 \ \text{lb} \qquad \textbf{Ans.}$$

Next, dismember the bracket and draw separate free-body diagrams of each member (Fig. 1-12). Members *AC* and *BC* are straight two-force members, and thus the forces in these members must act along the members. Pin *A* connects a

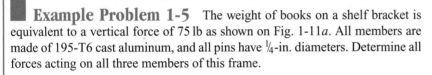

75 lb

8 in. 4 in.

A

5 in.

B

C

(a)

Figure 1-11

$\mathbf{T}_{AC}$

A

C

$\mathbf{T}_{AC}$

(a)

Figure 1-12(a)

75 lb

A_y

A_x

8 in.

B

5 in.

y

C

x

(b)

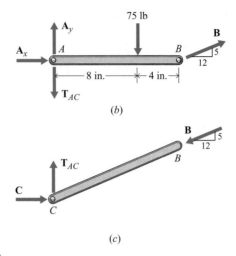

Figure 1-12(b, c)

support and two members. Since member AC is a two-force member, pin A will be left attached to member AB. The forces that act on pin A are the support reactions A_x and A_y and a vertical force due to the two-force member AC. Similarly, pin B connects two members, one of which is a two-force member. Therefore, pin B is left attached to member AB, and the only force on pin B is along the two-force member BC. Pin C connects a support and two members. Since both of the members are two-force members, pin C is arbitrarily left attached to member BC, and thus the force on pin C due to member AC is vertical. Then the equations of equilibrium can be written for member AB (Fig. 1-12b)

$$+\downarrow \Sigma M_B = 0: \qquad 4(75) + 12T_{AC} - 12(75.0) = 0$$
$$+\downarrow \Sigma M_A = 0: \qquad 12[(^5/_{13})B] - 8(75) = 0$$

from which

$$T_{AC} = 50.0 \text{ lb} \qquad B = 130.0 \text{ lb} \qquad \textbf{Ans.}$$

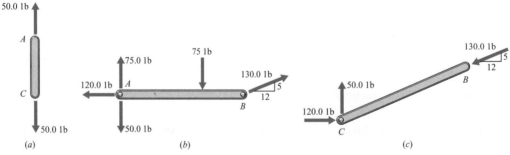

Figure 1-13

It is easily verified that these values also satisfy the equations of equilibrium for the other free-body diagrams. These forces are all shown on the "report diagrams" of Fig. 1-13.

Example Problem 1-6   Determine the reaction at support A of the pipe system shown in Fig. 1-14a. The 200-N force is parallel to the z-axis. Neglect the weights of the pipe and the wrench.

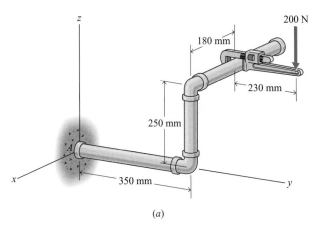

(a)

Figure 1-14(a)

SOLUTION

A free-body diagram of the pipe system is shown in Fig. 1-14b. The support at A is modeled as a rigid support that does not translate or rotate. There are three forces A_x, A_y, and A_z to prevent translation and three couples M_x, M_y, and M_z to prevent rotation. Couple M_x lies in the yz-plane, couple M_y lies in the xz-plane, and couple M_z lies in the xy-plane. Since there are six equations of equilibrium for a three-dimensional force system, all six unknowns can be found. Using Eqs. 1-3 and 1-4 yields

$\Sigma F_x = 0$: $A_x = 0$

$\Sigma F_y = 0$: $A_y = 0$

$\Sigma F_z = 0$: $A_z - 200 = 0$

$\Sigma M_x = 0$: $M_x - 200(0.350 + 0.230) = 0$

$\Sigma M_y = 0$: $M_y - 200(0.180) = 0$

$\Sigma M_z = 0$: $M_z = 0$

▶ The summation ΣM_x is the net tendency of all forces and moments to rotate the pipe about the x-axis through point A. For the 200-N force, this is just the magnitude of the force times the perpendicular distance between the line of action of the force and the x-axis through point A. For a more complicated force system, the tendency to rotate the body about the x-axis would be computed using the x-component of the vector product $\mathbf{r} \times \mathbf{F}$.

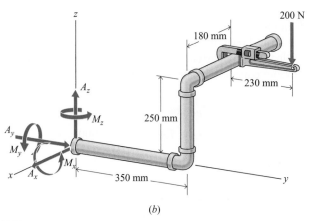

(b)

Figure 1-14(b)

Thus, the reaction at A is

$$A_x = 0 \text{ N} \qquad A_y = 0 \text{ N} \qquad A_z = 200 \text{ N} \qquad \textbf{Ans.}$$

$$M_x = 116.0 \text{ N} \cdot \text{m} \qquad M_y = 36.0 \text{ N} \cdot \text{m} \qquad M_z = 0 \text{ N} \cdot \text{m} \qquad \textbf{Ans.}$$

Since all the reactions are positive, they act in the directions shown on the free-body diagram of Fig. 1-14b.

Example Problem 1-7   A 1000-lb load is securely fastened to a hoisting cable as shown in Fig. 1-15a. The tension in the flexible cable does not change as it passes around the small frictionless pulley at the right support. The weight of the cable may be neglected. Plot the tensions in the two cables (T_{AB} and P) as a function of the sag distance d ($0 \le d \le 10$ ft). Determine the minimum sag d for which P is less than

(a) Twice the weight of the load.
(b) Four times the weight of the load.
(c) Eight times the weight of the load.

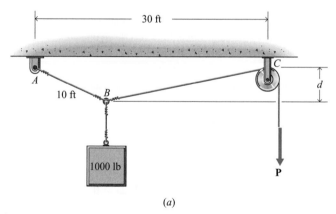

(a)

Figure 1-15(a)

SOLUTION
The ring B holds the wires together, and it will be isolated to generate the free-body diagram shown in Fig. 1-15b. The tension forces in the cables and the weight of the load are concurrent at the ring B. Writing the x- and y-components of the equilibrium equation for the free-body diagram of Fig. 1-15b results in

$$+ \rightarrow \Sigma F_x = 0: \qquad T_{BC} \cos \theta_C - T_{AB} \cos \theta_A = 0 \qquad (a)$$

$$+ \uparrow \Sigma F_y = 0: \qquad T_{AB} \sin \theta_A + T_{BC} \sin \theta_C - 1000 = 0 \qquad (b)$$

Solving Eq. (a) for T_{AB} gives

$$T_{AB} = \frac{T_{BC} \cos \theta_C}{\cos \theta_A} \qquad (c)$$

and substituting Eq. (*c*) into Eq. (*b*) gives

$$T_{BC}\frac{\sin\theta_C\cos\theta_A+\sin\theta_A\cos\theta_C}{\cos\theta_A}=1000 \qquad (d)$$

Before we can solve Eq. (*d*) for T_{BC}, we need to know how the angles θ_C and θ_A are related to the sag distance *d*. From the geometry of the triangles in Fig. 1-15*c*

$$\begin{aligned}\sin\theta_A &= d/10\\a &= 10\cos\theta_A\\b &= 30-a\\\tan\theta_C &= d/b\end{aligned} \qquad (e)$$

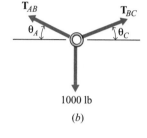

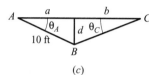

All that remains is to choose some values for *d* and to solve Eqs. (*c*)–(*e*) for the tensions. For example, when $d = 6$ ft, Eqs. (*e*) give

$$\begin{aligned}\theta_A &= \sin^{-1}\frac{6}{10}=36.8699°\\a &= 10\cos 36.8699° = 8\text{ ft}\\b &= 30-8=22\text{ ft}\\\theta_C &= \tan^{-1}\frac{6}{22}=15.2551°\end{aligned}$$

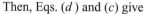

Figure 1-15(b, c)

Then, Eqs. (*d*) and (*c*) give

$$\begin{aligned}T_{BC} &= 1013.49\text{ lb}\\T_{AB} &= 1222.22\text{ lb}\end{aligned}$$

where $T_{BC} = P$ because the tension in the hoisting cable does not change as the cable goes around the small pulley. Figure 1-15*d* shows the results of repeating this process for various values of the sag distance *d* and graphing the results.

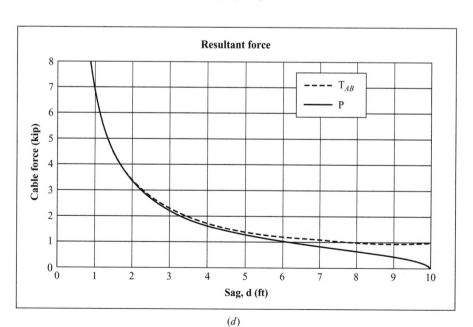

(*d*)

Figure 1-15(d)

When $d = 10$ ft, the load hangs directly below the support A, cable AB carries the entire load, and the hoisting cable is slack, $P = 0$ lb. As the load is raised (d gets smaller), the force in both cables increases. At $d = 3.28$ ft, the force in the hoisting cable is twice the load; at $d = 1.66$ ft, the force in the hoisting cable is four times the load; and at $d = 0.833$ ft, the force in the hoisting cable is eight times the weight of the load being lifted. As d goes to zero, the forces in the two cables both go to infinity.

PROBLEMS

Introductory Problems

1-1* A person is holding a 20-lb object as shown in Fig. P1-1. Determine the force **T** in the biceps muscle and the force **F** of the humerus against the ulna, in terms of the weight **W** of the forearm, which acts through G. For the position shown, both **T** and **F** act vertically.

Figure P1-1

1-2* A worker is using a hoist and cable to lift a 175-kg engine from a car as shown in Fig. P1-2. Determine the forces in the three cables attached to the ring.

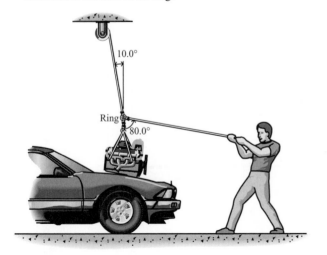

Figure P1-2

1-3 An 800-lb homogeneous cylinder is supported by two rollers as shown in Fig. P1-3. Determine the forces exerted by the rollers on the cylinder. All surfaces are smooth (frictionless).

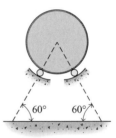

Figure P1-3

1-4* A curved slender bar is loaded and supported as shown in Fig. P1-4. Determine the reactions at supports A and B.

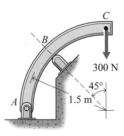

Figure P1-4

1-5 A curved slender bar is loaded and supported as shown in Fig. P1-5. Determine the reaction at support A.

Figure P1-5

1-6 Determine the forces in members BC, CD, and DE of the truss shown in Fig. P1-6.

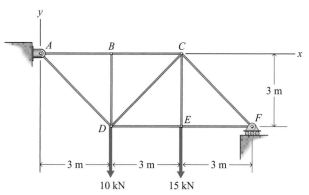

Figure P1-6

1-7* The lawn mower shown in Fig. P1-7 weighs 35 lb. Determine the force **P** required to move the mower at a constant velocity and the forces exerted on the front and rear wheels by the inclined surface.

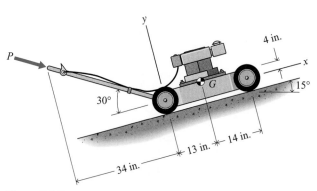

Figure P1-7

1-8 A human femur is modeled as shown in Fig. P1-8. The abductor muscle force is $M = 4060\,\text{N}$, and the femoral load is $J = 5210\,\text{N}$. Determine the force **P** and the couple **C**.

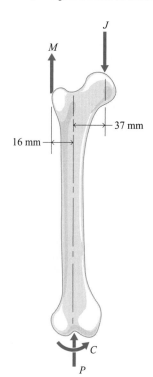

Figure P1-8

1-9* Determine the forces in members CD, CF, and FG of the bridge truss shown in Fig. P1-9.

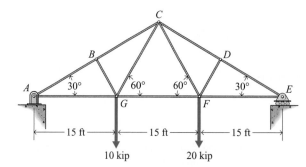

Figure P1-9

1-10* The coal wagon shown in Fig. P1-10 is used to haul coal from a mine. If the mass of the coal and wagon is 2000 kg, determine the force **P** required to move the wagon at a constant velocity and the forces exerted on the front and rear wheels by the inclined surface.

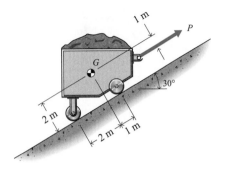

Figure P1-10

1-11 A 30-lb force **P** is applied to the brake pedal of an automobile as shown in Fig. P1-11. Determine the force **Q** applied to the brake cylinder and the reaction at support A.

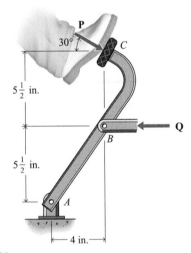

Figure P1-11

1-12* A beam is loaded and supported as shown in Fig. P1-12. Determine the reaction at support A.

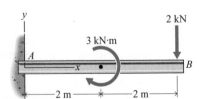

Figure P1-12

1-13 A beam is loaded and supported as shown in Fig. P1-13. Determine the reactions at supports A and B.

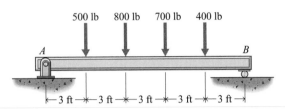

Figure P1-13

1-14 Pulleys A and B of the chain hoist shown in Fig. P1-14 are connected and rotate as a unit. The chain is continuous, and each of the pulleys contains slots that prevent the chain from slipping. Determine the force **F** required to hold a 450-kg block W in equilibrium if the radii of pulleys A and B are 90 mm and 100 mm, respectively.

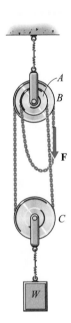

Figure P1-14

1-15* A bracket of negligible weight is used to support the distributed load shown in Fig. P1-15. Determine the reactions at the supports A and B.

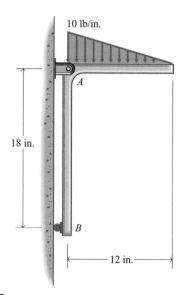

Figure P1-15

1-16 The wood plane shown in Fig. P1-16 moves with a constant velocity when subjected to the forces shown. Determine

a. The shearing force of the wood on the plane.
b. The normal force, and its location, of the wood on the plane.

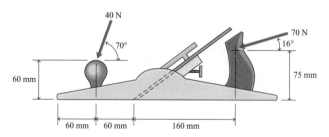

Figure P1-16

Intermediate Problems

1-17* Forces of 25 lb are applied to the handles of the pipe pliers shown in Fig. P1-17. Determine the force exerted on the pipe at D and the force exerted on handle DAB by the pin at A.

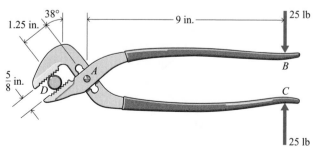

Figure P1-17

1-18* A pair of vise grip pliers is shown in Fig. P1-18. Determine the force $\mathbf{F}$ exerted on the block by the jaws of the pliers when a force $P = 100$ N is applied to the handles.

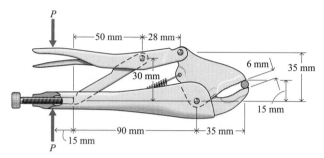

Figure P1-18

1-19 The Gambrel truss shown in Fig. P1-19 supports one side of a bridge; an identical truss supports the other side. Floor beams carry vehicle loads to the truss joints. Calculate the forces in members BC, BG, and CG when a truck weighing 7500 lb is stopped in the middle of the bridge as shown. The center of gravity of the truck is midway between the front and rear wheels.

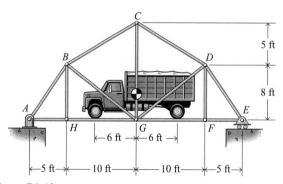

Figure P1-19

1-20* A transmission line truss supports a 5-kN load, as shown in Fig. P1-20. Determine the forces in members FG and CD.

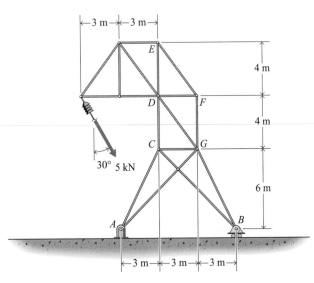

Figure P1-20

1-21 Three smooth homogeneous cylinders A, B, and C are stacked in a V-shaped trough as shown in Fig. P1-21. Cylinder A weighs 100 lb; cylinders B and C each weigh 200 lb. All cylinders have a 5-in diameter. Determine the minimum angle θ for equilibrium.

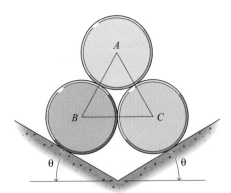

Figure P1-21

1-22 The mass of block A in Fig. P1-22 is 250 kg. Block A is supported by a small wheel that is free to roll on the continuous cable between supports B and C. The length of the cable is 42 m. Determine the distance x and the tension T in the cable when the system is in equilibrium.

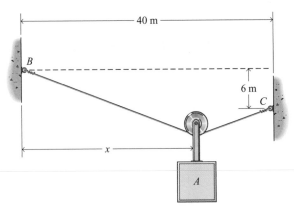

Figure P1-22

1-23* The wrecker truck of Fig. P1-23 has a weight of 15,000 lb and a center of gravity at G. The force exerted on the rear (drive) wheels by the ground consists of both a normal component B_y and a tangential component B_x while the force exerted on the front wheels consists of a normal force A_y only. Determine the maximum pull P that the wrecker can exert when $\theta = 30°$ if B_x cannot exceed $0.8B_y$ (because of friction considerations) and the wrecker does not tip over backwards (the front wheels remain in contact with the ground).

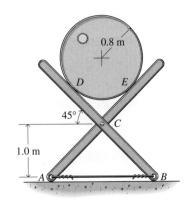

Figure P1-23

1-24 A drum of oil with a mass of 200 kg is supported by a pair of frames (the second frame is behind the one shown) as shown in Fig. P1-24. Determine all forces acting on member ACE.

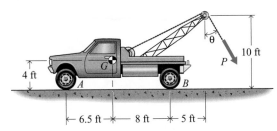

Figure P1-24

1-25* The hot-air balloon shown in Fig. P1-25 is tethered with three mooring cables. If the net lift of the balloon is 900 lb, determine the force exerted on the balloon by each of the three cables.

1-27 A force of 20 lb is required to pull the stopper *DE* in Fig. P1-27. Determine all forces acting on member *BCD*.

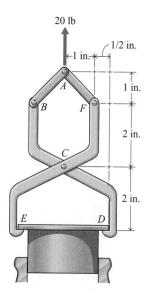

Figure P1-27

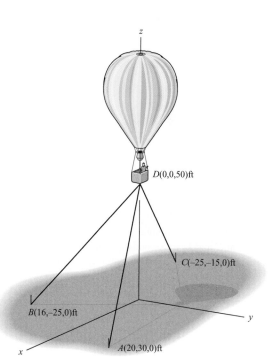

Figure P1-25

1-26* A 100-kg traffic light is supported by a system of cables as shown in Fig. P1-26. Determine the tensions in each of the three cables.

1-28 The front-wheel suspension of an automobile is shown in Fig. P1-28. The pavement exerts a vertical force of 2700 N on the tire. Determine the force in the spring and the forces at *A*, *B*, and *D*.

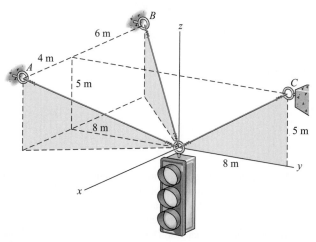

Figure P1-26

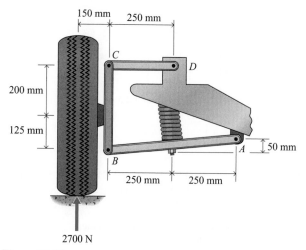

Figure P1-28

1-29* Determine all forces acting on member $ABCD$ of the frame of Fig. P1-29.

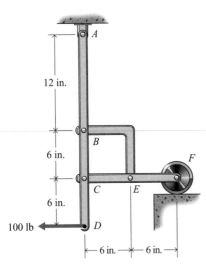

Figure P1-29

1-30 The flat roof of a building is supported by a series of parallel plane trusses spaced 2 m apart (only one such truss is shown in Fig. P1-30). Calculate the forces in all the members of a typical truss when water collects to a depth of 0.2 m as shown. The density of water is 1000 kg/m³.

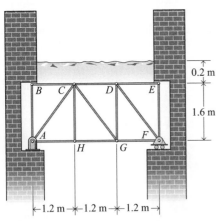

Figure P1-30

1-31 Two bodies W_1 and W_2 weighing 200 lb and 150 lb, respectively, rest on a cylinder and are connected by a rope as shown in Fig. P1-31. If all surfaces are smooth, determine

a. The reactions of the cylinder on the bodies.
b. The tension in the rope.
c. The angle θ.

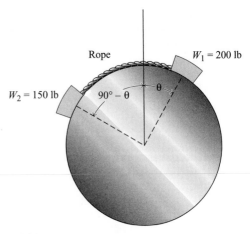

Figure P1-31

1-32* The homogeneous door shown in Fig. P1-32 has a mass of 60 kg and is held in the position shown by the rod AB. The rod is held in place by smooth horizontal pins at A and B. The hinges at C and D are smooth, and the hinge at C can support thrust along its axis. Determine all forces that act on the door.

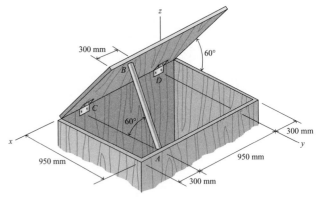

Figure P1-32

1-33 A farmer is using the hand winch shown in Fig. P1-33 to slowly raise a 40-lb bucket of water from a well. In the position shown, force **P** is vertical. The bearings at C and D exert only force reactions on the shaft. Bearing C can support thrust loading; bearing D cannot. Determine the magnitude of force **P** and the components of the bearing reactions.

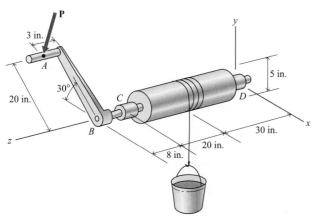

Figure P1-33

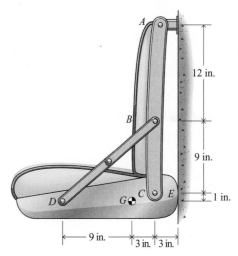

Figure P1-35

1-34*   A scissors jack for an automobile is shown in Fig. P1-34. The screw threads exert a force **F** on the blocks at joints A and B. Determine the force **P** exerted on the automobile if $F = 800$ N and $\theta = 15°$. Repeat for $\theta = 30°$ and $\theta = 45°$.

1-36   A frame is loaded and supported as shown in Fig. P1-36. Determine the reactions at supports A and C and all forces acting on member ADE.

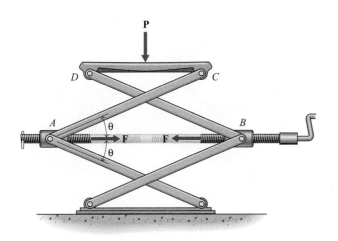

Figure P1-34

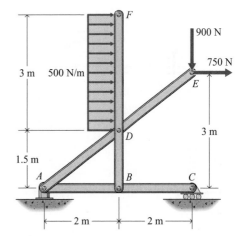

Figure P1-36

1-35*   The fold-down chair of Fig. P1-35 weighs 25 lb and has its center of gravity at G. Determine all forces acting on member ABC.

1-37   Forces of 50 lb are applied to the handles of the bolt cutter of Fig. P1-37. Determine

a.   All forces acting on the handle ABC.
b.   The force exerted on the bolt at E.

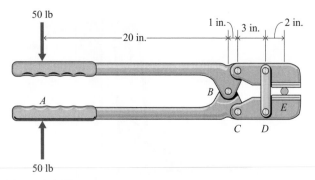

Figure P1-37

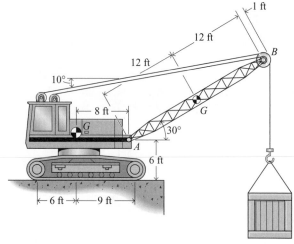

Figure P1-39

Challenging Problems

1-38* The garage door *ABCD* shown in Fig. P1-38 is being raised by a cable *DE*. The one-piece door is a homogeneous rectangular slab which has a mass of 100 kg. Frictionless rollers *B* and *C* run in tracks at each side of the door as shown. Determine the tension *T* in the cable and the forces **B** and **C** on the frictionless rollers when *d* = 1.875 m.

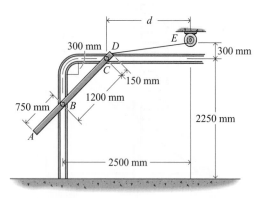

Figure P1-38

1-39* The crane and boom shown in Fig. P1-39 weigh 12,000 lb and 600 lb, respectively. When the boom is in the position shown, determine

a. The maximum load that can be lifted by the crane.

b. The tension in the cable used to raise and lower the boom when the load being lifted is 3600 lb.

c. The pin reaction at boom support *A* when the load being lifted is 3600 lb.

1-40 Figure P1-40 is a simplified sketch of the mechanism used to raise the bucket of a bulldozer. The bucket and its contents weigh 10 kN and have a center of gravity at *H*. Arm *ABCD* has a weight of 2 kN and a center of gravity at *B*; arm *DEFG* has a weight of 1 kN and a center of gravity at *E*. The weight of the hydraulic cylinders can be ignored. Calculate the force in the horizontal cylinders *CJ* and *EI* and all forces acting on arm *DEFG* for the position shown.

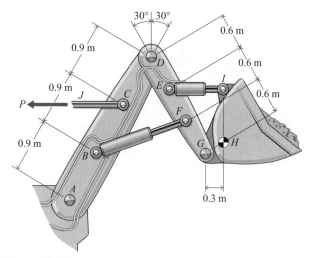

Figure P1-40

1-41 The mechanism of Fig. P1-41 is designed to keep its load level while raising it. A pin on the rim of the 4-ft-diameter pulley fits in a slot on arm *ABC*. Arms *ABC* and *DE* are each 4 ft long, and the package being lifted weighs 80 lb. The mechanism is raised by pulling on the rope that is wrapped around the pulley. Determine the force **P** applied to the rope and all forces acting on the arm *ABC* when the package has been lifted 4 ft, as shown.

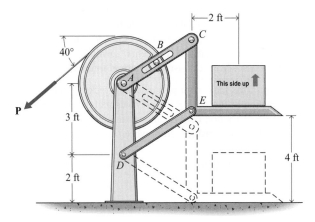

Figure P1-41

1-42* Bar AB of Fig. P1-42 has a uniform cross section, a mass of 25 kg, and a length of 1 m. Determine the angle θ for equilibrium.

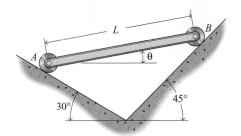

Figure P1-42

1-43 The homogeneous door shown in Fig. P1-43 has a mass of 25 kg and is supported in a horizontal position by two hinges and a bar. The hinges have been properly aligned; therefore, they exert only force reactions on the door. Assume that the hinge at B resists any force along the axis of the hinge pins. Determine the reactions at supports A, B, and D.

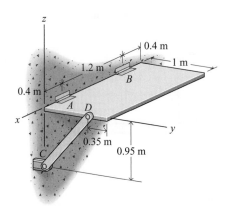

Figure P1-43

Computer Problems

1-44 A pair of steel pipes is stacked in a box as shown in Fig. P1-44. The masses and diameters of the smooth pipes are $m_A = 5$ kg, $m_B = 20$ kg, $d_A = 100$ mm, and $d_B = 200$ mm. Plot the two forces exerted on pipe A (by pipe B and by the side wall) as a function of the distance b between the walls of the box ($200\,\text{mm} \le b \le 300\,\text{mm}$). Determine the range of b for which

a. The force at the side wall is less than W_A, the weight of pipe A.
b. Neither of the two forces exceeds $2W_A$.
c. Neither of the two forces exceeds $4W_A$.

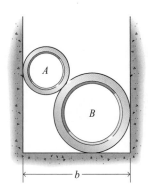

Figure P1-44

1-45 A worker positions a 250-lb crate by pulling on the rope BD as shown in Fig. P1-45. The 3-ft long rope BD is horizontal ($\theta = 0$) when the 5-ft long rope AB is vertical ($\phi = 0$).

a. What is the maximum distance $b_{\max}$ that the crate can be pulled to the side using this arrangement?
b. Calculate and plot the forces in ropes AB and BD as a function of the distance b for $0 \le b \le b_{\max}$.
c. How could the worker pull the crate to the side more than the $b_{\max}$ calculated in part a?

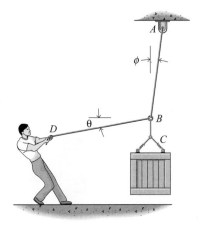

Figure P1-45

1-46 A 50-kg load is suspended from a pulley as shown in Fig. P1-46. Pulleys B and C are both frictionless and free to rotate, and the weight of the cable may be neglected. Plot the force P required for equilibrium as a function of the sag distance d ($0 \leq d \leq 1$ m). Determine the minimum sag d_{min} for which P is less than

a. Twice the weight of the load.
b. Four times the weight of the load.
c. Eight times the weight of the load.

1-48 The wrecker truck shown in Fig. P1-48 has a mass of 6800 kg and a center of gravity at G. The force exerted on the rear (drive) wheels by the ground consists of both a normal component B_y and a tangential component B_x, while the force exerted on the front wheels consists of a normal force A_y only.

a. Plot P, the maximum pull that the wrecker can exert, as a function of θ ($0° \leq \theta \leq 90°$) if B_x cannot exceed $0.8B_y$ (because of friction considerations) and the wrecker does not tip over backward (the front wheels remain in contact with the ground).
b. On the same graph, plot A_y, B_x, and B_y as functions of the angle θ.

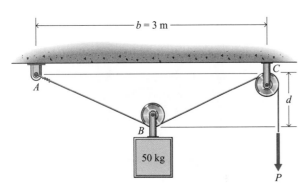

Figure P1-46

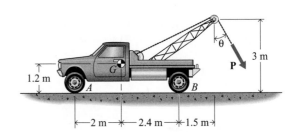

Figure P1-48

1-47 A 75-lb stop light is suspended between two poles as shown in Fig P1-47. Neglect the weight of the flexible cables and plot the tension in both cables as a function of the sag distance d ($0 \leq d \leq 8$ ft). Determine the minimum sag d_{min} for which both tensions are less than

a. 100 lb.
b. 250 lb.
c. 500 lb.

1-49 An overhead crane consists of an I-beam supported by a simple truss as shown in Fig. P1-49. If the uniform I-beam weighs 400 lb, plot the force in members BC, CF, and EF as a function of the position d ($0 \leq d \leq 8$ ft).

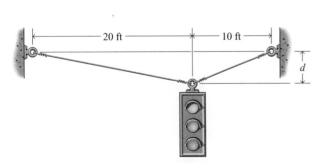

Figure P1-47

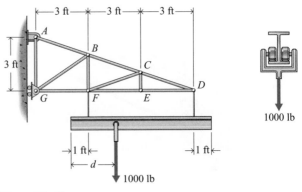

Figure P1-49

1-50 A light pole is braced using a lightweight flexible cable *DGH* as shown in Fig. P1-50. The uniform pole *ABCD* weighs 4200 N; the combined weight of the light fixture and arm *CEF* is 7500 N and acts at *E*; and the weight of the arm *BG* can be neglected. Assume that the arms *EC* and *BG* are braced to remain perpendicular to the pole *ABCD* and that the connection at *A* cannot provide any significant moment. Plot T_{DG} and T_{GH} (the tensions in the two parts of the cable) and F_{BG} (the force in the brace *BG*) as functions of *b*, the length of the brace, for $0.5\,\text{m} \le b \le 3\,\text{m}$.

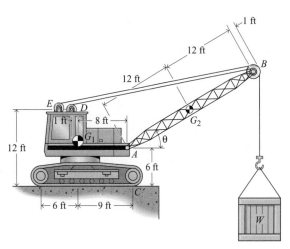

Figure P1-51

1-52 A group of workers proposes to raise a uniform 250-kg post *AB* to a vertical position using the rope and brace arrangement shown in Fig. P1-52*a*. Assume that the weight of the brace can be neglected and that end *A* acts as a frictionless pin for both the 6-m-long post *AB* and the 6-m-long brace *AC*.

a. Plot the rope force *P* and the force F_{AC} in the brace *AC* as functions of the angle θ ($0° \le \theta \le 90°$).
b. Repeat the problem if two braces are used as shown in Fig. P1-52*b*. Plot the rope force *P* and the brace forces F_{AC} and F_{AD} as functions of the angle θ ($0° \le \theta \le 90°$). Assume that when the force in brace *AD* becomes zero, the brace falls out of the way and from that point on the rope is attached to *C* instead of *D*.

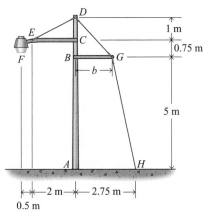

Figure P1-50

1-51 The crane and boom shown in Fig. P1-51 weigh 12,000 lb and 600 lb, respectively. The pulleys at *D* and *E* are small, and the cables attached to them are essentially parallel.

a. Plot *d*, the location of the resultant force of the ground on the crane relative to point *C*, as a function of the boom angle θ ($0° \le \theta \le 80°$) when the crane is lifting a 3600-lb load.
b. Plot $A/3600$ and $T_{BD}/3600$ as functions of the boom angle θ ($0° \le \theta \le 80°$) when the crane is lifting a 3600-lb load (in which *A* is the magnitude of the reaction force on the pin at *A*, T_{BD} is the tension in the cable raising the boom, and 3600 lb is the weight of the load being lifted).
c. It is desired that the resultant force on the tread always be at least 1 ft behind *C* to ensure that the crane is never in danger of tipping over. Plot W_{max}, the maximum load that may be lifted, as a function of the boom angle θ ($0° \le \theta \le 80°$). (Don't forget to check the tension in cable *BD*.)

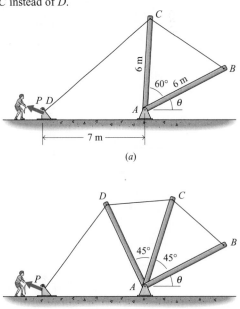

Figure P1-52

1-53 The hydraulic cylinder BC is used to tip the box of the dump truck shown in Fig. P1-53. If the combined weight of the box and the load is 22,000 lb and acts through the center of gravity G, plot

a. $C/22{,}000$, the force in the hydraulic cylinder divided by the weight of the truck box, as a function of the angle θ ($0° \leq \theta \leq 80°$).

b. $A/22{,}000$, the magnitude of the reaction force on the pin A divided by the weight of the truck box, as a function of the angle θ ($0° \leq \theta \leq 80°$).

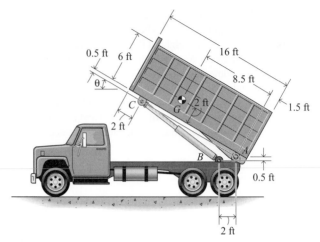

Figure P1-53

1-4 EQUILIBRIUM OF A DEFORMABLE BODY

All of the equilibrium problems considered thus far have assumed that the bodies are rigid. That is, the shape of the body and its orientation relative to its surroundings were assumed to be independent of the loads applied to the body. However, no body is perfectly rigid. Wires subjected to tension forces will stretch. Beams carrying loads will bend. Shafts subjected to torques will twist. In fact, one of the primary objectives of a course in mechanics of materials is to develop relationships between the loads applied to a nonrigid body and the deformation of the body.

If the wire or beam or shaft is very stiff, the amount of deformation will be very small and the deformation will have a negligible effect on the solution of the equilibrium equations. If the wire or beam or shaft is not very stiff, however, the deformation can affect the geometry of the problem used to write the equilibrium equations, which will in turn affect the solution of the equilibrium equations. The interaction between the loads acting on a body, the deformation of the body, and the geometry of the free-body diagram makes the solution of deformable body problems much more complex than the solution of rigid body problems. Frequently the solution of deformable body problems requires either a trial-and-error solution or a numerical solution or an iterative solution method.

Fortunately, most engineering structures and machines are designed "stiff," that is, they do not deform very much. For such problems, the solution of the equilibrium equations often ignores the deformation and treats the structure as though it were rigid. Example Problem 1-8 illustrates the difficulties encountered in the solution of deformable body problems and the errors that may result from neglecting the deformation in the solution of the equilibrium equations.

Example Problem 1-8   A 5000-lb weight W is to be supported by a very stiff (rigid) bar AB and a deformable (nonrigid) wire BC, as shown in Fig. 1-16a. The connections at A, B, and C are frictionless pins, the brackets at A and C are rigid, and the weights of the bar and the wire may be neglected. The

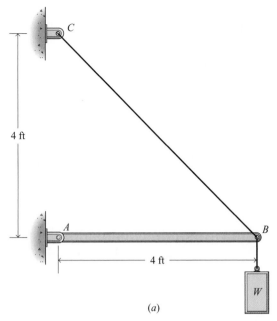

(a)

Figure 1-16(a)

relationship between the tension in the wire T_{BC} and the deformation (stretch) of the wire δ is given by $T_{BC} = k\delta$ where k depends on the cross-sectional area of the wire and the type of material from which the wire is made, and $\delta = L_f - L_i$, the difference between the final (deformed) length of the wire and the initial (undeformed) length of the wire. If Fig. 1-16a represents the unloaded configuration of the structure ($\delta = 0$), determine the tension in the wire after the load is applied for $k = 5000$ lb/in. Repeat for $k = 2500$ lb/in. and for $k = 1000$ lb/in.

SOLUTION

Since bar AB is rigid and the support bracket at A is fixed to the wall, bar AB must pivot about pin A due to the load W. Wire BC will increase in length by an amount δ since it is deformable. The loaded configuration of the structure is shown in Fig. 1-16b. The rotation of the bar can be described by the angle θ, and the deformation δ of the wire can be written as $\delta = L_f - L_i$.

A free-body diagram of the bar in the deformed configuration is shown in Fig. 1-16c. The equations of equilibrium give

$$+ \rightarrow \Sigma F_x = 0: \qquad A_x - T \cos(45 + \theta/2) = 0 \qquad (a)$$

$$+ \uparrow \Sigma F_y = 0: \qquad A_y + T \sin(45 + \theta/2) - W = 0 \qquad (b)$$

$$+ \downarrow \Sigma M_A = 0: \qquad [T \sin(45 + \theta/2)](4 \cos\theta)$$
$$-[T \cos(45 + \theta/2)](4 \sin\theta) \qquad (c')$$
$$-W(4 \cos\theta) = 0$$

Using the trigonometric relation

$$\sin(A - B) = \sin A \cos B - \cos A \sin B$$

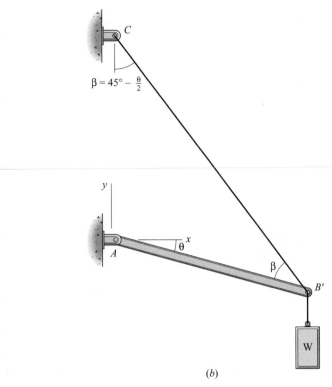

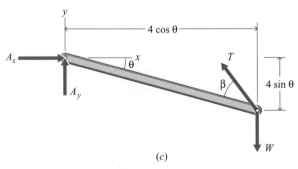

(b)

Figure 1-16(b)

Eq. (c') may be written

$$4T \sin (45 - \theta/2) - 4W \cos \theta = 0 \qquad (c)$$

Equations (a), (b), and (c) contain four unknowns A_x, A_y, T, and θ. Thus, there are four unknowns and three equations of equilibrium. We need an additional independent equation to solve the problem. The equations of equilibrium (a), (b), and (c) are necessary for equilibrium, but they are not sufficient to solve the problem. The additional equation comes from the relationship between the force in the wire and the deformation of the wire. Such a relationship will be developed in Chapter 4; we merely state the result here to complete this example.

(c)

Figure 1-16(c)

The force–deformation relation depends on the type of material from which the wire is made, and the relationship may be linear or nonlinear. Herein we limit our discussion to a linear relationship between force and deformation. Thus, we assume the behavior shown in Fig. 1-16d

$$T = k\delta \qquad\qquad (d)$$

where k is a material constant (the slope of a $T - \delta$ curve) that depends on the type of material from which the wire is made. For the wire,

$$L_i = \sqrt{(4)^2 + (4)^2} = 5.657 \text{ ft} = 67.88 \text{ in.}$$

Since $\delta = L_f - L_i = L_f - 67.88$ in., we need to determine the deformed length of the wire in order to find T. Using Fig. 1-16b and the law of cosines,

$$(B'C)^2 = L_f^2 = (48)^2 + (48)^2 - (2)(48)(48) \cos (90° + \theta)$$

from which

$$L_f = 67.88\sqrt{1 + \sin \theta} \text{ in.}$$

The deformation δ of the wire is

$$\begin{aligned} \delta = L_f - L_i &= 67.88\sqrt{1 + \sin \theta} - 67.88 \\ &= 67.88[\sqrt{1 + \sin \theta} - 1] \text{ in.} \end{aligned} \qquad (e)$$

The relationship between the applied load and the deformation of the structure is found by substituting Eq. (e) into Eq. (d). The result is then used in Eq. (c) to give

$$67.88\, k\, (\sqrt{1 + \sin \theta} - 1) \sin (45 - \theta/2) = W \cos \theta \qquad (f)$$

If $k = 5000$ lb/in., then Eq. (f) gives $\theta = 2.465°$; Eq. (c) gives

$$T = 7221 \text{ lb} \qquad\qquad \textbf{Ans.}$$

Equation (a) gives $A_x = 4995$ lb, and Eq. (b) gives $A_y = -215$ lb. This compares to the solution of Example Problem 1-1 (in which this structure was assumed to be rigid), which had $T = 7071$ lb, $A_x = 5000$ lb, and $A_y = 0$ lb. For this very stiff wire, the sag of the beam ($\theta = 2.465°$) would barely be noticeable and the error in the value of T from treating the structure as rigid would be only about 2 percent.
  If $k = 2500$ lb/in., then Eq. (f) gives $\theta = 5.097°$; Eq. (c) gives

$$T = 7379 \text{ lb} \qquad\qquad \textbf{Ans.}$$

Equation (a) gives $A_x = 4890$ lb; and Eq. (b) gives $A_y = -444$ lb.
  If $k = 1000$ lb/in., then Eq. (f) gives $\theta = 14.246°$; Eq. (c) gives

$$T = 7893 \text{ lb} \qquad\qquad \textbf{Ans.}$$

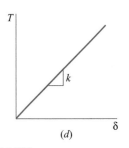

Figure 1-16(d)

▶ In Chapter 3 it will be shown that the constant of proportionality $k = \dfrac{EA}{L}$ where E is the modulus of elasticity of the material and L and A are the length and the cross-sectional area of the wire. For a steel wire $\frac{1}{8}$ in. in diameter, k = 5420 lb/in. or for a steel tie rod $\frac{1}{4}$ in. in diameter, $k = 21{,}690$ lb/in., little error is introduced by treating the structure as rigid. For an aluminum alloy wire 1/8 in. in diameter, however, $k = 1808$ lb/in., and the error introduced by treating the structure as rigid is about 6 percent. For more "stretchable" support materials, the error would be even greater.

Equation (*a*) gives $A_x = 4846$ lb, and Eq. (*b*) gives $A_y = -1230$ lb. For this less stiff wire, the sag of the beam is definitely noticeable and the error in the value of T from treating the structure as rigid would be over 10 percent.

The process illustrated in this example is typical of the solution of deformable body problems. Regardless of the type of structure or machine component or the type of loading, the solution process generally consists of these three steps

1. Equations of equilibrium, (*a*), (*b*), and (*c*)
2. Force-deformation relationship, Eq. (*d*)
3. Geometry of deformation, Eq. (*e*)

Since the equations of equilibrium must be applied to the forces acting on the deformed structure, the three sets of equations are often interdependent. It is this interdependence that makes the solution of deformable body problems more complex than the solution of rigid-body problems.

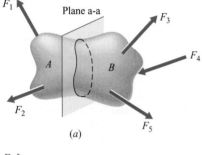

(*a*)

1-5 INTERNAL FORCES

In the study of mechanics of materials, it is necessary that we examine the internal forces that exist throughout the interior of a body. We consider an arbitrary body in equilibrium, as shown in Fig. 1-17*a*. The forces $\mathbf{F}_1$, $\mathbf{F}_2$, $\mathbf{F}_3$, $\mathbf{F}_4$, and $\mathbf{F}_5$ are applied loads and support reactions (found using the equations of equilibrium). We pass an imaginary "cutting plane" *a–a* (henceforth called a section) through the body, and separate the body into two parts *A* and *B*. Considering a free-body diagram of part *A* (Fig. 1-17*b*) we note that, in addition to the applied forces $\mathbf{F}_1$ and $\mathbf{F}_2$, the material of part *B* exerts forces on the material of part *A* over the section. These forces are internal to the body as a whole but are external for part A. The forces on the section are distributed over the surface in an unknown fashion. However, we can replace the distributed force system by a resultant force $\mathbf{R}$ and a resultant couple $\mathbf{C}$. In general, the couple $\mathbf{C}$ depends on where we place the force $\mathbf{R}$. In mechanics of materials we place $\mathbf{R}$ at the centroid C of the section, as shown in Fig. 1-17*c*. We use a double arrowhead to distinguish the couple $\mathbf{C}$ (vector) from the force $\mathbf{R}$ (vector). On the section, the distributed force system of Fig. 1-17*b* is statically equivalent to the force system $\mathbf{R}$ and $\mathbf{C}$ of Fig. 1-17*c*. The force system $\mathbf{R}$ and $\mathbf{C}$ will be referred to as an internal force system. We recognize that the internal force system depends on the orientation of the section.

Instead of part *A*, we could have considered part *B*, as shown in Fig. 1-17*d*. By applying Newton's third law to every pair of particles on the section for parts *A* and *B* of the body, we have that the distributed force systems over the section of parts *A* and *B* are equal in magnitude but opposite in sense, and thus the resultant force $\mathbf{R}$ and the resultant couple $\mathbf{C}$ on the two parts of the body are equal in magnitude but opposite in sense.

Because the body as a whole is in equilibrium, any portion of the body is also in equilibrium. Thus, using the equations of equilibrium and the force system shown in Fig. 1-17*c*,

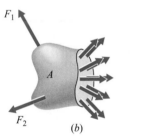

(*b*)

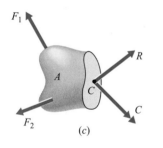

(*c*)

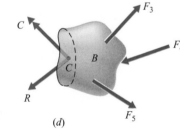

(*d*)

Figure 1-17

$$\Sigma \mathbf{F} = 0: \qquad \mathbf{F}_1 + \mathbf{F}_2 + \mathbf{R} = 0$$
$$\Sigma \mathbf{M}_C = 0: \qquad \mathbf{M}_1 + \mathbf{M}_2 + \mathbf{C} = 0$$

where $\mathbf{M}_1$ and $\mathbf{M}_2$ are the moments of forces $\mathbf{F}_1$ and $\mathbf{F}_2$, respectively, about the centroid of the section. We would find the same result using the free-body diagram of part B shown in Fig. 1-17d. Thus, we find the resultant of the internal force system using the equations of equilibrium. However, we cannot find the exact distribution of the internal forces until we learn how to determine the deformation of the body.

Experience indicates that materials behave differently to forces trying to pull atoms apart than to forces trying to slide atoms past each other. Therefore, it is standard practice to resolve the resultants $\mathbf{R}$ and $\mathbf{C}$ into components along and perpendicular to the section, as shown in Fig. 1-18. For convenience we select an xyz-coordinate system in which x is perpendicular to the section and y and z lie in the section. The component of $\mathbf{R}$ which is perpendicular to the section, R_x, is called a *normal* force; this force tends either to pull the body apart or to compress the body (Fig. 1-18a). The symbol P is often used to denote the normal force. The components of $\mathbf{R}$ that lie in the section are called *shear* forces; these forces tend to slide part A of the body relative to part B. The symbol V is often used to denote shear forces; hence, the forces V_y and V_z in Fig. 1-18a.

The component T of couple $\mathbf{C}$ shown in Fig. 1-18b tends to twist the body and is called a *twisting couple* (or twisting moment, or torque). The components M_y and M_z tend to bend the body and are called *bending couples* (or bending moments). Throughout this book we will examine the effects on a deformable body of the components of $\mathbf{R}$ and $\mathbf{C}$.

The section shown in Figs. 1-18a and b is called a positive section since the outward normal to the section points in a positive coordinate direction. The section shown in Figs. 1-18c and d is called a negative section since the outward normal to the section points in a negative coordinate direction.

A resultant force or couple component is defined as positive if the component is in a positive coordinate direction when acting on a positive section. Thus, all of the force and couple components shown in Figs. 1-18a and b are positive. If the internal forces exerted on part A of the body by part B are called positive, then the other half of the internal forces (exerted on part B of the body by part A) should also be called positive. Therefore, a force or couple component will also be defined as positive if the component is in the negative coordinate direction when acting on a negative section. Hence, all of the force and moment components shown in Figs. 1-18c and d are also positive.

The components of the internal force system can be found using the equations of equilibrium (Eqs. 1-3 and 1-4).

$$\Sigma F_x = 0 \qquad \Sigma F_y = 0 \qquad \Sigma F_z = 0 \tag{1-3}$$

$$\Sigma M_x = 0 \qquad \Sigma M_y = 0 \qquad \Sigma M_z = 0 \tag{1-4}$$

The equations of equilibrium should be applied to the body (or portion of the body) in its deformed state. However, as we saw in Example Problem 1-8, the support reactions (or, in this case internal forces) cannot be found until we know the relationship between the forces applied to the body and the deformation of the body. As we shall see in later chapters of this book, we can determine support reactions and internal forces using the undeformed configuration of a body. We accept this statement for now and illustrate the determination of internal forces with the following examples.

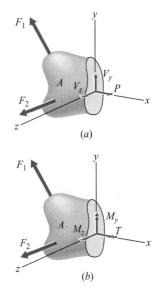
(a)

(b)

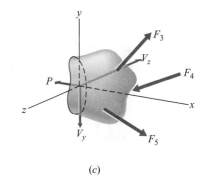
(c)

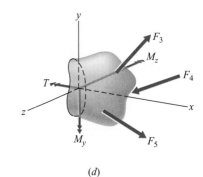
(d)

Figure 1-18

Example Problem 1-9   A post and bracket are used to support a pulley, as shown in Fig. 1-19a. A cable passing over the pulley supports a 2200-N force. Determine the internal forces on a section at the support at A.

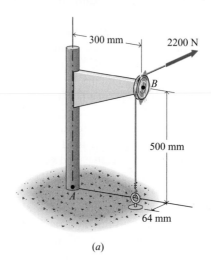

(a)

Figure 1-19(a)

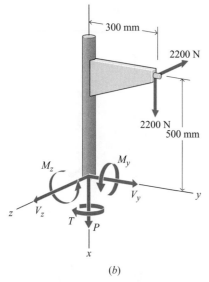

(b)

Figure 1-19(b)

SOLUTION

A section (perpendicular to the post) is passed through the post at the support at A. A free-body diagram of the part of the structure above the section is shown in Fig. 1-19b. The xyz-coordinate axes were arbitrarily selected, but the origin of the coordinate system is at the centroid of the section. Coordinate x is perpendicular to the section, whereas coordinates y and z lie in the section. Thus, the section on which the force and couple components act is a positive section, and the force and couple components shown in Figure 1-19b are positive. The couples T, M_y, and M_z are represented by curved arrows instead of vectors, as was done in Fig. 1-18b. The twisting couple T (torque) lies in the yz-plane; the bending couples M_y and M_z lie in the xz- and xy-planes, respectively. The equations of equilibrium (Eqs. 1-3 and 1-4) yield

$$\Sigma F_x = 0: \qquad P + 2200 = 0$$
$$\Sigma F_y = 0: \qquad V_y = 0$$
$$\Sigma F_z = 0: \qquad V_z - 2200 = 0$$
$$\Sigma M_x = 0: \qquad T - 2200(0.300) = 0$$
$$\Sigma M_y = 0: \qquad M_y - 2200(0.500) = 0$$
$$\Sigma M_z = 0: \qquad M_z - 2200(0.300) = 0$$

Solving for the internal force system gives

$$P = -2200\,\text{N} \qquad V_y = 0\,\text{N} \qquad V_z = 2200\,\text{N} \qquad \textbf{Ans.}$$
$$T = 660\,\text{N} \cdot \text{m} \qquad M_y = 1100\,\text{N} \cdot \text{m} \qquad M_z = 660\,\text{N} \cdot \text{m} \qquad \textbf{Ans.}$$

The negative sign for P indicates that the force is opposite to that shown in Fig. 1-19b, therefore, P is a compressive force.

Example Problem 1-10   The cantilever beam shown in Fig. 1-20a is subjected to both concentrated and distributed loads. Determine (a) the support reactions and (b) the internal forces on a section 4 m to the right of the support at A.

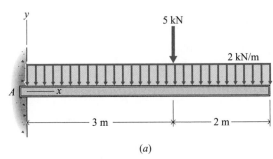

(a)

Figure 1-20(a)

SOLUTION

(a) A free-body diagram of the complete beam is shown in Fig. 1-20b. The support at A does not translate or rotate; thus, forces A_x and A_y and couple M_A may exist. The distributed load has been replaced by its resultant, $R = wl = 2\,\text{kN/m}(5\,\text{m}) = 10\,\text{kN}$, acting 2.5 m to the right of A. The equations of equilibrium give

$$+ \rightarrow \Sigma F_x = 0: \qquad A_x = 0$$

$$+ \uparrow \Sigma F_y = 0: \qquad A_y - 10 - 5 = 0$$

$$+ \downarrow \Sigma M_A = 0: \qquad M_A - 10(2.5) - 5(3) = 0$$

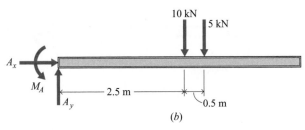

(b)

Figure 1-20(b)

Solving gives the support reactions

$$A_x = 0\,\text{kN} \qquad A_y = 15.00\,\text{kN} \qquad M_A = 40.0\,\text{kN} \cdot \text{m} \qquad \textbf{Ans.}$$

(b) A vertical section is passed through the beam 4 m to the right of A. A free-body diagram of the part of the beam to the left of the section is shown in Fig. 1-20c. The section as well as the force and couple components shown in Fig. 1-20c are positive. The distributed load is shown to illustrate that the forces acting on that part of the beam must be shown on the free-body diagram. The 10-kN resultant force on the free-body diagram of Fig. 1-20b cannot be used for the free-body diagram of the part of the beam shown in Fig. 1-20c. The free-body diagram of Fig. 1-20c is redrawn in Fig. 1-20d, where the distributed load on this part of the beam has been replaced by its resultant. The equations of equilibrium give

$$+ \rightarrow \Sigma F_x = 0: \quad P = 0$$

$$+ \uparrow \Sigma F_y = 0: \quad 15.00 - 8 - 5 + V_y = 0$$

$$+ \downarrow \Sigma M_z = 0: \quad 40.0 - 15.00(4) + 8(2) + 5(1) + M_z = 0$$

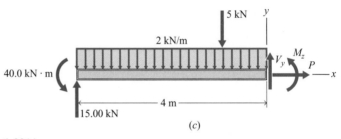

(c)

Figure 1-20(c)

The internal forces on the section are

$$P = 0\,\text{kN} \qquad V_y = -2.00\,\text{kN} \qquad M_z = -1.000\,\text{kN} \cdot \text{m} \qquad \textbf{Ans.}$$

The negative signs indicate that the directions of shear force V_y and bending couple M_z are opposite to those shown in Fig. 1-20d.

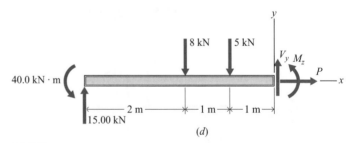

(d)

Figure 1-20(d)

▶ The internal force system acting on the portion of the beam to the right of the section is equal in magnitude and opposite in sense to the internal force system acting on the portion of the beam to the left of the section. Either portion of the beam can be used to determine the internal force system. Using the left portion of the beam, however, required that we first determine the forces and moments exerted on the beam by the wall. Generally speaking, the portion which is acted on by the fewest applied forces will be the easiest to use.

A more convenient way to solve part (b) is to use the free-body diagram of Fig. 1-20e, which shows the part of the beam to the right of the section (a negative section for the positive coordinate axes shown in Fig. 1-20d). The distributed load for this part of the beam has been replaced by its resultant; we note that the 5-kN concentrated load is not shown on the free-body diagram as it does not act on this part of the beam. Consistent with Newton's third law, the internal forces and moments in Fig. 1-20e are opposite in sense to the internal forces and moments in Fig. 1-20d. The forces P and V_y as well as the moment M_z in Fig. 1-20e are positive for this negative section. The equations of equilibrium give

$+ \rightarrow \Sigma F_x = 0: \qquad\qquad -P = 0$

$+ \uparrow \Sigma F_y = 0: \qquad\qquad -V_y - 2 = 0$

$+ \downarrow \Sigma M_z = 0: \qquad\qquad -M_z - 2(0.5) = 0$

The internal forces on the section are

$$P = 0 \, \text{kN} \qquad V_y = -2.00 \, \text{kN} \qquad M_z = -1.000 \, \text{kN} \cdot \text{m} \qquad \textbf{Ans.}$$

which are the same as the previously calculated internal forces. The internal forces are shown in Fig. 1-20f.

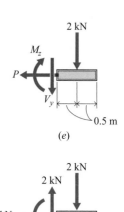

(e)

(f)

Figure 1-20(e, f)

Example Problem 1-11   A bag of potatoes is sitting on the chair in Fig. 1-21a. The force exerted by the potatoes on the frame at one side of the chair is equivalent to horizontal and vertical forces of 24 N and 84 N, respectively, at E and a force of 28 N perpendicular to member BH at G. Determine the internal forces on a section perpendicular to and midway between pins D and F.

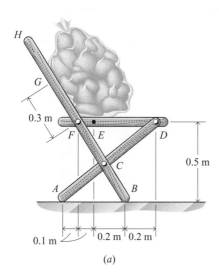

(a)

Figure 1-21(a)

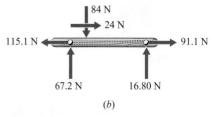

(b)

Figure 1-21(b)

SOLUTION
The forces acting on member DEF were found in Example Problem 1-4 and are shown in Fig. 1-21b. A section is passed midway between D and F. A free-body diagram of the portion of the member to the right of the section is shown in

Fig. 1-21c. Positive axes and internal forces are shown on the free-body diagram. The equations of equilibrium yield

$$+ \leftarrow \Sigma F_x = 0: \qquad P - 91.1 = 0 \qquad (a)$$

$$+ \downarrow \Sigma F_y = 0: \qquad V_y - 16.80 = 0 \qquad (b)$$

$$+ \downarrow \Sigma M_z = 0: \qquad M_z + 16.80(0.25) = 0 \qquad (c)$$

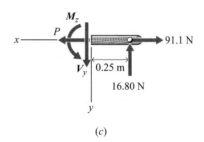

(c)

Figure 1-21(c)

The internal forces on the section are

$$P = 91.1\,\text{N} \qquad V_y = 16.80\,\text{N} \qquad M_z = -4.20\,\text{N} \cdot \text{m} \qquad \textbf{Ans.}$$

Students are encouraged to solve the problem using a free-body diagram of a portion of the member to the left of the section.

PROBLEMS

Introductory Problems

1-54* Three forces are applied along the centerline of a steel bar as shown in Fig. P1-54. Determine the internal forces on transverse cross sections in intervals AB, BC, and CD of the bar.

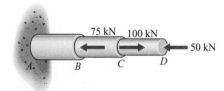

Figure P1-54

1-55* A human femur is modeled in Fig. P1-55. The abductor muscle force is $M = 90\,\text{lb}$, and the femoral load is $J = 120\,\text{lb}$. Determine the internal forces on section a–a.

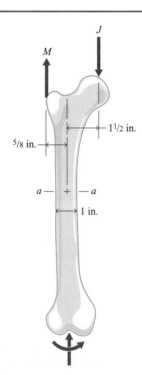

Figure P1-55

1-56 The man shown in Fig. P1-56 has a mass of 75 kg; the beam has a mass of 40 kg. The beam is in equilibrium with the man standing at the end and pulling on the cable. Determine the internal forces on a section perpendicular to and midway between

a. *A* and *B*.
b. *B* and *C*.

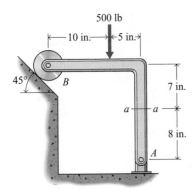

Figure P1-56

1-57* Determine the internal forces on section *a–a* in the angle bracket shown in Fig. P1-57.

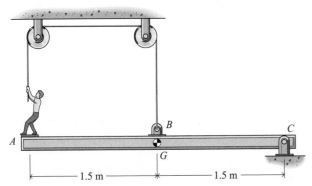

Figure P1-57

1-58 Determine the internal forces acting at the centroid of section *A–A* of the C-clamp shown in Fig. P1-58. The force **P** has a magnitude of 2000 N.

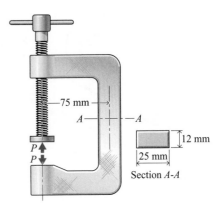

Figure P1-58

Intermediate Problems

1-59* Four sets of flexible cables, spaced at 120° intervals, are used to stabilize a 400-ft communications tower. The tower and one cable from each set are shown in Fig. P1-59. The weight of the tower is 40 lb/ft, and the communications equipment at the top weighs 2000 lb. Determine the axial forces (normal forces acting along the centerline) on transverse cross sections at points *A*, *B*, *C*, and *D* of the tower.

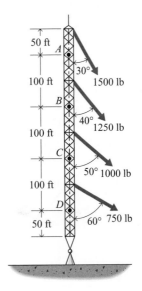

Figure P1-59

1-60* Determine the internal forces acting on section *a–a* in the bar rack shown in Fig. P1-60 if each bar has a mass of 50 kg.

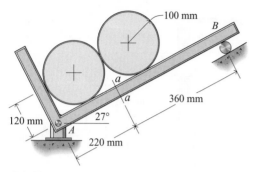

Figure P1-60

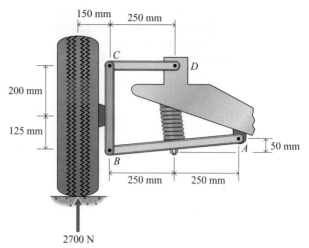

Figure P1-62

1-61 The reaction between a crutch and the ground is 35 lb, as shown in Fig. P1-61. Determine the internal forces acting on section *a–a*.

1-63 A pin-connected system of levers and bars is used as a toggle for a press as shown in Fig. P1-63. Three members are joined by pin *D*, as shown in the insert. Determine the internal forces on a section perpendicular to and midway between *D* and *E* when *P* = 1000 lb.

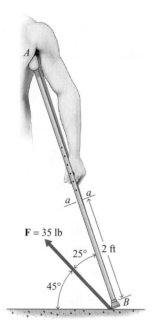

Figure P1-61

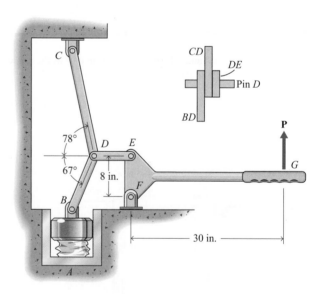

Figure P1-63

Challenging Problems

1-62 The front-wheel suspension of an automobile is shown in Fig. P1-62. The pavement exerts a vertical force of 2700 N on the tire. Determine the internal forces on a section perpendicular to and midway between *C* and *D*.

1-64* A steel shaft 120 mm in diameter is supported in flexible bearings at its ends. Two pulleys, each 500 mm in diameter, are keyed to the shaft. The pulleys carry belts that produce the forces shown in Fig. P1-64. Determine the internal forces on a vertical section through point *A*.

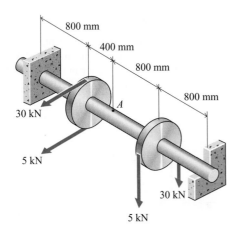

Figure P1-64

1-65* A device for lifting rectangular objects such as bricks and concrete blocks is shown in Fig. P1-65. The coefficients of friction at all vertical contact surfaces are $\mu_s = 0.4$ and $\mu_k = 0.3$. The device is to lift two blocks, each weighing 15 lb. Determine the internal forces on a vertical section 4 in. to the left of pin B.

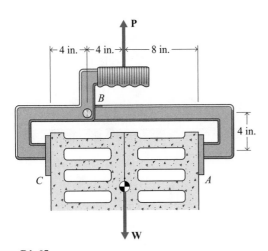

Figure P1-65

1-66 An automobile engine with a mass of 360 kg is supported by an engine hoist, as shown in Fig. P1-66. Determine the internal forces on section a–a.

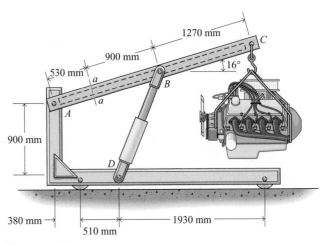

Figure P1-66

1-67 Determine the internal forces on section a–a of the pipe system shown in Fig. P1-67.

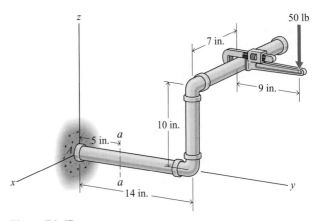

Figure P1-67

1-68* Determine the internal forces on section a–a in bar ABC of the three-bar frame shown in Fig. P1-68.

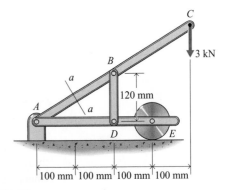

Figure P1-68

Computer Problems

1-69 The hook shown in Fig. P1-69 supports a 10-kip load. Plot P, V, and M, the internal forces and moment transmitted by a section of the hook, as a function of the angle θ ($0° \leq \theta \leq 150°$).

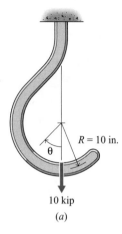

$R = 10$ in.

θ

10 kip

(a)

Figure P1-69

1-70 Forces of 100 N are being applied to the handles of the vise grip pliers shown in Fig. P1-70. Plot P, V, and M, the internal forces and moment transmitted by section a–a of the handle, as a function of the distance d (20 mm $\leq d \leq$ 30 mm).

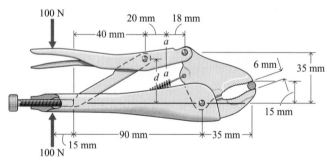

100 N

20 mm   18 mm

40 mm

a

6 mm   35 mm

d a

15 mm

90 mm   35 mm

15 mm

100 N

Figure P1-70

1-71 A 4000-lb cart rolls along a beam as shown in Fig. P1-71.

a. Show that the maximum bending moment in the beam occurs at the wheel that is closer to the middle of the beam.

b. Plot $|M|_{max}$, the maximum bending moment in the beam, as a function of the cart's position x (0 ft $\leq x \leq$ 15 ft).

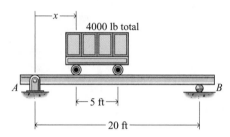

x

4000 lb total

A

B

5 ft

20 ft

Figure P1-71

1-72 A group of workers proposes to raise a uniform 250-kg post AB to a vertical position using the rope and brace arrangement shown in Fig. P1-72. Assume that the weight of the brace can be neglected and that end A acts as a frictionless pin for both the 6-m-long post AB and the 6-m-long brace AC. Plot P, V, and M, the internal forces and moment transmitted by section a–a of the post, as a function of the distance from A (0 m $\leq b \leq$ 6 m).

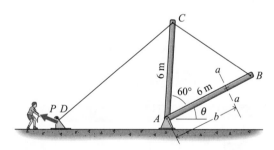

C

6 m

a

B

P D

60° 6 m

a

A

θ b

Figure P1-72

SUMMARY

Force is the action of one body on another. The primary objective of a course in mechanics of materials is the development of relationships between the forces applied to a body and the internal forces and deformations induced in the body.

Forces always exist in equal magnitude, opposite direction pairs. In order to study these forces, it is necessary to isolate the body from its surroundings or from other parts of the body at each point where forces are to be determined. The resultant of the internal force system is then found using the equations of equilibrium. The importance of drawing a correct free-body diagram cannot be overemphasized. The free-body diagram clearly establishes which body or portion of the body is being studied. A correct free-body diagram clearly identifies all forces (both known and unknown) that must be included in the equations of equilibrium.

A rigid body (a body that does not deform under the action of applied loads) is in equilibrium when the resultant of the system of forces acting on the body is zero. This condition is satisfied if the vector sum of all external forces acting on the body is zero and the vector sum of the moments of the external forces about any point O (on or off the body) is zero. Although no body is perfectly rigid, most engineering structures and machines are designed "stiff," that is, they do not deform very much. For such problems, the solution of the equilibrium equations often ignores the deformation and treats the structure as though it were rigid. However, we cannot find the exact distribution of the internal forces until we learn how to determine the deformation of the body.

The subject matter of this book forms the basis for the solution of three general types of problems encountered by engineers in the design and analysis of structures and machines:

- **Design**: Given a certain function to perform, of what materials should the machine or structure be constructed, and what should be the sizes and proportions of the various elements?
- **Checking**: Given the completed design, is it adequate—does it perform the function economically and without excessive deformation?
- **Rating**: Given a completed structure or machine, what is its actual load-carrying capacity?

Some knowledge of the physical and mechanical properties of materials is required in order to create a design, to properly evaluate a given design, or even to write the correct relation between an applied load and the resulting deformation of a loaded member. Essential information will be introduced as required.

Don't report final results with more accuracy than is justified by the data and don't round off numbers too much too soon. Although results should always be reported as accurately as possible, the final answer should not be more precise than the data from which it was derived. One of the tasks in all engineering work is to determine the accuracy of the given data and the expected accuracy of the final answer. Results should always reflect the accuracy of the given data. All data in Example Problems and Homework Problems will be assumed sufficiently accurate to justify rounding off the final answer to three or four significant figures.

REVIEW PROBLEMS

1-73* The collar A shown in Fig. P1-73 is free to slide on the smooth rod BC. Determine the forces exerted on the collar by the cable and by the rod when the force $F = 900$ lb is applied to the collar.

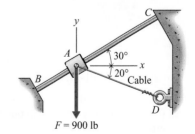

Figure P1-73

1-74*   A 500-kg mass is supported by a four-bar truss as shown in Fig. P1-74. Determine the force in each member of the truss.

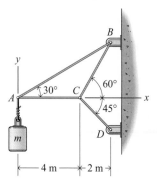

Figure P1-74

1-75   The electric motor shown in Fig. P1-75 weighs 25 lb. Due to friction between the belt and pulley, the belt forces have magnitudes of $T_1 = 21$ lb and $T_2 = 1$ lb. Determine the support reactions at A and B.

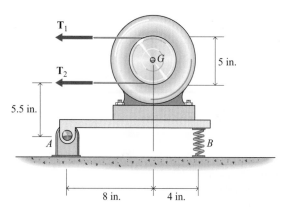

Figure P1-75

1-76*   Determine the force **P** required to push the 135-kg cylinder over the small block shown in Fig. P1-76.

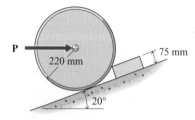

Figure P1-76

1-77   For the beam shown in Fig. P1-77, determine

a. The reactions at supports A and B.
b. The internal forces on a transverse cross section 10 ft to the right of support A.

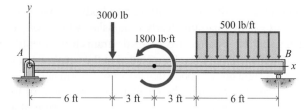

Figure P1-77

1-78   The jaws and bolts of the wood clamp in Fig. P1-78 are parallel. The bolts pass through swivel mounts so that no moments act on them. The clamp exerts forces of 300 N on each side of the board. Treat the forces on the boards as uniformly distributed over the contact areas and determine the forces in each of the bolts. Show on a sketch all forces acting on the upper jaw of the clamp.

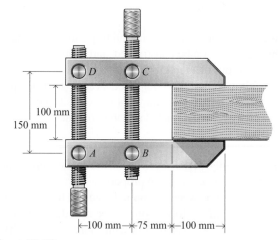

Figure P1-78

1-79*   A three-bar frame is loaded and supported as shown in Fig. P1-79. Determine the internal forces transmitted by

a. Section a–a in bar BEF.
b. Section b–b in bar $ABCD$.

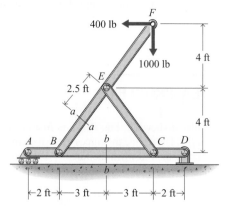

Figure P1-79

1-80 Forces of 5 N are applied to the handles of the paper punch of Fig. P1-80. Determine the force exerted on the paper at D and the force exerted on the pin at B by handle ABC.

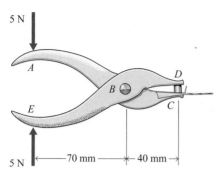

Figure P1-80

1-81* A shaft is loaded through a pulley and a lever (Fig. P1-81) that are fixed to the shaft. Friction between the belt and pulley prevents slipping of the belt. Determine the force **P** required for equilibrium and the reactions at supports A and B. The support at A is a ball bearing, and the support at B is a thrust bearing. The bearings exert only force reactions on the shaft.

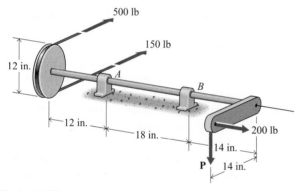

Figure P1-81

1-82* The masses of cartons 1, 2, and 3, which rest on the platform shown in Fig. P1-82, are 300 kg, 100 kg, and 200 kg, respectively. The mass of the platform is 500 kg. Determine the tensions in the three cables A, B, and C that support the platform.

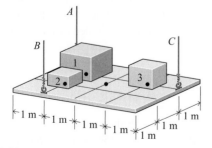

Figure P1-82

1-83 The clamp of Fig. P1-83 is used to hold two boards. If the clamping force is 300 lb, determine the internal forces on section a–a.

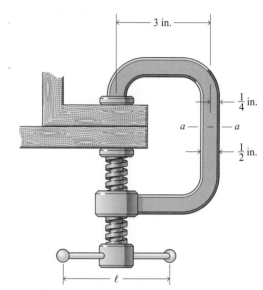

Figure P1-83

1-84 Two bars, a pulley, and a cable are used to support a block as shown in Fig. P1-84. The two bars have negligible weight. The mass of the pulley is 50 kg, and the mass of the block is 100 kg. Determine the internal forces on cross section a–a in bar AB.

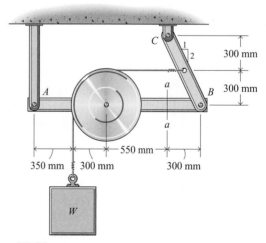

Figure P1-84

Chapter 2
Analysis of Stress: Concepts and Definitions

2-1 INTRODUCTION

Application of the equations of equilibrium is usually just the first step in the solution of engineering problems. Using these equations, an engineer can determine the forces exerted on a structure by its supports, the forces on bolts and rivets that connect parts of a machine, or the internal forces in cables or rods that either support the structure or are a part of the structure. A second and equally important step is determining the internal effect of the forces on the structure or machine. It is important, therefore, that all engineers understand the behavior of materials under the action of forces.

Safety and economy in a design are two considerations for which an engineer must accept responsibility. He or she must be able to calculate the intensity of the internal forces to which each part of a machine or structure is subjected and the deformation that each part experiences during the performance of its intended function. Then, by knowing the properties of the material from which the parts will be made, the engineer establishes the most effective size and shape of the individual parts and the appropriate means of connecting them.

In every subject area there are certain fundamental concepts of paramount importance to a satisfactory comprehension of the subject matter. For the subject mechanics of materials, a thorough mastery of the physical significance of stress and strain is paramount. The discussion of stress will be undertaken first; the study of strain will be taken up in Chapter 3.

2-2 NORMAL STRESS UNDER AXIAL LOADING

In the simplest qualitative terms, stress is the intensity of force. A body must be able to withstand the intensity of an internal force; if not, the body may rupture or deform excessively. Force intensity (stress) is force divided by the area over which the force is distributed. Thus,

$$\text{Stress} = \frac{\text{Force}}{\text{Area}} \qquad (2\text{-}1)$$

The forces shown in Fig. 2-1 are collinear with the centroidal axis of the eyebar and produce a tensile loading of the bar. These forces are called *axial* forces. When the eyebar is cut by a transverse plane, such as plane *a–a* of Fig. 2-1, a

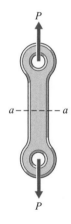

Figure 2-1

48

free-body diagram of the bottom half of the bar can be drawn as shown in Fig. 2-2. Equilibrium of this portion of the bar is obtained with a distribution of internal force that develops on the exposed cross section. This distribution of internal force has a resultant **F** that is normal to the exposed surface, is equal in magnitude to **P**, and has a line of action that is collinear with the line of action of **P**. An average intensity of internal force, which is also known as the average normal stress σ_{avg} on the cross section, can be computed as

$$\sigma_{\text{avg}} = \frac{F}{A} \tag{2-2}$$

where F is the magnitude of the internal force **F** and A is the cross-sectional area of the eyebar.

The Greek letter sigma (σ) is used to denote a normal stress in this book. A positive sign is used to indicate a tensile normal stress (member in tension), and a negative sign is used to indicate a compressive normal stress (member in compression). This sign convention is independent of the selection of a coordinate system.

Consider now a small area ΔA on the exposed cross section of the bar and let ΔF represent the magnitude of the resultant of the internal forces transmitted by this small area, as shown in Fig. 2-3. The average intensity of internal force being transmitted by area ΔA is obtained by dividing ΔF by ΔA. If the internal forces transmitted across the section are assumed to be continuously distributed, the area ΔA can be made smaller and smaller and will approach a point on the exposed surface in the limit. The corresponding force ΔF also becomes smaller and smaller. The stress at the point on the cross section to which ΔA converges is defined as

$$\sigma = \lim_{\Delta A \to 0} \frac{\Delta F}{\Delta A} \tag{2-3}$$

In general, the stress σ at a given point on a transverse cross section of an axially loaded bar will not be the same as the average stress computed by dividing the force F by the cross-sectional area A. For long, slender, axially loaded members such as those found in trusses and similar structures, however, it is generally assumed that the normal stresses are uniformly distributed except in the vicinity of the points of application of the loads. The subject of nonuniform stress distributions under axial loading will be discussed in a later chapter of this book.

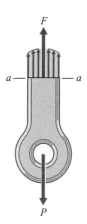

Figure 2-2

Figure 2-3

2-3 SHEARING STRESS IN CONNECTIONS

Loads applied to a structure or machine are generally transmitted to the individual members through connections which use rivets, bolts, pins, nails, or welds. In all of these connections, one of the most significant stresses induced is a shearing stress. The bolted and pinned connection shown in Fig. 2-4 will be used to introduce the concept of a shearing stress.

The method by which loads are transferred from one member of the connection to another is by means of a distribution of (internal) shearing force on a transverse cross section of the bolt or pin used to effect the connection. A free-body diagram of the left member of the connection of Fig. 2-4 is shown in

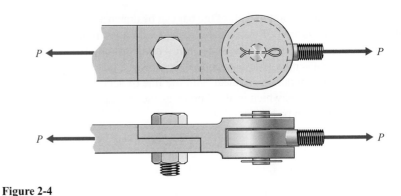

Figure 2-4

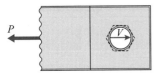

Figure 2-5

Figure 2-6

2.1

2.2

2.3

Fig. 2-5. In this diagram, a transverse cut has been made through the bolt, and the lower portion of the bolt remains in contact with the left member. The distribution of shearing force on the transverse cross section of the bolt has been replaced by a resultant shear force V. Since only one cross section of the bolt is used to effect load transfer between the members, the bolt is said to be in *single shear*; therefore, equilibrium requires that the resultant shear force V equal the applied load P. A free-body diagram for the threaded eyebar at the right end of the connection of Fig. 2-4 is shown in Fig. 2-6. In this diagram, two transverse cuts have been made through the bolt, and the middle portion of the bolt remains in contact with the eyebar. In this case, two transverse cross sections of the pin are used to effect load transfer between members of the connection and the pin is said to be in *double shear*. As a result, equilibrium requires that the resultant shear force V on each cross section of the pin equals one-half of the applied load P.

From the definition of stress given by Eq. 2-1, an average shearing stress on the transverse cross section of the bolt or pin can be computed as

$$\tau_{\text{avg}} = \frac{V}{A} \tag{2-4}$$

where V is the magnitude of the shear force $\mathbf{V}$ and A is the cross-sectional area of the bolt or pin.

The Greek letter tau (τ) is used to denote shearing stress in this book. A sign convention for shearing stress is presented in a later section of the book.

The stress at a point on the transverse cross section of the bolt or pin can be obtained by using the same type of limit process that was used to obtain Eq. 2-3 for the normal stress at a point. Thus,

$$\tau = \lim_{\Delta A \to 0} \frac{\Delta V}{\Delta A} \tag{2-5}$$

Unlike the normal stress in long, slender members, it can be shown that the shear stress τ cannot be uniformly distributed over the area. Therefore, the actual shear stress at any particular point and the maximum shear stress on a cross section will generally be different from the average shear stress calculated using Eq. 2-4. However, the design of simple connections is usually based on average stress considerations and this procedure is followed in this book.

Another type of shear loading is termed *punching shear*. Examples of this type of loading include the action of a punch in forming rivet holes in a metal plate, the tendency of building columns to punch through footings, and the tendency of a

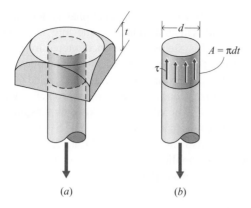

(a)   (b)

Figure 2-7

tensile axial load on a bolt to pull the shank of the bolt through the head (Fig. 2-7a). Under a punching shear load, the significant stress is the average shear stress on the surface described by the periphery of the punching member and the thickness of the punched member, for example, the shaded cylindrical area $A = \pi dt$ shown extending through the head of the bolt in Fig. 2-7b.

2-4 BEARING STRESS

Bearing stresses (compressive normal stresses) occur on the surface of contact between two interacting members. Actually, there are only two types of stress, normal and shear. Bearing stress is a just a name given to normal stress resulting from contact between two different bodies.

For the case of the connection shown in Fig. 2-4, bearing stresses occur on the surfaces of contact between the head of the bolt and the top plate and between the nut and the bottom plate. The force producing the stress is the axial tensile internal force **F** developed in the shank of the bolt as the nut is tightened. The area of interest for bearing stress calculations is the annular area $A = \frac{\pi}{4}(d_o^2 - d_i^2)$ of the bolt head or nut (see Fig. 2-8a) that is in contact with the plate. Thus, the average bearing stress σ_b is expressed as

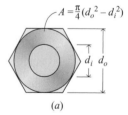

(a)

$$\sigma_b = \frac{F}{A} \tag{2-6}$$

Bearing stresses also develop on surfaces of contact where the shanks of bolts and pins are pressed against the sides of the hole through which they pass. Since the distribution of these forces is quite complicated, an average bearing stress σ_b is often used for design purposes. This stress is computed by dividing the force F transmitted across the surface of contact by the projected area $A = dt$ shown in Fig. 2-8b, instead of the actual contact area.

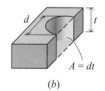

(b)

Figure 2-8

2-5 UNITS OF STRESS

Stress, being the intensity of internal force, has the dimensions of force per unit area (FL^{-2}). Until recently, the commonly used unit for stress in the United States was the pound per square inch (psi). Since metals can sustain stresses of several thousand pounds per square inch, the unit ksi (kip per square inch) is also frequently used (1 ksi = 1000 psi). With the advent of the International System of Units

(SI units), units of stress based on the international system are sometimes used in the United States and will undoubtedly come into wider use in the future. During the transition, both systems will be encountered by engineers; therefore, approximately one-half of the example problems and homework problems in this book are given using the U.S. customary system (pounds and inches) and the other half are given in SI units (newtons and meters). For problems with SI units, forces will be given in newtons (N) or kilonewtons (kN), dimensions in meters (m) or millimeters (mm), and masses in kilograms (kg). The SI unit for stress is a newton per square meter (N/m^2), also known as a pascal (Pa). Stress magnitudes normally encountered in engineering applications are expressed in meganewtons per square meter (MN/m^2) or megapascals (MPa).

Example Problem 2-1   A flat steel bar has axial loads (forces along the centroidal axis of the bar) applied at points A, B, C, and D, as shown in Fig. 2-9a. If the bar has a cross-sectional area of 3.00 in.^{2}, determine the normal stress in the bar

(a) On a cross section 20 in. to the right of point A.

(b) On a cross section 20 in. to the right of point B.

(c) On a cross section 20 in. to the right of point C.

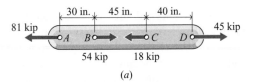

(a)

Figure 2-9(a)

SOLUTION
The sections on which the axial forces act are perpendicular to the axis of the bar and are located by the outward normal to the section. Since the applied forces are axial, the only internal forces acting on the sections are also axial; there are no shear forces or couples acting on the sections. The internal axial forces are assumed in the positive sense, as shown in the free-body diagrams of Fig. 2-9b.

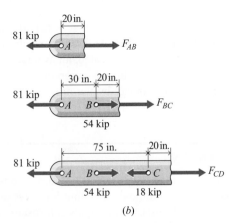

(b)

Figure 2-9(b)

The internal forces transmitted by cross sections in intervals AB, BC, and CD of the bar shown in Fig. 2-9a are obtained by using the free-body diagrams shown in Fig. 2-9b. Summing forces along the axis of the bar yields

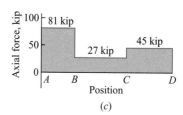

$$+ \rightarrow \Sigma F = F_{AB} - 81 = 0 \qquad F_{AB} = +81 \text{ kip} = 81 \text{ kip (T)}$$
$$+ \rightarrow \Sigma F = F_{BC} + 54 - 81 = 0 \qquad F_{BC} = +27 \text{ kip} = 27 \text{ kip (T)}$$
$$+ \rightarrow \Sigma F = F_{CD} + 54 - 18 - 81 = 0 \qquad F_{CD} = +45 \text{ kip} = 45 \text{ kip (T)}$$

(c)

Figure 2-9(c)

where (T) denotes a tensile force.

A pictorial representation of the distribution of internal force in bar $ABCD$ is shown in Fig. 2-9c. This type of representation is known as an *axial-force diagram* and has been shown to be useful in solving problems involving axial-force distributions. At the points where concentrated forces act, there are abrupt changes in the axial-force diagram. The forces in an axial-force diagram are internal forces. The required stresses (from Eq. 2-2) are

(a) $\qquad \sigma_{AB} = \dfrac{F_{AB}}{A} = \dfrac{+81}{3.00} = +27.0 \text{ ksi} = 27.0 \text{ ksi (T)}$ **Ans.**

(b) $\qquad \sigma_{BC} = \dfrac{F_{BC}}{A} = \dfrac{+27}{3.00} = +9.00 \text{ ksi} = 9.00 \text{ ksi (T)}$ **Ans.**

(c) $\qquad \sigma_{CD} = \dfrac{F_{CD}}{A} = \dfrac{+45}{3.00} = +15.00 \text{ ksi} = 15.00 \text{ ksi (T)}$ **Ans.**

▶ Since the force F_{AB} is the same everywhere in the 30-in. segment AB, the stress σ_{AB} will be the same everywhere in the segment AB. The stress is also assumed to be uniform across the member except near the points where the loads are applied. Determining how the stress changes from $\sigma_{AB} = 27.0$ ksi just to the left of B to $\sigma_{BC} = 9.00$ ksi just to the right of B is a very complicated problem which depends on many things, including the cross-sectional shape and the manner in which the 54-kip load is applied.

Example Problem 2-2 The round bar shown in Fig. 2-10a has steel, brass, and aluminum sections. Axial loads are applied at cross sections A, B, C, and D. If the allowable axial stresses are 125 MPa in the steel, 70 MPa in the brass, and 85 MPa in the aluminum, determine the diameters required for each of the sections. Assume that the allowable stresses are the same for tension (T) and compression (C).

▶ An allowable stress is the maximum permissible stress allowed in the design of a member. This will be discussed in more detail in Chapter 4.

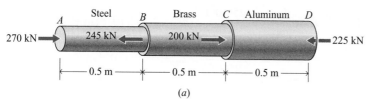

(a)

Figure 2-10(a)

SOLUTION

The internal forces transmitted by cross sections in intervals AB, BC, and CD of the bar shown in Fig. 2-10a are obtained by using the free-body diagrams shown in Fig. 2-10b. Summing forces along the axis of the bar yields

$$+ \rightarrow \Sigma F = F_s + 270 = 0 \qquad F_s = -270 \text{ kN} = 270 \text{ kN (C)}$$
$$+ \rightarrow \Sigma F = F_b + 270 - 245 = 0 \qquad F_b = -25 \text{ kN} = 25 \text{ kN (C)}$$
$$+ \rightarrow \Sigma F = F_a + 270 - 245 + 200 = 0 \qquad F_a = -225 \text{ kN} = 225 \text{ kN (C)}$$

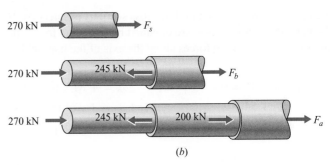

Figure 2-10(b)

An axial-force diagram for bar *ABCD* is shown in Fig. 2-10*c*. The cross-sectional areas of the bar required to limit the stresses to the specified values are obtained from Eq. 2-2. Thus,

▶ The analysis of this example ignores the problem of how the three different materials are to be fastened together. The actual stress in the vicinity of the concentrated loads may be considerably different than the average stress computed here. In addition, it can be shown that near abrupt changes in diameter or near holes through the section, the actual stress can be two to three times greater than the average stress at that section. This effect is called *stress concentration* and will be covered in later sections of this book and in structural design courses.

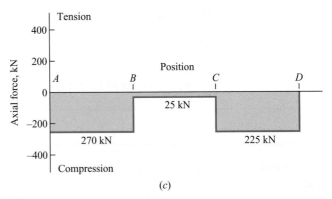

Figure 2-10(c)

(a) $A_s = \dfrac{\pi}{4}d_s^2 = \dfrac{F_s}{\sigma_s} = \dfrac{-270(10^3)}{-125(10^6)}$ $d_s = 52.44(10^{-3})\,\text{m} \cong 52.4\;\text{mm}$ **Ans.**

(b) $A_b = \dfrac{\pi}{4}d_b^2 = \dfrac{F_b}{\sigma_b} = \dfrac{-25(10^3)}{-70(10^6)}$ $d_b = 21.32(10^{-3})\,\text{m} \cong 21.3\;\text{mm}$ **Ans.**

(c) $A_a = \dfrac{\pi}{4}d_a^2 = \dfrac{F_a}{\sigma_a} = \dfrac{-225(10^3)}{-85(10^6)}$ $d_a = 58.05(10^{-3})\,\text{m} \cong 58.1\;\text{mm}$ **Ans.**

Example Problem 2-3 A brass tube with an outside diameter of 2.00 in. and a wall thickness of 0.375 in. is connected to a steel tube with an inside diameter of 2.00 in. and a wall thickness of 0.250 in. by using a 0.750-in-diameter pin as shown in Fig. 2-11*a*. Determine

(a) The shearing stress in the pin when the joint is carrying an axial load *P* of 10 kip.

(b) The length of joint required if the pin is replaced by a glued joint and the shearing stress in the glue must be limited to 250 psi.

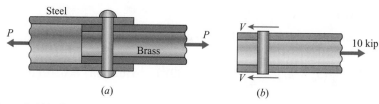

Steel

Brass

P P

V

V

10 kip

(a)   (b)

Figure 2-11(a–b)

SOLUTION

(a) A free-body diagram of the brass tube and pin is shown in Fig. 2-11b. Since the pin is in double shear, $A = 2(\frac{\pi}{4})(0.750)^2 = 0.8836$ in.^{2} Thus, from Eq. 2-4,

$$\tau = \frac{V}{A} = \frac{10}{0.8836} = 11.317 \text{ ksi} \cong 11.32 \text{ ksi} \qquad \textbf{Ans.}$$

▶ The maximum shear stress in the pin is about 50 percent greater than the average shear stress calculated here ($\tau = 11.32$ ksi). To compensate for this known underestimation of the shear stress, design engineers will make sure that the pin will be able to withstand shear stresses at least twice as great (a factor of safety of 2) as the average shear stress computed. Factors of safety will be discussed in Chapter 4 and in structural design courses.

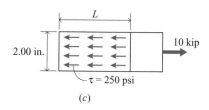

L

2.00 in.

10 kip

$\tau = 250$ psi

(c)

Figure 2-11(c)

(b) A free-body diagram of the brass tube and joint is shown in Fig. 2-11c. The shear stress acts over a length L of the outside circumference of the brass tube. For the glued joint $A = \pi dL = \pi(2.00)L = 2.00\,\pi L$ in.^{2} Thus, from Eq. 2-4,

$$\tau = \frac{V}{A} = \frac{10,000}{2.00\pi L} = 250 \text{ psi}$$

from which

$$L = 6.366 \text{ in.} \cong 6.37 \text{ in.} \qquad \textbf{Ans.}$$

■ **Example Problem 2-4**   The steel pipe column shown in Fig. 2-12 has an outside diameter of 150 mm and a wall thickness of 15 mm. The load imposed on the column by the timber beam is 150 kN. Determine

(a) The average bearing stress at the surface between the column and the steel bearing plate.

(b) The diameter of a circular bearing plate if the average bearing stress between the steel plate and the wood beam is not to exceed 3.25 MPa.

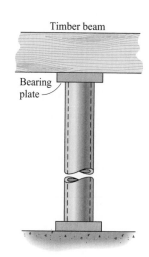

Timber beam

Bearing plate

Figure 2-12

SOLUTION

(a) The annular area between the steel column and the bearing plate is

$$A = \frac{\pi}{4}\left(d_o^2 - d_i^2\right) = \frac{\pi}{4}[(150)^2 - (120)^2] = 6362 \text{ mm}^2 = 6362(10^{-6}) \text{ m}^2$$

Thus, from Eq. 2-6,

$$\sigma_b = \frac{F}{A} = \frac{150(10^3)}{6362(10^{-6})} = 23.58(10^6) \text{ N/m}^2 \cong 23.6 \text{ MPa} \qquad \textbf{Ans.}$$

▶ Bearing stress is just the normal stress exerted on the surfaces of contact between two parts of a structure or machine element. Like the normal stress within the member itself, bearing stresses are usually assumed constant over the cross section of contact. However, if the bearing plate is not rigid, the stresses exerted on the beam by the outer edges of the bearing plate will likely be considerably smaller than the stresses exerted on the beam by the center of the bearing plate.

(b) The circular area between the bearing plate and the timber beam is

$$A = \frac{\pi}{4}d^2$$

Thus, from Eq. 2-6,

$$\sigma_b = \frac{F}{A} = \frac{150(10^3)}{(\pi/4)d^2} = 3.25(10^6)$$

from which

$$d = 242.4(10^{-3}) \text{ m} \cong 242 \text{ mm} \qquad \textbf{Ans.}$$

Example Problem 2-5 A vertical shaft is supported by a thrust collar and bearing plate as shown in Fig. 2-13a. The force imposed on the bearing plate by the collar is 50 kip. If the bearing stress between the collar and the bearing plate must not exceed 10 ksi, determine the minimum diameter collar that must be used. Assume that the bearing stress is uniformly distributed over the surface of the collar.

    If the collar is not rigid, the stress between the collar and the bearing plate will not be uniform. If the stress varies as shown in Fig. 2-13b (decreasing linearly from σ_{max} at the edge of the shaft to $\sigma_{max}/2$ at $r = 3$ in.), calculate and plot σ_{max}

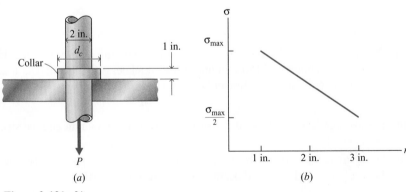

(a)

(b)

Figure 2-13(a, b)

versus the diameter d_c of the collar (2.5 in. $\leq d_c \leq 5.0$ in.). Now what minimum diameter collar must be used if the bearing stress must not exceed 10 ksi? What is the percent decrease in σ_{max} for a 3-in.-diameter collar compared to a 2.5-in.-diameter collar? For a 4.0-in.-diameter collar compared to a 3-in.-diameter collar?

SOLUTION
A free-body diagram of the shaft is shown in Fig. 2-13c. Summing forces in the vertical direction gives

$$+ \uparrow \Sigma F_y = 0: \qquad F - P = 0 \qquad F = P \qquad (a)$$

where the force on the collar is the sum of the bearing stresses $F = \int dF = \int \sigma \, dA$.

Rigid collar. If the bearing stresses are uniformly distributed over the annular ring of the collar, then

$$P = \sigma \int dA = \sigma A = \sigma \frac{\pi}{4} \left(d_c^2 - 2^2 \right)$$

Therefore, with $P = 50$ kip and $\sigma = 10$ ksi, the smallest diameter collar that can be used is

$$d_c = \sqrt{\frac{4(50)}{\pi(10)} + 4} = 3.22 \text{ in.} \qquad \textbf{Ans.}$$

Flexible collar. The equation for the linearly varying stress of Fig. 2-13b is

$$\sigma = \frac{\sigma_{max}}{4}(5 - r) \qquad (b)$$

Integrating the stresses of Eq. (b) over the annular ring of the collar gives

$$P = \int \sigma \, dA = \int_1^{r_c} \frac{\sigma_{max}}{4}(5 - r)(2\pi r \, dr)$$

$$= \frac{\pi \sigma_{max}}{4}\left[5(r_c^2 - 1) - \frac{2}{3}(r_c^3 - 1) \right]$$

Therefore, with $P = 50$ kip, the maximum stress on the collar is given by

$$\sigma_{max} = \frac{200}{\pi \left[5 \left(r_c^2 - 1 \right) - \frac{2}{3} \left(r_c^3 - 1 \right) \right]} \qquad (c)$$

For example, when $r_c = 1.25$ in. (and $d_c = 2.5$ in.)

$$\sigma_{max} = \frac{200}{\pi \left[5 \left(1.25^2 - 1 \right) - \frac{2}{3} \left(1.25^3 - 1 \right) \right]} = 29.242 \text{ ksi}$$

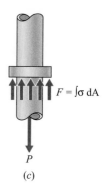

$F = \int \sigma \, dA$

P

(c)

Figure 2-13(c)

▶ The term $F = \int \sigma \, dA$ accounts for a normal stress that may vary over the contact area between the collar and bearing plate. If the collar is rigid, the stress σ is constant, and $F = \int \sigma \, dA$ reduces to Eq. (2-2).

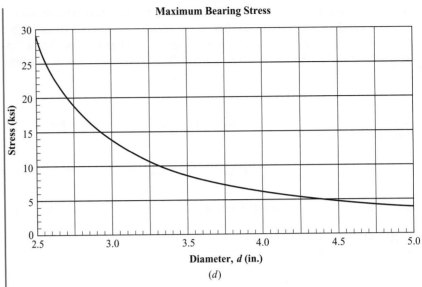

Figure 2-13(d)

Calculating Eq. (c) for values of d_c between 2.5 in. and 5.0 in. results in the graph of Fig. 2-13d.

From the graph of Fig. 2-13d, the smallest flexible collar for which $\sigma_{\max} \leq 10$ ksi is

$$d_c = 3.32 \text{ in.} \qquad \qquad \textbf{Ans.}$$

which is only 3.1% larger than the diameter of the smallest rigid collar. From Eq. (c) or from the graph of Fig. 2-13d, when $d_c = 3$ in., the maximum stress is $\sigma_{\max} = 13.642$ ksi and the percent decrease from when $d_c = 2.5$ in. is

$$\frac{29.242 - 13.642}{29.242}(100) = 53.4\% \qquad \qquad \textbf{Ans.}$$

The percent decrease in the maximum stress between $d_c = 3.0$ in., where the maximum stress is $\sigma_{\max} = 13.642$ ksi, and $d_c = 4.0$ in., where the maximum stress is $\sigma_{\max} = 6.161$ ksi, is

$$\frac{13.642 - 6.161}{13.642}(100) = 54.8\% \qquad \qquad \textbf{Ans.}$$

PROBLEMS

MecMovie Activities and Problems

MM2.1 Normal, shear, and bearing stress. Example; Try One. Simple pin connections used to illustrate various types of stress.

MM2.2 Normal stress–basic problems. Concept Checkpoints. Use normal stress concepts for four introductory problems.

MM2.3 Shear stress–basic problems. Concept Checkpoints. Use shear stress concepts for four introductory problems.

MM2.4 Load capacity of two-bar assembly. Example; Try One. Determine load capacity of two-bar structure given areas and allowable stresses for the two members.

MM2.5 Shear stress in pin support. Example: Concept Check-points. Determine shear stress in support.

MM2.6 Shear stress in bolted flanges. Example; Try One. Determine shear stress in bolted connection.

Introductory Problems

2-1* An aluminum tube with an outside diameter of 1.000 in. will be used to support a 10-kip load. If the axial stress in the member must be limited to 30 ksi (T) or (C), determine the wall thickness required for the tube.

2-2* Three steel bars with 25×15 -mm cross sections are welded to a gusset plate as shown in Fig. P2-2. Determine the normal stresses in the bars when the forces shown are being applied to the plate.

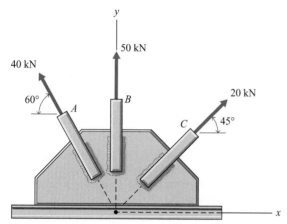

Figure P2-2

2-3 Loads are applied with rigid bearing plates to a system of steel bars, as shown in Fig. P2-3. The cross-sectional areas are 16 in.^{2} for bar AB, 4 in.^{2} for bar BC, and 12 in.^{2} for bar CD. Determine the normal stress on a typical cross section in each of the bars.

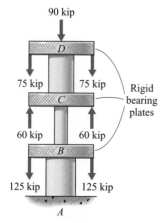

Figure P2-3

2-4* In order to hold a 130-kg crate in a stationary position, a worker exerts a force **P** at an angle θ on a rope as shown in Fig. P2-4. Determine the normal stress in the 15-mm diameter rope AB when $\theta = 20°$.

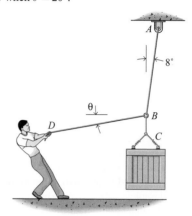

Figure P2-4

2-5 A device for determining the shearing strength of wood is shown in Fig. P2-5. The dimensions of the wood specimen are 6 in. wide by 8 in. high by 2 in. thick. If a load of 16,800 lb is required to fail the specimen, determine the shearing strength (the shear stress at the failure load) of the wood.

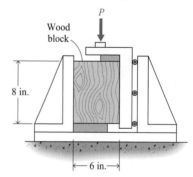

Figure P2-5

2-6 A lever is attached to the shaft of a steel gate valve with a square key, as shown in Fig. P2-6. If the shearing stress in the key must not exceed 125 MPa, determine the minimum dimension a that must be used if the key is 20 mm long.

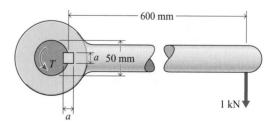

Figure P2-6

2-7* A coupling is used to connect a 2-in.-diameter plastic rod to a 1.5-in.-diameter rod, as shown in Fig. P2-7. If the average shearing stress in the adhesive must be limited to 500 psi, determine the minimum lengths L_1 and L_2 required for the joint when the applied axial load P is 8000 lb.

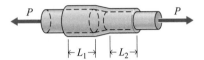

Figure P2-7

2-8 A vertical shaft is supported by a thrust collar and bearing plate as shown in Fig. P2-8. Determine the maximum axial load that can be applied to the shaft if the average punching shear stress in the collar and the average bearing stress between the collar and the plate are limited to 75 MPa and 100 MPa, respectively.

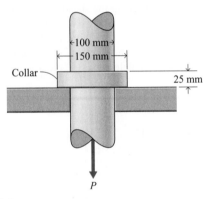

Figure P2-8

2-9* A 100-ton hydraulic punch press is used to punch holes in a 0.50-in.-thick steel plate, as illustrated schematically in Fig. P2-9. If the average punching shear resistance of the steel plate is 40 ksi, determine the maximum diameter hole that can be punched.

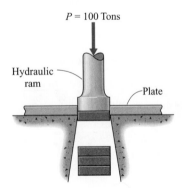

Figure P2-9

2-10* The weight W in the traction device shown in Fig. P2-10 has a mass of 45 kg. The continuous cord from A to W has a diameter of 10 mm. Determine the normal stress in segment BC of the cord.

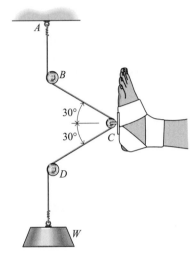

Figure P2-10

2-11 The 125–lb girl shown in Fig. P2-11 is doing a chin-up. For the position shown, the humerus is modeled as an axially loaded member. Determine the normal stress at section a–a if the section is modeled as

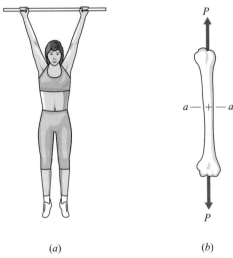

(a)   (b)

Figure P2-11

a. A solid circular cross section of 1-in. diameter.
b. A hollow cylinder of outside diameter 1-in. and inside diameter 0.6 in.

2-12 Pulleys 1 and 2 of the rope and pulley system shown in Fig. P2-12 are connected and rotate as a unit. The radii of pulleys 1 and 2 are 100 mm and 300 mm, respectively. Rope A is wrapped around pulley 1 and is fastened to pulley 1 at point A'. Rope B is wrapped around pulley 2 and is fastened to pulley 2 at point B'. Rope C is continuous over pulleys 3 and 4. Each rope has a diameter of 15 mm. Determine the normal stress in each segment of rope C if the weight W has a mass of 225 kg.

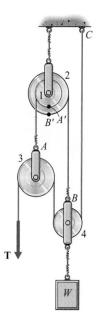

Figure P2-12

Intermediate Problems

2-13* Two flower pots, shown in Fig. P2-13, are supported with steel wires of equal diameter. Pot A weighs 10 lb and pot B weighs 8 lb. Determine the minimum required diameter of the wires if the normal stress in the wires must not exceed 18 ksi.

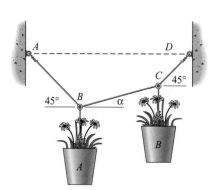

Figure P2-13

2-14* Member AD of the timber truss shown in Fig. P2-14 is framed into the 100×150 -mm bottom chord ABC as shown in the insert. Determine the dimension a that must be used if the average shearing stress parallel to the grain at the ends of chord ABC is not to exceed 2.25 MPa.

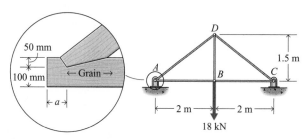

Figure P2-14

2-15 The inclined member AB of a timber truss is framed into a 4×6 -in. bottom chord, as shown in Fig. P2-15. Determine the axial compressive force in member AB when the average shearing stress parallel to the grain at the end of the bottom chord is 225 psi.

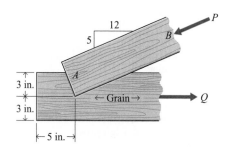

Figure P2-15

2-16* A scissors jack for an automobile is shown in Fig. P2-16. The screw threads exert a force **F** on the blocks at joints A and B.

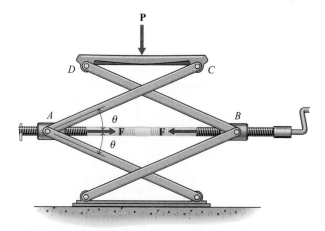

Figure P2-16

If $F = 700$ N, determine the shearing stress on a cross section of the 10-mm-diameter pin at C, which is in single shear when

a. $\theta = 15°$.
b. $\theta = 30°$.
c. $\theta = 45°$.

2-17 Forces of 25 lb are applied to the handles of the pipe pliers shown in Fig. P2-17. Determine the shearing stress on a cross section of the 0.25-in.-diameter pin at A.

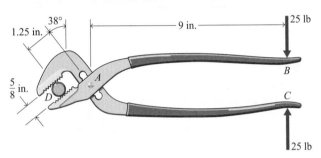

Figure P2-17

2-18 A pin-connected truss is loaded and supported as shown in Fig. P2-18. Determine

a. The normal stress in member DE if it has a cross-sectional area of 750 mm².
b. The minimum cross-sectional area for member BC if the normal stress is limited to 30 MPa.

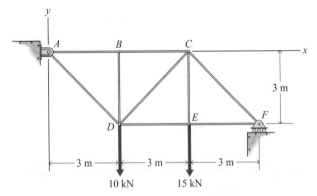

Figure P2-18

2-19* The fold-down chair of Fig. P2-19 weighs 30 lb and has its center of gravity at G. Determine the shearing stress on a

cross section of the 3/8-in.-diameter pin at B, which is in single shear.

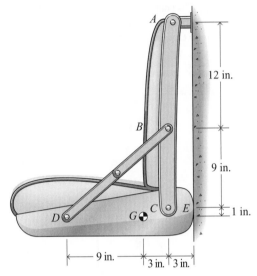

Figure P2-19

2-20 A pin-connected truss is loaded and supported as shown in Fig. P2-20. Determine

a. The normal stress in member CD if it has a cross-sectional area of 624 mm.²
b. The minimum cross-sectional area for member DF if the normal stress is limited to 25 MPa.

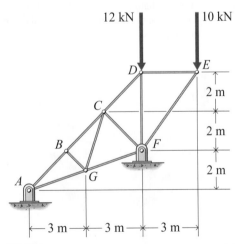

Figure P2-20

2-21   A force of 20 lb is required to pull the stopper *DE* in Fig. P2-21. Determine the shearing stress on the cross section of the 1/8-in.-diameter pin at *B*, which is in single shear.

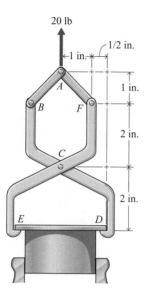

Figure P2-21

a. The normal stress on a section perpendicular to and midway between *D* and *E*. The cross-sectional area of member *DE* is 1.25 in.^{2}

b. The shearing stress on the cross section of the 0.25-in.-diameter pin at *E*, which is in double shear.

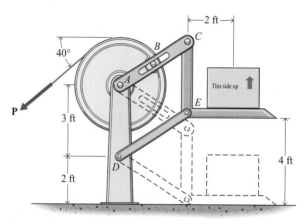

Figure P2-23

Challenging Problems

2-22*   A pair of vise grip pliers is shown in Fig. P2-22. When a force $P = 100$ N is being applied to the handles, determine

a. The average shearing stress in the pin at *A* if it has a 4-mm diameter and is in double shear.

b. The average shearing stress in the pin at *B* if it has a 5-mm diameter and is in double shear.

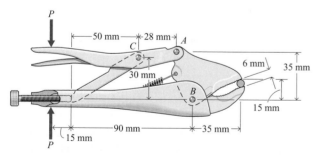

Figure P2-22

2-23*   The mechanism of Fig. P2-23 is designed to keep its load level while raising it. A pin on the rim of the 4-ft-diameter pulley fits in a slot on arm *ABC*. Arms *ABC* and *DE* are each 4 ft long, and the package being lifted weighs 80 lb. The mechanism is raised by pulling on the rope that is wrapped around the pulley. Determine for the position shown

2-24   Figure P2-24 is a simplified sketch of the mechanism used to raise the bucket of a bulldozer. The bucket and its contents weigh 10 kN and have a center of gravity at *H*. Arm *ABCD* has a weight of 2 kN and a center of gravity at *B*; arm *DEFG* has a weight of 1 kN and a center of gravity at *E*. The weight of the hydraulic cylinders can be ignored. Determine the required diameter of the pin at *D* if the shearing stress cannot exceed 120 MPa. The pin is in double shear.

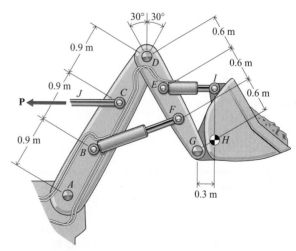

Figure P2-24

2-25* For the pin-connected structure of Fig. P2-25, determine the minimum diameter for the pin at joint D if the average shearing stress in the pin must be limited to 7500 psi.

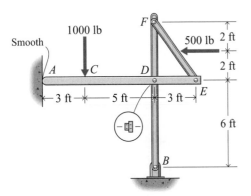

Figure P2-25

2-26 The front-wheel suspension of an automobile is shown in Fig. P2-26. The pavement exerts a vertical force of 2700 N on the tire. Determine the minimum diameters of the pins at A, B, C, and D if all pins are in double shear and the shear stress in any pin cannot exceed 125 MPa.

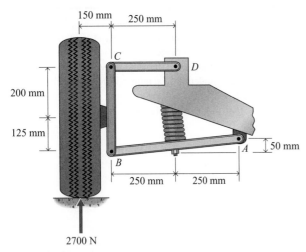

Figure P2-26

Computer Problems

2-27 A steel pipe will be used to support a 9000-lb load. If the wall thickness of the pipe is 10 percent of the pipe's outside diameter d_o, calculate and plot the normal stress in the pipe σ as a function of the diameter d_o (0.75 in. $\le d_o \le 3$ in.). If the axial stress in the pipe must be limited to 12 ksi, what is the smallest size standard steel pipe (see Appendix B) that could be used?

2-28 The steel pipe column shown in Fig. P2-28a has an outside diameter of 150 mm and a wall thickness of 15 mm. The load imposed on the column by the timber beam is 150 kN. If the bearing stress between the circular steel bearing plate and the wood beam is not to exceed 3.25 MPa, determine the minimum diameter bearing plate that must be used between the column and the beam. Assume that the bearing stress is uniformly distributed over the surface of the plate.

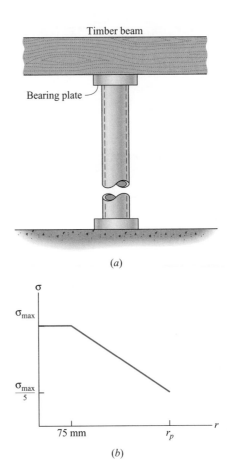

Figure P2-28

If the bearing plate is not rigid, the stress between the bearing plate and the wood beam will not be uniform. If the stress varies as shown in Fig. P2-28b (a uniform value of σ_{max} above the column and decreasing linearly to $\sigma_{max}/5$ at the outside edge r_p of the bearing plate), calculate and plot σ_{max} versus the radius r_p of the bearing plate (75 mm $\le r_p \le 500$ mm). Now what minimum diameter bearing plate must be used if the bearing stress must not exceed 3.25 MPa? What is the percent decrease in σ_{max} for a 400-mm-diameter bearing plate compared to a 150-mm-diameter bearing plate? For a 600-mm-diameter bearing plate compared to a 150-mm-diameter bearing plate?

2-29 A vertical shaft is supported by a thrust collar and bearing plate as shown in Fig. P2-29a. The force imposed on the bearing plate by the collar is 50 kip. If the bearing stress between the collar and the bearing plate must not exceed 10 ksi, determine the minimum diameter collar that must be used. Assume that the bearing stress is uniformly distributed over the surface of the collar.

2-30 The tie rod shown in Fig. P2-30a has a diameter of 40 mm and is used to resist the lateral pressure against the walls of a grain bin. The force imposed on the wall by the rod is 80 kN. If the bearing stress between the washer and the wall must not exceed 2.8 MPa, determine the minimum diameter washer that must be used between the head of the bolt and the grain bin wall. Assume that the bearing stress is uniformly distributed over the surface of the washer.

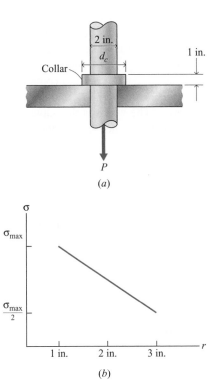

Figure P2-29

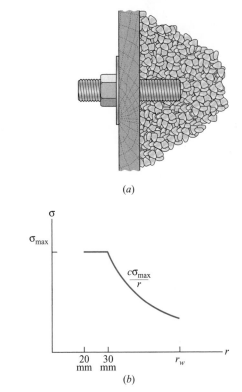

Figure P2-30

If the collar is not rigid, the stress between the collar and the bearing plate will not be uniform. If the stress varies as shown in Fig. P2-29b (decreasing linearly from σ_{max} at the edge of the shaft to $\sigma_{max}/2$ at $r = 3$ in.), calculate and plot σ_{max} versus the radius r_c of the collar (1 in. $\leq r_c \leq 2.5$ in.). Now what minimum diameter collar must be used if the bearing stress must not exceed 10 ksi? What is the percent decrease in σ_{max} for a 3.2-in.-diameter collar compared to a 2.4-in.-diameter collar? For a 4.0-in.-diameter collar? For a 5.0-in.-diameter collar?

If the washer is not rigid, the stress between the washer and the wall will not be uniform. If the stress varies as shown in Fig. P2-30b (a uniform value of σ_{max} under the 60-mm-diameter restraining nut and decreasing as $1/r$ to the outside edge r_w of the washer), calculate and plot σ_{max} versus the radius r_w of the washer (30 mm $\leq r_w \leq 200$ mm). Now what minimum diameter washer must be used if the bearing stress must not exceed 2.8 MPa? What is the percent decrease in σ_{max} for a 200-mm-diameter washer compared to no washer? For a 300-mm-diameter washer?

2-6   STRESSES ON AN INCLINED PLANE IN AN AXIALLY LOADED MEMBER

In Sections 2-2, 2-3, and 2-4, normal, shear, and bearing stresses for axially loaded members were introduced. Stresses on planes inclined to the axis of an axially

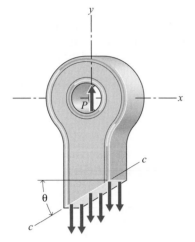

Figure 2-14

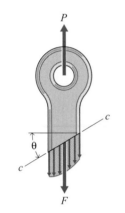

Figure 2-15

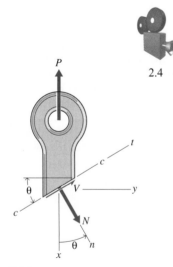

Figure 2-16

2.4

loaded bar will now be considered. When the eyebar shown in Fig. 2-1 is cut by an inclined plane, a free-body diagram of the upper portion of the bar would appear as shown in Fig. 2-14. Equilibrium of the upper portion of the bar is established by placing a distribution of internal force on the cut section, as shown in Fig. 2-15. The resultant F of this distribution of internal force is equal in magnitude to the applied load P and has a line of action that is coincident with the axis of the bar, as shown in Fig. 2-15. An average total stress S_{avg} on the inclined surface can be computed by using Eq. 2-2. This total stress conveys little information that is useful for design purposes, however, because experimental studies indicate that materials respond differently to forces that tend to pull surfaces apart than to forces that tend to slide surfaces relative to each other. Therefore, the resultant **F**, is usually resolved into normal and tangential (shear) components N and V, as shown in Fig. 2-16. It is these components that are used to compute the normal and shear stresses on the inclined surface using Eqs. 2-2 and 2-4. The axes and forces shown in Fig. 2-16 are all positive. The x-axis is the outward normal to a section perpendicular to the axis of the bar, and the n-axis is the outward normal to the inclined section. The angle θ is measured from a positive x-axis to a positive n-axis; a counterclockwise angle is positive. Positive y- and t-axes are located using the right-hand rule and a positive angle. The forces N and V shown in Fig. 2-16 are both positive (see Section 1-5). Also, the internal force on the x-section is positive.

From equilibrium, the normal and shear forces are given by $N = P \cos \theta$ and $V = -P \sin \theta$. The area on which these forces act is the area A_n of the inclined surface which is given by $A_n = A/\cos \theta$, where A is the cross-sectional area of the axially loaded member. Therefore,

$$\sigma_n = \frac{N}{A_n} = \frac{P \cos \theta}{A/\cos \theta} = \frac{P}{A} \cos^2 \theta = \frac{P}{2A}(1 + \cos 2\theta) \tag{2-7}$$

$$\tau_n = \frac{V}{A_n} = \frac{-P \sin \theta}{A/\cos \theta} = \frac{-P}{A} \sin \theta \cos \theta = \frac{-P}{2A} \sin 2\theta \tag{2-8}$$

In the preceding discussion, the assumption was made that the stresses were uniformly distributed over the inclinded surface.

Both the area of the inclined surface A_n and the values for the normal and shear forces N and V on the surface depend on the angle θ of the inclined plane with respect to the applied load; therefore, the normal and shear stresses σ_n and τ_n on the inclined plane also depend on the angle θ. This dependence of stress on both force and area means that stress is not a vector quantity; therefore, the laws of vector addition do not apply to stresses that act on different planes. This need not be cause for concern if, in the application of the equations of equilibrium (or motion), one always replaces a stress with a total force (stress multiplied by the appropriate area), thus reducing the problem to one involving ordinary force vectors. However, stresses that act on a single particular plane can be treated as vectors, since they all are associated with the same area.

A graph showing the magnitudes of σ_n and τ_n as a function of θ is shown in Fig. 2-17. These results indicate that σ_n is maximum when θ is 0° or 180°, that τ_n is maximum when θ is 45° or 135°, and also that $\tau_{max} = \sigma_{max}/2$. Therefore,

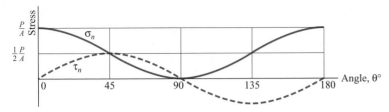

Figure 2-17

the magnitudes of the maximum normal and shearing stresses for axial tensile or compressive loading are

$$\sigma_{\max} = P/A \tag{2-9}$$

$$\tau_{\max} = P/2A \tag{2-10}$$

Note that the normal stress is either maximum or minimum on planes for which the shearing stress is zero. It can be shown that the shearing stress is always zero on the planes of maximum or minimum normal stress. The concepts of maximum and minimum normal stress and maximum shearing stress for more general cases will be treated in Sections 2-10, 2-11, and 2-12 of this chapter.

Laboratory experiments indicate that both normal and shearing stresses under axial loading are important because a brittle material loaded in tension will fail in tension on a transverse plane, whereas a ductile material loaded in tension will fail in shear on a 45° plane.

The plot of the magnitudes of the normal and shear stresses for axial loading, shown in Fig. 2-17, indicates that the sign of the shearing stress changes when θ is greater than 90°. The magnitude of the shearing stress for any angle θ, however, is the same as that for $90° + \theta$. The sign change merely indicates that the shear force V changes sense when $\theta > 90°$. Normal and shearing stresses on planes having aspects θ_1 and $90° + \theta_1$ are shown in Fig. 2-18.

The equality of shearing stresses on orthogonal planes can be demonstrated by applying the equations of equilibrium to the free-body diagram of a small rectangular block of thickness dz, shown in Fig. 2-19. If only a shearing force[1] $V_x = \tau_{yx} dx\, dz$ is applied to the top surface of the block, the equation $\Sigma F_x = 0$ will dictate the application of an oppositely directed force V_x to the bottom of the block, thus leaving the block subjected to a clockwise couple; this couple must be balanced by a counterclockwise couple composed of the oppositely directed forces V_y applied to the vertical faces of the block. Finally, application of the equation $\Sigma M_z = 0$ yields the following:

$$\tau_{yx}(dx\, dz)dy = \tau_{xy}(dy\, dz)dx$$

from which

$$\tau_{yx} = \tau_{xy} \tag{2-11}$$

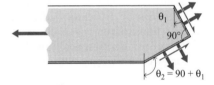

Figure 2-18

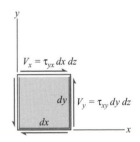

Figure 2-19

---

[1]The double subscript on the shearing stress is used to designate both the plane on which the stress acts and the direction of the stress. The first subscript indicates the plane (or rather the normal to the plane), and the second subscript indicates the direction of the stress. This will be discussed in more detail in Section 2–7.

Therefore, if a shearing stress exists at a point on any plane, there must also exist at this point a shearing stress of the same magnitude on an orthogonal plane. This statement is also valid when normal stresses are acting on the planes, because the normal stresses occur in collinear but oppositely directed pairs and thus have zero moment with respect to any axis.

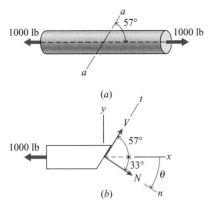

Figure 2-20

▶ Equations 2-7 and 2-8 give the normal and shear stresses on inclined surfaces in uniaxially loaded bars such as in this example. These equations must not be used for other, more general loading situations.

Example Problem 2-6   A plastic bar with a circular cross section of diameter 1.25 in. will be used to support an axial load of 1000 lb, as shown in Fig. 2-20a. Determine the normal and shearing stresses on section a–a.

SOLUTION
The positive coordinate axes (x–y and n–t) are shown in Fig. 2-20b. Since a positive angle θ is measured counterclockwise from positive x to positive n, the angle θ shown in Fig. 2-20b is negative ($\theta = -33°$). Using Eqs. 2-7 and 2-8,

$$\sigma_n = \frac{N}{A_n} = \frac{P}{2A}(1 + \cos 2\theta) = \frac{+1000}{2[\pi(1.25)^2/4]}[1 + \cos 2(-33°)]$$

$$= 573 \text{ psi} \qquad\qquad\qquad\qquad \textbf{Ans.}$$

$$\tau_n = \frac{V}{A_n} = \frac{-P}{2A}\sin 2\theta = \frac{-(+1000)}{2[\pi(1.25)^2/4]}\sin 2(-33°)$$

$$= +372 \text{ psi} \qquad\qquad\qquad\qquad \textbf{Ans.}$$

Alternatively, a positive angle of $+327°$ could have been used:

$$\sigma_n = \frac{P}{2A}(1 + \cos 2\theta) = \frac{+1000}{2[\pi(1.25)^2/4]}[1 + \cos 2(+327°)]$$

$$= 573 \text{ psi} \qquad\qquad\qquad\qquad \textbf{Ans.}$$

$$\tau_n = \frac{-P}{2A}\sin 2\theta = \frac{-(+1000)}{2[\pi(1.25)^2/4]}\sin 2(+327°)$$

$$= +372 \text{ psi} \qquad\qquad\qquad\qquad \textbf{Ans.}$$

Example Problem 2-7   The block shown in Fig. 2-21a has a 200 × 100-mm rectangular cross section. The normal stress on plane a–a is 12.00 MPa (C) when the load P is applied. If angle ϕ is 36°, determine

(a) The load P.
(b) The shearing stress on plane a–a.
(c) The magnitudes of the maximum normal and shearing stresses in the block.

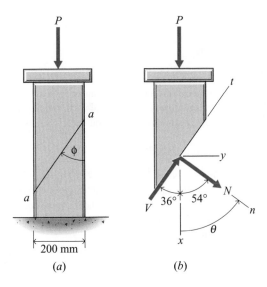

Figure 2-21

SOLUTION

(a) Positive values of coordinate axes and internal forces are shown in Fig. 2-21b. Using Eq. 2-7,

$$\sigma_n = \frac{N}{A_n} = \frac{P}{2A}(1 + \cos 2\theta)$$

$$-12(10^6) = \frac{P}{2(200)(100)(10^{-6})}[1 + \cos 2(+54°)]$$

$$P = -694.7(10^3) \text{ N} \cong 695 \text{ kN (C)} \qquad \textbf{Ans.}$$

(b) Using Eq. 2-8,

$$\tau_n = \frac{V}{A_n} = \frac{-P}{2A}\sin 2\theta = \frac{-(-694.7)(10^3)}{2(200)(100)(10^{-6})}\sin 2(+54°)$$

$$= +16.52(10^6) \frac{\text{N}}{\text{m}^2}$$

$$\tau_n = 16.52 \text{ MPa} \qquad \textbf{Ans.}$$

(c) From Eqs. 2-9 and 2-10,

$$\sigma_{max} = \frac{P}{A} = \frac{694.7(10^3)}{(200)(100)(10^{-6})} = 34.74(10^6) \frac{\text{N}}{\text{m}^2} \cong 34.7 \text{ MPa} \qquad \textbf{Ans.}$$

$$\tau_{max} = \frac{P}{2A} = \frac{694.7(10^3)}{2(200)(100)(10^{-6})} = 17.37(10^6) \frac{\text{N}}{\text{m}^2} = 17.37 \text{ MPa}$$

$$\textbf{Ans.}$$

▶ In uniaxial loading problems, the maximum normal stress always occurs on the transverse cross section ($\theta = 0°$) and the maximum shear stress ($\tau_{max} = \sigma_{max}/2$) occurs on a plane oriented at 45° to the axial direction. Like Eqs. 2-7 through 2-10, these results apply only to uniaxially loaded bars such as in this example; they must not be used for other, more general loading situations.

PROBLEMS

MecMovie Activities and Problems

MM2.7 Stresses on inclined plane. Example; Try One. Calculating normal and shear stresses acting on an inclined plane in an axial member.

MM2.8 Forces based on inclined plane stresses. Example; Try One. Given allowable normal and shear stresses and a specified inclined plane surface, determine the maximum axial load that can be applied.

Introductory Problems

2-31* An axial load P is applied to a timber block with a 4×4 -in. square cross section, as shown in Fig. P2-31. Determine the normal and shear stresses on the planes of the grain if $P = 5000$ lb.

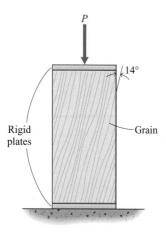

Figure P2-31

2-32* The steel bar shown in Fig. P2-32 will be used to carry an axial tensile load of 400 kN. If the thickness of the bar is 45 mm, determine the normal and shearing stresses on plane a–a.

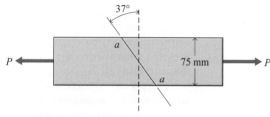

Figure P2-32

2-33 A structural steel bar with a 2×6-in. rectangular cross section is subjected to an axial tensile load of 270 kip. Determine the maximum normal and shear stresses in the bar.

2-34 A concrete cylinder 75 mm in diameter and 150 mm high failed along a plane making an angle of 57° with the horizontal when subjected to an axial vertical compressive load of 80 kN. Determine the normal and shear stresses on the failure plane.

Intermediate Problems

2-35* A steel bar with a 4×1-in. rectangular cross section is being used to transmit an axial tensile load, as shown in Fig. P2-35. Normal and shear stresses on plane a–a of the bar are 12 ksi tension and 9 ksi shear. Determine the angle θ and the applied load P.

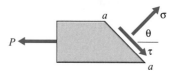

Figure P2-35

2-36* A steel bar with a butt-welded joint, as shown in Fig. P2-36, will be used to carry an axial tensile load of 400 kN. If the normal and shear stresses on the plane of the butt weld must be limited to 70 MPa and 45 MPa, respectively, determine the minimum thickness t required for the bar.

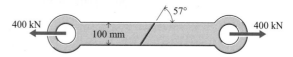

Figure P2-36

2-37 The shearing stress on plane a–a of the 4×8-in. rectangular block shown in Fig. P2-37 is 2 ksi when the axial load P is applied. If the angle ϕ is 35°, determine

a. The load P.
b. The normal stress on plane a–a.
c. The maximum normal and shearing stresses in the block.

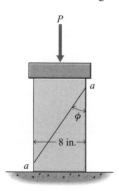

Figure P2-37

2-38 Determine the maximum axial load P that can be applied to the wood compression block shown in Fig. P2-38 if specifications require that the shear stress parallel to the grain not exceed 5.25 MPa, the compressive stress perpendicular to the grain not exceed 13.60 MPa, and the maximum shear stress in the block not exceed 8.75 MPa.

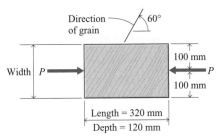

Figure P2-38

Challenging Problems

2-39* A steel eyebar 4×1-in. rectangular cross section has been designed to transmit an axial tensile load. The length of the eyebar must be increased by welding a new center section in the bar ($45° \leq \phi \leq 90°$) as shown in Fig. P2-39. The stresses in the weld material must be limited to 12 ksi in tension and 9 ksi in shear. Determine

a. The optimum angle ϕ for the joint.
b. The maximum safe load P for the redesigned member.

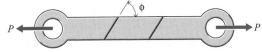

Figure P2-39

2-40* The bar shown in Fig. P2-40 has a 200×100-mm rectangular cross section. Determine

a. The normal and shearing stresses on plane a–a.
b. The maximum normal and shearing stresses in the bar.

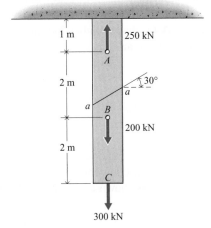

Figure P2-40

2-41 The two parts of the eyebar shown in Fig. P2-41 are connected with two 1/2-in.-diameter bolts (one on each side). Specifications for the bolts require that the axial tensile stress not exceed 12.0 ksi and that the shearing stress not exceed 8.0 ksi. Determine the maximum load P that can be applied to the eyebar without exceeding either specification.

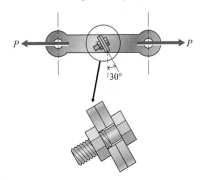

Figure P2-41

2-42 A tension member with a 50×100-mm rectangular cross section will be fabricated with an inclined glued joint ($45° \leq \phi \leq 90°$) at its midsection, as shown in Fig. P2-42. If the allowable stresses for the glue are 5 MPa in tension and 3 MPa in shear, determine

a. The optimum angle ϕ for the joint.
b. The maximum safe load P for the member.

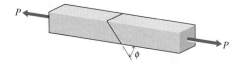

Figure P2-42

Computer Problems

2-43 Specifications for the $3 \times 3 \times 21$-in. rectangular block shown in Fig. P2-43 require that the normal and shearing

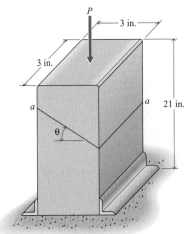

Figure P2-43

stresses on plane a–a not exceed 800 psi and 500 psi, respectively. If the plane a–a makes an angle $\theta = 37°$ with the horizontal, calculate and plot the ratios σ/σ_{max} and τ/τ_{max} as a function of the load $P\,(0 \leq P \leq 13\ \text{kip})$. What is the maximum load P_{max} that can be applied to the block? Which condition controls what the maximum load can be? Repeat for $\theta = 25°$. For what angle θ will the normal stress and the shear stress both reach their limiting values at the same time?

2-44 A steel eyebar with a 100×25-mm rectangular cross section has been designed to transmit an axial tensile load P. The length of the eyebar must be increased by welding a new center section in the bar, as shown in Fig. P2-44. If $P = 250\ \text{kN}$, calcu-

late and plot the normal stress σ_n and the shear stress τ_n in the weld material for weld angles $\phi\ (30° \leq \phi \leq 90°)$. If the stresses in the weld material must be limited to 80 MPa in tension and 60 MPa in shear, what ranges of ϕ would be acceptable for the joint? Repeat for $P = 305\ \text{kN}$ and for $P = 350\ \text{kN}$. Are weld angles $\phi < 30°$ reasonable? Why or why not?

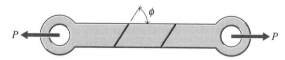

Figure P2-44

2-7 STRESS AT A GENERAL POINT IN AN ARBITRARILY LOADED MEMBER

In Sections 2-2 through 2-6 the concept of stress was introduced by considering the internal force distribution required to satisfy equilibrium in a portion of a bar under an axial load. The nature of the force distribution led to uniformly distributed normal and shearing stresses on transverse planes through the bar. In more complicated structural members or machine components, the stress distributions will not be uniform on arbitrary internal planes; therefore, a more general concept of the state of stress at a point is needed.

The nature of the internal force distribution at an arbitrary interior point O of a body of arbitrary shape that is in equilibrium under the action of a system of applied forces and support reactions $F_1, F_2, F_3, \ldots F_n$, can be studied by exposing an interior plane through O, as shown in Fig. 2-22a. The force distribution required on such an interior plane to maintain equilibrium of the isolated part of the body will not be uniform, in general; however, any distributed force acting on a small area ΔA surrounding a point of interest O can be replaced by a statically equivalent resultant force ΔF_n through O and a couple ΔM_n. The subscript n indicates that the resultant force and couple are associated with a particular plane through O—namely, the one having an outward normal in the n-direction at O. For any other plane through O the values of ΔF and ΔM could be different. Note that the line of action of ΔF_n or ΔM_n may not coincide with the direction of n. If the resultant force ΔF_n is divided by the area ΔA, an average force per unit area (average resultant stress) is obtained. As the area ΔA is made smaller and smaller, the force distribution becomes more and more uniform and the couple ΔM_n vanishes. In the limit $\Delta A \to 0$, a quantity known as the stress vector[2] or resultant stress, S_n, is obtained. Thus,

$$S_n = \lim_{\Delta A \to 0} \frac{\Delta F_n}{\Delta A}$$

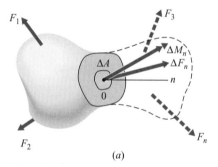

Figure 2-22(a)

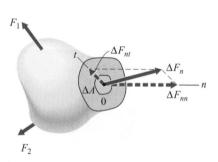

Figure 2-22(b)

In Section 2-6 it was pointed out that materials respond differently to components of the stress vector. In particular, components normal (n) and tangent (t) to the internal plane are important. As shown in Fig. 2-22b, the resultant

---

[2]The component of a tensor on a plane is a vector; therefore, on a particular plane, the stresses can be treated as vectors.

force ΔF_n can be resolved into components ΔF_{nn} normal to the plane and ΔF_{nt} tangent to the plane. A normal stress σ_n and a shearing stress τ_n are then defined as

$$\sigma_n = \lim_{\Delta A \to 0} \frac{\Delta F_{nn}}{\Delta A}$$

and

$$\tau_n = \lim_{\Delta A \to 0} \frac{\Delta F_{nt}}{\Delta A}$$

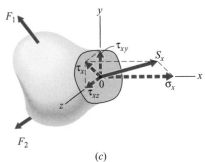

(c)

Figure 2-22(c)

For purposes of analysis it is convenient to reference stresses to some coordinate system. In a Cartesian coordinate system, the stresses on planes having outward normals in the x-, y-, and z-directions are usually chosen. Consider the plane having an outward normal in the x-direction (see Fig. 2-22c). In this case the normal and shear stresses on the plane will be σ_x and τ_x, respectively. Since τ_x, in general, will not coincide with the y- or z-axes, it must be resolved into the components τ_{xy} and τ_{xz}, as shown in Fig. 2-22c.

Unfortunately, the state of stress at a point in a material is not completely defined by these three components of the stress vector, since the stress vector itself depends on the orientation of the plane with which it is associated. An infinite number of planes can be passed through the point, resulting in an infinite number of stress vectors being associated with the point. Fortunately it can be shown (see Section 2-12) that the specification of stresses on three mutually perpendicular planes is sufficient to completely describe the state of stress at the point. The rectangular components of the stress vectors on planes (through point O) having outward normals in the positive coordinate directions are shown in Fig 2-23.

It is customary to show the stresses on positive and negative surfaces through a point using a small "element" such as is shown in Fig. 2-24. The six faces of the small element are denoted by the directions of their outward normals, so that the positive x-face is the one whose outward normal is in the direction of the positive x-axis. The coordinate axes x, y, and z are arranged as a right-hand system.

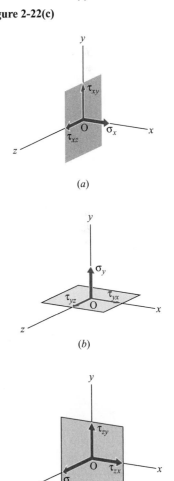

(a)

(b)

(c)

Figure 2-23

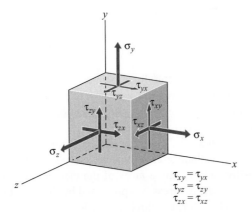

$$\tau_{xy} = \tau_{yx}$$
$$\tau_{yz} = \tau_{zy}$$
$$\tau_{zx} = \tau_{xz}$$

Figure 2-24

The sign convention for stresses is as follows:

- Normal stresses are indicated by the symbol σ and a single subscript to indicate the plane (actually the outward normal to the plane) on which the stress acts. Normal stresses are positive if they point in the direction of the outward normal. Thus, normal stresses are positive if tensile and negative if compressive.
- Shearing stresses are denoted by the symbol τ followed by two subscripts; the first subscript designates the normal to the plane on which the stress acts and the second designates the coordinate axis to which the stress is parallel. Thus, τ_{xz} is the shearing stress on a plane with outward normal in the x-direction. The stress acts parallel to the z-axis. A positive shearing stress points in the positive direction of the coordinate axis of the second subscript if it acts on a surface with an outward normal in the positive direction. Conversely, if the outward normal of the surface is in the negative direction, then the positive shearing stress points in the negative direction of the coordinate axis of the second subscript. The stresses shown on the element in Fig. 2-24 are all positive.

Summarizing, the resultant stress vector S_n is the result of the action of one part of the body on another part of the body, and it is a measure of the force intensity over a section of the body. The components of the stress vector, σ and τ, depend on the location of the point O as well as the orientation of the plane through the point. Of the nine components of stress, σ_x, σ_y, σ_z, τ_{xy}, τ_{yx}, τ_{yz}, τ_{zy}, τ_{zx}, and τ_{xz}, only six are independent because moment equilibrium requires that $\tau_{xy} = \tau_{yx}$, $\tau_{yz} = \tau_{zy}$, and $\tau_{zx} = \tau_{xz}$ (see Section 2-6).

2-8 TWO-DIMENSIONAL OR PLANE STRESS

Considerable insight into the nature of stress distributions can be gained by considering a state of stress known as two-dimensional or *plane stress*. For this case, two parallel faces of the small element shown in Fig. 2-24 are assumed to be free of stress. For purposes of analysis, let these faces be perpendicular to the z-axis. Thus,

$$\sigma_z = \tau_{zx} = \tau_{zy} = 0$$

From Eq. 2-11, however, this also implies that

2.5

$$\tau_{xz} = \tau_{yz} = 0$$

A state of plane stress occurs naturally at points on the outside surface of a body where the z-components of force are zero. A state of plane stress also occurs at points within thin plates where the z-dimension of the body is small and the z-components of force are zero.

For plane stress analysis, then, the only components of stress present are σ_x, σ_y, and $\tau_{xy} = \tau_{yx}$. A plane stress element is shown in Fig. 2-25. The stresses shown on the positive and negative faces of the element are shown as positive. For convenience, this state of stress is represented by the two-dimensional sketch shown in Fig. 2-26. However, the three-dimensional element of which the two-dimensional sketch is a plane projection should be kept in mind at all times.

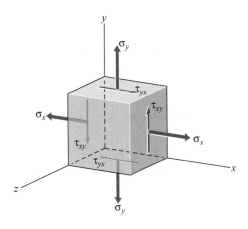

Figure 2-25

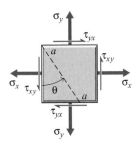

Figure 2-26

Normal and shearing stresses on an arbitrary plane, such as plane a–a in Fig. 2-26, can be obtained by using a free-body diagram and the equations of equilibrium as discussed in Section 2-9.

2-9 THE STRESS TRANSFORMATION EQUATIONS FOR PLANE STRESS

Equations relating the normal and shearing stresses σ_n and τ_{nt} on an arbitrary plane (whose normal is oriented at an angle θ with respect to a reference x-axis) through a point and the known stresses σ_x, σ_y, and $\tau_{xy} = \tau_{yx}$ on the reference planes can be developed using the free-body diagram and the equations of equilibrium. Consider the plane stress situation indicated in Fig. 2-27a (all stresses are positive), where the dotted line a–a represents any plane through the point (all planes are perpendicular to the plane of zero stress—the plane of the paper). Recall that a counterclockwise angle θ is positive, where θ is measured from the positive x-axis to the positive n-axis.

Figure 2-27b is a free-body diagram of a wedge-shaped element in which the areas of the faces are dA for the inclined face (plane a–a), $dA \cos \theta$ for the

2.6

2.7

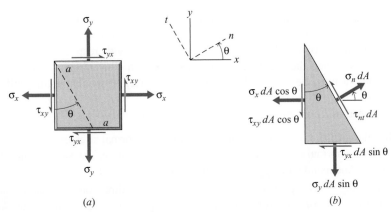

(a)   (b)

Figure 2-27

vertical face, and $dA \sin \theta$ for the horizontal face. The n-axis is perpendicular to the inclined face; the t-axis is parallel to the inclined face. The positive direction for the n-axis is outward from the surface (it is the outward normal), and the positive direction for the t-axis is $90°$ counterclockwise from the n-axis. The xy- and nt-axes shown in Fig. 2-27 are positive. The stresses shown in Fig. 2-27a are multiplied by the areas over which they act, resulting in the free-body diagram shown in Fig. 2-27b. Summing forces in the n-direction gives

$$+ \nearrow \Sigma F_n = 0 : \quad \begin{aligned} & \sigma_n dA - \sigma_x (dA \cos \theta) \cos \theta - \sigma_y (dA \sin \theta) \sin \theta \\ & - \tau_{yx} (dA \sin \theta) \cos \theta - \tau_{xy} (dA \cos \theta) \sin \theta = 0 \end{aligned}$$

from which, since $\tau_{yx} = \tau_{xy}$,

$$\sigma_n = \sigma_x \cos^2 \theta + \sigma_y \sin^2 \theta + 2\tau_{xy} \sin \theta \cos \theta \qquad (2\text{-}12a)$$

or, in terms of the double angle,

$$\begin{aligned} \sigma_n &= \frac{\sigma_x (1 + \cos 2\theta)}{2} + \frac{\sigma_y (1 - \cos 2\theta)}{2} + \frac{2\tau_{xy} (\sin 2\theta)}{2} \\ &= \frac{\sigma_x + \sigma_y}{2} + \frac{\sigma_x - \sigma_y}{2} \cos 2\theta + \tau_{xy} \sin 2\theta \end{aligned} \qquad (2\text{-}12b)$$

Summing forces in the t-direction gives

$$+ \nwarrow \Sigma F_t = 0 : \quad \begin{aligned} & \tau_{nt} dA + \sigma_x (dA \cos \theta) \sin \theta - \sigma_y (dA \sin \theta) \cos \theta \\ & - \tau_{xy} (dA \cos \theta) \cos \theta + \tau_{yx} (dA \sin \theta) \sin \theta = 0 \end{aligned}$$

from which

$$\tau_{nt} = -(\sigma_x - \sigma_y) \sin \theta \cos \theta + \tau_{xy} (\cos^2 \theta - \sin^2 \theta) \qquad (2\text{-}13a)$$

or, in terms of the double angle,

$$\tau_{nt} = -\frac{\sigma_x - \sigma_y}{2} \sin 2\theta + \tau_{xy} \cos 2\theta \qquad (2\text{-}13b)$$

Equations 2-12 and 2-13 provide a means for determining normal and shear stresses for plane stress on any plane whose outward normal is perpendicular to the z-axis and is oriented at an angle θ with respect to the reference x-axis. When these equations are used, the sign conventions used in their development must be rigorously followed; otherwise, erroneous results will be obtained.

These sign conventions can be summarized as follows:

1. Tensile normal stresses are positive; compressive normal stresses are negative. All of the normal stresses shown on Fig. 2-27 are positive. The sign of a normal stress is independent of the coordinate system being used.

2. A shearing stress is positive if it points in the positive direction of the coordinate axis of the second subscript when it is acting on a surface whose outward

normal is in a positive direction of the coordinate axis of the first subscript. Similarly, if the outward normal of the surface is in a negative direction, then a positive shearing stress points in the negative direction of the coordinate axis of the second subscript. All of the shearing stresses shown on Fig. 2-27 are positive. Shearing stresses pointing in the opposite directions would be negative. The sign of a shearing stress depends on the coordinate system being used.

3. An angle measured counterclockwise from the reference positive x-axis is positive. Conversely, angles measured clockwise from the reference x-axis are negative.

4. The (n, t, z) axes have the same order as the (x, y, z) axes. Both sets of axes form a right-hand coordinate system.

Example Problem 2-8   At a point in a structural member subjected to plane stress there are normal and shearing stresses on horizontal and vertical planes through the point, as shown in Fig. 2-28a. Use the stress transformation equations to determine:

2.8

(a) The normal and shearing stresses on plane a–b.

(b) The normal and shearing stresses on plane c–d, which is perpendicular to plane a–b.

(c) Show the stresses on planes a–b and c–d using a small element.

2.9

SOLUTION
First define the x–y and n–t (if needed) directions if they have not been specified. On the basis of the axes shown in Fig. 2-28b and the established sign conventions, σ_x is positive, whereas σ_y and τ_{xy} are negative. Thus, the given values for use in Eqs. 2-12 and 2-13 are

$$\sigma_x = +80\,\text{MPa} \qquad \sigma_y = -100\,\text{MPa}$$
$$\tau_{xy} = \tau_{yx} = -60\,\text{MPa}$$

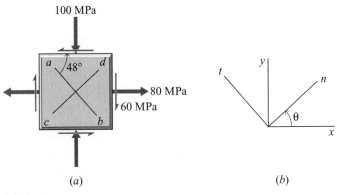

(a)

(b)

Figure 2-28(a–b)

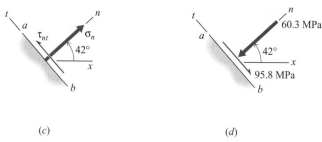

(c)   (d)

Figure 2-28(c–d)

(a) The angle θ for plane a–b is $+42°$, as shown in Fig. 2-28c. Thus, from Eqs. 2-12 and 2-13,

$$\sigma_n = \sigma_x \cos^2 \theta + \sigma_y \sin^2 \theta + 2\tau_{xy} \sin \theta \cos \theta$$
$$= 80 \cos^2(+42°) + (-100) \sin^2(+42°)$$
$$+ 2(-60) \sin(+42°) \cos(+42°)$$
$$= -60.26 \text{ MPa} \cong 60.3 \text{ MPa (C)} \qquad\qquad \textbf{Ans.}$$

$$\tau_{nt} = -(\sigma_x - \sigma_y) \sin \theta \cos \theta + \tau_{xy}(\cos^2 \theta - \sin^2 \theta)$$
$$= -[80 - (-100)] \sin(+42°) \cos(+42°)$$
$$+ (-60)[\cos^2(+42°) - \sin^2(+42°)]$$
$$= -95.78 \text{ MPa} \cong -95.8 \text{ MPa} \qquad\qquad \textbf{Ans.}$$

The results are shown in Fig. 2-28d.

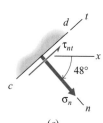

(e)

Figure 2-28(e)

(b) The angle θ for plane c–d is $-48°$ (or $+132°$), as shown in Fig. 2-28e. Thus, from Eqs. 2-12 and 2-13,

$$\sigma_n = \sigma_x \cos^2 \theta + \sigma_y \sin^2 \theta + 2\tau_{xy} \sin \theta \cos \theta$$
$$= 80 \cos^2(-48°) + (-100) \sin^2(-48°)$$
$$+ 2(-60) \sin(-48°) \cos(-48°)$$
$$= +40.26 \text{ MPa} \cong 40.3 \text{ MPa (T)} \qquad\qquad \textbf{Ans.}$$

$$\tau_{nt} = -(\sigma_x - \sigma_y) \sin \theta \cos \theta + \tau_{xy}(\cos^2 \theta - \sin^2 \theta)$$
$$= -[80 - (-100)] \sin(-48°) \cos(-48°)$$
$$+ (-60)[\cos^2(-48°) - \sin^2(-48°)]$$
$$= +95.78 \text{ MPa} \cong +95.8 \text{ MPa} \qquad\qquad \textbf{Ans.}$$

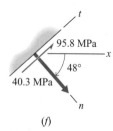

(f)

Figure 2-28(f)

The results are shown in Fig. 2-28f.

Since planes a–b and c–d are orthogonal, the shearing stresses on the two planes must be equal in magnitude to satisfy Eq. 2-11. Also, one of the stresses must tend to produce a clockwise rotation of the element while the other tends to produce a counterclockwise rotation. With the coordinate systems shown in Figs. 2-28d and 2-28f, this means that the shearing stresses calculated using Eq. 2-13 will have opposite signs on the two planes.

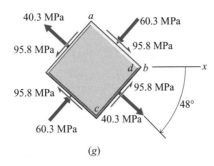

40.3 MPa
a
60.3 MPa
95.8 MPa
95.8 MPa
d
b
x
95.8 MPa
95.8 MPa
48°
c
40.3 MPa
60.3 MPa
(g)

Figure 2-28(g)

(c) The stresses acting on x–y planes through the point are shown in Fig. 2-28*a*. Since planes *a*–*b* and *c*–*d* through the point are perpendicular, the stresses on these planes, using the results of part (b), may be drawn on an element as shown in Fig. 2-28*g*.

Example Problem 2-9

At a point on the outside surface of a thin-walled pressure vessel (see Fig. 2-29*a*), stresses are 8000 psi (T) and zero shear on a horizontal plane and 4000 psi (C) and zero shear on a vertical plane, as shown in Figs. 2-29*b* and *c*. Determine the stresses at this point on plane *b*–*b* having a slope of 3 vertical to 4 horizontal.

SOLUTION

A small "element" is shown in Fig. 2-29*b*. There are no loads (or stresses) on the z-surface, and thus a state of plane stress exists. This state of stress is depicted in Fig. 2-29*c*. Figure 2-29*d* shows the x–y stresses, the n–t stresses on plane *b*–*b* (assumed positive), the positive coordinate axes, and the angle between the positive x- and n-axes ($\theta = -53.13°$). The given values of stresses for use in the stress transformation equations are

$$\sigma_x = -4000 \text{ psi}, \qquad \sigma_y = +8000 \text{ psi}, \qquad \text{and} \qquad \tau_{xy} = 0 \text{ psi}$$

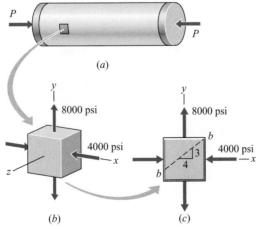

P
P
(a)
y
8000 psi
4000 psi
x
z
(b)
y
8000 psi
b
3
4
b
4000 psi
x
(c)

Figure 2-30(a–c)

y
8000 psi
t
τ_{nt}
4000 psi
x
σ_n
53.13°
n
(d)

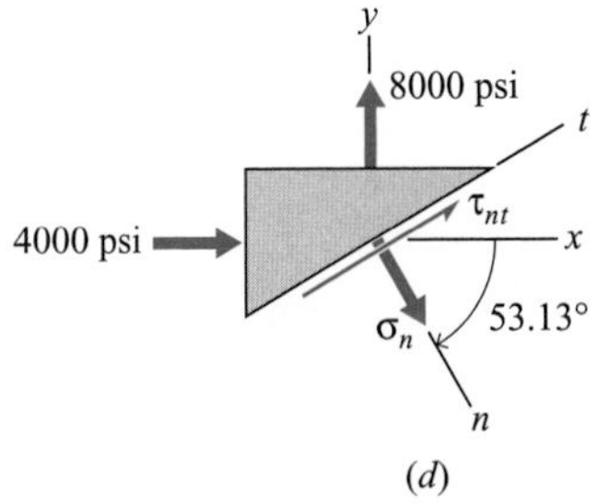

Figure 2-29(d)

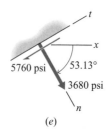

Figure 2-29(e)

Thus, from Eqs. 2-12a and 2-13a,

$$\sigma_n = \sigma_x \cos^2 \theta + \sigma_y \sin^2 \theta + 2\tau_{xy} \sin \theta \cos \theta$$
$$= (-4000) \cos^2(-53.13°) + (+8000) \sin^2(-53.13°) + 0$$
$$= +3680 \text{ psi} = 3680 \text{ psi (T)} \qquad \textbf{Ans.}$$

$$\tau_{nt} = -(\sigma_x - \sigma_y)\sin \theta \cos \theta + \tau_{xy}(\cos^2 \theta - \sin^2 \theta)$$
$$= -[(-4000) - (+8000)] \sin(-53.13°) \cos(-53.13°) + 0$$
$$= -5760 \text{ psi} \qquad \textbf{Ans.}$$

These results are shown in Fig. 2-29e.

Example Problem 2-10   The block shown in Fig. 2-30a has a 200 × 100-mm rectangular cross section. Determine the normal and shearing stresses on plane a–a when $\phi = 36°$ and $P = 600$ kN.

SOLUTION

Since the block is subjected to an axial load, the problem could be solved using Eqs. 2-7 and 2-8. Instead, the normal and shearing stresses on plane a–a will be found using the stress transformation equations, Eqs. 2-12a and 2-13a. The stress transformation equations are valid for any type of loading, if the state of stress is plane stress.

First, the x–y stresses are found. The block is sectioned as shown in Fig. 2-30b, and the positive axes are drawn. The normal stress σ_x is found using Eq. 2-2 as

$$\sigma_x = \frac{F}{A} = \frac{-600(10^3)}{(200)(100)(10^{-6})} = -30(10^6)\frac{\text{N}}{\text{m}^2}$$
$$= -30 \text{ MPa}$$

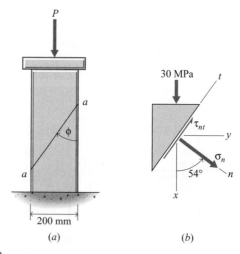

(a)   (b)

Figure 2-30(a–b)

The stresses σ_y and τ_{xy} are zero, as there are no loads to produce these stresses. The $x-y$ stresses for use in the stress transformation equations are $\sigma_x = -30$ MPa, $\sigma_y = 0$ MPa, and $\tau_{xy} = 0$ MPa. The stresses on plane $a-a$ (assumed positive), the $x-y$ stresses, positive coordinate axes, and the angle $\theta = +54°$ are shown in Fig. 2.30b. Equations 2-12a and 2-13a give

$$\sigma_n = \sigma_x \cos^2 \theta + \sigma_y \sin^2 \theta + 2\tau_{xy} \sin \theta \cos \theta$$
$$= (-30) \cos^2 (+54°) + 0 + 0$$
$$= -10.36 \text{ MPa} = 10.36 \text{ MPa (C)} \qquad \textbf{Ans.}$$
$$\tau_{nt} = -(\sigma_x - \sigma_y)\sin \theta \cos \theta + \tau_{xy} (\cos^2 \theta - \sin^2 \theta)$$
$$= -[(-30) - 0] \sin (+54°) \cos (+54°) + 0$$
$$= +14.27 \text{ MPa} \qquad \textbf{Ans.}$$

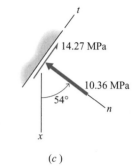

(c)

Figure 2-30(c)

These results are shown in Fig. 2-30c.

PROBLEMS

MecMovie Activities and Problems

MM2.9 Correct angle for stress transformations. Theory; Concept checkpoints. Easy method for finding the proper value of θ for use in the stress transformation equations.

MM2.10 Sign, sign, everywhere a sign. Game. Game focused on the correct sign conventions needed for the stress transformation equations.

Introductory Problems

2-45* At a point in a thin plate, there are normal stresses of 20 ksi (T) on a vertical plane and 10 ksi (C) on a horizontal plane, as shown in Fig. P2-45. Determine the normal and shear stresses at this point on the inclined plane $a-b$ shown in the figure.

2-46* At a point in a stressed body, there are normal stresses of 95 MPa (T) on a vertical plane and 125 MPa (T) on a horizontal plane, as shown in Fig. P2-46. Determine the normal and shear stresses at this point on the inclined plane $a-b$ shown in the figure.

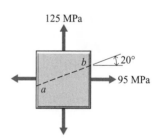

Figure P2-46

2-47 The stresses shown in Fig. P2-47 act at a point on the surface of a circular shaft that is subjected to a twisting moment M as shown. Determine the normal and shear stresses at this point on the inclined plane $a-b$ shown in the figure.

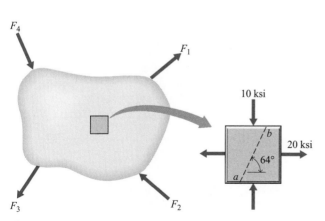

Figure P2-45

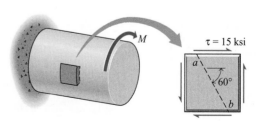

Figure P2-47

2-48* The stresses shown in Fig. P2-48 act at a point in a stressed body. Determine the normal and shear stresses at this point on the inclined plane a–b shown in the figure.

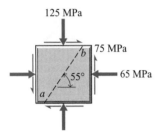

Figure P2-48

2-49 The stresses shown in Fig. P2-49 act at a point on the surface of a thin-walled pressure vessel that is subjected to an internal pressure, an axial load P, and a torque T. Determine the normal and shear stresses at this point on the inclined plane a–b shown in the figure.

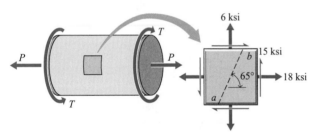

Figure P2-49

2-50 The stresses shown in Fig. P2-50 act at a point on the surface of a cantilever beam. Determine the normal and shear stresses at this point on the inclined plane a–b shown in the figure.

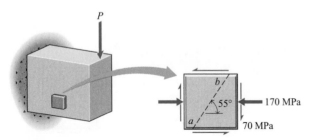

Figure P2-50

Intermediate Problems

2-51* An axial load P is applied to a timber block with a 4×4 -in. square cross section, as shown in Fig. P2-51. Determine the normal and shear stresses on the planes of the grain if $P = 5000$ lb. Use the stress transformation equations.

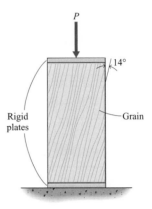

Figure P2-51

2-52* A steel bar with a butt-welded joint, as shown in Fig. P2-52, will be used to carry an axial tensile load of 400 kN. If the thickness of the bar is 40 mm, determine the normal and shear stress on the plane of the weld. Use the stress transformation equations.

Figure P2-52

2-53 Specifications for the $3 \times 3 \times 21$-in. block shown in Fig. P2-53 require that the normal and shear stresses on plane a–a not exceed 800 psi and 500 psi, respectively. Determine the maximum load P that can be applied without exceeding the specifications. Use the stress transformation equations.

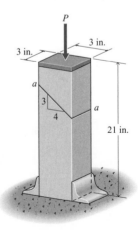

Figure P2-53

2-54 The shearing stress on plane a–b of the 100×200-mm rectangular block shown in Fig. P2-54 is 15 MPa when the axial load P is applied. If the angle ϕ is $35°$, determine

a. The load P.
b. The normal stress on plane a–b.

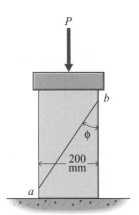

Figure P2-54

2-55* A timber block with a square cross section will be used to support a compressive load of 32 kip, as shown in Fig. P2-55. Determine the size of the block required if the shear stress parallel to the grain is not to exceed 800 psi and the compressive stress perpendicular to the grain is not to exceed 3500 psi. Use the stress transformation equations.

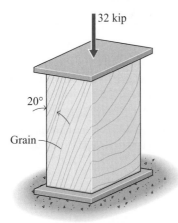

Figure P2-55

2-56 Determine the maximum axial load P that can be applied to the wood compression block shown in Fig. P2-56 if specifications require that the shear stress parallel to the grain not

exceed 5.25 MPa and the compressive stress perpendicular to the grain not exceed 13.60 MPa. Use the stress transformation equations.

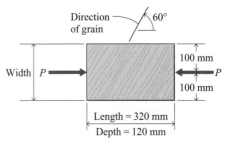

Figure P2-56

Challenging Problems

2-57* At a point in a machine component, the normal and shear stresses on an inclined plane are 4800 psi (T) and 1500 psi, respectively, as shown in Fig. P2-57. The normal stress on a vertical plane through the point is zero. Determine

a. The shear stresses on horizontal and vertical planes.
b. The normal stress on a horizontal plane.

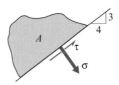

Figure P2-57

2-58* The stresses on horizontal and vertical planes at a point on the outside surface of a solid circular bar subjected to an axial load P and a torsional load T are shown in Fig. P2-58. The normal stress on the inclined plane a–b is 15 MPa (T). Determine

a. The normal stress σ_x on the vertical plane.
b. The magnitude and direction of the shear stress on the inclined plane a–b.

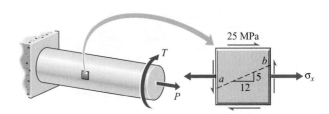

Figure P2-58

2-59 The stresses shown in Fig. P2-59a act at a point on the free surface of a stressed body. Determine

a. The normal and shear stresses at this point on the inclined plane a–b shown in the figure.
b. The normal stresses σ_n and σ_t and the shear stress τ_{nt} if they act on the stress element shown in Fig. P2-59b.

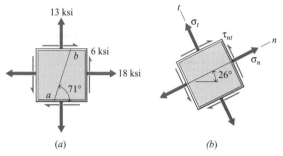

| (a) | (b) |

Figure P2-59

2-60* At a point in a stressed body, the stresses on two perpendicular planes are as shown in Fig. P2-60. Determine

a. The stresses on plane a–a.
b. The stresses on horizontal and vertical planes at the point.

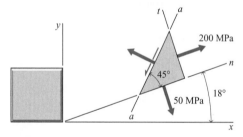

Figure P2-60

2-61 The stresses on horizontal and vertical planes at a point on the outside surface of a solid circular shaft subjected to an axial load P and a torque T are shown in Fig. P2-61. The normal stress on plane a–a at this point is 8000 psi (T). Determine

a. The magnitude of the shearing stresses τ_h and τ_v.
b. The magnitude and direction of the shearing stress on the inclined plane a–a.

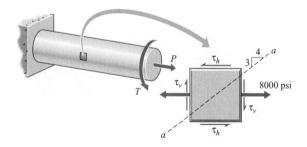

Figure P2-61

2-62 The stresses on horizontal and vertical planes at a point are shown in Fig. P2-62. The normal stress on the plane a–b is 15 MPa (T). Determine

a. The normal stress σ_x on the vertical plane.
b. The magnitude and direction of the shearing stress on the inclined plane a–b.

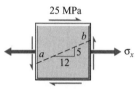

Figure P2-62

2-63* The thin-walled cylindrical pressure vessel shown in Fig. P2-63 was constructed by wrapping a thin steel plate into a helix that forms an angle of $\theta = 35°$ with respect to a transverse plane through the cylinder and butt-welding the resulting seam. In a thin-walled cylindrical pressure vessel, the normal stress σ_y on a horizontal plane through a point on the surface of the vessel is twice as large as the normal stress σ_x on a vertical plane through the point, and the shear stresses on both the horizontal and vertical planes are zero. If the stresses in the weld material on the plane of the weld must be limited to 10 ksi in tension and 7 ksi in shear, determine the maximum normal stress σ_x permitted in the vessel.

Figure P2-63

2-64 Known stresses at point A in a structural member (see Fig. P2-64) are 125-MPa tension and zero shear on plane b–b and 225-MPa compression on plane c–c. Determine

a. The stresses on a vertical plane through the point.
b. The stresses on a horizontal plane through the point.

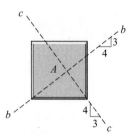

Figure P2-64

2-65 A steel bar with a 4×1-in. rectangular cross section is being used to transmit an axial tensile load, as shown in Fig. P2-65. Normal and shear stresses on plane a–b of the bar are 12-ksi tension and 9-ksi shear. Determine the angle θ and the applied load P.

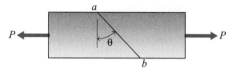

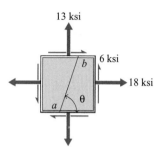

Figure P2-67

Figure P2-65

Computer Problems

2-66 The stresses on horizontal and vertical planes at a point on the outside surface of a solid circular shaft subjected to an axial load P and a torque T are shown in Fig. P2-66.

a. Calculate and graph the normal stress σ_n and the shearing stress τ_{nt} on the inclined plane a–b as a function of the angle θ ($0° \le \theta \le 180°$).
b. For what angle θ is the normal stress a maximum? A minimum? What is the value of the shear stress on these planes?
c. For what angle θ is the shear stress a maximum? A minimum? What is the value of the normal stress on these planes?

a. Calculate and graph the normal stress σ_n and the shearing stress τ_{nt} on the inclined plane a–b as a function of the angle θ ($0° \le \theta \le 180°$).
b. For what angle θ is the normal stress a maximum? A minimum? What is the value of the shear stress on these planes?
c. For what angle θ is the shear stress a maximum? A minimum? What is the value of the normal stress on these planes?

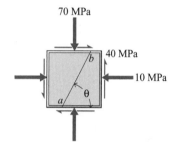

Figure P2-68

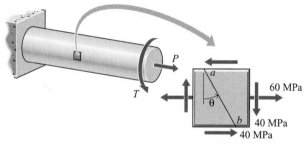

Figure P2-66

2-67 The stresses shown in Fig. P2-67 act at a point on the free surface of a stressed body.

2-68 The stresses shown in Fig. P2-68 act at a point on the free surface of a stressed body. Calculate the normal stress σ_n and the shearing stress τ_{nt} on the inclined plane a–b as a function of the angle θ ($0° \le \theta \le 180°$). For each angle θ, graph the negative of the shearing stress ($-\tau_{nt}$, vertical axis) as a function of the normal stress (σ_n, horizontal axis). On your graph, clearly identify the points associated with the angles $\theta = 0°$, 30°, 45°, 60°, 90°, 120°, 135°, 150°, and 180°.

2-10 PRINCIPAL STRESSES AND MAXIMUM SHEARING STRESS—PLANE STRESS

The transformation equations for plane stress ($\sigma_z = \tau_{zx} = \tau_{xz} = \tau_{zy} = \tau_{yz} = 0$), Eqs. 2.12 and 2-13, provide a means for determining the normal stress σ_n and the shearing stress τ_{nt} on different planes through a point in a stressed body. As an example, consider the state of stress shown on the stress element in Fig. 2-31a, which acts at a point on the free surface of a machine component or structural member. As the angle θ varies ($0° \le \theta \le 360°$), the normal stress σ_n and the

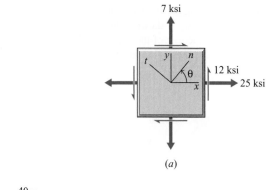

(a)

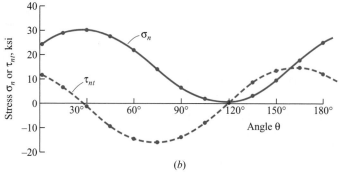

(b)

Figure 2-31

shearing stress τ_{nt} on the different planes vary as shown in Fig. 2-31b. Note that, for $\theta = 0°$, $\sigma_n = \sigma_x$ and $\tau_{nt} = \tau_{xy}$; and for $\theta = 90°$, $\sigma_n = \sigma_y$ and $\tau_{nt} = -\tau_{yx} = -\tau_{xy}$.

For design purposes, critical stresses at the point are usually the maximum tensile stress and the maximum shearing stress.

For a bar under axial load, the planes on which maximum normal stresses and maximum shearing stresses act are known from the results of Section 2-6. For more complicated forms of loading, these stresses can be determined by plotting curves similar to those shown in Fig. 2-31b for each different state of stress encountered, but this process is time-consuming and inefficient. Therefore, more general methods for finding the critical stresses have been developed.

The transformation equations for plane stress developed previously are as follows:

For normal stress σ_n,

$$\sigma_n = \frac{\sigma_x + \sigma_y}{2} + \frac{\sigma_x - \sigma_y}{2} \cos 2\theta + \tau_{xy} \sin 2\theta \qquad (2\text{-}12b)$$

For shear stress τ_{nt},

$$\tau_{nt} = -\frac{\sigma_x - \sigma_y}{2} \sin 2\theta + \tau_{xy} \cos 2\theta \qquad (2\text{-}13b)$$

Maximum and minimum values of σ_n occur at values of θ for which $d\sigma_n/d\theta$ is equal to zero. Differentiation of σ_n with respect to θ yields

$$\frac{d\sigma_n}{d\theta} = -(\sigma_x - \sigma_y) \sin 2\theta + 2\tau_{xy} \cos 2\theta \qquad (a)$$

Setting Eq. (*a*) equal to zero and solving gives

$$\tan 2\theta_p = \frac{2\tau_{xy}}{\sigma_x - \sigma_y} \qquad (2\text{-}14)$$

Note that the expression for $d\sigma_n/d\theta$ from Eq. (*a*) is numerically twice the value of the expression for τ_{nt} from Eq. 2-13*b*. Consequently, the shearing stress is zero on planes experiencing maximum and minimum values of normal stress. Planes free of shear stress are known as *principal planes*. Normal stresses occurring on principal planes are known as *principal stresses*. The values of θ_p from Eq. 2-14 give the orientations of two principal planes. A third principal plane for the plane stress state has an outward normal in the *z*-direction. For a given set of values of σ_x, σ_y, and τ_{xy}, there are two values of $2\theta_p$ differing by 180° and, consequently, two values of θ_p that are 90° apart. This proves that the principal planes are normal to each other.

When τ_{xy} and $(\sigma_x - \sigma_y)$ have the same sign, $\tan 2\theta_p$ is positive and one value of θ_p is between 0° and 90°, with the other value 180° greater, as shown in Fig. 2-32. Consequently, one value of θ_p is between 0° and 45°, and the other one is 90° greater. In the first case, both $\sin 2\theta_p$ and $\cos 2\theta_p$ are positive, and in the second case both are negative. When these functions of $2\theta_p$ are substituted into Eq. 2-12*b*, two in-plane principal stresses σ_{p1} and σ_{p2} are found to be

$$\sigma_{p1,p2} = \frac{\sigma_x + \sigma_y}{2} \pm \sqrt{\left(\frac{\sigma_x - \sigma_y}{2}\right)^2 + \tau_{xy}^2} \qquad (2\text{-}15)$$

Equation 2-15 gives the two principal stresses in the *xy*-plane, and the third one is $\sigma_{p3} = \sigma_z = 0$. Equation 2-14 gives the angles θ_p and $\theta_p + 90°$ between the *x*- (or *y*-) plane and the mutually perpendicular planes on which the principal stresses act. In order to determine which of the principal stresses (found using Eq. 2-15) acts on which of the principal planes (found using Eq. 2-14), substitute one of the values of θ_p into the stress transformation equation Eq. 2-12*a* (or Eq. 2-12*b*). Since θ_p is used, the calculated value of σ_n must be one of the principal stresses given by Eq. 2-15, and it acts on the surface whose normal points in the direction θ_p. Because principal planes are perpendicular, the other principal stress given by Eq. 2-15 acts on the surface whose normal points in the direction $\theta_p \pm 90°$. This procedure is illustrated in Example Problem 2-11.

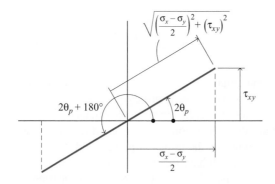

Figure 2-32

Note that if one or both of the principal stresses from Eq. 2-15 is negative, the algebraic maximum stress can have a smaller absolute value than the "minimum" stress.

The maximum in-plane shearing stress τ_p occurs on planes located by values of θ where $d\tau_{nt}/d\theta$ is equal to zero. Differentiation of Eq. 2-13b yields

$$\frac{d\tau_{nt}}{d\theta} = -(\sigma_x - \sigma_y)\cos 2\theta - 2\tau_{xy}\sin 2\theta$$

When $d\tau_{nt}/d\theta$ is equated to zero, the value of θ_τ is given by the expression

$$\tan 2\theta_\tau = -\frac{(\sigma_x - \sigma_y)}{2\tau_{xy}} \tag{2-16}$$

2.10

where θ_τ locates the planes of maximum in-plane shearing stress. Comparison of Eqs. 2-16 and 2-14 reveals that the two tangents are negative reciprocals. Therefore, the two angles $2\theta_p$ and $2\theta_\tau$ differ by 90°, and θ_p and θ_τ are 45° apart. This means that the planes on which the maximum in-plane shearing stresses occur are 45° from the principal planes. The maximum in-plane shearing stresses are found by substituting values of angle functions obtained from Eq. 2-16 in Eq. 2-13b. The results are

$$\tau_p = \pm\sqrt{\left(\frac{\sigma_x - \sigma_y}{2}\right)^2 + \tau_{xy}^2} \tag{2-17}$$

Equation 2-17 has the same magnitude as the second term of Eq. 2-15. Equation 2-16 gives two perpendicular planes of maximum in-plane shearing stress. The shearing stresses on these two planes have the same magnitude but opposite signs (Eq. 2-17). To determine which sign in Eq. 2-17 corresponds to each of the surfaces found using Eq. 2-16, substitute one value of θ_τ into the stress transformation equation for shearing stress (Eq. 2-13a or 2-13b). Since θ_τ is used, the calculated value of τ_{nt} must be one of the shear stresses given by Eq. 2-17, and it acts on the surface whose normal points in the direction θ_τ. This procedure is illustrated in Example Problem 2-11.

A useful relation between the principal stresses and the maximum in-plane shearing stress is obtained from Eqs. 2-15 and 2-17 by subtracting the values for the two in-plane principal stresses and substituting the value of the radical from Eq. 2-17. The result is

$$\tau_p = \frac{(\sigma_{p1} - \sigma_{p2})}{2} \tag{2-18}$$

or, in words, the maximum value of τ_{nt} (τ_p) is equal in magnitude to one-half the difference between the two in-plane principal stresses.

In general, when stresses act in three directions it can be shown (see Section 2-12 that there are three orthogonal planes on which the shearing stress is zero. These planes are known as the principal planes, and the stresses acting on them (the principal stresses) will have three values: one maximum, one minimum, and a third stress between the other two. The maximum shearing stress, τ_{max} on any plane that could be passed through the point, is one-half the difference between

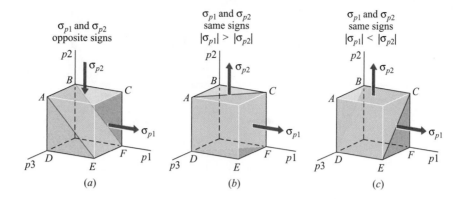

Figure 2-33

the maximum and minimum principal stresses and acts on planes that bisect the angles between the planes of the maximum and minimum normal stresses.

$$\tau_{max} = \frac{\sigma_{max} - \sigma_{min}}{2} \qquad (2\text{-}19)$$

When a state of plane stress exists, one of the principal stresses is zero. If the values of σ_{p1} and σ_{p2} from Eq. 2-15 have the same sign, then the third principal stress, σ_{p3} equals zero, will be either the maximum or the minimum normal stress. Thus, the maximum shearing stress may be

$$(\sigma_{p1} - \sigma_{p2})/2 \qquad (\sigma_{p1} - 0)/2 \quad \text{or} \quad (0 - \sigma_{p2})/2$$

depending on the relative magnitudes and signs of the principal stresses. These three possibilities are illustrated in Fig. 2-33, in which one of the two orthogonal planes on which the maximum shearing stress acts is shaded for each example.

The direction of the maximum shearing stress can be determined by drawing a wedge-shaped block with two sides parallel to the planes having the maximum and minimum principal stresses, and with the third side at an angle 45° with the other two sides. The direction of the maximum shearing stress must oppose the larger of the two principal stresses.

Another useful relation between the principal stresses and the normal stresses on the orthogonal planes shown in Fig. 2-34 is obtained by adding the values for the two principal stresses, as given by Eq. 2-15. Thus,

$$\sigma_{p1} + \sigma_{p2} = \sigma_x + \sigma_y \qquad (2\text{-}20)$$

or, in words, for plane stress, the sum of the normal stresses on any two orthogonal planes through a point in a body is a constant or invariant.

In the preceding discussion, "maximum" and "minimum" stresses were considered algebraic quantities, and it has already been pointed out that the minimum algebraic stress may have a larger magnitude than the maximum stress.

However, in the application to engineering problems (which includes the problems in this book), the term "maximum" will always refer to the largest absolute value (largest magnitude).

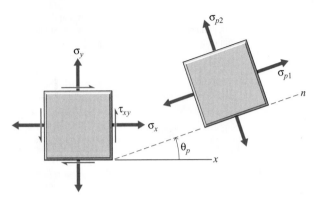

Figure 2-34

Applications of the formulas and procedures developed in this section are illustrated by the following examples.

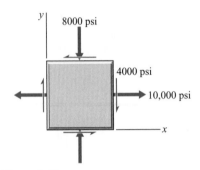

Figure 2-35

■ **Example Problem 2-11**   At a point in a structural member subjected to plane stress there are normal and shearing stresses on horizontal and vertical planes through the point, as shown on the element in Fig. 2-35.

(a) Determine the principal stresses and the maximum shearing stress at the point.
(b) Locate the planes on which these stresses act and show the stresses on the planes on which they act.
(c) Sketch a triangular stress element showing the principal stresses and the maximum shearing stress (and the associated normal stress).

SOLUTION

(a) On the basis of the axes shown in Fig. 2-35 and the established sign conventions, σ_x is positive, whereas σ_y and τ_{xy} are negative. For use in Eqs. 2-14 and 2-15, the given values are

$$\sigma_x = +10,000 \text{ psi} \quad \sigma_y = -8000 \text{ psi} \quad \tau_{xy} = -4000 \text{ psi}$$

When these values are substituted in Eq. 2-15, the principal stresses are found to be

$$\sigma_{p1,p2} = \frac{\sigma_x + \sigma_y}{2} \pm \sqrt{\left(\frac{\sigma_x - \sigma_x}{2}\right)^2 + \tau_{xy}^2}$$

$$= \frac{10,000 + (-8000)}{2} \pm \sqrt{\left(\frac{10,000 - (-8000)}{2}\right)^2 + (-4000)^2}$$

$$= 1000 \pm 9849$$

$$\sigma_{p1} = 1000 + 9849 = 10,849 \text{ psi} \cong 10,850 \text{ psi (T)} \qquad \textbf{Ans.}$$

$$\sigma_{p2} = 1000 - 9849 = -8849 \text{ psi} \cong 8850 \text{ psi (C)} \qquad \textbf{Ans.}$$

$$\sigma_{p3} = \sigma_z = 0 \qquad \textbf{Ans.}$$

The maximum in-plane shearing stress is given by Eq. 2-17, which is simply the last term in Eq. 2-15, and is $\tau_p = 9850$ psi. The maximum shearing stress is given by Eq. 2-19 as

$$\tau_{max} = \frac{\sigma_{max} - \sigma_{min}}{2}$$

$$= \frac{10,849 - (-8849)}{2} = 9849 \text{ psi} \cong 9850 \text{ psi} \qquad \textbf{Ans.}$$

Since σ_{p1} and σ_{p2} have opposite signs, the values of the maximum in-plane shearing stress (τ_p) and the maximum shearing stress (τ_{max}) are equal.

▶ The sum of the normal stresses on any two orthogonal planes is a constant for plane stress; $\sigma_x + \sigma_y = 10,000 + (-8000) = \sigma_{p1} + \sigma_{p2} = 10,850 + (-8850) = 2000$ psi. Also, the normal stress on the planes of maximum shear is equal to the average of the normal stresses $\sigma_n = 1000 = (\sigma_x + \sigma_y)/2 = (\sigma_{p1} + \sigma_{p2})/2$.

(b) When the given data are substituted in Eq. 2-14, the results are

$$\tan 2\theta_p = \frac{2\tau_{xy}}{\sigma_x - \sigma_y} = \frac{2(-4000)}{10,000 - (-8000)} = -0.4444$$

from which

$$2\theta_p = -23.96°, \quad 180° + (-23.96°) = +156.04°, \cdots$$

and

$$\theta_p = -11.98°, \ +78.02°, \cdots$$
$$\cong 11.98° \downarrow \text{ and } 78.0° \uparrow \qquad \textbf{Ans.}$$

Which principal stress is associated with which angle $(\theta_p = -11.98°$ or $= +78.02°)$ is determined by substituting the angles into the stress transformation equations. When $\theta = \theta_p = -11.98°$, Eq. 2-12a gives

$$\sigma_n = \sigma_x \cos^2 \theta + \sigma_y \sin^2 \theta + 2\tau_{xy} \sin \theta \cos \theta$$
$$= 10,000 \cos^2(-11.98°) + (-8000) \sin^2(-11.98°)$$
$$\qquad + 2(-4000) \sin(-11.98°) \cos(-11.98°)$$
$$= \sigma_{p1} = 10,849 \text{ psi} \cong 10,850 \text{ psi (T)}$$

The result is shown in Fig. 2-36a.

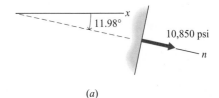

(a)

Figure 2-36(a)

When $\theta = \theta_p = +78.02°$, Eq. 2-12a gives

$$\sigma_n = \sigma_x \cos^2 \theta + \sigma_y \sin^2 \theta + 2\tau_{xy} \sin \theta \cos \theta$$
$$= 10,000 \cos^2(+78.02°) + (-8000) \sin^2(+78.02°)$$
$$\qquad + 2(-4000) \sin(+78.02°) \cos(+78.02°)$$
$$= \sigma_{p2} = -8849 \text{ psi} \cong 8850 \text{ psi (C)}$$

The result is shown in Fig. 2-36b.

The principal stresses shown on the principal planes in Figs. 2-36a and b are shown on a stress element in Fig. 2-36c. The maximum in-plane shear

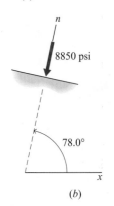

(b)

Figure 2-36(b)

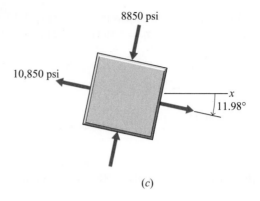

(c)

Figure 2-36(c)

stress occurs on a surface oriented at 45° to the principal directions. When $\theta = -11.98° + 45° = +33.02°$, Eqs. 2-12a and 2-13a give

$$\sigma_n = \sigma_x \cos^2 \theta + \sigma_y \sin^2 \theta + 2\tau_{xy} \sin \theta \cos \theta$$
$$= 10{,}000 \cos^2(+33.02°) + (-8000) \sin^2(+33.02°)$$
$$+ 2(-4000) \sin(+33.02°) \cos(+33.02°)$$
$$= +999.6 \text{ psi} \cong 1000 \text{ psi} \,(\text{T})$$

and

$$\tau_{nt} = -(\sigma_x - \sigma_y) \sin \theta \cos \theta + \tau_{xy}[\cos^2 \theta - \sin^2 \theta]$$
$$= -[10{,}000 - (-8000)] \sin(+33.02°) \cos(+33.02°)$$
$$+ (-4000)[\cos^2(+33.02°) - \sin^2(+33.02°)]$$
$$= -9849 \text{ psi} \cong -9850 \text{ psi}$$

These results are shown in Fig. 2-37.

Substituting $\theta = +33.02° + 90° = 123.02°$ into Eqs. 2-12a and 2-13a gives

$$\sigma_n = +1000 \text{ psi} \qquad \tau_{nt} = +9850 \text{ psi}$$

That is, the normal stresses on the perpendicular planes of maximum shearing stresses are equal both in magnitude and sign, whereas the shearing stresses have equal magnitude but opposite sign. These results are shown on the stress element in Fig. 2-38.

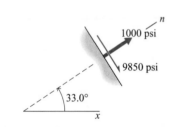

Figure 2-37

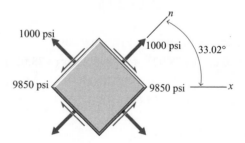

Figure 2-38

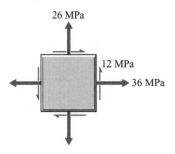

Figure P2-76

a. Determine the principal stresses and the maximum shearing stress at the point.
b. Show the stresses of part a on a three-dimensional triangular stress element.

2-77 Normal and shear stresses on horizontal and vertical planes through a point on the free surface of a structural member are shown in Fig. P2-77. Determine, and show on a triangular stress element, the principal stresses and maximum shear stress at the point.

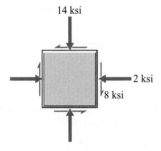

Figure P2-77

2-78 The stresses shown in Fig. P2-78 act on horizontal and vertical planes at a point on the free surface of a cantilever beam. Determine, and show on a triangular stress element, the principal stresses and the maximum shearing stress at the point.

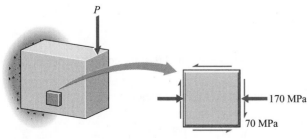

Figure P2-78

2-79 The stresses shown in Fig. P2-79 act on horizontal and vertical planes at a point on the free surface of a circular member that is subjected to a twisting moment (torque) M. Determine, and show on a triangular stress element, the principal stresses and the maximum shearing stress at the point.

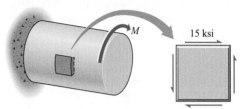

Figure P2-79

Challenging Problems

2-80* The stresses on the outside surface of a thin-walled pressure vessel are shown in Fig. P2-80. The vessel is constructed by wrapping a steel plate into a spiral and butt-welding the mating edges of the plate. If the butt-welded seams form an angle of $33°$ with the longitudinal axis of the vessel,

a. Determine the normal stress perpendicular to the weld a–b and the shear stress parallel to the weld and show the results on a sketch.
b. Determine the principal stresses and the maximum shearing stress at the point.
c. Show the results of part b on a triangular stress element.

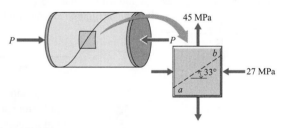

Figure P2-80

2-81* The stresses shown in Fig. P2-81 act at a point on the free surface of a stressed body. The principal stresses at the point are 20 ksi (C) and 12 ksi (T). Determine the unknown normal stresses on the horizontal and vertical planes and the angle θ_p between the x-axis and the maximum tensile stress at the point.

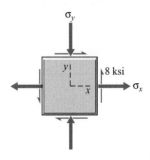

Figure P2-81

2-82 At a point on the free surface of a stressed body, a normal stress of 75 MPa (T) and an unknown negative shear stress exist on a horizontal plane, as shown in Fig. P2-82. One principal stress at the point is 200 MPa (T). The maximum in-plane shear stress at the point has a magnitude of 85 MPa. Determine the unknown stresses on the vertical plane, the unknown principal stress, and the maximum shear stress at the point.

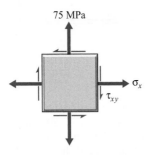

Figure P2-82

2-83 The principal compressive stress on a vertical plane through a point in a wooden block is equal to four times the principal compressive stress on a horizontal plane, as shown in Fig. P2-83. The plane of the grain is 30° clockwise from the vertical plane. If the normal and shear stresses on the plane of the grain must not exceed 300 psi (C) and 125 psi shear, determine the maximum allowable compressive stress on the horizontal plane.

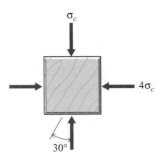

Figure P2-83

2-84* The stresses shown in Fig. P2-84 act at a point on the free surface of a stressed body.

a. Determine the normal and shear stresses at this point on the inclined plane a–b shown in the figure and show these stresses on a sketch.

b. Determine the principal stresses and the maximum shearing stress at the point and show these stresses on a triangular stress element.

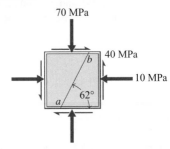

Figure P2-84

2-11 MOHR'S CIRCLE FOR PLANE STRESS

2.11

The German engineer Otto Mohr (1835–1918) developed a useful pictorial or graphic interpretation of the transformation equations for plane stress. This representation, commonly called Mohr's circle, involves the construction of a circle in such a manner that the coordinates of each point on the circle represent the normal and shearing stresses on one plane through the stressed point, and the angular position of the radius to the point gives the orientation of the plane. The proof that normal and shearing components of stress on an arbitrary plane through a point can be represented as points on a circle follows from Eqs. 2-12 and 2-13. Recall Eqs. 2-12b and 2-13b,

$$\sigma_n - \frac{\sigma_x + \sigma_y}{2} = \frac{\sigma_x - \sigma_y}{2} \cos 2\theta + \tau_{xy} \sin 2\theta$$

$$\tau_{nt} = -\frac{\sigma_x - \sigma_y}{2} \sin 2\theta + \tau_{xy} \cos 2\theta$$

Squaring both equations, adding, and simplifying yields

$$\left(\sigma_n - \frac{\sigma_x + \sigma_y}{2} \right)^2 + \tau_{nt}^2 = \left(\frac{\sigma_x - \sigma_y}{2} \right)^2 + \tau_{xy}^2$$

This is the equation of a circle in terms of the variables σ_n and τ_{nt}. The circle is centered on the σ-axis at a distance $(\sigma_x + \sigma_y)/2$ from the τ-axis, and the radius of the circle is given by

$$R = \sqrt{\left(\frac{\sigma_x - \sigma_y}{2}\right)^2 + \tau_{xy}^2}$$

Normal stresses are plotted as horizontal coordinates, with tensile stresses (positive) plotted to the right of the origin and compressive stresses (negative) plotted to the left. Shearing stresses are plotted as vertical coordinates, with those tending to produce a clockwise rotation of the stress element plotted above the σ-axis and those tending to produce a counterclockwise rotation of the stress element plotted below the σ-axis. The method for interpreting the sign to be associated with a particular shear stress value obtained from a Mohr's circle analysis will be illustrated in the discussion and example problems that follow.

2.12

Mohr's circle for any point subjected to plane stress can be drawn when stresses on two mutually perpendicular planes through the point are known. Consider, for example, the stress element of Fig. 2-41*a* with σ_x greater than σ_y and plot on Fig. 2-41*b* the points representing the given stresses. The coordinates of point V of Fig. 2-41*b* are the stresses on the vertical plane through the stressed point of Fig. 2-41*a* and point H is determined by the stresses on the horizontal plane through the point. Because $\tau_{yx} = \tau_{xy}$, point C, the center of the circle, is on the σ_n-axis.

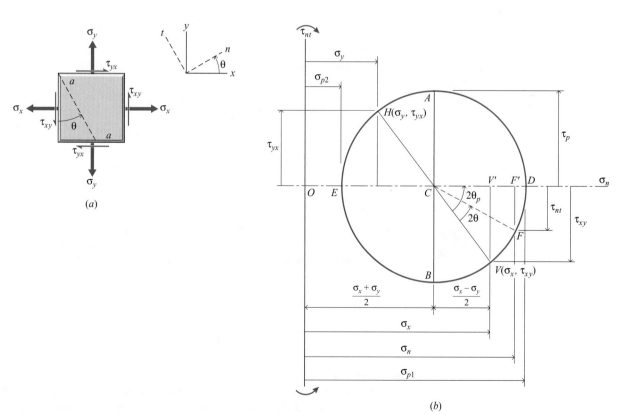

(a)

(b)

Figure 2-41

Line CV on Mohr's circle represents the plane (the vertical plane of Fig. 2-41a) through the stressed point from which the angle θ is measured. The coordinates of each point on the circle represent σ_n and τ_{nt} for one particular plane through the stressed point, the abscissa representing σ_n and the ordinate representing τ_{nt}. To demonstrate this statement, draw any radius CF in Fig. 2-41b at an angle 2θ counterclockwise from radius CV. From the figure, it is apparent that

$$OF' = OC + CF \cos(2\theta_p - 2\theta)$$

and since CF equals CV, the above equation reduces to

$$OF' = OC + CV \cos 2\theta_p \cos 2\theta + CV \sin 2\theta_p \sin 2\theta$$

Referring to Fig. 2-41b, note that

$$CV \cos 2\theta_p = CV' = (\sigma_x - \sigma_y)/2$$
$$CV \sin 2\theta_p = VV' = \tau_{xy}$$
$$OC = (\sigma_x + \sigma_y)/2 = \sigma_{\text{avg}}$$

Therefore,

$$OF' = OC + CV' \cos 2\theta + VV' \sin 2\theta$$
$$= \frac{\sigma_x + \sigma_y}{2} + \frac{\sigma_x - \sigma_y}{2} \cos 2\theta + \tau_{xy} \sin 2\theta$$

2.13

This expression is identical to Eq. 2-12b. Therefore, OF' is equal to σ_n. In a similar manner,

$$F'F = CF \sin(2\theta_p - 2\theta)$$
$$= CV \sin 2\theta_p \cos 2\theta - CV \cos 2\theta_p \sin 2\theta$$
$$= V'V \cos 2\theta - CV' \sin 2\theta$$
$$= \tau_{xy} \cos 2\theta - \frac{\sigma_x - \sigma_y}{2} \sin 2\theta$$

This expression is identical to Eq. 2-13b. Therefore, $F'F$ is equal to τ_{nt}.

2.14

Since the horizontal coordinate of each point on the circle represents the normal stress σ_n on some plane through the point, the maximum normal stress at the point is represented by OD, and its value is

$$\sigma_{p1} = OD = OC + CD = OC + CV$$
$$= \frac{\sigma_x + \sigma_y}{2} + \sqrt{\left(\frac{\sigma_x - \sigma_y}{2}\right)^2 + \tau_{xy}^2}$$

which agrees with Eq. 2-15.

Likewise, the vertical coordinate of each point on the circle represents the shearing stress τ_{nt} (called the *in-plane* shearing stress) on some plane through the point, which means that the maximum in-plane shearing stresses at the point are represented by CA and CB, and their value is

$$\tau_p = CA = CB = \sqrt{\left(\frac{\sigma_x - \sigma_y}{2}\right)^2 + \tau_{xy}^2}$$

which agrees with Eq. 2-17. If the two nonzero principal stresses have the same sign, the maximum shearing stress at the point will not be in the plane of the applied stresses.

The angle $2\theta_p$ from CV to CD is counterclockwise or positive, and its tangent is

$$\tan 2\theta_p = \frac{\tau_{xy}}{(\sigma_x - \sigma_y)/2}$$

which is Eq. 2-14. From the derivation of Eq. 2-15, the angle between the vertical plane and one of the principal planes was θ_p. In obtaining the same equation from Mohr's circle, the angle between the radii representing these same two planes is $2\theta_p$. In other words, all angles on Mohr's circle are twice the corresponding angles for the actual stressed body. The angle from the vertical plane to the horizontal plane in Fig. 2-41a is 90°, but in Fig. 2-41b, the angle between line CV (which represents the vertical plane) and line CH (which represents the horizontal plane) on Mohr's circle is 180°.

The results obtained from Mohr's circle have been shown to be identical with the equations derived from the free-body diagram of Fig. 2-27. Thus, Mohr's circle provides an extremely useful aid for both the visualization of and the solution of stresses on various planes through a point in a stressed body in terms of the stresses on two mutually perpendicular planes through the point. Although Mohr's circle can be drawn to scale and used to obtain values of stresses and angles by direct measurements on the figure, it is probably more useful as a pictorial aid to the analyst who is performing analytical determinations of stresses and their directions at the point.

2.15

When the state of stress at a point is specified by means of a sketch of a small element, the procedure for drawing and using Mohr's circle to obtain specific stress information can be briefly summarized as follows:

1. Choose a set of x–y reference axes.
2. Identify the stresses σ_x, σ_y and $\tau_{xy} = \tau_{yx}$ and list them with the proper sign.
3. Draw a set of σ_n–τ_{nt} coordinate axes with σ_n and τ_{nt} positive to the right and upward, respectively.
4. Plot the point $(\sigma_x, -\tau_{xy})$ and label it point V (vertical plane).
5. Plot the point (σ_y, τ_{yx}) and label it point H (horizontal plane).
6. Draw a line between V and H. This establishes the center C and the radius R of Mohr's circle.
7. Draw the circle.
8. An extension of the radius between C and V can be identified as the x-axis or the reference line for angle measurements (i.e., $\theta = 0°$).

By plotting points V and H as $(\sigma_x, -\tau_{xy})$ and (σ_y, τ_{yx}), respectively, shear stresses that tend to rotate the stress element clockwise will plot above the σ_n-axis, while those tending to rotate the element counterclockwise will plot below the σ_n-axis. The use of a negative sign with one of the shearing stresses (τ_{xy} or τ_{yx})

is required for plotting purposes, since for a given state of stress, the shearing stresses ($\tau_{xy} = \tau_{yx}$) have only one sign (both are positive or both are negative). The use of the negative sign at point V on Mohr's circle brings the direction of angular measurements 2θ on Mohr's circle into agreement with the direction of angular measurements θ on the stress element.

Once the circle has been drawn, the normal and shearing stresses on an arbitrary inclined plane a–a having an outward normal n that is oriented at an angle θ with respect to the reference x-axis (see Fig. 2-41a) can be obtained from the coordinates of point F (see Fig. 2-41b) on the circle that is located at angular position 2θ from the reference axis through point V. The coordinates of point F must be interpreted as stresses σ_n and $-\tau_{nt}$. Other points on Mohr's circle that provide stresses of interest are

1. Point D, which provides the principal stress σ_{p1}.

2. Point E, which provides the principal stress σ_{p2}.

3. Point A, which provides the maximum in-plane shearing stress $-\tau_p$ and the accompanying normal stress σ_{avg} that acts on the plane.

A negative sign must be used when interpreting shearing stresses τ_{nt} and τ_p obtained from the circle since a shearing stress tending to produce a clockwise rotation of the stress element is a negative shearing stress when a right-hand n–t coordinate system is used.

Problems of the type presented in Section 2-10 can readily be solved by this semigraphic method, as illustrated in the following examples.

Example Problem 2-13   At a point in a structural member subjected to plane stress there are normal and shearing stresses on horizontal and vertical planes through the point, as shown on the stress element in Fig. 2-42a. Determine and show on a sketch:

(a) The principal and maximum shearing stresses at the point.

(b) The normal and shearing stresses on plane a–a through the point.

SOLUTION

Mohr's circle (see Fig. 2-42b) is constructed from the given data by plotting point V (representing the stresses on the vertical plane) at $(8, -4)$ because the stresses on the vertical plane are 8 ksi (T) and 4 ksi (counterclockwise) shear. Likewise, point H (representing the stresses on the horizontal plane) has the coordinates $(-6, 4)$. Draw line HV, which is a diameter of Mohr's circle, and note that the center of the circle is at $(1, 0)$. The radius of the circle is

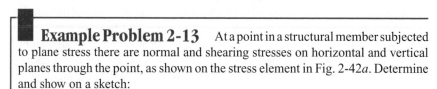

$$CV = \sqrt{\left(\frac{\sigma_x - \sigma_y}{2}\right)^2 + \tau_{xy}^2}$$

$$= \sqrt{7^2 + 4^2} = 8.062 \text{ ksi}$$

(a) The principal stresses and the maximum shearing stress at the point are

$$\sigma_{p1} = OD = 1 + 8.062 = +9.062 \text{ ksi} \cong 9.06 \text{ ksi (T)} \qquad \textbf{Ans.}$$

$$\sigma_{p2} = OE = 1 - 8.062 = -7.062 \text{ ksi} \cong 7.06 \text{ ksi (C)} \qquad \textbf{Ans.}$$

$$\sigma_{p3} = \sigma_z = 0 \qquad \textbf{Ans.}$$

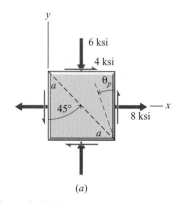

(a)

Figure 2-42(a)

▶ The point V on Mohr's circle represents the state of stress on the vertical surface of the stress element and can appear in any quadrant of the circle. If the shear stress on the vertical surface tends to rotate the stress element counterclockwise (as in this example), then the point V is plotted below the σ_n-axis. If the shear stress on the vertical surface tends to rotate the stress element clockwise, then the point V would be plotted above the σ_n-axis. If the normal stress on the vertical surface is positive, then the point V is plotted to the right of the τ_{nt}-axis (as in this example); if the normal stress is negative, it would be plotted to the left of the τ_{nt}-axis. If the normal stress on the vertical surface is more positive than the normal stress on the horizontal surface, then the point V is plotted to the right of the point H. If the normal stress on the vertical surface is less positive than the normal stress on the horizontal surface, then the point V would be plotted to the left of the point H.

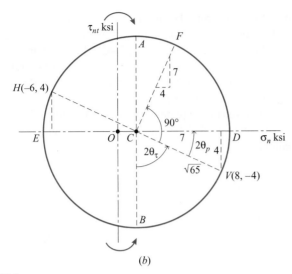

(b)

Figure 2-42(b)

Since σ_{p1} and σ_{p2} have opposite signs, the maximum shearing stress is

$$\tau_p = \tau_{max} = CA = CB = 8.062 \text{ ksi} \cong 8.06 \text{ ksi} \qquad \textbf{Ans.}$$

The principal planes are represented by lines CD and CE, where

$$\tan 2\theta_p = 4/7 = 0.5714$$

which gives

$$2\theta_p = +29.74° \quad \text{or} \quad \theta_p = 14.87° \nwarrow$$

Since the angle $2\theta_p$ is counterclockwise, the principal planes are counterclockwise from the vertical and horizontal planes of the stress block, as shown in Fig. 2-42c. Actually, the angle θ_p is measured from the x-axis to the outward normal to one of the principal planes. Since the outward normal and the plane are perpendicular, the angle between the x-plane and one of the principal planes is also θ_p. To determine which principal stress acts on which plane, note that as the radius of the circle rotates counterclockwise, the end of the radius CV moves from V to D, indicating that as the initially vertical plane rotates through 14.87°, the stresses change to 9.06 ksi (T) (normal) and 0 ksi (shear). Note also that the end H or radius CH moves to E, indicating that as the initially horizontal plane rotates through 14.87°, the stresses change to 7.06 ksi (C) (normal) and 0 ksi (shear). These stresses are shown on the sketch of Fig. 2-42c, in which the surfaces D and E are the principal planes represented by points D and E on Mohr's circle.

Point A on Mohr's circle represents the state of stress on a surface rotated 45° counterclockwise from the surface represented by point D on the circle. On this surface, the shear stress is the maximum it can be (in the xy-plane), $\tau_{nt} = \tau_p = \tau_{max} = 8.06$ ksi, and the normal stress is the average value $\sigma_n = (\sigma_x + \sigma_y)/2 = \sigma_C = 1.000$ ksi (the normal stress at the center of the circle). Likewise, point B represents the state of stress on a surface rotated

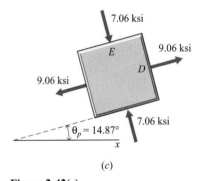

(c)

Figure 2-42(c)

▶ Positive θ is measured counterclockwise from the positive x-axis, whether on the stress element or on Mohr's circle. On a plane of the stress element a shear stress that rotates the element counterclockwise is plotted below the σ_n-axis on Mohr's circle, and a shear stress that rotates the element clockwise is plotted above the σ_n-axis of Mohr's circle.

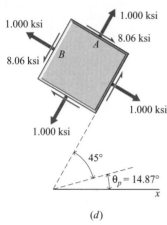

Figure 2-42(d)

(d)

45° counterclockwise from the surface represented by point E. The stresses on this surface are also $\sigma_n = \sigma_C = 1.000$ ksi and $\tau_{nt} = \tau_p = 8.06$ ksi. These stresses are shown on the sketch of Fig. 2-42d. Planes A and B in Fig. 2-42d are the surfaces represented by points A and B on Mohr's circle, Fig. 2-42b.

Note that any two orthogonal surfaces of Fig. 2-42c are sufficient to completely specify the principal stresses. Also, only one of the surfaces of Fig. 2-42d is required to completely specify the maximum shear stress. Therefore, these two separate sketches can be combined as shown on the triangular stress element in Fig. 2-42e.

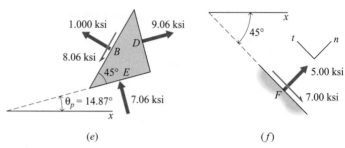

(e)   (f)

Figure 2-42(e–f)

(b) Plane a–a is 45° counterclockwise from the vertical plane; therefore, the corresponding radius of Mohr's circle is $2\theta = 90°$ counterclockwise from the line CV and is shown as CF on Fig. 2-42b. The coordinates of point F are seen to be (5, 7), which means that the stresses on plane a–a are 5.00 ksi (T) and 7.00 ksi shear in a direction to produce a clockwise rotation on a stress element having plane a–a as one of its faces. This shear stress would be defined as a negative shear stress because to produce a clockwise rotation, it must be directed in the negative t-direction associated with plane a–a (see Fig. 2-42f).

Example Problem 2-14
At a point in a structural member subjected to plane stress there are normal and shearing stresses on horizontal and vertical planes through the point, as shown in Fig. 2-43a. Determine, and show on a sketch, the principal stresses and the maximum shearing stress at the point.

SOLUTION
Mohr's circle (see Fig. 2-43b) is constructed from the given data by plotting point V (representing the stresses on the vertical plane) at (72, 24) and point H (representing the stresses on the horizontal plane) at (36, −24). Line HV between the two points is a diameter of Mohr's circle. The circle is centered at point C (54, 0) and has a radius CV equal to 30 MPa. The principal stresses at the point are

$$\sigma_{p1} = OD = 54 + 30 = 84.0 \text{ MPa (T)} \qquad \textbf{Ans.}$$

$$\sigma_{p2} = OE = 54 - 30 = 24.0 \text{ MPa (T)} \qquad \textbf{Ans.}$$

$$\sigma_{p3} = \sigma_z = 0 \qquad \textbf{Ans.}$$

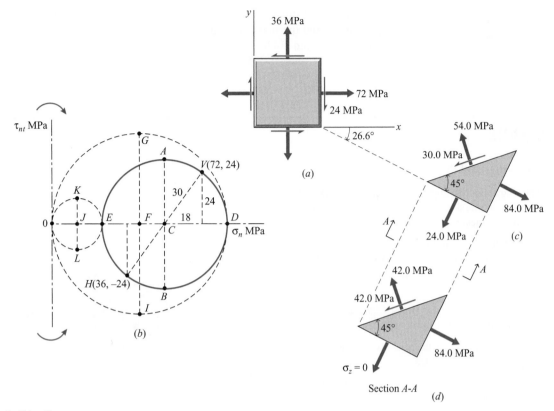

Figure 2-43(a–d)

The principal planes are represented by lines CD and CE, where

$$\tan 2\theta_p = -24/18 = -1.3333$$

which gives

$$2\,\theta_p = -53.13° = 53.13° \downarrow$$
$$\theta_p = -26.57° = 26.57° \downarrow$$

Since σ_{p1} and σ_{p2} have the same sign, the maximum in-plane shearing stress τ_p is not the maximum shearing stress τ_{max} at the point. The maximum shearing stress at the point is represented on Mohr's circle by drawing an additional circle (shown dashed in Fig. 2-43b) which has line OD as a diameter. This circle is centered at point F (42, 0) and has a radius FG equal to 42 MPa. This circle represents the combinations of normal and shearing stresses existing on planes obtained by rotating the element about the principal axis associated with the principal stress σ_{p2}. A third Mohr's circle (shown dotted in Fig. 2-43b) has line OE as a diameter. This circle is centered at point J (12, 0) and has a radius JK equal to 12 MPa. This circle represents the combinations of normal and shearing stresses existing on planes obtained by rotating the element about the principal axis associated with the principal stress σ_{p1}. Thus,

$$\tau_p = CA = 30.0 \text{ MPa}$$
$$\tau_{max} = FG = 42.0 \text{ MPa} \qquad \qquad \textbf{Ans.}$$

The principal stresses σ_{p1}, σ_{p2}, and $\sigma_z = \sigma_{p3} = 0$, the maximum in-plane shearing stress τ_p, and the maximum shearing stress τ_{max} at the point are all shown on Figs. 2-43c and 2-43d.

PROBLEMS

MecMovie Activities and Problems

MM2.12   Sketching stress transformation results. Learning tool. Constructing appropriate sketches showing orientation of principal and maximum shear stresses.

MM2.13   Coach Mohr's Circle of Stress. Theory; Interactive example; Game. Learn to construct and use Mohr's circle to determine principal stresses including the proper orientation of the principal stress planes.

MM2.14   Absolute maximum shear stress. Example; Try one. Investigate a three-dimensional stress state at a point.

MM2.15   Mohr's circle game—plane stress. Game. Recognize correctly constructed Mohr's circles.

MM2.16   Mohr's circle game—principal/max shear. Game. Given a Mohr's circle, recognize the corresponding stress element for the principal stress state and the maximum in-plane shear stress state.

Introductory Problems

2-85*   The stresses shown in Fig. P2-85 act at a point on the free surface of a stressed body. Using Mohr's circle, determine, and show on a sketch, the normal and shear stresses at this point on plane *a–a*.

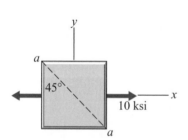

Figure P2-85

2-86*   At a point in a structural member subjected to plane stress there are stresses on horizontal and vertical planes through the point, as shown in Fig. P2-86. Use Mohr's circle to determine

the normal and shear stresses at this point on plane *a–a*, and show these stresses on a sketch.

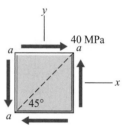

Figure P2-86

2-87   The stresses shown in Fig. P2-87 act at a point on the free surface of a stressed body. Use Mohr's circle to determine the normal and shear stresses at this point on the inclined plane *a–b* shown in the figure. Show these stresses on a sketch.

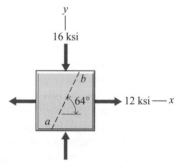

Figure P2-87

2-88   The stresses shown in Fig. P2-88 act at a point on the free surface of a stressed body. Use Mohr's circle to determine the normal and shear stresses at this point on the inclined plane *a–b* shown in the figure. Show these stresses on a sketch.

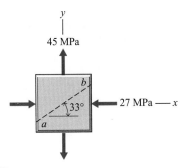

Figure P2-88

Intermediate Problems

2-89* At a point in a structural member subjected to plane stress there are stresses on horizontal and vertical planes through the point, as shown in Fig. P2-89. Use Mohr's circle to determine the principal stresses and the maximum shear stress at the point. Show these stresses on the planes on which they act.

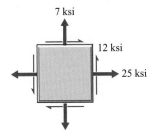

Figure P2-89

2-90* At a point in a structural member subjected to plane stress there are stresses on horizontal and vertical planes through the point, as shown in Fig. P2-90. Use Mohr's circle to determine the principal stresses and the maximum shear stress at the point. Show these stresses on the planes on which they act.

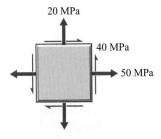

Figure P2-90

2-91 At a point in a structural member subjected to plane stress there are stresses on horizontal and vertical planes through the point, as shown in Fig. P2-91. Use Mohr's circle to determine the principal stresses and the maximum shear stress at the point. Show these stresses on the planes on which they act.

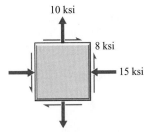

Figure P2-91

2-92* The stresses shown in Fig. P2-92 act at a point on the free surface of a machine component. Use Mohr's circle to determine the principal stresses and the maximum shear stress at the point. Show these stresses on the planes on which they act.

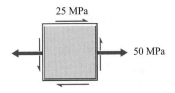

Figure P2-92

2-93 The stresses shown in Fig. P2-93 act at a point on the free surface of a thin-walled pressure vessel. Use Mohr's circle to determine the principal stresses and the maximum shear stress at the point. Show these stresses on the planes on which they act.

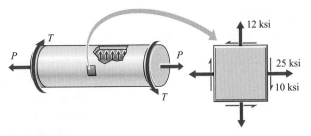

Figure P2-93

Challenging Problems

2-94* At a point in a structural member subjected to plane stress there are stresses on horizontal and vertical planes through the point, as shown in Fig. P2-94. Using Mohr's circle,

a. Determine the principal stresses and the maximum shear stress at the point. Show these stresses on a triangular stress element.

b. Determine the normal and shear stresses on the inclined plane *a–b* shown in the figure. Show the results on a sketch of the plane on which these stresses act.

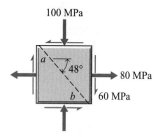

Figure P2-94

2-95* The stresses on horizontal and vertical planes at a point on the outside surface of a solid circular shaft subjected to an axial load *P* and a torque *T* are shown in Fig. P2-95. Using Mohr's circle,

a. Determine the principal stresses and the maximum shear stress at the point. Show these stresses on a triangular stress element.

b. Determine the normal and shear stresses on the inclined plane *a–b* shown in the figure. Show the results on a sketch of the plane on which these stresses act.

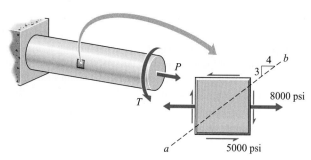

Figure P2-95

2-96 At a point in a structural member subjected to plane stress there are normal and shear stresses on horizontal and vertical planes through the point, as shown in Fig. P2-96. Using Mohr's circle,

a. Determine the principal stresses and the maximum shear stress at the point. Show these stresses on a triangular stress element.

b. Determine the normal and shear stresses on the inclined plane *a–b* shown in the figure. Show the results on a sketch of the plane on which these stresses act.

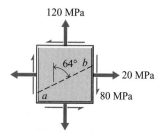

Figure P2-96

2-97 At a point in a machine component subjected to plane stress there are normal and shear stresses on horizontal and vertical planes through the point, as shown in Fig. P2-97. Using Mohr's circle,

a. Determine the principal stresses and the maximum shear stress at the point. Show these stresses on a triangular stress element.

b. Determine the normal and shear stresses on the inclined plane *a–b* shown in the figure. Show the results on a sketch of the plane on which these stresses act.

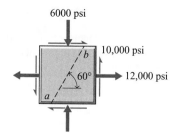

Figure P2-97

2-12 GENERAL STATE OF STRESS AT A POINT

The general state of stress at a point was previously illustrated in Fig. 2-24. Expressions for the stresses on any oblique plane through the point in terms of stresses on the reference planes can be developed with the aid of the free-body diagram of Fig. 2-44, where the *n*-axis is normal to the oblique (shaded) face. The areas of

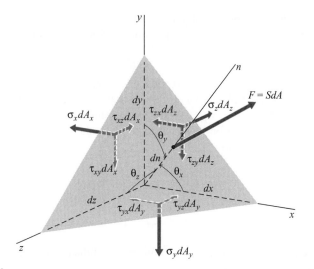

Figure 2-44

the faces of the element are dA for the oblique face and $dA \cos \theta_x$, $dA \cos \theta_y$, and $dA \cos \theta_z$ for the x-, y-, and z-faces, respectively.[3]

The resultant force F on the oblique face is $S\,dA$, where S is the resultant stress (stress vector—see Section 2-7) on the area and is equal to $\sqrt{\sigma_n^2 + \tau_{nt}^2}$. The forces on the x-, y-, and z-faces are shown as three components, the magnitude of each being the product of the area and the appropriate stress. If we use l, m, and n for $\cos \theta_x$, $\cos \theta_y$, and $\cos \theta_z$, respectively, the three force equations of equilibrium in the x-, y-, and z-directions are

$$F_x = S_x dA = \sigma_x dA\,l + \tau_{yx}dA\,m + \tau_{zx}dA\,n$$
$$F_y = S_y dA = \sigma_y dA\,m + \tau_{zy}dA\,n + \tau_{xy}dA\,l$$
$$F_z = S_z dA = \sigma_z dA\,n + \tau_{xz}dA\,l + \tau_{yz}dA\,m$$

where l, m, and n are direction cosines. Thus, the three orthogonal components of the resultant stress are

$$S_x = \sigma_x l + \tau_{yx}m + \tau_{zx}n$$
$$S_y = \tau_{xy}l + \sigma_y m + \tau_{zy}n \qquad (a)$$
$$S_z = \tau_{xz}l + \tau_{yz}m + \sigma_z n$$

The normal component σ_n of the resultant stress S equals $S_x l + S_y m + S_z n$; therefore, from Eq. (a), the following equation for the normal stress on any oblique plane through the point is obtained:

$$\sigma_n = \sigma_x l^2 + \sigma_y m^2 + \sigma_z n^2 + 2\tau_{xy}lm + 2\tau_{yz}mn + 2\tau_{zx}nl \qquad (2\text{-}21)$$

---

[3] These relationships can be established by considering the volume of the tetrahedron shown in Fig. 2-44. Thus, $V = \frac{1}{3}dn\,dA = \frac{1}{3}dx\,dA_x = \frac{1}{3}dy\,dA_y = \frac{1}{3}dz\,dA_z$. But $dn = dx \cos \theta_x = dy \cos \theta_y = dz \cos \theta_z$; therefore, $dA_x = dA \cos \theta_x$, $dA_y = dA \cos \theta_y$, $dA_z = dA \cos \theta_z$.

The shearing stress τ_{nt} on the oblique plane can be obtained from the relation $S^2 = \sigma_n^2 + \tau_{nt}^2$. For a given problem the values of S and σ_n will be obtained from Eqs. (a) and 2-21.

A principal plane was previously defined as a plane on which the shear stress τ_{nt} is zero. The normal stress σ_n on such a plane was defined as a principal stress σ_p. If the oblique plane of Fig. 2-44 is a principal plane, then $S = \sigma_p$ and $S_x = \sigma_p l$, $S_y = \sigma_p m$, $S_z = \sigma_p n$. When these components are substituted into Eq. (a), the equations can be rewritten to produce the following homogeneous linear equations in l, m, and n:

$$(\sigma_p - \sigma_x)l - \tau_{yx}m - \tau_{zx}n = 0$$
$$(\sigma_p - \sigma_y)m - \tau_{zy}n - \tau_{xy}l = 0 \qquad (b)$$
$$(\sigma_p - \sigma_z)n - \tau_{xz}l - \tau_{yz}m = 0$$

This set of equations has a nontrivial solution only if the determinant of the coefficients of l, m, and n is equal to zero. Thus,

$$\begin{vmatrix} (\sigma_p - \sigma_x) & -\tau_{yx} & -\tau_{zx} \\ -\tau_{xy} & (\sigma_p - \sigma_y) & -\tau_{zy} \\ -\tau_{xz} & -\tau_{yz} & (\sigma_p - \sigma_z) \end{vmatrix} = 0$$

Expansion of the determinant yields the following cubic equation for determining the principal stresses:

$$\sigma_p^3 - (\sigma_x + \sigma_y + \sigma_z)\sigma_p^2 + (\sigma_x\sigma_y + \sigma_y\sigma_z + \sigma_z\sigma_x - \tau_{xy}^2 - \tau_{yz}^2 - \tau_{zx}^2)\sigma_p$$
$$- (\sigma_x\sigma_y\sigma_z - \sigma_x\tau_{yz}^2 - \sigma_y\tau_{zx}^2 - \sigma_z\tau_{xy}^2 + 2\tau_{xy}\tau_{yz}\tau_{zx}) = 0 \qquad \text{(2-22)}$$

For given values of σ_x, σ_y, ..., τ_{zx}, Eq. 2-22 gives three values for the principal stresses σ_{p1}, σ_{p2}, σ_{p3}. By substituting these values for σ_p, in turn, into Eq. (b) and using the relation

$$l^2 + m^2 + n^2 = 1 \qquad (c)$$

three sets of direction cosines may be determined for the normals to the three principal planes. The foregoing discussion verifies the existence of three mutually perpendicular principal planes for the most general state of stress.

In developing equations for maximum and minimum normal stresses, the special case will be considered in which $\tau_{xy} = \tau_{yz} = \tau_{zx} = 0$. No loss in generality is introduced by considering this special case since it involves only a reorientation of the reference x-, y-, z-axes to coincide with the principal directions. Since the x-, y-, z-planes are now principal planes, the stresses σ_x, σ_y, σ_z become σ_{p1}, σ_{p2}, and σ_{p3}. Solving Eq. (a) for the direction cosines yields

$$l = S_x/\sigma_{p1} \qquad m = S_y/\sigma_{p2} \qquad n = S_z/\sigma_{p3}$$

By substituting these values into Eq. (c), the following equation is obtained:

$$\frac{S_x^2}{\sigma_{p1}^2} + \frac{S_y^2}{\sigma_{p2}^2} + \frac{S_z^2}{\sigma_{p3}^2} = 1 \qquad (d)$$

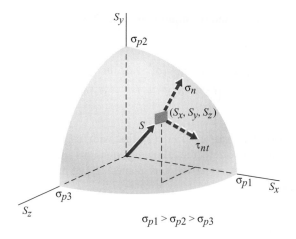

Figure 2-45

The plot of Eq. (*d*) is the ellipsoid shown in Fig. 2-45. It can be observed that the magnitude of σ_n is everywhere less than that of S (since $S^2 = \sigma_n^2 + \tau_{nt}^2$) except at the intercepts, where S is σ_{p1}, σ_{p2}, or σ_{p3}. Therefore, it can be concluded that two of the principal stresses (σ_{p1} and σ_{p3} of Fig. 2-45) are the maximum and minimum normal stresses at the point. The third principal stress is intermediate in value and has no particular significance. The discussion above demonstrates that the set of principal stresses includes the maximum and minimum normal stresses at the point.

Continuing with the special case where the given stresses σ_x, σ_y, and σ_z are principal stresses, we can develop equations for the maximum shearing stress at the point. The resultant stress on the oblique plane is given by the expression

$$S^2 = S_x^2 + S_y^2 + S_z^2$$

Substitution of values for S_x, S_y, and S_z from Eq. (*a*), with zero shearing stresses, yields the expression

$$S^2 = \sigma_x^2 l^2 + \sigma_y^2 m^2 + \sigma_z^2 n^2 \qquad (e)$$

Also, from Eq. 2-21,

$$\sigma_n^2 = (\sigma_x l^2 + \sigma_y m^2 + \sigma_z n^2)^2 \qquad (f)$$

Since $S^2 = \sigma_n^2 + \tau_{nt}^2$, an expression for the shearing stress τ_{nt} on the oblique plane is obtained from Eqs. (*e*) and (*f*) as

$$\tau_{nt} = \sqrt{\sigma_x^2 l^2 + \sigma_y^2 m^2 + \sigma_z^2 n^2 - (\sigma_x l^2 + \sigma_y m^2 + \sigma_z n^2)^2} \qquad (2\text{-}23)$$

The planes on which maximum and minimum shearing stresses occur can be obtained from Eq. 2-23 by differentiating with respect to the direction cosines *l*, *m*, and *n*. One of the direction cosines in Eq. 2-23 (*n*, for example) can be eliminated by solving Eq. (*c*) for n^2 and substituting into Eq. 2-23. Thus,

$$\tau_{nt} = \{(\sigma_x^2 - \sigma_z^2)l^2 + (\sigma_y^2 - \sigma_z^2)m^2 \\ + \sigma_z^2 - [(\sigma_x - \sigma_z)l^2 + (\sigma_y - \sigma_z)m^2 + \sigma_z]^2\}^{1/2} \tag{g}$$

By taking the partial derivatives of Eq. (*g*), first with respect to *l* and then with respect to *m* and equating to zero, the following equations are obtained for determining the direction cosines associated with planes having maximum and minimum shearing stresses:

$$l\left[\frac{1}{2}(\sigma_x - \sigma_z) - (\sigma_x - \sigma_z)l^2 - (\sigma_y - \sigma_z)m^2\right] = 0 \tag{h}$$

$$m\left[\frac{1}{2}(\sigma_y - \sigma_z) - (\sigma_x - \sigma_z)l^2 - (\sigma_y - \sigma_z)m^2\right] = 0 \tag{i}$$

One solution of these equations is obviously $l = m = 0$. Then from Eq. (*c*), $n = \pm 1$. But this is just the principal surface whose normal is in the *z*-direction, and on which the shear stresses are their minimum values, $\tau_{zx} = \tau_{zy} = 0$.

Solutions different from zero are also possible for this set of equations. Consider first that $m = 0$; then from Eq. (*h*), $l = \pm\sqrt{1/2}$ and from Eq. (*c*), $n = \pm\sqrt{1/2}$. This is the surface whose normal is 45° relative to both the *x*- and *z*-axes and perpendicular to the *y*-axis. This surface has the largest shear stress of all surfaces having a normal perpendicular to the *y*-axis. Also if $l = 0$, then from Eq. (*i*), $m = \pm\sqrt{1/2}$ and from Eq. (*c*), $n = \pm\sqrt{1/2}$. This is the surface whose normal is 45° relative to both the *y*- and *z*-axes and perpendicular to the *x*-axis. This surface has the largest shear stress of all surfaces having a normal perpendicular to the *x*-axis. Repeating the above procedure by eliminating *l* and *m* in turn from Eq. (*g*) yields other values for the direction cosines that make the shearing stresses maximum or minimum. All the possible solutions are listed in Table 2-1. In the last line of the table the planes corresponding to the direction cosines in the column above are shown shaded. Note that in each case only one of the two possible planes is shown.

Table 2-1   Direction Cosines for Shearing Stresses

	Minimum			Maximum		
	1	2	3	4	5	6
l	± 1	0	0	$\pm\sqrt{1/2}$	$\pm\sqrt{1/2}$	0
m	0	± 1	0	$\pm\sqrt{1/2}$	0	$\pm\sqrt{1/2}$
n	0	0	± 1	0	$\pm\sqrt{1/2}$	$\pm\sqrt{1/2}$

Substituting values for the direction cosines from column 4 of Table 2-1 into Eq. 2-23, with σ_x, σ_y, σ_z replaced with σ_{p1}, σ_{p2}, σ_{p3}, yields

$$
\tau_{max} = \sqrt{\frac{1}{2}\sigma_{p1}^2 + \frac{1}{2}\sigma_{p2}^2 - \left(\frac{1}{2}\sigma_{p1} + \frac{1}{2}\sigma_{p2}\right)^2}
$$

$$
= \frac{\sigma_{p1} - \sigma_{p2}}{2}
$$

Similarly, using the values of the cosines from columns 5 and 6 gives

$$
\tau_{max} = \frac{\sigma_{p1} - \sigma_{p3}}{2} \quad \text{and} \quad \tau_{max} = \frac{\sigma_{p2} - \sigma_{p3}}{2}
$$

Of these three possible results, the largest magnitude will be the maximum stress; hence, the expression for the maximum shearing stress is

$$
\tau_{max} = \frac{\sigma_{max} - \sigma_{min}}{2} \tag{2-24}
$$

which verifies the statement in Section 2-10 regarding maximum shearing stress.

Example Problem 2-15   At a point in a stressed body, the known stresses are

$$
\sigma_x = 14 \text{ ksi (T)} \qquad \sigma_y = 12 \text{ ksi (T)} \qquad \sigma_z = 10 \text{ ksi (T)}
$$
$$
\tau_{xy} = 8 \text{ ksi} \qquad \tau_{yz} = -10 \text{ ksi} \qquad \tau_{zx} = 6 \text{ ksi}
$$

Determine

(a) The normal and shearing stresses on a plane whose outward normal is oriented at angles of 61.3°, 53.1°, and 50.2° with the x-, y-, and z-axes, respectively.

(b) The principal stresses and the maximum shearing stress at the point.

SOLUTION

(a) The direction cosines for a plane whose outward normal is oriented at angles of 61.3°, 53.1°, and 50.2° with the x-, y-, and z-axes, respectively, are

$$
l = \cos 61.3° = 0.4802
$$
$$
m = \cos 53.1° = 0.6004
$$
$$
n = \cos 50.2° = 0.6401
$$

The three orthogonal components of the resultant stress S on the plane are

$$S_x = \sigma_x l + \tau_{yx} m + \tau_{zx} n$$
$$= 14(0.4802) + 8(0.6004) + 6(0.6401) = 15.367 \text{ ksi}$$
$$S_y = \tau_{xy} l + \sigma_y m + \tau_{zy} n$$
$$= 8(0.4802) + 12(0.6004) - 10(0.6401) = 4.645 \text{ ksi}$$
$$S_z = \tau_{xz} l + \tau_{yz} m + \sigma_z n$$
$$= 6(0.4802) - 10(0.6004) + 10(0.6401) = 3.278 \text{ ksi}$$

▶ Stresses are not vectors and generally cannot be manipulated as vectors. However, the total force on the surface having area dA and normal in the n-direction has a magnitude of $S \, dA$. The magnitudes of the normal and tangential components of this force are $\sigma_n \, dA$ and $\tau_{nt} \, dA$, respectively. Therefore, dividing the relationship between the force vector components, $(S \, dA)^2 = (\sigma_n \, dA)^2 + (\tau_{nt} \, dA)^2$ by dA^2 and rearranging gives the relationship between the stress components $\tau_{nt}^2 = S^2 - \sigma_n^2$ or $\tau_{nt} = \sqrt{S^2 - \sigma_n^2}$.

Once the three orthogonal components of the resultant stress S on the plane are known, the resultant stress S, the normal stress σ_n, and the shear stress τ_{nt} on the plane can be determined by using the equations

$$S = \sqrt{S_x^2 + S_y^2 + S_z^2}$$
$$= \sqrt{(15.367)^2 + (4.645)^2 + (3.278)^2} = 16.385 \text{ ksi}$$

$$\sigma_n = S_x l + S_y m + S_z n$$
$$= 15.367(0.4802) + 4.645(0.6004) + 3.278(0.6401) \qquad \textbf{Ans.}$$
$$= 12.266 \text{ ksi} \cong 12.27 \text{ ksi} \ (T)$$

$$\tau_{nt} = \sqrt{S^2 - \sigma_n^2} = \sqrt{(16.385)^2 - (12.266)^2} \qquad \textbf{Ans.}$$
$$= 10.863 \text{ ksi} \cong 10.86 \text{ ksi} \ (T)$$

(b) The three principal stresses and the maximum shearing stress at the point are determined by using Eqs. 2-22 and 2-24. Substituting the given values of stress at the point into Eq. 2-22 yields

$$\sigma_p^3 - 36\sigma_p^2 + 228\sigma_p + 1752 = 0$$

which has the solution

$$\sigma_{p1} = \sigma_{\max} = 22.08 \text{ ksi} \cong 22.1 \text{ ksi} \ (T) \qquad \textbf{Ans.}$$
$$\sigma_{p2} = \sigma_{\text{int}} = 18.263 \text{ ksi} \cong 18.26 \text{ ksi} \ (T) \qquad \textbf{Ans.}$$
$$\sigma_{p3} = \sigma_{\min} = -4.344 \text{ ksi} \cong 4.34 \text{ ksi} \ (C) \qquad \textbf{Ans.}$$

Finally, Eq. 2-24 yields

$$\tau_{\max} = \frac{\sigma_{\max} - \sigma_{\min}}{2} = \frac{22.08 - (-4.344)}{2}$$
$$= 13.212 \text{ ksi} \cong 13.21 \text{ ksi} \qquad \textbf{Ans.}$$

■ **Example Problem 2-16**   At a point in a stressed body, the known stresses are

$$\sigma_x = 60 \text{ MPa (T)} \qquad \sigma_y = 40 \text{ MPa (T)} \qquad \sigma_z = 20 \text{ MPa (T)}$$
$$\tau_{xy} = +40 \text{ MPa} \qquad \tau_{yz} = +20 \text{ MPa} \qquad \tau_{xz} = +30 \text{ MPa}$$

Determine

(a) The principal stresses and the maximum shearing stress at the point.

(b) The orientation of the plane on which the maximum tensile stress acts.

SOLUTION

(a) The three principal stresses and the maximum shearing stress at the point are determined by using Eqs. 2-22 and 2-24. Substituting the given values of stress at the point into Eq. 2-22 yields

$$\sigma_p^3 - 120\sigma_p^2 + 1500\sigma_p - 4000 = 0$$

which has the solution

$$\sigma_{p1} = \sigma_{max} = 106.23 \text{ MPa} \cong 106.2 \text{ MPa (T)} \qquad \textbf{Ans.}$$
$$\sigma_{p2} = \sigma_{int} = 10.000 \text{ MPa} \cong 10.00 \text{ MPa (T)} \qquad \textbf{Ans.}$$
$$\sigma_{p3} = \sigma_{min} = 3.765 \text{ MPa} \cong 3.77 \text{ MPa (T)} \qquad \textbf{Ans.}$$

Equation 2-24 yields

$$\tau_{max} = \frac{\sigma_{max} - \sigma_{min}}{2} = \frac{106.23 - 3.765}{2}$$
$$= 51.23 \text{ MPa} \cong 51.2 \text{ MPa} \qquad \textbf{Ans.}$$

(b) The orientation of the plane on which the maximum tensile stress acts is obtained by substituting the stress value $\sigma_{p1} = \sigma_{max} = 106.23$ MPa into Eqs. (*b*) and solving for *l*, *m*, and *n* after noting that $l^2 + m^2 + n^2 = 1$. Thus,

$$(46.23)l - 40m - 30n = 0 \qquad (1)$$
$$(66.23)m - 20n - 40l = 0 \qquad (2)$$
$$(86.23)n - 30l - 20m = 0 \qquad (3)$$

From Eqs. 1 and 2,

$$m = 0.7623l$$

From Eqs. 2 and 3,

$$n = 0.5247l$$

▶ These three homogeneous equations are not independent because the determinate of their coefficients is zero (by construction). Therefore, these three equations cannot give unique values for the three direction cosines, *l*, *m*, and *n*; they can only give relationships between the direction cosines. The fourth equation, $l^2 + m^2 + n^2 = 1$, must be used along with these three equations in order to get values for the three direction cosines.

Therefore,

$$l^2 + (0.7623l)^2 + (0.5247l)^2 = 1$$

which yields

$$l = 0.7339 \qquad \theta_{1x} = 42.79° \cong 42.8° \qquad \textbf{Ans.}$$
$$m = 0.5595 \qquad \theta_{1y} = 55.98° \cong 56.0° \qquad \textbf{Ans.}$$
$$n = 0.3850 \qquad \theta_{1z} = 67.36° \cong 67.4° \qquad \textbf{Ans.}$$

PROBLEMS

Introductory Problems

2-98* At a point in a stressed body, the known stresses are $\sigma_x = 40$ MPa (T), $\sigma_y = 20$ MPa (C), $\sigma_z = 20$ MPa (T), $\tau_{xy} = +40$ MPa, $\tau_{yz} = 0$ MPa, and $\tau_{zx} = +30$ MPa. Determine the normal and shear stresses on a plane whose outward normal is oriented at angles of 40°, 75°, and 54° with the x-, y-, and z-axes, respectively.

2-99* At a point in a stressed body, the known stresses are $\sigma_x = 14$ ksi (T), $\sigma_y = 12$ ksi (T), $\sigma_z = 10$ ksi (T), $\tau_{xy} = +4$ ksi, $\tau_{yz} = -4$ ksi, and $\tau_{zx} = 0$ ksi. Determine the normal and shear stresses on a plane whose outward normal is oriented at angles of 40°, 60°, and 66.2° with the x-, y-, and z-axes, respectively.

2-100 At a point in a stressed body, the known stresses are $\sigma_x = 60$ MPa (T), $\sigma_y = 90$ MPa (T), $\sigma_z = 60$ MPa (T), $\tau_{xy} = +120$ MPa, $\tau_{yz} = +75$ MPa, and $\tau_{zx} = +90$ MPa. Determine the normal and shear stresses on a plane whose outward normal is oriented at angles of 60°, 70°, and 37.3° with the x-, y-, and z-axes, respectively.

2-101* At a point in a stressed body, the known stresses are $\sigma_x = 0$ ksi, $\sigma_y = 0$ ksi, $\sigma_z = 0$ ksi, $\tau_{xy} = +6$ ksi, $\tau_{yz} = +10$ ksi, and $\tau_{zx} = +8$ ksi. Determine the normal and shear stresses on a plane whose outward normal makes equal angles with the x-, y-, and z-axes.

2-102 At a point in a stressed body, the known stresses are $\sigma_x = 72$ MPa (T), $\sigma_y = 32$ MPa (C), $\sigma_z = 0$ MPa, $\tau_{xy} = +21$ MPa, $\tau_{yz} = 0$ MPa, and $\tau_{zx} = +21$ MPa. Determine the normal and shear stresses on a plane whose outward normal makes equal angles with the x-, y-, and z-axes.

Intermediate Problems

2-103* At a point in a stressed body, the known stresses are $\sigma_x = 12$ ksi (T), $\sigma_y = 10$ ksi (C), $\sigma_z = 8$ ksi (T), $\tau_{xy} = +8$ ksi, $\tau_{yz} = -10$ ksi, and $\tau_{zx} = +12$ ksi. Determine the principal stresses and the maximum shearing stress at the point.

2-104* At a point in a stressed body, the known stresses are $\sigma_x = 40$ MPa (T), $\sigma_y = 20$ MPa (C), $\sigma_z = 20$ MPa (T), $\tau_{xy} = +40$ MPa, $\tau_{yz} = 0$ MPa, and $\tau_{zx} = +30$ MPa. Determine the principal stresses and the maximum shearing stress at the point.

2-105 At a point in a stressed body the known stresses are $\sigma_x = 14$ ksi (T), $\sigma_y = 12$ ksi (T), $\sigma_z = 10$ ksi (T), $\tau_{xy} = +4$ ksi, $\tau_{yz} = -4$ ksi, and $\tau_{zx} = 0$ ksi. Determine the principal stresses and the maximum shearing stress at the point.

2-106* At a point in a stressed body, the known stresses are $\sigma_x = 60$ MPa (T), $\sigma_y = 90$ MPa (T), $\sigma_z = 60$ MPa (T), $\tau_{xy} = +120$ MPa, $\tau_{yz} = +75$ MPa, and $\tau_{zx} = +90$ MPa. Determine the principal stresses and the maximum shearing stress at the point.

2-107 At a point in a stressed body, the known stresses are $\sigma_x = 0$ ksi, $\sigma_y = 0$ ksi, $\sigma_z = 0$ ksi, $\tau_{xy} = +6$ ksi, $\tau_{yz} = +10$ ksi, and $\tau_{zx} = +8$ ksi. Determine the principal stresses and the maximum shearing stress at the point.

2-108 At a point in a stressed body, the known stresses are $\sigma_x = 72$ MPa (T), $\sigma_y = 32$ MPa (C), $\sigma_z = 0$ MPa, $\tau_{xy} = +21$ MPa, $\tau_{yz} = 0$ MPa, and $\tau_{zx} = +21$ MPa. Determine the principal stresses and the maximum shearing stress at the point.

Challenging Problems

2-109* At a point in a stressed body, the known stresses are $\sigma_x = 18$ ksi (C), $\sigma_y = 15$ ksi (C), $\sigma_z = 12$ ksi (C), $\tau_{xy} = -15$ ksi, $\tau_{yz} = +12$ ksi, and $\tau_{zx} = -9$ ksi. Determine

a. The principal stresses and the maximum shearing stress at the point.

b. The orientation of the plane on which the maximum compressive stress acts.

2-110* At a point in a stressed body, the known stresses are $\sigma_x = 75$ MPa (T), $\sigma_y = 35$ MPa (T), $\sigma_z = 55$ MPa (T), $\tau_{xy} = +45$ MPa, $\tau_{yz} = +28$ MPa, and $\tau_{zx} = +36$ MPa. Determine

a. The principal stresses and the maximum shearing stress at the point.

b. The orientation of the plane on which the maximum tensile stress acts.

2-111 At a point in a stressed body, the known stresses are $\sigma_x = 18$ ksi (T), $\sigma_y = 12$ ksi (T), $\sigma_z = 6$ ksi (T), $\tau_{xy} = +12$ ksi, $\tau_{yz} = -6$ ksi, and $\tau_{zx} = +9$ ksi. Determine

a. The principal stresses and the maximum shearing stress at the point.

b. The orientation of the plane on which the maximum tensile stress acts.

2-112 At a point in a stressed body, the known stresses are $\sigma_x = 100$ MPa (T), $\sigma_y = 100$ MPa (C), $\sigma_z = 80$ MPa (T), $\tau_{xy} = +50$ MPa, $\tau_{yz} = -70$ MPa, and $\tau_{zx} = -64$ MPa. Determine

a. The principal stresses and the maximum shearing stress at the point.

b. The orientation of the plane on which the maximum compressive stress acts.

SUMMARY

Application of the equations of equilibrium is usually just the first step in the solution of engineering problems. Using these equations, an engineer can determine the forces exerted on a structure by its supports, the forces on bolts and rivets that connect parts of a machine, or the internal forces in cables or rods that either support the structure or are a part of the structure. A second and equally important step is determining the internal effect of the forces on the structure or machine.

In the simplest qualitative terms, stress is the intensity of internal force. A body must be able to withstand the intensity of internal force; if not, the body may rupture or deform excessively. Force intensity (stress) is force divided by the area over which the force is distributed

$$\text{Stress} = \frac{\text{Force}}{\text{Area}} \tag{2-1}$$

Experimental studies indicate that materials respond differently to forces that tend to pull surfaces apart than to forces that tend to slide surfaces relative to each other. Therefore, the resultant internal force is usually resolved into normal and tangential (shear) components. These are the components that are used to compute the stresses on an internal surface. The ratio of the normal force N and the area on which it acts is the normal stress σ. Positive normal stresses are tensile normal stresses—they tend to stretch the material. The ratio of the tangential (shear) component of force V and the area on which it acts is the shear stress τ. The positive direction for shear stress is in a direction $90°$ counterclockwise from the outward normal direction to the surface.

In general, the stress at a given point on a transverse cross section of an axially loaded bar will not be the same as the average stress computed by dividing the total force F by the total cross-sectional area A. For long, slender, axially loaded members such as those found in trusses and similar structures, however, it is generally assumed that the normal stresses are uniformly distributed except in the vicinity of the points of application of the loads.

It can be shown that the shear stress cannot be uniformly distributed over the area. Therefore, the actual shear stress at any particular point and the maximum shear stress on a cross section will generally be different than the average shear stress. However, the design of simple connections is usually based on average stress considerations.

The stress at a point depends on the orientation of the surface (the area A) used to compute it. For example, in an axially loaded member the normal stress and the shear stress on a surface oriented at an angle θ to the axis are given by

$\sigma_n = \frac{P}{2A}(1 + \cos 2\theta)$ and $\tau_n = \frac{-P}{2A} \sin 2\theta$, respectively. That is, the normal stress is a maximum ($\sigma_{max} = P/A$) and the shear stress is zero on a transverse surface ($\theta = 0°$), and the shear stress is a maximum ($\tau_{max} = P/2A = \sigma_{max}/2$) on a surface oriented at an angle of $\theta = 45°$ to the axial direction.

Although the normal stress and the shear stress at a point both depend on the orientation of the surface on which they act, the state of stress at a point is completely determined by the normal stress and shear stress on three mutually orthogonal surfaces through the point. For two-dimensional or plane stress (for which $\sigma_z = \tau_{zx} = \tau_{zy} = \tau_{xz} = \tau_{yz} = 0$), the normal and shear stresses on a plane oriented at an angle θ relative to the x-axis are given by

$$\sigma_n = \frac{\sigma_x + \sigma_y}{2} + \frac{\sigma_x - \sigma_y}{2}\cos 2\theta + \tau_{xy} \sin 2\theta \tag{2-12b}$$

$$\tau_{nt} = -\frac{\sigma_x - \sigma_y}{2} \sin 2\theta + \tau_{xy} \cos 2\theta \tag{2-13b}$$

respectively.

For design purposes, the critical stresses at a point are usually the maximum normal stress and the maximum shearing stress. The maximum (and minimum) values of normal stress are called the principal stresses and always occur on planes free of shear stress, which are called principal planes. At every point in a stressed body, there exist three principal stresses acting on mutually orthogonal planes. In plane stress situations, two principal stresses are in the xy-plane

$$\sigma_{p1,p2} = \frac{\sigma_x + \sigma_y}{2} \pm \sqrt{\left(\frac{\sigma_x - \sigma_y}{2}\right)^2 + \tau_{xy}^2} \tag{2-15}$$

and the third principal stress is $\sigma_{p3} = \sigma_z = 0$. The principal directions are oriented at an angle θ_p given by

$$\tan 2\theta_p = \frac{2\tau_{xy}}{\sigma_x - \sigma_y} \tag{2-14}$$

relative to the x-axis.

The maximum in-plane shearing stress can be found from the principal stresses

$$\tau_p = \frac{\sigma_{p1} - \sigma_{p2}}{2} \tag{2-18}$$

The maximum shear stress at the point may be $(\sigma_{p1} - \sigma_{p2})/2$, $(\sigma_{p1} - 0)/2$, or $(0 - \sigma_{p2})/2$, depending on the relative magnitudes and signs of the principal stresses. The planes associated with the maximum shear stress bisect the angles between the planes experiencing maximum and minimum normal stresses.

The transformation equations for plane stress have a simple graphical representation called Mohr's circle. Normal stresses are plotted on the horizontal axis, with tensile stresses (positive) plotted to the right of the origin and compressive stresses (negative) plotted to the left of the origin. Shearing stresses are plotted on the vertical axis with those tending to produce a clockwise rotation of the stress element plotted above the σ-axis and those tending to produce a counterclockwise rotation of the stress element plotted below the σ-axis. Each point on Mohr's circle represents the state of stress on some surface through the given point. The angular separation of two points on Mohr's circle is in the same direction but twice as large as the physical angular separation of the two surfaces represented by the points.

Since Mohr's circle is simply a graphical representation of the transformation equations, every problem that can be solved using Mohr's circle can also be solved using the stress transformation equations. Mohr's circle is simply an alternative (graphical) method of representing and working with the stress transformation equations. Although Mohr's circle can be drawn to scale and used to obtain values of stresses and angles by direct measurements on the figure, it is more useful as a pictorial aid to the analyst who is performing analytical determinations of stresses and their directions at the point.

REVIEW PROBLEMS

2-113* A farmer is extracting a post from the ground using the structure shown in Fig. P2-113. If the force required to remove the post is 2000 lb, determine the normal stresses in the 0.25-in.-diameter cables *CE* and *AB*.

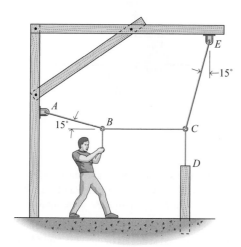

Figure P2-113

2-114* Two tie rods are used to support a 75-kN load as shown in Fig. P2-114. Determine

a. The minimum cross-sectional area required for each of the rods if the normal stress in each rod must be limited to 75 MPa.

b. The minimum diameters required for the pins at *A* and *C* if the shear stress in each pin must be limited to 100 MPa. Both pins are in single shear.

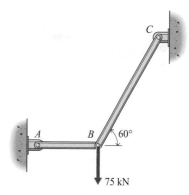

Figure P2-114

2-115 A pin-connected truss is loaded and supported as shown in Fig. P2-115. Determine

a. The normal stress in member *AC* if it has a cross-sectional area of 1.477 in.^{2}

b. The minimum cross-sectional area for member *CD* if the axial stress must be limited to 3500 psi.

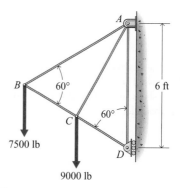

Figure P2-115

2-116* Determine the maximum axial load P that can be applied to the 150×180-mm wood compression block shown in Fig. P2-116 if the shear stress parallel to the grain must not exceed 1.40 MPa and the normal stress perpendicular to the grain must not exceed 12 MPa.

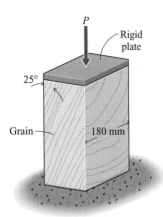

Figure P2-116

2-117 The stresses shown in Fig. P2-117 act at a point on the free surface of a stressed body. Determine the normal and shear stresses at this point on the inclined plane a–b shown in the figure. Show these stresses on a sketch of the plane on which they act.

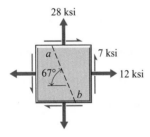

Figure P2-117

2-118 The stresses shown in Fig. P2-118 act at a point on the free surface of a stressed body. Determine the principal stresses and the maximum shear stress at the point. Show these stresses on a triangular stress element.

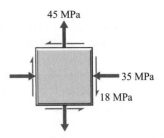

Figure P2-118

2-119* At a point in a structural member subjected to plane stress there are normal and shear stresses on horizontal and vertical planes through the point, as shown in Fig. P2-119. Using Mohr's circle,

a. Determine the principal stresses and the maximum shear stress at the point. Show these stresses on a triangular stress element.
b. Determine the normal and shear stresses on the inclined plane a–b shown in the figure. Show these stresses on a sketch of the plane on which they act.

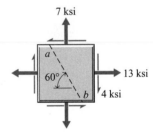

Figure P2-119

2-120 At a point in a stressed body, the known stresses are $\sigma_x = 53$ MPa (T), $\sigma_y = 28$ MPa (C), $\sigma_z = 36$ MPa (T), $\tau_{xy} = +24$ MPa, $\tau_{yz} = -18$ MPa, and $\tau_{zx} = +46$ MPa. Determine

a. The normal and shear stresses on a plane whose outward normal is oriented at angles of $40°$, $75°$, and $54°$ with the x-, y-, and z-axes, respectively.
b. The principal stresses and the maximum shear stress at the point.
c. The orientation of the plane on which the maximum tensile stress acts.

2-121 Demonstrate that Eq. 2-22 reduces to Eq. 2-15 for the state of plane stress.

Chapter 3
Analysis of Strain: Concepts and Definitions

3-1 INTRODUCTION

Relationships were developed in Chapter 2 between forces and stresses and between stresses on planes having different orientations at a point using equilibrium considerations. No assumptions involving deformations or materials used in fabricating the body were made; therefore, the results are valid for an idealized rigid body or for a real deformable body. In the design of structural elements or machine components, the deformations experienced by the body, as a result of the applied loads, often represent as important a design consideration as the stresses. For this reason, the nature of the deformations experienced by a real deformable body as a result of internal force or stress distributions will be studied, and methods to measure or compute deformations will be established.

3-2 DISPLACEMENT, DEFORMATION, AND STRAIN

3-2-1 Displacement
When a system of loads is applied to a machine component or structural element, individual points of the body generally move. This movement of a point with respect to some convenient reference system of axes is a vector quantity known as a *displacement*. For example, consider a body where displacements are restricted to the x–y plane. In Fig. 3-1 the solid lines represent a body before loads are applied, and the dashed lines represent the body after loads are applied and the body has deformed. Line AB of initial length L_i has deformed into a line of final length L_f. Point A is displaced to point A'. The vector from A to A' is called the *displacement* of point A. For this two-dimensional example, the scalar components of the displacement vector are u_A in the x-direction and v_A in the y-direction. If $L_i = L_f$, the body is rigid; and if $L_i \neq L_f$, the body has deformed. In some instances, displacements are associated with a translation and/or a rotation of the body as a whole and neither the size nor the shape of the body is changed. A study of displacements in which neither the size nor the shape of the body is changed is the concern of courses in rigid-body mechanics. When displacements induced by applied loads cause the size and/or shape of a body to be altered, individual points of the body move relative to one another. The change in any dimension associated with these relative displacements is known as a *deformation* and will be designated by the Greek letter delta (δ).

121

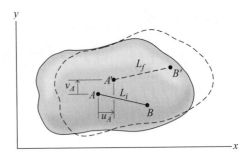

Figure 3-1

3-2-2 Deformation

Deformation may be related to force or stress or to a change in temperature. The deformation (δ_{AB}) of line AB in Fig. 3-1 is $\delta_{AB} = L_f - L_i$. Two rods of identical material and identical cross-sectional area subjected to different loads (Fig. 3-2a) can have the same deformation if the second rod is half as long as the first. Similarly, if two rods of identical material and identical cross-sectional area are subjected to identical loads (Fig. 3-2b), the deformation in a 2-L-long rod will be twice as large as the deformation in a 1-L-long rod. Therefore, a quantitative measure of the intensity of the deformation is needed, just as stress is used to measure the intensity of an internal force (force per unit area).

3-2-3 Strain

Strain (deformation per unit length) is the quantity used to measure the intensity of a deformation just as stress (force per unit area) is used to measure the intensity of an internal force. In Chapter 2, two types of stresses were defined: normal stresses and shearing stresses. This same classification is used for strains. *Normal strain*, designated by the Greek letter epsilon (ϵ), measures the change in size (elongation or contraction of an arbitrary line segment) of a body during deformation. *Shearing strain*, designated by the Greek letter gamma (γ), measures the change in shape (change in angle between two lines that are orthogonal in the undeformed state) of a body during a deformation. The deformation or strain may be the result of a stress, of a change in temperature, or of other physical phenomena such as grain growth or shrinkage. In this book only strains resulting from changes in temperature or stress are considered.

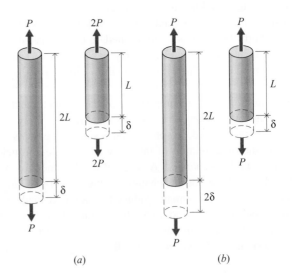

(a)   (b)

Figure 3-2

3-2-4 Average Axial Strain

The change in length (or width) of a simple bar under an axial load (see Fig. 3-3a) can be used to illustrate the idea of a normal strain. The average axial strain (a normal strain, hereafter called axial strain) ϵ_{avg} over the length of the bar is obtained by dividing the axial deformation δ_n by the original length of the bar (the length L)

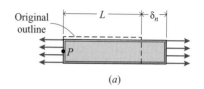

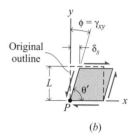

$$\epsilon_{avg} = \frac{\delta_n}{L} \qquad (3\text{-}1)$$

That is, the axial strain is the deformation δ_n in the direction of the length divided by the length L.

3-2-5 Axial Strain at a Point

In those cases in which the deformation is nonuniform along the length of the bar (a long bar hanging under its own weight, for example), the average axial strain given by Eq. 3-1 may be significantly different from the axial strain at an arbitrary point P along the bar. The axial strain at a point can be determined by making the length over which the axial deformation is measured smaller and smaller. In the limit as $\Delta L \to 0$, a quantity defined as the axial strain at the point, $\epsilon(P)$, is obtained. This limit process is indicated by the expression

Figure 3-3

$$\epsilon(P) = \lim_{\Delta L \to 0} \frac{\Delta \delta_n}{\Delta L} = \frac{d\delta_n}{dL} \qquad (3\text{-}2)$$

3-2-6 Shearing Strain

In a similar manner a deformation involving a change in shape can be used to illustrate a shearing strain. The average shearing strain γ_{avg} is obtained by dividing the deformation δ_s in a direction normal to the length L by the length (Fig. 3-3b)

$$\gamma_{avg} = \frac{\delta_s}{L} = \tan \phi \qquad (3\text{-}3)$$

Since δ_s/L is usually very small (typically $\delta_s/L < 0.001$), $\sin \phi \cong \tan \phi \cong \phi$ where ϕ is measured in radians. Therefore, $\gamma_{avg} = \phi = \delta_s/L$ is the decrease in the angle between two reference lines that are orthogonal in the undeformed state. Again, for those cases in which the deformation is nonuniform, the shearing strain at a point, $\gamma_{xy}(P)$, associated with two orthogonal reference lines x and y is obtained by measuring the shearing deformation as the size of the element is made smaller and smaller. In the limit as $\Delta L \to 0$

$$\gamma_{xy}(P) = \lim_{\Delta L \to 0} \frac{\Delta \delta_s}{\Delta L} = \frac{d\delta_s}{dL} \qquad (3\text{-}4a)$$

The angle γ_{xy} is difficult to observe and even more difficult to measure. An equivalent expression for shearing strain that is sometimes useful for calculations is

$$\gamma_{xy}(P) = \frac{\pi}{2} - \theta' \qquad (3\text{-}4b)$$

In this expression, θ' is the angle in the deformed state between the two initially orthogonal reference lines.

3-2-7 Units of Strain

Equations 3-1 through 3-4 indicate that both normal and shearing strains are dimensionless quantities; however, normal strains are frequently expressed in units of inch per inch (in./in.) or micro-inch per inch (μin./in.), while shearing strains are expressed in radians or microradians. [The symbol μ is frequently used to indicate micro- (10^{-6}).]

3-2-8 Sign Convention for Strains

From the definition of normal strain given by Eq. 3-1 or 3-2 it is evident that normal strain is positive when a line elongates and negative when the line contracts. In general, if the axial stress is tensile, the axial deformation will be an elongation. Therefore, positive normal strains are referred to as *tensile strains*. The reverse will be true for compressive axial stresses; therefore, negative normal strains are referred to as *compressive strains*. From Eq. 3-4 it is evident that shearing strains will be positive if the angle between reference lines decreases. If the angle increases, the shearing strain is negative. Positive and negative shearing strains are not given special names. Normal and shearing strains for most engineering materials in the elastic range (see Section 4-2) seldom exceed values of 0.2 percent (0.002 mm/mm or 0.002 in./in. or 0.002 rad or just 0.002).

> **Example Problem 3-1**   A 1.00-in.-diameter steel bar is 8 ft long. The diameter is reduced to 1/2 in. in a 2-ft central portion of the bar. When an axial load is applied to the ends of the bar, the axial strain in the central portion of the bar is 960 μin./in., the total elongation of the bar is 0.04032 in., and the diameter of the central portion shrinks to 0.49986 in. Determine

(a) The elongation of the central portion of the bar.

(b) The axial strain in the end portions of the bar.

(c) The diametral strain in the central portion of the bar.

SOLUTION

▶ The central portion of the bar has an undeformed length of 2 ft. The end portions have an undeformed length of 8 ft − 2 ft = 6 ft.

(a) The elongation of the central portion of the bar is obtained by using Eq. 3-1. Thus,

$$\delta_C = \epsilon_{\text{avg}} L = 960(10^{-6})(2)(12)$$
$$= 0.02304 \text{ in.} \cong 0.0230 \text{ in.} \qquad \textbf{Ans.}$$

(b) The elongation of the end portions of the bar is

$$\delta_E = \delta_{\text{total}} - \delta_C = 0.04032 - 0.02304 = 0.01728 \text{ in.}$$

The axial strain in the end portions of the bar is obtained by using Eq. 3-1 as follows:

$$\epsilon_E = \frac{\delta_E}{L} = \frac{0.01728}{6(12)} = 240(10^{-6}) = 240 \,\mu\text{in./in.} \qquad \textbf{Ans.}$$

(c) The strain in the diametral direction is given by Eq. 3-1. Thus, if D represents the diameter of the bar in the central portion,

$$\epsilon_D = \frac{\delta_D}{L} = \frac{0.49986 - 0.5}{0.5} = -280(10^{-6}) = -280\,\mu\text{in./in.} \qquad \textbf{Ans.}$$

The negative sign for ϵ_D indicates that the diameter decreases in length.

Example Problem 3-2

The shear force V shown in Fig. 3-4a produces an average shearing strain γ_{avg} of 1000 μrad in the block of material. Determine the horizontal movement of point A (Fig. 3-4b) resulting from application of the shear force V.

SOLUTION

The horizontal movement of point A is obtained by using Eq. 3-3. Thus,

$$\delta_A = \gamma_{\text{avg}}L = 1000(10^{-6})(10)$$

$$= 0.0100 \text{ mm} = 10.00\,\mu\text{m} \qquad \textbf{Ans.}$$

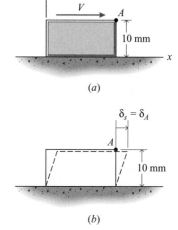

(a)

(b)

Figure 3-4

Example Problem 3-3

A rigid bar $FEBC$ is supported by two steel deformable bars AB and DE, as shown in Fig. 3-5a. There is no strain in the vertical bars before the load P is applied. After the load P is applied, the axial strain in bar DE is 0.0006 in./in. Determine

(a) The axial strain in bar AB.
(b) The axial strain in bar AB if there is a 0.001-in. clearance in the connection at B before the load is applied.

SOLUTION

Since the bar $FEBC$ is rigid, it pivots about the pin at F. The initial and final configurations of bar $FEBC$ are shown in Fig. 3-5b (greatly exaggerated). End B of bar AB moves along a circular arc of radius BF. The displacement of B (the vector BB') has components $-u_B$ and $-v_B$, both being negative as they are in the negative coordinate directions. The strain in member AB is given by Eq. 3-1, in which $\delta_n = \delta_{AB}$, $\delta_{AB} = L_f - L_i$, $L_i = AB$, and $L_f = AB'$. The axial deformation in bar AB is

$$\delta_{AB} = \sqrt{(L + v_B)^2 + u_B^2} - L$$

which may be written

$$\delta_{AB}^2 + 2\delta_{AB}L + L^2 = L^2 + 2v_B L + u_B^2$$

For small displacements, the terms involving the squares of the displacements (δ^2_{AB}, v^2_B, and u^2_B) may be neglected compared to the remaining nonsquared terms ($2\delta_{AB}L$ and $2v_B L$). Therefore

$$\delta_{AB} \cong v_B$$

In a similar manner

$$\delta_{DE} \cong v_E$$

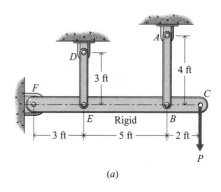

(a)

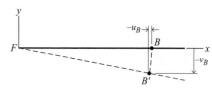

(b)

Figure 3-5(a,b)

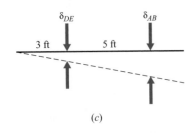

(c)

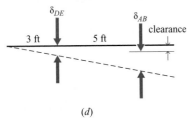

(d)

Figure 3-5(c,d)

▶ By similar triangles, the vertical displacement of the ends of bars AB and DE are related by $\dfrac{a}{8} = \dfrac{b}{3}$ where $b = \delta_{DE}$ and $a = \delta_{AB}$ + clearance. Therefore, δ_{AB} + clearance = $\dfrac{8}{3}\delta_{DE}$.

Thus, for small displacements, the axial deformation in either bar is equal to the component of the displacement of one end of the bar (relative to the other end) taken in the direction of the undeformed orientation of the bar.

The change in length of bar DE is given by Eq. 3-1 as

$$\delta_{DE} = \epsilon_{DE}L_{DE} = 0.0006(3)$$
$$= 0.0018 \text{ ft} = 0.0216 \text{ in.}$$

The relationship between the deformations in bars AB and DE is controlled by the rigid bar $FEBC$ because it does not deform but simply rotates about the support, as shown in Figs. 3-5c and 3-5d. Sketches of the type shown in Figs. 3-5c and 3-5d are referred to as deformation diagrams.

(a) From the geometry of Fig. 3-5c, the change in length of bar AB is

$$\delta_{AB} = \frac{8}{3}\delta_{DE} = \frac{8}{3}(0.0216) = 0.0576 \text{ in.}$$

The strain in bar AB is given by Eq. 3-1 as

$$\epsilon_{AB} = \frac{\delta_{AB}}{L_{AB}} = \frac{0.0576}{4(12)} = 0.001200 \text{ in./in} = 1200 \ \mu\text{in./in.} \qquad \textbf{Ans.}$$

(b) From the geometry of Fig. 3-5d, the relationship between the deformations in bars AB and DE is

$$\delta_{AB} + \text{clearance} = \frac{8}{3}\delta_{DE}$$

Thus

$$\delta_{AB} = \frac{8}{3}\delta_{DE} - \text{clearance} = \frac{8}{3}(0.0216) - 0.001 = 0.0566 \text{ in.}$$

The strain in bar AB is given by Eq. 3-1 as

$$\epsilon_{AB} = \frac{\delta_{AB}}{L_{AB}} = \frac{0.0566}{4(12)} = 0.001179 \text{ in./in.} = 1179 \ \mu\text{in./in.} \qquad \textbf{Ans.}$$

PROBLEMS

MecMovie Activities and Problems

MM3.1 Normal strain in rod assembly. Example; Try One. Determine axial strain.

MM3.2 Normal strain in pinned rod assembly. Example; Try One. Determine axial strain.

MM3.3 Normal strain - basic problems. Concept checkpoints. Use normal strain concepts for four introductory problems.

Introductory Problems

3-1* A 25-ft length of steel wire is subjected to a tensile load that produces a change in length of 0.625 in. Determine the axial strain in the wire.

3-2* Compression tests of concrete indicate that concrete fails when the axial compressive strain is 1200 μm/m. Determine the maximum change in length that a 200-mm-diameter $\times$ 400-mm-long concrete test specimen can tolerate before failure occurs.

3-3 A 1/2-in.-diameter structural steel rod was loaded in tension to fracture. The 8.00-in. gage length of the rod was marked off in 1.00-in. lengths before loading. After the rod broke, the strain in the 2.00-in. length containing the fracture was found to be 0.450 in./in., and the total elongation of the other 6 in. was found to be 1.50 in. Determine the average strain for the 8.00-in. gage length.

3-4 A structural steel bar was loaded in tension to fracture. The 200-mm gage length of the bar was marked off in 25-mm lengths before loading. After the rod broke, the 25-mm segments were found to have lengthened to 30.0, 30.5, 31.5, 34.0, 44.5, 32.0, 31.0, and 30.0 mm, consecutively. Determine

a. The average strain over the 200-mm gage length.
b. The maximum average strain over any 50-mm length.

Intermediate Problems

3-5* The 0.5 × 2 × 4-in. rubber mounts shown in Fig. P3-5 are used to isolate the vibrational motion of a machine from its supports. Determine the average shearing strain in the rubber mounts if the rigid frame displaces 0.01 in. vertically relative to the support.

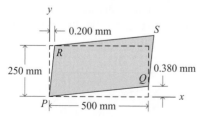

Figure P3-5

3-6* A thin rectangular plate is uniformly deformed as shown in Fig. P3-6. Determine the shearing strain γ_{xy} at P.

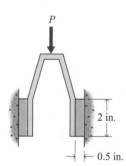

Figure P3-6

3-7 Mutually perpendicular axes in an unstressed member were found to be oriented at 89.92° when the member was stressed. Determine the shearing strain associated with these axes in the stressed member.

3-8 A thin rectangular plate $ABCD$ is uniformly deformed into the dashed line shown in Fig. P3-8. Determine

a. The average normal strain of side AB of the plate.
b. The average shearing strain γ_{xy}.

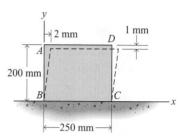

Figure P3-8

Challenging Problems

3-9* A rigid steel plate A is supported by three rods as shown in Fig. P3-9. There is no strain in the rods before the load P is applied. After load P is applied, the axial strain in rod DE is 800 μin./in. Determine

a. The axial strain in rods BC.
b. The axial strain in rods BC if there is a 0.006-in. clearance in the connections between plate A and rod BC before the load is applied.

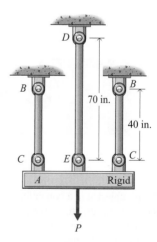

Figure P3-9

3-10* The load P produces an axial strain in the steel post D of Fig. P3-10 of 0.0075 m/m. Determine

a. The axial strain in the aluminum rod CE.
b. The axial strain in the aluminum rod CE if there is a 0.10-mm clearance in the connection at E in addition to the

0.09-mm clearance between B and D before the load P is applied.

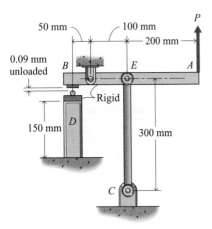

Figure P3-10

3-11 The sanding-drum mandrel shown in Fig. P3-11 is made for use with a hand drill. The mandrel is made from a rubber-like material that expands when the nut is tightened to secure the sanding drum placed over the outside surface. If the diameter D of the mandrel increases from 2.00 in. to 2.15 in. as the nut is tightened, determine

a. The average normal strain along a diameter of the mandrel.
b. The circumferential strain at the outside surface of the mandrel.

Figure P3-11

3-12* A steel rod is subjected to a nonuniform heating that produces an extensional (axial) strain that is proportional to the square of the distance from the unheated end ($\epsilon = kx^2$). If the strain is 1250 μm/m at the midpoint of a 3.00-m-long rod, determine

a. The change in length of the rod.
b. The average axial strain over the length L of the rod.
c. The maximum axial strain in the rod.

3-13 The load P produces an axial strain in the brass post B of Fig. P3-13 of 0.0014 in./in. Determine the axial strain in the aluminum alloy rod AD.

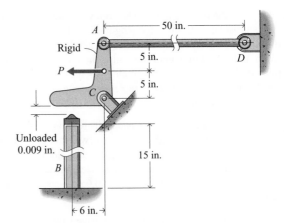

Figure P3-13

3-14 A steel sleeve is connected to a steel shaft with a flexible rubber insert, as shown in Fig. P3-14. The insert has an inside diameter of 85 mm and an outside diameter of 110 mm. When the unit is subjected to a torque T, the shaft rotates $1.5°$ with respect to the sleeve. Assume that radial lines in the unstressed state remain straight as the rubber deforms. Determine the shearing strain $\gamma_{r\theta}$ in the rubber insert

a. At the inside surface.
b. At the outside surface.

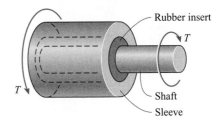

Figure P3-14

3-15* A thin triangular plate is uniformly deformed as shown in Fig. P3-15. Determine the shearing strain at P associated with the two edges (PQ and PR) that were orthogonal in the undeformed plate.

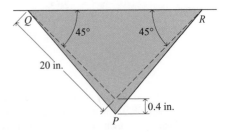

Figure P3-15

3-3 THE STATE OF STRAIN AT A POINT

The material of Section 3-2 serves to convey the concept of strain as a unit deformation, but it is inadequate for other than one-directional loading. The extension of the concept to biaxial loading is essential because of the important role played by strain in experimental methods of stress evaluation. In many practical engineering problems involving the design of structural or machine elements, the configuration and loading are too complicated to permit stress determination solely by mathematical analysis; hence, this technique is supplemented by laboratory measurements.

Strains can be measured by several methods, but, except for the simplest cases, stresses cannot be obtained directly. Therefore, the usual procedure used in experimental stress analysis is to measure the strains and calculate the state of stress by using the stress-strain equations presented later in Section 4-3.

The complete state of strain at an arbitrary point P in a body under load can be determined by considering the deformation associated with a small volume of material surrounding the point. For convenience, the volume is normally assumed to have the shape of a rectangular parallelepiped with its faces oriented perpendicular to the reference axes x, y, and z in the undeformed state, as shown in Fig. 3-6a. Since the element of volume is very small, deformations are assumed to be uniform; therefore, parallel planes remain plane and parallel and straight lines remain straight in the deformed element, as shown in Fig. 3-6b. The final size of the deformed element is determined by the lengths of the three edges dx', dy', and dz'. The distorted shape of the element is determined by the angles θ'_{xy}, θ'_{yz}, and θ'_{zx} between faces.

The Cartesian components of strain at the point can be expressed in terms of the deformations by using the definitions of normal and shearing strain presented in Section 3-2. These are the strain components associated with the Cartesian components of stress discussed in Section 2-7 and shown in Fig. 2-24. Thus,

$$\epsilon_x = \frac{dx' - dx}{dx} = \frac{d\delta_x}{dx} \qquad \gamma_{xy} = \frac{\pi}{2} - \theta'_{xy}$$

$$\epsilon_y = \frac{dy' - dy}{dy} = \frac{d\delta_y}{dy} \qquad \gamma_{yz} = \frac{\pi}{2} - \theta'_{yz} \qquad (3\text{-}5a)$$

$$\epsilon_z = \frac{dz' - dz}{dz} = \frac{d\delta_z}{dz} \qquad \gamma_{zx} = \frac{\pi}{2} - \theta'_{zx}$$

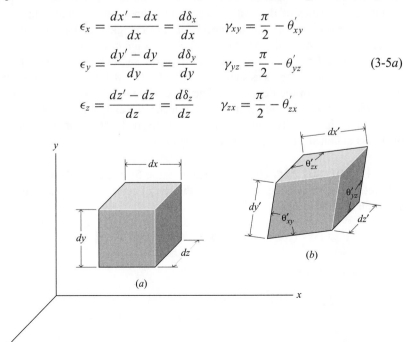

(a)

(b)

Figure 3-6

In a similar manner, the normal strain component associated with a line oriented in an arbitrary n direction and the shearing strain component associated with two arbitrary orthogonal lines oriented in the n and t directions in the undeformed element are given by

$$\epsilon_n = \frac{dn' - dn}{dn} = \frac{d\delta_n}{dn} \qquad \gamma_{nt} = \frac{\pi}{2} - \theta'_{nt} \qquad (3\text{-}5b)$$

Alternative forms of Eqs. 3-5 that will be useful in later developments are

$$
\begin{aligned}
dx' &= (1 + \epsilon_x)\,dx & \theta'_{xy} &= \frac{\pi}{2} - \gamma_{xy} \\
dy' &= (1 + \epsilon_y)\,dy & \theta'_{yz} &= \frac{\pi}{2} - \gamma_{yz} \\
dz' &= (1 + \epsilon_z)\,dz & \theta'_{zx} &= \frac{\pi}{2} - \gamma_{zx} \\
dn' &= (1 + \epsilon_n)\,dn & \theta'_{nt} &= \frac{\pi}{2} - \gamma_{nt}
\end{aligned}
\qquad (3\text{-}6)
$$

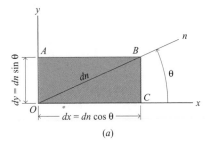

(a)

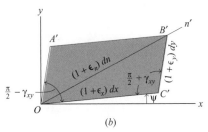

(b)

Figure 3-7

3-4 THE STRAIN TRANSFORMATION EQUATIONS FOR PLANE STRAIN

The method of relating the components of strain associated with a Cartesian coordinate system to the normal and shearing strains associated with other orthogonal directions will be illustrated by considering the two-dimensional or plane strain case. If the x–y plane is taken as the reference plane, then for conditions of plane strain $\epsilon_z = \gamma_{zx} = \gamma_{zy} = 0$.[1]

Consider Fig. 3-7a in which the shaded rectangle represents a small unstrained element of material having the configuration of a rectangular parallelepiped. The sides of the rectangle are along directions for which the strains ϵ_x, ϵ_y, and γ_{xy} are known. The proportion of the rectangle is chosen so that the diagonal OB points in the direction n for which the strain ϵ_n is to be determined. The angle θ is considered positive when measured counterclockwise from the positive x-axis to the positive n-axis as shown in Fig. 3-7a. This definition of positive angle θ is consistent with the definition of positive angle θ for stress transformation. When the body is subjected to a system of loads, the element assumes the shape indicated in Fig. 3-7b. The dimensions of the deformed element are given in terms of strains obtained from Eqs. 3-6, as shown in Fig. 3-7b.

3-4-1 Normal Strain ϵ_n
An expression for the normal strain ϵ_n in the n-direction can be obtained by applying the law of cosines to the triangle $OC'B'$ shown in Fig. 3-7b. Thus,

$$(OB')^2 = (OC')^2 + (C'B')^2 - 2(OC')(C'B')\cos\left(\frac{\pi}{2} + \gamma_{xy}\right)$$

or in terms of strains

$$
\begin{aligned}
[(1 + \epsilon_n)dn]^2 = {}& [(1 + \epsilon_x)dx]^2 + [(1 + \epsilon_y)dy]^2 \\
& - 2[(1 + \epsilon_x)dx][(1 + \epsilon_y)dy][-\sin\gamma_{xy}]
\end{aligned}
\qquad (a)
$$

---

[1]Note that although there is no strain in the z-direction, $\epsilon_z = 0$, there must be a stress in the z-direction, $\sigma_z \neq 0$. That is, a stress is required in the z-direction to prevent deformation in the z-direction.

Substituting $dx = dn \cos \theta$ and $dy = dn \sin \theta$ (see Fig. 3-7a) into Eq. (a) yields

$$(1 + \epsilon_n)^2 (dn)^2 = (1 + \epsilon_x)^2 (dn)^2 (\cos^2 \theta) + (1 + \epsilon_y)^2 (dn)^2 (\sin^2 \theta)$$
$$+ 2(dn)^2 (\sin \theta)(\cos \theta)(1 + \epsilon_x)(1 + \epsilon_y)(\sin \gamma_{xy}) \qquad (b)$$

Since the strains are small, it follows that $\epsilon^2 \ll \epsilon$, $\sin \gamma \cong \gamma$, and so forth; hence, all second-degree terms such as ϵ^2, $\gamma\epsilon$, and the like can be neglected as Eq. (b) is expanded to become

$$1 + 2\epsilon_n = (1 + 2\epsilon_x) \cos^2 \theta + (1 + 2\epsilon_y) \sin^2 \theta + 2\gamma_{xy} \sin \theta \cos \theta$$

which reduces to

$$\epsilon_n = \epsilon_x \cos^2 \theta + \epsilon_y \sin^2 \theta + \gamma_{xy} \sin \theta \cos \theta \qquad (3\text{-}7a)$$

or in terms of the double angle

$$\epsilon_n = \frac{\epsilon_x + \epsilon_y}{2} + \frac{\epsilon_x - \epsilon_y}{2} \cos 2\theta + \frac{\gamma_{xy}}{2} \sin 2\theta \qquad (3\text{-}7b)$$

Equations 3-7a and 3-7b are called the strain transformation equations for normal strain when the state of strain is two-dimensional or plane strain. Note the similarity between the normal strain transformation equations (Eqs. 3-7a and 3-7b) and the normal stress transformation equations (Eqs. 2-12a and 2-12b).

3-4-2 Shearing Strain γ_{nt}

The shearing strain γ_{nt} measures the amount by which the right angle between the n- and t-directions decreases as the material deforms. As the material deforms, the n-direction rotates counterclockwise through an angle ϕ_n, as shown in Fig. 3-8. Applying the law of sines to triangle $OC'B'$,

$$\frac{OB'}{\sin \angle OC'B'} = \frac{B'C'}{\sin \angle B'OC'}$$

which gives

$$B'C'\sin \angle OC'B' = OB'\sin \angle B'OC'$$

or in terms of the strains,

$$(1 + \epsilon_y)dy \sin \left(\frac{\pi}{2} + \gamma_{xy} \right) = (1 + \epsilon_n) \, dn \sin [\theta + (\phi_n - \psi)] \qquad (c)$$

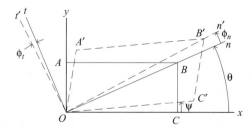

Figure 3-8

Since the strains (ϵ_x, ϵ_y, γ_{xy}, and ϵ_n) and the angles ϕ_n and ψ are all small (but the angle θ is not necessarily small),

$$\sin\left(\frac{\pi}{2} + \gamma_{xy}\right) = \cos\gamma_{xy} \cong 1$$

$$\sin[\theta + (\phi_n - \psi)] = \sin\theta\,\cos(\phi_n - \psi) + \cos\theta\,\sin(\phi_n - \psi)$$
$$\cong \sin\theta + (\phi_n - \psi)\cos\theta$$

and Eq. (c) can be written

$$(1 + \epsilon_y)dy \cong (1 + \epsilon_n)dn\,[\sin\theta + (\phi_n - \psi)\cos\theta] \qquad (d)$$

where $dy = dn\,\sin\theta$ (see Fig. 3-7a). Therefore, Eq. (d) can be reduced to

$$(\epsilon_y - \epsilon_n)\sin\theta \cong (\phi_n - \psi)\cos\theta + \epsilon_n(\phi_n - \psi)\cos\theta$$
$$\cong (\phi_n - \psi)\cos\theta \qquad (e)$$

since ϵ_n, ϕ_n, and ψ are all small. Substituting Eq. (3-7a) into Eq. (e) and solving for ϕ_n yields

$$\phi_n = -(\epsilon_x - \epsilon_y)\sin\theta\cos\theta - \gamma_{xy}\sin^2\theta + \psi \qquad (f)$$

Equation (f) gives the counterclockwise rotation of a line which makes an angle of θ with the x-axis initially. Thinking of ϕ_n as $\phi(\theta)$, the counterclockwise rotation of the t-axis can be written

$$\phi_t = \phi\left(\theta + \frac{\pi}{2}\right)$$
$$= -(\epsilon_x - \epsilon_y)\sin\left(\theta + \frac{\pi}{2}\right)\cos\left(\theta + \frac{\pi}{2}\right) - \gamma_{xy}\sin^2\left(\theta + \frac{\pi}{2}\right) + \psi$$
$$= (\epsilon_x - \epsilon_y)\cos\theta\sin\theta - \gamma_{xy}\cos^2\theta + \psi \qquad (g)$$

Finally, the shearing strain γ_{nt} is the decrease in the right angle between the orthogonal n- and t-directions or the difference between the rotations ϕ_n and ϕ_t. Therefore,

$$\gamma_{nt} = \phi_n - \phi_t$$
$$= -2(\epsilon_x - \epsilon_y)\sin\theta\cos\theta + \gamma_{xy}(\cos^2\theta - \sin^2\theta) \qquad (3\text{-}8a)$$

or in terms of the double angle

$$\gamma_{nt} = -(\epsilon_x - \epsilon_y)\sin 2\theta + \gamma_{xy}\cos 2\theta \qquad (3\text{-}8b)$$

Equations 3-8a and 3-8b are the strain transformation equations for shearing strain when the state of strain is two-dimensional or plane strain. Note the similarity between the shear strain transformation equations (Eqs. 3-8a and 3-8b) and the shear stress transformation equations (Eqs. 2-13a and 2-13b).

Equations 3-7 and 3-8 provide a means for determining the normal strain ϵ_n associated with a line oriented in an arbitrary n-direction in the x–y plane and the shearing strain γ_{nt} associated with any two orthogonal lines oriented in the n- and t-directions in the x–y plane when the strains ϵ_x, ϵ_y, and γ_{xy} associated with the coordinate directions are known. When these equations are used, the sign conventions used in their development must be rigorously followed. The sign conventions used are as follows:

1. Tensile strains are positive; compressive strains are negative.
2. Shearing strains that decrease the angle between the two lines at the origin of coordinates are positive.
3. Angles measured counterclockwise from the reference x-axis are positive.
4. The (n, t, z) axes have the same order as the (x, y, z) axes. Both sets of axes form a right-hand coordinate system.

■ Example Problem 3-4 The strain components at a point are $\epsilon_x = +800\mu$, $\epsilon_y = -1000\mu$, and $\gamma_{xy} = -600\mu$. Determine the strain components ϵ_n, ϵ_t, and γ_{nt} if the xy- and nt-axes are oriented as shown in Fig. 3-9.

SOLUTION
The n-axis is located at an angle $\theta_n = -30°$ with respect to the x-axis; therefore, the strain ϵ_n is given by Eq. 3-7a as

$$\epsilon_n = \epsilon_x \cos^2 \theta_n + \epsilon_y \sin^2 \theta_n + \gamma_{xy} \sin \theta_n \cos \theta_n$$
$$= 800 \cos^2(-30°) + (-1000) \sin^2(-30°) + (-600) \sin(-30°) \cos(-30°)$$
$$= 609.8\mu \cong 610\mu \qquad \textbf{Ans.}$$

The t-axis is located at an angle $\theta_t = +60°$ with respect to the x-axis; therefore, the strain ϵ_t is given by Eq. 3-7a as

$$\epsilon_t = \epsilon_x \cos^2 \theta_t + \epsilon_y \sin^2 \theta_t + \gamma_{xy} \sin \theta_t \cos \theta_t$$
$$= 800 \cos^2(+60°) + (-1000) \sin^2(+60°) + (-600) \sin(+60°) \cos(+60°)$$
$$= -809.8\mu \cong -810\mu \qquad \textbf{Ans.}$$

In a similar manner, the shearing strain γ_{nt} is given by Eq. 3-8a as

$$\gamma_{nt} = -2(\epsilon_x - \epsilon_y) \sin \theta_n \cos \theta_n + \gamma_{xy} (\cos^2 \theta_n - \sin^2 \theta_n)$$
$$= -2[800 - (-1000)] \sin(-30°) \cos(-30°)$$
$$\qquad + (-600)[\cos^2(-30°) - \sin^2(-30°)]$$
$$= 1258.8 \ \mu\text{rad} \cong 1259 \ \mu\text{rad} \qquad \textbf{Ans.}$$

Therefore, a line element in the n-direction has increased in length, a line element in the t-direction has decreased in length, and the angle at the origin of the nt-axes is now less than 90°.

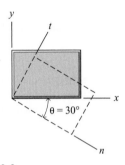

Figure 3-9

3.1

3.2

▶ As a check of the small angle assumptions, note that when $\gamma_{nt} = 0.0012588$ rad, $\sin \gamma_{nt} = 0.00125900 \cong \gamma_{nt}$ and $\cos \gamma_{nt} = 0.9999992 \cong 1$.

3.3

PROBLEMS

Introductory Problems

3-16* The thin rectangular plate shown in Fig. P3-16 is uniformly deformed such that $\epsilon_x = -2000 \ \mu\text{m/m}$, $\epsilon_y = -1500 \ \mu\text{m/m}$, and $\gamma_{xy} = +1250 \ \mu\text{rad}$. Determine the normal strain ϵ_n in the plate.

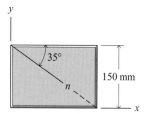

Figure P3-16

3-17* The thin rectangular plate shown in Fig. P3-17 is uniformly deformed such that $\epsilon_x = +880 \ \mu\text{in./in.}$, $\epsilon_y = +960 \ \mu\text{in./in.}$, and $\gamma_{xy} = -750 \ \mu\text{rad}$. Determine

a. The normal strain ϵ_{AC} along diagonal AC of the plate.
b. The normal strain ϵ_{BD} along diagonal BD of the plate.

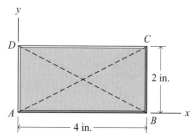

Figure P3-17

3-18 The thin square plate shown in Fig. P3-18 is uniformly deformed such that $\epsilon_x = +1750 \ \mu\text{m/m}$, $\epsilon_y = -2200 \ \mu\text{m/m}$, and $\gamma_{xy} = -800 \ \mu\text{rad}$. Determine

a. The normal strains ϵ_n and ϵ_t in the plate.
b. The shearing strain γ_{nt} in the plate.

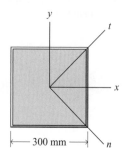

Figure P3-18

Intermediate Problems

3-19* A thin square plate 30 in. on a side is uniformly deformed into the rectangle indicated by the dashed lines shown in Fig. P3-19. Determine

a. The normal strains ϵ_x and ϵ_y and the shearing strain γ_{xy}.
b. The normal strain ϵ_n.

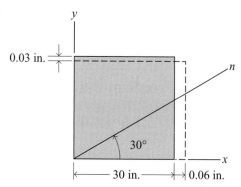

Figure P3-19

3-20* A plate undergoing plane strain is deformed into the dashed shape shown in Fig. P3-20. Determine the strains ϵ_x, ϵ_y, and γ_{xy}.

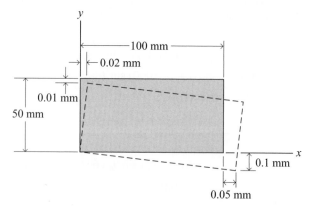

Figure P3-20

3-21 The strain components at a point in a body undergoing plane strain are $\epsilon_x = -800 \ \mu\text{in./in.}$, $\epsilon_y = +640 \ \mu\text{in./in.}$, and $\gamma_{xy} = -960 \ \mu\text{rad}$. Determine the strain components ϵ_n, ϵ_t, and γ_{nt}

when the nt-axes are oriented at $\theta = 42°$ counterclockwise from the xy-axes as shown in Fig. P3-21.

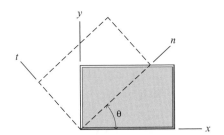

Figure P3-21

3-22 The strain components at a point in a body undergoing plane strain are $\epsilon_x = +720\ \mu\text{m/m}$, $\epsilon_y = -480\ \mu\text{m/m}$, and $\gamma_{xy} = +360\ \mu\text{rad}$. Determine the strain components ϵ_n, ϵ_t, and γ_{nt} when the nt-axes are oriented at $\theta = 30°$ clockwise from the xy-axes as shown in Fig. P3-22.

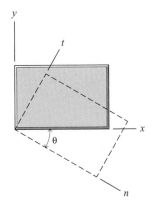

Figure P3-22

Challenging Problems

3-23 Using the small strain approximation, show that Eq. (*b*) of Section 3-4 reduces to Eq. 3-7*a* of Section 3-4.

3-24* The thin rectangular plate shown in Fig. P3-24 is uniformly deformed such that $\epsilon_{AB} = -1200\ \mu\text{m/m}$, $\epsilon_{BD} = +750\ \mu\text{m/m}$, and $\epsilon_{AD} = -600\ \mu\text{m/m}$. Determine the normal strain ϵ_y and the shearing strain γ_{xy} in the plate.

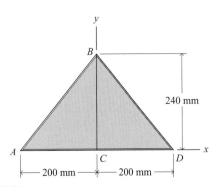

Figure P3-24

3-25* The thin rectangular plate shown in Fig. P3-25 is uniformly deformed such that $\epsilon_n = +1575\ \mu\text{in./in.}$, $\epsilon_t = +1350\ \mu\text{in./in.}$, and $\epsilon_x = +1250\ \mu\text{in./in.}$. Determine

a. The shearing strain γ_{nt} in the plate.
b. The normal strain ϵ_y in the plate.

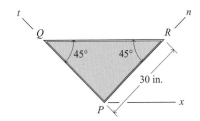

Figure P3-25

3-26 The thin rectangular plate shown in Fig. P3-26 is uniformly deformed such that $\epsilon_x = +1950\ \mu\text{m/m}$, $\epsilon_y = -1625\ \mu\text{m/m}$, and $\epsilon_n = -1275\ \mu\text{m/m}$. Determine

a. The shearing strain γ_{xy} in the plate.
b. The normal strain ϵ_{QR} along the diagonal QR of the plate.

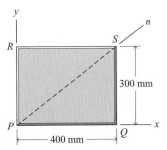

Figure P3-26

3-5 PRINCIPAL STRAINS AND MAXIMUM SHEAR STRAIN

The similarity between Eqs. 3-7 and 3-8 for plane strain and Eqs. 2-12 and 2-13 for plane stress indicates that all the equations developed for plane stress can be applied to plane strain by substituting ϵ_x for σ_x, ϵ_y for σ_y, and $\gamma_{xy}/2$ for τ_{xy}. Thus, from Eqs. 2-14, 2-15, and 2-17, expressions are obtained for determining in-plane principal directions, in-plane principal strains, and the maximum in-plane shear strain. Thus,

$$\tan 2\theta_p = \frac{\gamma_{xy}}{\epsilon_x - \epsilon_y} \tag{3-9}$$

$$\epsilon_{p1}, \epsilon_{p2} = \frac{\epsilon_x + \epsilon_y}{2} \pm \sqrt{\left(\frac{\epsilon_x - \epsilon_y}{2}\right)^2 + \left(\frac{\gamma_{xy}}{2}\right)^2} \tag{3-10}$$

$$\gamma_p = 2\sqrt{\left(\frac{\epsilon_x - \epsilon_y}{2}\right)^2 + \left(\frac{\gamma_{xy}}{2}\right)^2} \tag{3-11}$$

In the previous equations, normal strains that are tensile and shear strains that decrease the angle between the faces of the element at the origin of coordinates (see Fig. 3-7) are positive.

When a state of plane strain exists, Eq. 3-10 gives the two in-plane principal strains while the third principal strain is $\epsilon_{p3} = \epsilon_z = 0$.[2] An examination of Eqs. 3-10 and 3-11 indicates that the maximum in-plane shear strain is the difference between the in-plane principal strains, but this may not be the maximum shear strain at the point. The maximum shear strain at the point may be $(\epsilon_{p1} - \epsilon_{p2})$, $(\epsilon_{p1} - 0)$, or $(0 - \epsilon_{p2})$, depending on the relative magnitudes and signs of the principal strains. The lines associated with the maximum shear strain bisect the angles between lines experiencing maximum and minimum normal strains. The three possibilities are illustrated in Fig. 3-10.

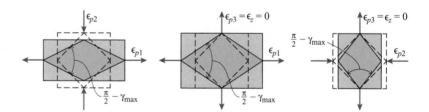

Figure 3-10

Example Problem 3-5   The strain components at a point in a body under a state of plane strain are $\epsilon_x = +1200\mu$, $\epsilon_y = -600\mu$, and $\gamma_{xy} = +900\mu$. Determine the principal strains and the maximum shear strain at the point. Show the principal strain deformations and the maximum shear strain distortion on a sketch.

---

[2]Note that although there is no strain in the z-direction, $\epsilon_{p3} = \epsilon_z = 0$, there must be a stress in the z-direction, $\sigma_{p3} = \sigma_z \neq 0$. That is, a stress is required in the z-direction to prevent deformation in the z-direction.

SOLUTION

The in-plane principal strains ϵ_{p1} and ϵ_{p2} are given by Eq. 3-10 as

$$\epsilon_{p1}, \epsilon_{p2} = \frac{\epsilon_x + \epsilon_y}{2} \pm \sqrt{\left(\frac{\epsilon_x - \epsilon_y}{2}\right)^2 + \left(\frac{\gamma_{xy}}{2}\right)^2}$$

$$(\epsilon_{p1}, \epsilon_{p2})(10^6) = \frac{1200 + (-600)}{2} \pm \sqrt{\left(\frac{1200 - (-600)}{2}\right)^2 + \left(\frac{900}{2}\right)^2}$$

$$= 300 \pm 1006.2$$

$$\epsilon_{p1} = (300 + 1006.2)(10^{-6}) = 1306.2(10^{-6}) \cong 1306\mu \qquad \textbf{Ans.}$$

$$\epsilon_{p2} = (300 - 1006.2)(10^{-6}) = -706.2(10^{-6}) \cong -706\mu \qquad \textbf{Ans.}$$

$$\epsilon_{p3} = \epsilon_z = 0 \qquad \textbf{Ans.}$$

The in-plane principal strain directions are given by Eq. 3-9 as

$$\tan 2\theta_p = \frac{\gamma_{xy}}{\varepsilon_x - \varepsilon_y} = \frac{900}{(1200) - (-600)}$$

from which $2\theta_p = 26.57°$ and $2\theta_p = 26.57° + 180° = 206.57°$ or

$$\theta_p = 13.285° \quad \text{and} \quad \theta_p = 103.285°$$

Which of the angles θ_p is associated with which of the in-plane principal strains is found by substituting one of the values of θ_p into the strain transformation equation (Eq. 3-7a). Using $\theta = \theta_p = 13.285°$ gives

$$\epsilon_n = \epsilon_x \cos^2 \theta + \epsilon_y \sin^2 \theta + \gamma_{xy} \sin \theta \cos \theta$$
$$= (1200 \, \mu) \cos^2(13.285°) + (-600\mu) \sin^2(13.285°)$$
$$+ (900 \, \mu) \sin (13.285°) \cos (13.285°)$$
$$= 1306.2 \, \mu = \epsilon_{p1}$$

The principal strains ϵ_{p1} and ϵ_{p2} are perpendicular. There is no shearing strain associated with the principal strain directions. A sketch showing the principal strains is shown in Fig. 3-11a.

The maximum in-plane shear strain γ_p is given by Eq. 3-11 as

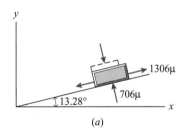

$$\gamma_p = 2\sqrt{\left(\frac{\epsilon_x - \epsilon_y}{2}\right)^2 + \left(\frac{\gamma_{xy}}{2}\right)^2}$$

$$\gamma_p(10^6) = 2\sqrt{\left(\frac{1200 - (-600)}{2}\right)^2 + \left(\frac{900}{2}\right)^2} = 2012$$

$$\gamma_p = 2012(10^{-6}) \cong 2010\mu \qquad \textbf{Ans.}$$

Since ϵ_{p1} and ϵ_{p2} have opposite signs,

$$\gamma_{\max} = \gamma_p = 2010\mu \qquad \textbf{Ans.}$$

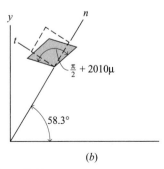

Figure 3-11

The angle θ_γ associated with the nt-axes of γ_{max} can be found using Eq. 2-16 and substituting ϵ for σ and $\gamma/2$ for τ, which results in

$$\tan 2\theta_\gamma = \frac{-(\varepsilon_x - \varepsilon_y)}{\gamma_{xy}} = \frac{-(1200) - (-600)}{900}$$

from which

$$\theta_\gamma = -31.715° \quad \text{and} \quad \theta_\gamma = 58.285°$$

Substituting $\theta = \theta_\gamma = 58.285°$ into the shear strain transformation equation (Eq. 3-8a) gives

$$\begin{aligned}
\gamma_{nt} &= -2(\epsilon_x - \epsilon_y)\sin\theta\cos\theta + \gamma_{xy}(\cos^2\theta - \sin^2\theta) \\
&= -2[(1200\mu) - (-600\mu)]\sin(58.285°)\cos(58.285°) \\
&\quad + (900\mu)[\cos^2(58.285°) - \sin^2(58.285°)] \\
&= -2012.5\mu \cong -2010\mu = \gamma_{max}
\end{aligned}$$

Figure 3-11b is a sketch showing γ_{max}. Since γ_{max} is negative, the 90° angle between the n- and t-directions increases.

Example Problem 3-6

The strain components at a point in a body under a state of plane strain are $\epsilon_x = +720\mu$, $\epsilon_y = +520\mu$, and $\gamma_{xy} = +480\mu$. Determine the principal strains and the maximum shear strain at the point. Show the principal strain deformations and the maximum shear strain distortion on a sketch.

SOLUTION

The in-plane principal strains ϵ_{p1} and ϵ_{p2} are given by Eq. 3-10 as

$$\epsilon_{p1}, \epsilon_{p2} = \frac{\epsilon_x + \epsilon_y}{2} \pm \sqrt{\left(\frac{\epsilon_x - \epsilon_y}{2}\right)^2 + \left(\frac{\gamma_{xy}}{2}\right)^2}$$

$$(\epsilon_{p1}, \epsilon_{p2})(10^6) = \frac{720 + 520}{2} \pm \sqrt{\left(\frac{720 - 520}{2}\right)^2 + \left(\frac{480}{2}\right)^2}$$

$$= 620 \pm 260$$

$$\epsilon_{p1} = (620 + 260)(10^{-6}) = 880(10^{-6}) = 880\mu \qquad \textbf{Ans.}$$
$$\epsilon_{p2} = (620 - 260)(10^{-6}) = 360(10^{-6}) = 360\mu \qquad \textbf{Ans.}$$
$$\epsilon_{p3} = \epsilon_z = 0 \qquad \textbf{Ans.}$$

The maximum in-plane shear strain γ_p is given by Eq. 3-11 as

$$\gamma_p = 2\sqrt{\left(\frac{\epsilon_x - \epsilon_y}{2}\right)^2 + \left(\frac{\gamma_{xy}}{2}\right)^2}$$

$$\gamma_p(10^6) = 2\sqrt{\left(\frac{720 - 520}{2}\right)^2 + \left(\frac{480}{2}\right)^2} = 520$$

$$\gamma_p = 520(10^{-6}) = 520\mu \qquad\qquad \textbf{Ans.}$$

Since ϵ_{p1} and ϵ_{p2} have the same signs,

$$\gamma_{\max} = \epsilon_{p1} - \epsilon_{p3} = 880\mu - 0 = 880\mu \qquad\qquad \textbf{Ans.}$$

The in-plane principal strain directions are given by Eq. 3-9 as

$$\tan 2\theta_p = \frac{\gamma_{xy}}{\epsilon_x - \epsilon_y} = \frac{480}{720 - 520} = 2.4000$$

$$2\theta_p = 67.38° \qquad \theta_p = 33.69° \cong 33.7° \qquad\qquad \textbf{Ans.}$$

The required sketch is given in Fig. 3-12.

▶ The sides of the small dotted square inside the larger dashed square are rotated 45° relative to the principal directions. These directions experience the maximum shearing strain possible in the x–y plane. As the material deforms, the small dotted square deforms into a diamond in which two of the angles are smaller than 90° by γ_p and the other two are larger than 90° by γ_p. Since the difference $\epsilon_{p1} - \epsilon_{p3}$ is larger than the difference $\epsilon_{p1} - \epsilon_{p2}$, the maximum shear deformation in the 13-plane (shown in the lower part of Fig. 3-12) is greater than the maximum in-plane shear strain γ_p.

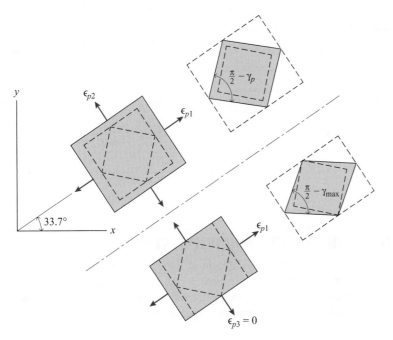

Figure 3-12

PROBLEMS

In Problems 3-27 through 3-38 the strain components ϵ_x, ϵ_y, and γ_{xy} are given for a point in a body subjected to plane strain. Determine the principal strains and the maximum shearing strain at the point. Show the principal strain deformations and the maximum shearing strain distortion on a sketch.

Introductory Problems

Problem	ϵ_x	ϵ_y	γ_{xy}
3-27*	$+600\mu$	-200μ	-480μ
3-28*	$+960\mu$	-320μ	$+500\mu$
3-29	$+900\mu$	-300μ	$+480\mu$
3-30	-900μ	$+600\mu$	-420μ
3-31*	$+750\mu$	-1000μ	$+360\mu$
3-32	-750μ	$+410\mu$	-250μ

Intermediate Problems

Problem	ϵ_x	ϵ_y	γ_{xy}
3-33*	$+720\mu$	$+520\mu$	$+480\mu$
3-34*	-540μ	-980μ	$+560\mu$
3-35	$+864\mu$	$+432\mu$	$+288\mu$
3-36*	$+900\mu$	$+650\mu$	$+300\mu$
3-37	-325μ	-625μ	$+680\mu$
3-38	-900μ	-650μ	-600μ

In Problems 3-39 through 3-46 certain strains and angles are given for a point in a body subjected to plane strain. Determine the unknown quantities for each problem and prepare a sketch showing the principal strain deformations and the maximum shearing strain distortions. In some problems there may be more than one possible value of θ_p depending on the sign of γ_{xy}.

Challenging Problems

Problem	ϵ_x	ϵ_y	γ_{xy}	ϵ_{p1}	ϵ_{p2}	γ_{max}	θ_p
3-39*	$+480\mu$	-1200μ			-1400μ		
3-40*	$+300\mu$	-800μ		$+1500\mu$			
3-41	$+800\mu$			$+1280\mu$		$+2400\mu$	
3-42	-450μ				-780μ	$+960\mu$	
3-43*			-1800μ	$+225\mu$			$-30°$
3-44			$+840\mu$	$+1100\mu$			$+20°$
3-45		-750μ	-750μ		-1500μ		
3-46*		$+750\mu$	-750μ	$+1000\mu$			

3-6 MOHR'S CIRCLE FOR PLANE STRAIN

The pictorial or graphic representation of Eqs. 2-12 and 2-13, known as Mohr's circle for stress, can be used with Eqs. 3-7 and 3-8 to yield a Mohr's circle for strain. The equation for the strain circle obtained from the equation for the stress circle by using a change in variables ϵ for σ and $\gamma/2$ for τ is

$$\left(\epsilon_n - \frac{\epsilon_x + \epsilon_y}{2}\right)^2 + \left(\frac{\gamma_{nt}}{2}\right)^2 = \left(\frac{\epsilon_x - \epsilon_y}{2}\right)^2 + \left(\frac{\gamma_{xy}}{2}\right)^2$$

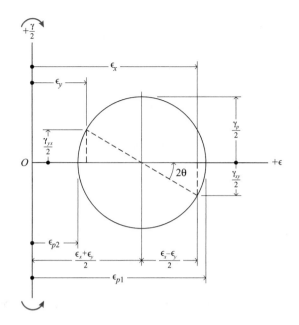

Figure 3-13

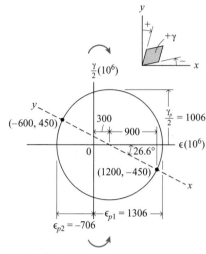

Figure 3-14

The variables in this equation are ϵ_n and $\gamma_{nt}/2$. The circle is centered on the ϵ axis at a distance $(\epsilon_x + \epsilon_y)/2$ from the origin and has a radius

$$R = \sqrt{\left(\frac{\epsilon_x - \epsilon_y}{2}\right)^2 + \left(\frac{\gamma_{xy}}{2}\right)^2}$$

Mohr's circle for the strains of Fig. 3-7 (with $\epsilon_x > \epsilon_y$) is shown in Fig. 3-13. It is apparent that the sign convention for shear strain needs to be extended to cover the construction of Mohr's circle. Observe for a positive shear strain (indicated in Fig. 3-7) that the edge of the element parallel to the x-axis tends to rotate counterclockwise while the edge parallel to the y-axis tends to rotate clockwise. For Mohr's circle construction, the clockwise rotation will be designated positive and the counterclockwise rotation will be designated negative. This is consistent with the sign convention for shear stresses given in Section 2-7. A Mohr's circle solution for Example Problem 3-5 is shown in Fig. 3-14.

PROBLEMS

In Problems 3-47 through 3-58 certain strains and angles are given for a point in a body subjected to plane strain. Use Mohr's circle to determine the unknown quantities for each problem and prepare a sketch showing the angle θ_p, the principal strain deformations, and the maximum shearing strain distortions. In some problems there may be more than one possible value of θ_p, depending on the sign of γ_{xy}.

MecMovie Activities and Problems

MM3.4 Coach Mohr's Circle of Strain. Theory; Interactive Example; Game. Learn to construct and use Mohr's circle to determine principal strains including the proper orientation of the principal strain planes.

Introductory Problems

Problem	ϵ_x	ϵ_y	γ_{xy}	ϵ_{p1}	ϵ_{p2}	γ_p	γ_{max}	θ_p
3-47*				$+400\mu$	-600μ			$+18.43°$
3-48*				$+945\mu$	-785μ			$+16.85°$
3-49				$+708\mu$	-104μ			$-34.10°$
3-50				-114μ	-903μ			$+19.26°$

Intermediate Problems

Problem	ϵ_x	ϵ_y	γ_{xy}	ϵ_{p1}	ϵ_{p2}	γ_p	γ_{max}	θ_p
3-51*	$+950\mu$	-225μ	-275μ					
3-52*	$+900\mu$	-333μ	$+982\mu$					
3-53	$+750\mu$	$+390\mu$	-900μ					
3-54	$+600\mu$	$+480\mu$	$+480\mu$					

Challenging Problems

Problem	ϵ_x	ϵ_y	γ_{xy}	ϵ_{p1}	ϵ_{p2}	γ_p	γ_{max}	θ_p
3-55*	-680μ	$+320\mu$		$+414\mu$				
3-56*	$+450\mu$	$+150\mu$		$+780\mu$				
3-57	$+360\mu$	$+750\mu$			$+120\mu$			
3-58	-300μ	$+600\mu$			-450μ			

3-7 STRAIN MEASUREMENT AND ROSETTE ANALYSIS

In most experimental work involving strain measurement, the strains are measured on a free surface of a member where a state of plane stress exists. If the outward normal to the surface is taken as the z-axis, then $\sigma_z = \tau_{zx} = \tau_{zy} = 0$. Since this state of stress offers no restraint to out-of-plane deformation, a normal strain ϵ_z develops in addition to the in-plane strains ϵ_x, ϵ_y, and γ_{xy}. The shear strains γ_{zx} and γ_{zy} remain zero; therefore, the normal strain ϵ_z is a principal strain. In Section 3-5, expressions were developed for the plane strain case relating the in-plane principal strains (ϵ_{p1} and ϵ_{p2}) and their orientations to the in-plane strains ϵ_x, ϵ_y, and γ_{xy}. For the plane stress case, which involves the normal strain ϵ_z in addition to the in-plane strains, similar expressions are needed.

As an illustration of the effects of an out-of-plane displacement on the deformation (change in length) of a line segment originally located in the x–y plane, consider line AB of Fig. 3-15. As a result of the loads imposed on the member, the line AB is displaced and extended into the line $A'B'$. The displacements associated with point A' are u in the x-direction, v in the y-direction, and w in the z-direction. Point B' displaces $u + du$ in the x-direction, $v + dv$ in the y-direction, and $w + dw$ in the z-direction. The deformation δ_{AB} is obtained from the original length of the line, and the displacements du, dv, and dw of point B' with respect to A'. Thus,

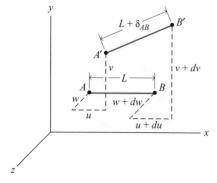

Figure 3-15

$$(A'B')^2 = (L + \delta_{AB})^2 = (L + du)^2 + (dv)^2 + (dw)^2$$

and after squaring the terms on both sides, the result is

$$L^2 + 2L\delta_{AB} + \delta_{AB}^2 = L^2 + 2L(du) + (du)^2 + (dv)^2 + (dw)^2$$

If the deformations are small, the second-degree terms can be neglected; hence,

$$\delta_{AB} = du$$

This indicates that the normal strain along AB (δ_{AB} divided by L) is not affected by the presence of the out-of-plane displacements. In fact, none of the in-plane strains is affected; therefore, Eqs. 3-7 and 3-8 are valid not only for the plane strain case but also for the plane stress case present when strain measurements are made on a free surface.

Electrical resistance strain gages have been developed to provide accurate measurements of normal strain. The gage may be an etched foil conductor mounted on epoxy or polyimide backing (see Fig. 3-16). The foil gage is cemented to the material for which the strain is to be determined. As the material is strained, the wires are lengthened or shortened; this changes the electrical resistance of the gage. The change in resistance is measured by means of a Wheatstone bridge, which may be calibrated to directly read strain.

Shear strains are more difficult to measure directly than normal strains. The electrical resistance strain gages are sensitive only to normal strains and cannot respond to shear strains. Instead, shear strains are often obtained by measuring normal strains in two or three different directions. The shearing strain γ_{xy} can be computed from the normal strain data by using Eq. 3-7a. For example, consider the most general case of three arbitrary normal strain measurements, as shown in

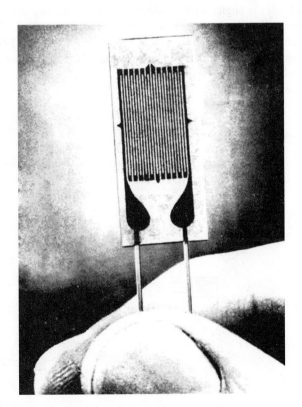

Figure 3-16

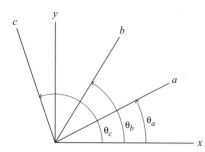

Figure 3-17

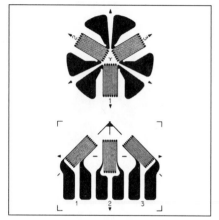

Figure 3-18

Fig. 3-17. From Eq. 3-7a,

$$\epsilon_a = \epsilon_x \cos^2 \theta_a + \epsilon_y \sin^2 \theta_a + \gamma_{xy} \sin \theta_a \cos \theta_a$$

$$\epsilon_b = \epsilon_x \cos^2 \theta_b + \epsilon_y \sin^2 \theta_b + \gamma_{xy} \sin \theta_b \cos \theta_b \qquad (3\text{-}12)$$

$$\epsilon_c = \epsilon_x \cos^2 \theta_c + \epsilon_y \sin^2 \theta_c + \gamma_{xy} \sin \theta_c \cos \theta_c$$

From the measured values of ϵ_a, ϵ_b, and ϵ_c and a knowledge of the gage orientations θ_a, θ_b, and θ_c with respect to the reference x-axis, the values of ϵ_x, ϵ_y, and γ_{xy} can be determined by simultaneous solution of the three equations. In practice the angles θ_a, θ_b, and θ_c are selected to simplify the calculations. Multiple-element strain gages used for this type of measurement are known as *strain rosettes*. Two rosette configurations that are marketed commercially are shown in Fig. 3-18.

It should be noted that the strain rosettes shown in Fig. 3-19 are all equivalent. The choice of which to use is often determined by the geometry of the machine part and the point at which the strains are to be determined. The rosettes shown in Figs. 3-19a and 3-19b could be used to determine strains at a point on the free surface of a shaft, pressure vessel, or other type of machine component. The point of interest would be located at the center of the triangular arrangement of gages for the rosette shown in Fig. 3-19a or at the intersection of the three gage lines for the rosette shown in Fig. 3-19b. Neither of these rosettes could be used to determine the strains at a point near the edge of a hole or other type of boundary in a machine component. Since the intersections of the gage lines for the rosettes shown in Figs. 3-19c and 3-19d are outside the regions occupied by the gages, they can be used to determine the strains at a point near the edge of a hole or other type of boundary in a machine component.

In this book the angles used to identify the normal strain directions of the various elements of a rosette will always be measured counterclockwise from the reference x-axis. Once ϵ_x, ϵ_y, and γ_{xy} have been determined, Eqs. 3-9, 3-10, and 3-11 can be used to determine the in-plane principal strains, their orientations, and the maximum in-plane shear strain at the point.

In Section 4-3, it will be shown that for plane stress

$$\epsilon_z = \epsilon_{p3} = -\frac{\nu}{1 - \nu}(\epsilon_x + \epsilon_y) \qquad (3\text{-}13)$$

where ν is Poisson's ratio (a property of the material used in fabricating the member), which is defined in Section 4-2. For the case of plane stress, this out-of-plane

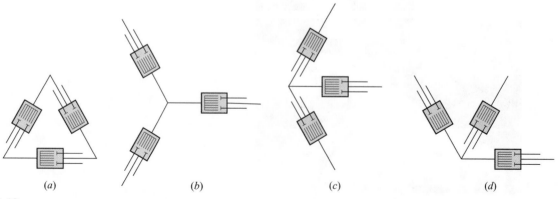

(a)   (b)   (c)   (d)

Figure 3-19

principal strain is important because the maximum shear strain at the point may be $(\epsilon_{p1} - \epsilon_{p2})$, $(\epsilon_{p1} - \epsilon_{p3})$, or $(\epsilon_{p3} - \epsilon_{p2})$, depending on the relative magnitudes and signs of the principal strains at the point.

Strain-measuring transducers such as electrical resistance strain gages measure the average normal strain under the sensing foil element of the gage (along some gage length) and not the strain at a point. So long as the gage length is kept small, errors associated with such measurements can be kept within acceptable limits. The following example illustrates the application of Eqs. 3-12 to principal strain and maximum shear strain determinations under conditions of plane stress.

Example Problem 3-7   A strain rosette, composed of three electrical resistance strain gages making angles of 0°, 60°, and 120° with the x-axis (see Fig. 3-20) was mounted on the free surface of a material for which Poisson's ratio is 1/3. Under load, the following strains were measured:

$$\epsilon_a = +1000\mu \qquad \epsilon_b = +750\mu \qquad \epsilon_c = -650\mu$$

Determine the principal strains and the maximum shear strain. Show the directions of the in-plane principal strains on a sketch.

SOLUTION
The given data are substituted in Eqs. 3-12 to yield

$\epsilon_a = \epsilon_0 = \epsilon_x = +1000\mu$ (measured)
$\epsilon_c = \epsilon_{60} = -650\mu = 1000\mu \cos^2 60° + \epsilon_y \sin^2 60° + \gamma_{xy} \sin 60° \cos 60°$
$\epsilon_b = \epsilon_{120} = +750\mu = 1000\mu \cos^2 120° + \epsilon_y \sin^2 120° + \gamma_{xy} \sin 120° \cos 120°$

from which

$$\epsilon_y = -266.7\mu \quad \text{and} \quad \gamma_{xy} = -1616.6\mu$$

The in-plane principal strains ϵ_{p1} and ϵ_{p2} are given by Eq. 3-10 as

$$\epsilon_{p1}, \epsilon_{p2} = \frac{\epsilon_x + \epsilon_y}{2} \pm \sqrt{\left(\frac{\epsilon_x - \epsilon_y}{2}\right)^2 + \left(\frac{\gamma_{xy}}{2}\right)^2}$$

$$(\epsilon_{p1}, \epsilon_{p2})(10^6) = \frac{1000 - 266.7}{2} \pm \sqrt{\left(\frac{1000 + 266.7}{2}\right)^2 + \left(\frac{-1616.6}{2}\right)^2}$$

$$= 366.7 \pm 1026.9$$

$$\epsilon_{p1} = (366.7 + 1026.9)(10^{-6}) = 1393.6(10^{-6}) \cong 1394\mu \qquad \textbf{Ans.}$$

$$\epsilon_{p2} = (366.7 - 1026.9)(10^{-6}) = -660.2(10^{-6}) \cong -660\mu \qquad \textbf{Ans.}$$

The third principal strain $\epsilon_{p3} = \epsilon_z$ is given by Eq. 3-13 as

$$\epsilon_z = \epsilon_{p3} = -\frac{v}{1-v}(\epsilon_x + \epsilon_y)$$
$$= -\frac{1/3}{1 - (1/3)}(1000\mu - 266.7\mu)$$
$$= -366.7\mu \cong -367\mu \qquad \textbf{Ans.}$$

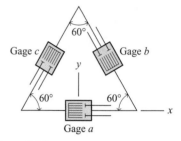

Figure 3-20

▶ Note that gage b could be considered oriented 120° above the x-axis or 60° below the x-axis because $\cos^2 120° = \cos^2(-60°)$, $\sin^2 120° = \sin^2(-60°)$, and $\cos 120° \sin 120° = \cos(-60°) \sin(-60°)$. Similarly, gage c could be considered oriented 60° above the x-axis or 120° below the x-axis.

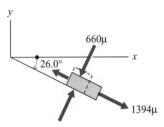

Figure 3-21

Since ϵ_{p3} is less compressive than the in-plane principal strain ϵ_{p2}, the maximum shear strain at the point is the maximum in-plane shear strain γ_p. Thus,

$$\gamma_{\max} = \gamma_p = \epsilon_{p1} - \epsilon_{p2}$$
$$= 1393.6\mu + 660.2\mu = 2053.8\mu \cong 2050\mu \qquad \textbf{Ans.}$$

The in-plane principal strain directions are given by Eq. 3-9 as

$$\tan 2\theta_p = \frac{\gamma_{xy}}{\epsilon_x - \epsilon_y} = \frac{-1616.6}{1000 + 266.7} = -1.2763$$
$$2\theta_p = -51.92° \qquad \theta_p = -25.96° \cong -26.0° \qquad \textbf{Ans.}$$

The required sketch is given in Fig. 3-21.

PROBLEMS

MecMovie Activities and Problems

MM3.5 Strain measurement with rosettes. Example; Try One. Calculating strain states with rosette strain data.

Introductory Problems

3-59* At a point on the free surface of a steel ($\nu = 0.30$) machine part, the strain rosette shown in Fig. P3-59 was used to obtain the following normal strain data: $\epsilon_a = +750$ μin./in., $\epsilon_b = -125$ μin./in., and $\epsilon_c = -250$ μin./in. Determine

a. The strain components ϵ_x, ϵ_y, and γ_{xy} at the point.
b. The principal strains and the maximum shearing strain at the point. A sketch is not required.

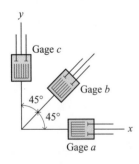

Figure P3-59

a. The strain components ϵ_x, ϵ_y, and γ_{xy} at the point.
b. The principal strains and the maximum shearing strain at the point. A sketch is not required.

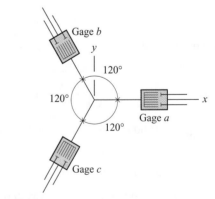

Figure P3-60

3-60* At a point on the free surface of a steel ($\nu = 0.30$) machine part, the strain rosette shown in Fig. P3-60 was used to obtain the following normal strain data: $\epsilon_a = -555$ μm/m, $\epsilon_b = +925$ μm/m, and $\epsilon_c = +740$ μm/m. Determine

3-61 The strain rosette shown in Fig. P3-61 is attached to a point on the free surface of an aluminum ($\nu = 0.33$) machine part. The following normal strain data were taken: $\epsilon_a = +800$ μin./in., $\epsilon_b = +950$ μin./in., and $\epsilon_c = +600$ μin./in. Determine

a. The strain components ϵ_x, ϵ_y, and γ_{xy} at the point.
b. The principal strains and the maximum shearing strain at the point. A sketch is not required.

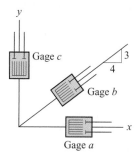

Figure P3-61

Intermediate Problems

3-62* At a point on the free surface of an aluminum alloy ($v = 0.33$) machine part, the strain rosette shown in Fig. P3-62 was used to obtain the following normal strain data: $\epsilon_a = +780$ μm/m, $\epsilon_b = +345$ μm/m, and $\epsilon_c = -332$ μm/m.

a. Determine the strain components ϵ_x, ϵ_y, and γ_{xy} at the point.
b. Determine the principal strains and the maximum shearing strain at the point. A sketch is not required.

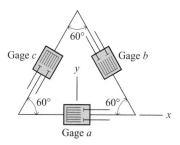

Figure P3-62

3-63* At a point on the outside surface of a steel ($v = 0.30$) thin-walled pressure vessel the strain rosette shown in Fig. P3-63 indicates the normal strains $\epsilon_a = +36$ μin./in., $\epsilon_b = +310$ μin./in., and $\epsilon_c = +150$ μin./in. Gages a and c are oriented in the axial and hoop directions of the vessel, respectively. Determine the principal strains and the maximum shearing strain at this point. Show the principal strain deformations and maximum shear strain distortion on a sketch.

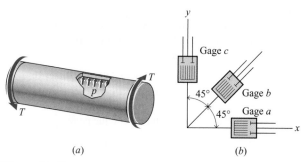

(a)   (b)

Figure P3-63

3-64 The strain rosette shown in Fig. P3-64 was used to obtain normal strain data at a point on the free surface of a 2024-T4 aluminum alloy ($v = 0.30$) machine part. The gage readings were $\epsilon_a = +525$ μm/m, $\epsilon_b = +450$ μm/m, and $\epsilon_c = +1425$ μm/m. Determine the principal strains at this point, and show the principal strain deformations on a sketch.

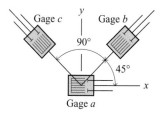

Figure P3-64

Challenging Problems

3-65* At a point on the free surface of a steel ($v = 0.30$) machine part, the strain rosette shown in Fig. P3-65 was used to obtain the following normal strain data: $\epsilon_a = +875$ μin./in., $\epsilon_b = +700$ μin./in., and $\epsilon_c = +350$ μin./in.

a. Determine the principal strains and the maximum shearing strain at the point. Prepare a sketch showing all of these strains.
b. Determine the normal strain in the n-direction at the point.

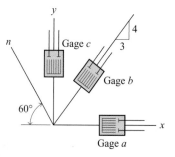

Figure P3-65

3-66   At a point on the free surface of an aluminum alloy ($v = 0.33$) machine part, the strain rosette shown in Fig. P3-66 was used to obtain the following normal strain data: $\epsilon_a = +875$ μm/m, $\epsilon_b = +700$ μm/m, and $\epsilon_c = -650$ μm/m.

a. Determine the principal strains and the maximum shearing strain at the point. Prepare a sketch showing all of these strains.

b. Determine the shearing strain γ_{nt} at the point.

3-67*   At a point on the free surface of an aluminum alloy ($v = 0.33$) machine part, the strain rosette shown in Fig. P3-67 was used to obtain the following normal strain data: $\epsilon_a = +800$ μin./in., $\epsilon_b = +950$ μin./in., and $\epsilon_c = +600$ μin./in.

a. Determine the principal strains and the maximum shearing strain at the point. Prepare a sketch showing all of these strains.

b. Determine the normal strain in the n-direction at the point.

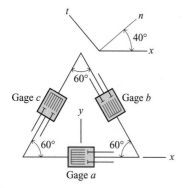

Figure P3-66

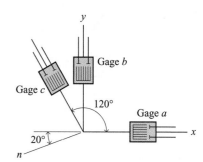

Figure P3-67

SUMMARY

When a system of loads is applied to a machine component or structural element, individual points of the body generally move. A study of displacements in which neither the size nor the shape of the body is changed is the concern of courses in rigid-body mechanics. When displacements induced by applied loads cause the size and/or shape of a body to be altered, individual points of the body move relative to one another. In the design of structural elements or machine components, the deformations experienced by the body as a result of the applied loads often represent as important a design consideration as the stresses.

Strain (deformation per unit length) is the quantity used to measure the intensity of deformation, just as stress (force per unit area) is used to measure the intensity of an internal force. When the deformation δ is in the same direction as the gage length L, the strain is called *normal strain*

$$\epsilon_{\text{avg}} = \frac{\delta_n}{L} \tag{3-1}$$

When the deformation δ is perpendicular to the gage length L, the strain is called *shear strain*

$$\gamma_{\text{avg}} = \frac{\delta_s}{L} = \tan \phi \tag{3-3}$$

Since δ_s/L is usually very small (typically $\delta_s/L < 0.001$), $\sin \phi \cong \tan \phi \cong \phi$, where ϕ is measured in radians, and $\gamma_{\text{avg}} = \phi = \delta_s/L$ is the decrease in the angle between two reference lines that are orthogonal in the undeformed state.

Although the normal strain and the shear strain at a point both depend on the orientation of the gage length L, the state of strain at a point is completely determined by the normal strain and shear strain in three mutually orthogonal directions at the point. For *plane strain* (for which $\epsilon_z = \gamma_{zx} = \gamma_{zy} = 0$, $\sigma_z \neq 0$), the normal and shear strains for a pair of axes oriented at an angle θ relative to the x-axis are given by

$$\epsilon_n = \epsilon_x \cos^2 \theta + \epsilon_y \sin^2 \theta + \gamma_{xy} \sin \theta \cos \theta \qquad (3\text{-}7a)$$

$$\gamma_{nt} = -2(\epsilon_x - \epsilon_y) \sin \theta \cos \theta + \gamma_{xy}(\cos^2 \theta - \sin^2 \theta) \qquad (3\text{-}8a)$$

respectively.

For design purposes, the critical strains at a point are usually the maximum normal strain and the maximum shearing strain. The maximum (and minimum) values of normal strain are called the *principal strains* and always occur for sets of axes that are free of shear strain and which are called *principal directions*. At every point in a stressed body, there exist three principal strains acting in mutually orthogonal directions. In plane strain situations, two principal strains are in the x–y plane

$$\epsilon_{p1,p2} = \frac{\epsilon_x + \epsilon_y}{2} \pm \sqrt{\left(\frac{\epsilon_x - \epsilon_y}{2}\right)^2 + \left(\frac{\gamma_{xy}}{2}\right)^2} \qquad (3\text{-}10)$$

and the third principal strain is $\epsilon_{p3} = \epsilon_z = 0$. The principal directions are oriented at an angle θ_p given by

$$\tan 2\theta_p = \frac{\gamma_{xy}}{\epsilon_x - \epsilon_y} \qquad (3\text{-}9)$$

relative to the xy-axes.

The maximum in-plane shearing strain occurs for lines oriented at $45°$ to the principal directions and can be found from the principal strains

$$\gamma_p = \epsilon_{p1} - \epsilon_{p2}$$

The maximum shear strain at the point may be $(\epsilon_{p1} - \epsilon_{p2})$, $(\epsilon_{p1} - 0)$, or $(0 - \epsilon_{p2})$, depending on the relative magnitudes and signs of the principal strains. The lines associated with the maximum shear strain bisect the angles between the lines experiencing maximum and minimum normal strains.

All of the transformation equations developed for plane stress in Chapter 2 can be applied to plane strain by substituting ϵ for σ and $\gamma/2$ for τ. Likewise, the pictorial or graphic representation of Eqs. 2-12 and 2-13, known as Mohr's circle for stress, can be used with Eqs. 3-7 and 3-8 to yield a similar Mohr's circle for strain. The angle relationships on Mohr's circle for strain are identical to the angle relationships on Mohr's circle for stress.

Strains can be measured by several methods, but except for the simplest cases, stresses cannot be obtained directly. Therefore, the usual procedure used in experimental stress analysis is to measure the strains and calculate the state of stress by using the stress-strain equations presented later in Section 4-3. In most experimental work involving strain measurement, the strains are measured on a free surface of a member where a state of plane stress exists. However, Eqs. 3-7 and 3-8 are also valid for the plane stress case present when strain measurements are made on a free surface.

Electrical resistance strain gages have been developed to provide accurate measurements of normal strain. Shear strains are more difficult to measure directly than normal strains and are often obtained by measuring the normal strains in two or three different directions using a strain rosette. The shearing strain γ_{xy} can then be computed from the normal strain data by using Eqs. 3-12.

In Section 4-3, it will be shown that for plane stress

$$\epsilon_z = \epsilon_{p3} = -\frac{\nu}{1-\nu}(\epsilon_x + \epsilon_y) \tag{3-13}$$

where ν is Poisson's ratio (a property of the material used in fabricating the member), which is defined in Section 4-2. For the case of plane stress, this out-of-plane principal strain is important because the maximum shear strain at the point may be $(\epsilon_{p1} - \epsilon_{p2})$, $(\epsilon_{p1} - \epsilon_{p3})$, or $(\epsilon_{p3} - \epsilon_{p2})$, depending on the relative magnitudes and signs of the principal strains at the point.

REVIEW PROBLEMS

3-68*   A rigid bar AD is supported by two rods as shown in Fig. P3-68. There is no strain in the vertical bars before the load P is applied. After the load P is applied, the axial strain in rod BF is 400 μm/m. Determine the axial strain in rod CE.

a. The axial strain in bar EF.
b. The axial strain in bar EF if there is a 0.005-in. clearance in the connection between bars EF and CD.

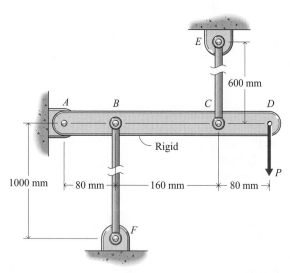

Figure P3-68

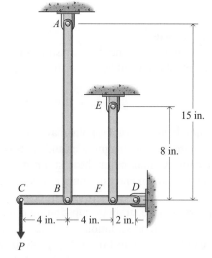

Figure P3-69

3-69*   The rigid bar CD of Fig. P3-69 is horizontal under no load, and bars EF and AB are unstrained. When the load P is applied, the axial strain in bar AB is found to be 0.0015 in./in. Determine

3-70   A brake block has the shape of a circular ring segment (Fig. P3-70a). During application of the brake, the outer surface of the block rotates with respect to the inner surface, as shown in Fig. P3-70b. Determine, in terms of ϕ, R_1, and

R_2, the shearing strains at point A and B in the block (associated with the radial and circumferential directions). Note that radial lines before deformation are assumed to be straight lines after deformation. Solve by assuming that the angle ϕ is small.

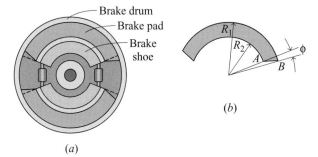

(a)

Figure P3-70

3-71* The thin square plate shown in Fig. P3-71 is uniformly deformed such that $\epsilon_x = +3200\ \mu\text{in./in.}$, $\epsilon_y = +1500\ \mu\text{in./in.}$, and $\gamma_{xy} = +1000\ \mu\text{rad}$. Determine

a. The normal strain ϵ_n in the plate.
b. The shearing strain γ_{nt} in the plate.

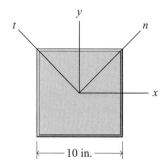

Figure P3-71

3-72 The thin rectangular plate shown in Fig. P3-72 is uniformly deformed such that $\epsilon_x = +1500\ \mu\text{m/m}$, $\epsilon_y = -1250\ \mu\text{m/m}$, and $\gamma_{xy} = +1000\ \mu\text{rad}$. Determine the normal strain ϵ_{BD} along the diagonal BD of the plate.

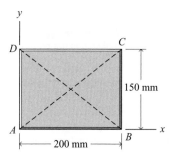

Figure P3-72

3-73 The strain components at a point in a body under a state of plane strain are $\epsilon_x = +1000\ \mu\text{in./in.}$, $\epsilon_y = -800\ \mu\text{in./in.}$, and $\gamma_{xy} = -800\ \mu\text{rad}$. Determine the principal strains and the maximum shearing strain at the point. Show the principal strain deformations and the maximum shearing strain distortion on a sketch.

3-74* The strain components at a point are $\epsilon_x = -600\ \mu\text{m/m}$, $\epsilon_y = +1200\ \mu\text{m/m}$, and $\gamma_{xy} = +2000\ \mu\text{rad}$.

a. Sketch Mohr's circle for the state of strain at this point.
b. Determine the principal strains and the maximum shearing strain at the point.
c. Show the principal strain deformations and the maximum shearing strain distortion on a sketch.

3-75 The strain rosette shown in Fig. P3-75 was used to obtain the following strain data at a point on the free surface of a steel $(v = 0.30)$ machine part: $\epsilon_a = +600\ \mu\text{in./in.}$, $\epsilon_b = +500\ \mu\text{in./in.}$, and $\epsilon_c = -200\ \mu\text{in./in.}$ Determine the principal strains and the maximum shearing strain at the point. A sketch is not required.

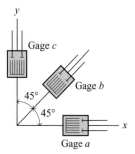

Figure P3-75

3-76* The strain components at a point are $\epsilon_x = -800\ \mu\text{m/m}$, $\epsilon_y = +640\ \mu\text{m/m}$, and $\gamma_{xy} = -960\ \mu\text{rad}$.

a. Sketch Mohr's circle for the state of strain at this point.
b. Determine the principal strains and the maximum shearing strain at the point. A sketch is not required.

3-77   At a point on the free surface of an aluminum alloy ($v = 0.33$) machine component, the three strain gages shown in Fig. P3-77 were used to obtain the following strain data: $\epsilon_a = +800\ \mu\text{in./in.}$, $\epsilon_b = +960\ \mu\text{in./in.}$, and $\epsilon_c = +800\ \mu\text{in./in.}$ Determine the principal strains and the maximum shearing strain at the point. Show the principal strain deformations and the maximum shearing strain distortion on a sketch.

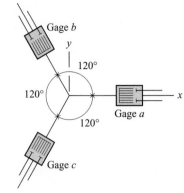

Figure P3-77

Chapter 4
Material Properties and Stress-Strain Relationships

4-1 INTRODUCTION

The satisfactory performance of a structure frequently is determined by the amount of deformation or distortion that can be permitted. A deformation of a few thousandths of an inch might make a boring machine useless, whereas the hook on a drag line might deflect several inches without impairing its usefulness. It is often necessary to relate loads and temperature changes on a structure to the deformations produced by the loads and temperature changes. Experience has shown that the deformations caused by loads and by temperature effects are essentially independent of each other. The deformations due to the two effects may be computed separately and added together to get the total deformation.

4-2 STRESS-STRAIN DIAGRAMS

The relationship between loads and deformation in a structure can be obtained by plotting diagrams showing loads and deflections for each member and each type of loading in a structure. However, the relationship between load and deformation depends on the dimensions of the members as well as on the type of material from which the members are made. For example, the graph of Fig. 4-1a (one-dimensional loading) shows the relationship between the force required to stretch three bars of the same material but of different lengths and cross-sectional areas and the resulting deformations of the bars. It is not clear from this graph that these three curves all describe the same material behavior. However, if these curves are redrawn plotting stress versus deformation as in Fig. 4-1b, the data for the first and third bars form a single line. If the curves are redrawn plotting stress versus strain as in Fig. 4-1c, the data for all three curves form a single line. That is, curves showing the relationship between stress and strain (such as Fig. 4-1c) are independent of the size and shape of the member and depend only on the type of material from which the members are made. Such diagrams are called *stress-strain diagrams*.

4-2-1 The Tensile Test
Data for stress-strain diagrams are obtained by applying an axial load to a test specimen and measuring the load and deformation simultaneously. A testing machine (Fig. 4-2) is used to strain the specimen and to measure the load required to produce the deformation. The stress is obtained by

153

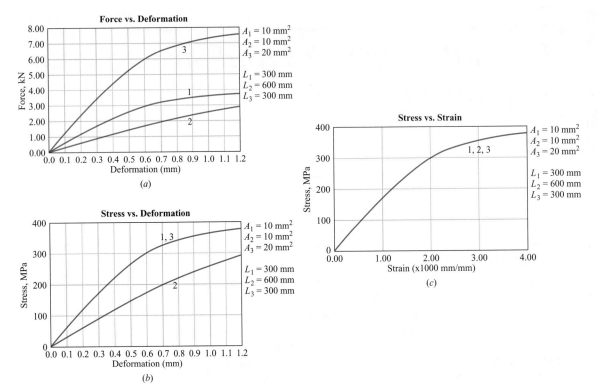

Figure 4-1

Figure 4-2 Hydraulic testing machine set up for a tension test. (Courtesy of MTS Systems Corporation.)

dividing the load by the initial cross-sectional area of the specimen. The area will change somewhat during the loading, and the stress obtained using the initial area is obviously not the exact stress occurring at higher loads. However, it is the stress most commonly used in designing structures. Stress obtained by dividing the load by the actual area is frequently called the *true stress* and is useful in explaining the fundamental behavior of materials.

4-2-2 Strain Measurement
Strains are small in materials used in engineering structures, often less than 0.001, and their accurate determination requires special measuring equipment. Normal strain is obtained by measuring the deformation δ in a length L and dividing δ by L. Instruments for measuring the deformation δ are called *strain gages* or *extensometers* and obtain the desired accuracy by multiplying levers, dial indicators, beams of light, or other means. The electrical resistance strain gage is widely used for this type of measurement.

True strain, like true stress, is computed on the basis of the actual length of the test specimen during the test and is used primarily to study the fundamental properties of materials. The difference between *nominal stress and strain*, computed from initial dimensions of the specimen, and true stress and strain is negligible for stresses usually encountered in engineering structures, but sometimes the difference becomes important with larger stresses and strains. A more complete discussion of the experimental determination of stress and strain will be found in various books on experimental stress analysis.[1]

4-2-3 Example Stress-Strain Diagrams
Figures 4-3*a, b*, and *c* show tensile stress-strain diagrams for structural steel (a low-carbon steel), for a magnesium alloy, and for a gray cast iron, respectively. These diagrams will be used to explain a number of properties useful in the study of mechanics of materials.

Although some of the relationships that follow have a basis in theory (for example, the linear relationship between stress and strain for small strain), others are purely empirical fits to experimental data. In either case, the values of specific constants for various materials must be experimentally determined.

4-2-4 Modulus of Elasticity
The initial portion of the stress-strain diagram for most materials used in engineering structures (see Figs. 4-3*a* and *b*) is a straight line. The stress-strain diagrams for some materials, such as gray cast iron (see Fig. 4-3*c*) and concrete, show a slight curve even at very small stresses, but it is common practice to draw a straight line to average the data for the first part of the diagram and neglect the curvature. The proportionality of load to deflection was first recorded by Robert Hooke who observed in 1678, "ut tensio sic vis" (as the stretch so the force); this relationship is frequently referred to as Hooke's law,

$$\sigma = E\epsilon \qquad (4\text{-}1a)$$

in which the constant of proportionality, E, is the slope of the straight-line portion of the stress-strain diagram. *It is important to realize that Hooke's law (Eq. 4-1a)*

---

[1] *Experimental Stress Analysis*, 3rd ed. J. W. Dally and W. F. Riley, McGraw-Hill, New York, 1991.

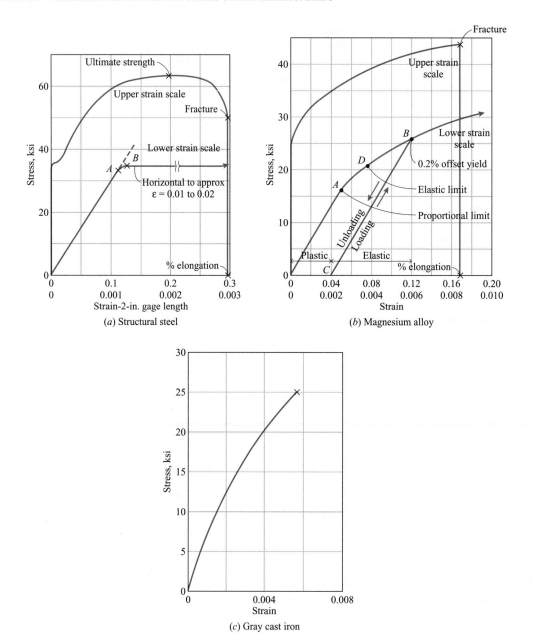

Figure 4-3

only describes the initial linear portion of the stress-strain diagram and is valid only for bars loaded in uniaxial extension as in the testing machine of Fig. 4-2. A modified version of Hooke's law which is valid for materials being stretched in two or three directions at the same time will be derived later in Section 4-3.

Thomas Young, in 1807, suggested what amounts to using the ratio of stress to strain to measure the stiffness of a material. This ratio is called *Young's modulus* or the *modulus of elasticity* and is the slope of the straight-line portion of the stress-strain diagram. Young's modulus is written as

$$E = \sigma / \epsilon \quad \text{(normal stress-strain)} \qquad (4\text{-}1b)$$

A similar modulus called the *shear modulus* or the *modulus of rigidity* relates the shearing stress τ and the shearing strain γ.

$$G = \tau/\gamma \qquad \text{(shear stress-strain)} \qquad (4\text{-}1c)$$

The maximum stress for which stress and strain are proportional is called the *proportional limit* and is indicated by the ordinates at points A on Fig. 4-3a or b. The exact point of the proportional limit is difficult to determine from the stress-strain curve.

For points on the stress-strain curve beyond the proportional limit (such as point C on Fig. 4-4), other quantities such as the tangent modulus and the secant modulus are used as measures of the stiffness of a material. The *tangent modulus* E_t is defined as the slope of the stress-strain diagram at a particular stress level. Thus, the tangent modulus is a function of the stress (or strain) for stresses greater than the proportional limit. For stresses less than the proportional limit, the tangent modulus is the same as Young's modulus. The *secant modulus* E_s is the ratio of the stress to the strain at any point on the diagram. Young's modulus E, the tangent modulus E_t, and the secant modulus E_s are all illustrated in Fig. 4-4.

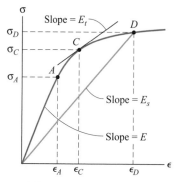

Figure 4-4

4-2-5 Elastic Limit

The action is said to be *elastic* if the strain resulting from loading disappears when the load is removed. The *elastic limit* (point D in Fig. 4-3b) is the maximum stress for which the material acts elastically. For stresses above the elastic limit, some deformation (strain) remains when the load is removed (and the stress goes to zero). For most materials it is found that the stress-strain diagram for unloading (see line BC in Fig. 4-3b) is approximately parallel to the loading portion (see line OA in Fig. 4-3b). If the specimen is again loaded, the stress-strain diagram will usually follow the unloading curve until it reaches a stress a little less than the maximum stress attained during the initial loading, at which time it will start to curve in the direction of the initial loading curve. As indicated in Fig. 4-3b, the proportional limit for the second loading (point B) is greater than that for the initial loading. This phenomenon is called *strain hardening* or *work hardening*.

When the stress exceeds the elastic limit (or proportional limit for practical purposes), it is found that a portion of the deformation remains after the load is removed. The deformation remaining after an applied load is removed is called *plastic deformation* (OC in Fig. 4-3b). Plastic deformation which depends only on the amount of load (stress) and is independent of the time duration of the applied load is known as *slip*. Plastic deformation that continues to increase under a constant stress is called *creep*. In many instances creep continues until fracture occurs; however, in other instances the rate of creep decreases and approaches zero as a limit. Some materials are much more susceptible to creep than others, but most materials used in engineering exhibit creep at elevated temperatures. The total strain is thus made up of elastic strain, possibly combined with plastic strain that results from slip, creep, or both. When the load is removed, the elastic portion of the strain is recovered, but the plastic part (slip and creep) remains as permanent set.[2]

---

[2]In some instances a portion of the strain that remains immediately after the stress is removed may disappear after a period of time. This reduction of strain is sometimes called *recovery*.

4-2-6 Yield Point
A precise value for the proportional limit is difficult to obtain when the transition of the stress-strain diagram from a straight line to a curve is gradual. For this reason, other measures of stress that can be used as a practical elastic limit are required. The yield point and the yield strength for a specified offset are used for this purpose.

The *yield point* is the stress at which there is an appreciable increase in strain with no increase in stress, with the limitation that, if straining is continued, eventually the stress will again increase. This latter specification indicates that there is a kink or "knee" in the stress-strain diagram, as indicated in Fig. 4-3*a*. The yield point is easily determined without the aid of strain-measuring equipment because the load indicated by the testing machine ceases to rise (or may even drop) at the yield point. Unfortunately, few materials possess this property, the most common examples being low-carbon steels.

4-2-7 Yield Strength
The *yield strength* is defined as the stress that will induce a specified permanent set, usually 0.05 to 0.3 percent (which is equivalent to a strain of 0.0005 to 0.003) with 0.2 percent the most commonly used value. The yield strength can be conveniently determined from a stress-strain diagram by laying off the specified offset (permanent set) on the strain axis (*OC* in Fig. 4-3*b*) and drawing a line *CB* parallel to *OA*. The stress indicated by the intersection of *CB* and the stress-strain diagram is the yield strength for the specified offset (0.2 percent in Fig. 4-3*b*).

4-2-8 Ultimate Strength
The maximum stress (based on the original area) developed in a material before rupture is called the *ultimate strength* of the material (Fig. 4-3*a*), and the term may be modified as the ultimate tensile, compressive, or shearing strength of the material. Ductile materials undergo considerable plastic tensile or shearing deformation before rupture. When the ultimate strength of a ductile material is reached, the cross-sectional area of the test specimen starts to decrease or neck down (see Fig. 4-5), and the resultant load that can be carried by the specimen decreases. Thus, the stress based on the original area decreases beyond the ultimate strength of the material (Fig. 4-3*a*), although the true stress continues to increase until rupture.

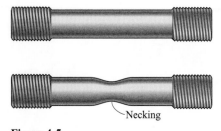

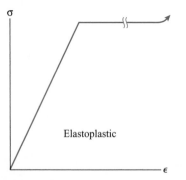

Necking

Figure 4-5

4-2-9 Elastoplastic Materials
Most engineering structures are designed so that the stresses are less than the proportional limit; therefore, Young's modulus provides a simple and convenient relationship between stress and strain. When the stress exceeds the proportional limit, no simple relation exists between stress and strain. Various empirical equations have been proposed relating the stress and strain beyond the proportional limit. A stress-strain diagram similar to the one shown in Fig. 4-6 (elastoplastic) is frequently assumed for mild steel or other materials with similar properties in order to simplify calculations. For mild steel the plastic strain that occurs at the yield point with no increase in stress is 16–20 times the elastic strain at the proportional limit.

σ

Elastoplastic

ε

Figure 4-6

4-2-10 Ductility
Strength and stiffness are not the only properties of interest to a design engineer. Another important property is *ductility*, defined as the capacity for plastic deformation in tension or shear. This property controls the amount of cold forming to which a material may be subjected. The forming of automobile bodies and the manufacture of fencing and other wire products all

require ductile materials. Ductility is also an important property of materials used for fabricated structures. Under static loading, the presence of large stresses in the region of rivet holes or welds may be ignored, since ductility permits considerable plastic action to take place in the region of high stress, with a resulting redistribution of stress and the establishment of equilibrium. Two commonly used quantitative indices of ductility are the ultimate elongation (expressed as a percent elongation or the gage length at rupture) and the reduction of cross-sectional area at the section where rupture occurs (expressed as a percentage of the original area).

4-2-11 Creep Limit

The property indicating the resistance of a material to failure by creep is known as the *creep limit* and is defined as the maximum stress for which the plastic strain will not exceed a specified amount during a specified time interval at a specified temperature. The creep limit is important when designing parts to be fabricated with polymeric materials (commonly known as plastics) and when designing metal parts that will be subjected to high temperatures and sustained loads (for example, the turbine blades in a turbojet engine).

4-2-12 Poisson's Ratio

A material loaded in one direction will undergo strains perpendicular to the direction of the load in addition to those parallel to the load. The ratio of the lateral or perpendicular strain (ϵ_{lat} or ϵ_t) to the longitudinal or axial strain (ϵ_{long} or ϵ_a) is called *Poisson's ratio* after Simeon D. Poisson, who identified the constant in 1811. The symbol ν is used for Poisson's ratio, which is given by the equation[3]

$$\nu = -\frac{\epsilon_{lat}}{\epsilon_{long}} = -\frac{\epsilon_t}{\epsilon_a} \tag{4-2}$$

The ratio $\nu = -\epsilon_t/\epsilon_a$ is valid only for a uniaxial state of stress. Since the lateral strain and the axial strain always have opposite signs, the negative sign in Eq. 4-2 ensures that ν will have a positive value. Like the modulus of elasticity E and the shear modulus G, Poisson's ratio is a property of the material. It will be shown in Section 4-3 that ν is related to E and G by the formula

$$E = 2(1 + \nu)G \tag{4-3}$$

Therefore, Poisson's ratio is a constant for stresses below the proportional limit and has a value between $1/4$ and $1/3$ for most metals.

4-2-13 Effect of Composition

The alloy content of a material affects the stress-strain behavior of the material. For example, Fig. 4-7 shows the effects of various alloy content on the stress-strain curves for steels of different strength levels ranging from a very hard, strong, brittle steel (A) to a relatively soft, ductile steel (E). The alloy content does not affect the modulus of elasticity but does affect the elastic limit, the ultimate strength, the fracture strength, and the ductility of the steel.

---

[3]Equation 4-2 relating Poisson's ratio and the lateral and axial strains is only valid for a uniaxial state of stress such as in the tension test described in this section. If the material is stressed in the lateral as well as the axial direction, then the lateral and axial strains will be independent quantities, and their ratio could be positive, negative, or even zero.

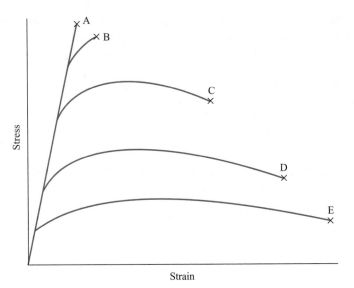

Figure 4-7

4-2-14 Effect of Temperature   Temperature also affects the stress-strain behavior of a material. The stress-strain diagrams shown in Fig. 4-3 were for room temperature. Figure 4-8 shows the effect of temperature on the tensile stress-strain diagram for a class 40 gray iron. The ductility of the material increases as the temperature increases, whereas the ultimate strength decreases as the temperature increases.

4-2-15 Effect of Tension or Compression   The stress-strain behavior of some materials depends on whether the axial load is tension or

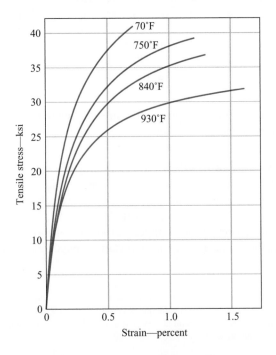

Figure 4-8

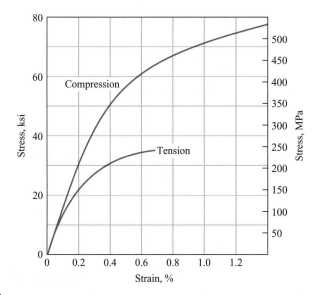

Figure 4-9

compression. For ductile materials the tension and compression behavior are usually assumed to be the same. For brittle materials the stress-strain curve obtained in a tension test differs from the curve obtained in a compression test. For example, Fig. 4-9 shows the tension and compression stress-strain curves for a class 35 gray iron. The linear range of stress-strain behavior for the compression curve is larger than the linear range for the tension curve. The ultimate strength in compression is also greater than the ultimate strength in tension.

The properties discussed in this section are primarily concerned with static or continuous loading or with slowly varying loading. Properties of a few typical materials may be found in Appendix B, Average Properties of Selected Materials.

Example Problem 4-1   A 100-kip axial load is applied to a $1 \times 4 \times 90$-in. rectangular bar. When loaded, the 4-in. side measures 3.9986 in., and the length has increased 0.09 in. Determine Poisson's ratio, the modulus of elasticity, and the modulus of rigidity of the material.

SOLUTION
The lateral and longitudinal strains and the axial stress for the bar are

$$\delta_{\text{lat}} = 3.9986 - 4 = -0.0014 \text{ in.}$$

$$\epsilon_{\text{lat}} = \frac{\delta_{\text{lat}}}{L} = \frac{-0.0014}{4} = -0.00035$$

$$\epsilon_{\text{long}} = \frac{\delta_{\text{long}}}{L} = \frac{0.09}{90} = 0.00100$$

$$\sigma = \frac{P}{A} = \frac{100}{4(1)} = 25 \text{ ksi}$$

Poisson's ratio ν is obtained by using Eq. 4-2,

$$\nu = -\frac{\epsilon_{\text{lat}}}{\epsilon_{\text{long}}} = -\frac{-0.00035}{0.00100} = 0.35 \qquad \textbf{Ans.}$$

The modulus of elasticity E is obtained by using Eq. 4-1b,

$$E = \frac{\sigma}{\epsilon} = \frac{25}{0.00100} = 25{,}000 \text{ ksi} \qquad \textbf{Ans.}$$

The modulus of rigidity is obtained by using Eq. 4-3,

$$G = \frac{E}{2(1+v)} = \frac{25{,}000}{2(1+0.35)} = 9260 \text{ ksi} \qquad \textbf{Ans.}$$

PROBLEMS

MecMovie Activities and Problems

MM4.1 Hooke's Law – basic problems. Concept checkpoints. Four basic problems requiring the use of Hooke's Law.

Introductory Problems

4-1* A 1.50-in.-diameter rod 20 ft long elongates 0.48 in. under a load of 53 kip. The diameter of the rod decreases 0.001 in. during the loading. Determine the modulus of elasticity, Poisson's ratio, and the modulus of rigidity for the material.

4-2* At the proportional limit, a 200-mm gage length of a 15-mm-diameter alloy bar has elongated 0.90 mm and the diameter has been reduced 0.022 mm. The total axial load carried was 62.6 kN. Determine the modulus of elasticity, Poisson's ratio, and the proportional limit for the material.

4-3 A 1/4 × 2-in. flat alloy bar elongates 0.08 in. in a length of 5 ft under a total axial load of 10,000 lb. The proportional limit of the material is 35,000 psi. Determine

a. The axial stress in the bar.
b. The modulus of elasticity for the material.
c. The total change in each lateral dimension if Poisson's ratio for the material is 0.25.

Intermediate Problems

4-4* A tensile test specimen having a diameter of 5.64 mm and a gage length of 50 mm was tested to fracture. Stress and strain values, which were calculated from load and deformation data obtained during the test, are shown in Fig. P4-4. Determine

a. The modulus of elasticity.
b. The proportional limit.
c. The ultimate strength.
d. The yield strength (0.05% offset).
e. The yield strength (0.2% offset).

f. The fracture stress.
g. The true fracture stress if Poisson's ratio $v = 0.30$ remains constant.
h. The tangent modulus at a stress level of 400 MPa.
i. The secant modulus at a stress level of 400 MPa.

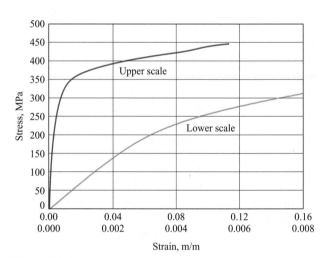

Figure P4-4

4-5 A tensile test specimen having a diameter of 0.250 in. and a gage length of 2.000 in. was tested to fracture. Stress and strain values, which were calculated from load and deformation data obtained during the test, are shown in Fig. P4-5. Determine

a. The modulus of elasticity.
b. The proportional limit.
c. The ultimate strength.
d. The yield strength (0.05% offset).
e. The yield strength (0.2% offset).
f. The fracture stress.
g. The true fracture stress if the final diameter of the specimen at the location of the fracture was 0.212 in.

h. The tangent modulus at a stress level of 56 ksi.
i. The secant modulus at a stress level of 56 ksi.

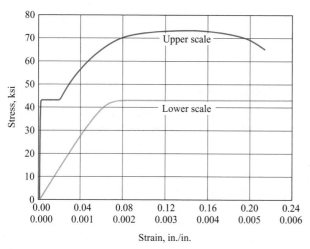

Figure P4-5

4-6 A tensile test specimen having a diameter of 5.64 mm and a gage length of 50 mm was tested to fracture. Stress and strain values, which were calculated from load and deformation data obtained during the test, are shown in Fig. P4-6. Determine

a. The modulus of elasticity.
b. The proportional limit.
c. The ultimate strength.
d. The yield strength (0.05% offset).
e. The yield strength (0.2% offset).
f. The fracture stress.
g. The true fracture stress if the final diameter of the specimen at the location of the fracture was 4.75 mm.
h. The tangent modulus at a stress level of 440 MPa.
i. The secant modulus at a stress level of 440 MPa.

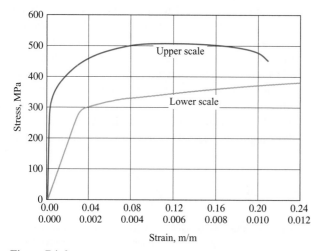

Figure P4-6

Challenging Problems

4-7* A tensile test specimen having a diameter of 0.505 in. and a gage length of 2.00 in. was tested to fracture. Load and deformation data obtained during the test were as follows:

Load (lb)	Change in Length (in.)	Load (lb)	Change in Length (in.)
0	0	12,600	0.0600
2,200	0.0008	13,200	0.0800
4,300	0.0016	13,900	0.1200
6,400	0.0024	14,300	0.1600
8,200	0.0032	14,500	0.2000
8,600	0.0040	14,600	0.2400
8,800	0.0048	14,500	0.2800
9,200	0.0064	14,400	0.3200
9,500	0.0080	14,300	0.3600
9,600	0.0096	13,800	0.4000
10,600	0.0200	13,000	Fracture
11,800	0.0400		

Determine

a. The modulus of elasticity.
b. The proportional limit.
c. The ultimate strength.
d. The yield strength (0.05% offset).
e. The yield strength (0.2% offset).
f. The fracture stress.
g. The true fracture stress if the final diameter of the specimen at the location of the fracture was 0.425 in.
h. The tangent modulus at a stress level of 46,000 psi.
i. The secant modulus at a stress level of 46,000 psi.

4-8 A tensile test specimen having a diameter of 11.28 mm and a gage length of 50 mm was tested to fracture. Load and deformation data obtained during the test were as follows:

Load (kN)	Change in Length (mm)	Load (kN)	Change in Length (mm)
0	0	43.8	1.50
7.6	0.02	45.8	2.00
14.9	0.04	48.3	3.00
22.2	0.06	49.7	4.00
28.5	0.08	50.4	5.00
29.9	0.10	50.7	6.00
30.6	0.12	50.4	7.00
32.0	0.16	50.0	8.00
33.0	0.20	49.7	9.00
33.3	0.24	47.9	10.00
36.8	0.50	45.1	Fracture
41.0	1.00		

Determine

a. The modulus of elasticity.
b. The proportional limit.
c. The ultimate strength.
d. The yield strength (0.05% offset).

e. The yield strength (0.2% offset).
f. The fracture stress.
g. The true fracture stress if the final diameter of the specimen at the location of the fracture was 9.50 mm.
h. The tangent modulus at a stress level of 315 MPa.
i. The secant modulus at a stress level of 315 MPa.

4-3 GENERALIZED HOOKE'S LAW

Hooke's law (see Eq. 4-1) can be extended to include the biaxial (see Fig. 2–26 and triaxial (see Fig. 2-24) states of stress often encountered in engineering practice. Consider Fig. 4-10, which shows a differential element of material subjected to a biaxial state of normal stress. Shear stresses have not been shown on the faces of the element because they produce distortion of the element (angle changes) but do not produce changes in the lengths of the sides of the element which would contribute to the normal strains. The deformations of the element in the directions of the normal stresses, for a combined loading, can be determined by computing the deformations resulting from the individual stresses separately and adding the values obtained algebraically. This procedure is based on the principle of superposition, which states that the effects of separate loadings can be added algebraically if two conditions are satisfied:

1. Each effect is linearly related to the load that produced it.

2. The effect of the first load does not significantly change the effect of the second load.

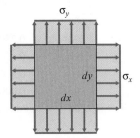

Figure 4-10

4.1

Condition 1 is satisfied if the stresses do not exceed the proportional limit for the material. Condition 2 is satisfied if the deformations are small so that the small changes in the areas of the faces of the element do not produce significant changes in the stresses.

The deformations of the element of Fig. 4-10, associated with the stresses σ_x and σ_y, are shown in Fig. 4-11. The shaded square in Fig. 4-11a indicates the original or unstrained configuration of the element. Under the action of the tensile stress σ_x (uniaxial state of stress), the element experiences a tensile strain of σ_x/E (Eq. 4-1a) in the x-direction and a compressive strain of $-v\sigma_x/E$ (Eq. 4-2) in the y-direction. These strains cause the element to stretch an amount $(\sigma_x/E)dx$ in the x-direction and to contract an amount $(v\sigma_x/E)dy$ in the y-direction to the configuration indicated in Fig. 4-11b (the deformations are greatly exaggerated). Then, under the action of the tensile stress σ_y superimposed on the stress σ_x, the element experiences a tensile strain of σ_y/E in the y-direction and a compressive strain of $v\sigma_y/E$ in the x-direction. These strains cause the element to stretch an amount $(\sigma_y/E)dy$ in the y-direction and to contract an amount $(v\sigma_y/E)dx$ in the x-direction, as shown in Fig. 4-11c. If the material is isotropic,[4] Young's modulus has the same value for all directions and the final deformation in the x-direction is (Fig. 4-11d),

$$d\delta_x = \epsilon_x dx = \frac{\sigma_x}{E}dx - v\frac{\sigma_y}{E}dx$$

---

[4]In an isotropic material, material properties such as the modulus of elasticity and Poisson's ratio are independent of direction within the material. Examples of nonisotropic materials are fiber-reinforced materials, wood, and many crystalline materials.

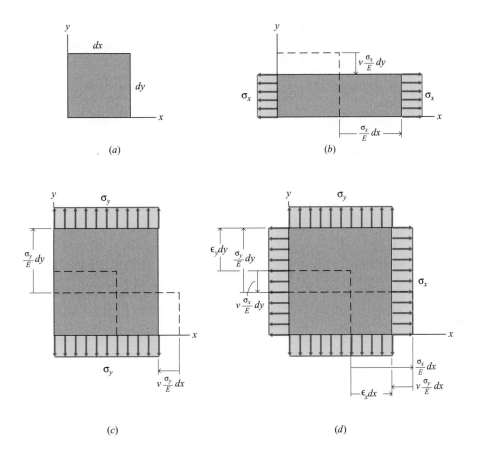

Figure 4-11

Proceeding in a similar fashion for the deformations in the y- and z-directions, the three normal strains are

$$\epsilon_x = \frac{1}{E}(\sigma_x - v\sigma_y)$$

$$\epsilon_y = \frac{1}{E}(\sigma_y - v\sigma_x)$$

$$\epsilon_z = -\frac{v}{E}(\sigma_x + \sigma_y)$$

(4-4)

Note that $\epsilon_z \neq 0$ even though $\sigma_z = 0$. That is, when the material is pulled in the x- and y-directions, it will shrink in the z-direction unless prevented by a stress in the z-direction. Note that Eqs. 4-4 are valid for a state of plane stress in which $\sigma_z = \tau_{zx} = \tau_{zy} = 0$.

The analysis above is readily extended to triaxial principal stresses, and the expressions for strain become

$$\epsilon_x = \frac{1}{E}[\sigma_x - v(\sigma_y + \sigma_z)]$$

$$\epsilon_y = \frac{1}{E}[\sigma_y - v(\sigma_x + \sigma_z)]$$

$$\epsilon_z = \frac{1}{E}[\sigma_z - v(\sigma_x + \sigma_y)]$$

(4-5)

In these expressions, tensile stresses and strains are considered positive, and compressive stresses and strains are considered negative.

When Eqs. 4-4 are solved for the stresses in terms of the strains, they give

$$\sigma_x = \frac{E}{1 - v^2}(\epsilon_x + v\epsilon_y)$$
$$\sigma_y = \frac{E}{1 - v^2}(\epsilon_y + v\epsilon_x) \tag{4-6}$$

and $\sigma_z = 0$ by assumption. Equations 4-6 can be used to calculate normal stresses from measured or computed normal strains. When Eqs. 4-5 are solved for stresses in terms of strains, they give

$$\sigma_x = \frac{E}{(1 + v)(1 - 2v)}[(1 - v)\epsilon_x + v(\epsilon_y + \epsilon_z)]$$
$$\sigma_y = \frac{E}{(1 + v)(1 - 2v)}[(1 - v)\epsilon_y + v(\epsilon_z + \epsilon_x)] \tag{4-7}$$
$$\sigma_z = \frac{E}{(1 + v)(1 - 2v)}[(1 - v)\epsilon_z + v(\epsilon_x + \epsilon_y)]$$

In Chapter 3, the expression

$$\epsilon_z = \epsilon_{p_3} = -\frac{v}{1 - v}(\epsilon_x + \epsilon_y) \tag{3-13}$$

was stated without proof for the case of plane stress. From Eqs. 4-4 and 4-6, which represent Hooke's law for the case of plane stress,

$$\epsilon_z = -\frac{v}{E}(\sigma_x + \sigma_y) = -\frac{v}{E}\left[\frac{E}{1 - v^2}\right][(\epsilon_x + v\epsilon_y) + (\epsilon_y + v\epsilon_x)]$$
$$= -\frac{v}{1 - v}(\epsilon_x + \epsilon_y) \tag{3-13}$$

This out-of-plane principal strain is important since the maximum shear strain at the point may be $(\epsilon_{p1} - \epsilon_{p2})$, $(\epsilon_{p1} - \epsilon_{p3})$, or $(\epsilon_{p3} - \epsilon_{p2})$, depending on the relative magnitudes and signs of the principal strains at the point.

Torsion test specimens are used to study material behavior under pure shear, and it is observed that a shear stress produces only a single corresponding shear strain (Chapter 6). Thus, Hooke's law extended to shear stresses (for linear elastic material behavior) is simply

$$\tau = G\gamma \tag{4-8}$$

where G is the shear modulus or modulus of rigidity. Equations 4-7 and 4-8 seem to indicate that three elastic constants E, v, and G are required to determine the deformations (and strains) in a material resulting from an arbitrary state of stress. In fact, only two of these constants need to be determined experimentally for a given material.

The relationship between the elastic constants E, v, and G can be determined by considering the stresses and strains produced by an axial tensile load in a bar

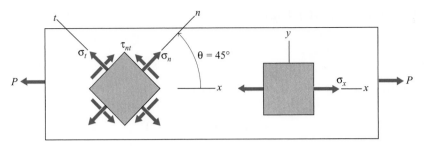

Figure 4-12

of the material (Fig. 4-12). The shearing stresses for the xy coordinate axes (σ_x, $\sigma_y = \tau_{xy} = 0$) and the nt coordinate axes (σ_n, σ_t, τ_{nt}) are related through the stress transformation equation (Eq. 2-13a),

$$
\begin{aligned}
\tau_{nt} &= -(\sigma_x - \sigma_y) \sin\theta\cos\theta + \tau_{xy}(\cos^2\theta - \sin^2\theta) \\
&= -(\sigma_x - 0)\sin(+45°)\cos(+45°) + 0 \\
&= -\frac{\sigma_x}{2}
\end{aligned}
\tag{a}
$$

The relationship between the shearing strains for the xy and the nt coordinate axes is determined using the strain transformation equation (Eq. 3.8a) as

$$
\begin{aligned}
\gamma_{nt} &= -2(\epsilon_x - \epsilon_y)\sin\theta\cos\theta + \gamma_{xy}(\cos^2\theta - \sin^2\theta) \\
&= -2(\epsilon_x - \epsilon_y)\sin(+45°)\cos(+45°) + 0 \\
&= -(\epsilon_x - \epsilon_y)
\end{aligned}
\tag{b}
$$

Since the bar shown in Fig. 4-12 is subjected to an axial force, the relationship between ϵ_x and ϵ_y is found using Eq. 4-2; that is,

$$
v = \frac{-\epsilon_t}{\epsilon_a} = \frac{-\epsilon_y}{\epsilon_x}
$$

from which

$$
\epsilon_y = -v\epsilon_x
\tag{c}
$$

Substituting Eq. (c) into Eq. (b) gives

$$
\gamma_{nt} = -\epsilon_x(1 + v)
\tag{d}
$$

Equation 4-1c may be used to give a relationship between shearing stress and shearing strain, resulting in

$$
\tau_{nt} = G\gamma_{nt}
\tag{e}
$$

Since γ_{nt} is negative [Eq. (d)] the shearing stress τ_{nt} is also negative, giving

$$
\tau_{nt} = G\gamma_{nt} = -G\epsilon_x(1 + v)
\tag{f}
$$

Recall Hooke's law (Eq. 4-1a)

$$\epsilon_x = \frac{\sigma_x}{E} \tag{g}$$

Finally, combining Eqs. (a), (f), and (g) gives

$$\frac{-\sigma_x}{2} = -G\left(\frac{\sigma_x}{E}\right)(1+v)$$

or

$$G = \frac{E}{2(1+v)} \tag{4-9}$$

Equation 4-9 shows that, for homogeneous, isotropic materials, the material properties E, v, and G are related; there are only two independent material properties. Values of E and G for selected materials are given in Appendix B, Tables B-17 (in U.S. customary units) and B-18 (in SI units).

   If Eq. 4-9 is substituted into Eq. 4-8, an alternate form of generalized Hooke's law for shearing stress and strain in isotropic materials is obtained. Thus,

$$\tau_{xy} = G\gamma_{xy} = \frac{E}{2(1+v)}\gamma_{xy}$$

$$\tau_{yz} = G\gamma_{yz} = \frac{E}{2(1+v)}\gamma_{yz} \tag{4-10}$$

$$\tau_{zx} = G\gamma_{zx} = \frac{E}{2(1+v)}\gamma_{zx}$$

   Equations 4-4 through 4-10 are widely used for experimental determinations of stress. The following Example Problems illustrate the method of application.

Example Problem 4-2   At a point on the surface of an alloy steel ($E = 210$ GPa and $v = 0.30$) machine part subjected to a biaxial state of stress, the measured strains were $\epsilon_x = +1394$ μm/m, $\epsilon_y = -660$ μm/m, and $\gamma_{xy} = 2054$ μrad. Determine

(a) The stress components σ_x, σ_y, and τ_{xy} at the point.
(b) The principal stresses and the maximum shear stress at the point. Locate the planes on which these stresses act and show the stresses on a triangular element.

SOLUTION

(a) The normal stresses σ_x and σ_y are obtained by using Eqs. 4-6. Thus,

$$\sigma_x = \frac{E}{1-v^2}(\epsilon_x + v\epsilon_y)$$

$$= \frac{210(10^9)}{1-(0.30)^2}[1394 + 0.30(-660)](10^{-6})$$

$$= +276.0(10^6)\,\text{N/m}^2 = 276\,\text{MPa (T)} \qquad \textbf{Ans.}$$

$$\sigma_y = \frac{E}{1 - v^2}(\epsilon_y + v\epsilon_x)$$

$$= \frac{210(10^9)}{1 - (0.30)^2}[-660 + 0.30(1394)](10^{-6})$$

$$= -55.80(10^6)\,\text{N/m}^2 = 55.8\,\text{MPa (C)} \qquad \textbf{Ans.}$$

The shear stress τ_{xy} is obtained by using Eqs. 4-10. Thus,

$$\tau_{xy} = \frac{E}{2(1 + v)}\gamma_{xy}$$

$$= \frac{210(10^9)}{2(1 + 0.30)}(2054)(10^{-6})$$

$$= 165.90(10^6)\,\text{N/m}^2 = 165.9\ \text{MPa} \qquad \textbf{Ans.}$$

(b) When the values for σ_x, σ_y, and τ_{xy} are substituted in Eq. 2-15, the principal stresses are found to be

$$\sigma_{p1,p2} = \frac{\sigma_x + \sigma_y}{2} \pm \sqrt{\left(\frac{\sigma_x - \sigma_y}{2}\right)^2 + \tau_{xy}^2}$$

$$= \frac{276.0 - 55.80}{2} \pm \sqrt{\left(\frac{276.0 + 55.80}{2}\right)^2 + (165.90)^2}$$

$$= 110.10 \pm 234.62$$

$$\sigma_{p1} = 110.10 + 234.62 = 344.72\,\text{MPa} \cong 345\,\text{MPa (T)} \qquad \textbf{Ans.}$$
$$\sigma_{p2} = 110.10 - 234.62 = -124.52\,\text{MPa} \cong 124.5\,\text{MPa (C)} \qquad \textbf{Ans.}$$

A state of plane stress exists on the surface of the machine part; therefore,

$$\sigma_{p3} = \sigma_z = 0 \qquad \textbf{Ans.}$$

Since σ_{p1} and σ_{p2} have opposite signs, the maximum shearing stress is given by Eq. 2-18 as

▶ The direction of $\tau_{max} = 235$ MPa on Fig. 4-13 is to oppose the larger of the principal stresses $\sigma_{p1} = 345$ MPa and $\sigma_{p2} = 124.5$ MPa. As observed in Example Problems 2-11 and 2-12, the normal stress on the surface of maximum shear stress is the average of the principal stresses $\sigma_n = [344.72 + (-124.52)]/2 = 110.1$ MPa.

$$\tau_{max} = \tau_p = \frac{1}{2}(\sigma_{p1} - \sigma_{p2})$$

$$= \frac{1}{2}(344.72 + 124.52) = 234.62\,\text{MPa} \cong 235\,\text{MPa} \qquad \textbf{Ans.}$$

The in-plane principal stress directions are given by Eq. 2-14 as

$$\tan 2\theta_p = \frac{2\tau_{xy}}{\sigma_x - \sigma_y} = \frac{2(165.90)}{276.0 + 55.80} = 1.0000$$

$$2\theta_p = 45.00° \qquad \theta_p = 22.5° \qquad \textbf{Ans.}$$

The required sketch is shown in Fig. 4-13.

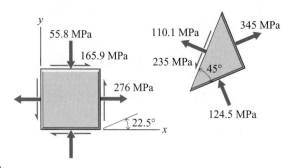

Figure 4-13

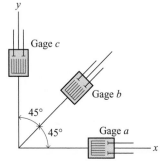

Figure 4-14

Example Problem 4-3 At a point on the free surface of a steel ($E = 30{,}000$ ksi and $v = 0.30$) machine part, the strain rosette shown in Fig. 4-14 was used to obtain the following normal strain data: $\epsilon_a = +650\ \mu\text{in./in.}$, $\epsilon_b = +475\ \mu\text{in./in.}$, and $\epsilon_c = -250\ \mu\text{in./in.}$

(a) Determine the stress components σ_x, σ_y, and τ_{xy} at the point.

(b) Determine the principal strains and the maximum shearing strain at the point. Prepare a sketch showing all of these strains.

(c) Determine the principal stresses and the maximum shearing stress at the point. Prepare a sketch showing all of these stresses.

SOLUTION

(a) The measured strain values are

$$\epsilon_a = \epsilon_x = +650\mu, \quad \epsilon_b = +475\mu\ (\theta = 45^\circ), \text{ and } \epsilon_c = \epsilon_y = -250\mu$$

The shear strain γ_{xy} is obtained from these measured data by using Eq. 3-7a. Thus,

$$\begin{aligned}
\epsilon_b &= \epsilon_x \cos^2 \theta_b + \epsilon_y \sin^2 \theta_b + \gamma_{xy} \sin \theta_b \cos \theta_b \\
&= 650\mu \cos^2(45^\circ) - 250\mu \sin^2(45^\circ) + \gamma_{xy} \sin(45^\circ) \cos(45^\circ) \\
&= +475\mu
\end{aligned}$$

from which $\gamma_{xy} = 550\mu$.

The normal stresses σ_x and σ_y are obtained by using Eqs. 4-6. Thus,

$$\sigma_x = \frac{E}{1 - v^2}(\epsilon_x + v\epsilon_y) = \frac{30{,}000}{1 - (0.30)^2}[650 + 0.30(-250)](10^{-6})$$
$$= +18.956\ \text{ksi} \cong 18.96\ \text{ksi (T)} \qquad \textbf{Ans.}$$

$$\sigma_y = \frac{E}{1 - v^2}(\epsilon_y + v\epsilon_x) = \frac{30{,}000}{1 - (0.30)^2}[-250 + 0.30(650)](10^{-6})$$
$$= -1.8132\ \text{ksi} \cong 1.813\ \text{ksi (C)} \qquad \textbf{Ans.}$$

The shear stress τ_{xy} is obtained by using Eqs. 4-10. Thus,

$$\tau_{xy} = \frac{E}{2(1+v)}\gamma_{xy} = \frac{30,000}{2(1+0.30)}(550)(10^{-6})$$
$$= 6.346 \, \text{ksi} \cong 6.35 \, \text{ksi} \qquad \textbf{Ans.}$$

(b) The in-plane principal strains ϵ_{p1} and ϵ_{p2} are obtained from the strain data by using Eq. 3-10. Thus,

$$\epsilon_{p_1}, \epsilon_{p_2} = \frac{\epsilon_x + \epsilon_y}{2} \pm \sqrt{\left(\frac{\epsilon_x - \epsilon_y}{2}\right)^2 + \left(\frac{\gamma_{xy}}{2}\right)^2}$$

$$= \frac{650\mu - 250\mu}{2} \pm \sqrt{\left(\frac{650\mu + 250\mu}{2}\right)^2 + \left(\frac{550\mu}{2}\right)^2}$$

$$= 200\mu \pm 527.4\mu$$

$$\epsilon_{p_1} = 200\mu + 527.4\mu = 727.4\mu \cong 727\mu \qquad \textbf{Ans.}$$

$$\epsilon_{p_2} = 200\mu - 527.4\mu = -327.4\mu \cong -327\mu \qquad \textbf{Ans.}$$

The third principal strain $\epsilon_{p3} = \epsilon_z$ is obtained by using Eq. 3-13. Thus,

$$\epsilon_{p3} = \epsilon_z = -\frac{v}{1-v}(\epsilon_x + \epsilon_y)$$

$$= -\frac{0.30}{1-0.30}(650\mu - 250\mu) = -171.4\mu \qquad \textbf{Ans.}$$

Since ϵ_{p3} is less compressive than the in-plane principal strain ϵ_{p2}, the maximum shear strain at the point is the maximum in-plane shear strain γ_p. Thus,

$$\gamma_{\text{max}} = \gamma_p = \epsilon_{p_1} - \epsilon_{p_2}$$
$$= 727.4\mu + 327.4\mu = 1054.8\mu \cong 1055\mu \qquad \textbf{Ans.}$$

The in-plane principal strain directions are given by Eq. 3-9 as

$$\tan 2\theta_p = \frac{\gamma_{xy}}{\epsilon_x - \epsilon_y} = \frac{550\mu}{650\mu + 250\mu} = 0.61111$$
$$2\theta_p = 31.429° \qquad \theta_p = 15.714° \cong 15.71° \qquad \textbf{Ans.}$$

The required sketch is shown in Fig. 4-15.

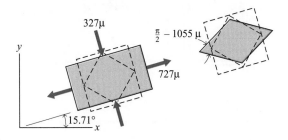

Figure 4-15

▶ Note that Eqs. 4-6 relate the stresses σ and the strains ϵ for any pair of orthogonal directions in the plane—the x- and y-directions or the n- and t-directions or the 1- and 2-directions.

(c) Once the principal strains ϵ_{p1} and ϵ_{p2} are known, Eqs. 4-6 can be used to determine the principal stresses σ_{p1} and σ_{p2}. Thus,

$$\sigma_{p_1} = \frac{E}{1 - v^2}(\epsilon_{p_1} + v\epsilon_{p_2})$$

$$= \frac{30,000}{1 - (0.30)^2}[727.4 + 0.30(-327.4)](10^{-6})$$

$$= +20.742 \text{ ksi} \cong 20.7 \text{ ksi (T)} \qquad \textbf{Ans.}$$

$$\sigma_{p_2} = \frac{E}{1 - v^2}(\epsilon_{p_2} + v\epsilon_{p_1})$$

$$= \frac{30,000}{1 - (0.30)^2}[-327.4 + 0.30(727.4)](10^{-6})$$

$$= -3.599 \text{ ksi} \cong 3.60 \text{ ksi (C)} \qquad \textbf{Ans.}$$

Since a state of plane stress exists on the surface of the machine part, the third principal stress is

$$\sigma_{p_3} = \sigma_z = 0 \qquad \textbf{Ans.}$$

The maximum shear stress is obtained using Eq. 4-10 and the maximum shear strain. Thus,

$$\tau_{\max} = \frac{E}{2(1 + v)}\gamma_{\max} = \frac{30,000}{2(1 + 0.30)}(1054.8)(10^{-6})$$

$$= 12.171 \text{ ksi} \cong 12.17 \text{ ksi} \qquad \textbf{Ans.}$$

The two in-plane principal stresses have opposite signs; therefore, the maximum shear stress can also be obtained as

$$\tau_{\max} = \tau_p = \frac{1}{2}(\sigma_{p_1} - \sigma_{p_2})$$

$$= \frac{1}{2}(20.742 + 3.599) = 12.171 \text{ ksi} \cong 12.17 \text{ ksi} \qquad \textbf{Ans.}$$

▶ As observed in Example Problems 2-11 and 2-12, the normal stress on the surface of maximum shear stress is the average of the principal stresses $\sigma_n = [20.742 + (-3.599)]/2 = 8.57$ ksi.

The required sketch is shown in Fig. 4-16.

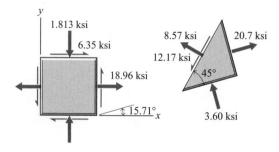

Figure 4-16

Example Problem 4-4 A $10 \times 10 \times 10$-in. block of steel ($E = 30,000$ ksi and $v = 0.30$) is loaded with a uniformly distributed pressure of $30,000$ psi on the four faces having outward normals in the x- and y-directions. Rigid frictionless constraints limit the deformation of the block in the z-direction to $+0.002$ in. Determine the normal stress σ_z that develops as the pressure is applied.

SOLUTION

Figure 4-17 depicts the stresses acting on the block. For simplicity, the constraint (which is perpendicular to the z-axis) is not shown. The uniformly distributed pressure causes the stresses $\sigma_x = -30,000$ psi and $\sigma_y = -30,000$ psi. Since the constraint is rigid and frictionless, there are no shearing stresses on the z-faces of the block ($\tau_{zx} = \tau_{zy} = 0$). The normal stress σ_z is developed when the deformation in the z-direction becomes 0.002 in. Thus, the stresses and deformation in the z-direction give

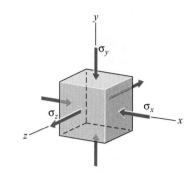

Figure 4-17

$$\sigma_x = -30,000\,\text{psi}$$

$$\sigma_y = -30,000\,\text{psi}$$

$$\sigma_z = \text{unknown}$$

$$\delta_z = +0.002\,\text{in.}$$

The third of Eqs. 4-5 gives

$$\epsilon_z = \frac{1}{E}[\sigma_z - v(\sigma_x + \sigma_y)]$$

But the normal strain ϵ_z is related to the given value of deformation $\delta_z = +0.002$ in. by Eq. 3-1,

$$\epsilon_z = \frac{\delta_z}{L}$$

where L is the original length of the block in the z-direction. Combining the two expressions for ϵ_z gives

$$\frac{\delta_z}{L} = \frac{1}{E}[\sigma_z - v(\sigma_x + \sigma_y)]$$

or

$$\frac{0.002}{10} = \frac{1}{30(10^6)}[\sigma_z - (0.3)(-30,000 - 30,000)]$$

and solving for σ_z gives

$$\sigma_z = -12,000\,\text{psi} = 12.00\,\text{ksi (C)} \qquad \text{**Ans.**}$$

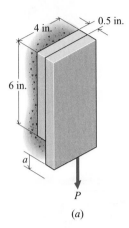

(a)

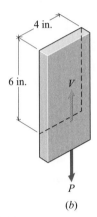

(b)

Figure 4-18

Example Problem 4-5

A $0.5 \times 6 \times 4$-in. piece of rubber is attached to a wall and a piece of steel (Fig. 4-18a). When the force P is 30 lb, the rigid steel plate displaces downward $a = 0.0003$ in. Determine the shear modulus of the rubber.

SOLUTION

The shear modulus is given by Eq. 4-10 as

$$G = \tau/\gamma$$

To determine the shear stress, consider the free-body diagram of the steel plate shown in Fig. 4-18b, where V is the resultant shear force acting over the 4×6-in. area of the rubber. For vertical equilibrium, $V = P = 30$ lb. Therefore,

$$\tau = \frac{V}{A_s} = \frac{30}{(4)(6)} = 1.25 \, \text{psi}$$

The shear strain is found using Eq. 3-3,

$$\gamma = \frac{\delta_s}{L} = \frac{0.0003}{0.5} = 0.0006$$

Therefore,

$$G = \frac{\tau}{\gamma} = \frac{1.25}{0.0006} = 2080 \, \text{psi} \qquad \textbf{Ans.}$$

PROBLEMS

MecMovie Activities and Problems

MM4.2 Principal stresses from rosette data. Example; Try one. Using strain gage rosette data to compute the normal and shear strains, the normal and shear stresses, and the principal stresses in the x-y plane.

Introductory Problems

In Problems 4-9 through 4-12 the strain components ϵ_x, ϵ_y, and γ_{xy} are given for a point on the surface of a machine component. Determine the stresses σ_x, σ_y, and τ_{xy} at the point.

Problem	ϵ_x	ϵ_y	γ_{xy}	E	ν
4-9*	$+900\mu$	-300μ	-400μ	10,000 ksi	0.30
4-10*	$+1175\mu$	-1250μ	$+850\mu$	190 GPa	0.25
4-11	$+500\mu$	$+250\mu$	$+150\mu$	15,000 ksi	0.34
4-12	$+1000\mu$	$+400\mu$	$+800\mu$	210 GPa	0.25

Intermediate Problems

4-13* Determine the state of strain that corresponds to the following state of stress at a point in a steel ($E = 30,000$ ksi and $\nu = 0.30$) machine part: $\sigma_x = 15,000$ psi, $\sigma_y = 5000$ psi, $\sigma_z = 7500$ psi, $\tau_{xy} = 5500$ psi, $\tau_{yz} = 4750$ psi, and $\tau_{zx} = 3200$ psi.

4-14* Determine the state of strain that corresponds to the following state of stress at a point in an aluminum ($E = 73$ GPa and $\nu = 0.33$) machine part: $\sigma_x = 120$ MPa, $\sigma_y = -85$ MPa, $\sigma_z = 45$ MPa, $\tau_{xy} = 35$ MPa, $\tau_{yz} = 48$ MPa, and $\tau_{zx} = 76$ MPa.

4-15 The $0.5 \times 2 \times 4$-in. rubber mounts shown in Fig. P4-15 are used to isolate the vibrational motion of a machine from its supports. The shear modulus of the rubber is 3000 psi. Determine the force P required to displace the rigid frame 0.001 in. vertically.

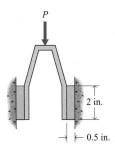

Figure P4-15

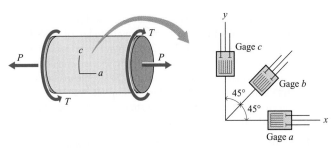

Figure P4-18

4-16* At a point on the free surface of an aluminum ($E = 73$ GPa and $\nu = 0.33$) machine part, the strain rosette shown in Fig. P4-16 was used to obtain the following strain data: $\epsilon_a = +875\ \mu\text{m/m}$, $\epsilon_b = +700\ \mu\text{m/m}$, and $\epsilon_c = -650\ \mu\text{m/m}$. Determine the stress components σ_x, σ_y, and τ_{xy}.

4-19* The stresses on the free surface of a bar subjected to an axial force P and a torque T are shown in Fig. P4-19. The bar is made of steel ($E = 30,000$ ksi and $\nu = 0.30$). Determine the strain components ϵ_x, ϵ_y, and γ_{xy}.

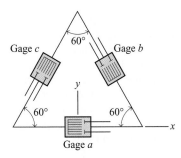

Figure P4-16

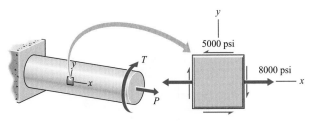

Figure P4-19

4-17 The strain rosette shown in Fig. P4-17 was used to measure the normal strains on the free surface of a structural steel ($E = 30,000$ ksi and $\nu = 0.30$) member. The measured strains were $\epsilon_a = +650\ \mu\text{in./in.}$, $\epsilon_b = +475\ \mu\text{in./in.}$, and $\epsilon_c = -250\ \mu\text{in./in.}$ Determine the stress components σ_x, σ_y, and τ_{xy}.

Challenging Problems

4-20* The strain rosette shown in Fig. P4-20 was used to obtain the following normal strain data at a point on the free surface of a steel ($E = 200$ GPa and $\nu = 0.30$) machine part: $\epsilon_a = -555\ \mu\text{m/m}$, $\epsilon_b = +925\ \mu\text{m/m}$, and $\epsilon_c = +740\ \mu\text{m/m}$.

a. Determine the stress components σ_x, σ_y, and τ_{xy} at the point.
b. Determine the principal stresses and the maximum shear stress at the point. Prepare a sketch showing these stresses on a triangular element.

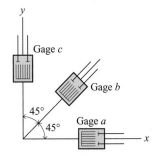

Figure P4-17

4-18 A steel ($E = 200$ GPa and $\nu = 0.30$) thin-walled pressure vessel is subjected to an internal pressure, an axial force P, and a torque T, as shown in Fig. P4-18. A strain rosette mounted on the outside surface of the vessel measured the strains: $\epsilon_a = +540\ \mu\text{m/m}$, $\epsilon_b = +930\ \mu\text{m/m}$, and $\epsilon_c = +20\ \mu\text{m/m}$, where a is in the longitudinal direction and c is in the hoop direction. Determine the stress components σ_x, σ_y, and τ_{xy}.

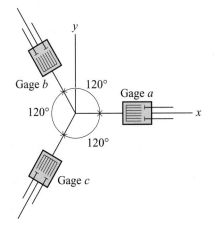

Figure P4-20

4-21* At a point on the free surface of an alloy steel ($E = 30,000$ ksi and $v = 0.30$) machine part, normal strains of $+1000$ μin./in., $+2000$ μin./in., and $+1200$ μin./in. were measured at angles of $0°$, $60°$, and $120°$, respectively. Design considerations limit the maximum normal stress to 74 ksi, the maximum shearing stress to 40 ksi, the maximum normal strain to 2200 μin./in., and the maximum shearing strain to 2500 μrad. What is your evaluation of the design?

4-22 A $10 \times 10 \times 25.4$-mm block of rubber-like material ($E = 1.4$ GPa and $v = 0.40$) is to be pushed into a lubricated $10 \times 10 \times 25$-mm flat-bottomed hole in a rigid material, as shown in Fig. P4-22. Determine the load P required to push the block into the hole until its top surface is flush with the top of the hole.

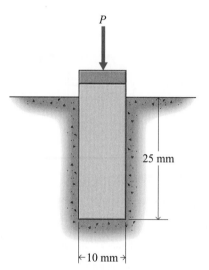

Figure P4-22

4-23 The strain rosette shown in Fig. P4-23 was used to measure the normal strains at a point on the free surface of an aluminum ($E = 10,600$ ksi and $v = 0.33$) member. The measured strains were $\epsilon_a = +875$ μin./in., $\epsilon_b = +700$ μin./in., and $\epsilon_c = -350$ μin./in.

a. Determine the principal strains and the maximum shearing strain at the point.
b. Use the results of part a to determine the principal stresses and the maximum shear stress at the point. A sketch is not required.

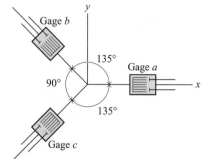

Figure P4-23

4-24 A thick-walled cylindrical pressure vessel will be used to store gas under a pressure of 100 MPa. During initial pressurization of the vessel, axial and hoop components of strain were measured on the inside and outside surfaces. On the inside surface the axial strain was $+500$ μm/m and the hoop strain was $+750$ μm/m. On the outside surface the axial strain was $+500$ μm/m and the hoop strain was $+100$ μm/m. Determine the axial and hoop components of stress associated with these strains if $E = 200$ GPa and $v = 0.30$.

4-25* A steel sleeve is connected to a steel shaft with a flexible rubber insert, as shown in Fig. P4-25. The rubber insert has an outside diameter of 4.25 in. and an inside diameter of 3.25 in. The length of the insert is 8 in.; the shear modulus is 2000 psi. For a force $P = 300$ lb, determine the deflection of the shaft with respect to the sleeve.

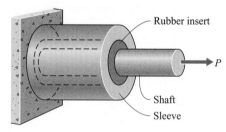

Figure P4-25

4-4 THERMAL STRAIN

Most engineering materials when unrestrained expand when heated and contract when cooled. The thermal strain due to a one-degree ($1°$) change in temperature is designated by α and is known as the *coefficient of thermal expansion*.

The thermal strain (of an unrestrained body) due to a temperature change of ΔT degrees is

$$\epsilon_T = \alpha \Delta T \tag{4-11}$$

Like the material constants described in the last section, the value of α for various materials must be determined experimentally. The coefficient of thermal expansion is approximately constant for a large range of temperatures (in general, the coefficient increases with an increase of temperature). For a homogeneous,[5] isotropic material, the coefficient applies to all dimensions (all directions). Values of the coefficient of expansion for several materials are included in Appendix B.

4-4-1 Total Strains

Strains caused by temperature changes and strains caused by applied loads are essentially independent. The total strain in any direction is the sum of the strain caused by the applied loads and the strain caused by the temperature change

$$\epsilon_{total} = \epsilon_\sigma + \epsilon_T \qquad (4\text{-}12a)$$

For a member loaded only in the x-direction, the total axial strain (strain in the load direction) is given by[6]

$$\begin{aligned} \epsilon_{axial} &= \epsilon_\sigma + \epsilon_T \\ &= \left(\frac{\sigma_x}{E} - \frac{\nu\sigma_y}{E} - \frac{\nu\sigma_z}{E} \right) + \alpha\,\Delta T \qquad (4\text{-}12b) \\ &= \frac{\sigma}{E} + \alpha\,\Delta T \end{aligned}$$

and the total lateral or transverse strain (strain in the direction perpendicular to the load) is given by

$$\begin{aligned} \epsilon_{trans} &= \epsilon_\sigma + \epsilon_T \\ &= \left(\frac{\sigma_y}{E} - \frac{\nu\sigma_x}{E} - \frac{\nu\sigma_z}{E} \right) + \alpha\,\Delta T \qquad (4\text{-}12c) \\ &= \frac{-\nu\sigma}{E} + \alpha\,\Delta T \end{aligned}$$

Since homogeneous, isotropic materials, when unrestrained, expand uniformly in all directions when heated (and contract uniformly when cooled), neither the shape of the body nor the shear stresses and shear strains are affected by temperature changes.

■ **Example Problem 4-6** The aluminum [$E = 70$ GPa, $\alpha = 22.5(10^{-6})/°C$] block shown in Fig. 4-19 rests on a smooth, horizontal surface. When the body is subjected to a temperature change of $\Delta T = +20°C$, determine

(a) The thermal strains ϵ_{Tx}, ϵ_{Ty}, and ϵ_{Tz}.
(b) The deformations in the coordinate directions δ_x, δ_y, and δ_z.
(c) The shearing strain γ_{xy}.

---

[5] In a homogeneous material, material properties such as modulus of elasticity and Poisson's ratio do not vary from point to point. Examples of nonhomogeneous materials are concrete (which consists of sand and rocks held together by cement) and particle board (which consists of sawdust and wood chips held together by glue).

[6] Assuming the deformation remains in the linearly elastic range so that Hooke's law (Eq. 4-5) applies.

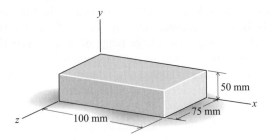

Figure 4-19

SOLUTION

(a) The thermal strain is given by Eq. 4-11, $\epsilon_T = \alpha\,\Delta T$, and is the same for each coordinate direction. Thus,

$$\epsilon_T = \epsilon_{Tx} = \epsilon_{Ty} = \epsilon_{Tz} = \alpha\,\Delta T = 22.5(10^{-6})(20)$$
$$= 450(10^{-6})\,\text{m/m} = 450\,\mu\text{m/m} \qquad\qquad \textbf{Ans.}$$

(b) Deformations may be calculated using Eq. 3-1, $\delta = \epsilon L$. Since the block is not constrained and there are no applied forces, there are no stresses, $\sigma_x = \sigma_y = \sigma_z = 0$. Therefore, the load portion of the strain is zero ($\epsilon_\sigma = 0$) and the total strain is the same as the thermal strain, $\epsilon_{\text{total}} = \epsilon_\sigma + \epsilon_T = \epsilon_T$. Although the total strain is the same for each of the coordinate directions, the initial lengths of the member are different for each of the coordinate directions. Therefore, the deformations in the coordinate directions are

$$\delta_x = \epsilon_{Tx}L = 450(10^{-6})(100) = 0.0450\,\text{mm} \qquad \textbf{Ans.}$$
$$\delta_y = \epsilon_{Ty}L = 450(10^{-6})(50) = 0.0225\,\text{mm} \qquad \textbf{Ans.}$$
$$\delta_z = \epsilon_{Tz}L = 450(10^{-6})(75) = 0.0338\,\text{mm} \qquad \textbf{Ans.}$$

(c) Since the body is not constrained and there are no applied forces, there are no stresses ($\sigma_x = \sigma_y = \sigma_z = \tau_{xy} = \tau_{yz} = \tau_{zx} = 0$) and the total strain is the same as the thermal strain, $\epsilon_{\text{total}} = \epsilon_\sigma + \epsilon_T = \epsilon_T$. However, the thermal strain is the same in all directions. Therefore, original 90° angles do not change. That is, there are no shearing strains.

$$\gamma_{xy} = 0 \qquad\qquad \textbf{Ans.}$$

Example Problem 4-7   A 1/2-in.-diameter steel [$E = 30{,}000$ ksi, $\nu = 0.30$, $\alpha = 6.5\,(10^{-6})/°\text{F}$] rod has an initial length of 6 ft. Determine

(a) The change in length of the rod after a tensile load of 5000 lb is applied to the rod, and the temperature of the rod decreases 50°F.

(b) The change in diameter of the rod for the conditions given in part (a).

SOLUTION

Since the rod is loaded in only one direction, Eq. 4-12b can be used for the axial direction and Eq. 4-12c for the transverse (diametral) direction.

(a) Strain ϵ is the ratio of change in length δ and initial length L, and stress σ is the ratio of force P and area A. Therefore, Eq. 4-12b can be written

$$\frac{\delta}{L} = \frac{P}{AE} + \alpha \, \Delta T$$

and the rod will stretch

$$\delta = \left(\frac{P}{AE} + \alpha \Delta T\right) L$$

$$= \left[\frac{5000}{30(10^6)(\pi/4)(1/2)^2} + 6.5(10^{-6})(-50)\right](6)$$

$$= 0.003143 \text{ ft} \cong 0.0377 \text{ in.} \qquad \textbf{Ans.}$$

▶ It is important that the terms in Eq. 4-12 be compatible. That is, a positive strain ϵ is a stretch (positive δ) and is caused by a tensile stress (positive σ) and a temperature increase (positive ΔT). Since the force is tensile, it is entered as a positive 5000 lb. Because the temperature decreases, it is entered as a negative 50°F.

(b) In the diametral direction the strain is still the ratio of the change in length and the initial length, but here length refers to the diameter. Therefore, Eq. 4-12c can be written

$$\frac{\delta_d}{d} = \frac{-vP}{AE} + \alpha \Delta T$$

and the diameter of the rod will stretch

$$\delta_d = \left(\frac{-vP}{AE} + \alpha \, \Delta T\right) d$$

$$= \left[\frac{-(0.3)(5000)}{30(10^6)(\pi/4)(1/2)^2} + 6.5(10^{-6})(-50)\right](1/2)$$

$$= -0.000290 \text{ in.} \qquad \textbf{Ans.}$$

▶ Since strain $\epsilon = \delta/L$ is dimensionless, the deformation δ and the gage length L must both have the same units. They can both be in feet as in part (a) of this problem or both in inches as in part (b).

PROBLEMS

Introductory Problems

4-26* A cast iron pipe has an inside diameter of 70 mm and outside diameter of 105 mm. The length of the pipe is 2.5 m. The coefficient of thermal expansion for cast iron is $\alpha = 12.1(10^{-6})/°C$. Determine the dimension changes caused by

a. An increase in temperature of 70°C.
b. A decrease in temperature of 85°C.

4-27* A 1 × 2-in. rectangular bar of steel [$E = 30,000$ ksi and $\alpha = 6.6(10^{-6})/°F$] has a length of 4 ft. The bar is subjected to an axial load $P = 3000$ lb and a temperature increase of $\Delta T°$F. Determine the temperature increase ΔT if the elongation of the bar is 0.05 in.

4-28 An airplane has a wing span of 40 m. Determine the change in length of the aluminum alloy [$\alpha = 22.5(10^{-6})/°C$] wing span if the plane leaves the ground at a temperature of 40°C and climbs to an altitude where the temperature is −40°C.

4-29 A large cement kiln has a length of 225 ft and a diameter of 12 ft. Determine the change in length and diameter of the structural steel shell [$\alpha = 6.5(10^{-6})/°F$] caused by an increase in temperature of 250°F.

Intermediate Problems

4-30* A bronze [$\alpha_B = 16.9(10^{-6})/°C$] sleeve with an inside diameter of 99.8 mm is to be placed over a solid steel [$\alpha_S = 11.9(10^{-6})/°C$] cylinder that has an outside diameter of 100 mm. If the temperatures of the cylinder and sleeve remain equal, how much must the temperature be increased in order for the bronze sleeve to slip over the steel cylinder?

4-31* Determine the movement of the pointer of Fig. P4-31 with respect to the scale zero when the temperature increases 80°F.

The coefficients of thermal expansion are $6.6(10^{-6})/°F$ for the steel and $12.5(10^{-6})/°F$ for the aluminum.

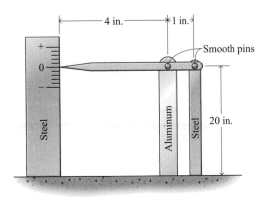

Figure P4-31

4-32 Determine the horizontal movement of point A in Fig. P4-32 due to a temperature increase of 75°C. Assume member AE has an insignificant coefficient of thermal expansion. The coefficients of thermal expansion are $11.9(10^{-6})/°C$ for the steel and $22.5(10^{-6})/°C$ for the aluminum alloy.

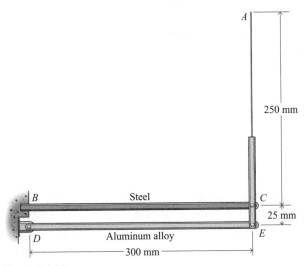

Figure P4-32

4-33 An aluminum [$E = 10,000$ ksi, $v = 0.33$, and $\alpha = 12.5(10^{-6})/°F$] rod of 0.25-in. diameter and 2-ft length is subjected to an axial compressive load of 4000 lb and a temperature change of +60°F. Determine the change in diameter of the rod.

Challenging Problems

4-34* A 25-mm-diameter aluminum [$E = 73$ GPa, $v = 0.33$, and $\alpha = 22.5(10^{-6})/°C$] rod hangs vertically while suspended from one end. A 2500-kg mass is attached at the other end. After the load is applied, the temperature decreases 50°C. Determine

a. The axial stress in the rod.
b. The axial strain in the rod.
c. The change in diameter of the rod.

4-35 A steel [$E = 30,000$ ksi and $\alpha = 6.5(10^{-6})/°F$] surveyor's tape 1/2 in. wide × 1/32 in. thick is exactly 100 ft long at 72°F and under a pull of 10 lb. What correction should be introduced if the tape is used to make a 100-ft measurement at a temperature of 100°F and under a pull of 25 lb?

4-36 The stepped bar shown in Fig. P4-36 is subjected to an axial load $P = 25$ kN and a temperature change $\Delta T = +20°C$. Segment AB is steel [$E = 200$ GPa and $\alpha = 12(10^{-6})/°C$] with a 50-mm diameter and length 200 mm; segment BC is aluminum [$E = 70$ GPa and $\alpha = 22.5(10^{-6})/°C$) with a diameter of 25 mm and length 150 mm. Determine

a. The deformation of segment AB.
b. The deformation of segment BC.
c. The deflection of a cross section at C.

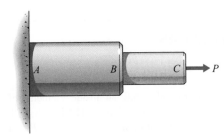

Figure P4-36

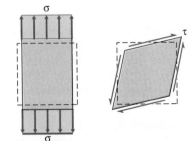

Figure 4-20

4-5 STRESS-STRAIN EQUATIONS FOR ORTHOTROPIC MATERIALS

The generalized Hooke's law for homogeneous, isotropic materials (Eqs. 4-5 and 4-10) is found in Section 4-3. Recall that a body is homogeneous and isotropic if the material properties are neither a function of position within the body nor a function of direction within the body, respectively. Such a body has three material properties (modulus of elasticity, modulus of rigidity or shear modulus, and Poisson's ratio), only two of which are independent. For an isotropic material (Fig. 4-20), a normal tensile stress causes an elongation in the direction of the stress and a contraction in the direction perpendicular to the stress. Shear stresses cause only shearing deformations. These types of deformations exist for isotropic materials regardless of the direction of the stress.

For some materials, wood for example, three material properties are not sufficient to describe the material behavior. Wood, however, has three mutually perpendicular planes of material symmetry, one parallel to the grain, one tangential to the grain, and one radial. Bodies having material properties that are different in three mutually perpendicular directions at a point within a body but that have three mutually perpendicular planes of material symmetry are known as *orthotropic materials*. The material properties are a function of orientation within the body.

Orthotropic materials have three natural axes that are mutually perpendicular (such as for the wood previously discussed). Figure 4-21 shows an orthotropic material subjected to an applied stress in the direction of a natural axis. As for the isotropic material, the orthotropic material elongates in the direction of the stress and contracts in the perpendicular direction. The magnitudes of the elongations and contractions are not the same for the two materials, since the material properties for the orthotropic material are dependent on the orientation of the applied stress. If the stress in Fig. 4-21 were applied in the perpendicular direction, the elongation and contraction would have different values from those shown in Fig. 4-21. On the other hand, the elongations and contractions of the isotropic material of Fig. 4-20 are independent of the direction of the applied load.

An applied stress that is not in the direction of one of the natural axes of an orthotropic material behaves as shown in Fig. 4-22. A normal stress produces an elongation, a contraction, and a shearing deformation. A shear stress also produces an elongation, a contraction, and a shearing deformation. Thus, depending on the direction of the stress, there may exist a coupling between elongation, contraction, and shearing deformation. No coupling exists for an isotropic material, regardless of the direction of the stress. The mechanical behavior of orthotropic materials (or anisotropic materials—no planes of material symmetry) is more complex than for isotropic materials. A complete description of such behavior may be found in books dealing with composite materials.

In this book, discussion will be limited to orthotropic materials subjected to plane stress. For example, consider a thin piece of material (such as an epoxy) reinforced by unidirectional fibers (such as graphite), as shown in Fig. 4-23. This material is orthotropic. Axes 1, 2, and 3 are in the principal material directions (natural axes), and planes 1–2, 2–3, and 1–3 are planes of material symmetry. For plane stress, $\sigma_3 = \tau_{31} = \tau_{32} = 0$. To determine the strain-stress relations, assume that the material behavior is linear elastic and use the principle of superposition.

First, consider that the orthotropic specimen is subjected to uniaxial loading in the 1-direction, as shown in Fig. 4-24a. The stress and strains are

$$\sigma_1 = \frac{P}{A} \qquad \epsilon_1 = \frac{\sigma_1}{E_1} \qquad v_{12} = -\frac{\epsilon_2}{\epsilon_1} \qquad (a)$$

where v_{12} is the Poisson's ratio for loading in the 1-direction and E_1 is the modulus of elasticity for the same loading (Fig. 4-24b).

Next, consider uniaxial loading in the 2-direction, as shown in Fig. 4-25a. The stress and strains are

$$\sigma_2 = \frac{P}{A} \qquad \epsilon_2 = \frac{\sigma_2}{E_2} \qquad v_{21} = -\frac{\epsilon_1}{\epsilon_2} \qquad (b)$$

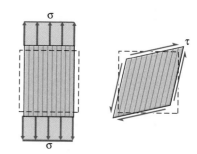

Figure 4-21

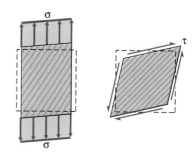

Figure 4-22

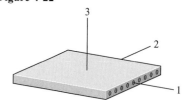

Figure 4-23

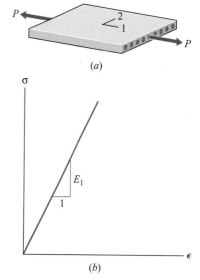

Figure 4-24

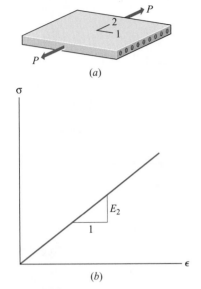

(a)

(b)

Figure 4-25

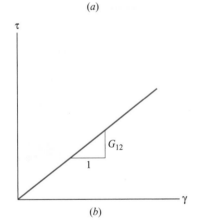

(a)

(b)

Figure 4-26

where v_{21} is the Poisson's ratio and E_2 is the modulus of elasticity for loading in the 2-direction (Fig. 4-25b).

Finally, the material is subjected to the shear stresses shown in Fig. 4-26a. The shearing stress is

$$\gamma_{12} = \frac{\tau_{12}}{G_{12}} \qquad (c)$$

where G_{12} is the shear modulus in the 1-2 plane (Fig. 4-26b).

Using Eqs. (a) and (b) and the principle of superposition, the strains are

$$\epsilon_1 = \frac{\sigma_1}{E_1} - v_{21}\epsilon_2 = \frac{\sigma_1}{E_1} - v_{21}\frac{\sigma_2}{E_2} \qquad (d)$$

$$\epsilon_2 = -v_{12}\epsilon_1 + \frac{\sigma_2}{E_2} = \frac{\sigma_2}{E_2} - v_{12}\frac{\sigma_1}{E_1} \qquad (e)$$

Summarizing, the stress-strain equations for the unidirectionally reinforced material of Fig. 4-23 for plane stress are

$$\epsilon_1 = \frac{\sigma_1}{E_1} - v_{21}\frac{\sigma_2}{E_2}$$

$$\epsilon_2 = \frac{\sigma_2}{E_2} - v_{12}\frac{\sigma_1}{E_1} \qquad (4\text{-}13)$$

$$\gamma_{12} = \frac{\tau_{12}}{G_{12}}$$

Equations 4-13 can be solved for the stresses, giving the stress-strain equations:

$$\sigma_1 = \frac{E_1}{1 - v_{12}v_{21}}(\epsilon_1 + v_{21}\epsilon_2)$$

$$\sigma_2 = \frac{E_2}{1 - v_{12}v_{21}}(\epsilon_2 + v_{12}\epsilon_1) \qquad (4\text{-}14)$$

$$\tau_{12} = G_{12}\gamma_{12}$$

Of the five material properties in Eqs. 4-14, only four are independent,[7] since

$$\frac{v_{12}}{E_1} = \frac{v_{21}}{E_2} \qquad (4\text{-}15)$$

The stress-strain equations (Eqs. 4-14) only apply when the loading is in the principal material directions. For loadings in other directions, refer to the composite materials book by Jones.[7]

If the material is isotropic ($E_1 = E_2 = E$; $v_{12} = v_{21} = v$; and $G_{12} = G$), Eqs. 4-14 reduce to Eqs. 4-6 and 4-8. For isotropic materials, recall that there are only two independent material properties.

Some typical material properties for a unidirectional composite material are shown in Table 4-1.[8] The Poisson's ratio v_{21} can be found by using Eq. 4-15.

[7]*Mechanics of Composite Materials*, R. M. Jones, Scripta Book Co., Washington, D.C., 1975.
[8]*Introduction to Composite Materials*, S. W. Tsai and H. T. Hahn, Technomic Publishing Co., Inc., Westport, CT, 1980.

Table 4-1   **Material Properties for Two Unidirectional Composites**

Type	Material	E_1 GPa (ksi)	E_2 GPa (ksi)	v_{12}	G_{12} GPa (ksi)
T300/5208	Graphite/ Epoxy	181 (26,300)	10.3 (1494)	0.28	7.17 (1040)
Scotchply 1002	Glass/ Epoxy	38.6 (5600)	8.27 (1199)	0.26	4.14 (600)

As previously stated, wood is an orthotropic material. In this book, wood members will be subjected to either axial loads (tension or compression) or bending loads (Chapter 7). The isotropic material properties listed in Appendix B are sufficient to solve problems involving these loading situations.

Example Problem 4-8   A unidirectional T300/5208 graphite/epoxy composite material is loaded in the principal material directions with stresses $\sigma_1 = 50$ ksi, $\sigma_2 = 6$ ksi, and $\tau_{12} = 2$ ksi. Determine the normal and shear strains in the principal material directions.

SOLUTION
The material properties, from Table 4-1, are $E_1 = 26,300$ ksi, $E_2 = 1494$ ksi, $G_{12} = 1040$ ksi, and $v_{12} = 0.28$. The Poisson's ratio v_{21} is found using Eq. 4-15 and is

$$v_{21} = \frac{E_2}{E_1}v_{12} = \frac{1494}{26,300}(0.28) = 0.01591$$

The strains are found using Eqs. 4-13. Thus,

$$\epsilon_1 = \frac{\sigma_1}{E_1} - v_{21}\frac{\sigma_2}{E_2} = \frac{50}{26,300} - 0.01591\frac{6}{1494} = 0.0018372 \cong 1837\mu \qquad \textbf{Ans.}$$

$$\epsilon_2 = \frac{\sigma_2}{E_2} - v_{12}\frac{\sigma_1}{E_1} = \frac{6}{1494} - 0.28\frac{50}{26,300} = 0.003484 \cong 3480\mu \qquad \textbf{Ans.}$$

$$\gamma_{12} = \frac{\tau_{12}}{G_{12}} = \frac{2}{1040} = 0.0019231 \cong 1923\mu \qquad \textbf{Ans.}$$

PROBLEMS

Introductory Problems

4-37*   A thin plate of T300/5208 unidirectional composite material is subjected to the strains $\epsilon_1 = +2000$ μin./in., $\epsilon_2 = +4000$ μin./in., and $\gamma_{12} = +1500$ μrad. Determine the normal and shear stresses in the principal material directions.

4-38*   A unidirectional composite plate of Scotchply 1002 Glass/Epoxy is subjected to the stresses $\sigma_1 = 30$ MPa, $\sigma_2 = -2$ MPa, and $\tau_{12} = 0.3$ MPa. Determine the normal and shear strains in the principal material directions.

4-39   A thin plate of T300/5208 unidirectional composite material is subjected to the stresses $\sigma_1 = 40$ ksi, $\sigma_2 = -10$ ksi, and $\tau_{12} = 2$ ksi. Determine the normal and shear strains in the principal material directions.

4-40   The material properties for a Boron/Epoxy unidirectional composite are $E_1 = 200$ GPa, $E_2 = 20$ GPa, $v_{12} = 0.23$, and $G_{12} = 6$ GPa. Determine the stresses in the principal material directions if the strains are $\epsilon_1 = +1000$ μm/m, $\epsilon_2 = +500$ μm/m, and $\gamma_{12} = +300$ μrad.

Intermediate Problems

4-41* A thin 10×10-in. square plate of T300/5208 unidirectional composite material is subjected to the stresses shown in Fig. P4-41. The y-axis is in the fiber direction. Determine the change in length of each side of the plate.

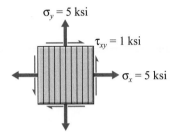

$\sigma_y = 5$ ksi

$\tau_{xy} = 1$ ksi

$\sigma_x = 5$ ksi

Figure P4-41

4-42 A plate of unidirectional Scotchply 1002 Glass/Epoxy has the dimensions shown in Fig. P4-42. The fibers are in the x-direction. For the stresses shown on Fig. P4-42, determine the change in length of each side of the plate.

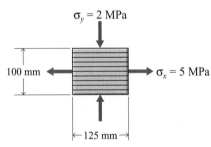

$\sigma_y = 2$ MPa

100 mm

$\sigma_x = 5$ MPa

$\leftarrow$ 125 mm $\rightarrow$

Figure P4-42

Challenging Problems

4-43 The unidirectional composite material shown in Fig. P4-43 is subjected to a uniform stress σ_x. Both the fibers and the matrix are isotropic materials with material properties E_f, v_f, G_f and E_m, v_m, G_m, respectively. The fibers and matrix are perfectly bonded so that $\epsilon_x = \epsilon_f = \epsilon_m$. The cross-sectional areas of the composite, the fibers, and the matrix are A, A_f, A_m, respectively. If both the fibers and the matrix behave in a linear elastic manner, show that the longitudinal modulus of the composite E_x is given by

$$E_x = \frac{E_f A_f + E_m A_m}{A}$$

In addition, if the volume fractions of the fiber and the matrix are $V_f = A_f/A$ and $V_m = A_m/A$, respectively, show that the above equation can be written

$$E_x = V_f E_f + V_m E_m$$

σ_x

x

Figure P4-43

4-44 For the unidirectional composite material of Problem 4-43, show that

$$\frac{P_f}{P} = \frac{E_f V_f}{E_f V_f + E_m(1 - V_f)}$$

where P is the total force carried by the composite and P_f is the force carried by the fibers. Prepare a plot of P_f/P (percent) versus E_f/E_m. Construct the plot for $1\% \leq P_f/P \leq 100\%$ and $0.1 \leq E_f/E_m \leq 100$ and for fiber volume fractions $0.1 \leq V_f \leq 0.9$.

SUMMARY

In 1678, Robert Hooke observed that the stretch of a structural member acted on by an axial load was directly proportional to the magnitude of the load applied to the member. When stress is plotted as a function of strain (rather than load as a function of deformation), the resulting curves are independent of the size and shape of the member and depend only on the type of material from which the member is made. The relationship between the load acting on a structural member and the deformation of the member must be determined experimentally. Data for stress-strain diagrams are obtained by applying an axial load to a test specimen and measuring the load and deformation simultaneously. The initial portion of the stress-strain diagram for most materials used in engineering structures is a straight line and is represented by Hooke's law,

$$\sigma = E\epsilon \tag{4-1a}$$

where the constant of proportionality (modulus of elasticity, E) must be determined from the experimental data. Although the stress-strain diagram for some materials such as gray cast iron and concrete show a slight curve even at very small stresses, it is common practice to draw a straight line to average the data for the first part of the diagram and neglect the curvature. It is important to realize that Hooke's law (Eq. 4-1a) only describes the initial linear portion of the stress-strain diagram and is valid only for uniaxially loaded bars.

A body loaded in one direction will undergo strains perpendicular to the direction of the load in addition to those parallel to the load. The ratio of the lateral or perpendicular strain to the longitudinal or axial strain is called Poisson's ratio:

$$v = -\frac{\epsilon_{\text{lat}}}{\epsilon_{\text{long}}} = -\frac{\epsilon_t}{\epsilon_a} \tag{4-2}$$

Because the lateral strain and the axial strain always have opposite signs, the negative sign in Eq. 4-2 ensures that v will have a positive value. Like the modulus of elasticity E and the shear modulus G, Poisson's ratio is a property of the material. Poisson's ratio v is related to E and G by the formula

$$E = 2(1 + v)G \tag{4-3}$$

Therefore, Poisson's ratio is a constant for stresses below the proportional limit and has a value between $1/4$ and $1/3$ for most metals. Equation 4-2 relating Poisson's ratio and the lateral and axial strains is only valid for a uniaxial state of stress such as in the tension test.

Hooke's law (Eq. 4-1a) can be extended to biaxial and triaxial states of stress often encountered in engineering practice by using the principal of superposition. The results for a biaxial state of stress are

$$\epsilon_x = \frac{1}{E}\left(\sigma_x - v\sigma_y\right)$$

$$\epsilon_y = \frac{1}{E}\left(\sigma_y - v\sigma_x\right) \tag{4-4}$$

$$\epsilon_z = -\frac{v}{E}\left(\sigma_x + \sigma_y\right)$$

Solving Eqs. 4-4 for the stresses in terms of the strains gives

$$\sigma_x = \frac{E}{1 - v^2}(\epsilon_x + v\epsilon_y)$$

$$\sigma_y = \frac{E}{1 - v^2}(\epsilon_y + v\epsilon_x) \tag{4-6}$$

Equations 4-6 can be used to calculate normal stresses from measured or computed normal strains.

Torsion test specimens are used to study material behavior under pure shear, and it is observed that a shearing stress produces only a single corresponding shear strain. Thus, Hooke's law extended to shearing stresses is simply

$$\tau = G\gamma \tag{4-8}$$

If unrestrained, most engineering materials expand when heated and contract when cooled. The thermal strain of an unrestrained body due to a temperature change of ΔT degrees is

$$\epsilon_T = \alpha\,\Delta T \tag{4-11}$$

where α is known as the coefficient of thermal expansion, and it is approximately constant for a large range of temperatures. For a homogeneous, isotropic material, the coefficient applies to all dimensions. Like the constants E and G, the value for α for various materials must be determined experimentally.

Strains caused by temperature changes and strains caused by applied loads are essentially independent. The total normal strain in a body acted upon by both temperature changes and axially applied loads is given by

$$\epsilon_{\text{total}} = \epsilon_\sigma + \epsilon_T \tag{4-12a}$$

Because homogeneous, isotropic materials expand uniformly in all directions when heated (and contract uniformly when cooled), neither the shape of the body nor the shearing stresses and shearing strains are affected by temperature changes if the body is unrestrained.

For an isotropic material, a normal tensile stress causes an elongation in the direction of the stress and a contraction in the direction perpendicular to the stress. Shear stresses cause only shearing deformations. For isotropic materials, the deformations depend only on the magnitude of the stress and not on the direction of the stress. In an isotropic material, the material properties are not a function of direction within the body.

Bodies that have three mutually perpendicular planes of material symmetry and that have material properties that are different in those three mutually perpendicular directions at a point within a body are known as orthotropic materials. In wood, for example, the elastic modulus for stresses along the grain is different from the elastic modulus for stresses across the grain of the wood. Since the material properties are a function of orientation within the body, the mechanical behavior of orthotopic materials is more complex than for isotropic materials. In orthotropic materials, a normal stress can produce a shearing deformation as well as an elongation or a contraction; a shear stress can produce an elongation or a contraction as well as a shearing deformation.

REVIEW PROBLEMS

4-45*   Stress-strain diagrams for a steel alloy at two temperatures are shown in Fig. P4-45. For each temperature, determine

a. The modulus of elasticity.
b. The yield strength (0.2 percent offset).

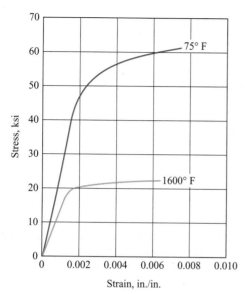

Figure P4-45

4-46* The stresses shown in Fig. P4-46 act at a point on the free surface of an aluminum alloy ($E = 70$ GPa and $v = 0.33$) machine component. Determine the normal strain ϵ_n that would be indicated by the strain gage shown in the figure.

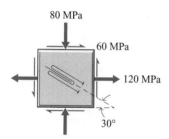

Figure P4-46

4-47 The stresses shown in Fig. P4-47a act at a point on the free surface of a steel ($E = 30,000$ ksi and $v = 0.30$) machine component. Determine the normal strains ϵ_a, ϵ_b, and ϵ_c that would be indicated by the strain rosette shown in Fig. P4-47b.

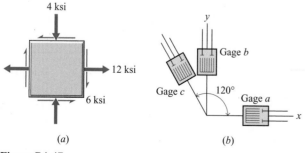

Figure P4-47

4-48* A $10 \times 25 \times 50$-mm block of 0.4 percent C hot-rolled steel ($E = 210$ GPa and $G = 80$ GPa) is placed between two rigid, frictionless walls and subjected to the stress shown in Fig. P4-48. Determine the change in length of the 10-mm side of the block.

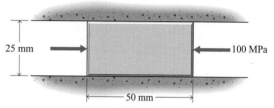

Figure P4-48

4-49 Two thin blocks of structural steel ($E = 29,000$ ksi and $G = 11,000$ ksi) are subjected to the stresses shown in Fig. P4-49. The 2×2-in. block shown in Fig. P4-49a is subjected to a state of biaxial stress (σ_y may be either tension or compression), and the 2×3-in. block shown in Fig. P4-49b is subjected to a uniaxial state of stress. If the x-component of deformation δ_x is to be the same for the two blocks, determine the value of σ_y.

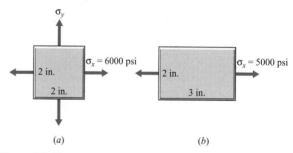

Figure P4-49

4-50 Stress-strain diagrams for two steel alloys are shown in Fig. P4-50. For each alloy, determine

a. The modulus of elasticity.
b. The yield strength (0.2 percent offset).

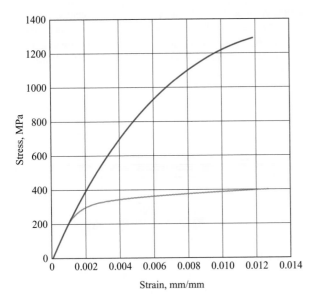

Figure P4-50

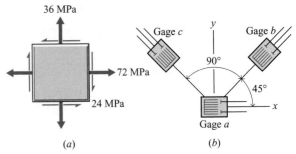

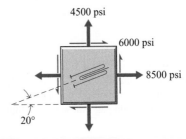

Figure P4-52

4-53 The stresses shown in Fig. P4-53 act at a point on the free surface of a steel ($E = 30,000$ ksi and $v = 0.30$) machine component. Determine the normal strain ϵ_n that would be indicated by the strain gage shown in the figure.

Figure P4-53

4-51* Steel [$E = 29,000$ ksi and $\alpha = 6.6(10^{-6})/^\circ$F] rails 55 ft long are separated 0.125 in. when the temperature is 60°F. Determine

a. The temperature at which the rails will just touch.
b. The gap between the rails when the temperature drops to 10°F.

4-52 The stresses shown in Fig. P4-52a act at a point on the free surface of a brass ($E = 100$ GPa and $v = 0.28$) machine component. Determine the normal strains ϵ_a, ϵ_b, and ϵ_c that would be indicated by the strain rosette shown in Fig. P4-52b.

4-54* At a point on the free surface of an aluminum alloy ($E = 73$ GPa and $v = 0.33$) machine part, the measured strains are $\epsilon_x = +825$ μm/m, $\epsilon_y = +950$ μm/m, and $\gamma_{xy} = +680$ μrad.

a. Determine the stresses σ_x, σ_y, and τ_{xy} at the point.
b. Determine the principal stresses and the maximum shearing stress at the point. Show these stresses on a triangular element.

Chapter 5
Axial Loading Applications and Pressure Vessels

5-1 INTRODUCTION

The problem of determining internal forces and deformations at all points within a body subjected to external forces is extremely difficult when the loading or geometry of the body is complicated. A refined analytical method of analysis that attempts to obtain general solutions to such problems is known as the *theory of elasticity*. The number of problems solved by such methods has been limited; therefore, practical solutions to most design problems are obtained by what has become known as the mechanics of materials approach. With this approach, real structural elements are analyzed as idealized models subjected to simplified loadings and restraints. The resulting solutions are approximate, because they consider only effects that significantly affect the magnitudes of stresses, strains, and deformations.

In Chapters 2 and 3, the concepts of stress and strain were developed. A discussion of material behavior in Chapter 4 led to the development of equations relating stress to strain. In the remaining chapters of the book, the stresses and deformations produced in a wide variety of structural members by axial, torsional, and flexural loadings will be considered. The mechanics of material analyses, as presented here, are somewhat less rigorous than the theory of elasticity approach, but experience indicates that the results obtained are quite satisfactory for most engineering problems.

5-2 DEFORMATION OF AXIALLY LOADED MEMBERS

Uniform Member: When a straight bar of uniform cross section is axially loaded by forces applied at the ends, the axial strain along the length of the bar is assumed to have a constant value,[1] and the elongation (or contraction) of the bar resulting from the axial load P may be expressed as $\delta = \epsilon L$ (by the definition of average axial strain). If Hooke's law (Eq. 4-1a) applies, the axial deformation may be expressed

---

[1] The forces at the ends of such members must be equal in magnitude, opposite in direction, and directed along the axis of the member. Furthermore, the internal forces at any position along the member must be the same as the forces at the ends of the member and also must act along the axis of the member.

in terms of either stress or load as

$$\delta = \epsilon L = \frac{\sigma L}{E} \tag{5-1}$$

$$\delta = \frac{PL}{EA} \tag{5-2}$$

The first form will be convenient in elastic problems in which the limiting axial stress and axial deformation are both specified and either the maximum allowable load or the required size (cross-sectional area) of the member are to be determined. The stress corresponding to the specified deformation can be obtained from Eq. 5-1 and compared to the allowable stress. The smaller of the two values can then be used to compute the allowable load or the required cross-sectional area. In general, Eq. 5-1 is preferred when the problem involves the determination or comparison of stresses.

Multiple Loads/Sizes: Equation 5-2, which gives the elongation (or contraction) δ occurring over some length L, applies only to uniform members for which P, A, and E are constant over the entire length L. If a bar is subjected to a number of axial loads at different points along the bar, or if the bar consists of parts having different cross-sectional areas or of parts composed of different materials (Fig. 5-1a), then the change in length of each part can be computed by using Eq. 5-2. The changes in length of the various parts of the bar can then be added algebraically to give the total change in length of the complete bar

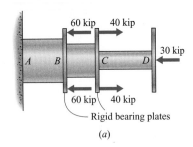

60 kip   40 kip

30 kip

A   B   C   D

60 kip   40 kip

Rigid bearing plates

(a)

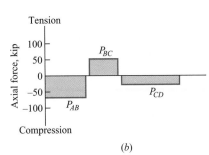

Tension

Axial force, kip

100

50

0

−50

−100

P_{BC}

P_{CD}

P_{AB}

Compression

(b)

Figure 5-1

$$\delta = \sum_{i=1}^{n} \delta_i = \sum_{i=1}^{n} \frac{P_i L_i}{E_i A_i} \tag{5-3}$$

where A_i and E_i are both constant on segment i and the force P_i is the internal force in segment i of the bar and is usually different from the forces applied at the ends of the segment. These forces must be calculated from equilibrium of the segment and are often shown on an axial-force diagram such as Fig. 5-1b.

5-2-1 Nonuniform Deformation
For cases in which the axial force or the cross-sectional area varies continuously along the length of the bar (Fig. 5-2), Eq. 5-2 is not valid. The axial strain at a point for the case of nonuniform deformation was defined in Section 3-2 as $\epsilon = d\delta/dL$. Thus, the increment of deformation associated with a differential element of length $dL = dx$ may be expressed as $d\delta = \epsilon\, dx$. If Hooke's law applies, the strain may again be expressed as $\epsilon = \sigma/E$, where $\sigma = P_x/A_x$. The subscripts indicate that both the applied load P_x and the cross-sectional area A_x may be functions of position x along the bar. Thus,

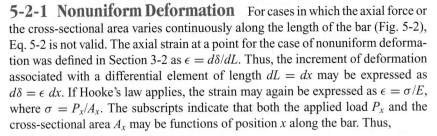

$$d\delta = \frac{P_x}{EA_x}\, dx \tag{a}$$

Integrating Eq. (a) yields the following expression for the total elongation (or contraction) of the bar:

$$\delta = \int_0^L d\delta = \int_0^L \frac{P_x}{EA_x}\, dx \tag{5-4}$$

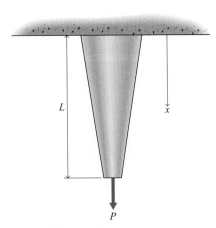

L   x

P

Figure 5-2

Equation 5-4 gives acceptable results for tapered bars, provided the angle between the sides of the bar does not exceed 20°.

Example Problem 5-1   The compression member shown in Fig. 5-3a consists of a solid aluminum bar A, which has an outside diameter of 100 mm, a brass tube B, which has an outside diameter of 150 mm and an inside diameter of 100 mm, and a steel pipe C, which has an outside diameter of 200 mm and an inside diameter of 125 mm. The moduli of elasticity of the aluminum, brass, and steel are 73, 100, and 210 GPa, respectively. Determine the overall shortening of the member under the action of the indicated loads.

SOLUTION

The forces transmitted by cross sections in parts A, B, and C of the axially loaded member shown in Fig. 5-3a are obtained by using the free-body diagrams shown in Fig. 5-3b. Summing forces along the axis of the bar yields

▶ The internal axial forces P_A, P_B, and P_C are all drawn as tension forces on the free-body diagrams of Fig. 5-3b. If the forces evaluate to be positive, then they will be plotted as positive (tension) forces on the axial-force diagram of Fig. 5-3c and they will cause the respective segment of the bar to stretch (δ will be positive). If the forces evaluate to be negative, then they will be plotted as negative (compression) forces and will cause the respective segment of the bar to shrink (δ will be negative).

$$+\uparrow \sum F = 0: \qquad -P_A - 650 = 0$$
$$P_A = -650 \text{ KN} = 650 \text{ kN (C)}$$

$$+\uparrow \sum F = 0: \qquad -P_B - 650 - 850 = 0$$
$$P_B = -1500 \text{ kN} = 1500 \text{ kN (C)}$$

$$+\uparrow \sum F = 0: \qquad -P_C - 650 - 850 - 1500 = 0$$
$$P_C = -3000 \text{ kN} = 3000 \text{ kN (C)}$$

A pictorial representation of the distribution of axial, or internal, force in the member is shown in Fig. 5-3c. The cross-sectional areas of the aluminum, brass, and steel are

$$A_A = \frac{\pi}{4}d^2 = \frac{\pi}{4}(100)^2 = 7854 \text{ mm}^2 = 0.007854 \text{ m}^2$$

$$A_B = \frac{\pi}{4}(d_o^2 - d_i^2) = \frac{\pi}{4}(150^2 - 100^2) = 9817 \text{ mm}^2 = 0.009817 \text{ m}^2$$

$$A_C = \frac{\pi}{4}(d_o^2 - d_i^2) = \frac{\pi}{4}(200^2 - 125^2) = 19,144 \text{ mm}^2 = 0.019144 \text{ m}^2$$

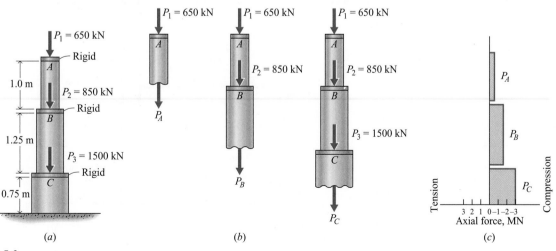

(a)   (b)   (c)

Figure 5-3

▶ Since the cross-sectional areas and the internal axial forces are different in segments A, B, and C of the bar, the deformations δ_A, δ_B, and δ_C must be computed separately and then added together to get the total deformation of the bar.

The changes in length of the different parts are obtained by using Eq. 5-2. Thus,

$$\delta_A = \frac{P_A L_A}{E_A A_A} = \frac{-650(10^3)(1.0)}{73(10^9)(0.007854)} = -1.1337(10^{-3}) \text{ m} = -1.1337 \text{ mm}$$

$$\delta_B = \frac{P_B L_B}{E_B A_B} = \frac{-1500(10^3)(1.25)}{100(10^9)(0.009817)} = -1.9100(10^{-3}) \text{ m} = -1.9100 \text{ mm}$$

$$\delta_C = \frac{P_C L_C}{E_C A_C} = \frac{-3000(10^3)(0.75)}{210(10^9)(0.019144)} = -0.5597(10^{-3}) \text{ m} = -0.5597 \text{ mm}$$

The total change in length of the complete bar is given by Eq. 5-3 as

$$\delta_{\text{total}} = \delta_A + \delta_B + \delta_C$$
$$= -1.1337 - 1.9100 - 0.5597 = -3.6034 \text{ mm} \cong -3.60 \text{ mm} \qquad \textbf{Ans.}$$

The negative sign indicates that the complete bar decreases in length.

◼ **Example Problem 5-2** The rigid yokes B and C of Fig. 5-4a are securely fastened to the 2-in. square steel ($E = 30{,}000$ ksi) bar AD. Determine

(a) The maximum normal stress in the bar.

(b) The change in length of segment AB.

(c) The change in length of segment BC.

(d) The change in length of the complete bar.

SOLUTION

Since bar AD is subjected to a number of axial loads applied at different points along the bar, the different sections AB, BC, and CD of the bar will transmit different levels of load. An axial-force diagram, such as the one shown in Fig. 5-4b, provides a pictorial representation of the levels of internal force in each of the sections and serves as an aid for stress and deformation calculations.

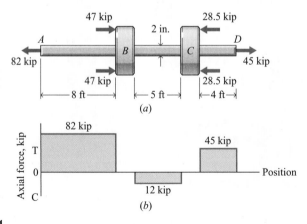

Figure 5-4

The axial-force diagram is easily constructed by isolating different portions of the bar with imaginary cuts and computing the internal force that must be transmitted by the cross section exposed by the cut to maintain equilibrium of the isolated portion of the bar. For example, an imaginary cut anywhere in section AB will expose a cross section that must transmit 82 kip to maintain equilibrium of the portions of the bar on either side of the cut.

(a) Examination of the axial-force diagram indicates that the maximum load transmitted by any section of the bar is 82 kip; therefore, since the bar has a uniform cross section,

$$\sigma_{max} = \frac{P_{max}}{A} = \frac{82}{4} = 20.5 \text{ ksi (T)} \qquad \textbf{Ans.}$$

(b) Since different parts of the bar are transmitting different levels of load, Eq. 5-2 must be used to determine the changes in length associated with each of the different parts of the bar. For segment AB,

$$\delta_{AB} = \frac{P_{AB}L_{AB}}{E_{AB}A_{AB}} = \frac{+82(8)(12)}{30,000(4)} = +0.06560 \text{ in.} \cong +0.0656 \text{ in.} \qquad \textbf{Ans.}$$

The positive sign indicates that segment AB of the bar increases in length.

(c) For segment BC,

$$\delta_{BC} = \frac{P_{BC}L_{BC}}{E_{BC}A_{BC}} = \frac{-12(5)(12)}{30,000(4)} = -0.00600 \text{ in.} \qquad \textbf{Ans.}$$

The negative sign indicates that segment BC of the bar decreases in length.

(d) For segment CD,

$$\delta_{CD} = \frac{P_{CD}L_{CD}}{E_{CD}A_{CD}} = \frac{+45(4)(12)}{30,000(4)} = +0.01800 \text{ in.}$$

The positive sign indicates that segment CD of the bar increases in length. The deformations of the individual segments δ_{AB}, δ_{BC}, and δ_{CD} are then added algebraically to give the change in length of the complete bar. Thus,

$$\delta_{AD} = \sum \frac{PL}{EA} = \delta_{AB} + \delta_{BC} + \delta_{CD}$$
$$= +0.06560 - 0.00600 + 0.01800$$
$$= +0.07760 \text{ in.} \cong +0.0776 \text{ in.} \qquad \textbf{Ans.}$$

The positive sign indicates that the complete bar increases in length.

5.1

5.2

Example Problem 5-3 A homogeneous bar of uniform cross section A hangs vertically while suspended from one end, as shown in Fig. 5-5a. Determine

(a) The elongation of the bar due to its own weight W in terms of $W, L, A,$ and E.

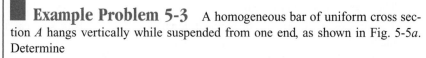

L

(a)

Figure 5-5(a)

(b) The elongation of the bar if the bar is also subjected to an axial tensile force P at its lower end.

SOLUTION

(a) A free-body diagram of a segment of the bar, Fig. 5-5b, shows that the axial force is a function of x, the distance from the free end of the bar. Thus, Eq. 5-4 is applicable. The weight of the segment of the bar shown in Fig. 5-5b is $W_x = \gamma V_x = \gamma Ax$, where γ is the specific weight of the material of which the bar is made and V_x is the volume of the bar segment. From Eq. 5-4,

$$\delta = \int_0^L \frac{P_x}{EA_x}\, dx = \frac{1}{EA}\int_0^L \gamma Ax\, dx = \frac{\gamma}{E}\int_0^L x\, dx$$

where γ, A, and E are each constant. The elongation of the bar is

$$\delta = \frac{\gamma}{E}\int_0^L x\, dx = \frac{\gamma x^2}{2E}\Big]_0^L = \frac{\gamma L^2}{2E}$$

The weight of the bar is $W = \gamma AL$, from which $\gamma = W/AL$. Thus, the elongation of the bar is

$$\delta = \frac{\gamma L^2}{2E} = \frac{W}{AL}\left[\frac{L^2}{2E}\right] = \frac{WL}{2AE} \qquad \textbf{Ans.}$$

(b) When the bar is subjected to its own weight and a concentrated force P at the free end, the elongation would be found using the method of superposition (Section 4-3). That is, the elongations found using Eqs. 5-2 and 5-4 separately would be added algebraically to find the elongation due to the combined effects of the weight of the bar and the concentrated force. When the bar is subjected to the axial force, the elongation of the bar is given by Eq. 5-2,

$$\delta = \frac{PL}{EA}$$

Due to the weight of the bar and the axial force, the elongation is

$$\delta = \frac{WL}{2EA} + \frac{PL}{EA} = \frac{L}{EA}\left(\frac{W}{2} + P\right) \qquad \textbf{Ans.}$$

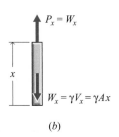

$P_x = W_x$

x

$W_x = \gamma V_x = \gamma Ax$

(b)

Figure 5-5(b)

Example Problem 5-4   A bar of steel ($E = 200$ GPa and $v = 0.30$) with a 30×30-mm square cross section is subjected to an axial compressive load P of 180 kN, as shown in Fig. 5-6a. Determine

(a) The change in length of the bar.

(b) The stresses σ_x, σ_y, and τ_{xy} on element A, which is on the outside surface of the bar.

(c) The strains ϵ_x, ϵ_y, and γ_{xy}.

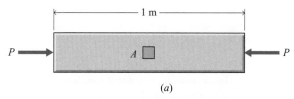

Figure 5-6(a)

SOLUTION

(a) Since P, A, and E are all constant over the length L of the bar, the change in length of the bar may be found using Eq. 5-2. Thus,

$$\delta = \frac{PL}{EA} = \frac{-180(10^3)(1)}{200(10^9)(0.030)^2} = -1.000(10^{-3})\text{ m} = -1.000\text{ mm} \quad \textbf{Ans.}$$

(b) The element at A is subjected to a state of plane stress, as shown in Fig. 5-6b. Since the bar is loaded by an axial force, the only nonzero stress is σ_x. Furthermore, σ_x is compressive since P is a compressive force and is

$$\sigma_x = \frac{P}{A} = \frac{-180(10^3)}{(0.030)^2} = -200(10^6)\text{ N/m}^2$$
$$= -200\text{ MPa} = 200\text{ MPa (C)} \quad \textbf{Ans.}$$

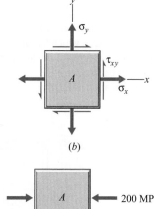

The stresses on the element at A are shown in Fig. 5-6c.

(c) Since the bar is axially loaded, the normal strains could be calculated using Eqs. 3-1 and 4-2. An alternative approach is to use Eq. 4-4 for plane stress. The required normal strains are ϵ_x and ϵ_y; thus

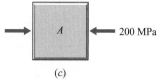

Figure 5-6(b–c)

$$\epsilon_x = \frac{1}{E}\left[\sigma_x - \nu\sigma_y\right]$$
$$= \frac{1}{200(10^9)}\left[-200(10^6) - (0.30)(0)\right]$$
$$= -0.001000 = -1000\ \mu\text{m/m} \quad \textbf{Ans.}$$
$$\epsilon_y = \frac{1}{E}\left[\sigma_y - \nu\sigma_x\right]$$
$$= \frac{1}{200(10^9)}\left[(0) - (0.30)(-200)(10^6)\right]$$
$$= +0.000300 = +300\ \mu\text{m/m} \quad \textbf{Ans.}$$

Since the xy-axes remain perpendicular after the bar is loaded, the shear strain γ_{xy} is zero. Alternatively, using Eq. 4-1c,

$$\gamma_{xy} = \frac{\tau_{xy}}{G} = \frac{0}{G} = 0\ \mu\text{rad} \quad \textbf{Ans.}$$

PROBLEMS

MecMovie Activities and Problems

MM5.1 Axial deformation – basic problems. Concept check-points. Use the axial deformation equation for three introductory problems.

MM5.2 Compound axial members – basic problems. Concept checkpoints. Apply the axial deformation concept to compound axial members.

Introductory Problems

5-1* The tension member of Fig. P5-1 consists of a steel ($E = 30,000$ ksi) pipe A, which has an outside diameter of 6 in. and an inside diameter of 4.5 in., and a solid aluminum alloy ($E = 10,600$ ksi) bar B, which has a diameter of 4 in. Determine the overall elongation of the member.

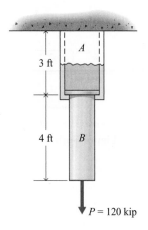

Figure P5-1

5-2* A flat steel ($E = 200$ GPa) bar 100 mm wide and 25 mm thick is subjected to the axial forces shown in Fig. P5-2. Determine the change in length of

a. Segment AB of the bar.
b. Segment BC of the bar.
c. The complete bar.

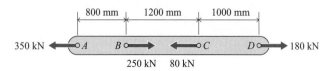

Figure P5-2

5-3 The steel ($E = 29,000$ ksi) pipe column shown in Fig. P5-3 has an outside diameter of 6 in., a wall thickness of 0.6 in., and a length of 24 in. The axial load imposed on the column by the timber beam is 30,000 lb. Determine

a. The normal stress in the column.
b. The change in length of the column.
c. The average axial normal strain in the column.

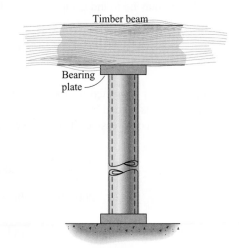

Figure P5-3

5-4* An aluminum alloy ($E = 73$ GPa) bar is loaded and supported as shown in Fig. P5-4. The diameters of the top and bottom sections of the bar are 25 mm and 15 mm, respectively. Determine the deflection

a. Of cross section a–a.
b. Of cross section b–b.

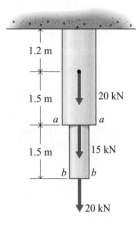

Figure P5-4

5-5 An 8-ft-long steel ($E = 30,000$ ksi) bar has a 1.5-in. diameter over one-half its length and a 1.0-in. diameter over the other half (Fig. P5-5). For the axial loads shown, determine

a. The change in length of the bar.
b. The change in length of a bar of uniform cross section having the same weight as the bar of part a.

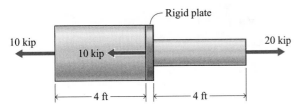

Figure P5-5

5-6 A structural tension member of aluminum alloy ($E = 70$ GPa) has a rectangular cross section 25 mm × 75 mm and is 2 m long. Determine the maximum axial load that may be applied if the normal stress is not to exceed 100 MPa and the total elongation is not to exceed 4 mm.

5-7* The roof and second floor of a building are supported by the column shown in Fig. P5-7. The column is a structural steel (see Appendix B for properties) section having a cross-sectional area of 9 in^2. The roof and floor subject the column to the axial forces shown. Determine

a. The amount that the first floor will settle.
b. The amount that the roof will settle.

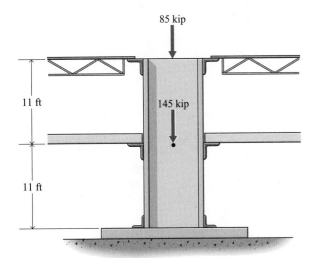

Figure P5-7

5-8 A steel ($E = 200$ GPa) rod, which has a diameter of 30 mm and a length of 1.0 m, is attached to the end of a Monel

($E = 180$ GPa) tube, which has an internal diameter of 40 mm, a wall thickness of 10 mm, and a length of 2.0 m, as shown in Fig. P5-8. Determine the load required to stretch the assembly 3.00 mm.

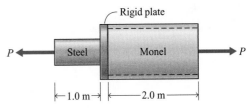

Figure P5-8

Intermediate Problems

5-9* A flat 1 × 2-in. bar of 6061-T6 aluminum alloy ($E = 10,000$ ksi) is subjected to the axial loads shown in Fig. P5-9.

a. Determine the change in length of the bar.
b. Determine the xy components of stress at points A, B, and C on the outside surface of the bar. Show these stresses on stress elements for these points.

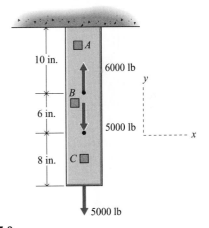

Figure P5-9

5-10* A hollow structural steel ($E = 200$ GPa) tube A with an outside diameter of 60 mm and an inside diameter of 50 mm is fastened to a 2014-T4 aluminum ($E = 73$ GPa) bar B that has a 50-mm diameter over one-half its length and a 25-mm diameter over the other half. The bar is loaded and supported as shown in Fig. P5-10. Determine

a. The change in length of the steel tube.
b. The overall change in length of the member.
c. The average axial strain in the steel tube.
d. The maximum normal stress in the member.

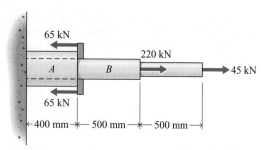

Figure P5-10

c. The average axial strain in the circular segment of the member.
d. The change in diameter of the circular segment of the member.

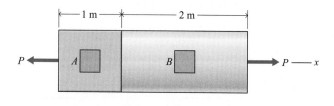

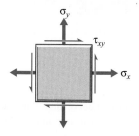

Figure P5-12

5-11 The tension member of Fig. P5-11 consists of a structural steel ($E = 29,000$ ksi and $v = 0.30$) pipe A, which has an outside diameter of 6 in. and an inside diameter of 4.5 in., and a solid 2014-T4 aluminum alloy ($E = 10,600$ ksi and $v = 0.33$) bar B, which has a diameter of 4 in. Determine

a. The change in length of the steel pipe.
b. The overall deflection of the member.
c. The maximum normal and shearing stresses in the aluminum bar.
d. The change in diameter of segment B of the member.

5-13* A 3/4-in.-diameter × 3-ft-long structural steel ($E = 29,000$ ksi, $v = 0.30$) rod is supporting an axial tensile load P of 5 kip, as shown in Fig. P5-13. Determine

a. The elongation of the bar.
b. The stresses σ_x, σ_y, and τ_{xy} on the element at A, which is on the outside surface of the rod.
c. The components of strain ϵ_x, ϵ_y, and γ_{xy} at point A.

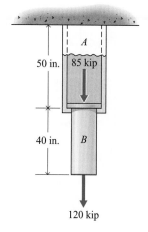

Figure P5-11

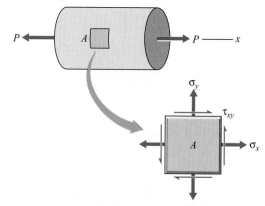

Figure P5-13

5-12* A 2024-T4 aluminum alloy ($E = 73$ GPa and $v = 0.33$) bar 3 m long has a 25-mm square cross section over 1 m of its length and a 25-mm-diameter circular cross section over the other 2 m of its length, as shown in Fig. P5-12. The bar is supporting an axial tensile load P of 50 kN. Determine

a. The elongation of the bar.
b. The stresses σ_x, σ_y, and τ_{xy} on the elements at A and B, which are on the outside surface of the assembly.

5-14 The roof and second floor of a building are supported by the column shown in Fig. P5-14. The column is a structural steel ($E = 200$ GPa and $G = 76$ GPa) W305 × 74 wide-flange section with a cross-sectional area of 9485 mm². The roof and floor subject the column to the axial forces shown. Determine

a. The change in length of the column.
b. The stresses σ_x, σ_y, and τ_{xy} on the elements A and B, which are on the outside surface of the web of the column.
c. The components of strain ϵ_x, ϵ_y, and γ_{xy} at points A and B.

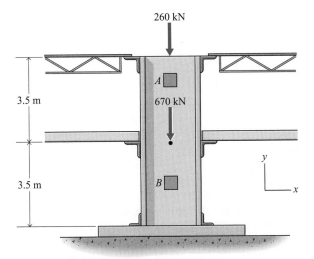

Figure P5-14

5-15 A hollow brass ($E = 15,000$ ksi and $G = 5600$ ksi) tube A with a 4-in. outside diameter and a 2-in. inside diameter is fastened to a solid 2-in.-diameter steel ($E = 30,000$ ksi) rod B, as shown in Fig. P5-15. Determine

a. The deflection of cross section a–a.
b. The deflection of cross section b–b.
c. The horizontal and vertical components of stress on an element on the outside surface of the tube and rod.
d. The change in both the inside and outside diameters of the tube.

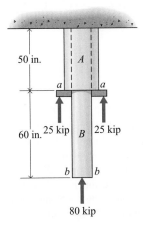

Figure P5-15

Challenging Problems

5-16* An aluminum alloy ($E = 73$ GPa) tube A with an outside diameter of 75 mm is used to support a 25-mm-diameter steel ($E = 200$ GPa) rod B, as shown in Fig. P5-16. Determine the minimum thickness t required for the tube if the maximum deflection of the loaded end of the rod must be limited to 0.40 mm.

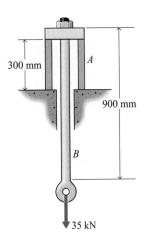

Figure P5-16

5-17* A structural steel ($E = 29,000$ ksi and $\gamma = 0.284$ lb/in³) bar of rectangular cross section consists of uniform and tapered sections as shown in Fig. P5-17. The width of the tapered section varies linearly from 2 in. at the bottom to 5 in. at the top. The bar has a constant thickness of 1/2 in. Determine the elongation of the bar resulting from application of the 30-kip load P and the weight of the bar.

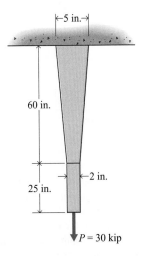

Figure P5-17

5-18 Determine the elongation, due to its own weight, of the homogeneous bar of Fig. P5-18. Express the results in terms of L, E, and the specific weight γ of the material. The taper of the bar is slight enough for the assumption of a uniform axial stress distribution over a cross section to be valid.

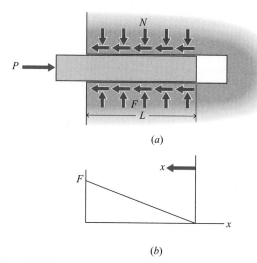

(a)

(b)

Figure P5-20

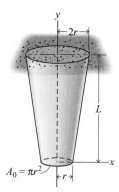

Figure P5-18

5-19* A homogeneous bar of specific weight γ, uniform cross section A, and length L is suspended at one end and hangs vertically, as shown in Fig. P5-19. The stress-strain relationship for the rod is given by $\sigma = K\epsilon^{1/2}$, where K is a constant. Determine the elongation of the rod.

5-21 Determine the change in length of the homogeneous conical bar of Fig. P5-21 due to its own weight. Express the results in terms of L, E, and the specific weight γ of the material. The taper of the bar is slight enough for the assumption of a uniform axial stress distribution over a cross section to be valid.

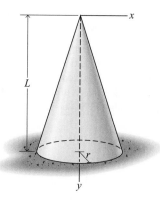

Figure P5-21

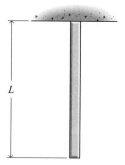

Figure P5-19

5-20 A uniform circular member of cross sectional area A is pressed into a slightly smaller circular hole, as shown in Fig. P5-20a. As a result, normal and frictional forces are developed over the surface of the cylindrical shaft. The friction force varies linearly with depth L, as shown in Fig. P5-20b. Determine the shortening of the member.

5-22* An aluminum alloy ($E = 70$ GPa) bar of circular cross section consists of uniform and tapered sections as shown in Fig. P5-22. The diameter of the tapered section varies linearly from 50 mm at the top to 20 mm at the bottom. The uniform section at the top is hollow and has a wall thickness of 10 mm. Determine the elongation of the bar resulting from application of the 75-kN load. Neglect the weight of the bar.

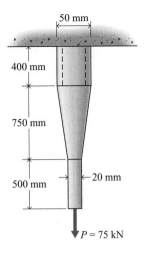

Figure P5-22

Computer Problems

5-23 A 3-in.-diameter structural steel ($E = 30,000$ ksi and $v = 0.30$) shaft is pressed into a slightly smaller circular hole as shown in Fig. P5-23. Assume that the normal stresses σ_n and the shearing stresses τ are uniformly distributed over the surface of the cylindrical shaft. If the magnitudes of the normal and shearing stresses are 600 psi and 300 psi, respectively, compute and plot

a. The force P required to insert the shaft as a function of the length L that has been inserted ($0 \leq L \leq 15$ in.).
b. The axial stress $\sigma_x(x)$ in the shaft as a function of the distance x from the surface of the hole when 15 in. of the shaft has been inserted ($0 \leq x \leq 15$ in.).
c. The axial deformation $\delta_x(x)$ of the shaft as a function of the distance x from the surface of the hole when 15 in. of the shaft has been inserted ($0 \leq x \leq 15$ in.).

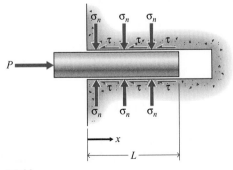

Figure P5-23

5-24 The 300-mm-diameter timber ($E = 13$ GPa and $v = 0.30$) pile shown in Fig. P5-24 is being extracted from the ground. Assume that the horizontal normal stresses σ_n and the vertical shearing stresses τ (which are a function of the type of soil surrounding the pile) can be approximated by the expressions

$$\sigma_n(x) = \gamma x(1 - \sin\phi) \qquad \tau(x) = \sigma_n \tan \phi$$

where $\gamma = 400$ N/m³ is the specific weight of the surrounding soil, $\phi = 28°$ is the friction angle of the soil, and x is the distance from the ground surface. Compute and plot

a. The axial stress $\sigma_x(x)$ in the pile as a function of the distance x from the ground surface when the entire 8-m length of the pile is in the ground ($0 \leq x \leq 8$ m).
b. The axial deformation $\delta_x(x)$ of the pile as a function of the distance x from the ground surface when the entire 8-m length of the pile is in the ground ($0 \leq x \leq 8$ m).
c. The force P required to extract the pile from the ground as a function of the length L of the pile that remains in the ground ($0 \leq L \leq 8$ m).

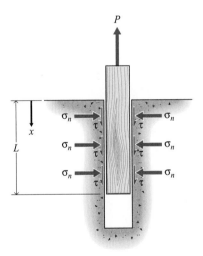

Figure P5-24

5-3   DEFORMATIONS IN A SYSTEM OF AXIALLY LOADED BARS

It is sometimes necessary to determine axial deformations and strains in a loaded system of pin-connected deformable bars (two-force members). The problem is

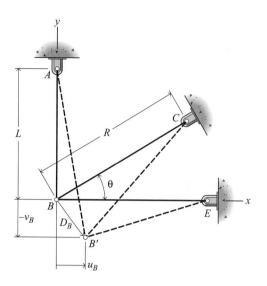

Figure 5-7

approached through a study of the geometry of the deformed system, from which the axial deformation δ of the various bars in the system are obtained. Suppose, for example, one is interested in the axial deformations of bars AB, BC, and BE of Fig. 5-7, in which the solid lines represent the unstrained (unloaded) configuration of the system and the dashed lines represent the configuration due to a force (not shown) applied at B. The displacement of point B is the vector $\mathbf{D}_B$. The scalar components of this displacement are u_B and $-v_B$ (since the v component is in the negative y-direction) in the x- and y-directions, respectively. The change in length of member AB is

$$\delta_{AB} = L_f - L_i$$

where the subscripts f and i denote final and initial length of member AB, respectively. Thus, the change in length of AB (the deformation) is

$$\delta_{AB} = \sqrt{(L + v_B)^2 + u_B^2} - L$$

Transposing the last term and squaring both sides give

$$\delta_{AB}^2 + 2L\delta_{AB} + L^2 = L^2 + 2Lv_B + v_B^2 + u_B^2$$

If the displacement components are small (the usual case for stiff materials and elastic action), the terms involving the squares of the displacements may be neglected; hence,

$$\delta_{AB} \cong v_B$$

In a similar manner,

$$\delta_{BE} \cong u_B$$

The axial deformation in bar BC is

$$\delta_{BC} = \sqrt{(R \cos \theta - u_B)^2 + (R \sin \theta + v_B)^2} - R$$

Transposing the last term and squaring both sides gives

$$\delta_{BC}^2 + 2R\delta_{BC} + R^2 = R^2 \cos^2 \theta - 2Ru_B \cos \theta + u_B^2$$
$$+ R^2 \sin^2 \theta + 2Rv_B \sin \theta + v_B^2$$

Neglecting small second-degree terms and noting that $\sin^2 \theta + \cos^2 \theta = 1$, one obtains

$$\delta_{BC} \cong v_B \sin \theta - u_B \cos \theta$$

or in terms of the deformations of the other two bars,

$$\delta_{BC} \cong \delta_{AB} \sin \theta - \delta_{BE} \cos \theta$$

The geometric interpretation of this equation is indicated by the shaded right triangles of Fig. 5-8.

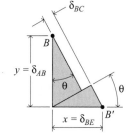

Figure 5-8

The general conclusion that may be drawn from the above discussion is that, *for small displacements, the axial deformation in any bar may be assumed equal to the component of the displacement of one end of the bar (relative to the other end) taken in the direction of the unstrained orientation of the bar.* Any rigid members of the system will change orientation or position but will not be deformed in any manner. For example, if bar BE of Fig. 5-7 were rigid and subjected to a small downward rotation, point B could be assumed to be displaced vertically through a distance y, and δ_{BC} would be equal to $y \sin \theta$.

Example Problem 5-5

A tie rod and a pipe strut are used to support a 50-kN load, as shown in Fig. 5-9a. The cross-sectional areas are 650 mm² for tie rod AB and 925 mm² for pipe strut BC. Both members are made of structural steel that has a modulus of elasticity of 200 GPa. Determine

(a) The normal stresses in tie rod AB and pipe strut BC.

(b) The lengthening or shortening of tie rod AB and pipe strut BC.

(c) The horizontal and vertical components of the displacement of point B.

(d) The angles through which members AB and BC rotate.

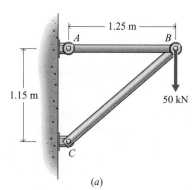

Figure 5-9(a)

SOLUTION

(a) The forces in members AB and BC can be determined by using the free-body diagram of joint B shown in Fig. 5-9b. Thus,

$$\xrightarrow{+} \Sigma F_x = 0: \qquad -F_{AB} + F_{BC} \cos 42.61° = 0$$
$$+\uparrow \Sigma F_y = 0: \qquad F_{BC} \sin 42.61° - 50 = 0$$

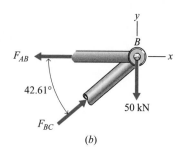

(b)

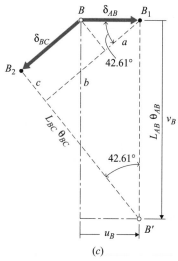

(c)

Figure 5-9(b–c)

▶ Imagine that the pin connecting the bars at B is removed for a moment. As the tie rod AB stretches, it wants to push the joint from point B to point B_1. Similarly, as the pipe strut BC shrinks, it wants to pull the joint from point B to point B_2. Then, the tie rod AB must rotate clockwise about pin A and the joint moves downward from B_1 to B', the pipe strut rotates clockwise about pin C and the joint moves downward from B_2 to B', and the pin can be replaced in the joint. The displacement of point B is the vector from B to B'.

from which

$$F_{AB} = +54.36 \text{ kN} = 54.36 \text{ kN (T)}$$
$$F_{BC} = -73.85 \text{ kN} = 73.85 \text{ kN (C)}$$

The normal stresses in members AB and BC are determined by using Eq. 2-2. Thus,

$$\sigma_{AB} = \frac{F_{AB}}{A_{AB}} = \frac{+54.36(10^3)}{650(10^{-6})}$$
$$= +83.63(10^6) \text{ N/m}^2 \cong 83.6 \text{ MPa (T)} \qquad \textbf{Ans.}$$

$$\sigma_{BC} = \frac{F_{BC}}{A_{BC}} = \frac{-73.85(10^3)}{925(10^{-6})}$$
$$= -79.84(10^6) \text{ N/m}^2 \cong 79.8 \text{ MPa (C)} \qquad \textbf{Ans.}$$

(b) The changes in length of the members are determined by using Eq. 5-1. Thus,

$$\delta_{AB} = \frac{\sigma_{AB}L_{AB}}{E} = \frac{+83.63(10^6)(1.25)}{200(10^9)}$$
$$= +0.5227(10^{-3}) \text{ m} \cong +0.523 \text{ mm} \qquad \textbf{Ans.}$$

$$\delta_{BC} = \frac{\sigma_{BC}L_{BC}}{E} = \frac{-79.84(10^6)(1.699)}{200(10^9)}$$
$$= -0.6782(10^{-3}) \text{ m} \cong -0.678 \text{ mm} \qquad \textbf{Ans.}$$

(c) The horizontal and vertical displacements u_B and v_B of point B are indicated on the deformation diagram shown in Fig. 5-9c. The deformations have been greatly exaggerated in this diagram and the arcs through which the members rotate have been replaced by straight lines drawn perpendicular to the unloaded positions of the members. From the diagram it is observed that

$$u_B = \delta_{AB} = +0.5227 \text{ mm} \cong 0.523 \text{ mm} \qquad \textbf{Ans.}$$

and pin B moves to the right a distance, $u_B = 0.523$ mm. Also,

$$a = \delta_{AB} \cos 42.61° = 0.5227 \cos 42.61° = 0.3847 \text{ mm}$$
$$b = |\delta_{BC}| = 0.6782 \text{ mm}$$
$$\sin 42.61° = \frac{\delta_{BC} + a}{v_B} = \frac{0.6782 + 0.3847}{v_B} = \frac{1.0629}{v_B}$$
$$v_B = \frac{1.0629}{\sin 42.61°} = 1.5700 \text{ mm} \cong 1.570 \text{ mm} \qquad \textbf{Ans.}$$

That is, pin B moves down a distance $v_B = 1.570$ mm.

(d) As pin B moves from B to B', bar AB rotates clockwise about pin A through an angle θ_{AB} to AB' (see the deformation diagram shown in Fig. 5-9c). Assuming that the angle is very small, $\tan \theta_{AB} \cong \theta_{AB}$ (where θ_{AB} is in radians), and

$$\theta_{AB} \cong \tan \theta_{AB} = \frac{v_B}{L_{AB}} = \frac{1.570}{1250} = 0.001256 \text{ rad} \cong 0.0720° \qquad \textbf{Ans.}$$

Similarly, as pin B moves from B to B', bar BC rotates clockwise about pin C through an angle θ_{BC} to CB'. Again assuming that the angle is very small, $\tan \theta_{BC} \cong \theta_{BC}$, and

$$\begin{aligned}
\theta_{BC} \cong \tan \theta_{BC} &= \frac{c + v_B \cos 42.61°}{L_{BC}} \\
&= \frac{0.5227 \sin 42.61° + 1.5700 \cos 42.61°}{\sqrt{1150^2 + 1250^2}} \\
&= 0.000889 \text{ rad} \cong 0.0509° \qquad \textbf{Ans.}
\end{aligned}$$

Note that the forces F_{AB} and F_{BC} were found using a free-body diagram drawn in the undeformed instead of the deformed configuration. The implications of using free-body diagrams in the undeformed configuration will be discussed in Section 5-4.

Example Problem 5-6   A rigid bar CD is loaded and supported as shown in Fig. 5-10a. Bars A and B are unstressed before the load P is applied. Bar A is made of stainless steel ($E = 190$ GPa) and has a cross-sectional area of 750 mm^2. Bar B is made of an aluminum alloy ($E = 73$ GPa) and has a cross-sectional area of 1250 mm^2. After the load P is applied, the strain in bar B is found to be 1200 μm/m. Detertmine

(a) The vertical components of the displacements of pins F and E and of point D.
(b) The change in length of member A.

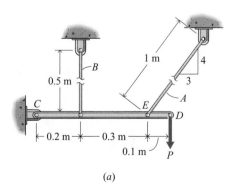

(a)

Figure 5-10(a)

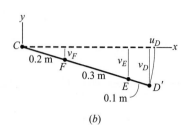

(b)

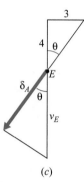

(c)

Figure 5-10(b–c)

SOLUTION

(a) The deformation of member B is found using Eq. 3-1,

$$\delta_B = \epsilon_B L_B = 1200(10^{-6})(0.5) = 600(10^{-6}) \text{ m}$$

Since bar CD is rigid, it rotates about the pin at C as shown in Fig. 5-10b. Point D moves in an arc of a circle with radius CD. The vector $\overline{DD'}$ is the displacement of D, and u_D and v_D are the x- and y-components of this displacement, respectively. The vertical components of the displacements of points E and F are v_E and v_F, respectively. The component of displacement v_F is

$$v_F \cong \delta_B = 600(10^{-6}) \text{ m} = 0.600 \text{ mm} \qquad \textbf{Ans.}$$

Using similar triangles,

$$\frac{v_D}{6} = \frac{v_E}{5} = \frac{v_F}{2}$$

Therefore,

$$v_D = 3v_F = 3(0.600) \text{ mm} = 1.800 \text{ mm} \qquad \textbf{Ans.}$$

and

$$v_E = 2.5v_F = 2.5(0.600) \text{ mm} = 1.500 \text{ mm} \qquad \textbf{Ans.}$$

(b) Since the displacements are small, the arc through which member A has rotated has been replaced by a straight line drawn perpendicular to the undeformed position of member A, as shown in Fig. 5-10c. The change in length of member A is

$$\delta_A = v_E \cos \theta = 1.500(4/5) = 1.200 \text{ mm} \qquad \textbf{Ans.}$$

PROBLEMS

Introductory Problems

5-25* Two tie rods are used to support a load $P = 16$ kip as shown in Fig. P5-25. Rod AB is made of an aluminium alloy with a modulus of elasticity of 10,600 ksi, a length of 80 in., and a cross-sectional area of 0.6 in.^{2} Rod BC is made of structural steel with a modulus of elasticity of 29,000 ksi, a length of 160 in., and a cross-sectional area of 1.25 in.^{2} Determine

a. The elongation of each rod.
b. The horizontal and vertical displacements of pin B.

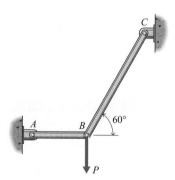

Figure P5-25

5-26* A tie rod and a strut are used to support a 50-kN load as shown in Fig. P5-26. Tie rod *AB* is made of a titanium alloy ($E = 96$ GPa) and has a cross-sectional area of 450 mm². Strut *BC* is made of Monel ($E = 180$ GPa) and has a cross-sectional area of 1450 mm². Determine

a. The lengthening or shortening of the rod and strut.
b. The horizontal and vertical displacements of pin *B*.

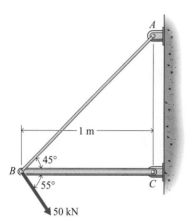

Figure P5-26

5-27 Two structural steel ($E = 29,000$ ksi) cables are used to support the 220-lb traffic light shown in Fig. P5-27. Each cable has a cross-sectional area of 0.015 in.² Determine

a. The normal stresses in each cable.
b. The change in length of each cable.
c. The vertical displacement of point *C*.

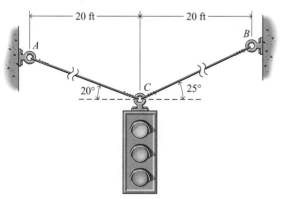

Figure P5-27

Intermediate Problems

5-28* A tie rod *AB* and a pipe strut *AC* are used to support a 100-kN load as shown in Fig. P5-28. Both members are made

of structural steel ($E = 200$ GPa). The cross-sectional areas are 620 mm² for *AB* and 1000 mm² for *AC*. Determine

a. The normal stresses in the tie rod and pipe strut
b. The changes in length of the tie rod and pipe strut.
c. The horizontal and vertical displacement of point *A*.

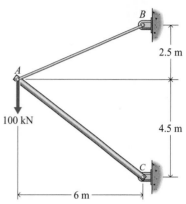

Figure P5-28

5-29 Two tie rods are used to support a 10-kip load as shown in Fig. P5-29. Rod *AC*, which is made of an aluminum alloy with a modulus of elasticity of 10,600 ksi and a yield strength of 41 ksi, is 10 ft long and has a cross-sectional area of 0.326 in.² Rod *BC*, which is made of structural steel with a modulus of elasticity of 29,000 ksi and a yield strength of 36 ksi, is 15 ft long and has a cross-sectional area of 0.508 in.² Determine

a. The elongation of each rod.
b. The horizontal and vertical displacements of pin *C*.

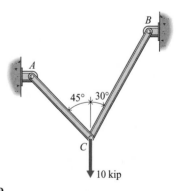

Figure P5-29

5-30 A rigid bar *CD* is loaded and supported as shown in Fig. P5-30. Bars *A* and *B* are unstressed before the load *P* is applied. Bar *A* is made of steel ($E = 200$ GPa) and has a cross-sectional area of 1250 mm². Bar *B* is made of brass

($E = 100$ GPa) and has a cross-sectional area of 940 mm². After the load P is applied, the strain in bar B is found to be 1500 μm/m. Determine

a. The vertical displacement of pin C.
b. The load P.

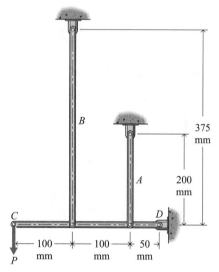

Figure P5-30

Challenging Problems

5-31* The three bars shown in Fig. P5-31 will be used to support a load P of 100 kip. All of the bars have the same cross-sectional area of 2.5 in.² and all are 6 ft long. Bar A is made of Monel ($E = 26,000$ ksi), bar B is made of a magnesium alloy ($E = 6500$ ksi), and bar C is made of structural steel ($E = 29,000$ ksi). A strain gage mounted on member C indicates that the axial strain is 680 μin./in. Determine the change in length of each member and the force P.

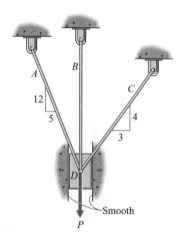

Figure P5-31

5-32 A pin-connected structure is loaded and supported as shown in Fig. P5-32. Member CD is rigid and is horizontal before load P is applied. Bar A is made of structural steel ($E = 200$ GPa), and bar B is made of an aluminum alloy ($E = 73$ GPa). The axial strain in member A is 625 μm/m. Determine the vertical displacement of the pin used to apply the load.

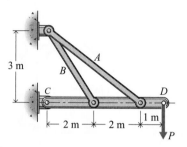

Figure P5-32

5-33 Two struts are used to support a 30-kip load as shown in Fig. P5-33. Strut AB is made of 2024-T4 aluminum alloy ($E = 10,600$ ksi) and has a cross-sectional area of 1.25 in.² Strut BC is made of structural steel ($E = 29,000$ ksi) and has a cross-sectional area of 2.50 in.² Determine

a. The maximum normal and shearing stresses in strut AB.
b. The maximum normal and shearing stresses in strut BC.
c. The lengthening or shortening of both struts.
d. The horizontal and vertical displacements of pin B.

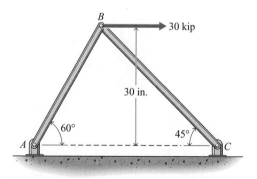

Figure P5-33

5-4 STATICALLY INDETERMINATE AXIALLY LOADED MEMBERS

In many simple structures (and mechanical systems) constructed with axially loaded members, it is possible to find the reactions at supports and the forces in the individual members by drawing free-body diagrams and solving equilibrium equations. Such structures (and systems) are referred to as being statically determinate.

For many other structures (and mechanical systems), the equations of equilibrium are not sufficient for the determination of axial forces in the members and reactions at the supports; these structures (and systems) are referred to as being statically indeterminate. Problems of this type can be analyzed by supplementing the equilibrium equations with additional equations involving the geometry of the deformations in the members of the structure or system. The following outline of procedure will be helpful in the analysis of problems involving statically indeterminate situations.

1. Draw a free-body diagram.
2. Note the number of unknowns involved (magnitudes and positions).
3. Recognize the type of force system on the free-body diagram and note the number of independent equations of equilibrium available for this system.
4. If the number of unknowns exceeds the number of equilibrium equations, a deformation equation must be written for each extra unknown.
5. When the number of independent equilibrium equations and deformation equations equals the number of unknowns, the equations can be solved simultaneously. Deformations and forces must be related in order to solve the equations simultaneously.

Hooke's law (Eq. 4-1) and the definitions of stress (Eq. 2-1) and strain (Eq. 3-1) can be used to relate deformations and forces when all stresses are less than the corresponding proportional limits of the materials used in the fabrication of the members. If some of the stresses exceed the proportional limits of the materials, stress-strain diagrams can be used to relate the loads and deformations. In this section, the problems will be limited to the region of elastic action of the materials. Problems involving inelastic behavior of materials are discussed in Section 5-7. It is recommended that a displacement diagram be drawn showing deformations to assist in obtaining the correct deformation equation. The displacement diagram should be as simple as possible (a line diagram), with the deformations indicated with exaggerated magnitudes and clearly dimensioned. Note that an equilibrium equation and the corresponding deformation equation must be *compatible*; that is, when a tensile force is assumed for a member in the free-body diagram, a tensile deformation must be indicated for the same member in the deformation diagram. If the diagrams are compatible, a negative result will indicate that the assumption was wrong; however, the magnitude of the result will be correct.

In most engineering applications, a body is assumed to be rigid when the equations of equilibrium are used to determine support reactions even though it is a fact that the body deforms when the loads are applied. For example, consider the lever/cable system shown in Fig. 5-11a and assume that lever ABC is rigid and that the weights of the lever and cable are negligible with respect to the applied loads. The cable is fastened to and wraps around the circular sector at the left end of the

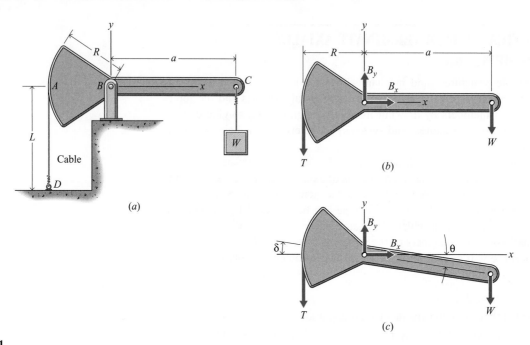

Figure 5-11

lever. The no-load position of the system is shown in Fig. 5-11*a*. Equilibrium of the system under two conditions will be investigated: (1) the cable is rigid and (2) the cable deforms.

5-4-1 Condition (1)

When both the cable and the lever are assumed to be rigid, the free-body diagram for the lever is as shown in Fig. 5-11*b*. The moment equilibrium equation for lever *ABC* is

$$+\!\downarrow \Sigma M_B = 0: \qquad\qquad TR - Wa = 0$$

Thus, the tension T in the rigid cable is

$$T = \frac{Wa}{R} \qquad\qquad (a)$$

5-4-2 Condition (2)

When the cable is assumed to be deformable, the free-body diagram for the lever after the cable deforms (deformed state of equilibrium) is as shown in Fig. 5-11*c*. The moment equilibrium equation for the lever *ABC* is then

$$+\!\downarrow \Sigma M_B = 0: \qquad\qquad TR - W(a \cos \theta) = 0$$

and the tension in the deformed cable is

$$T = \frac{Wa}{R} \cos \theta \qquad\qquad (b)$$

Equation (*b*) cannot be solved for T, since θ is unknown. Since the remaining equations of equilibrium do not provide the additional information needed to solve for T, the problem is statically indeterminate. Statically indeterminate problems are solved by using the equilibrium equations [in this case, Eq. (*b*) above] together with equations obtained from the deformation of the member.

The deformation of the cable is given by Eq. 5-2 as

$$\delta = \frac{TL}{EA} \qquad (c)$$

where E is the modulus of elasticity and A is the cross-sectional area of the cable. Combining Eqs. (*b*) and (*c*) gives

$$\frac{\delta EA}{L} = \frac{Wa}{R} \cos \theta \qquad (d)$$

which has two unknowns, δ and θ. However, the cable wraps around a circular sector on the lever; therefore,

$$\delta = R\theta \qquad (e)$$

Substituting Eq. (*e*) into Eq. (*d*), and rearranging yield

$$R^2 EA\theta = WaL \cos \theta \qquad (f)$$

Equation (*f*) can be solved for θ (if the geometric parameters R, A, a, and L, the material property E, and the weight W are known) by using trial and error, by using numerical methods, or by plotting both sides of Eq. (*f*) versus θ and locating the intersection of the two curves. Once θ is known, Eq. (*b*) is used to find the tension in the cable.

Even though Eq. (*a*) for the rigid cable and Eq. (*b*) for the deformable cable are similar, the computations required to find tension T in Eq. (*b*) are somewhat lengthy. Computational difficulties aside, are the results obtained by using Eq. (*a*) significantly different from those obtained by using Eq. (*b*) in typical engineering situations? To help answer this question, consider the next several examples.

Example 1: The cable is rigid.
If $W = 100$ lb, $a = 30$ in., and $R = 15$ in.
Equation (*a*) yields: $T = 200$ lb

Example 2: The cable is a 3/32-in.-diameter steel ($E = 29{,}000$ ksi) wire.
If $W = 100$ lb, $a = 30$ in., $R = 15$ in., and $L = 45$ in.
Equation (*f*) yields: $\theta = 0.002997$ rad $= 0.1717°$
Equation (*b*) yields: $T = 199.999$ lb
The percent difference in T in the two examples is

$$\%D = \frac{200 - 199.999}{199.999}(100) = 0.0005\%$$

This error is acceptable for practical engineering problems. A check on the normal stress in the wire ($\sigma = P/A$) yields $\sigma = 29.0$ ksi. This stress level is within the linear range of the stress-strain behavior of the steels used to produce wire products. The normal stress must be in the linear range for Eq. (*c*) to be valid.

Example 3:   The cable is a 3/32-in.-diameter aluminum ($E = 10{,}600$ ksi) wire.
If $W = 100$ lb, $a = 30$ in., $R = 15$ in., and $L = 45$ in.
Equation (f) yields: $\theta = 0.00820$ rad $= 0.4698°$
Equation (b) yields: $T = 199.993$ lb
The percent difference in T for Examples 1 and 3 is
$\% D = 0.0035 \%$.

A check on the normal stress in the aluminum wire yields $\sigma = 29.0$ ksi, which is again within the linear range of the stress-strain behavior of aluminum so that Eq. (c) is valid. Examples 2 and 3 indicate that the tension in the wire changes very little when the stiffness of the wire is changed by a factor of almost 3 and that both values are essentially the same as that obtained using the rigid wire assumption (Example 1).

In Sections 5-2 and 5-3, a rigid body was used when calculating support reactions and internal forces; these forces were then used to determine stresses and deformations. Equilibrium requirements should be satisfied when a body is in the deformed configuration. However, the previous example illustrates that forces may be determined, within engineering accuracy, using the equilibrium equations and a free-body diagram in the undeformed configuration. These forces may then be used to determine stresses and deformations with sufficient accuracy for most engineering applications.

In all the following problems and examples, the loading members, pins, and supports are assumed to be rigid, and the mechanism is so constructed that the force system is coplanar. The procedure outlined above is illustrated in the following examples.

Example Problem 5-7   Nine 25-mm-diameter steel ($E = 200$ GPa)

reinforcing bars are used in the short concrete ($E = 30$ GPa) pier shown in Fig. 5-12a. An axial load P of 650 kN is applied to the pier through a rigid capping plate. Determine

(a) The stresses in the concrete and in the steel bars.
(b) The shortening of the pier.

SOLUTION

▶ Note that both forces are drawn on the free-body diagram of Fig. 5-12b as *pushing* on the rigid capping plate. The capping plate will exert equal pushing forces back on the concrete pier and the steel reinforcing bars. Thus, the forces P_C and P_R are both compressive forces; they will lead to compressive stresses σ_C and σ_R, and they will lead to shortening of the pier and the rods (as drawn on the deformation diagram of Fig. 5-12c).

(a) A free-body diagram of the rigid capping plate is shown in Fig. 5-12b. The free-body diagram contains two unknown forces: the resultant force P_C exerted by the concrete and the resultant force P_R exerted by the rods. Since only one equation of equilibrium, $\Sigma F_y = 0$, is available, the problem is statically indeterminate. The additional equation needed to obtain a solution to the problem is obtained from the deformation diagram shown in Fig. 5-12c. As the load P is applied to the rigid capping plate, it moves downward an amount δ, which represents the deflection (the vertical component of the displacement) experienced by both the steel rods and the concrete. The relationship between load and deflection for axial loading is given by Eq. 5-2. Thus, the two equations needed to solve the problem are as follows.

Equilibrium equation,

$$+\uparrow \ \Sigma F_y = 0: \qquad P_R + P_C - P = P_R + P_C - 650(10^3) = 0 \qquad (a)$$

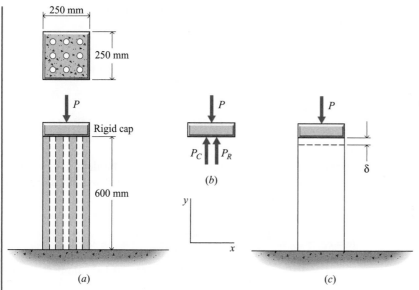

Figure 5-12

Deformation equation,

$$\delta_R = \delta_C$$
$$\frac{P_R L_R}{E_R A_R} = \frac{P_C L_C}{E_C A_C}$$

The total cross-sectional area A_R for the nine steel rods is

$$A_R = 9\left(\frac{\pi}{4}\right)(25)^2 = 4418 \text{ mm}^2$$

The cross-sectional area A_C for the concrete is

$$A_C = (250)^2 - A_R = (250)^2 - 4418 = 58{,}080 \text{ mm}^2$$

▶ The area of all nine rods is required since P_R is the resultant force acting on all nine rods. Similarly, A_C is the total area of the concrete over which P_C acts (the 250-mm square minus the area of the rods).

The deformation equation then yields

$$\frac{P_R(0.600)}{200(10^9)(4418)(10^{-6})} = \frac{P_C(0.600)}{30(10^9)(58{,}080)(10^{-6})}$$

from which

$$P_R = 0.5071 P_C \qquad\qquad (b)$$

Solving Eqs. (a) and (b) simultaneously yields

$$P_R = 218.7(10^3) \text{ N} \cong 219 \text{ kN (C)}$$
$$P_C = 431.3(10^3) \text{ N} \cong 431 \text{ kN (C)}$$

The normal stresses in the rods and in the concrete are obtained by using Eq. 2-2. Thus,

$$\sigma_R = \frac{P_R}{A_R} = \frac{218.7(10^3)}{4418(10^{-6})} = 49.50(10^6) \text{ N/m}^2 \cong 49.5 \text{ MPa (C)} \qquad \textbf{Ans.}$$

$$\sigma_C = \frac{P_C}{A_C} = \frac{431.3(10^3)}{58,080(10^{-6})} = 7.426(10^6) \text{ N/m}^2 \cong 7.43 \text{ MPa (C)} \qquad \textbf{Ans.}$$

(b) The shortening of the pier is obtained from either the deformation of the rods or the deformation of the concrete since they are equal. Thus, from the deformation of the rods,

$$\delta = \delta_C = \delta_R = \frac{\sigma_R L_R}{E_R} = \frac{49.50(10^6)(0.600)}{200(10^9)}$$

$$= 0.1485(10^{-3}) \text{ m} = 0.1485 \text{ mm} \qquad \textbf{Ans.}$$

■ **Example Problem 5-8** A rigid plate C is used to transfer a 20-kip load P to a steel ($E = 30,000$ ksi) rod A and to an aluminum alloy ($E = 10,000$ ksi) pipe B, as shown in Fig. 5-13a. The supports at the top of the rod and bottom of the pipe are rigid, and there are no stresses in the rod or pipe before the load P is applied. The cross-sectional areas of rod A and pipe B are 0.800 in.^{2} and 3.00 in.^{2}, respectively. Determine

(a) The normal stresses in rod A and pipe B.

(b) The displacement of plate C.

SOLUTION

(a) A free-body diagram of plate C and portions of rod A and pipe B is shown in Fig. 5-13b. The free-body diagram contains two unknown forces, P_A and P_B. Since only one equation of equilibrium, $\Sigma F_y = 0$, is available, the problem is statically indeterminate. The additional equation needed to obtain a solution to the problem is obtained from the deformation diagram shown in Fig 5-13c. As the load P is applied to plate C, it moves downward an amount δ, which represents the deflection experienced by both rod A and pipe B. The relationship between load and deflection for axial loading is given by Eq. 5-1. Thus, the two equations needed to solve the problem are as follows. Equilibrium equation,

▶ When the 20-kip load is applied to the rigid plate C, the plate will move down a distance δ; the steel rod A will stretch an amount $\delta_A = \delta = P_A L_A / E_A A_A = \sigma_A L_A / E_A$ (where P_A is a tension force and σ_A is a tension stress); and the aluminum pipe B will shrink an equal amount $\delta_B = \delta = P_B L_B / E_B A_B = \sigma_B L_B / E_B$ (where P_B is a compression force and σ_B is a compression stress). Since P_A is a tension force and P_B is a compression force, P_A is shown as *pulling* on the rod and P_B is shown as *pushing* on the pipe in the free-body diagram of Fig. 5-13b.

$$+\uparrow \ \Sigma F_y = 0 : \qquad\qquad P_A + P_B - 20 = 0$$

or in terms of stresses

$$0.800\sigma_A + 3.00\sigma_B = 20 \qquad\qquad (a)$$

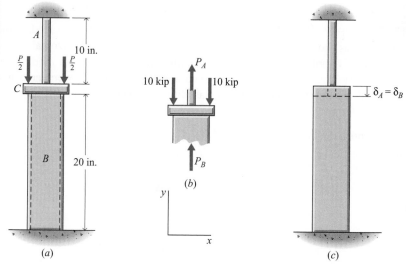

Figure 5-13

Deformation equation,

$$\delta_A = \delta_B$$

$$\frac{\sigma_A L_A}{E_A} = \frac{\sigma_B L_B}{E_B}$$

$$\frac{\sigma_A(10)}{30,000} = \frac{\sigma_B(20)}{10,000}$$

from which

$$\sigma_A = 6\sigma_B \qquad (b)$$

Solving Eqs. (a) and (b) simultaneously yields

$$\sigma_A = 15.384 \text{ ksi} \cong 15.38 \text{ ksi (T)} \qquad \textbf{Ans.}$$

$$\sigma_B = 2.564 \text{ ksi} \cong 2.56 \text{ ksi (C)} \qquad \textbf{Ans.}$$

▶ Note that Eqs. (a) and (b) could just as easily have been written and solved in terms of the forces P_A and P_B rather than in terms of the stresses σ_A and σ_B. Stresses were selected since the problem asked for the stresses but did not ask for the forces.

(b) The displacement of plate C is the same as the deflection of rod A or the deflection of pipe B. Thus, from Eq. 5-1,

$$\delta_C = \delta_A = \delta_B = \frac{\sigma_A(10)}{30,000}$$

$$= \frac{15.384(10)}{30,000}$$

$$= 0.005128 \text{ in.} \cong 0.00513 \text{ in. downward} \qquad \textbf{Ans.}$$

Example Problem 5-9   A pin-connected structure is loaded and supported, as shown in Fig. 5-14a. Member *CD* is rigid and is horizontal before the load *P* is applied. Member *A* is an aluminum alloy bar with a modulus of elasticity of 75 GPa and a cross-sectional area of 1000 mm². Member *B* is a structural steel bar with a modulus of elasticity of 200 GPa and a cross-sectional area of 500 mm². Determine

(a) The normal stresses in bars *A* and *B*.

(b) The vertical component of the displacement of point *D*.

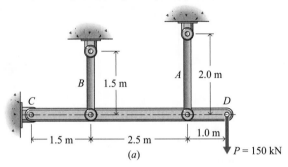

(a)

Figure 5-14(a)

SOLUTION

(a) A free-body diagram of member *CD* and portions of members *A* and *B* is shown in Fig. 5-14b. The free body-diagram contains four unknown forces: C_x, C_y, F_A, and F_B; therefore, since only three equilibrium equations are available, the problem is statically indeterminate. As the load *P* is applied to member *CD*, it will tend to rotate clockwise about pin *C* and produce deformations in members *A* and *B*, as shown in Fig. 5-14c. The extensions shown in Fig. 5-14c are compatible with the tensile forces shown in members *A* and *B* in Fig. 5-14b. The unknown reaction at *C* is not needed to complete the solution of the problem and can be eliminated from further consideration by summing moments about pin *C*. The equilibrium and deformation equations needed to solve for F_A and F_B are

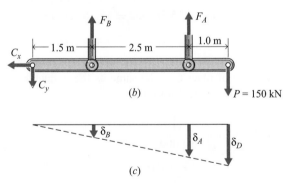

(b)

(c)

Figure 5-14(b–c)

Equilibrium equation,

$$+\uparrow \Sigma M_C = 0: \qquad\qquad P(5) - F_A(4) - F_B(1.5) = 0$$

from which

$$4F_A + 1.5F_B = 150(10^3)(5) = 750(10^3) \text{ N} \qquad (a)$$

Deformation equation (using the principle of similar triangles),

$$\frac{\delta_A}{4} = \frac{\delta_B}{1.5}$$

$$\frac{F_A L_A}{4E_A A_A} = \frac{F_B L_B}{1.5E_B A_B}$$

$$\frac{F_A(2)}{4(75)(10^9)(1000)(10^{-6})} = \frac{F_B(1.5)}{1.5(200)(10^9)(500)(10^{-6})}$$

or

$$F_A = 1.5F_B \qquad (b)$$

Solving Eqs. (a) and (b) simultaneously yields

$$F_A = 150.0(10^3) \text{ N} = 150.0 \text{ kN}$$
$$F_B = 100.0(10^3) \text{ N} = 100.0 \text{ kN}$$

The normal stresses in the two bars are

$$\sigma_A = \frac{F_A}{A_A} = \frac{150.0(10^3)}{1000(10^{-6})} = 150.0(10^6) \text{ N/m}^2 = 150.0 \text{ MPa (T)} \qquad \textbf{Ans.}$$

$$\sigma_B = \frac{F_B}{A_B} = \frac{100.0(10^3)}{500(10^{-6})} = 200.0(10^6) \text{ N/m}^2 = 200 \text{ MPa (T)} \qquad \textbf{Ans.}$$

▶ Note that when point D moves down 5.00 mm, the rigid bar rotates clockwise 0.001 rad $\cong 0.0573°$. This slight angle will have a negligible effect on the free-body diagram and the equilibrium equation.

(b) Since bar CD rotates as a rigid body, the vertical component of the displacement of point D is (again using the principle of similar triangles)

$$\delta_D = \frac{5}{4}\delta_A = \frac{5(150.0)(10^3)(2)}{4(75)(10^9)(1000)(10^{-6})}$$

$$= 5.000(10^{-3}) \text{ m} = 5.00 \text{ mm downward} \qquad \textbf{Ans.}$$

5.3

Example Problem 5-10 A 1/2-in.-diameter alloy-steel bolt ($E = 30,000$ ksi) passes through a cold-rolled brass sleeve ($E = 15,000$ ksi) as shown in Fig. 5-15a. The cross-sectional area of the sleeve is 0.375 in.^{2} Determine the normal stresses produced in the bolt and sleeve by tightening the nut 1/4 turn (0.020 in.).

5.4

SOLUTION
A free-body diagram of the nut and parts of the bolt and sleeve is shown in Fig. 5-15b. The free-body diagram contains two unknown forces F_B and F_S. Since the only equilibrium equation available is $\Sigma F_x = 0$, the problem is statically

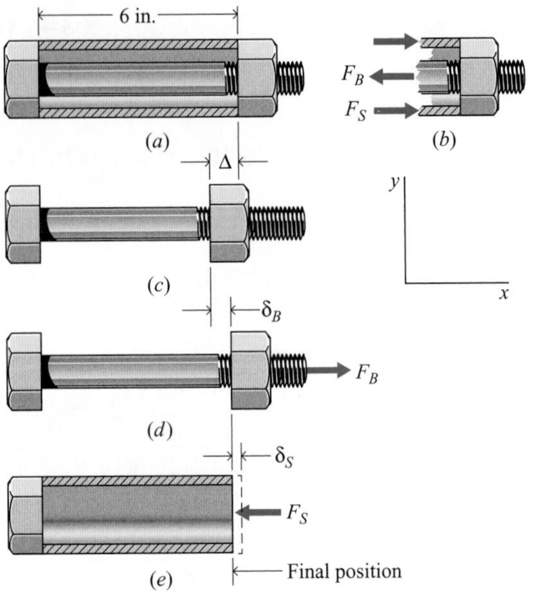

Figure 5-15

indeterminate. The additional equation needed to obtain a solution to the problem is obtained from deformation considerations. As the nut is turned it would move a distance $\Delta = 0.020$ in., as shown in Fig. 5-15c, if the sleeve were not present; however, the sleeve is present and the movement is resisted. As a result, tensile stresses develop in the bolt and compressive stresses develop in the sleeve. These stresses produce the extension δ_B of the bolt and contraction δ_S of the sleeve shown in Figs. 5-15d and e. The deformation equation obtained from the final positions of the nut and sleeve is $\delta_S + \delta_B = \Delta$. The two equations needed to solve the problem are:

Equilibrium equation,

$$\xrightarrow{+}\Sigma F_x = 0: \qquad\qquad F_S - F_B = 0$$

or in terms of stresses

$$0.375\sigma_S = \frac{\pi}{4}\left(\frac{1}{2}\right)^2 \sigma_B$$

from which

$$\sigma_S = 0.5236\,\sigma_B \qquad\qquad (a)$$

▶ Although a free-body diagram of the bolt alone or of the nut alone would work just as well for purposes of writing the equilibrium equation, it would not be as good for purposes of writing the deformation equation. Since it is desired to relate the forces in the sleeve and in the bolt to the deformations that these forces cause, the free-body diagram should show these forces. These forces are made visible by cutting a section through the bolt and sleeve as shown in the free-body diagram of Fig. 5-15b. On Fig. 5-15b it is clear that the force F_B represents a tension force in the steel bolt while the force F_S represents a compression force in the aluminum sleeve.

Deformation equation,

$$\delta_B + \delta_S = \Delta$$

$$\frac{\sigma_B L_B}{E_B} + \frac{\sigma_S L_S}{E_S} = \Delta$$

$$\frac{\sigma_B(6)}{30,000} + \frac{\sigma_S(6)}{15,000} = 0.020 \qquad\qquad (b)$$

Solving Eqs. (a) and (b) simultaneously yields

$$\sigma_B = 48.84 \text{ ksi} \cong 48.8 \text{ ksi (T)} \qquad\qquad \textbf{Ans.}$$

$$\sigma_S = 25.58 \text{ ksi} \cong 25.6 \text{ ksi (C)} \qquad\qquad \textbf{Ans.}$$

▶ If the bolt were rigid, then the sleeve would shrink by the amount that the nut was tightened, $\delta_S = \Delta$. However, the bolt is not rigid, so the bolt will stretch an amount δ_B, and the shrink of the sleeve is reduced by the amount that the bolt stretches, $\delta_S = \Delta - \delta_B$.

PROBLEMS

MecMovie Activities and Problems

MM5.3 Rod and post in series. Example; Try one. Determine normal stresses and deflections in an axial structure consisting of two rods connected end-to-end and subjected to a concentrated load at the connection.

MM5.4 Coaxial tube and core. Example; Try one. Determine normal stresses for a tube and a core rod subjected to a concentrated load.

MM5.5 Rigid bar and two axial members. Example; Try one. Determine normal stresses in two axial members that are connected to a pinned rigid bar. Also, determine deflection of rigid bar.

MM5.6 Rigid bar with two opposing members. Example; Try one. Determine normal stresses in two axial members that are connected but on opposite sides of a pinned rigid bar. Also, determine deflection of rigid bar.

Introductory Problems

5-34* A hollow brass ($E = 100$ GPa) tube A with an outside diameter of 100 mm and an inside diameter of 50 mm is fastened to a 50-mm-diameter steel ($E = 200$ GPa) rod B, as shown in Fig. P5-34. The supports at the top and bottom of the assembly and the collar C used to apply the 500-kN load P are rigid. Determine

a. The normal stresses in each of the members.
b. The deflection of the collar C.

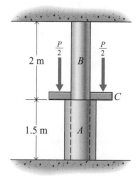

Figure P5-34

5-35* The $7.5 \times 7.5 \times 20$-in. oak ($E = 1800$ ksi) block shown in Fig. P5-35 was reinforced by bolting two $2 \times 7.5 \times 20$-in. steel ($E = 29,000$ ksi) plates to opposite sides of the block. If the load P is 700 kip, determine

a. The normal stress in each member of the assembly.
b. The shortening of the block when the load of part a is applied.

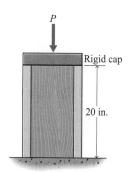

Figure P5-35

5-36 A 3-mm-diameter cord ($E = 7$ GPa) that is covered with a 0.5-mm-thick plastic sheath ($E = 14$ GPa) is subjected to an axial tensile load P, as shown in Fig. P5-36. The load is transferred to the cord and sheath by rigid blocks attached to the ends of the assembly. The length of the cord–sheath assembly is 500 mm, and the load P is 90 N. Determine the forces carried by the cord and sheath.

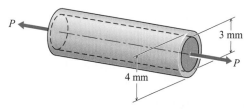

Figure P5-36

5-37* A load P will be supported by a structure consisting of a rigid bar A, two aluminum alloy ($E = 10,600$ ksi) bars B, and a stainless-steel ($E = 28,000$ ksi) bar C, as shown in Fig. P5-37. Each bar has a cross-sectional area of 2.00 in.^{2} If the bars are unstressed before the load P is applied, determine the normal stresses in the bars after a 40-kip load is applied.

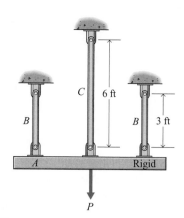

Figure P5-37

5-38   Member *CD* of the structure shown in Fig. P5-38 is rigid and is subjected to a load *P* of 5 kN. Members *A* and *B* are steel ($E = 200$ GPa) wires and each has a cross-sectional area of 80 mm². Determine

a. The forces in each wire.
b. The vertical component of the displacement of point *C*.

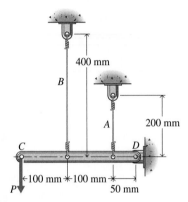

Figure P5-38

5-39   Each part of the stepped circular shaft shown in Fig. P5-39 is aluminum ($E = 10,600$ ksi and $v = 0.33$). If segment *AB* has a diameter of 2.25 in., segment *BC* has a diameter of 3.20 in., and the load *P* is 150 kip, determine

a. The deflection of a cross section at *B*.
b. The axial strain in part *AB*.
c. The change in diameter of part *AB*.

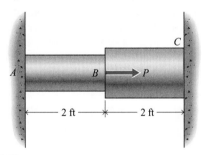

Figure P5-39

5-40*   The rigid 4500-kg homogeneous slab shown in Fig. P5-40 is supported by two cables. Cable *A* is aluminum ($E = 70$ GPa), and cable *B* is steel ($E = 200$ GPa). Determine the ratio of the diameters of the two cables if the slab is to remain horizontal.

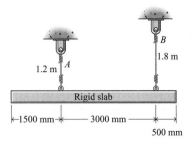

Figure P5-40

Intermediate Problems

5-41*   A hollow steel ($E = 30,000$ ksi) tube *A* with an outside diameter of 2.5 in. and an inside diameter of 2.0 in. is fastened to an aluminum ($E = 10,000$ ksi) bar *B* that has a 2-in. diameter over one-half of its length and a 1-in. diameter over the other half. The assembly is attached to unyielding supports at the left and right ends and is loaded as shown in Fig. P5-41. Determine

a. The normal stresses in all parts of the bar.
b. The deflection of cross section *a–a*.

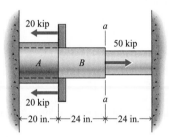

Figure P5-41

5-42*   The $100 \times 100 \times 500$-mm aluminum ($E = 73$ GPa) block is reinforced by bolting a $25 \times 100 \times 500$-mm steel ($E = 200$ GPa) plate on one side of the block as shown in Fig. P5-42. The structure is compressed between two rigid plates by a force $P = 30$ kN. If the rigid plates are to remain horizontal, determine the location *x* of the load *P*.

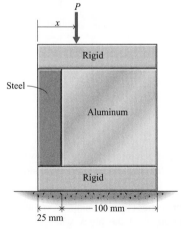

Figure P5-42

5-43 The assembly shown in Fig. P5-43 consists of a steel bar A ($E = 30,000$ ksi and $A = 1.25$ in.^{2}), a rigid bearing plate C that is securely fastened to bar A, and a bronze bar B ($E = 15,000$ ksi and $A = 3.75$ in.^{2}). A clearance of 0.015 in. exists between the bearing plate C and bar B before the assembly is loaded. After a load P of 95 kip is applied to the bearing plate, determine

a. The normal stresses in bars A and B.
b. The vertical displacement of the bearing plate C.

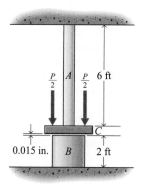

Figure P5-43

5-44* A 150-mm-diameter × 200-mm-long polymer ($E = 2.10$ GPa) cylinder will be attached to a 45-mm-diameter × 400-mm-long brass ($E = 100$ GPa) rod by using the flange type of connection shown in Fig. P5-44. A 0.15-mm clearance exists between the parts as a result of a machining error. If the bolts are inserted and tightened, determine

a. The normal stresses produced in each of the members.
b. The final position of the flange–polymer interface after assembly with respect to the left support.

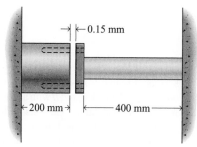

Figure P5-44

5-45* A rigid bar AD is supported by two steel ($E = 29,000$ ksi) wires of the same cross-sectional area of 0.3 in.^{2}, as shown in

Fig. P5-45. If the force P is 1000 lb, determine the change in length of each wire.

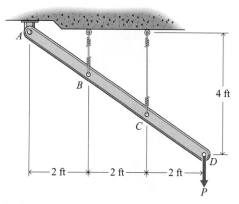

Figure P5-45

5-46 The rigid bar CDE, shown in Fig. P5-46, is horizontal before the load P is applied. Tie rod A is a hot-rolled steel ($E = 200$ GPa) bar with a length of 450 mm and a cross-sectional area of 300 mm^2. Post B is an oak timber ($E = 12$ GPa) with a length of 375 mm and a cross-sectional area of 4500 mm^2. After the 225-kN load P is applied, determine

a. The normal stresses in bar A and post B.
b. The shearing stress in the 20-mm-diameter pin at C, which is in double shear.
c. The vertical displacement of point D.

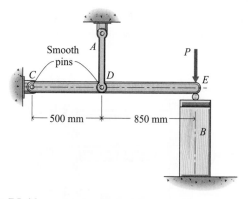

Figure P5-46

5-47 Five 1-in.-diameter steel ($E = 29,000$ ksi) reinforcing bars will be used in a 3-ft-long concrete ($E = 4500$ ksi) pier with a square cross section, as shown in Fig. P5-47. The allowable strengths in compression for steel and concrete are 18 ksi and

1.4 ksi, respectively. Determine the minimum size of pier required to support a 200-kip axial load.

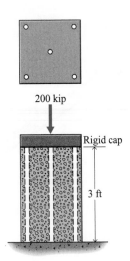

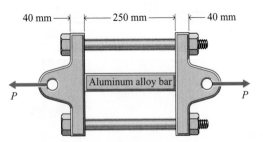

Figure P5-47

Challenging Problems

5-48* The two faces of the clamp shown in Fig. P5-48 are 250 mm apart when the two stainless-steel ($E = 190$ GPa) bolts connecting them are unstretched. A force P is applied to separate the faces of the clamp so that an aluminum alloy ($E = 73$ GPa) bar with a length of 251 mm can be inserted as shown. Each of the bolts has a cross-sectional area of 120 mm^2, and the bar has a cross-sectional area of 625 mm^2. After the load P is removed, determine

a. The axial stresses in the bolts and in the bar.
b. The change in length of the aluminum alloy bar.

Figure P5-48

5-49* A 1/2-in.-diameter alloy-steel ($E = 30,000$ ksi) bolt passes through a cold-rolled brass ($E = 15,000$ ksi) sleeve as shown in Fig. P5-49. The cross-sectional area of the sleeve is

0.375 in.^{2} Determine the normal stresses produced in the bolt and sleeves by tightening the nut 1/4 turn (0.020 in.).

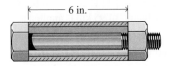

Figure P5-49

5-50 Bar BF of Fig. P5-50 is made of steel ($E = 210$ GPa), and bar CE is made of aluminum alloy ($E = 73$ GPa). The cross-sectional areas are 1200 mm^2 for bar BF and 900 mm^2 for bar CE. As a result of a misalignment of the pin holes at A, B, and C, a force of 50 kN upward must be applied at D, after pins A and B are in place, to permit insertion of pin C. Determine

a. The normal stress in bar CE when the force P is removed with all pins in place.
b. The vertical component of the displacement of pin D from its no-load position.

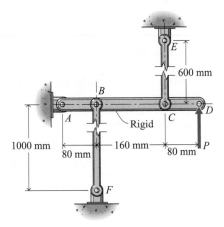

Figure P5-50

5-51 The mechanism of Fig. P5-51 consists of structural steel ($E = 29,000$ ksi) rod A with a cross-sectional area of 0.50 in.^{2}, a cold-rolled brass ($E = 15,000$ ksi) rod B with a cross-sectional area of 1.20 in.^{2}, and a rigid bar C. The nuts at the top ends of rods A and B are initially tightened to the point where all slack is removed from the mechanism but the bars remain free of stress. Determine

a. The axial stress induced in rod A by advancing the nut at the top of rod B one turn (0.10 in.).
b. The vertical displacement of the nut at the top end of rod A.

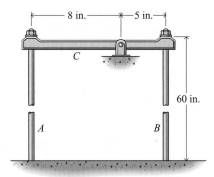

Figure P5-51

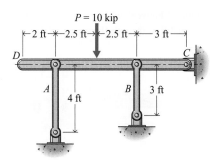

Figure P5-53

5-52* A solid circular aluminum ($E = 73$ GPa and $G = 28$ GPa) bar of constant diameter 30 mm is subjected to the axial loads $P_1 = 6$ kN and $P_2 = 3$ kN as shown in Fig. P5-52. Determine

a. The normal stresses in segments AB, BC, and CD of the bar.
b. The maximum shear stress in the bar.
c. The change in length of segment BC.
d. The change in diameter of segment BC.
e. The axial strain in segment AB.

5-54 The bronze ($E = 100$ GPa) post D of Fig. P5-54 has a cross-sectional area of 2500 mm^2, and the high-strength steel ($E = 200$ GPa) bar C has a cross-sectional area of 600 mm^2. Bar AB and the bearing block on post D are to be considered rigid. The clearance between post D and bar AB is 0.09 mm before the load P is applied. If the axial stresses are not to exceed 215 MPa for the steel and 95 MPa for the bronze, determine the maximum load that can be applied.

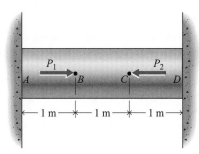

Figure P5-52

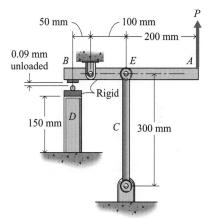

Figure P5-54

Computer Problems

5-55 A high-strength steel bolt ($E = 30,000$ ksi and $A = 0.785$ in.^{2}) passes through a brass sleeve ($E = 15,000$ ksi and $A = 1.767$ in.^{2}), as shown in Fig. P5-55. As the nut is tightened, it advances a distance of 0.125 in. along the bolt for each complete turn of the nut. Compute and plot

a. The axial stresses σ_s (in the steel bolt) and σ_b (in the brass sleeve) as functions of the angle of twist θ of the nut ($0° \le \theta \le 180°$).
b. The elongations δ_s (of the steel bolt) and δ_b (of the brass sleeve) as functions of the angle of twist θ of the nut ($0° \le \theta \le 180°$).

5-53* A pin-connected structure is loaded and supported as shown in Fig. P5-53. Member CD is rigid and is horizontal before the load P is applied. Member A is an aluminum alloy bar with a modulus of elasticity of 10,600 ksi and a cross-sectional area of 2.25 in.^{2} Member B is a stainless-steel bar with a modulus of elasticity of 28,000 ksi and a cross-sectional area of 1.75 in.^{2} After the load P is applied to the structure, determine

a. The normal stresses in bars A and B.
b. The maximum shearing stresses in bars A and B.
c. The shearing stress in the 0.5 in.-diameter pin at C, which is in double shear.
d. The vertical displacement of point D.

c. The distance L between the two washers as a function of the angle of twist θ of the nut ($0° \le \theta \le 180°$).

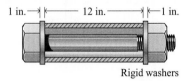

Figure P5-55

5-56 The mechanism of Fig. P5-56 consists of a structural steel ($E = 200$ GPa) rod A with a cross-sectional area of 350 mm², a cold-rolled brass ($E = 100$ GPa) rod B with a cross-sectional area of 750 mm², and a rigid bar C. The nuts at the top ends of rods A and B are initially tightened to the point where all slack is removed from the mechanism but the bars remain free of stress. If a nut advances 2.5 mm with each full turn (360°), calculate and plot

a. The axial stresses σ_A and σ_B in the rods as functions of the angle θ through which the nut at the top end of rod B is rotated ($0° \le \theta \le 180°$).
b. The vertical displacement δ_{nut} of the nut at the top end of rod A as a function of the angle θ through which the nut at the top end of rod B is rotated ($0° \le \theta \le 180°$).
c. The rotation angle ϕ (deg) of arm C as a function of the angle θ through which the nut at the top end of rod B is rotated ($0° \le \theta \le 180°$).

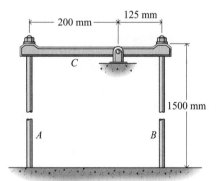

Figure P5-56

5-57 Initially, the arms of the crank C shown in Fig. P5-57 are horizontal and vertical; there is a 0.009-in. gap between the horizontal arm and the brass ($E = 15,000$ ksi and $A = 12$ in.²) post B; and the aluminum ($E = 10,000$ ksi and $A = 2$ in.²) rod A is horizontal. If the crank C is rigid, calculate and plot

a. The axial stresses σ_A (in the aluminum rod) and σ_B (in the brass post) as functions of the force P (0 kip $\le P \le$ 30 kip).

b. The changes in length δ_A (of the aluminum rod) and δ_B (of the brass post) as functions of the force P (0 kip $\le P \le$ 30 kip).
c. The rotation angle θ (deg) of the crank C as a function of the force P (0 kip $\le P \le$ 30 kip).

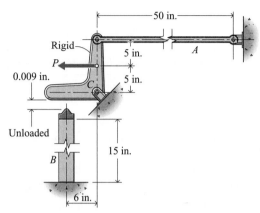

Figure P5-57

5-58 Initially, the bar AB shown in Fig. P5-58 is horizontal; there is a 0.09-mm gap between the horizontal arm and the bronze ($E = 100$ GPa and $A = 2500$ mm²) post D; and the aluminum ($E = 73$ GPa and $A = 600$ mm²) rod C is vertical. If the bar AB is rigid, calculate and plot

a. The axial stresses σ_C (in the aluminum rod) and σ_D (in the bronze post) as functions of the force P (0 kN $\le P \le$ 150 kN).
b. The changes in length δ_C (of the aluminum rod) and δ_D (of the bronze post) as functions of the force P (0 kN $\le P \le$ 150 kN).
c. The rotation angle θ (deg) of the bar AB as a function of the force P (0 kN $\le P \le$ 150 kN).

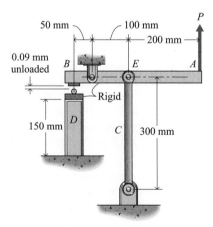

Figure P5-58

5-5 THERMAL EFFECTS

When a temperature change takes place while a member is restrained (free movement restricted or prevented), stresses (referred to as thermal stresses) are induced in the member. For example, the bar AB of Fig. 5-16a is securely fastened to rigid supports at both ends. Since the ends of the bar are fixed, the total deformation of the bar must be zero

$$\delta_{\text{total}} = \delta_T + \delta_\sigma = \epsilon_T L + \epsilon_\sigma L$$
$$0 = \alpha \Delta T L + \frac{\sigma}{E} L$$

The term δ_T is the deformation due to a temperature change, and δ_σ is the deformation due to an axial load. If the temperature of the bar increases (ΔT positive), then the induced stress must be negative and the wall must push on the ends of the rod. If the temperature of the bar decreases (ΔT negative), then the induced stress must be positive and the wall must pull on the ends of the rod.

That is, if end B were not attached to the wall and the temperature dropped, end B would move to B', a distance $|\delta_T| = |\epsilon_T L| = |\alpha \Delta T L|$, as indicated in Fig. 5-16b. Therefore, for the total deformation of the bar to be zero, the wall at B must apply a force $P = \sigma A$ (Fig. 5-16c) of sufficient magnitude to move end B through a distance $\delta_P = \epsilon_\sigma L = (\sigma/E)L$ so that the length of the bar is again L, the distance between the walls. Since the walls do not move, $|\delta_T| = \delta_P$, or

$$\delta_P - |\delta_T| = \delta_P + \delta_T = 0$$

and thus the total deformation of the bar is zero.

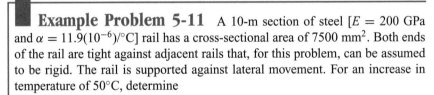

(a)

(b)

(c)

Figure 5-16

Example Problem 5-11   A 10-m section of steel [$E = 200$ GPa and $\alpha = 11.9(10^{-6})/°C$] rail has a cross-sectional area of 7500 mm^2. Both ends of the rail are tight against adjacent rails that, for this problem, can be assumed to be rigid. The rail is supported against lateral movement. For an increase in temperature of 50°C, determine

(a) The normal stress in the rail.

(b) The internal force on a cross section of the rail.

5.5

5.6

SOLUTION

The rail is modeled as shown in Fig. 5-16a. Since the temperature increases, the deformations shown in Figs. 5-16b and 5-16c are reversed, but the magnitudes of δ_T and δ_σ are equal.

(a) The change in length of the rail resulting from the temperature change is given by modifying Eq. 4-11 as

$$\delta_T = \epsilon_T L = \alpha L \,\Delta T = 11.9(10^{-6})(10)(50) = 5.950(10^{-3})\,\text{m} = 5.95\,\text{mm}$$

The stress required to resist a change in length of 5.95 mm is given by Eq. 5-1 as

$$\sigma = \frac{E\delta}{L} = \frac{200(10^9)(5.95)(10^{-3})}{10}$$
$$= 119.0(10^6)\,\text{N/m}^2 = 119.0\,\text{MPa (C)}\qquad\textbf{Ans.}$$

(b) The internal force on a cross section of the rail is

$$F = \sigma A = 119.0(10^6)(7500)(10^{-6})$$
$$= 892.5(10^3) \text{ N} \cong 893 \text{ kN (C)} \qquad \textbf{Ans.}$$

Example Problem 5-12   After a load $P = 150$ kN is applied to the pin-connected structure shown in Fig. 5-17a, the temperature increases 100°C. The thermal coefficients of expansion are $22(10^{-6})$/°C for the aluminum alloy rod A and $12(10^{-6})$/°C for the steel rod B. The moduli of elasticity of aluminum and steel are 75 GPa and 200 GPa, respectively. The cross-sectional area of members A and B are 1000 mm² and 500 mm², respectively. If member CD is rigid, determine

(a) The normal stresses in bars A and B.
(b) The vertical component of the displacement of point D.

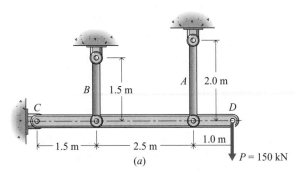

Figure 5-17(a)

SOLUTION

(a) A free-body diagram for bar CD is shown in Fig. 5-17b. The deformations in bars A and B are functions of both load and temperature change, as shown in Fig. 5-17c. Thus,

$$\delta_A = \delta_{AP} + \delta_{AT}$$
$$\delta_B = \delta_{BP} + \delta_{BT} \qquad (a)$$

where the subscripts P and T refer to load and temperature, respectively. The equilibrium and deformation equations needed to solve for F_A and F_B are Equilibrium equation,

$$+\uparrow \Sigma M_C = 0: \qquad P(5) - F_A(4) - F_B(1.5) = 0$$

from which

$$4F_A + 1.5F_B = 750(10^3) \text{ N} \qquad (b)$$

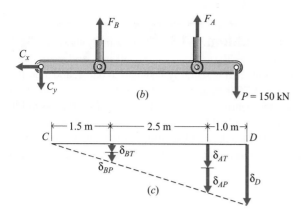

Figure 5-17(b–c)

▶ As the rigid bar *CD* rotates about pin *C*, the pins *B*, *A*, and *D* move on circular arcs about pin *C*. If the angle of rotation is small, these motions can be approximated as vertical (perpendicular to the bar *CD*) displacements, as shown in Fig. 5-17c. From similar triangles, $\delta_B/1.5 = \delta_A/4.0 = \delta_D/5$.

Deformation equation,

$$\frac{\delta_A}{4} = \frac{\delta_B}{1.5}$$

$$\frac{F_A L_A}{4 E_A A_A} + \frac{\alpha_A L_A (\Delta T)}{4} = \frac{F_B L_B}{1.5 E_B A_B} + \frac{\alpha_B L_B (\Delta T)}{1.5}$$

$$\frac{F_A (2.0)}{4(75)(10^9)(1000)(10^{-6})} + \frac{22(10^{-6})(2.0)(100)}{4}$$

$$= \frac{F_B (1.5)}{1.5(200)(10^9)(500)(10^{-6})} + \frac{12(10^{-6})(1.5)(100)}{1.5}$$

from which

$$F_A = 1.5 F_B + 15(10^3) \qquad (c)$$

Solving Eqs. (b) and (c) simultaneously yields

$$F_A = 153.00(10^3) \text{ N} = 153.00 \text{ kN}$$
$$F_B = 92.00(10^3) \text{ N} = 92.00 \text{ kN}$$

The stresses in the two bars are

$$\sigma_A = \frac{F_A}{A_A} = \frac{153.00(10^3)}{1000(10^{-6})} = 153.0(10)^6 \text{ N/m}^2 = 153.0 \text{ MPa (T)} \qquad \textbf{Ans.}$$

$$\sigma_B = \frac{F_B}{A_B} = \frac{92.00(10^3)}{500(10^{-6})} = 184.0(10^6) \text{ N/m}^2 = 184.0 \text{ MPa (T)} \qquad \textbf{Ans.}$$

(b) Since bar *CD* rotates as a rigid body, the vertical component of the displacement of point *D* is

$$\delta_D = \frac{5}{4}\delta_A = \frac{5 F_A L_A}{4 E_A A_A} + \frac{5}{4}\alpha_A L_A(\Delta T)$$

$$= \frac{5(153.00)(10^3)(2.0)}{4(75)(10^9)(1000)(10^{-6})} + \frac{5}{4}(22)(10^{-6})(2.0)(100)$$

$$= 10.60(10^{-3}) \text{ m} = 10.60 \text{ mm} \downarrow \qquad \textbf{Ans.}$$

5.7

Example Problem 5-13 The assembly shown in Fig. 5-18a consists of a steel rod A (E_A = 30,000 ksi, A_A = 2.50 in.², and α_A = 6.6 × 10⁻⁶/°F), a rigid bearing plate C that is securely fastened to bar A, and a bronze bar B (E_B = 15,000 ksi, A_B = 3.75 in.², and α_B = 9.4 × 10⁻⁶/°F). A clearance of 0.015 in. exists between the bearing plate C and bar B before the assembly is loaded. If a load P = 5 kip is applied to the bearing plate and then the temperature of the assembly is slowly raised, calculate and plot the stresses σ_A in the steel rod and σ_B in the bronze bar as a function of the temperature increase ΔT for 0°F < ΔT < 50°F.

SOLUTION

The first step is to draw a free-body diagram. When the force P is applied to the bearing plate C, we expect the plate to be pushed down (causing a tensile force in A) and press against the bar B (causing a compressive force in B). The free-body diagram is drawn accordingly (Fig. 5-18b). The only equation of equilibrium that gives any useful information is the sum of forces in the vertical direction

$$+\uparrow \Sigma F_y = 0: \qquad T_A + P_B - 5 = 0 \qquad T_A + P_B = 5 \text{ kip}$$
$$2.5\sigma_A + 3.75\sigma_B = 5000 \text{ lb} \qquad (a)$$

Therefore, the problem is statically indeterminate and we have to write a compatibility equation relating the deformations to solve for the stresses.

Since T_A is a tensile force, the stress deformation $(\sigma L/E)_A$ will represent a stretch of the rod A. When the temperature of A increases, the stretch of A will be even greater. Therefore, the total stretch of A is the sum of the stress deformation (a stretch) and the temperature deformation (also a stretch)

$$\delta_A = \left(\frac{\sigma L}{E} + \alpha \Delta TL\right)_A \qquad (b)$$

Likewise, since P_B is a compressive force, the stress deformation $(\sigma L/E)_B$ will represent a shrink of the bar B. When the temperature increases, bar B will stretch, thus reducing the shrink caused by the force P_B. Therefore, the total shrink of B is the difference of the stress deformation (a shrink) and the temperature deformation (a stretch)

$$\delta_B = \left(\frac{\sigma L}{E} - \alpha \Delta TL\right)_B \qquad (c)$$

The deformation diagram (Fig. 5-18c) relates the stretch δ_A, the shrink δ_B, and the initial gap of 0.015 in.

$$\delta_A = \delta_B + 0.015 \qquad (d)$$

Combining Eqs. (b), (c), and (d) gives

$$\left(\frac{\sigma L}{E} + \alpha \Delta TL\right)_A = \left(\frac{\sigma L}{E} - \alpha \Delta TL\right)_B + 0.015 \qquad (e)$$

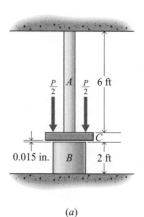

(a)

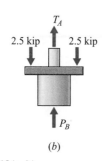

(b)

Figure 5-18(a–b)

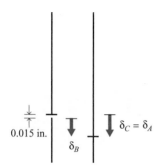

(c)

Figure 5-18(c)

or, putting in numbers,

$$\left[\frac{\sigma_A(72)}{(30)(10^6)} + (6.6)(10^{-6})\Delta T(72)\right]$$
$$= \left[\frac{\sigma_B(24)}{(15)(10^6)} - (9.4)(10^{-6})\Delta T(24)\right] + 0.015 \tag{f}$$

Multiplying through by 30×10^6 and rearranging gives

$$72\sigma_A - 48\sigma_B = 450{,}000 - 21{,}024\Delta T \tag{g}$$

Finally, solving Eqs. (a) and (g) simultaneously gives the stresses

$$\sigma_A = 4942.308 - 202.154\Delta T$$
$$\sigma_B = 134.769\Delta T - 1961.539 \tag{h}$$

which shows that both σ_A and σ_B are linear functions of the temperature increase ΔT. However, this says that for a temperature increase of less than about 14.5°F, the stress in bar B is negative or in tension (opposite what was initially assumed). But there is nothing pulling on bar B that could put it in tension, and this solution cannot be valid. (Actually, this solution would apply if the plate C were pulled down to the bar B, welded to it, and then released. The tension stress above would be the tension in the weld until the temperature increase exceeded about 14.5°F.) Therefore, for a temperature increase of less than about 14.5°F, the assumption that the applied force closes the gap and the plate pushes on the bar B is not correct.

Starting over with a new free-body diagram (Fig. 5-18d) gives the equilibrium equation

$$+\uparrow \Sigma F_y = 0: \qquad T_A - 5 = 0 \qquad T_A = 5 \text{ kip}$$
$$2.5\sigma_A = 5000 \text{ lb} \tag{i}$$

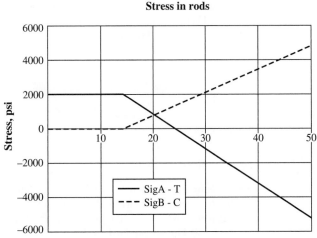

Figure 5-18(d)

Stress in rods

Stress, psi — Temperature increase, deg. F

— SigA - T
- - - SigB - C

(e)

Figure 5-18(e)

So now the stresses in the two bars are (for $\Delta T < 14.5°$)

$$\sigma_A = 2000 \text{ psi}$$
$$\sigma_B = 0 \text{ psi} \qquad\qquad (j)$$

For temperature increases greater than $\Delta T > 14.5°$, the stresses are given by Eq. (h). The stresses are shown in the graph of Fig. 5-18e. Positive stresses represent tension in bar A and compression in bar B (as assumed on the free-body diagrams), and negative stresses represent compression in bar A and tension in bar B (opposite what was assumed on the free-body diagrams).

PROBLEMS

MecMovie Activities and Problems

MM5.7 Rod and post with temperature change. Example; Try one. Determine normal stresses in end-to-end axial members due to temperature increase.

MM5.8 Coaxial bars with temperature. Examples; Try one. Determine internal forces and normal strains in axial bars connected by pins. Also, determine shear stress in connecting pin.

Introductory Problems

5-59* A 3-in.-diameter × 80-in.-long aluminum alloy [$E = 10,600$ ksi and $\alpha = 12.5 \ (10^{-6})/°$F] bar is stress free after being attached to rigid supports, as shown in Fig. P5-59. If the temperature drops 100°F, determine

a. The normal stress in the bar.
b. The change in length of the bar.
c. The normal stress in the bar if the rigid support at B is removed.
d. The change in length of the bar for the conditions stated in part c.

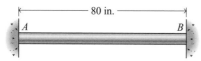

Figure P5-59

5-60* A 6-m-long × 50-mm-diameter rod of aluminum alloy [$E = 70$ GPa, $v = 0.346$, and $\alpha = 22.5 \ (10^{-6})/°$C] is attached at the ends to supports that yield to permit a change in length of 1.00 mm in the rod when stressed. When the temperature is 35°C, there is no stress in the rod. After the temperature of the rod drops to −20°C, determine

a. The normal stress in the rod.
b. The change in diameter of the rod.

5-61 A bar consists of 3-in.-diameter aluminum alloy [$E = 10,600$ ksi, $v = 0.33$, and $\alpha = 12.5 \ (10^{-6})/°$F] and 4-in.-diameter steel [$E = 30,000$ ksi, $v = 0.30$, and $\alpha = 6.6(10^{-6})/°$F] parts, as

shown in Fig. P5-61. If end supports are rigid and the bar is stress free at 0°F, determine

a. The normal stress in both parts of the bar at 80°F.
b. The change in diameter of the steel part of the bar.
c. The displacement of the rigid plate.

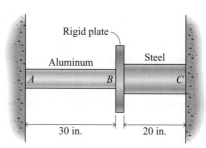

Figure P5-61

5-62 A steel tie rod containing a rigid turnbuckle (Fig. P5-62) has its end attached to rigid walls. During the summer, when the temperature is 30°C, the turnbuckle is tightened to produce a stress in the rod of 15 MPa. Determine the normal stress in the rod in the winter when the temperature is −10°C. Use $E = 200$ GPa and $\alpha = 11.9 \ (10^{-6})/°$C.

Figure P5-62

Intermediate Problems

5-63* Nine 3/4-in.-diameter steel ($E = 30,000$ ksi) reinforcing bars were used when the short concrete ($E = 4500$ ksi) pier shown in Fig. P5-63 was constructed. After a load P of 150 kip was applied to the pier, the temperature increased 100°F. The coefficients of thermal expansion for steel and concrete are 6.6(10^{-6})/°F and 6.0(10^{-6})/°F, respectively. Determine

a. The normal stresses in the concrete and in the steel bars after the temperature increases.
b. The change in length of the pier resulting from the combined effects of the temperature change and the load.

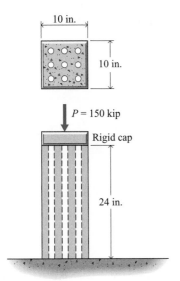

Figure P5-63

5-64* The 50-mm-diameter circular steel [$E = 200$ GPa, and $\alpha = 11.9\,(10^{-6})/°C$] bar shown in Fig. P5-64 is subjected to an axial load of $P = 100$ kN. After the load is applied, the temperature decreases 20°C. Determine the normal stress in segments AB and BC of the bar.

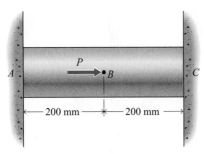

Figure P5-64

5-65 A rigid block of negligible weight is suspended by three wires, as shown in Fig. P5-65. Wires B are steel [$E = 29,000$ ksi, $\alpha = 6.6(10^{-6})/°F$, and $A = 0.25$ in.^{2}], and wire C is aluminum [$E = 10,600$ ksi, $\alpha = 12.5(10^{-6})/°F$, and $A = 0.50$ in.^{2}]. If the temperature rises 50°F, determine the normal stress in each wire.

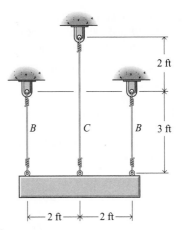

Figure P5-65

5-66* Bar B of the pin-connected system of Fig. P5-66 is made of an aluminum alloy [$E = 70$ GPa, $A = 300$ mm^2, and $\alpha = 22.5\,(10^{-6})/°C$], and bar A is made of a hardened carbon steel [$E = 210$ GPa, $A = 1200$ mm^2, and $\alpha = 11.9\,(10^{-6})/°C$]. Bar CDE is to be considered rigid. When the system is unloaded at 40°C, bars A and B are unstressed. After the load P is applied, the temperature of both bars decreases to 15°C. Determine

a. The normal stresses in bars A and B.
b. The maximum shearing stresses in bars A and B.
c. The shearing stress in the 20-mm-diameter pin at C, which is in double shear.
d. The vertical component of the displacement of pin E.

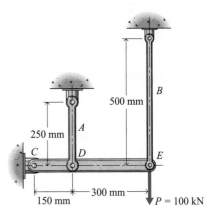

Figure P5-66

5-67 A rigid block weighing 5000 lb is suspended by three wires, as shown in Fig. P5-65. Wires B are steel [$E = 29,000$ ksi, $\alpha = 6.6(10^{-6})/°F$, and $A = 0.25$ in.^{2}], and wire C is aluminum [$E = 10,600$ ksi, $\alpha = 12.5(10^{-6})/°F$, and $A = 0.50$ in.^{2}]. If the temperature rises 50°F, determine the normal stress in each wire.

5-68 The pin-connected structure shown in Fig. P5-68 consists of a rigid bar $ABCD$, a steel [$E = 210$ GPa, and $\alpha = 11.9(10^{-6})/°C$] bar BF, and an aluminum alloy [$E = 73$ GPa and $\alpha = 22.5(10^{-6})/°C]$ bar CE. The cross-sectional areas are 1200 mm^2 for bar BF and 900 mm^2 for bar CE. The bars are un-stressed when the structure is assembled at 40°C. Determine

a. The normal stresses in the bars after the temperature is re-duced to −20°C.
b. The shearing stresses in the 30-mm-diameter pins at A, B, and C. Pin B is in double shear, and pins A and C are in single shear.
c. The vertical component of the displacement of pin D from it original position.

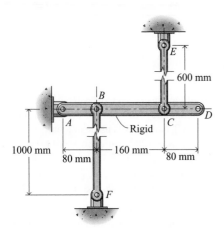

Figure P5-68

5-69*   A 200-lb block W is suspended by an aluminum $[E = 10,600$ ksi, $\alpha = 12.5(10^{-6})/°F$, and $A = 0.15$ in.$^2]$ wire AB, as shown in Fig. P5-69. There is an initial clearance of 0.08 in. between the block and the floor. Determine the normal stress in the wire

a. For a temperature increase of 25°F.
b. For a temperature increase of 60°F.

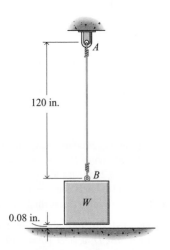

Figure P5-69

Challenging Problems

5-70*   A solid circular bar of aluminum $[E = 74$ GPa and $\alpha = 12.5 (10^{-6})/°C]$ has the shape shown in Fig. P5-70. The bar is securely fastened to rigid supports at each end and is essentially stress free at 20°C. Assuming that plane sections perpendicular to the longitudinal axis of the bar remain plane, determine the stress at section x $(0 \leq x \leq 1$ m) when the temperature is 70°C.

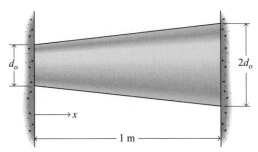

Figure P5-70

5-71*   The pin-connected structure shown in Fig. P5-71 consists of a cold-rolled bronze $[E = 15,000$ ksi, $\alpha = 9.4(10^{-6})/°F]$ bar A, which has a cross-sectional area of 3.00 in.^{2}, and two 0.2 per-cent C hardened steel $[E = 30,000$ ksi and $\alpha = 6.6 (10^{-6})/°F]$ bars B, which have cross-sectional areas of 2.50 in.^{2} If the tem-perature of bar A decreases 50°F and the temperature of bars B increases 30°F after the 200-kip load is applied, determine

a. The normal stresses in the bars.
b. The displacement of pin C.

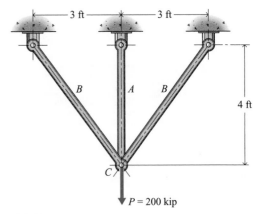

Figure P5-71

5-72   The two faces of the clamp of Fig. P5-72 are 250 mm apart when the two stainless-steel $[E = 190$ GPa, A = 115 mm^2 (each), and $\alpha = 17.3 (10^{-6})/°C]$ bolts connecting them are unstretched. A force P is applied to separate the faces of the clamp so that an aluminum alloy [$E = 73$ GPa, $A = 625$ mm², and $\alpha = 22.5(10^{-6})/°$C] bar with a length of 250.5 mm can be inserted as shown. After the load P is removed, the temperature is raised 100°C. Determine the normal stresses in the bolts and in the bar, and the distance between the faces of the clamp.

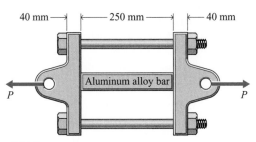

40 mm ——→ |←—— 250 mm ——→| |←— 40 mm

Aluminum alloy bar

P P

Figure P5-72

5-73 The three bars shown in Fig. P5-73 are 0.5 in. thick, 1 in. wide, and 10 in. long. Bars S are steel [$E = 29,000$ ksi and $\alpha = 6.6(10^{-6})/°$F], and bar A is aluminum [$E = 10,600$ ksi and $\alpha = 12.5(10^{-6})/°$F]. The assembly is held together by single rivets of 0.5-in. diameter at each end, and is stress free at 20°C. After the temperature drops 40°C, determine

a. The normal stress in the rivets.
b. The shear stress in the rivets.

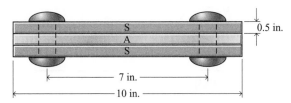

|← 7 in. →|
0.5 in.
S
A
S
|← 10 in. →|

Figure P5-73

5-74 A prismatic bar [$E = 70$ GPa and $\alpha = 22.5\ (10^{-6})/°$C], free of stress at room temperature, is fastened to rigid walls at its ends. One end of the bar is heated to 100°C while the other end is maintained at room temperature. The change in temperature ΔT along the bar is proportional to the square of the distance from the unheated end. Determine the normal stress in the bar after the change in temperature.

Computer Problems

5-75 An aluminum [$E = 10,000$ ksi, $\alpha = 12.5\ (10^{-6})/°$F, and $A = 1.400$ in.²] bolt passes through a steel [$E = 30,000$ ksi, $\alpha = 6.6\ (10^{-6})/°$F, and $A = 0.400$ in.²] sleeve, as shown in Fig. P5-75. Initially, the nut is tightened against the washer at room

temperature until the bolt has a tensile force of 3500 lb, and then the temperature of the assembly is slowly raised. Calculate and plot

a. The stress σ_{al} in the aluminum bolt and the stress σ_{st} in the steel sleeve as a function of the temperature increase ΔT (0° $\leq \Delta T \leq 100°$F).
b. The change in length of the aluminum bolt δ_{al} and the change in length of the steel sleeve δ_{st} as a function of the temperature increase ΔT (0° $\leq \Delta T \leq 100°$F).

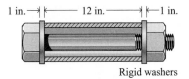

1 in. →| |←—— 12 in. ——→| |←1 in.

Rigid washers

Figure P5-75

5-76 A pin-connected structure is loaded and supported as shown in Fig. P5-76. Member AB is rigid; C is a steel [$E = 200$ GPa, $A = 600$ mm², and $\alpha = 11.9\ (10^{-6})/°$C] rod; and D is an aluminum alloy [$E = 73$ GPa, $A = 2500$ mm², and $\alpha = 22.5(10^{-6})/°$C] post. Initially, bar AB is horizontal, rod C is vertical, and there is a 0.09-mm gap between the horizontal arm and the post D. If a 35-kN force P is applied to the right end of the bar AB and the temperature of the system is slowly raised, calculate and plot

a. The axial stresses σ_C (in the steel rod) and σ_D (in the aluminum post) as functions of the temperature increase ΔT (0°C $\leq \Delta T \leq 40°$C).
b. The changes in length δ_C (of the steel rod) and δ_D (of the aluminum post) as functions of the temperature increase ΔT (0°C $\leq \Delta T \leq 40°$C).
c. The rotation angle θ (deg.) of the bar AB as a function of the temperature increase ΔT (0°C $\leq \Delta T \leq 40°$C).

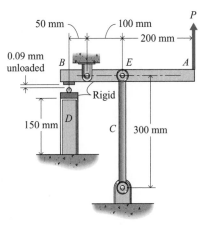

50 mm   100 mm P
200 mm →
0.09 mm unloaded
B E A
Rigid
150 mm D
C 300 mm

Figure P5-76

5-77 A pin-connected structure is loaded and supported as shown in Fig. P5-77. Member CD is rigid; A is a stainless steel [$E = 28,000$ ksi, $\alpha = 6.6(10^{-6})/°$F, and $A = 1.75$ in.^{2}] bar; and B is an aluminum alloy [$E = 10,600$ ksi, $\alpha = 12.5 (10^{-6})/°$F, and $A = 2.25$ in.^{2}] bar. Initially, CD is horizontal and bars A and B are unstressed. If the temperature of the system is slowly raised, calculate and plot

a. The axial stresses σ_A (in the steel bar) and σ_B (in the aluminum bar) as functions of the temperature increase ΔT $(0° \leq \Delta T \leq 100°$F).
b. The changes in length δ_A (of the steel bar) and δ_B (of the aluminum bar) as functions of the temperature increase ΔT $(0° \leq \Delta T \leq 100°$F).

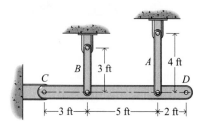

Figure P5-77

5-78 Member $ABCD$ of the pin-connected structure shown in Fig. P5-78 is rigid; bar BF is made of steel [$E = 210$ GPa, $A = 1200$ mm^2, and $\alpha = 11.9(10^{-6})/°$C]; and bar CE is made of an aluminum alloy [$E = 73$ GPa, $A = 900$ mm^2, and $\alpha =$

$22.5(10^{-6})/°$C]. As a result of a misalignment of the pin holes at A, B, and C, bar CE must be heated $80°$C (after pins A and B are in place) to permit insertion of pin C. Calculate and plot

a. The axial stresses σ_{BF} (in the steel bar) and σ_{CE} (in the aluminum bar) as functions of the temperature decrease (as bar CE cools back down to room temperature) ΔT $(-80°$C $\leq \Delta T \leq 0°$C).
b. The changes in length δ_{BF} (in the steel bar) and δ_{CE} (in the aluminum bar) as functions of the temperature decrease (as bar CE cools back down to room temperature) ΔT $(-80°$C $\leq \Delta T \leq 0°$C).

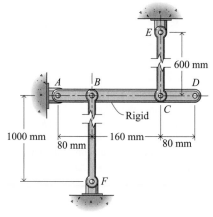

Figure P5-78

5-6 STRESS CONCENTRATIONS

In the foregoing section, it was assumed that the average normal stress given by the expression $\sigma = P/A$ is the significant or critical stress. For many problems, this is true: for other problems, however, the maximum normal stress on a given section may be considerably greater than the average normal stress. For certain combinations of loading and material, the maximum rather than the average is the important stress. If there exists in the structural or machine element a discontinuity that interrupts the stress path (called a *stress trajectory*),[2] the stress at the discontinuity may be considerably greater than the nominal (average, in the case of centric loading) stress on the section; thus, there is a stress concentration at the discontinuity. This is illustrated in Fig. 5-19, in which a type of discontinuity is shown in the upper part of Figs. 5-19a, b, and c and the approximate distribution of normal stress on a transverse plane is shown in the accompanying lower figure. The ratio of the maximum stress to the nominal stress on the section is known

---

[2]A *stress trajectory* is a line everywhere parallel to the maximum normal stress.

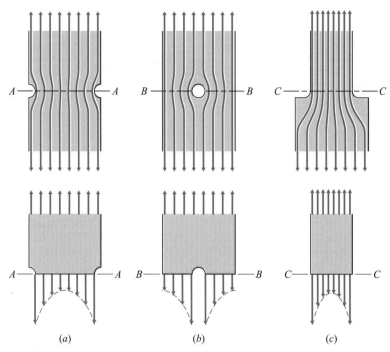

Figure 5-19

as the *stress concentration factor*. Thus, the expression for the maximum normal stress in a centrically loaded member becomes

$$\sigma = K\frac{P}{A} \qquad (5\text{-}5)$$

where A is either the gross area or the net area (area at the reduced section) depending on the value used for K, the stress concentration factor. Curves, similar to the ones shown in Fig. 5-20, can be found in numerous design handbooks. It is important that the user of such curves (or tables of factors) ascertain whether the factors are based on the gross or net section. The factors K_t shown in Figs. 5-20a, b, and c are based on the net section. Sometimes the solution of a problem is expedited by use of the factor K_g based on the gross section, and for this purpose conversion expressions are given with the various curves.

A classic example of the solution of a problem involving a localized redistribution of stress occurs in the case of a small circular hole in a wide plate under uniform unidirectional tension.[3] The theory of elasticity solution is expressed in terms of a radial stress σ_r, a tangential stress σ_θ, and a shearing stress $\tau_{r\theta}$,

---

[3]This solution was obtained by G. Kirsch; see Z. Ver. deut. Ing., Vol. 42, 1898. See also *Elasticity in Engineering Mechanics*, A. P. Boresi and P. P. Lynn, Prentice-Hall, Englewood Cliffs, N.J., 1974, pp. 304–309.

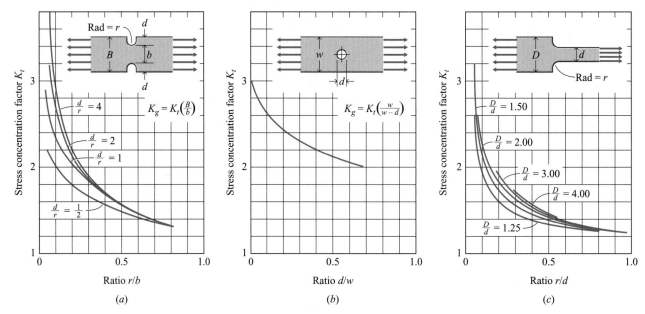

Figure 5-20 Stress concentration factors for grooves, holes, and fillets.

as shown in Fig. 5-21. The equations are

$$\sigma_r = \frac{\sigma}{2}\left(1 - \frac{a^2}{r^2}\right) - \frac{\sigma}{2}\left(1 - \frac{4a^2}{r^2} + \frac{3a^4}{r^4}\right)\cos 2\theta$$

$$\sigma_\theta = \frac{\sigma}{2}\left(1 + \frac{a^2}{r^2}\right) + \frac{\sigma}{2}\left(1 + \frac{3a^4}{r^4}\right)\cos 2\theta \qquad (5\text{-}6)$$

$$\tau_{r\theta} = \frac{\sigma}{2}\left(1 + \frac{2a^2}{r^2} + \frac{3a^4}{r^4}\right)\sin 2\theta$$

On the boundary of the hole (at $r = a$), these equations reduce to

$$\sigma_r = 0$$
$$\sigma_\theta = \sigma(1 + 2\cos 2\theta)$$
$$\tau_{r\theta} = 0$$

At $\theta = 0°$, the tangential stress σ_θ equals 3σ, where σ is the uniform tensile stress in the plate in regions far removed from the hole. Thus, the stress concentration factor associated with this type of discontinuity is 3.

The localized nature of a stress concentration can be evaluated by considering the distribution of tangential stress σ_θ along the x-axis ($\theta = 0°$) in Fig. 5-21. Here

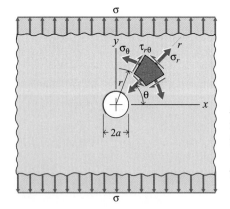

Figure 5-21

$$\sigma_\theta = \frac{\sigma}{2}\left(2 + \frac{a^2}{r^2} + \frac{3a^4}{r^4}\right)$$

At a distance $r = 3a$ (one hole diameter from the hole boundary), this equation yields $\sigma_\theta = 1.074\sigma$. Thus, the stress that began as three times the nominal stress at the boundary of the hole has decayed to a value only 7 percent greater than the nominal stress at a distance of one diameter from the hole. This rapid decay is typical of the redistribution of stress in the neighborhood of a discontinuity.

Stress concentration is not significant in the case of static loading of a ductile material (defined in Section 4-2) because the material will yield inelastically in the region of high stress and, with the accompanying redistribution of stress, equilibrium may be established and no harm done. However, if the load is an impact or repeated load, instead of the action described above, the material may fracture. Also, if the material is brittle, even a static load may cause fracture. Therefore, in the case of impact or repeated loading on any material or static loading on a brittle material, the presence of stress concentration should not be ignored. Before we leave the subject of stress concentration, it should be noted that in regions of support and load application, the stress distribution varies from the nominal value (defined as the stress obtained from elementary theories of stress distribution—uniform for centric loading). This fact was discussed in 1864 by Barre de Saint-Venant (1797–1886), a French mathematician. Saint-Venant observed that although localized distortions in such regions produced stress distributions different from the theoretical distributions, these localized effects disappeared at some distance (the implication being that the distance is not of great magnitude[4]) from such locations. This statement is known as *Saint-Venant's principle* and is constantly used in engineering design.

▮ **Example Problem 5-14** The machine part shown in Fig. 5-22 is 20 mm thick and is made of 0.4 percent carbon hot-rolled steel. Determine the maximum safe load P if the maximum normal stress is not to exceed 144 MPa.

SOLUTION

The maximum normal stress in the machine part will occur either in the fillet between the two sections or on the boundary of the hole. At the fillet,

$$D/d = 90/60 = 1.5$$
$$r/d = 15/60 = 0.25$$

From Fig. 5-20c,

$$K_t \cong 1.62$$

Thus, from Eq. 5-5,

$$P = \sigma A_t/K_t = 144(10^6)(60)(20)(10^{-6})/1.62$$
$$= 106.67(10^3) \text{ N} \cong 106.7 \text{ kN}$$

At the hole,

$$d/w = 27/90 = 0.3$$

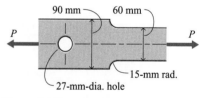

Figure 5-22

▶ Recall that the graphs in Fig. 5-20 are all based on the net area (the smallest area that carries the stress at the discontinuity). Therefore, the correct area to use at the fillet is $A_t = (60 \times 10^{-3})(20 \times 10^{-3})$ m². Similarly, the correct area to use at the hole is $A_t = [(90-27) \times 10^{-3})](20 \times 10^{-3})$ m².

---

[4]For example, it can be shown mathematically that the localized effect of a concentrated load on a beam may disappear at a section slightly greater than the depth of the beam away from the load. See *Theory of Elasticity*, S. Timoshenko and J. N. Goodier, 3rd ed., McGraw-Hill, New York, 1970.

From Fig. 5-20b,

$$K_t \cong 2.30$$

Thus, from Eq. 5-5,

$$P = \sigma A_t / K_t = 144(10^6)(90 - 27)(20)(10^{-6})/2.30$$
$$= 78.89(10^3) \text{ N} \cong 78.9 \text{ kN}$$

Therefore,

$$P_{\text{max}} = 78.9 \text{ kN} \qquad \textbf{Ans.}$$

PROBLEMS

Introductory Problems

5-79* The machine part shown in Fig. P5-79 is 1/4 in. thick and is made of SAE 4340 heat-treated steel. Determine the maximum safe load P if the maximum normal stress is not to exceed 66 ksi.

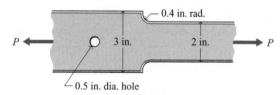

Figure P5-79

5-80 The machine part shown in Fig. P5-80 is 10 mm thick and is made of cold-rolled 18-8 stainless steel. Determine the maximum safe load P if the maximum normal stress is not to exceed 760 MPa.

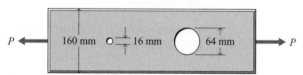

Figure P5-80

Intermediate Problems

5-81 A 1/8-in. thick × 4-in.-wide steel bar is transmitting an axial tensile load of 500 lb. After the load is applied, a 1/64-in.-diameter hole is drilled through the bar as shown in Fig. P5-81.

a. Determine the stress at point A (on the edge of the hole) in the bar before and after the hole is drilled. Use Fig. 5-20b.
b. Determine the stress at point A using Eq. 5-6.
c. Repeat parts a and b if the diameter of the hole is increased to 1 in.
d. Repeat parts a and b if the diameter of the hole is increased to 2 in.

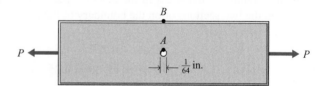

Figure P5-81

5-82* The machine part shown in Fig. P5-82 is 20 mm thick, is made of cold-rolled red brass, and is subjected to a tensile load P of 100 kN. Determine the minimum radius r that can be used between the two sections if the maximum normal stress is not to exceed 205 MPa.

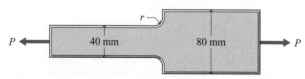

Figure P5-82

Challenging Problems

5-83* The 1/2-in.-thick bar with semicircular ($d/r = 1$) edge grooves, shown in Fig. P5-83, is made of structural steel and will be subjected to an axial tensile load P of 10 kip. Determine the minimum safe width B for the bar if the maximum normal stress is limited to 20 ksi.

5-84 The 25-mm-thick aluminum ($E = 73$ GPa and $G = 28$ GPa) bar shown in Fig. P5-84 is subjected to an axial load P of 180 kN. Determine the components of strain ϵ_x, ϵ_y, and γ_{xy} at point A on the edge of the hole.

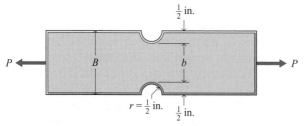

Figure P5-83

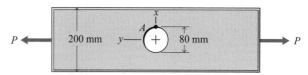

Figure P5-84

5-7 INELASTIC BEHAVIOR OF AXIALLY LOADED MEMBERS

Preceding sections in this chapter of the book covered elastic analyses of axially loaded members. Stress was proportional to strain (Hooke's law applies), and the maximum normal and shearing stresses in the members were not permitted to exceed the corresponding proportional limits (or yield strengths) of the materials. In some design situations, the restriction on elastic behavior is not required and a limited amount of inelastic action can be permitted. The theory of inelastic behavior in axially loaded members is introduced in this chapter.

The stress-strain diagram used to represent the behavior of a specific material was introduced in Section 4-2. Typical diagrams for structural steel and a magnesium alloy were presented in Figs. 4-3a and b. These diagrams were used to define a number of material properties.

The useful small-deformation region of the stress-strain diagram for structural steel can be approximated by the idealized diagram shown in Fig. 5-23. The elastic portion of the diagram is a straight line whose slope is the modulus of elasticity E of the material. The inelastic portion of the diagram is a straight line of zero slope beginning at the yield stress. A material for which the slope is zero in the initial portion of the inelastic region is called an *elastoplastic* material. The stress-strain diagram for structural steel, shown in Fig. 4-3a, indicates that the yield stress is approximately 36 ksi, and the strain at yield is approximately 0.00124, which gives a modulus of elasticity of approximately 29,000 ksi. Figure 4-3a indicates that plastic flow continues with no increase in stress beyond the yield stress until the strain reaches a value of approximately 0.015. Beyond this point (which is beyond the useful small-deformation region of the material) strain hardening begins to occur and an increase in the stress level is required to produce additional strain.

The useful small-deformation region of the stress-strain diagram for an aluminum alloy can be approximated by the idealized diagram shown in Fig. 5-24. The elastic portion of the diagram is a straight line. The inelastic portion of the diagram is a straight line of different slope beginning at the yield stress. A material for which the slope in the initial portion of the inelastic region is not zero is called a *strain-hardening* material. The data presented in Fig. 5-24 indicate that the yield stress for the aluminum alloy are approximately 42 ksi,

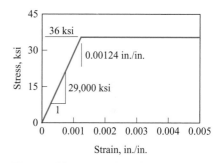

Figure 5-23 Structural steel.

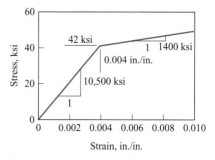

Figure 5-24 Aluminum alloy.

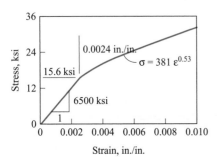

Figure 5-25 Magnesium alloy.

and the strain at yield is approximately 0.004. Thus, the modulus of elasticity for the aluminum alloy is approximately 10,500 ksi. The slope of the diagram in the initial portion of the inelastic region is 1400 ksi; therefore, the increment of stress required to produce a specified increment of strain in the inelastic region is less than it is in the elastic region. It is obvious from the diagram that a strain-hardening material does not permit an increase in strain without an increase in stress.

The useful small-deformation region of the stress-strain diagram for a magnesium alloy can be approximated by the idealized diagram shown in Fig. 5-25. The elastic portion of the diagram is a straight line. The inelastic portion of the diagram cannot be represented adequately by a straight line of different slope beginning at the yield stress; therefore, this nonlinear region is defined by an appropriate mathematical function. The data presented in Fig. 5-25 indicate that the yield stress for the magnesium alloy is approximately 15.6 ksi and the strain at yield is approximately 0.0024. Thus, the modulus of elasticity for the magnesium alloy is approximately 6500 ksi. It is obvious from the diagram that this magnesium alloy is a strain-hardening material that does not permit an increase in strain without an increase in stress.

A statically determinate, axially loaded member will deform elastically until stresses in the member reach the yield point of the material. Once the yield stress for the member is exceeded, the strain in the member must be obtained by referring to the stress-strain diagram for the material. Failure of such members is assumed when a prescribed level of strain is reached.

The elastic analysis of statically indeterminate, axially loaded members was discussed in Section 5-4. The procedure outlined in that section for solving statically indeterminate problems consisted of the following steps:

1. Draw a free-body diagram.
2. Identify the unknown forces.
3. List the independent equations of equilibrium.
4. Write the required number of deformation equations.
5. Solve the equilibrium and deformation equations for the unknown forces.

When the stresses in some members extend into the inelastic range, stress-strain diagrams such as those shown in Figs. 5-23, 5-24, and 5-25 must be used to relate the loads and the deflections before the equations can be solved for the unknown forces. For the case of an elastoplastic material (Fig. 5-23), the force in the member will have a constant value after the yield stress is reached. Once the forces in these members are known, the remaining forces can be determined by using statics methods. For the case of strain-hardening materials (Figs. 5-24 and 5-25), trial-and-error solutions are required because the stress in each of the members depends on the deflection imposed on the member.

The procedure for analyzing axially loaded members with stresses extending into the inelastic range is illustrated in the following example.

Example Problem 5-15   The rigid plate C in Fig. 5-26a is fastened to the 0.50-in.-diameter steel rod A and to the aluminum alloy pipe B. The other ends of A and B are fastened to rigid supports. When the force P is zero, there are no stresses in A and B. Rod A is made of low-carbon steel (assumed to be elastoplastic) with a proportional limit and yield point of 40 ksi and a modulus

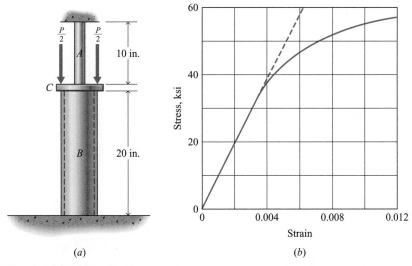

(a)   (b)

Figure 5-26(a–b)

of elasticity of 30,000 ksi. Pipe B has a cross-sectional area of 2.00 in.^{2} and is made of an aluminum alloy with the stress-strain diagram shown in Fig. 5-26b. A load of 30 kip is applied to plate C as shown. Determine the normal stresses in rod A and pipe B and the displacement of plate C.

SOLUTION
Figure 5-26c is a free-body diagram of plate C and portions of members A and B. The free-body diagram contains two unknown forces. Only one equation of equilibrium is available; therefore, the problem is statically indeterminate. As the load is applied to plate C, it moves downward an amount δ, which represents the total deformation in members A and B (see Fig. 5-26d). This observation provides a displacement equation that can be used with the equilibrium equation

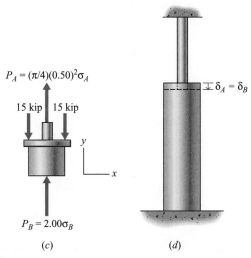

(c)   (d)

Figure 5-26(c–d)

to solve the problem. The two equations are equilibrium

$$+\uparrow \Sigma F_y = 0: \qquad P_A + P_B - 30 = 0$$
$$(\pi/4)(0.50)^2\sigma_A + 2.00\sigma_B - 30 = 0$$
$$0.19635\sigma_A + 2.00\sigma_B - 30 = 0 \qquad\qquad (a)$$

and compatibility

$$\delta_A = \delta_B$$

which can be rewritten using $\delta = \epsilon L$ as

$$\epsilon_A(10) = \epsilon_B(20)$$

In order to solve these equations, the stresses and strains must be related. If the stresses are less than the corresponding proportional limits, Hooke's law can be used. If Hooke's law ($\sigma = E\epsilon$) is assumed to be valid, the compatibility equation becomes

$$\frac{10\sigma_A}{30(10^3)} = \frac{20\sigma_B}{10(10^3)}$$

where $E_B = 10(10^3)$ ksi is the slope of the linear portion of the stress-strain diagram shown in Fig. 5-26b. Thus,

$$\sigma_A = 6\sigma_B \qquad\qquad (b)$$

The simultaneous solution of Eqs. (a) and (b) gives

$$\sigma_A = 56.64 \text{ ksi} \cong 56.6 \text{ ksi (T)}$$
$$\sigma_B = 9.440 \text{ ksi} \cong 9.44 \text{ ksi (C)}$$

These results indicate that bar A is stressed beyond the proportional limit of the steel (40 ksi) and that Hooke's law does not apply. Since the material is assumed to be perfectly plastic beyond the proportional limit, the stress in bar A must be 40 ksi. When this value is substituted in equilibrium equation, Eq. (a), the equation becomes

$$+\uparrow \Sigma F_y = 0: \qquad 0.19635\sigma_A + 2.00\sigma_B - 30 = 0$$
$$0.19635(40) + 2.00\sigma_B - 30 = 0$$

▶ If the stress in B had been more than the proportional limit of the aluminum, it would have been necessary to determine the strain from the curve in Fig. 5-26b. If the material in bar A were not perfectly plastic, a trial-and-error solution would be required, since the stresses in both A and B depend on the movement of C.

from which

$$\sigma_B = 11.073 \text{ ksi} \cong 11.07 \text{ ksi (C)} \qquad\qquad \textbf{Ans.}$$
$$\sigma_A = 40.00 \text{ ksi (T)} \qquad\qquad \textbf{Ans.}$$

The stress in B is less than the proportional limit for aluminum ($\cong 35$ ksi from Fig. 5-26b); therefore, Hooke's law can be used to determine the movement of the plate, which is

$$\delta = \delta_B = \frac{\sigma_B L_B}{E_B} = \frac{11.073(20)}{10(10^3)} = 0.02214 \text{ in.} \cong 0.0221 \text{ in.} \downarrow \qquad \textbf{Ans.}$$

PROBLEMS

Introductory Problems

5-85* Member ABC of Fig. P5-85 is rigid and bar CD is un-stressed before the load P is applied. Bar CD has a cross-sectional area of 1.00 in.^{2} and is made of an aluminum alloy with the idealized stress-strain diagram shown in Fig. 5-24. Determine

a. The normal stress in bar CD when $P = 28{,}000$ lb.
b. The normal stress in bar CD when $P = 35{,}000$ lb.
c. The change in length of bar CD when $P = 35{,}000$ lb.

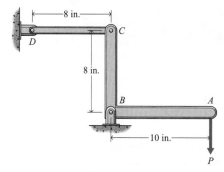

Figure P5-85

5-86* A 20-mm-diameter aluminum alloy bar is subjected to the axial loads shown in Fig. P5-86a. The idealized stress-strain diagram for the aluminum alloy is shown in Fig. P5-86b. Determine

a. The normal stress in each segment of the bar for $P = 20$ kN.
b. The normal stress in each segment of the bar for $P = 65$ kN.

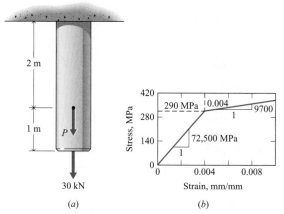

(a)   (b)

Figure P5-86

5-87 The stepped bar shown in Fig. P5-87 has a 0.75-in. diameter over half its length and a 1.25-in. diameter over the other half. The bar is made of an aluminum alloy with the idealized stress-strain diagram shown in Fig. 5-24. Determine

a. The change in length of the bar when $P = 10{,}000$ lb.
b. The change in length of the bar when $P = 20{,}000$ lb.

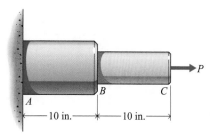

Figure P5-87

Intermediate Problems

5-88* The rigid bar CD of Fig. P5-88 is horizontal, and bars A and B are unstressed before load P is applied. Bar A has a cross-sectional area of 500 mm^2 and is made of an aluminum alloy that has a proportional limit of 330 MPa and a modulus of elasticity of 73 GPa. Bar B has a cross-sectional area of 750 mm^2 and is made of a low-carbon steel (elastoplastic) that has a proportional limit and yield point of 275 MPa and a modulus of elasticity of 210 GPa. Determine

a. The normal stresses in bars A and B after load P is applied.
b. The shearing stress in the 30-mm-diameter pin at C, which is in double shear.
c. The maximum shearing stresses in bars A and B.
d. The vertical component of the displacement of pin D.

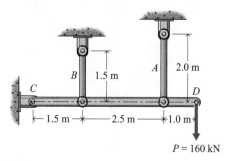

Figure P5-88

5-89* The rigid, weightless bar shown in Fig. P5-89 is supported by two steel (yield strength = 100 ksi, $E = 29{,}000$ ksi, and $A = 1.6$ in.^{2}) cables and an aluminum (yield strength = 70 ksi, $E = 10{,}600$ ksi, and $A = 3.2$ in.^{2}) cable. Determine the largest load P that can be applied without any cable exceeding its yield strength.

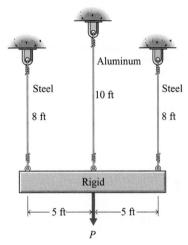

Figure P5-89

5-90 Bar A of Fig. P5-90 is made of an aluminum alloy that has a proportional limit of 380 MPa and a modulus of elasticity of 72 GPa. Bars B are made of structural steel (elastoplastic) that has a proportional limit and yield point of 250 MPa and a modulus of elasticity of 200 GPa. All bars have cross-sectional areas of 1500 mm^2. For $L = 900$ mm, determine

a. The normal stresses in bars A and B after a 1110-kN load P is applied.
b. The vertical displacement (deflection) of pin C produced by the 1110-kN load.

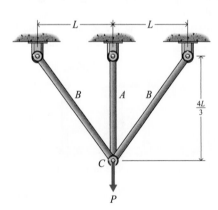

Figure P5-90

Challenging Problems

5-91* Bar A of Fig. P5-91 is made of an aluminum alloy with a proportional limit of 55 ksi and a modulus of elasticity of 10,500 ksi. Bar B is made of structural steel (elastoplastic) with a proportional limit and yield point of 36 ksi and a modulus of elasticity of 29,000 ksi. Both bars have cross-sectional areas of 1.5 in.^{2} Determine

a. The normal stresses in bars A and B after a 50-kip load P is applied.
b. The shearing stress in the 1.00-in.-diameter pin at C, which is in double shear.

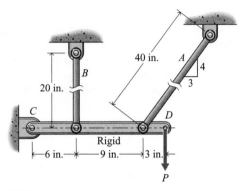

Figure P5-91

5-92 An aluminum circular rod is securely fastened to a steel tube, and the assembly is subjected to an axial load, as shown in Fig. P5-92a. The cross-sectional area of each member of the assembly is 315 mm^2. The stress-strain diagram for the two materials is shown in Fig. P5-92b. Determine the displacement of the lower end of the assembly when $P = 530$ kN.

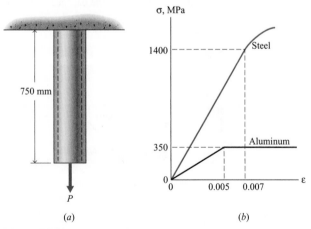

Figure P5-92

5-93 The rigid bar CD of Fig. P5-93 is horizontal, and bars A and B are unstressed before the load P is applied. Bar A has a cross-sectional area of 2.00 in.^{2} and is made of an aluminum alloy that has the stress-strain diagram shown in Fig. 5-26b. Bar B has a cross-sectional area of 2.50 in.^{2} and is made of a low-carbon steel (elastoplastic) that has a proportional limit and yield point of 36 ksi and a modulus of elasticity of 30,000 ksi. Determine

a. The normal stress in each of the bars after a 40-kip load P is applied.
b. The normal stress in each of the bars after a 60-kip load P is applied.
c. The vertical displacement (deflection) of C after a 65-kip load P is applied.
d. The axial strain in bar A when $P = 40$ kip.
e. The axial strain in bars A and B when $P = 65$ kip.

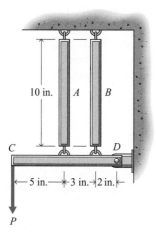

Figure P5-93

Computer Problems

5-94 The rigid bar CDE of Fig. P5-94 is horizontal, and bars A and B are unstressed before the load P is applied. Bar A has a cross-sectional area of 500 mm^2 and is made of an aluminum alloy that has a proportional limit of 330 MPa and a modulus of elasticity of 73 GPa. Bar B has a cross-sectional area of 750 mm^2 and is made of a low-carbon steel (elastoplastic) that has a proportional limit and a yield point of 275 MPa and a modulus of elasticity of 210 GPa. Calculate and plot

a. The axial stresses σ_A (in the aluminum bar) and σ_B (in the steel bar) as functions of the load P ($0 \text{ kN} \leq P \leq 260$ kN).
b. The changes in length δ_A (of the aluminum bar) and δ_B (of the steel bar) as functions of the load P ($0 \text{ kN} \leq P \leq 260$ kN).

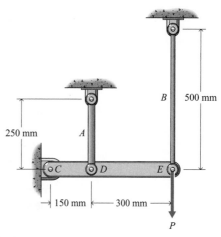

Figure P5-94

5-95 A 3/4-in.-diameter bolt passes through a sleeve, as shown in Fig. P5-95a. As the nut is tightened, it advances a distance of 0.125 in. along the bolt for each complete turn of the nut. The bolt is made of low-carbon steel (assumed to be elastoplastic) with a proportional limit and a yield point of 40 ksi and a modulus of elasticity of 30,000 ksi. The sleeve has a cross-sectional area of 0.40 in.^{2} and is made of an aluminum alloy with the stress-strain diagram shown in Fig. P5-95b. Calculate and plot

a. The axial stresses σ_{st} (in the steel bolt) and σ_{al} (in the aluminum sleeve) as functions of the angle of twist θ of the nut ($0° \leq \theta \leq 210°$).
b. The changes in length δ_{st} (of the steel bolt) and δ_{al} (of the aluminum sleeve) as functions of the angle of twist θ of the nut ($0° \leq \theta \leq 210°$).
c. The distance L between the two washers as functions of the angle of twist θ of the nut ($0° \leq \theta \leq 210°$).

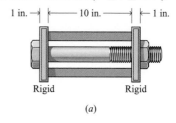

(a)

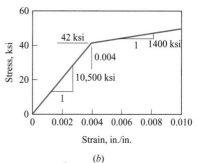

(b)

Figure P5-95

5-8 THIN-WALLED PRESSURE VESSELS

A pressure vessel is described as thin walled when the ratio of the wall thickness to the radius of the vessel is so small that the distribution of normal stress on a plane perpendicular to the surface of the vessel is essentially uniform throughout the thickness of the vessel. Actually, this stress varies from a maximum value at the inside surface to a minimum value at the outside surface of the vessel, but it can be shown that if the ratio of the wall thickness to the inner radius of the vessel is less than 0.1, the maximum normal stress is less than 5 percent greater than the average. Boilers, gas storage tanks, pipelines, metal tires, and hoops are normally analyzed as thin-walled elements. Gun barrels, certain high-pressure vessels in the chemical processing industry, and cylinders and piping for heavy hydraulic presses need to be treated as thick-walled vessels.

Problems involving thin-walled vessels subjected to liquid (or gas) pressure p are readily solved with the aid of free-body diagrams of sections of the vessels together with the fluid contained therein. In the following subsections, spherical, cylindrical, and other thin shells of revolution are considered.

5.8

1. **Spherical Pressure Vessels**. A typical thin-walled spherical pressure vessel used for gas storage is shown in Fig. 5-27. If the weights of the gas and vessel are negligible (a common situation), symmetry of loading and geometry requires that stresses on sections that pass through the center of the sphere be equal. Thus, on the small triangular element shown in Fig. 5-28a,

$$\sigma_x = \sigma_y = \sigma_n$$

Figure 5-27 Hortonsphere for gas storage in Superior, Wisconsin. (Courtesy of Chicago Bridge and Iron Co., Chicago, Ill.)

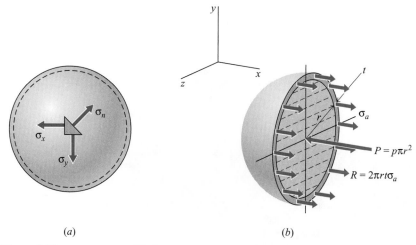

(a)   (b)

Figure 5-28

Furthermore, there are no shearing stresses on any of these planes because there are no loads to induce them. The normal stress component in a sphere is known as a *meridional* or *axial* stress and is commonly denoted as σ_m or σ_a.

The free-body diagram shown in Fig. 5-28b can be used to evaluate the stress $\sigma_x = \sigma_y = \sigma_n = \sigma_a$ in terms of the pressure p, and the radius r and thickness t of the spherical vessel. The force R is the resultant of the internal forces that act on the cross-sectional area of the sphere, which is exposed by passing a plane through the center of the sphere. The force P is the resultant of the fluid forces acting on the fluid remaining within the hemisphere. From a summation of forces in the x-direction,

$$R - P = 0$$

or

$$2\pi r t \sigma_a = p\pi r^2$$

from which

$$\sigma_a = \frac{pr}{2t} \tag{5-7}$$

2. **Cylindrical Pressure Vessels**. A typical thin-walled cylindrical pressure vessel used for gas storage is shown in Fig. 5-29. Normal stresses, such as those shown on the small element of Fig. 5-30a, are easy to evaluate by using appropriate free-body diagrams. The normal stress component on a transverse plane is known as an axial or meridional stress and is commonly denoted as σ_a or σ_m. The normal stress component on a longitudinal plane is known as a *hoop, tangential,* or *circumferential* stress and is denoted as σ_h, σ_t, or σ_c. There are no shearing stresses on transverse or longitudinal planes.

5.9

The free-body diagram used for axial stress determination is similar to Fig. 5-28b, which was used for the sphere, and the results are the same. The free-body diagram used for the hoop stress determination is shown in Fig. 5-30b.

Figure 5-29 Cylindrical tanks for gas storage in Northlake, Illinois. (Courtesy of Chicago Bridge and Iron Co., Chicago Ill.)

The force P_x is the resultant of the fluid forces acting on the fluid remaining within the portion of the cylinder isolated by the longitudinal plane and two transverse planes. The forces Q are the resultant of the internal forces on the cross-sectional area exposed by the longitudinal plane containing the axis of the cylinder. From a summation of forces in the x-direction,

$$2Q - P_x = 0$$

or

$$2\sigma_h Lt = p2rL$$

from which

$$\sigma_h = \frac{pr}{t} \tag{5-8a}$$

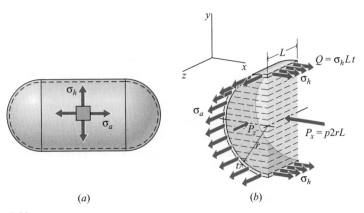

(a)

(b)

Figure 5-30

Also (see Eq. 5-7),

$$\sigma_a = \frac{pr}{2t} \qquad (5\text{-}8b)$$

The previous analysis of the stresses in a cylindrical vessel subjected to uniform internal pressure indicates that the stress on a longitudinal plane is twice the stress on a transverse plane. Consequently, a longitudinal joint needs to be twice as strong as a transverse (or girth) joint.

3. **Thin Shells of Revolution.** The discussion in the two previous subsections was limited to thin-walled cylindrical and spherical vessels under uniform internal pressure. The theory can be extended to include other shapes and other loading conditions. Consider, for example, the thin shell of revolution shown in Fig. 5-31. Such shells are generated by rotating a plane curve, called the *meridian*, about an axis lying in the plane of the curve. Shapes that can be formed in this manner include the sphere, hemisphere, torus (doughnut), cylinder, cone, and ellipsoid. In shells of revolution, the two unknown principal stresses are a meridional stress σ_m that acts on a plane perpendicular to the meridian and a tangential stress σ_t that acts on a plane perpendicular to a parallel. Both of these stresses are shown on the small element of Fig. 5-31. The two stresses can be evaluated by using two equilibrium equations.

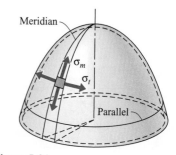

Figure 5-31

The small element of Fig. 5-31 is shown enlarged in Fig. 5-32. The element has a uniform thickness, is subjected to an internal pressure p, and has different curvatures in two orthogonal directions. The resultant forces on the various surfaces are shown on the diagram. Summing forces in the n-direction gives

$$P - 2F_m \sin d\theta_m - 2F_t \sin d\theta_t = 0$$

from which

$$p(2r_t d\theta_t)(2r_m d\theta_m) = 2\sigma_m(2tr_t d\theta_t) \sin d\theta_m + 2\sigma_t(2tr_m d\theta_m) \sin d\theta_t$$

and since, for small angles, $\sin d\theta \cong d\theta$, the above equation becomes

$$\frac{\sigma_m}{r_m} + \frac{\sigma_t}{r_t} = \frac{p}{t} \qquad (5\text{-}9)$$

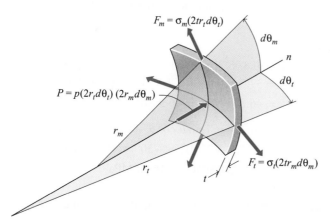

Figure 5-32

In Eq. 5-9, r_m is the radius of curvature of the meridian, σ_m is the meridional stress, r_t is the curvature of the parallel, σ_t is the circumferential stress, and t is the thickness of the wall of the thin shell of revolution.

Since Eq. 5-9 contains two unknown stresses, an additional independent equation is needed. Such an equation can be obtained by considering equilibrium of a portion of the vessel above or below the parallel that passes through the point of interest. Application of Eq. 5-9 is illustrated in Example Problem 5-17.

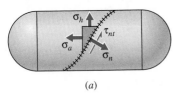

(a)

Figure 5-33(a)

Example Problem 5-16   A cylindrical pressure vessel with an inside diameter of 1.50 m is constructed by wrapping a 15-mm-thick steel plate into a spiral and butt-welding the mating edges of the plate, as shown in Fig. 5-33a. The butt-welded seams form an angle of 30° with a transverse plane through the cylinder. Determine the normal stress σ_n perpendicular to the weld and the shearing stress τ_{nt} parallel to the weld when the internal pressure in the vessel is 1500 kPa.

SOLUTION
The hoop stress σ_h and the axial stress σ_a in the cylinder can be determined by using Eqs. 5-8a and b. Thus,

$$\sigma_h = \frac{pr}{t} = \frac{1500(10^3)(0.75)}{0.015} = 75.0(10^6) \text{ N/m}^2 = 75.0 \text{ MPa}$$

$$\sigma_a = \frac{pr}{2t} = \frac{1500(10^3)(0.75)}{2(0.015)} = 37.5(10^6) \text{ N/m}^2 = 37.5 \text{ MPa}$$

The normal stress σ_n perpendicular to the weld and the shearing stress τ_{nt} parallel to the weld can be determined by using the stress transformation equations, Eqs. 2-12 and 2-13. The stresses for use in these equations are

$$\sigma_x = \sigma_a = +37.5 \text{ MPa} \qquad \sigma_y = \sigma_h = +75.0 \text{ MPa} \qquad \tau_{xy} = 0$$

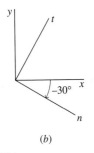

(b)

Figure 5-33(b)

► The longitudinal and hoop directions are principal directions, and these surfaces are free of shear stress. However, the weld seam (and any other surface not aligned with the axial or hoop directions) will be subjected to both a normal stress and a shear stress.

The angle θ for the plane parallel to the weld is $-30°$, as shown in Fig. 5-33b. Thus, from Eqs. 2-12 and 2-13,

$$\sigma_n = \sigma_x \cos^2 \theta + \sigma_y \sin^2 \theta + 2\tau_{xy} \sin \theta \cos \theta$$
$$= 37.5 \cos^2(-30°) + 75.0 \sin^2(-30°) + 0$$
$$= 46.875 \text{ MPa} \cong 46.9 \text{ MPa (T)} \qquad \textbf{Ans.}$$

$$\tau_{nt} = -(\sigma_x - \sigma_y) \sin \theta \cos \theta + \tau_{xy}(\cos^2 \theta - \sin^2 \theta)$$
$$= -(37.5 - 75.0) \sin(-30°) \cos(-30°) + 0$$
$$= -16.238 \text{ MPa} \cong -16.24 \text{ MPa} \qquad \textbf{Ans.}$$

The minus sign indicates that the direction of the shearing stress τ_{nt} is opposite to that shown in Fig. 5-33a.

Example Problem 5-17   A pressure vessel of 1/4-in. steel plate has the shape of a paraboloid closed by a thick, flat plate, as shown in Fig. 5-34a. The equation of the generating parabola is $y = x^2/4$, where x and y are in inches. Determine the meridional and tangential stresses σ_m and σ_t in the shell at a point 16 in. above the bottom of the vessel due to an internal gas pressure of 250 psi.

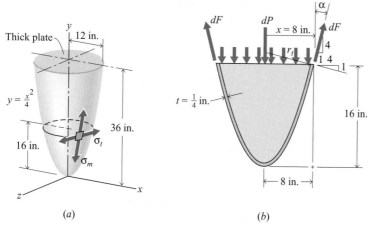

(a)   (b)

Figure 5-34

SOLUTION

Determining σ_m

The meridional stress σ_m, which must be tangent to the shell, can be determined with the aid of the free-body diagram shown in Fig. 5-34b. From the equation of the parabola, the radius x and the slope of the shell dy/dx at $y = 16$ in. are determined to be 8 in. and 4/1, respectively.

Summing forces in the y-direction gives

$$-\int_{A_P} dP + \int_{A_\sigma} dF \cos\alpha = 0$$

$$-p\pi x^2 + \sigma_m 2\pi x t \cos\alpha = 0$$

Substituting the given data yields

$$-250(\pi)(8^2) + \sigma_m(2\pi)(8)(1/4)(4/\sqrt{17}) = 0$$

from which

$$\sigma_m = 4123 \text{ psi} \cong 4120 \text{ psi (T)} \qquad\qquad \textbf{Ans.}$$

Determining σ_t

In order to find σ_t from Eq. 5-9, the radii r_m and r_t at the point must be determined. The radius of curvature r_m of the shell in the xy plane is determined from the expression

$$r_m = \frac{[1 + (dy/dx)^2]^{1.5}}{d^2y/dx^2} = \frac{(1 + 4^2)^{1.5}}{1/2} = 140.19 \text{ in.}$$

and the perpendicular radius r_t is found from the geometry of Fig. 5-34b as

$$r_t = 8(\sqrt{17}/4) = 8.246 \text{ in.}$$

Then, from Eq. 5-9

$$\frac{p}{t} = \frac{\sigma_m}{r_m} + \frac{\sigma_t}{r_t}$$

$$\frac{250}{1/4} = \frac{4123}{140.19} + \frac{\sigma_t}{8.246}$$

from which

$$\sigma_t = 8003 \text{ psi} \cong 8000 \text{ psi (T)} \qquad \textbf{Ans.}$$

PROBLEMS

MecMovie Activities and Problems

MM5.9 Cylindrical pressure vessels – weld stresses. Example; Try one. Example finding normal and shear stresses produced on a weld seam as a result of internal pressure in a cylindrical tank.

MM5.10 Strain gage to measure tank pressure. Example; Try one. Application of Hooke's Law for biaxial stress and strain transformation equations to determine tank pressure in a cylinder.

MM5.11 Critical strain reading for pressure tank. Example; Try one. Use generalized Hooke's Law and 3D Mohr's circle to find strain corresponding to a specified shear stress.

Introductory Problems

5-96* Determine the maximum normal stress in a 300-mm-diameter basketball that has a 2-mm wall thickness after it has been inflated to a pressure of 100 kPa.

5-97* A steel pipe with an inside diameter of 10 in. will be used to transmit steam under a pressure of 800 psi. If the hoop stress in the pipe must be limited to 10 ksi because of a longitudinal weld in the pipe, determine the minimum satisfactory thickness for the pipe.

5-98 A cylindrical propane tank, similar to the ones shown in Fig. 5-29, has an outside diameter of 3.25 m and a wall thickness of 22 mm. If the allowable hoop stress is 100 MPa and the allowable axial stress is 45 MPa, determine the maximum internal pressure that can be applied to the tank.

5-99 A spherical gas storage tank, similar to the one shown in Fig. 5-27, has a diameter of 35 ft and a wall thickness of 7/8 in.

Determine the maximum normal stress in the tank if the gas pressure is 100 psi.

Intermediate Problems

5-100* A steel boiler 1 m in diameter is welded using a spiral seam that makes an angle of 30° with respect to a transverse plane of the boiler, as shown in Fig. P5-100. For an internal pressure of 950 kPa and a wall thickness of 50 mm, determine

a. The normal stress perpendicular to the weld.
b. The shearing stress parallel to the weld.

Figure P5-100

5-101* A cylindrical pressure vessel with an internal diameter of 6 ft and a wall thickness of 0.5 in. is subjected to an internal pressure of 200 psi. The vessel is made of steel having a modulus of elasticity of 29,000 ksi and a shear modulus of 11,000 ksi. Determine

a. The axial and hoop stresses in the vessel.
b. The axial and hoop strains in the vessel.

5-102 The cylindrical tank shown in Fig. P5-102 is 20 m in diameter, is made of structural steel ($E = 200$ GPa), and will be used to store stove oil with a density of 850 kg/m³. Determine the minimum wall thickness required if the maximum normal stress is not to exceed 80 MPa.

Figure P5-102

5-103 A standpipe 12 ft in diameter and 50 ft tall is being constructed for use as a storage tank for water with a specific weight of 62.4 lb/ft³ (Fig. P5-102). The wall thickness of the vessel is 0.5 in.

a. Determine the axial and hoop stresses at the lower end of the vessel. Neglect stress concentrations.
b. Determine the axial and hoop stresses at a point 25 ft above the bottom of the vessel.

Challenging Problems

5-104* A cylindrical pressure vessel is fabricated by butt-welding 20-mm-thick steel ($E = 200$ GPa and $G = 76$ GPa) plate with a spiral seam, as shown in Fig. P5-104. The pressure in the tank is 2800 kPa. Determine

a. The stresses parallel and perpendicular to the weld.
b. The axial and hoop strains.
c. The maximum shearing stress at a point on the outside surface of the vessel.
d. The maximum shearing stress at a point on the inside surface of the vessel.

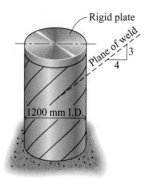

Figure P5-104

5-105* The strains measured on the outside surface of the cylindrical pressure vessel shown in Fig. P5-105 are $\epsilon_1 = +619$ μin./in. and $\epsilon_2 = +330$ μin./in. The angle $\theta = 30°$. The outside diameter of the vessel is 20 in., and the wall thickness is

1/8 in. The vessel is made of 0.4 percent carbon hot-rolled steel ($E = 30,000$ ksi and $G = 11,600$ ksi). Determine

a. The stresses σ_1 and σ_2 in the vessel.
b. The internal pressure applied to the vessel.

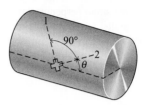

Figure P5-105

5-106 An internal pressure p is applied to the thin-walled toroidal shell (pressurized doughnut) shown in Fig. P5-106. Determine the axial and hoop stresses σ_a and σ_h at points A and B on the horizontal plane of symmetry of the shell in terms of p, R, r, and t. *Hint:* the hoop stress σ_h at point A can be determined by using a free body consisting of one quarter of the shell, such as the part shown cross-hatched in Fig. P5-106.

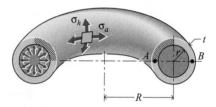

Figure P5-106

5-107 A hemispherical tank of radius r and thickness t is supported by a flange, as shown in the cross section of Fig. P5-107. The tank is filled with a fluid having a specific weight γ. Determine, in terms of γ, r, and t, the meridional and tangential stresses σ_m and σ_t at a depth $y = r/2$.

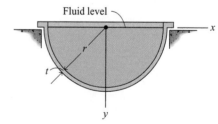

Figure P5-107

Computer Problems

5-108 A 1200-mm-diameter cylindrical pressure tank is fabricated by butt-welding 20-mm-thick plate with a spiral seam as shown in Fig. P5-108. If the seam angle is $\theta = 37°$, compute and plot the normal stress perpendicular to the weld and the shearing stress parallel to the weld as functions of the internal pressure p (10 kPa $\leq p \leq$ 2800 kPa).

5-109 A 4-ft-diameter cylindrical pressure tank is fabricated by butt-welding 3/4-in.-thick plate with a spiral seam as shown in Fig. P5-108. The pressure in the tank is 200 psi. Compute and plot the normal stress perpendicular to the weld and the shearing stress parallel to the weld as functions of the seam angle θ ($0° \leq \theta \leq 60°$).

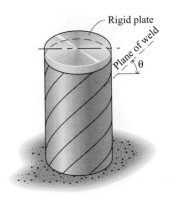

Figure P5-108

5-9 COMBINED EFFECTS—AXIAL AND PRESSURE LOADS

Previously, in Chapter 4, we found that an axial load produces a constant normal stress on a transverse plane; the axial stress is given by $\sigma = F/A$. A pressure load acting on a thin-walled cylindrical pressure vessel produces normal stresses on both transverse and longitudinal planes; the stresses are $\sigma_a = pr/2t$ and $\sigma_h = pr/t$, respectively (Section 5-8).

Consider an axial load and a pressure load acting on a thin-walled pressure vessel simultaneously, as shown in Fig. 5-35a, where P is the axial load and p is the pressure load. Both of these loads produce a normal stress on the x-plane (a transverse plane) while only the pressure load produces a normal stress on the y-plane (a longitudinal plane). For a point on the outside surface of the thin-walled vessel, the stresses are shown on a three-dimensional element in Fig. 5-35b. The

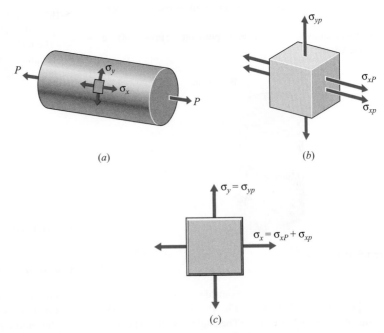

Figure 5-35

stress σ_{xP} is due to the axial load, and the stresses $\sigma_{xp}\,(=\sigma_a)$ and $\sigma_{yp}\,(=\sigma_h)$ are due to the pressure. The x-components may be added since both are tensile stresses and act on the same plane at the same point. Since a point on the outside of the vessel is a free surface, $\sigma_z = 0$; a plane stress element showing the stresses is shown in Fig. 5-35c, where $\sigma_x = \sigma_{xP} + \sigma_{xp}$. There are no shear stresses on the element shown in Fig. 5-35c as there are no loads to produce them. Thus, the xy-components of stress (Fig. 5-35c) are also principal stresses. Example Problem 5–18 illustrates the procedures discussed.

Example Problem 5-18   The cylindrical pressure tank shown in Fig. 5-36a has an inside diameter of 4 ft and a wall thickness of 3/4 in. The pressure in the tank is 400 psi. An additional axial load of 30,000 lb is applied to the top end of the tank through a rigid bearing plate. Determine

(a) The stresses σ_x, σ_y, and τ_{xy} on a stress element at point A, which is on the outside surface of the tank.

(b) The normal and shearing stresses on an inclined plane oriented at $+30°$ from the x-axis.

SOLUTION

(a) The stresses on element A are due to the combined effects of internal pressure and axial load. The internal pressure produces tensile normal (hoop and axial) stresses $\sigma_h = \sigma_x$ and $\sigma_a = \sigma_y$. The axial load produces a compressive normal stress σ_y. There are no shearing stresses on transverse or longitudinal planes. Since element A is on the outside surface of the tank, it is subjected to a state of plane stress.

Stresses due to internal pressure:

$$\sigma_x = \sigma_h = \frac{pr}{t} = \frac{400(2)(12)}{0.75} = 12{,}800 \text{ psi (T)}$$

$$\sigma_y = \sigma_a = \frac{pr}{2t} = \frac{400(2)(12)}{2(0.75)} = 6400 \text{ psi (T)}$$

Stresses due to axial load:

$$\sigma_y = \frac{F}{A} = \frac{-30{,}000}{(\pi/4)(49.5^2 - 48^2)} = -261.2 \text{ psi} = 261.2 \text{ psi (C)}$$

The stresses σ_x and σ_y at point A, obtained by superimposing the results for the two forms of loading, are

$$\sigma_x = 12{,}800 \text{ psi (T)} \qquad\qquad \textbf{Ans.}$$

$$\sigma_y = 6400 - 261.2 = 6138.8 \text{ psi} \cong 6140 \text{ psi (T)} \qquad\qquad \textbf{Ans.}$$

These results are shown on the stress element of Fig. 5-36b.

(b) The stresses on the inclined plane, shown in Fig. 5-36c, are found using the stress transformation equation, Eqs. 2-12a and 2-13a. The stresses and

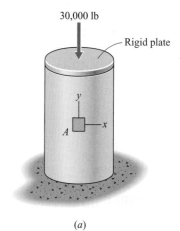

30,000 lb

Rigid plate

A

(a)

Figure 5-36(a)

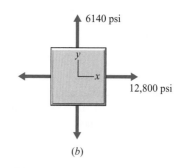

6140 psi

12,800 psi

(b)

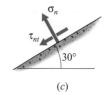

σ_n

τ_{nt}

30°

(c)

Figure 5-36(b–c)

angle θ to be used in the stress transformation equations are

$$\sigma_x = +12,800 \text{ psi} \qquad \tau_{xy} = 0 \text{ psi}$$
$$\sigma_y = +6140 \text{ psi} \qquad \theta = +120°$$

Thus,

$$\sigma_n = \sigma_x \cos^2 \theta + \sigma_y \sin^2 \theta + 2\tau_{xy} \sin \theta \cos \theta$$
$$= 12,800 \cos^2 (120°) + 6140 \sin^2 (120°) + 0$$
$$= 7805 \text{ psi} \cong 7810 \text{ psi} \qquad \text{Ans.}$$
$$\tau_{nt} = -(\sigma_x - \sigma_y) \sin \theta \cos \theta + \tau_{xy}(\cos^2 \theta - \sin^2 \theta)$$
$$= -(12,800 - 6140) \cos (120°) \sin (120°) + 0$$
$$= 2884 \text{ psi} \cong 2880 \text{ psi} \qquad \text{Ans.}$$

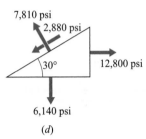

7,810 psi
2,880 psi
30°   12,800 psi
6,140 psi
(d)

Figure 5-36(d)

The xy and nt stresses are shown on the triangular element of Fig. 5-36d.

PROBLEMS

Introductory Problems

5-110* The thin-walled cylindrical pressure vessel shown in Fig. P5-110 is subjected to the axial load P of 40 kN and an internal pressure p of 2 MPa. The vessel has an inside diameter of 1000 mm and a wall thickness of 20 mm. Determine the stresses σ_x, σ_y, and τ_{xy}, and show them on a rectangular stress element.

Figure P5-110

5-111 The pressure vessel shown in Fig. P5-111 has an inside diameter of 3 ft and a wall thickness of 0.375 in. The internal pressure is 300 psi. Determine the maximum load P that may be applied if the maximum axial normal stress cannot exceed 18 ksi.

Figure P5-111

Intermediate Problems

5-112 A cylindrical pressure vessel is fabricated by butt-welding 20-mm-thick plate with a spiral seam, as shown in Fig. P5-112.

The pressure in the tank is 2800 kPa, and an axial load of 130 kN is applied to the end of the tank through a rigid bearing plate. Determine

a. The normal stress perpendicular to the weld.
b. The shearing stress parallel to the weld.
c. The maximum shearing stress at a point on the outside surface of the vessel.
d. The maximum shearing stress at a point on the inside surface of the vessel.

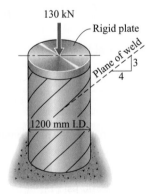

130 kN
Rigid plate
Plane of weld
3
4
1200 mm I.D.

Figure P5-112

5-113* The thin-walled cylindrical pressure vessel shown in Fig. P5-113 has an inside diameter of 40 in. and a wall thickness of 0.4 in. The axial force P is 10,000 lb. The normal stress at point A on plane B–B (which is perpendicular to the surface of the plate at A) is 11,000 psi. Determine the air pressure in the tank.

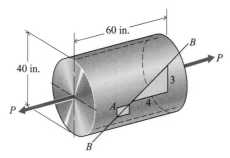

Figure P5-113

Challenging Problems

5-114* A thin-walled cylindrical pressure vessel is made of steel ($E = 200$ GPa and $G = 76$ GPa), has an inside diameter of 1000 mm, has a wall thickness of 20 mm, and is subjected to an axial tensile load of 50 kN and an internal pressure of 2.5 kPa. Determine the axial and hoop strains at a point on the outside surface of the vessel.

5-115 A thin-walled pressure vessel, similar to the one shown in Fig. P5-111, has an inside diameter of 4 ft and a wall thickness of 0.25 in. The internal pressure is 100 psi, and the axial tensile load is 5000 lb. Determine the maximum shear stress at points on the outside and inside surfaces of the vessel.

Computer Problems

5-116 A 1200-mm-diameter cylindrical pressure tank is fabricated by butt-welding 20-mm-thick plate with a spiral seam as shown in Fig. P5-116. An axial load of 130 kN is applied to the end of the tank through a rigid bearing plate. If the seam angle is $\theta = 37°$, compute and plot the normal stress perpendicular to the weld and the shearing stress parallel to the weld as functions of the internal pressure p (10 kPa $\leq p \leq$ 2800 kPa).

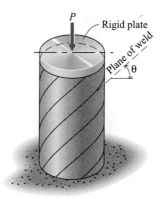

Figure P5-116

5-117 A 4-ft-diameter cylindrical pressure tank is fabricated by butt-welding 3/4-in.-thick plate with a spiral seam as shown in Fig. P5-116. The pressure in the tank is 200 psi, and an axial load of 30,000 lb is applied to the end of the tank through a rigid bearing plate. Compute and plot the normal stress perpendicular to the weld and the shearing stress parallel to the weld as functions of the seam angle θ ($0° \leq \theta \leq 60°$).

5-10 THICK-WALLED CYLINDRICAL PRESSURE VESSELS

The problem of determining the tangential stress σ_t and the radial stress σ_r at any point in the wall of a thick-walled cylinder, in terms of the pressures applied to the cylinder, was solved by the French elastician G. Lame in 1833. The results can be applied to a wide variety of design situations involving cylindrical pressure vessels, hydraulic cylinders, piping systems, and shrink- and press-fit applications.

Consider a thick-walled cylinder having an inner radius a and an outer radius b, as shown in Fig. 5-37a. The cylinder is subjected to an internal pressure p_i and an external pressure p_o. For purposes of analysis, the thick-walled cylinder can be considered to consist of a series of thin rings. A typical ring located at a radial distance ρ from the axis of the cylinder and having a thickness $d\rho$ is shown by the dashed lines in Fig. 5-37a. As a result of the internal and external pressure loadings, a radial stress σ_r would develop at the interface between rings located at radial position ρ, while a slightly different radial stress $(\sigma_r + d\sigma_r)$ would develop at radial position $(\rho + d\rho)$. These stresses would be uniformly distributed over the inner and outer surfaces of the ring, as shown in Fig. 5-37b. Shearing stresses

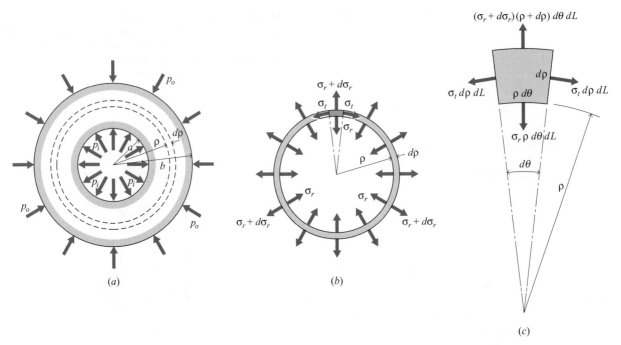

Figure 5-37

would not develop on the inner and outer surfaces of the ring, since the pressure loadings do not tend to force the rings to rotate with respect to one another.

In Section 5-8, which dealt with thin-walled pressure vessels, it was shown that a tangential or hoop component of stress develops when a pressure difference exists between the inner and outer surfaces of a thin shell or ring. The planes on which these tangential stresses act can be exposed by considering only a small part of a ring, as shown shaded in Fig. 5-37b. Since the ring is assumed to be thin, the tangential stress σ_t can be considered to be uniformly distributed through the thickness of the ring. A relationship between radial stress σ_r and tangential stress σ_t can be obtained from equilibrium considerations. A free-body diagram of a small part of a ring, such as the one shown in Fig. 5-37c for the shaded part of Fig. 5-37b, is useful for this determination. The axial stress σ_a, which may be present in the cylinder, has been omitted from this diagram since it does not contribute to equilibrium in the radial or tangential directions. The free-body diagram is assumed to have a length dL along the axis of the cylinder.

From a summation of forces in the radial direction,

$$(\sigma_r + d\sigma_r)(\rho + d\rho)\, d\theta\, dL - \sigma_r \rho\, d\theta dL - 2\sigma_t\, d\rho\, dL \sin\frac{d\theta}{2} = 0 \qquad (a)$$

By neglecting higher-order terms and noting that for small angles $\sin d\theta/2 \cong d\theta/2$, Eq. ($a$) can be reduced to

$$\rho\frac{d\sigma_r}{d\rho} + \sigma_r - \sigma_t = 0 \qquad (b)$$

Equation (b) cannot be integrated directly, since both σ_r and σ_t are functions of radial position ρ. In previous instances, when such statically indeterminate situations were encountered, the problem was solved by considering deformations of the structure.

For the case of a thick-walled cylinder, the axial strain ϵ_a at any point in the wall of the cylinder can be expressed in terms of σ_a, σ_r, and σ_t by using the generalized Hooke's law (see Eq. 4-5). Thus,

$$\epsilon_a = [\sigma_a - \nu(\sigma_r + \sigma_t)]/E \tag{c}$$

The assumption normally made concerning strains in the thick-walled cylinder, which has been verified by careful measurement, is that the axial strain is uniform. This means that plane transverse cross sections before loading remain plane and parallel after the internal and external pressures are applied. So far as the axial stress σ_a is concerned, two cases are of interest in a wide variety of design applications. In the first case, the axial loads induced by the pressure are not carried by the walls of the cylinder ($\sigma_a = 0$). This situation arises in gun barrels and in many types of hydraulic cylinders where pistons carry the axial loads. In the second case, the walls of the cylinder carry the loads. This situation occurs in pressure vessels with various types of end closures or heads. In this second case, in regions of the cylinder away from the ends, it has been found that the axial stress is uniformly distributed over the cross section. Hence ϵ_a, σ_a, E, and ν are constant for the two cases being considered; therefore, it follows from Eq. (c) that

$$\sigma_r + \sigma_t = (\sigma_a - E\epsilon_a)/\nu = 2C_1 \tag{d}$$

The constant is taken as $2C_1$ for convenience in the following derivation.

When the value for σ_t from Eq. (d) is substituted into Eq. (b), this latter equation may be written as

$$\rho\frac{d\sigma_r}{d\rho} + 2\sigma_r = 2C_1 \tag{e}$$

If Eq. (e) is multiplied by ρ, the terms before the equal sign can be expressed as $\frac{d}{d\rho}(\rho^2\sigma_r)$, and thus Eq. (e) becomes

$$\frac{d}{d\rho}(\rho^2\sigma_r) = 2C_1\rho$$

Integrating yields

$$\rho^2\sigma_r = C_1\rho^2 + C_2$$

where C_2 is a constant of integration. Thus,

$$\sigma_r = C_1 + \frac{C_2}{\rho^2} \tag{f}$$

The tangential stress σ_t is then obtained from Eq. (d) as

$$\sigma_t = C_1 - \frac{C_2}{\rho^2} \tag{g}$$

Values for the constants C_1 and C_2 in Eqs. (f) and (g) can be determined by using the known values for the pressures at the inside and outside surfaces of the cylinder. These values, commonly referred to as *boundary conditions,* are

$$\sigma_r = -p_i \quad \text{at} \quad \rho = a$$
$$\sigma_r = -p_o \quad \text{at} \quad \rho = b$$

The minus signs indicate that the pressures (normally considered as positive quantities) produce compressive normal stresses at the surfaces on which they are applied. Substituting the boundary conditions into Eq. (f) yields

$$C_1 = \frac{a^2 p_i - b^2 p_o}{b^2 - a^2}$$

$$C_2 = -\frac{a^2 b^2 (p_i - p_o)}{b^2 - a^2}$$

The desired expressions for σ_r and σ_t are obtained by substituting these values for C_1 and C_2 into Eqs (f) and (g). Thus,

$$\sigma_r = \frac{a^2 p_i - b^2 p_o}{b^2 - a^2} - \frac{a^2 b^2 (p_i - p_o)}{(b^2 - a^2)\rho^2}$$
$$\sigma_t = \frac{a^2 p_i - b^2 p_o}{b^2 - a^2} + \frac{a^2 b^2 (p_i - p_o)}{(b^2 - a^2)\rho^2}$$

(5-10)

Radial and circumferential deformations δ_r and δ_t play important roles in many practical problems. The change in circumference δ_t of the thin ring shown in Fig. 5-37a, when the pressures p_i and p_o are applied to the cylinder, may be expressed in terms of the radial displacement δ_r of a point on the ring as

$$\delta_t = 2\pi \delta_r$$

The circumferential deformation δ_t may also be expressed in terms of the tangential strain ϵ_t as

$$\delta_t = \epsilon_t c$$

where $c = 2\pi\rho$ is the circumference of the ring. Thus,

$$\delta_r = \epsilon_t \rho$$

For many applications, the axial stress $\sigma_a = 0$. The tangential strain ϵ_t can then be expressed in terms of the radial stress σ_r and the tangential stress σ_t by using the generalized Hooke's law. Thus,

$$\epsilon_t = (\sigma_t - v\sigma_r)/E$$

The radial displacement of a point in the wall is then obtained in terms of the radial and tangential stresses present at the point as

$$\delta_r = (\sigma_t - v\sigma_r)\rho/E \qquad (5\text{-}11)$$

As mentioned previously, Eqs. 5-10 and 5-11 can be used to compute stresses and deformations in a wide variety of design situations involving pressure vessels, hydraulic cylinders, and so on. Reduced forms of these equations are used with sufficient frequency to warrant consideration of the following special cases.

Case 1   (Internal Pressure Only):

If the loading is limited to an internal pressure p_i ($p_o = 0$), Eqs. 5-10 reduce to

$$\sigma_r = \frac{a^2 p_i}{b^2 - a^2}\left(1 - \frac{b^2}{\rho^2}\right)$$
$$\sigma_t = \frac{a^2 p_i}{b^2 - a^2}\left(1 + \frac{b^2}{\rho^2}\right)$$

(5-12)

Examination of these equations indicates that σ_r is always a compressive stress, while σ_t is always a tensile stress. In addition, σ_t is always larger than σ_r and is maximum at the inside surface of the cylinder. Substituting the values of σ_r and σ_t from the previous two equations into Eq. 5-11 yields the deformation equation applicable for this special case ($p_o = 0$ and $\sigma_a = 0$). Thus,

$$\delta_r = \frac{a^2 p_i}{(b^2 - a^2)E\rho}[(1 - v)\rho^2 + (1 + v)b^2]$$

(5-13)

Case 2   (External Pressure Only):

If the loading is limited to an external pressure p_o ($p_i = 0$), Eqs. 5-10 reduce to

$$\sigma_r = -\frac{b^2 p_o}{b^2 - a^2}\left(1 - \frac{a^2}{\rho^2}\right)$$
$$\sigma_t = -\frac{b^2 p_o}{b^2 - a^2}\left(1 + \frac{a^2}{\rho^2}\right)$$

(5-14)

In this case, both σ_r and σ_t are always compressive. The tangential stress is always larger than the radial stress and assumes its maximum value at the inner surface of the cylinder. Substituting the values of σ_r and σ_t from the previous two equations into Eq. 5-11 yields the deformation equation applicable for this special case ($p_i = 0$ and $\sigma_a = 0$). Thus,

$$\delta_r = -\frac{b^2 p_o}{(b^2 - a^2)E\rho}[(1 - v)\rho^2 + (1 + v)a^2]$$

(5-15)

Case 3   (External Pressure on a Solid Circular Cylinder):

Radial and tangential stresses σ_r and σ_t and the radial displacement δ_r for this special case can be obtained from the expression developed for case 2 by letting the hole in the cylinder vanish ($a = 0$). Thus,

$$\sigma_r = -p_o$$
$$\sigma_t = -p_o$$

(5-16)

and

$$\delta_r = -\frac{1-\nu}{E}p_o\rho \tag{5-17}$$

The minus sign in these equations indicates that the stresses are both compressive and the radius of the cylinder is reduced when the external pressure p_o is applied. For this case the stresses are independent of radial position ρ and have a constant magnitude equal to the applied pressure.

Example Problem 5-19 A steel ($E = 30,000$ ksi and $\nu = 0.30$) cylinder with an inside diameter of 8 in. and an outside diameter of 16 in. is subjected to an internal pressure of 15,000 psi, as shown in Fig. 5-38a. Axial loads induced by the pressure are not carried by the walls of the cylinder ($\sigma_a = 0$). Determine

(a) The maximum tensile stress in the cylinder.

(b) The maximum shearing stress in the cylinder.

(c) The increase in the inside diameter as the pressure is applied.

(d) The increase in the outside diameter as the pressure is applied.

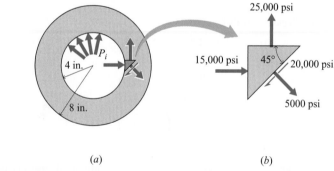

(a)   (b)

Figure 5-38(a–b)

SOLUTION

(a) The maximum tensile stress in a thick-walled cylinder subjected to an internal pressure is the tangential stress at the inside surface of the cylinder and is given by Eq. 5-12 as

$$\sigma_t = \frac{a^2 p_i}{b^2 - a^2}\left(1 + \frac{b^2}{\rho^2}\right)$$

$$= \frac{4^2(15,000)}{8^2 - 4^2}\left(1 + \frac{8^2}{4^2}\right) = 25,000 \text{ psi (T)} \qquad \textbf{Ans.}$$

(b) The maximum shearing stress also occurs at a point on the inside surface of the cylinder on a plane inclined 45° with respect to a radial line, as shown in Fig. 5-38b. The minimum normal stress at the point is $\sigma_r = -p_i =$

$-15{,}000$ psi, since the axial stress $\sigma_a = 0$. Therefore, from Eq. 2-19,

$$\tau_{max} = \frac{\sigma_{max} - \sigma_{min}}{2} = \frac{25{,}000 - (-15{,}000)}{2} = 20{,}000 \text{ psi} \qquad \textbf{Ans.}$$

A Mohr's circle representation of these results is shown in Fig. 5-38c.

(c) The radial displacement of a point in the wall of a cylinder subjected to internal pressure ($p_o = 0$ and $\sigma_a = 0$) is given by Eq. 5-13 as

$$\delta_r = \frac{a^2 p_i}{(b^2 - a^2)E\rho}[(1 - \nu)\rho^2 + (1 + \nu)b^2]$$

On the inside surface where $\rho = 4$ in.,

$$\delta_{ri} = \frac{4^2(15{,}000)}{(8^2 - 4^2)(30{,}000{,}000)(4)}[(1 - 0.30)(4^2) + (1 + 0.30)(8^2)]$$
$$= 0.003933 \text{ in.}$$

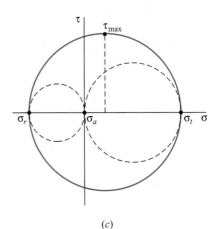

(c)

Figure 5-38(c)

Therefore,

$$\Delta D_i = 2\delta_{ri} = 2(0.003933) = 0.007866 \text{ in.} \cong 0.00787 \text{ in.} \qquad \textbf{Ans.}$$

(d) On the outside surface where $\rho = 8$ in.,

$$\delta_{ro} = \frac{4^2(15{,}000)}{(8^2 - 4^2)(30{,}000{,}000)(8)}[(1 - 0.30)(8^2) + (1 + 0.30)(8^2)]$$
$$= 0.002667 \text{ in.}$$

Therefore,

$$\Delta D_o = 2\delta_{ro} = 2(0.002667) = 0.005334 \text{ in.} \cong 0.00533 \text{ in.} \qquad \textbf{Ans.}$$

PROBLEMS

Introductory Problems

5-118* A thick-walled cylindrical pressure vessel with an inside diameter of 250 mm and an outside diameter of 400 mm is subjected to an internal pressure of 75 MPa. Determine

a. The tangential stress σ_t at a point on the inside surface.
b. The tangential stress σ_t at a point on the outside surface.
c. The maximum shearing stress in the cylinder.

5-119 Demonstrate that for a cylindrical vessel with a shell thickness $(b-a)$ of one-tenth the radius a, subjected only to internal pressure, the error involved in computing the hoop tension σ_t by the method of Section 5-9 instead of Eq. 5-12 is approximately 5 percent.

5-120 A thick-walled cylindrical pressure vessel with an inside diameter of 200 mm and an outside diameter of 300 mm is made of 0.2% C hardened steel, which has a yield strength of 430 MPa. Determine the maximum internal pressure that may be applied to the vessel if the yield strength must not be exceeded.

5-121* A thick-walled cylindrical pressure vessel with an inside diameter of 4 in. and an outside diameter of 8 in. has a design specification that limits the maximum shearing stress to 24,000 psi. Determine the maximum internal pressure that may be applied to the vessel.

Intermediate Problems

5-122 A steel cylinder with an inside diameter of 200 mm and an outside diameter of 300 mm is subjected to an internal pressure of 75 MPa. Determine

  a. The maximum tensile stress in the cylinder.
  b. The maximum shearing stress at a point in the cylinder midway between the inside and outside surfaces.

5-123* A steel cylinder with an inside diameter of 8 in. and an outside diameter of 14 in. is subjected to an internal pressure of 25,000 psi. Determine

  a. The maximum shearing stress in the cylinder.
  b. The radial and tangential stresses at a point in the cylinder midway between the inside and outside surfaces.

Challenging Problems

5-124* A steel cylinder with an inside diameter of 50 mm and an outside diameter of 250 mm is subjected to an internal pressure of 85 MPa. Determine the circumferential force per unit length carried by a circular element having an inside diameter of 100 mm and a thickness of 25 mm.

5-125 A navel gun ($E = 30,000$ ksi and $G = 11,600$ ksi) with a bore of 16 in. develops a maximum internal pressure of 20,000 psi. Determine

  a. The minimum outside diameter required if the maximum tensile stress permitted in the gun barrel is 36,000 psi.
  b. The change in inside diameter when the gun is fired if the outside diameter of part a is used.

5-11 DESIGN

A designer must select a material and properly proportion a member to perform a specified function without failure. *Failure* is defined as the state or condition in which a member or structure no longer functions as intended. To accomplish the design task, one must anticipate the type of failure (*failure mode*) that may occur. Once the failure mode has been determined, the significant material property that controls failure is established. Design computations are performed using mathematical relationships between load and stress or load and deformation.

5-11-1 Modes of Failure   The *mode of failure* of a member depends on many factors: the type of material, the manner of loading, the rate of loading, and environmental conditions. Discussion in this book will be limited to members subjected to static or slowly applied loads at room temperature. Furthermore, failure modes are limited to *elastic failure*, which occurs as a result of excessive elastic deformation; *yielding* (sometimes referred to as slip failure), characterized by excessive plastic deformation; and *failure by fracture* (complete separation of the material).

5-11-2 Significant Material Property   Associated with each mode of failure is a significant material property. When a structure is designed to avoid elastic failure, the modulus of elasticity is the significant material property. Since yielding is characterized by excessive plastic deformation, the significant material property is the yield strength. Failure by fracture may be due to sudden fracture of a brittle material, fracture of a material with cracks or flaws, or fracture due to repeated loading. In this book, fracture failure will be limited to sudden fracture, where the significant material property is the ultimate strength.

5-11-3 Mathematical Analysis   A failure criterion is needed to perform design computations. The failure criterion may be based on a probabilistic model, an allowable stress model, or a load resistance factor design (LRFD) model. Only the allowable stress model (called *allowable stress design*, ASD) will be discussed in this book. Once the mode of failure is established, the ASD model states

that the design is satisfactory so long as

$$\text{Strength} \geq \text{Stress} \tag{5-18}$$

in which "strength" is the significant material property and "stress" refers to the computed stress in the member. For example, stress $= \sigma = F/A$ for an axially loaded member.

5-11-4 Factor of Safety

Most design problems involve many unknown variables. The load that the structure or machine must carry is usually estimated. The actual load may vary considerably from the estimate, especially when loads at some future time must be considered. Since testing usually damages a material, the properties of a material used in a structure cannot be evaluated directly but are normally determined by testing specimens of a similar material. Furthermore, the actual stresses that will exist in a structure are unknown because the calculations are based on assumptions about the distribution of stresses in the material. Because of these and other unknown variables it is customary to write Eq. 5-18 as

$$\text{Strength} \geq (\text{Factor of Safety})(\text{Stress}) \tag{5-19}$$

where the *factor of safety* (FS) takes into account the imponderables.

As an aid to understanding the use of Eq. 5-19, consider an axially loaded rod that is to be designed (find the required diameter) so that the material does not yield. Then, since the mode of failure is yielding, the significant material property is the yield strength σ_y, that is, strength equals σ_y. Then, Eq. 5-19 becomes

$$\sigma_y \geq (\text{FS})(F/A)$$

or

$$\sigma_y \geq (\text{FS})\left(\frac{F}{\pi d^2/4}\right)$$

Solving for the diameter gives

$$d \geq \sqrt{4(\text{FS})(F)/(\pi \sigma_y)}$$

For a given material (σ_y known), a given factor of safety, and a given load, the minimum required diameter would be

$$d_{\min} = \sqrt{4(\text{FS})(F)/(\pi \sigma_y)}$$

The following Example Problems illustrate the use of the design principles previously discussed.

Example Problem 5-20 An axially loaded circular bar is subjected to a load of 6500 lb. The bar is made of structural steel, and the factor of safety is to be 1.5. Determine the minimum diameter bar required if yielding is to be avoided.

SOLUTION

Since the mode of failure is yielding, the significant material property (strength) is the yield strength. Using Table B-17 in Appendix B, the yield strength of structural steel is $36(10^3)$ psi. Equation (5-19) then gives

$$\sigma_y \geq (\text{FS})(F/A)$$

$$\sigma_y \geq (\text{FS})\left(\frac{F}{\pi d^2/4}\right)$$

$$d^2 \geq \frac{4F(\text{FS})}{\pi \sigma_y}$$

$$d^2 \geq \frac{4(6500)(1.5)}{\pi(36)(10^3)}$$

Therefore,

$$d \geq 0.587 \text{ in.} \qquad \qquad \textbf{Ans.}$$

The minimum required diameter is 0.587 in. If rods are commercially available in increments of 1/8 in., a rod of diameter 5/8 in. would be selected.

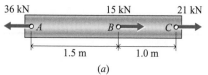

(a)

Figure 5-39(a)

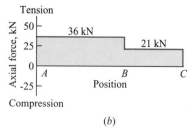

(b)

Figure 5-39(b)

■ **Example Problem 5-21**   An axially loaded circular bar made of structural steel has a constant cross-sectional area and is subjected to the forces shown in Fig. 5-39a. The factor of safety, based on failure by yielding, is to be 1.8. Determine the minimum permissible diameter of the bar required to support the loads.

SOLUTION

The forces transmitted by sections AB and BC are obtained from free-body diagrams of portions of the bar isolated by using cutting planes to the right of pin A and to the left of pin C and drawing the axial force diagram shown in Fig. 5-39b. Thus, the maximum load transmitted by any cross section is $F_{AB} = 36$ kN. Since the criterion for failure is yielding, the significant material property is the yield strength. From Table B-18 in Appendix B, $\sigma_y = 250$ MPa. Proceeding as in the previous example,

$$d^2 \geq \frac{4F(\text{FS})}{\pi \sigma_y}$$

where $F = F_{AB}$ is the largest internal force in the constant diameter bar. Substituting the numerical values,

$$d^2 \geq \frac{4(36)(10^3)(1.8)}{\pi(250)(10^6)}$$

$$d \geq 0.01817 \text{ m}$$

Therefore,

$$d_{\min} = 18.17 \text{ mm} \qquad \qquad \textbf{Ans.}$$

Example Problem 5-22   A 40-lb light is supported at the midpoint of a 10-ft length of wire that is made of 0.2% C hardened steel, as shown in Fig. 5-40a. For reasons of safety, a factor of safety of 3 based on the yield strength of the wire is specified. Spools of wire are available with diameters of 10, 20, 30, 40, and 50 mil (1 mil = 0.001 in.). What spool size would you select for suspending the light?

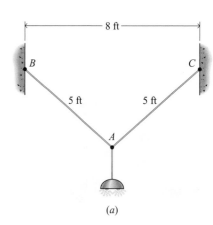

(a)

SOLUTION
The forces transmitted by cables AB and AC are obtained form the free-body diagram of joint A of the cable system shown in Fig. 5-40b. The wire angle is

$$\theta = \cos^{-1}\frac{4}{5} = 36.87°$$

From the horizontal component of the equilibrium equation

$$+ \rightarrow \Sigma F_x = 0: \qquad T_{AC}\cos\theta - T_{AB}\cos\theta = 0 \qquad (a)$$

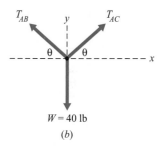

(b)

we get that the two tension forces must be equal, $T_{AC} = T_{AB}$. Then, from the vertical component of the equilibrium equation

$$+ \uparrow \Sigma F_y = 0: \qquad T_{AC}\sin 36.87° + T_{AB}\sin 36.87° - 40 = 0 \qquad (b)$$ **Figure 5-40**

Substituting $T_{AC} = T_{AB}$ into Eq. (b) gives

$$T_{AC} = T_{AB} = 33.33 \text{ lb}$$

For 0.2% C hardened steel (see Table B-17 in Appendix B), $\sigma_y = 62$ ksi. Therefore, proceeding as in the previous examples,

$$d^2 \geq \frac{4F(\text{FS})}{\pi\sigma_y}$$

$$d^2 \geq \frac{4(33.33)(3)}{\pi(62)(10^3)}$$

$$d \geq 0.0453 \text{ in.}$$

The required spool size is 50 (use 50-mil wire).   **Ans.**

Example Problem 5-23   The short post shown in Fig. 5-41a is subjected to an axial compressive load $P = 150$ kip. The load is applied to the post through a rigid steel plate. The core of the post is annealed bronze, and the outer segment of the post is composed of two symmetrically placed plates of 2024-T4 aluminum. Each of the aluminum plates is one-fourth as thick as the bronze core. Determine the minimum thickness t required if the factor of safety based on failure by yielding is 1.5.

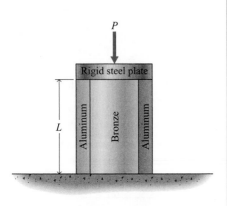

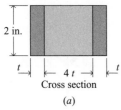

2 in.

Cross section

(a)

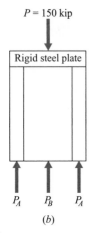

(b)

Figure 5-41

SOLUTION

Since the criterion for failure is yielding, the significant property for each material is the yield strength. From Table B-17 in Appendix B, the modulus of elasticity and the yield strength are $E_B = 15,000$ ksi and $\sigma_y = 20$ ksi for the bronze and $E_A = 10,600$ ksi and $\sigma_y = 48$ ksi for the aluminum. The free-body diagram (Fig. 5-41b) has two unknown forces: P_A, the force in the aluminum plates, and P_B, the force in the bronze core. Since only one equation of equilibrium is available,

$$+ \uparrow \Sigma F_y = 0: \qquad 2P_A + P_B - 150 = 0$$
$$2P_A + P_B = 150 \text{ kip} \qquad (a)$$

the problem is statically indeterminate. As the rigid steel plate pushes down on the top of the post, the bronze core and the two aluminum plates will all shorten the same amount. Therefore, the deformation equation is $\delta_A = \delta_B$ which gives

$$\left(\frac{PL}{EA}\right)_A = \left(\frac{PL}{EA}\right)_B$$

$$\frac{(2P_A)L}{(10,600)[2(2t)]} = \frac{P_B L}{(15,000)8t}$$

or

$$P_A = 0.17667\, P_B \qquad (b)$$

Solving Eqs. (a) and (b) yields

$$P_A = 19.582 \text{ kip}$$
$$P_B = 110.84 \text{ kip}$$

The failure criterion is

$$\sigma_y \geq (\text{FS})(P/A)$$
$$A \geq \frac{P(\text{FS})}{\sigma_y}$$

Applying the failure criterion to each member of the structure yields

$$2t \geq \frac{19.582(1.5)}{48} \qquad t \geq 0.306 \text{ in.}$$

for the aluminum and

$$8t \geq \frac{110.84(1.5)}{20} \qquad t \geq 1.039 \text{ in.}$$

for the bronze. Therefore, the minimum thickness is

$$t_{\min} = 1.039 \text{ in.} \qquad \textbf{Ans.}$$

PROBLEMS

MecMovie Activities and Problems

MM5.12 Beam strut structure factors of safety. Example; Try one. Evaluate safety factors for various facets of a simple structure.

MM5.13 Design bolts for splice plate structure. Example; Try one. Use factor of safety concept in sizing bolts for a tension connection.

MM5.14 Load capacity of beam-strut structure. Example; Try one. Determine distributed load capacity of structure given normal and shear stress limits.

Introductory Problems

5-126* A short standard-weight steel pipe (see Appendix B) is used to support an axial compressive load of 100 kN. If yielding ($\sigma_y = 250$ MPa) should not occur and the factor of safety is to be 1.6, determine the smallest nominal diameter pipe that may be used to support the load.

5-127* A short column made of structural steel is used to support the floor beams of a building, as shown in Fig. P5-127. Each floor beam (A and B) transmits a force of 40 kip to the column. The column has the shape of a wide-flange (W) section (see Appendix B). The factor of safety based on failure by yielding is 3.0. Select the lightest wide-flange section that will support the given loads.

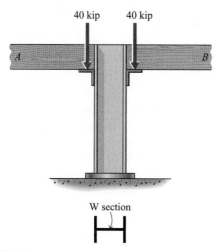

Figure P5-127

5-128 The two structural steel (see Appendix B) rods A and B shown in Fig. P5-128 are used to support a mass $m = 2000$ kg. If failure is by yielding and a factor of safety of 1.75 is specified, determine the diameters of the rods (to the nearest 1 mm) that must be used to support the mass. Both rods are to have the same diameter.

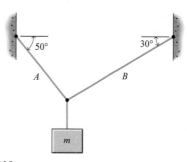

Figure P5-128

Intermediate Problems

5-129* The machine component shown in Fig. P5-129 is made of hot-rolled Monel. The forces at B are applied to the component with a rigid collar that is firmly attached to the component. If the mode of failure is yielding and the factor of safety is 1.5, determine the minimum permissible diameter of each segment of the machine component.

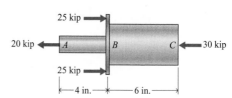

Figure P5-129

5-130* An axial load $P = 1000$ kN is applied to the rigid steel bearing plate on the top of the short column shown in Fig. P5-130. The outside segment of the column is made of structural steel. The inside core is made of fairly high-strength concrete. Both segments are square. The failure modes are yielding for the steel and fracture for the concrete. The factor of safety is to be 1.4. If the area of the concrete is to be 10 times the area of the steel, determine the required dimensions.

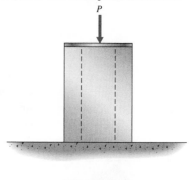

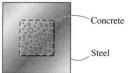

Figure P5-130

5-131 Four axial forces are applied to the 1-in. thick, 0.4% C hot-rolled steel bar as shown in Fig. P5-131. The factor of safety for failure by yielding is 1.75. Determine the minimum width w of the constant cross-sectional area bar.

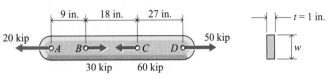

Figure P5-131

Challenging Problems

5-132 The two parts of the eyebar shown in Fig. P5-132 are connected by two bolts (one on each side of the eyebar). The bolts are made of a grade of steel with a tensile yield strength of 1035 MPa and a shear yield strength of 620 MPa. The eyebar is subjected to the forces $P = 85$ kN. Determine the minimum bolt diameter required to safely support the forces if the mode of failure is yielding and the factor of safety is 1.5.

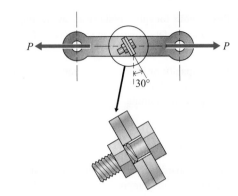

Figure P5-132

SUMMARY

When a straight bar of uniform cross section is axially loaded by forces applied at its ends, the axial strain along the length of the bar is assumed to have a constant value, and the elongation (or contraction) of the bar resulting from the axial load P may be expressed as $\delta = \epsilon L$ (by the definition of average axial strain). If Hooke's law (Eq. 4-1a) applies (if the stresses are less than the proportional limits of the materials used in the fabrication of the members), the axial deformation may be expressed in terms of either stress or load as

$$\delta = \epsilon L = \frac{\sigma L}{E} = \frac{PL}{EA} \qquad (5\text{-}1,2)$$

where P, σ, A, and E are all constant over the entire length L. If the stress exceeds the proportional limit of the materials, stress-strain diagrams can be used to relate the load and deformation.

If a bar is subjected to a number of axial loads at different points along the bar, or if the bar consists of parts having different cross-sectional areas or of parts composed of different materials, then the change in length of each part can be computed by using Eq. 5-1 or 5-2. The changes in length of the various parts of the bar can then be added algebraically to give the total change in length of the complete bar

$$\delta = \sum_{i=1}^{n} \delta_i = \sum_{i=1}^{n} \frac{P_i L_i}{E_i A_i} \qquad (5\text{-}3)$$

where A_i and E_i are both constant on segment i and the force P_i is the internal force in segment i of the bar and is usually different than the forces applied at the ends of the segment. These forces must be calculated from equilibrium of the segment and are often shown on an axial force diagram.

It is sometimes necessary to determine axial deformations and strains in a loaded system of pin-connected deformable bars (two-force members). The solution of such problems involves a static analysis (free-body diagrams and equilibrium equations) combined with a study of the geometry of the deformed system. For small displacements, the axial deformation in any bar may be assumed equal to the component of the displacement of one end of the bar (relative to the other end) taken in the direction of the unstrained orientation of the bar. If the statics and the deformation portions of the problem can be separated and solved separately, the problem is called statically determinate. If the statics and deformation portions cannot be separated and solved separately, the problem is called statically indeterminate. In either case, a displacement diagram showing deformations should be drawn to assist in obtaining the correct deformation equation. The free-body diagrams, equilibrium equations, and deformation diagrams must be compatible. That is, when a tensile force is assumed for a member in the free-body diagram, a tensile deformation (stretch) must be indicated for the same member in the deformation diagram. If the diagrams are compatible, a negative result will indicate that the assumption was wrong; however, the magnitude of the result will be correct.

When a bar is subjected to a temperature change in addition to a stress, the total deformation of the bar is the sum of the deformation due to stress plus the deformation due to the temperature change

$$\delta_{\text{total}} = \delta_\sigma + \delta_T = \frac{\sigma L}{E} + \alpha \Delta T L$$

where α is the coefficient of thermal expansion. A tensile stress ($\sigma > 0$) and a temperature increase ($\Delta T > 0$) both cause a stretch of the bar ($\delta > 0$).

The stress at a discontinuity in a structural or machine element may be considerably greater than the nominal or average stress on the section. The ratio of the maximum stress at a discontinuity to the nominal stress on the section is called the stress concentration factor. Thus, the maximum normal stress at a discontinuity in a centrically loaded member is

$$\sigma = K\frac{P}{A} \qquad (5\text{-}5)$$

Graphs or tables for K, the stress concentration factor, can be found in numerous design handbooks, and they may be based on either the gross area or the net area (area at the reduced section). It is important when using such stress concentration graphs or tables to ascertain whether the factors are based on the gross or net section.

Stress concentration is a very localized effect. For example, the stress on the boundary of a hole in a large plate under uniform unidirectional tension is 3 times the nominal stress—the stress in regions far removed from the hole. However, at a distance of one hole diameter from the edge of the hole, the stress is only about 7 percent greater than the nominal stress.

Stress concentration is not significant in the case of static loading of a ductile material because the material will yield inelastically in the region of high stress. As a result of the accompanying redistribution of stress, equilibrium may be established and no harm done. However, if the load is an impact or repeated load, the material may fracture. Also, if the material is brittle, even a static load may cause fracture.

A pressure vessel is described as thin-walled when the ratio of the wall thickness to the radius of the vessel is so small that the distribution of normal stress on a plane perpendicular to the surface of the vessel is essentially uniform throughout the thickness of the vessel. If the ratio of the wall thickness to the inner radius of the vessel is less than about 0.1, the maximum normal stress is less than 5 percent greater than the average. In a spherical, thin-walled pressure vessel, the normal stress on a section that passes through the center of the sphere is called a meridional or axial stress and is given by

$$\sigma_a = \frac{pr}{2t} \tag{5-7}$$

where r and t are the radius and wall thickness of the pressure vessel, and p is the internal pressure. There are no shearing stresses on any of these planes since there are no loads to induce them. In a cylindrical, thin-walled pressure vessel, the stress on a longitudinal plane is called a hoop or circumferential stress and is given by

$$\sigma_h = \frac{pr}{t} \tag{5-8a}$$

The normal stress on a transverse plane is called an axial stress and has the same value as in a spherical pressure vessel

$$\sigma_a = \frac{pr}{2t} \tag{5-8b}$$

Although there are no shearing stresses on transverse or longitudinal planes, there will be both normal and shear stresses on any other plane such as a spiral weld surface. If a pressure vessel is subjected to an axial load in addition to an internal pressure, the normal stresses may be found by superimposing the normal stresses due to internal pressure and the normal stresses due to the axial load.

In a thick-walled pressure vessel, normal stresses are not uniform throughout the thickness of the vessel. In addition, for a thick-walled pressure vessel, the radial component of normal stress cannot be ignored. In a thick-walled cylindrical vessel, the radial normal stress and the tangential (hoop) normal stress vary with radial position ρ according to

$$\begin{aligned}
\sigma_r &= \frac{a^2 p_i - b^2 p_o}{b^2 - a^2} - \frac{a^2 b^2 (p_i - p_o)}{(b^2 - a^2)\rho^2} \\
\sigma_t &= \frac{a^2 p_i - b^2 p_o}{b^2 - a^2} + \frac{a^2 b^2 (p_i - p_o)}{(b^2 - a^2)\rho^2}
\end{aligned} \tag{5-10}$$

where p_i is the internal pressure (at $\rho = a$) and p_o is the external pressure (at $\rho = b$).

In order to select a material and properly proportion a member to perform a specified function without failure, a designer must anticipate the type of failure that may occur. Once the type of failure has been determined, the significant material property that controls failure is established. This property is then divided by the factor of safety to determine the allowable stress to be used in the design computations. In this book's context of design, attention will be focused on elastic failure, failure by yielding, and failure by fracture.

REVIEW PROBLEMS

5-133* An alloy steel ($E = 30,000$ ksi) bar is loaded and supported as shown in Fig. P5-133. The loading collar at B is free to slide on section BC. The diameters of sections $AB, BC,$ and CD are 2.50 in., 1.50 in., and 1.00 in., respectively. The lengths of all three segments are 15 in. Determine the normal stresses in each section and the overall change in length of the bar.

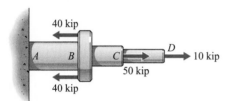

Figure P5-133

5-134* A tension member consists of a 50-mm-diameter brass ($E = 100$ GPa) bar connected to a 32-mm-diameter stainless-steel ($E = 190$ GPa) bar, as shown in Fig. P5-134. Determine the maximum load P that can be applied to the bar if the normal stress in the brass must be limited to 200 MPa, the normal stress in the steel must be limited to 500 MPa, and the total elongation of the bar must be limited to 5.60 mm.

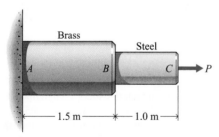

Figure P5-134

5-135 Two rigid bars (AB and BC) and a 1/2-in.-diameter structural steel ($E = 30,000$ ksi) tie rod AC are used to support a 3000-lb load P, as shown in Fig. P5-135. Determine the normal stress in the tie rod and the change in length of the 30-in. reduced section of the tie rod as the load P is applied.

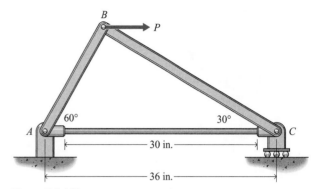

Figure P5-135

5-136* A rigid bar ABC is supported by two links, as shown in Fig. P5-136. Link BD is made of an aluminum alloy ($E = 73$ GPa) and has a cross-sectional area of 1250 mm^2. Link CE is made of structural steel ($E = 200$ GPa) and has a cross-sectional area of 750 mm^2. Determine the normal stress in each of the links and the deflection of point A as the 50-kN load P is applied.

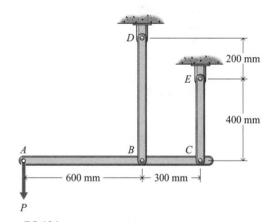

Figure P5-136

5-137 Bar A of Fig. P5-137 is a steel ($E = 30,000$ ksi) rod that has a cross-sectional area of 1.24 in.^{2} Member B is a brass ($E = 15,000$ ksi) post that has a cross-sectional area of 4.00 in.^{2} Determine the maximum permissible value for the load P if the allowable normal stresses are 30 ksi for the steel and 20 ksi for the brass.

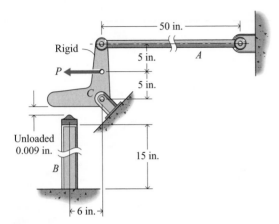

Figure P5-137

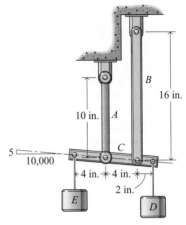

Figure P5-139

5-138 Bar C of Fig. P5-138 is an aluminum alloy ($E = 73$ GPa) rod that has a cross-sectional area of 625 mm². Member D is a wood ($E = 12$ GPa) post that has a cross-sectional area of 2500 mm². Determine the maximum permissible value for the load P if the allowable normal stresses are 100 MPa for the aluminum and 30 MPa for the wood.

5-140 A 90-mm-diameter brass ($E = 100$ GPa) bar is securely fastened to a 50-mm-diameter steel ($E = 200$ GPa) bar. The ends of the composite bar are then attached to rigid supports, as shown in Fig. P5-140. Determine the stresses in the brass and the steel after a temperature drop of 70°C occurs. The thermal coefficients of expansion for the brass and steel are $17.6(10^{-6})$/°C and $11.9(10^{-6})$/°C, respectively.

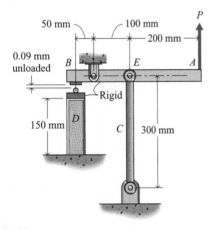

Figure P5-138

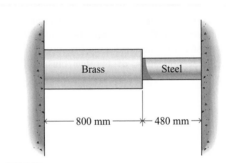

Figure P5-140

5-139* The pin-connected structure shown in Fig. P5-139 occupies the position shown when unloaded. When the loads $D = 16$ kip and $E = 8$ kip are applied to the structure, the rigid bar C must become horizontal. Bar A is made of an aluminum alloy ($E = 10,600$ ksi), and bar B is made of bronze ($E = 15,000$ ksi). If the normal stresses in the bars must be limited to 20 ksi in the aluminum alloy and 15 ksi in the bronze, determine

a. The minimum cross-sectional areas that will be satisfactory for the bars.
b. The changes in length of rods A and B.

5-141 Three bars, each 50 mm wide × 25 mm thick × 4 m long, are connected and loaded as shown in Fig. P5-141. Bar A is made of Monel, which has a proportional limit of 400 MPa and a modulus of elasticity of 180 GPa. Bar B is made of a magnesium alloy that has a proportional limit of 100 MPa and a modulus of elasticity of 40 GPa. Bar C is made of structural steel (elastoplastic) that has a proportional limit and yield point of 240 MPa and a modulus of elasticity of 200 GPa. Determine

a. The normal stress in each of the bars after a 650-kN load P is applied.
b. The vertical displacement (deflection) of pin D produced by the 650-kN load.

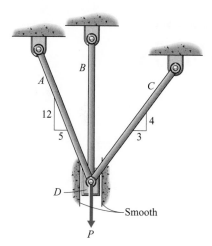

Figure P5-141

5-142* The rigid bar AB of Fig. P5-142 is horizontal, and bar C and post D are unstressed before the load P is applied. Bar C has a cross-sectional area of 600 mm² and is made of a low-carbon steel (elastoplastic) that has a proportional limit and yield point of 240 MPa and a modulus of elasticity of 200 GPa. Post D has a cross-sectional area of 2000 mm² and is made of cold-rolled brass, which has a proportional limit of 410 MPa and a modulus of elasticity of 100 GPa. Determine

a. The normal stresses in bar C and post D after a 100-kN load P is applied.
b. The vertical displacement (deflection) of point A produced by the 100-kN load.

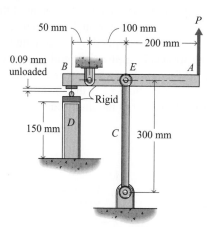

Figure P5-142

5-143* The conical water tank shown in Fig. P5-143 was fabricated from 1/8-in.-thick steel plate. When the tank is completely full of water (specific weight $\gamma = 62.4 \text{ lb/ft}^3$), determine the axial and hoop stresses σ_a and σ_h at a point in the wall 8 ft below the apex of the cone.

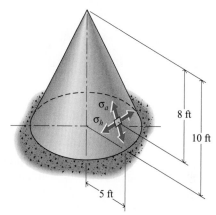

Figure P5-143

5-144 A thin-walled cylindrical pressure vessel has an outside diameter of 2 m and a wall thickness of 10 mm. The vessel is made of steel with a modulus of elasticity of 200 GPa and a Poisson's ratio of 0.30. During proof testing of the vessel, an axial strain of 300 μm/m is recorded. Determine

a. The internal pressure applied to the vessel.
b. The axial and hoop stresses in the vessel.
c. The maximum shearing stress in the vessel.
d. The hoop strain present when the axial strain was measured.

5-145 A gun barrel with an inside diameter of 3.00 in. and an outside diameter of 7.00 in. is made of steel having a yield strength of 50 ksi. Determine the maximum internal pressure that may be applied to the gun barrel before yielding occurs.

5-146* A hydraulic cylinder with an inside diameter of 200 mm and an outside diameter of 450 mm is made of steel ($E = 210$ GPa and $v = 0.30$). For an internal pressure of 125 MPa, determine

a. The maximum tensile stress in the cylinder.
b. The change in internal diameter of the cylinder.

Chapter 6
Torsional Loading of Shafts

6-1 INTRODUCTION

The problem of transmitting a torque (a couple) from one plane to a parallel plane is frequently encountered in the design of machinery. The simplest device for accomplishing this function is a circular shaft such as that connecting an electric motor with a pump, compressor, or other machine. A modified free-body diagram (the weight and bearing reactions are not shown because they contribute no useful information to the torsion problem) of a shaft used to transmit a torque from a driving motor A to a coupling B is shown in Fig. 6-1. The resultant of the electromagnetic forces applied to armature A of the motor is a couple resisted by the resultant of the bolt forces (another couple) acting on the flange coupling B. The circular shaft transmits the torque from the armature to the coupling. Typical torsion problems involve determinations of significant stresses in and deformations of shafts.

A segment of the shaft between transverse planes a–a and b–b of Fig. 6-1 will be studied. The complicated stress distributions at the locations of the torque-applying devices are beyond the scope of this elementary treatment of the torsion problem. A free-body diagram of the segment of the shaft between sections a–a and b–b is shown in Fig. 6-2 with the torque applied by the armature indicated on the left end as T. The resisting torque T_r at the right end of the segment is the resultant of the moment of the differential forces dF acting on the transverse plane b–b. The force dF is equal to $\tau_\rho dA$ where τ_ρ is the shearing stress on the transverse plane at a distance ρ from the center of the shaft and dA is a differential

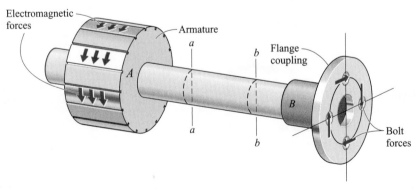

Figure 6-1

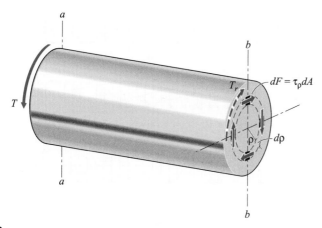

Figure 6-2

area. For circular sections, the shearing stress on any transverse plane is always perpendicular to the radius to the point. The resisting torque is statically equivalent to the sum of the torques produced by dF

$$T_r = \int_{\text{area}} \rho \, dF = \int_{\text{area}} \rho \, \tau_\rho \, dA \qquad (6\text{-}1)$$

The law of variation of the shearing stress on the transverse plane (τ as a function of radial position ρ) must be known before the integral of Eq. 6-1 can be evaluated. Thus, the problem of determining the relationship between torque and shearing stress is statically indeterminate. Recalling the procedures developed in Chapter 5, the solution of a statically indeterminate problem requires the use of the equation of equilibrium, an analysis of deformation, and the relationship between stress and strain.

In 1784 C. A. Coulomb, a French engineer, developed (experimentally) a relationship between applied torque and angle of twist for circular bars.[1] In a paper published in 1820,[1] A. Duleau, another French engineer, derived the same relationship analytically by making the assumption that a plane section before twisting remains plane after twisting and a diameter remains straight. Visual examination of twisted models indicates that these assumptions are correct for circular sections either solid or hollow (provided the hollow section is circular and symmetrical with respect to the axis of the shaft), but incorrect for any other shape. Compare, for example, the distortions of rubber models with circular and rectangular cross sections shown in Fig. 6-3. Figure 6-3b shows the circular shaft after loading and illustrates that plane sections remain plane. For the rectangular shaft, plane sections before loading (Fig. 6-3c) become warped after loading (Fig. 6-3d).

6-2 TORSIONAL SHEARING STRAIN

If a plane transverse cross section of a circular shaft before twisting remains plane after twisting and a diameter of the section remains straight, the distortion of the shaft of Fig. 6-2 will be as indicated in Fig. 6-4a, where points B and D on a

---

[1] From *History of Strength of Materials*, S. P. Timoshenko, McGraw-Hill, New York, 1953.

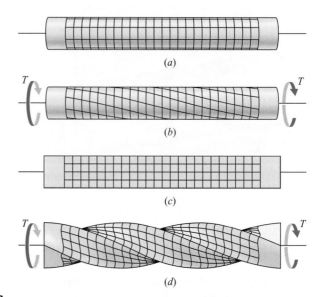

Figure 6-3

common radius in a plane move to points B' and D' in the same plane and still on the same radius. The angle θ is called the *angle of twist*. The surface ABB' of Fig. 6-4a is shown in plan view in Fig. 6-4b, in which a differential element of the material at B (Fig. 6-4c) is distorted at B' due to shearing stress (Fig. 6-4d). Clearly, the angle ϕ of Fig. 6-4b is the same as the shearing strain γ_c of Fig. 6-4d. Similar figures could be drawn for the surface EDD'. It is recommended that the reader review the concept of shearing strain in Section 3-2.

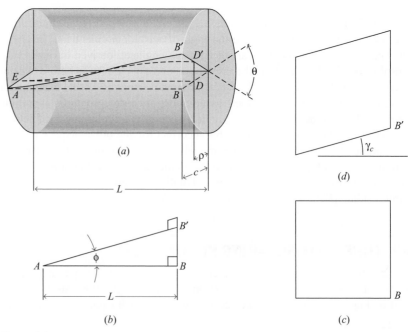

Figure 6-4

At this point the assumption is made that all longitudinal elements (AB, ED, etc.) have the same length L (which limits the results to straight shafts of constant diameter). From Fig. 6-4, the shearing strain γ_ρ at a distance ρ from the center of the shaft and γ_c at the surface of the shaft ($\rho = c$) are related to the angle of twist θ by

$$\tan \gamma_c = \frac{BB'}{AB} = \frac{c\theta}{L}$$

and

$$\tan \gamma_\rho = \frac{DD'}{ED} = \frac{\rho\theta}{L}$$

or, if the strain is small ($\tan \gamma \cong \sin \gamma \cong \gamma$, γ in radians),

$$\gamma_c = \frac{c\theta}{L} \tag{6-2a}$$

and

$$\gamma_\rho = \frac{\rho\theta}{L} \tag{6-2b}$$

Combining Eqs. 6-2a and 6-2b gives

$$\theta = \frac{\gamma_c L}{c} = \frac{\gamma_\rho L}{\rho}$$

which indicates that the shearing strain

$$\gamma_\rho = \frac{\gamma_c}{c}\rho \tag{6-3}$$

is zero at the center of the shaft and increases linearly with respect to the distance ρ from the axis of the shaft. Equation 6-3 is the result of the deformation analysis of a circular shaft subjected to torsional loading. This equation can be combined with Eq. 6-1 once the relationship between shearing stress τ and shearing strain γ is known.

Up to this point, no assumption has been made about the relationship between stress and strain or about the type of material of which the shaft is made. Therefore, Eq. 6-3 is valid for elastic or inelastic action and for homogeneous or heterogeneous materials, provided the strains are not too large ($\tan \gamma \cong \gamma$). Problems in this book will be assumed to satisfy this requirement.

6-3 TORSIONAL SHEARING STRESS—THE ELASTIC TORSION FORMULA

If the assumption is now made that Hooke's law applies (the accompanying limitation is that the stresses must be below the proportional limit of the material), the shearing stress τ is related to the shearing strain γ by the expression $\tau = G\gamma$

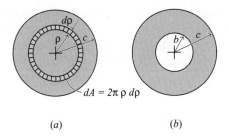

$$dA = 2\pi\,\rho\,d\rho$$

(a)   (b)

Figure 6-5

6.1

(Eq. 4-1c). Then, multiplying Eq. 6-3 by the shear modulus (modulus of rigidity) G gives

$$\tau_\rho = \frac{\tau_c}{c}\rho \tag{6-4}$$

When Eq. 6-4 is substituted into Eq. 6-1, the result is

$$T_r = \frac{\tau_c}{c}\int \rho^2 dA = \frac{\tau_\rho}{\rho}\int \rho^2 dA \tag{a}$$

The integral in Eq. (a) is called the *polar second moment of area*[2] and is given the symbol J. For a solid circular shaft (see Fig. 6-5a), the polar second moment J is

$$J = \int \rho^2 dA = \int_0^c \rho^2 (2\pi\,\rho\,d\rho) = \frac{\pi c^4}{2} \tag{6-5a}$$

For a circular annulus (see Fig. 6-5b), the polar second moment J is

$$\begin{aligned} J = \int \rho^2 dA &= \int_b^c \rho^2 (2\pi\,\rho\,d\rho) \\ &= \frac{\pi c^4}{2} - \frac{\pi b^4}{2} = \frac{\pi}{2}\left(r_o^4 - r_i^4\right) \end{aligned} \tag{6-5b}$$

where r_o and r_i are the outer and inner radii, respectively, of the circular annulus. In terms of the polar second moment J, Eq. (a) can be written as

$$T_r = \frac{\tau_c J}{c} = \frac{\tau_\rho J}{\rho} \tag{b}$$

or solving for the unknown shearing stress τ

$$\tau_\rho = \frac{T_r \rho}{J} \qquad \text{and} \qquad \tau_c = \frac{T_r c}{J} \tag{6-6}$$

---

[2]Integrals of the type $\int x^2\,dA$ arise often in mechanics and are given the general name second moments of area. Second moments of area are discussed further in Chapter 7, where they are used to relate stresses to internal forces and moments in beams. Second moments of area are sometimes (improperly) called moments of inertia since they are closely related to the moment of inertia integral $\int \rho^2\,dm$, which arises in dynamics.

Like the shearing strain γ_ρ, the shearing stress τ_ρ is zero at the center of the shaft and increases linearly with respect to the distance ρ from the axis of the shaft. Both the shearing strain γ and the shearing stress τ are maximum when $\rho = c$. Equation 6-6 is known as the *elastic torsion formula*, in which τ_ρ is the shearing stress on a transverse plane at a distance ρ from the axis of the shaft, and T is the resisting torque (the torque produced on the transverse plane by the shearing stresses). Equation 6-6 is valid for both solid and hollow circular shafts. The resisting torque T_r is generally different than the external torques applied to various points along the shaft and must be obtained from a free-body diagram and an equilibrium equation. The procedure for calculating the resisting torque is illustrated in Example Problem 6-1. Note that Eq. 6-6 applies only for linearly elastic action in homogeneous and isotropic materials since Hooke's law $\tau = G\gamma$ was used in its development.

6-4 TORSIONAL DISPLACEMENTS

Frequently the amount of twist in a shaft is of paramount importance. Therefore, determination of the angle of twist is a common problem for the machine designer. The fundamental approach to such problems is provided by the following equations:

$$\gamma_\rho = \rho\frac{\theta}{L} \quad \text{or} \quad \gamma_\rho = \rho\frac{d\theta}{dL} \tag{6-2}$$

$$\tau_\rho = \frac{T_r\rho}{J} \quad \text{or} \quad \tau_c = \frac{T_r c}{J} \tag{6-6}$$

$$G = \frac{\tau}{\gamma} \tag{4-1}$$

The second form of Eq. 6-2 is used when the torque or the cross section varies as a function of position along the length of the shaft. Equation 6-2 is valid for both elastic and inelastic action. Equation 6-6 is the elastic torsion formula that provides the shearing stress τ_ρ on a transverse plane at a distance ρ from the axis of the shaft. Equation 4-1 is Hooke's law for shearing stress. The last two expressions are limited to stresses below the proportional limit of the material (elastic action). The three equations can be combined to give several different relationships; for example,

$$\theta = \frac{\gamma_\rho L}{\rho} = \frac{\tau_\rho L}{G\rho} \tag{6-7a}$$

or

$$\theta = \frac{T_r L}{GJ} \tag{6-7b}$$

The angle of twist determined from the above expressions is for a segment of shaft having a length L, a constant diameter (therefore, $J = $ constant), constant material properties ($G = $ constant), and carrying a constant resisting torque T_r.

The resisting torque T_r is the torque produced on the transverse plane by the shear stresses and is generally different from the external torques applied to the shaft at various sections by gears, pulleys, or couplings. Ideally, the length of the shaft should not include sections too near (within about one-half shaft diameter of) places where mechanical devices are attached. For practical purposes, however, it is customary to neglect local distortions at all connections and to compute angles as though there were no discontinuities.

If T_r, G, or J is not constant along the length of the shaft, Eq. 6-7b takes the form

$$\theta = \sum_{i=1}^{n} \frac{T_{ri} L_i}{G_i J_i} \tag{6-7c}$$

where each term in the summation is for a length L where T_r, G, and J are constant. If T_r, G, or J is a function of x (the distance along the length of the shaft), the angle of twist is found using

$$\theta = \int_0^L \frac{T_r dx}{GJ} \tag{6-7d}$$

6.2

Up to this point, no mention has been made of a sign convention for the internal torque or for the shear stress and the angle of twist in a shaft undergoing torsion. In Eq. 6-6 the signs of the internal torque and the shear stress are often ignored. Shear stresses in one direction are no better or worse than shear stresses in another direction. In Eq. 6-7c, if the angles of twist $\theta_i = \dfrac{T_{ri} L_i}{G_i J_i}$ of each segment L_i are in the same direction, they are simply added together. However, if the angles of twist of the different segments are not all in the same direction, it is necessary to decide which angles of twist to call positive and which to call negative.

A common sign convention used in torsion is that the internal resisting torque and the angle of twist are considered positive when the vectors representing them point outward from the internal section. Recall that the direction of a torque (moment) vector or a rotation vector is the axis about which the moment or the rotation occurs. The sense of the moment or rotation is counterclockwise when looking back along the vector toward the internal section. All of the torques and rotations shown in Figs. 6-6a and c are positive.

The problem with this sign convention (or with any other sign convention) is that it is not always consistent with the positive direction for shear stress as defined in Chapter 2. That is, a positive torque acting on a vertical shaft as shown in Fig. 6-6a results in a positive shear stress on an element next to the internal transverse section as shown in Fig. 6-6b. However, a positive torque acting on a horizontal shaft as shown in Fig. 6-6c results in a negative shear stress on an element next to the internal transverse section as shown in Fig. 6-6d.

In the rest of this chapter, the sign of the internal resisting torque will be ignored when calculating shear stress (Eq. 6-6), and the direction of the shear stress will be obtained from the direction of the torque.[3] In torque diagrams and in the calculation of angles of twist, internal resisting torques and angles of twist will be considered positive when the vectors representing them point outward from the internal section.

---

[3] When the torsional shear stress is combined with normal stresses and other shear stresses in combined loading situations, it will be necessary to know the direction of the torsional shear stress.

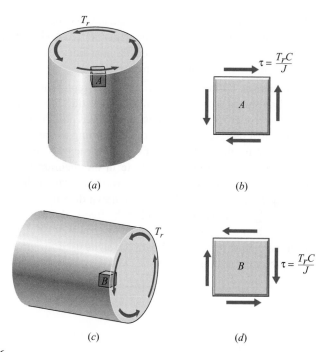

$$\tau = \frac{T_r C}{J}$$

$$\tau = \frac{T_r C}{J}$$

(a) (b)

(c) (d)

Figure 6-6

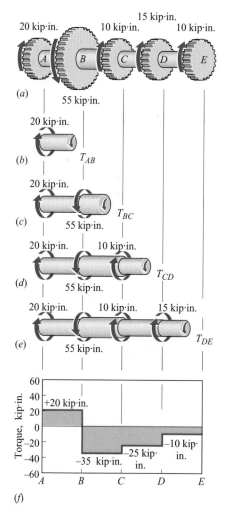

(a)

(b)

(c)

(d)

(e)

(f)

Figure 6-7

■ **Example Problem 6-1** A steel shaft is used to transmit torque from a motor to operating units in a factory. The torque is input at gear B (see Fig. 6-7a) and is removed at gears A, C, D, and E.

(a) Determine the torques transmitted by transverse cross sections in intervals AB, BC, CD, and DE of the shaft.

(b) Draw a torque diagram for the shaft.

SOLUTION

(a) The torques transmitted by transverse cross sections, or resisting torques, in intervals AB, BC, CD, and DE of the shaft shown in Fig. 6-7a are obtained by using the four free-body diagrams shown in Figs. 6-7b, c, d, and e. The internal resisting torques are labeled T_{AB}, T_{BC}, T_{CD}, and T_{DE} and are drawn in the positive direction on the free-body diagrams. The moment equilibrium equation $\Sigma M = 0$ about the axis of the shaft yields

$+\downarrow\Sigma M = 0$: $\qquad T_{AB} - 20 = 0 \quad T_{AB} = +20.0\,\text{kip}\cdot\text{in.}$ **Ans.**

$+\downarrow\Sigma M = 0$: $\qquad T_{BC} - 20 + 55 = 0 \quad T_{BC} = -35.0\,\text{kip}\cdot\text{in.}$ **Ans.**

$+\downarrow\Sigma M = 0$: $\quad T_{CD} - 20 + 55 - 10 = 0 \quad T_{CD} = -25.0\,\text{kip}\cdot\text{in.}$ **Ans.**

$+\downarrow\Sigma M = 0$: $T_{DE} - 20 + 55 - 10 - 15 = 0 \quad T_{DE} = -10.00\,\text{kip}\cdot\text{in.}$ **Ans.**

For all of the above calculations, a free-body diagram of the part of the bar to the left of the transverse section has been used. A free-body diagram of the part of the bar to the right of the section would have yielded identical results. In fact, for the determination of T_{CD} and T_{DE}, the free-body diagram to the right of the section would have been more efficient since fewer torques would have appeared on the free-body diagrams.

▶ It is the torques T_{AB}, T_{BC}, T_{CD}, and T_{DE} shown on the torque diagram (not the torques applied to the gears) that are used in the elastic torsion formula (Eqs. 6-6 and 6-7).

(b) A torque diagram is a graph in which abscissas represent distances along the shaft and ordinates represent the internal resisting torques at the corresponding transverse cross sections. Positive torques point outward from the cross section when represented as a vector according to the right-hand rule. A torque diagram for the shaft of Fig. 6-7a, constructed by using the results from part (a), is shown in Fig. 6-7f. Note in the diagram that the abrupt changes in torque are equal to the applied torques at gears A, B, C, D, and E. Thus, the torque diagram could have been drawn directly below the sketch of the shaft of Fig. 6-7a, without the aid of the free-body diagrams shown in Figs. 6-7b, c, d, and e, by using the applied torques at gears A, B, C, D, and E. However, care must be exercised with the signs used for torques because Fig. 6-7f represents resisting torques.

6.3

6.4

Example Problem 6-2   A hollow steel shaft with an outside diameter of 400 mm and an inside diameter of 300 mm is subjected to a torque of 300 kN · m, as shown in Fig. 6-8. The modulus of rigidity G (shear modulus) for the steel is 80 GPa. Determine

(a) The maximum shearing stress in the shaft.

(b) The shearing stress on a transverse cross section at the inside surface of the shaft.

(c) The magnitude of the angle of twist in a 2-m length.

SOLUTION

Equations for shearing stress and angle of twist in a circular shaft subjected to a torque contain the polar second moment J of the cross section, which is given by Eq. 6-5b as

$$J = \frac{\pi}{2}\left(r_o^4 - r_i^4\right) = \frac{\pi}{2}(200^4 - 150^4)$$
$$= 1718.1(10^6)\,\text{mm}^4 = 1718.1(10^{-6})\,\text{m}^4$$

(a) The resisting torque on all cross sections of the shaft is $T_r = 300$ kN · m. The maximum shearing stress occurs on a transverse cross section at the outer surface of the shaft and is given by Eq. 6-6 as

$$\tau_c = \frac{T_r c}{J} = \frac{300(10^3)(200)(10^{-3})}{1718.1(10^{-6})}$$
$$= 34.92(10^6)\,\text{N/m}^2 \cong 34.9\,\text{MPa} \qquad \textbf{Ans.}$$

The shear stress on an element at the outside surface is shown in Fig. 6-9.

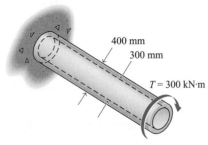

400 mm

300 mm

$T = 300$ kN·m

Figure 6-8

(b) The shearing stress on a transverse cross section at the inner surface of the shaft is given by Eq. 6-6 as

$$\tau_\rho = \frac{T_r \rho}{J} = \frac{300(10^3)(150)(10^{-3})}{1718.1(10^{-6})}$$

$$= 29.19(10^6) \, \text{N/m}^2 \cong 26.2 \, \text{MPa} \qquad \textbf{Ans.}$$

Figure 6-9

(c) The angle of twist in a 2-m length is given by Eq. 6-7b as

$$\theta = \frac{T_r L}{GJ} = \frac{300(10^3)(2)}{80(10^9)(1718.1)(10^{-6})}$$

$$= 0.004365 \, \text{rad} \cong 0.00437 \, \text{rad} \qquad \textbf{Ans.}$$

▶ Remember that the angle θ in Eq. 6-7 is in radians since the small angle approximation $\tan \theta \cong \theta$ was used in its development. In this case, $\tan 0.004365 \, \text{rad} = 0.004365$, and the small angle approximation is certainly appropriate.

Example Problem 6-3 A solid steel shaft 14 ft long has a diameter of 6 in. for 9 ft of its length and a diameter of 4 in. for the remaining 5 ft. The shaft is attached to a wall at its left end and is in equilibrium when subjected to the two torques shown in Fig. 6-10a. The modulus of rigidity (shear modulus) of the steel is 12,000 ksi. Determine

6.5

(a) The maximum shearing stress in the shaft.
(b) The rotation of end B of the 6-in. segment with respect to end A.
(c) The rotation of end C of the 4-in. segment with respect to end B.
(d) The rotation of end C with respect to end A.

6.6

6.7

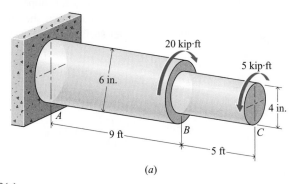

(a)

Figure 6-10(a)

SOLUTION
In general, free-body diagrams should be drawn to evaluate the resisting torque in each section of the shaft. Such diagrams are shown in Figs. 6-10b and c in which T_{AB} and T_{BC} are the internal resisting torques in segments AB and BC and are drawn in the positive direction. From the free-body diagram of Fig. 6-10b,

$$\Sigma M_x = 0: \qquad T_{AB} + 20 - 5 = 0 \qquad T_{AB} = -15 \, \text{kip} \cdot \text{ft}$$

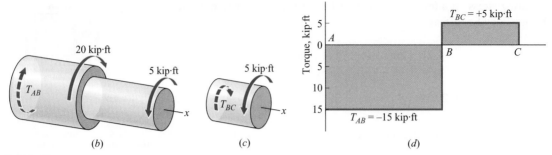

Figure 6-10(b–d)

From the free-body diagram of Fig. 6-10c,

$$\Sigma M_x = 0: \qquad\qquad T_{BC} - 5 = 0 \qquad\qquad T_{BC} = +5\,\text{kip}\cdot\text{ft}$$

A torque diagram, such as the one shown in Fig. 6-10d, provides a pictorial representation of the levels of torque being transmitted by each of the sections and serves as an aid for stress and deformation calculations.

Equations for the shearing stress and angle of twist in a circular shaft subjected to a torque contain the polar second moment J of the cross section, which is given by Eq. 6-5a as

$$J_{BC} = \frac{\pi}{2}c^4 = \frac{\pi}{2}(2^4) = 25.13\ \text{in.}^4$$

$$J_{AB} = \frac{\pi}{2}c^4 = \frac{\pi}{2}(3^4) = 127.23\ \text{in.}^4$$

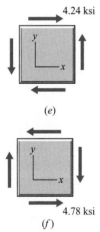

(e)

(f)

Figure 6-10(e–f)

(a) The location of the maximum shearing stress is not apparent; hence, the stress must be checked at both sections. The maximum shearing stress on a transverse cross section occurs at the outer surface of the shaft and is given by Eq. 6-6 as

$$\tau_{AB} = \frac{T_{AB}c_{AB}}{J_{AB}} = \frac{15(12)(3)}{127.23} = 4.244\,\text{ksi}$$

$$\tau_{BC} = \frac{T_{BC}c_{BC}}{J_{BC}} = \frac{5(12)(2)}{25.13} = 4.775\,\text{ksi}$$

Therefore,

$$\tau_{\text{max}} = \tau_{BC} = 4.775\,\text{ksi} \cong 4.78\,\text{ksi} \qquad\qquad\qquad \textbf{Ans.}$$

The shear stresses τ_{AB} and τ_{BC} are shown in Figs. 6-10e and f, respectively. The direction of the shear stresses is the same as the direction of the internal torques. For use in the stress transformation equations, τ_{AB} would be a positive shear stress and τ_{BC} would be a negative shear stress.

(b) As the resisting torque of $-15\,\text{kip}\cdot\text{ft}$ is transmitted from section to section in segment AB of the shaft, the section at B twists relative to the section at A by an amount $\theta_{B/A}$, as shown on the angle of twist diagram of Fig. 6-10g.

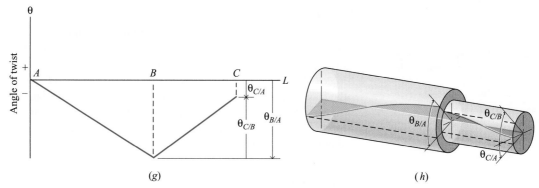

(g)   (h)

Figure 6-10(g–h)

The slope of the angle of twist diagram $\theta_{B/A}/L_{AB}$ is constant since the term T/JG (in Eq. 6-7b) is constant. The rotation of the section at B (angle of twist in the 6-in. section) is given by Eq. 6-7b as

$$\theta_{B/A} = \frac{T_{AB}L_{AB}}{G_{AB}J_{AB}} = \frac{-15(12)(9)(12)}{12,000(127.23)}$$

$$= -0.012733 \text{ rad} \cong 0.01273 \text{ rad} -\uparrow - \qquad \textbf{Ans.}$$

(c) Similarly for segment BC,

$$\theta_{C/B} = \frac{T_{BC}L_{BC}}{G_{BC}J_{BC}} = \frac{+5(12)(5)(12)}{12,000(25.13)}$$

$$= +0.011938 \text{ rad} \cong 0.01194 \text{ rad} -\downarrow - \qquad \textbf{Ans.}$$

▶ Equation 6-7 can be used to calculate the angle of twist of any section of the shaft relative to any other section of the shaft so long as T, G, and J are all constant on the section.

(d) If there were no resisting torque being transmitted by segment BC, it would rotate as a rigid body through angle $\theta_{B/A}$. However, the resisting torque of 5 kip · ft causes the section at C to rotate relative to the section at B by an amount $\theta_{C/B}$ as segment BC deforms, as shown in Fig. 6-10g. The resultant of the deformations in the two segments of the shaft is

$$\theta_{C/A} = \theta_{C/B} + \theta_{B/A}$$
$$= (+0.011938) + (-0.012733)$$
$$= -0.000795 \text{ rad}$$
$$= 0.000795 \text{ rad} -\uparrow- \qquad \textbf{Ans.}$$

The deformations for the entire shaft are pictorially shown in Fig. 6-10h.

Example Problem 6-4   Two 1.50-in.-diameter steel ($G = 12,000$ ksi) shafts are connected with gears, as shown in Fig. 6-11a. The diameters of gears B and C are 10 in. and 6 in., respectively. If an input torque of $T_A = 750$ lb · ft is applied at section A of shaft AB, determine

6.8

(a) The maximum shearing stress on a cross section of shaft CD.

(b) The rotation of section A of shaft AB with respect to its no-load position.

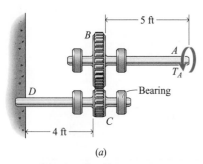

(a)

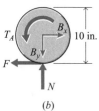

(b)

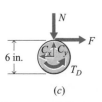

(c)

Figure 6-11(a–c)

SOLUTION

(a) The torque at section D of shaft CD required for equilibrium of the system can be determined from equilibrium considerations for the two shafts. As an aid for these considerations, free-body diagrams for gears B and C are shown in Figs. 6-11b and c, respectively. The forces B_x, B_y, C_x, and C_y are the forces at the frictionless bearings, and do not affect the torque calculations. The input torque T_A in shaft AB is transferred to shaft CD by means of the gear tooth force F shown in the two diagrams. Thus, from a summation of moments about the axis of each of the shafts:

For shaft AB, $\qquad\qquad\qquad T_A - r_B F = 0 \qquad\qquad (a)$

For shaft CD, $\qquad\qquad\qquad T_D - r_C F = 0 \qquad\qquad (b)$

Since the force F in Eqs. (a) and (b) must be equal,

$$T_D = (r_C/r_B)T_A = (3/5)(750) = 450 \, \text{lb} \cdot \text{ft}$$

The magnitude of the resisting torque on all cross sections of the shaft CD is $T_r = 450 \, \text{lb} \cdot \text{ft}$. The maximum shearing stress on a transverse cross section of shaft CD occurs at the outside surface of the shaft and is given by Eq. 6-6 as

$$\tau_c = \frac{T_{CD}c_{CD}}{J_{CD}} = \frac{450(12)(0.75)}{(\pi/2)(0.75)^4}$$

$$= 8149 \, \text{psi} \cong 8150 \, \text{psi} \qquad\qquad \textbf{Ans.}$$

(b) The quantities required for the determination for the rotation of section A of shaft AB with respect to its no-load position are illustrated on the angle of twist diagram shown in Fig. 6-11d. The rotation of the section at C relative to the section at D in shaft CD is given by Eq. 6-7b as

$$\theta_{C/D} = \frac{T_{CD}L_{CD}}{G_{CD}J_{CD}} = \frac{-450(12)(4)(12)}{12{,}000{,}000(\pi/2)(0.75)^4} = 0.04346 \, \text{rad} - \uparrow -$$

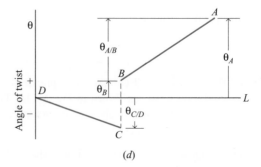

(d)

Figure 6-11(d)

The teeth on gears B and C must move through the same arc length but in opposite directions. Therefore,

$$s = r_B\theta_B = r_C\theta_{C/D}$$

from which

$$\theta_B = (r_C/r_B)(\theta_{C/D}) = (3/5)(0.04346) = 0.02608\,\text{rad} \,-\!\downarrow\!-$$

The magnitude of the resisting torque on all cross sections of the shaft AB is $T_r = 750\,\text{lb}\cdot\text{ft}$, and the rotation of the section at A relative to the section at B in shaft AB is given by Eq. 6-7b as

$$\theta_{A/B} = \frac{T_{AB}L_{AB}}{G_{AB}J_{AB}} = \frac{750(12)(5)(12)}{12,000,000(\pi/2)(0.75)^4} = 0.09054\,\text{rad} \,-\!\downarrow\!-$$

Finally,

$$\theta_A = \theta_B + \theta_{A/B}$$
$$= 0.02608 + 0.09054 = 0.11662\,\text{rad} = 6.68° \,-\!\downarrow\!- \qquad \textbf{Ans.}$$

▶ The torque T_A causes the entire shaft AB to rotate counterclockwise (when viewed from end A), and section A rotates farther than does the gear B. However, when gear B rotates counterclockwise, gear C rotates clockwise (again viewed from end C). Thus, the angle of twist diagram (Fig. 6-11d) goes from a negative angle of twist at C to a positive angle of twist at B.

Example Problem 6-5

The solid circular tapered shaft of Fig. 6-12 is subjected to end torques applied in transverse planes. Determine the magnitude of the angle of twist in terms of T, L, G, and r. Assume elastic action and a slight taper.

SOLUTION

Note that Eq. 6-2 was developed assuming that plane cross sections remain plane and that all longitudinal elements have the same length. Neither of these assumptions is strictly valid for the tapered shaft; but if the taper is slight, the error involved is negligible. Therefore, from Eqs. 6-2, 6-6, and 4-1,

$$d\theta = \frac{\gamma}{\rho}dx \qquad \tau = \frac{T_r\rho}{J} = \frac{T\rho}{\pi\rho^4/2} = \frac{2T}{\pi\rho^3} \qquad \gamma = \frac{\tau}{G}$$

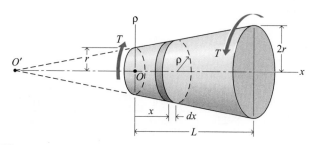

Figure 6-12

Therefore,

$$d\theta = \frac{\tau}{G\rho}dx = \frac{2T}{G\pi\rho^4}dx$$

The radius ρ can be expressed as a function of x; thus,

$$\rho = r + \frac{2r - r}{L}x = \frac{r}{L}(L + x)$$

Substituting this value for ρ into the expression for $d\theta$ gives

$$d\theta = \frac{2TL^4}{G\pi r^4(L + x)^4}dx$$

Integrating to obtain angle θ yields

$$\theta = \frac{2TL^4}{G\pi r^4}\int_0^L \frac{dx}{(L + x)^4} = -\frac{2TL^4}{3G\pi r^4}\left(\frac{1}{8L^3} - \frac{1}{L^3}\right) = \frac{7TL}{12G\pi r^4} \qquad \textbf{Ans.}$$

Alternatively, the origin of coordinates can be placed at a distance L to the left of O in Fig. 6-12 (point O'). The function for ρ then becomes

$$\rho = \frac{r}{L}x$$

and

$$\theta = \frac{2TL^4}{G\pi r^4}\int_L^{2L} \frac{dx}{x^4} = \frac{7TL}{12G\pi r^4} \qquad \textbf{Ans.}$$

PROBLEMS

MecMovies Activities and Problems

MM6.1   Torsion concepts. Concept checkpoints. Basic torsion problems involving internal torques, shear stress, and angles of twist.

MM6.2   Gear basics. Theory; Concept checkpoints. Basic gear relationships for torque, rotation angle, rotation speed, and power transmission.

MM6.3   Gear trains: torque and shear stress. Concept checkpoints. Basic calculations involving two shafts connected by gears.

MM6.4   Gear trains: torque and shear stress. Concept checkpoints. Basic calculations involving three shafts connected by gears.

MM6.5   Gear Trains: angles of twist. Concept checkpoints.

Introductory Problems

6-1*   For the steel shaft shown in Fig. P6-1,

a. Determine the torques transmitted by transverse sections in intervals *AB, BC, CD*, and *DE* of the shaft.

b. Draw a torque diagram for the shaft.

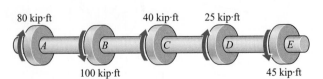

Figure P6-1

6-2* For the steel shaft shown in Fig. P6-2,

a. Draw a torque diagram for the shaft.

b. Determine the maximum torque transmitted by any transverse cross section of the shaft.

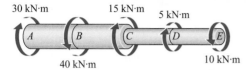

Figure P6-2

6-3 For the steel shaft shown in Fig. P6-3,

a. Draw a torque diagram for the shaft.

b. Determine the maximum torque transmitted by any transverse cross section of the shaft.

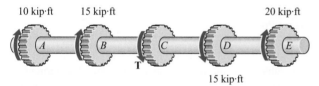

Figure P6-3

6-4* The motor shown in Fig. P6-4 supplies a torque of 500 N · m to the shaft *BCDE*. The torques removed at gears *C*, *D*, and *E* are 100 N · m, 150 N · m, and 250 N · m, respectively.

a. Determine the torques transmitted by cross sections in intervals *BC*, *CD*, and *DE* of the shaft.

b. Draw a torque diagram for the shaft.

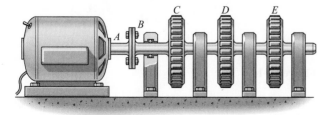

Figure P6-4

6-5 A solid circular steel shaft 2 in. in diameter is subjected to a torque of 18,000 lb · in. The modulus of rigidity (shear modulus) for the steel is 12,000 ksi. Determine

a. The maximum shearing stress in the shaft.

b. The magnitude of the angle of twist in a 6-ft length of the shaft.

6-6 A hollow steel shaft has an outside diameter of 120 mm and an inside diameter of 80 mm. The shaft is subjected to a torque of 28 kN · m. The modulus of rigidity (shear modulus) for the steel is 80 GPa. Determine

a. The shearing stress on a transverse cross section at the outside surface of the shaft.

b. The shearing stress on a transverse cross section at the inside surface of the shaft.

c. The magnitude of the angle of twist in a 2.0-m length of the shaft.

d. The magnitude of the angle of twist in a 2.0-m-long solid shaft that has the same weight as the hollow shaft.

6-7* Specifications for a solid circular aluminum alloy ($G = 4000$ ksi) rod 6.5 ft long require that it be adequate to resist a torque of 2200 lb · ft without twisting more than 5° or exceeding a shearing stress of 14.5 ksi. What minimum diameter is required?

6-8 A torque of 10 kN · m is supplied to the steel ($G = 80$ GPa) factory drive shaft of Fig. P6-8 by a belt that drives pulley *A*. A torque of 6 kN · m is taken off by pulley *B* and the remainder by pulley *C*. Shafts *AB* and *AC* are 2.25 m long and 1.60 m long, respectively. If the diameter of shaft *AB* is 80 mm and the diameter of shaft *AC* is 65 mm, determine

a. The maximum shearing stress in each of the shafts.

b. The angle of twist of pulley *B* with respect to pulley *A*.

c. The angle of twist of pulley *C* with respect to pulley *B*.

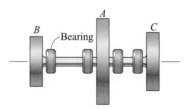

Figure P6-8

Intermediate Problems

6-9* The shaft shown in Fig. P6-9 consists of a brass ($G = 5600$ ksi) tube *AB* that is securely connected to a solid stainless-steel ($G = 12,500$ ksi) bar *BC*. Tube *AB* has an outside diameter of 5 in. and an inside diameter of 2.5 in. Bar *BC* has a diameter of 3.5 in. Torques T_1 and T_2 are 70 kip · in. and 30 kip · in., respectively, in the directions shown. Determine

a. The maximum shearing stress in the shaft.

b. The rotation of a section at *C* with respect to its no-load position.

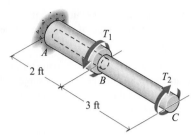

Figure P6-9

6-10* The solid circular steel ($G = 80$ GPa) shaft of Fig. P6-10 has a diameter of 100 mm. If the gears are spaced at 1.50-m intervals, determine

a. The maximum shearing stress in the shaft.
b. The rotation of a section at D with respect to a section at B.
c. The rotation of a section at E with respect to a section at A.

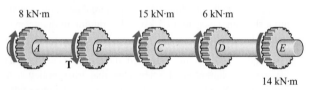

Figure P6-10

6-11 The hollow circular steel ($G = 12,000$ ksi) shaft of Fig. P6-11 is in equilibrium under the torques indicated. Determine

a. The maximum shearing stress in the shaft.
b. The rotation of a section at D with respect to a section at B.
c. The rotation of a section at D with respect to a section at A.

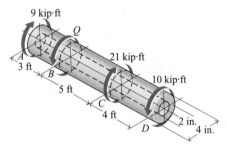

Figure P6-11

6-12* A motor supplies a torque of 5.5 kN · m to the constant diameter steel ($G = 80$ GPa) line shaft shown in Fig. P6-12. Three machines are driven by the gears B, C, and D on the shaft, and they require torques of 3.0 kN · m, 1.5 kN · m, and 1.0 kN · m, respectively. Determine

a. The minimum diameter required if the maximum shearing stress in the shaft is limited to 100 MPa.
b. The rotation of gear D with respect to the coupling at A if the coupling and gears are spaced at 2-m intervals and the shaft diameter is 75 mm.

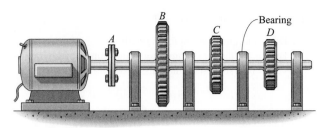

Figure P6-12

6-13 A solid circular aluminum alloy ($G = 4000$ ksi) shaft with diameters of 2.5 in. and 1.75 in. is subjected to a torque T, as shown in Fig. P6-13. The allowable shearing stress is 8000 psi, and the maximum allowable angle of twist in the 7-ft length is 0.04 rad. Determine the maximum allowable value of T.

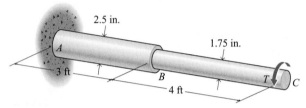

Figure P6-13

6-14 A stepped steel ($G = 80$ GPa) shaft has the dimensions and is subjected to the torques shown in Fig. P6-14. Determine

a. The maximum shearing stress on a section 3 m from the left end of the shaft.
b. The rotation of a section 2 m from the left end of the shaft with respect to its no-load position.
c. The rotation of the section at the right end of the shaft with respect to its no-load position.

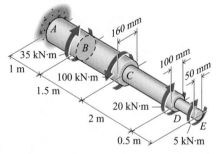

Figure P6-14

6-15* A torque T is applied to the right end of shaft AB of Fig. P6-15. The mean diameter of bevel gear C is twice that of bevel gear B. Both shafts are made of steel ($G = 12,000$ ksi). Shaft AB has a diameter of 1.5 in., and shaft CD has a diameter of 2.0 in. If the maximum shearing stress in either shaft must not exceed 15 ksi, determine

a. The maximum permissible torque T.
b. The rotation of a section at A relative to its no-load position.

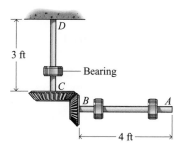

Figure P6-15

6-16 The solid circular shaft and the hollow tube shown in Fig. P6-16 are both attached to a rigid circular plate at the left end. A torque $T_A = 2$ kN · m applied to the right end of the shaft is resisted by a torque T_B at the right end of the tube. The shaft is made of steel ($G = 80$ GPa), and the tube is made of an aluminum alloy ($G = 28$ GPa). If the shaft has a diameter of 50 mm and the tube has an outside diameter of 80 mm, determine

a. The maximum inside diameter that can be used for the tube if the maximum shearing stress in the tube must be limited to 50 MPa.
b. The maximum inside diameter that can be used for the tube if the rotation of the right end of the shaft with respect to the right end of the tube must be limited to 0.25 rad.

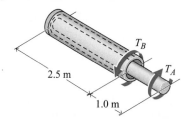

Figure P6-16

Challenging Problems

6-17* A torque of 30 lb · ft is applied through gear A to the left end of the gear train shown in Fig. P6-17. The diameters of gears B and C are 5 in. and 2 in., respectively. If the maximum shearing stresses in the aluminum alloy ($G = 3800$ ksi) shafts AB and CD are limited to 12 ksi, determine

a. The minimum permissible diameter for shaft AB.
b. The minimum permissible diameter for shaft CD.
c. The maximum length for shaft CD if the rotation of a section at D with respect to a section at C must not exceed 0.5 rad.

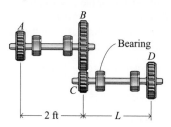

Figure P6-17

6-18* The motor shown in Fig. P6-18 supplies a torque of 45 kN · m to shaft AB. Two machines are powered by gears D and E. The torque delivered by gear E to the machine is 8 kN · m. Shafts AB and CDE are made of steel ($G = 80$ GPa) and have 150-mm and 80-mm diameters, respectively. If the diameters of gears B and C are 450 mm and 150 mm, respectively, determine

a. The maximum shearing stress in shaft AB.
b. The maximum shearing stress in shaft CDE.
c. The rotation of gear E relative to gear D.

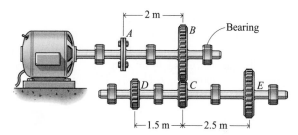

Figure P6-18

6-19 Torque is applied to the steel ($G = 11,600$ ksi) shaft shown in Fig. P6-19 through gear C and is removed through gears A and B. If the torque applied to gear C by the motor is 8800 lb · ft and the torque removed through gear B is 5200 lb · ft, determine

a. The minimum permissible diameter for each section of the shaft if the maximum shearing stresses must not exceed 18 ksi.
b. The minimum permissible uniform diameter for a shaft with $L_1 = 5$ ft and $L_2 = 4$ ft if the rotation of gear A relative to gear C must be less than 0.15 rad.

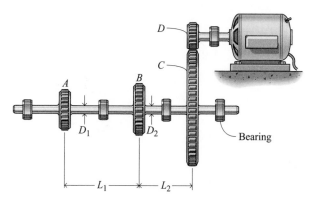

Figure P6-19

6-20* The solid circular tapered shaft of Fig. P6-20 is subjected to a constant torque T. Determine the angle of twist in terms of T, L, G, r, and m.

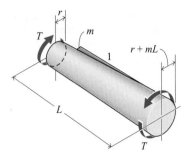

Figure P6-20

6-21 The solid cylindrical shaft of Fig. P6-21 is subjected to a uniformly distributed torque of q. Determine the rotation of the left end of the shaft caused by the distributed torque q. Express your answer in terms of q, L, G, and c. The dimensions of q are moment per unit of length.

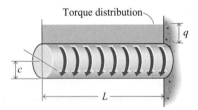

Figure P6-21

6-22 The tapered circular shaft of Fig. P6-22 has an axial hole of constant diameter throughout its length. Determine the angle of twist due to a constant torque T in terms of T, L, G, R, and r.

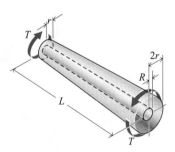

Figure P6-22

6-23* The hollow tapered shaft of Fig. P6-23 has a constant wall thickness t. Determine the angle of twist for a constant torque T in terms of T, L, G, t, and r. Note that when t is small, the approximate expression for the polar second moment of area ($J = r^2 A$ where A is the cross-sectional area of the shaft) may be used.

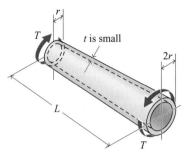

Figure P6-23

6-24 The solid cylindrical shaft of Fig. P6-24 is subjected to a distributed torque that varies linearly from zero at the left end to q at the right end. Determine the rotation of the left end of the shaft caused by the distributed torque q. Express your answer in terms of q, L, G, and c. The dimensions of q are moment per unit of length.

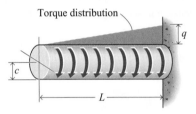

Figure P6-24

Computer Problems

6-25 A hollow circular steel ($G = 11,000$ ksi) shaft 3 ft long is being designed to transmit a torque T of 3000 lb · ft. The outer radius r_o of the shaft can vary, but the cross-sectional area A of the shaft must remain constant ($A = 3$ in.^{2}). Calculate and plot

a. The angle of twist θ for the 3-ft length as a function of the outer radius r_o (1 in. $\leq r_o \leq 4$ in.).
b. The maximum shearing stress τ_c in the shaft as a function of the outer radius r_o (1 in. $\leq r_o \leq 4$ in.).

6-26 A hollow circular brass ($G = 40$ GPa) shaft 2 m long is being designed to transmit a torque T of 7500 N · m. The outer radius r_o of the shaft must be fixed ($r_o = 50$ mm); however, the inner radius r_i of the shaft can vary (0 mm $\leq r_i \leq 45$ mm). Calculate and plot

a. The angle of twist θ for the 2-m length as a function of the radius ratio r_i/r_o ($0 \leq r_i/r_o \leq 0.9$).
b. The maximum shearing stress τ_c in the shaft as a function of the radius ratio r_i/r_o ($0 \leq r_i / r_o \leq 0.9$).

6-27 A hollow circular steel ($G = 11,000$ ksi) shaft 3 ft long is being designed to transmit a torque T of 3000 lb · ft. The wall thickness of the shaft is 0.25 in.

a. Calculate and plot the angle of twist θ for the 3-ft length as a function of the outer radius r_o (0.5 in. $\leq r_o \leq 3$ in.).

b. Calculate and plot the maximum shearing stress τ_c in the shaft as a function of the outer radius r_o (0.5 in. $\leq r_o \leq$ 3 in.).

c. Since increasing the outside radius of the shaft increases its weight and cost, what do you think would be a reasonable minimum value for the outside radius?

6-28   The shaft of Fig. P6-28 is turned out of aluminum ($G =$ 28 GPa). Section AC is 1.5 m long and 100 mm in diameter (AB is hollow, and BC is solid); section CD is 0.50 m long and 75 mm in diameter.

a. If the diameter of the hole from A to B is 75 mm, calculate and plot the angle of twist $\theta_{D/A}$ as a function of the distance L_{AB} (0 m $\leq L_{AB} \leq$ 1.4 m).

b. Repeat for a 90-mm-diameter hole.

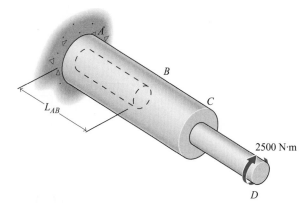

Figure P6-28

6-5   STRESSES ON OBLIQUE PLANES

At this point it is necessary to ascertain whether the transverse plane is a plane of maximum shearing stress and whether there are other significant stresses induced by torsion. For this study, the stresses at point A in the shaft of Fig. 6-13a will be analyzed. Figure 6-13b shows a differential element taken from the shaft at A and the stresses acting on transverse and longitudinal planes. The shearing stress τ_{xy} can be determined by means of the torsion formula.[4] The equality of shearing stresses on orthogonal planes was previously discussed in Section 2-6 (see Eq. 2-11).

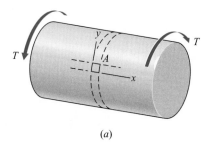

(a)

$$\tau_{yx} = \tau_{xy} \tag{6-8}$$

Therefore, if a shearing stress exists at a point on any plane, a shearing stress of the same magnitude must also exist at this point on an orthogonal plane. This statement is also valid when normal stresses are acting on the planes, since the normal stresses occur in collinear but oppositely directed pairs, and thus they have zero moment with respect to any axis.

The normal stress σ_n and the shear stress τ_{nt} on the inclined plane of Fig. 6-13c can be found by using the stress transformation equations (Eqs. 2-12 and 2-13). The stresses and angle for use in these equations are

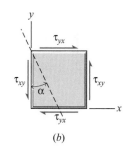

(b)

$$\sigma_x = 0 \qquad \sigma_y = 0 \qquad \tau_{xy} = \tau_{yx} \qquad \theta = \alpha$$

Equations 2-12 and 2-13 then give

$$\begin{aligned}\sigma_n &= \sigma_x \cos^2 \theta + \sigma_y \sin^2 \theta + 2\tau_{xy} \sin \theta \cos \theta \\ &= 0 + 0 + 2\tau_{xy} \sin \alpha \cos \alpha = 2\tau_{xy} \sin \alpha \cos \alpha\end{aligned} \tag{a}$$

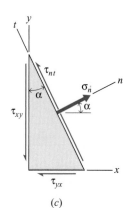

(c)

<hr/>

[4]The torque T is generated by a shear stress $\tau_{xy} = Tc/J$. For the applied torque of Fig. 6-13a, the shear stress will act in the direction shown on Fig. 6-13b (which is in the positive sense as defined in Section 2-8). The double subscript naming convention for the shear stress was described in Section 2-7.

Figure 6-13

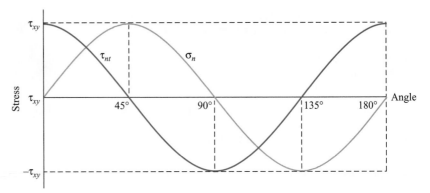

Figure 6-14

and

$$\tau_{nt} = -(\sigma_x - \sigma_y) \sin \theta \cos \theta + \tau_{xy}(\cos^2 \theta - \sin^2 \theta)$$
$$= 0 + \tau_{xy}(\cos^2 \alpha - \sin^2 \alpha) = \tau_{xy}(\cos^2 \alpha - \sin^2 \alpha) \tag{b}$$

Expressing Eqs. (*a*) and (*b*) in terms of the double angle 2α yields

$$\sigma_n = \tau_{xy} \sin 2\alpha \tag{6-9}$$
$$\tau_{nt} = \tau_{xy} \cos 2\alpha \tag{6-10}$$

In Eqs. 6-9 and 6-10, σ_n is the normal stress on the inclined plane and τ_{nt} is the shearing stress on the same plane. The shearing stress τ_{xy} is found using the elastic torsion formula (Eq. 6-6). At a given point of the circular shaft τ_{xy} is constant, and thus Eqs. 6-9 and 6-10 show that the stresses σ_n and τ_{nt} are functions of the angle of the inclined plane α. The results obtained from Eqs. 6-9 and 6-10 are shown on the graph of Fig. 6-14, from which it is apparent that the maximum shearing stress occurs on both transverse ($\alpha = 0$) and longitudinal ($\alpha = 90°$) planes. The graph also shows that maximum normal stresses occur on planes oriented at 45° with the axis of the bar and perpendicular to the surface of the bar. On one of these planes ($\alpha = 45°$ on Fig. 6-14), the normal stress is tension, and on the other ($\alpha = 135°$), the normal stress is compression. Furthermore, all of these maximum stresses have the same magnitude; hence, the elastic torsion formula gives the magnitude of both the maximum normal stress and the maximum shearing stress at a point in a circular shaft subjected to pure torsion (the only loading is a torque).

Any of the stresses discussed previously may be significant in a given problem. Compare, for example, the failures shown in Fig. 6-15. In Fig. 6-15*a*, the steel rear axle of a truck split longitudinally. One would also expect this type of failure to occur in a shaft of wood with the grain running longitudinally. In Fig. 6-15*b*, the compressive stress caused the thin-walled aluminum alloy tube to buckle along one 45° plane, while the tensile stress caused tearing on the other 45° plane. Buckling of thin-walled tubes subjected to torsional loading is a matter of paramount concern to the designer. In Fig. 6-15*c*, the tensile stresses caused the gray cast iron bar to fail in tension—typical of any brittle material subjected to torsion. In Fig. 6-15*d*, the low-carbon steel bar failed in shear on a plane that is almost transverse—a typical failure for a ductile material. The reason the fracture in Fig. 6-15*d* did not occur on a transverse plane is that under the large plastic twisting deformation before

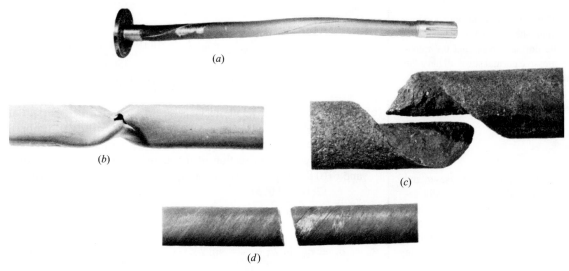

Figure 6-15

rupture (note the spiral lines indicating elements originally parallel to the axis of the bar), longitudinal elements were subjected to both torsion and axial tensile loading because the grips of the testing machine would not permit the bar to shorten as the elements were twisted into spirals. This axial tensile stress changes the plane of maximum shearing stress from a transverse to an oblique plane (resulting in a warped surface of rupture).[5] This will be discussed in later sections of this book.

Example Problem 6-6   A cylindrical tube is fabricated by butt-welding a 6-mm-thick steel plate along a spiral seam, as shown in Fig. 6-16. If the maximum compressive stress in the tube must be limited to 80 MPa, determine

(a) The maximum torque T that can be applied to the tube.

(b) The factor of safety with respect to failure by fracture for the weld, when a torque of 12 kN · m is applied, if the ultimate strengths of the weld metal are 205 MPa in shear and 345 MPa in tension.

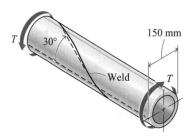

Figure 6-16

SOLUTION

(a) For the cylindrical tube,

$$J = \frac{\pi}{2}(75^4 - 69^4) = 14.096(10^6)\,\text{mm}^4 = 14.096(10^{-6})\,\text{m}^4$$

The magnitude of the maximum compressive stress in the tube is given by Eq. 6-6 (see Fig. 6-14) as

$$\sigma_{\max} = \tau_c = \frac{T_r c}{J} = 80\,\text{MPa} = 80(10^6)\,\text{N/m}^2$$

---

[5]The tensile stress is not entirely due to the grips because the plastic deformation of the outer elements of the bar is considerably greater than that of the inner elements. This results in a spiral tensile stress in the outer elements and a similar compressive stress in the inner elements.

▶ Recall that for a shaft in pure torsion, the maximum shearing stress, the maximum tensile normal stress, and the maximum compressive normal stress all have the same value:

$$\tau_c = \sigma_{\max T} = \sigma_{\max C} = T_r c / J$$

▶ In pure torsion the normal stresses σ_x and σ_y are both zero, and the stress transformation equations (Eqs. 2-12 and 2-13) reduce to Eqs. 6-9 and 6-10 only for the case of pure torsion. Equations 6-9 and 6-10 must not be used for axially loaded problems or any other more general loading situations.

Thus,

$$T_{\max} = T_r = \frac{\sigma_{\max} J}{c} = \frac{80(10^6)(14.096)(10^{-6})}{75(10^{-3})}$$
$$= 15.036(10^3) \text{ N} \cdot \text{m} \cong 15.04 \text{ kN} \cdot \text{m} \qquad \textbf{Ans.}$$

(b) The normal stress σ_n and the shear stress τ_{nt} on the weld surface are given by Eqs. 6-9 and 6-10. The clockwise torque $T = 12 \text{ kN} \cdot \text{m}$ on the right end of the shaft causes a negative resisting torque $T_r = -12 \text{ kN} \cdot \text{m}$, to be felt at each cross section of the shaft. But for the purpose of Eqs. 6-9 and 6-10 this negative resisting torque is generated by a positive shear stress

$$\tau_{xy} = \frac{12(10^3)(0.075)}{14.096(10^{-6})}.$$

$$\sigma_n = \tau_{xy} \sin 2\alpha = \frac{T_r c}{J} \sin 2\alpha = \frac{12(10^3)(75)(10^{-3})}{14.096(10^{-6})} \sin 2(60°)$$
$$= 55.29(10^6) \text{ N/m}^2 = 55.29 \text{ MPa (T)}$$

$$\tau_{nt} = \tau_{xy} \cos 2\alpha = \frac{T_r c}{J} \cos 2\alpha = \frac{12(10^3)(75)(10^{-3})}{14.096(10^{-6})} \cos 2(60°)$$
$$= -31.92(10^6) \text{ N/m}^2 = -31.92 \text{ MPa}$$

The minus sign indicates that the direction of τ_{nt} is opposite to that shown on Fig. 6-13c.

The factor of safety with respect to failure by fracture (normal stress) for the weld is

$$FS_\sigma = \frac{\sigma_{\text{ult}}}{\sigma_n} = \frac{345}{55.29} = 6.24$$

The factor of safety with respect to failure by fracture (shear stress) for the weld is

$$FS_\tau = \frac{\tau_{\text{ult}}}{\tau_{nt}} = \frac{205}{31.92} = 6.42$$

Therefore, the overall factor of safety with respect to failure for the weld surface is the smaller of FS_σ and FS_τ, or

$$FS = FS_\sigma = 6.24 \qquad \textbf{Ans.}$$

PROBLEMS

Introductory Problems

6-29*   A solid circular steel ($G = 12,000$ ksi) shaft with diameters as shown in Fig. P6-29 is subjected to a torque $T = 15$ kip $\cdot$ in. Determine

a. The maximum tensile stress in section AB of the shaft.
b. The maximum compressive stress in section BC of the shaft.
c. The rotation of a section at C with respect to its no-load position.

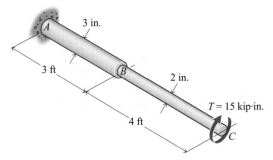

Figure P6-29

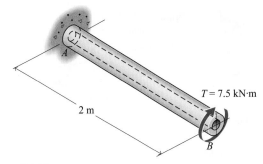

Figure P6-32

6-30* Determine the maximum torque that can be resisted by a hollow circular shaft having an inside diameter of 50 mm and an outside diameter of 90 mm without exceeding a normal stress of 75 MPa (T) or a shearing stress of 80 MPa.

6-31 A cylindrical tube is fabricated by butt-welding a 0.075-in.-thick plate with a spiral steam, as shown in Fig. P6-31. A torque T is applied to the tube through a rigid plate. If the outside diameter of the tube is 1.50 in. and $T = 1000$ lb · in., determine

a. The normal stress perpendicular to the weld and the shearing stress parallel to the weld.
b. The maximum tensile and compressive stresses in the tube.

Intermediate Problems

6-33* The motor shown in Fig. P6-33 develops a torque $T = 1500$ lb · ft; the output torques from gears C and D are equal. The mean diameters of gears A and B are 12 in. and 4 in., respectively. If the diameter of the motor shaft is 2 in. and the diameter of the power shaft is 1.25 in., determine

a. The torque in the power shaft between gears B and C.
b. The torque in the power shaft between gears C and D.
c. The maximum tensile and compressive stresses in each shaft.

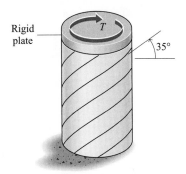

Figure P6-31

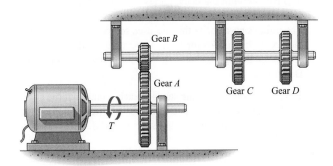

Figure P6-33

6-32 The hollow circular steel ($G = 80$ GPa) shaft shown in Fig. P6-32 has an outside diameter of 120 mm and an inside diameter of 60 mm. Determine

a. The maximum compressive stress in the shaft.
b. The maximum compressive stress in the shaft after the inside diameter is increased to 100 mm.
c. The rotation of end B with respect to its no-load position for the conditions of parts a and b.

6-34* A solid circular structural steel (see Appendix B for properties) shaft is securely fastened to a solid cold-rolled bronze shaft, as indicated in Fig. P6-34. For the steel, the allowable normal stress is 125 MPa and the allowable shearing stress is 75 MPa; for the bronze, the allowable normal stress is 260 MPa and the allowable shearing stress is 150 MPa. The allowable angle of twist in the 3.5-m length is 2.5°. Determine the maximum permissible value of the torque T.

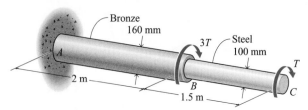

Figure P6-34

6-35 The hollow circular steel ($G = 12,000$ ksi) shaft of Fig. P6-35 is in equilibrium under the torques indicated. Determine

a. The maximum compressive stress in the shaft at a point near the outside surface of the shaft.
b. The maximum compressive stress in the shaft at a point near the inside surface of the shaft.
c. The rotation of end D with respect to end A.

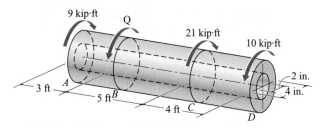

Figure P6-35

6-36 Five 600-mm-diameter pulleys are keyed to a 40-mm-diameter solid steel ($G = 76$ GPa) shaft, as shown in Fig. P6-36. The pulleys carry belts that are used to drive machinery in a factory. Belt tensions for normal operating conditions are indicated on the figure. Each segment of the shaft is 1.5 m long. Determine

a. The maximum shearing stress in each segment of the shaft.
b. The maximum tensile and compressive stresses in the shaft.
c. The rotation of end E with respect to end A.

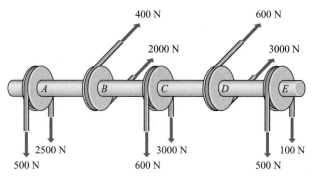

Figure P6-36

Challenging Problem

6-37 When the two torques shown in Fig. P6-37 are applied to the steel ($G = 12,000$ ksi) shaft, point A moves 0.172 in. in the direction indicated by torque T_1. Determine

a. The torque T_1.
b. The maximum tensile stress in section BC of the shaft.
c. The maximum compressive stress in section CD of the shaft.

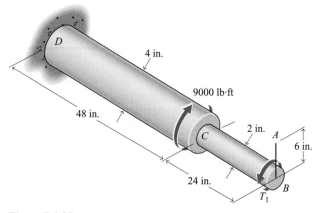

Figure P6-37

6-6 POWER TRANSMISSION

6.9

One of the most common uses for the circular shaft is the transmission of power; therefore, no discussion of torsion would be adequate without including this topic. Power is the time rate of doing work, and the basic relationship for work done by a constant torque is $W_k = T\phi$ where W_k is work and ϕ is the angular displacement of the shaft in radians. The time derivative of this expression gives

6.10

$$\text{Power} = \frac{dW_k}{dt} = T\frac{d\phi}{dt} = T\omega \tag{6-11}$$

where dW_k/dt is power in lb · ft per minute (or similar units), T is a constant torque in lb · ft, and ω is the angular velocity of the shaft (assumed constant) in rad/min.

All units, of course, may be changed to any other consistent set of units. Since the angular velocity is usually given in revolutions per minute (rpm), the conversion of revolutions to radians will often be necessary. Also, in the English system of units, power is usually given in units of horsepower, and the relation 1 hp = 33,000 lb · ft/min will be found useful. In the International (SI) system of units, power is given in watts (N · m/s). Solution of a power transmission problem is illustrated in the following example problem.

6.11

Example Problem 6-7 A diesel engine for a small commercial boat operates at 200 rpm and delivers 800 hp through a gearbox with a ratio of 4 to 1 to the propeller as shown in Fig. 6-17. Both the shaft from the engine to the gearbox and the propeller shaft are to be solid and made of heat-treated alloy steel. Determine the minimum permissible diameters for the two shafts if the allowable shearing stress is 20 ksi and the angle of twist in a 10-ft length of the propeller shaft is not to exceed 4°. Neglect power loss in the gearbox and assume (incorrectly because of thrust stresses) that the propeller shaft is subjected to pure torsion.

SOLUTION
The first step is the determination of the torques to which the shafts are to be subjected. By means of the expression, power = $T\omega$, the torques are obtained as follows:

$$800(33,000) = T_1(200)(2\pi)$$

from which

$$T_1 = 21,010 \text{ lb} \cdot \text{ft}$$

which is the torque at the crank shaft of the engine. Because the propeller shaft speed is four times that of the crank shaft and power loss in the gearbox is to be neglected, the torque on the propeller shaft is one-fourth that on the crank shaft and is equal to 5252 lb · ft. The torsion formula can be used to determine the shaft sizes necessary to satisfy the stress specification. For the main shaft,

$$\frac{J}{c} = \frac{T}{\tau} = \frac{21,010(12)}{20(10^3)} = \frac{(\pi/2)c_1^4}{c_1}$$

$$c_1^3 = 8.024 \quad \text{and} \quad c_1 = 2.002 \text{ in.}$$

or the shaft from the engine to the gearbox should be

$$d = 2c_1 = 2(2.002) = 4.004 \text{ in.} \cong 4.00 \text{ in.} \qquad \textbf{Ans.}$$

The torque on the propeller shaft is one-fourth that on the main shaft, and this is the only change in the expression for c_1^3; therefore,

$$c_2^3 = 8.024/4 \qquad c_2 = 1.2612 \text{ in.}$$

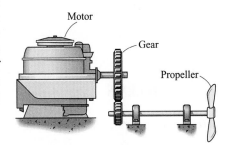

Motor
Gear
Propeller

Figure 6-17

▶ The angular velocity ω in Eq. 6-11 must be in either *radians per second* or *radians per minute*; 1 *revolution* equals 2π *radians*.

▶ Note that specifying the power that the shaft must transmit is just another way of specifying the torque that the shaft must withstand. After determining the torque from the given power, the application of the shearing stress equation (Eq. 6-6) and the angle of twist equation (Eq. 6-7) is the same as in previous examples.

The size of the propeller shaft needed to satisfy the distortion specification is found using Eq. 6-7*b*.

$$\theta = \frac{T_r L}{JG}$$

$$\frac{\pi}{180}(4) = \frac{5252(12)(10)(12)}{(\pi c_2^4/2)(12)(10^6)}$$

$$c_2^4 = 5.747 \qquad c_2 = 1.5483 > 1.2612$$

Therefore, the propeller shaft must be

$$d = 2c_2 = 2(1.5483) = 3.0966 \text{ in.} \cong 3.10 \text{ in.} \qquad \textbf{Ans.}$$

PROBLEMS

MecMovie Activities and Problems

MM6.6 Gear trains: power transmission (two shafts). Concept checkpoints. Basic calculations involving power transmission in two shafts connected by gears.

MM6.7 Gear trains: power transmission (three shafts). Concept checkpoints. Basic calculations involving power transmission in three shafts connected by gears.

Introductory Problems

6-38* The shaft of a diesel engine is being designed to transmit 240 kW at 180 rpm. Determine the minimum diameter required if the maximum shearing stress in the shaft is not to exceed 80 MPa.

6-39 A steel ($G = 12,000$ ksi) shaft with a 4-in. diameter must not twist more than 0.06 rad in a 20-ft length. Determine the maximum power that the shaft can transmit at 270 rpm.

Intermediate Problems

6-40* A 3-m-long hollow steel ($G = 80$ GPa) shaft has an outside diameter of 100 mm and an inside diameter of 60 mm. The maximum shearing stress in the shaft is 80 MPa, and the angular velocity is 200 rpm. Determine

a. The power being transmitted by the shaft.
b. The magnitude of the angle of twist in the shaft.

6-41* The hydraulic turbines in a water-power plant rotate at 60 rpm and are rated at 20,000 hp. The 30-in.-diameter shaft between the turbine and the generator is made of steel ($G = 12,000$ ksi) and is 20 ft long. Determine

a. The maximum shearing stress in the shaft at rated load.
b. The angle of twist in the 20-ft length at rated load.

6-42 A solid circular steel ($G = 80$ GPa) shaft 1.5 m long transmits 200 kW at a speed of 400 rpm. If the allowable shearing stress is 70 MPa and the allowable angle of twist is 0.045 rad, determine

a. The minimum permissible diameter for the shaft.
b. The speed at which this power can be delivered if the shearing stress is not to exceed 50 MPa in a shaft with a diameter of 75 mm.

6-43 The engine of an automobile supplies 162 hp at 3800 rpm to the drive shaft. If the maximum shearing stress in the drive shaft must be limited to 5 ksi, determine

a. The minimum diameter required for a solid drive shaft.
b. The maximum inside diameter permitted for a hollow drive shaft if the outside diameter is 3 in.
c. The percentage savings in weight realized if the hollow shaft is used instead of the solid shaft.

6-44 A hollow shaft of aluminum alloy ($G = 28$ GPa) is to transmit 1200 kW at 1800 rpm. The shearing stress is not to exceed 100 MPa, and the angle of twist is not to exceed 0.20 rad in a 3-m length. Determine the minimum permissible outside diameter if the inside diameter is to be three-fourths of the outside diameter.

6-45* A motor delivers 350 hp at 1800 rpm to a gearbox, which reduces the speed to 200 rpm to drive a ball mill. If the maximum shearing stress in the steel shafts ($G = 12,000$ ksi) is not to exceed 15 ksi and the angle of twist in a 10-ft length is not to exceed 0.10 rad, determine the minimum permissible diameter for each of the shafts.

Challenging Problems

6-46* A motor supplies 200 kW at 250 rpm to gear *A* of the factory drive shaft shown in Fig. P6-46. Gears *B* and *C* transfer

125 kW and 75 kW, respectively, to operating machinery in the factory. For an allowable shearing stress of 75 MPa, determine

a. The minimum permissible diameter d_1 for shaft AB.
b. The minimum permissible diameter d_2 for shaft BC.
c. The rotation of gear C with respect to gear A if both shafts are made of steel ($G = 80$ GPa) and have diameters of 75 mm.

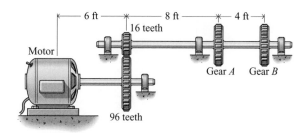

Figure P6-47

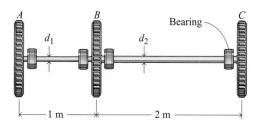

Figure P6-46

6-47 The motor shown in Fig. P6-47 develops 100 hp at a speed of 360 rpm. Gears A and B deliver 40 hp and 60 hp, respectively, to operating units in a factory. If the maximum shearing stress in the shafts must be limited to 12 ksi, determine

a. The minimum satisfactory diameter for the motor shaft.
b. The minimum satisfactory diameter for the power shaft.

6-48 A motor provides 180 kW of power at 400 rpm to the drive shafts shown in Fig. P6-48. The maximum shearing stress in the three solid steel ($G = 80$ GPa) shafts must not exceed 70 MPa. Gears A, B, and C supply 40 kW, 60 kW, and 80 kW, respectively, to operating units in the plant. Determine

a. The minimum satisfactory diameter for shaft D.
b. The minimum satisfactory diameter for shaft E.
c. The minimum satisfactory diameter for shaft F.

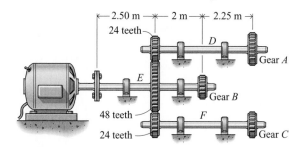

Figure P6-48

6-7 STATICALLY INDETERMINATE MEMBERS

All problems discussed in the preceding sections of this chapter were statically determinate; therefore, only the equations of equilibrium were required to determine the resisting torque at any section. Occasionally, torsionally loaded members are constructed and loaded such that the member is statically indeterminate (the number of independent equilibrium equations is less than the number of unknowns). When this occurs, distortion equations, which involve angles of twist, must be written until the total number of equations agrees with the number of unknowns to be determined. A simplified angle of twist diagram will often be of assistance in obtaining the correct equations. The following examples illustrate the procedures to be followed in solving statically indeterminate torsion problems.

6.12

Example Problem 6-8 The circular shaft AC of Fig. 6-18a is fixed to rigid walls at A and C. The solid section AB is made of annealed bronze ($G_{AB} = 45$ GPa), and the hollow section BC is made of aluminum alloy ($G_{BC} = 28$ GPa). There is no stress in the shaft before the 30-kN · m torque is applied. Determine the maximum shearing stresses in both the bronze and aluminum portions of the shaft after the torque is applied.

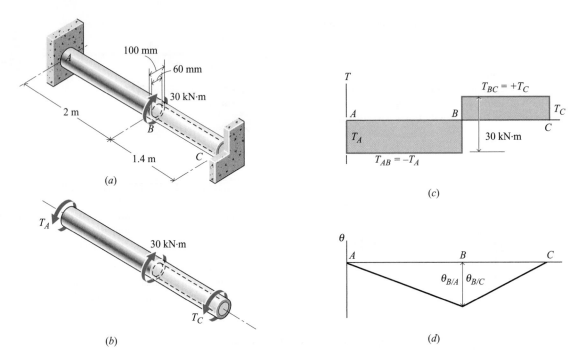

Figure 6-18

▶ The torque diagram (Fig. 6-18c) is calculated by drawing free-body diagrams and solving the equations of equilibrium for sections of the shaft to the left of B and to the right of B. Alternatively, the torque diagram can be drawn by observing that it jumps T_A at the left end of the shaft; it jumps 30 kN · m (in the opposite direction) at section B; and it jumps T_C (in the same direction as T_A) at section C.

▶ An alternative interpretation of Eq. (b) is that the rotations of the two segments must have equal magnitude of opposite sense. That is, $\theta_{B/A} = -\theta_{C/B}$.

SOLUTION

A free-body diagram of the shaft is shown in Fig. 6-18b. The torques T_A and T_C at the supports are unknown. A summation of moments about the axis of the shaft gives

$$T_A + T_C = 30(10^3)\,\text{N} \cdot \text{m} \qquad (a)$$

This is the only independent equation of equilibrium relating the two unknown torques T_A and T_C; therefore, the problem is statically indeterminate. A second equation can be obtained from the deformation of the shaft since the rotation of the two ends of the shaft are not independent.

The torque diagram for the shaft (shown in Fig. 6-18c) represents the results of applying the equilibrium equations to free-body diagrams of portions of the shaft. The torque in every section of the shaft between A and B is $T_{AB} = -T_A$, and the torque in every section of the shaft between B and C is $T_{BC} = +T_C$. Where the torque is negative the shaft rotates in a negative sense, and where the torque is positive the shaft rotates in a positive sense as shown in the rotation diagram of Fig. 6-18d. However, because the two ends of the shaft are attached to the walls, the total rotation of the shaft must be zero:

$$\theta_{\text{total}} = \theta_{B/A} + \theta_{C/B} = 0 \qquad (b)$$

Since Eq. (a) is expressed in terms of T_A and T_C, the convenient form of Eq. (b) for use here is Eq. 6-7b. Substituting Eq. 6-7b into Eq. (b) and using $T_{AB} = -T_A$ and $T_{BC} = +T_C$ gives

$$\frac{T_A L_{AB}}{G_{AB}\,J_{AB}} = \frac{T_C L_{BC}}{G_{BC}\,J_{BC}} \qquad (c)$$

For the two segments of the shaft, the polar second moments of area are

$$J_{AB} = (\pi/2)(50^4) = 9.817(10^6)\,\text{mm}^4 = 9.817(10^{-6})\,\text{m}^4$$
$$J_{BC} = (\pi/2)(50^4 - 30^4) = 8.545(10^6)\,\text{mm}^4 = 8.545(10^{-6})\,\text{m}^4$$

▶ It is the torques in the torque diagram (Fig. 6-18c) that are used in the angle of twist equation. That is, the torque in the left segment of the shaft is $T_{AB} = -T_A$ and the torque in the right segment of the shaft is $T_{BC} = +T_C$.

Therefore,

$$\frac{T_A(2)}{45(10^9)(9.817)(10^{-6})} = \frac{T_C(1.4)}{28(10^9)(8.545)(10^{-6})}$$

from which

$$T_C = 0.7737\,T_A \qquad\qquad (d)$$

Solving Eqs. (a) and (d) simultaneously yields

$$T_A = 16{,}914\,\text{N} \cdot \text{m} = 16.914\,\text{kN} \cdot \text{m}$$
$$T_C = 13{,}086\,\text{N} \cdot \text{m} = 13.086\,\text{kN} \cdot \text{m}$$

The stresses in the two portions of the shaft can then be obtained by using Eq. 6-6. Thus,

$$\tau_{AB} = \frac{T_A c_{AB}}{J_{AB}} = \frac{16.914(10^3)(50)(10^{-3})}{9.817(10^{-6})}$$
$$= 86.14(10^6)\,\text{N/m}^2 \cong 86.1\,\text{MPa} \qquad\qquad \textbf{Ans.}$$

$$\tau_{BC} = \frac{T_C c_{BC}}{J_{BC}} = \frac{13.086(10^3)(50)(10^{-3})}{8.545(10^{-6})}$$
$$= 76.57(10^6)\,\text{N/m}^2 \cong 76.6\,\text{MPa} \qquad\qquad \textbf{Ans.}$$

The stresses on the outside surface of the shaft, τ_{AB} and τ_{BC}, are shown in Figs. 6-19a and b, respectively.

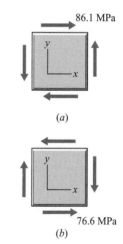

86.1 MPa

(a)

76.6 MPa

(b)

Figure 6-19

Example Problem 6-9   A hollow circular aluminum alloy ($G_a = 4000$ ksi) cylinder has a steel ($G_s = 11{,}600$ ksi) core, as shown in Fig. 6-20a. The steel and aluminum parts are securely connected at the ends. If the allowable stresses in the steel and aluminum must be limited to 14 ksi and 10 ksi, respectively, determine

6.13

(a) The maximum torque T that can be applied to the right end of the composite shaft.

(b) The magnitude of the rotation of the right end of the composite shaft when the torque of part (a) is applied.

6.14

SOLUTION

A free-body diagram of the shaft is shown in Fig. 6-20b. Since torques and stresses will be related by the torsion formula, which is limited to cross sections of homogeneous material, two unknown torques, the torque in the aluminum T_a

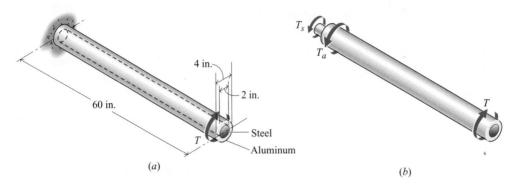

4 in.

2 in.

60 in.

Steel

Aluminum

T

T_s

T_a

T

(a)

(b)

Figure 6-20

and the torque in the steel T_s, have been placed on the left end of the shaft. Summing moments with respect to the axis of the shaft yields

$$T_a + T_s = T \qquad (a)$$

Since Eq. (a) is the only independent equation of equilibrium, the problem is statically indeterminate. A second equation can be obtained from the deformation of the shaft. The fact that the shaft is nonhomogeneous does not invalidate the assumptions of plane cross sections remaining plane and diameters remaining straight. As a result, strains remain proportional to the distance from the axis of the shaft; however, stresses are not proportional to the radii throughout the entire cross section since G is not single-valued. The steel and aluminum parts of the shaft experience the same angle of twist because of the secure connections at the ends. Thus,

$$\theta_s = \theta_a$$

Since maximum shearing stresses are specified, the convenient form of the angle of twist equation for use in this example is Eq. 6-7a. Thus,

$$\frac{\tau_s L_s}{G_s c_s} = \frac{\tau_a L_a}{G_a c_a}$$

$$\frac{\tau_s(60)}{11.6(10^6)(1)} = \frac{\tau_a(60)}{4.0(10^6)(2)}$$

from which

$$\tau_s = 1.45 \, \tau_a \qquad (b)$$

It is obvious from Eq. (b) that the shearing stress is in the steel controls; therefore,

$$\tau_s = 14 \, \text{ksi} \qquad \tau_a = 14/1.45 = 9.655 \, \text{ksi} < 10 \, \text{ksi}$$

(a) Once the maximum shearing stresses in the steel and aluminum portions of the shaft are known, Eq. 6-6 can be used to determine the torques transmitted by the two parts of the shaft. Thus,

$$T_s = \frac{\tau_s J_s}{c_s} = \frac{14,000(\pi/2)(1^4)}{1} = 21,990 \text{ lb} \cdot \text{in.}$$

$$T_a = \frac{\tau_a J_a}{c_a} = \frac{9655(\pi/2)(2^4 - 1^4)}{2} = 113,750 \text{ lb} \cdot \text{in.}$$

From Eq. (a),

$$T = T_a + T_s = 113,750 + 21,990$$
$$= 135,740 \text{ lb} \cdot \text{in.} \cong 135.7 \text{ kip} \cdot \text{in.} \qquad \textbf{Ans.}$$

(b) The rotation of the right end of the shaft with respect to its no-load position can be determined by using either Eq. 6-7a or Eq. 6-7b. If Eq. 6-7a is used,

$$\theta = \theta_a = \theta_s = \frac{\tau_s L_s}{G_s c_s} = \frac{14,000(60)}{11.6(10^6)(1)} = 0.0724 \text{ rad} \qquad \textbf{Ans.}$$

Example Problem 6-10   The torsional assembly shown in Fig. 6-21a consists of a solid bronze ($G_B = 45$ GPa) shaft CD and a hollow aluminum alloy ($G_A = 28$ GPa) shaft EF that has a steel ($G_S = 80$ GPa) core. The ends C and F are fixed to rigid walls, and the steel core of shaft EF is connected to the flange at E so that the aluminum and steel parts act as a unit. The two flanges D and E are bolted together, and the bolt clearance permits flange D to rotate through 0.03 rad before EF carries any of the load. Determine the maximum shearing stress in each of the shaft materials when the torque $T = 54$ kN $\cdot$ m is applied to flange D.

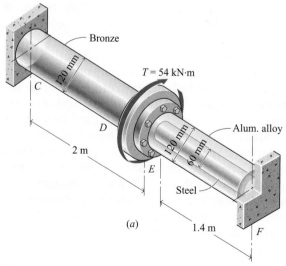

(a)

Figure 6-21(a)

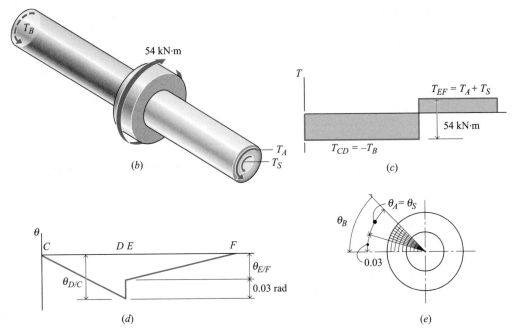

Figure 6-21(b–e)

▶ The polar diagram (Fig. 6-21e) shows the rotation of the coupling as viewed along the shaft from end F. A radial horizontal line on the coupling D will rotate clockwise through an angle θ_B. For the first 0.03 radians that the coupling D rotates, the bolts slip in the holes, and the coupling E does not move. As the coupling D continues to rotate, it pulls the coupling E with it and causes a rotation in both the steel and the aluminum shafts $\theta_S = \theta_A$.

SOLUTION

A free-body diagram for the assembly is shown in Fig. 6-21b. An unknown torque T_B is shown at the left support, and two unknown torques T_A and T_S are shown at the right support. Summing moments with respect to the axis of the shaft, as shown in the torque diagram of Fig. 6-21c, gives

$$T_B + T_A + T_S = 54(10^3)\,\text{N} \cdot \text{m} \qquad (a)$$

Equation (a) is the only independent equilibrium equation that can be written relating the three unknown torques T_A, T_B, and T_S. Since there are three unknown torques and only one equilibrium equation, two deformation equations are needed to solve the problem.

Two different types of angle of twist diagrams are shown in Figs. 6-21d and e. The torque diagram shown in Fig. 6-21c represents the results of applying the equilibrium equations to free-body diagrams of portions of the shaft. The torque in every section of the shaft between C and D is the negative of the torque that the left support exerts on the brass shaft ($T_{CD} = -T_B$), and the torque in every section of the shaft between E and F is the sum of the torques that the right support exerts on the aluminum and steel shafts ($T_{EF} = T_A + T_S$). Where the torque is negative the shaft rotates in a negative sense, and where the torque is positive the shaft rotates in a positive sense as shown in the rotation diagram of Fig. 6-21d. (The same quantities are shown in a polar form of representation in Fig. 6-21e.)

However, because the two ends of the shaft are attached to the walls, the total rotation of the shaft must be zero

$$\theta_{\text{total}} = \theta_{D/C} + \theta_{E/D} + \theta_{F/E} = 0 \qquad (b)$$

in which $\theta_{E/D} = +0.03$ rad is the rotation that occurs between the two parts of the coupling. In addition, since the aluminum and steel parts of the shaft act as a single unit, they must rotate an identical amount

$$(\theta_{F/E})_A = (\theta_{F/E})_S \tag{c}$$

Equations (a), (b), and (c) can be written in terms of the same three unknowns (torque, angle, or stress) and solved simultaneously. Since maximum stresses are required, Eqs. (a), (b), and (c) will be written in terms of the maximum stress in each material by using Eqs. 6-6 and 6-7a in which the polar second moments of area are

$$J_A = (\pi/2)(60^4 - 30^4) = 19.085(10^6)\,\text{mm}^4 = 19.085(10^{-6})\,\text{m}^4$$
$$J_B = (\pi/2)(60^4) = 20.36(10^6)\,\text{mm}^4 = 20.36(10^{-6})\,\text{m}^4$$
$$J_S = (\pi/2)(30^4) = 1.2723(10^6)\,\text{mm}^4 = 1.2723(10^{-6})\,\text{m}^4$$

Thus, Eq. (a) can be written

$$\frac{\tau_B J_B}{c_B} + \frac{\tau_A J_A}{c_A} + \frac{\tau_S J_S}{c_S} = 54(10^3)$$

$$\frac{\tau_B(20.36)(10^{-6})}{60(10^{-3})} + \frac{\tau_A(19.085)(10^{-6})}{60(10^{-3})} + \frac{\tau_S(1.2723)(10^{-6})}{30(10^{-3})} = 54(10^3)$$

from which

$$16\tau_B + 15\tau_A + 2\tau_S = 2.546(10^9) \tag{d}$$

Similarly, Eq. (b) can be written

$$\frac{-\tau_B L_B}{G_B c_B} + \frac{\tau_A L_A}{G_A c_A} + 0.03 = 0$$

$$\frac{-\tau_B(2)}{(45)(10^9)(60)(10^{-3})} + \frac{\tau_A(1.4)}{(28)(10^9)(60)(10^{-3})} + 0.03 = 0$$

from which

$$8\tau_B = 9\tau_A + 324(10^6) \tag{e}$$

Finally, Eq. (c) can be written

$$\frac{\tau_A L_A}{G_A c_A} = \frac{\tau_S L_S}{G_S c_S}$$

$$\frac{\tau_A(1.4)}{28(10^9)(60)(10^{-3})} = \frac{\tau_S(1.4)}{80(10^9)(30)(10^{-3})}$$

from which

$$10\tau_A = 7\tau_S \tag{f}$$

Solving Eqs. (d), (e), and (f) simultaneously yields

$$\tau_A = 52.93(10^6)\,\text{N/m}^2 \cong 52.9\,\text{MPa} \qquad \textbf{Ans.}$$

$$\tau_B = 100.0(10^6)\,\text{N/m}^2 = 100.0\,\text{MPa} \qquad \textbf{Ans.}$$

$$\tau_S = 75.62(10^6)\,\text{N/m}^2 \cong 75.6\,\text{MPa} \qquad \textbf{Ans.}$$

PROBLEMS

MecMovie Activities and Problems

MM6.8 Shear stresses in coaxial shafts. Example; Try one. Determine internal torques and shear stresses, and shaft rotation angle in two coaxial shafts

MM6.9 Shear stresses in end-to-end shafts. Example; Try one. Determine internal torques, shear stresses, and rotation angles for a compound torsion member.

MM6.10 Maximum torque for composite shaft. Example; Try one. Determine the maximum torque that can be applied to a compound torsion member given allowable shear stresses.

Introductory Problems

6-49* The 2-in.-diameter steel ($G = 12{,}000$ ksi) shaft shown in Fig. P6-49 is fixed to rigid walls at both ends. When a torque of 3500 lb · ft is applied as shown, determine

a. The maximum shearing stress in the shaft.
b. The angle of rotation of the section where the torque is applied with respect to its no-load position.

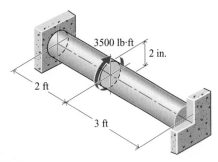

Figure P6-49

6-50* A steel ($G = 80$ GPa) tube with an inside diameter of 100 mm and an outside diameter of 125 mm is encased in a Monel ($G = 65$ GPa) tube with an inside diameter of 125 mm and an outside diameter of 175 mm, as shown in Fig. P6-50. The tubes are connected at the ends to form a composite shaft. The shaft is subjected to a torque of 10 kN · m. Determine

a. The maximum shearing stress in each material.
b. The angle of twist in a 2-m length.

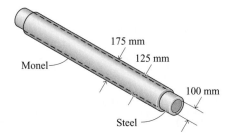

Figure P6-50

6-51 A 3-in.-diameter cold-rolled steel ($G = 11{,}600$ ksi) shaft, for which the maximum allowable shearing stress is 15 ksi, exhibited severe corrosion in a certain installation. It is proposed to replace the shaft with one in which an aluminum alloy ($G = 4000$ ksi) tube 1/4 in. thick is bonded to the outer surface of the cold-rolled steel shaft to produce a composite shaft. If the maximum allowable shearing stress in the aluminum alloy shell is 12 ksi, determine

a. The maximum torque that the original shaft can transmit.
b. The maximum torque that the replacement shaft can transmit.

6-52* A composite shaft consists of a bronze ($G = 45$ GPa) sleeve with an outside diameter of 80 mm and an inside diameter of 60 mm over a solid aluminum ($G = 28$ GPa) rod with an outside diameter of 60 mm. If the allowable shearing stress in the bronze is 150 MPa, determine

a. The maximum torque T that can be transmitted by the composite shaft.
b. The maximum shearing stress in the aluminum rod when the maximum torque is being transmitted.

6-53 Two 3-in.-diameter solid circular steel ($G = 11{,}600$ ksi) and bronze ($G = 6500$ ksi) shafts are rigidly connected and supported as shown in Fig. P6-53. A torque T is applied at the junction of the two shafts as indicated. The allowable shearing stresses are 18 ksi for the steel and 6 ksi for the bronze. Determine the maximum torque T that can be applied.

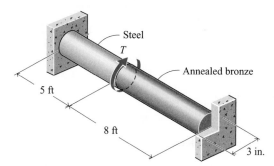

Figure P6-53

Intermediate Problems

6-54* A composite shaft, as shown in Fig. P6-54, consists of a solid brass ($G = 39$ GPa) core with an outside diameter of 40 mm covered by a steel ($G = 80$ GPa) tube with an inside diameter of 40 mm and a wall thickness of 20 mm, which is in turn covered by an aluminum alloy ($G = 28$ GPa) sleeve with an inside diameter of 80 mm and a wall thickness of 10 mm. The three materials are bonded so that they act as a unit. Determine

a. The maximum shearing stress in each material when the assembly is transmitting a torque of 15 kN · m.
b. The angle of twist in a 3-m length when the assembly is transmitting a torque of 10 kN · m.

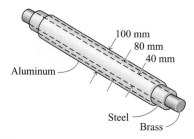

Figure P6-54

6-55* The composite shaft shown in Fig. P6-55 is used as a torsional spring. The solid circular polymer ($G = 150$ ksi) portion of the shaft is encased in and firmly attached to a steel ($G = 12,000$ ksi) sleeve for part of its length. If a torque T of 1000 lb · in. is being transmitted by the composite shaft, determine

a. The rotation of a cross section at C.
b. The rotation of a cross section at C if the steel shell is assumed to be rigid.
c. The percent error introduced by assuming the steel shell to be rigid.

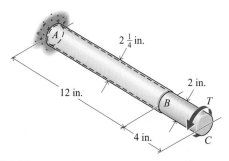

Figure P6-55

6-56 A hollow steel ($G = 80$ GPa) tube with an outside diameter of 100 mm and an inside diameter of 50 mm is covered with a Monel ($G = 65$ GPa) tube that has an outside diameter of 125 mm and an inside diameter of 102 mm. The tubes are connected at the ends to form a composite shaft. If the allowable shearing stress in the steel is 70 MPa and the allowable shearing stress in the Monel is 85 MPa, determine

a. The maximum torque T that the composite shaft can transmit.
b. The angle of twist in a 2.5-m length when the composite shaft is transmitting the maximum torque.

6-57* A solid aluminum alloy ($G = 4000$ ksi) rod with an outside diameter of 2 in. is used as a shaft. A hollow steel ($G = 12,000$ ksi) tube with an inside diameter of 2 in. is placed over the rod to increase the torque-transmitting capacity of the shaft. The tube and the rod are attached at the ends to form a composite shaft. Determine the minimum tube thickness required to permit the torque-transmitting capacity of the shaft to be increased by 50 percent.

6-58 A composite shaft consists of a bronze ($G = 45$ GPa) shell that has an outside diameter of 100 mm bonded to a solid steel ($G = 80$ GPa) core. Determine the diameter of the steel core when the torque resisted by the steel core is equal to the torque resisted by the bronze shell.

6-59 The steel ($G = 12,000$ ksi) shaft shown in Fig. P6-59 is attached to rigid walls at both ends. The right 10 ft of the shaft is hollow, having an inside diameter of 2 in. Determine

a. The maximum shearing stress in the shaft.
b. The angle of rotation of the section where the torque is applied with respect to its no-load position.

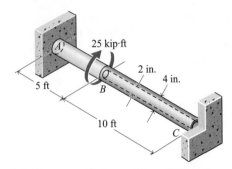

Figure P6-59

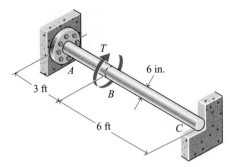

Figure P6-61

6-60* A disk and two circular shafts are connected and supported between rigid walls, as shown in Fig. P6-60. Shaft AB is made of brass ($G = 39$ GPa) and has a diameter of 75 mm and a length of 300 mm. Shaft BC is made of Monel ($G = 65$ GPa) and has a diameter of 60 mm and a length of 450 mm. If a torque of 15 kN · m is applied to the disk, determine

a. The maximum shearing stress in each of the shafts.
b. The angle of rotation of the disk with respect to its no-load position.
c. The maximum tensile and compressive stresses in the shaft.

Challenging Problems

6-62* The torsional assembly of Fig. P6-62 consists of an aluminum alloy ($G = 28$ GPa) segment AB securely connected to a steel ($G = 80$ GPa) segment BCD by means of a flange coupling with four bolts. The diameters of both segments are 75 mm, the cross-sectional area of each bolt is 150 mm^2, and the bolts are located 75 mm from the center of the shaft. If the shearing stress in the bolts must be limited to 60 MPa, determine

a. The maximum torque T that can be applied at section C.
b. The maximum shearing stress in the steel.

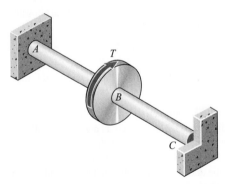

Figure P6-60

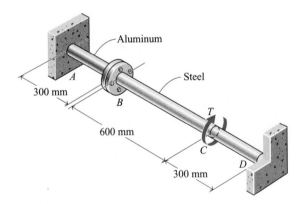

Figure P6-62

6-61 The solid steel ($G = 12,000$ ksi) shaft shown in Fig. P6-61 is fixed to the wall at C. The bolt holes in the flange at A have an angular misalignment of 0.0018 rad with respect to the holes in the wall. Determine

a. The torque, applied at B, required to align the bolt holes.
b. The maximum shearing stress in the shaft after the bolts are inserted and tightened and the torque at B is removed.
c. The maximum torque that can be applied at section B after the bolts are tightened if the maximum shearing stress in the shaft is not to exceed 10 ksi.

6-63* The steel ($G = 12,000$ ksi) shaft shown in Fig. P6-63 will be used to transmit a torque of 1000 lb · in. The hollow portion AE of the shaft is connected to the solid portion BF with two pins at C and D as shown. If the average shearing stress in the pins must be limited to 25 ksi, determine the minimum satisfactory diameters for each of the pins.

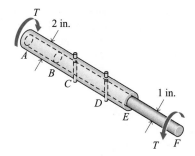

Figure P6-63

6-64 A stainless-steel ($G = 86$ GPa) shaft 2.5 m long extends through and is attached to a hollow brass ($G = 39$ GPa) shaft 1.5 m long, as shown in Fig. P6-64. Both shafts are fixed at the wall. When the two torques shown are applied to the shaft, determine

a. The maximum shearing stress in the steel.
b. The maximum shearing stress in the brass.
c. The maximum compressive stress in the brass.
d. The rotation of the right end of the shaft.

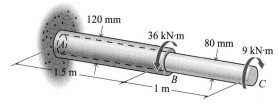

Figure P6-64

6-65 The 4-in-diameter shaft shown in Fig. P6-65 is composed of brass ($G = 6000$ ksi) and steel ($G = 12,000$ ksi) segments. Determine the maximum permissible magnitude for the torque T, applied at C, if the allowable shearing stresses are 5 ksi for the brass and 12 ksi for the steel.

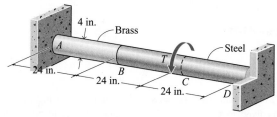

Figure P6-65

6-66* The circular shaft shown in Fig. P6-66 consists of a steel ($G = 80$ GPa) segment ABC securely connected to a bronze

($G = 40$ GPa) segment CD. Ends A and D of the shaft are fastened securely to rigid supports. Determine

a. The maximum shearing stress in the bronze segment.
b. The maximum shearing stress in the steel segment.
c. The rotation of a section at B with respect to its no-load position.

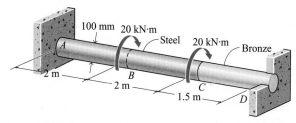

Figure P6-66

6-67 The shaft shown in Fig. P6-67 consists of a 6-ft-long hollow steel ($G = 12,000$ ksi) section AB and a 4-ft-long solid aluminum alloy ($G = 4000$ ksi) section CD. The torque T of 40 kip · ft is applied initially only to the steel section AB. Section CD is then connected and the torque T is released. When the torque is released, the connection slips 0.010 rad before the aluminum section takes any load. Determine

a. The maximum shearing stress in the aluminum alloy.
b. The maximum shearing stress in the steel after the torque T is released.
c. The final rotation of the collar at B with respect to its no-load position.

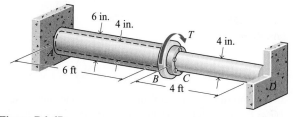

Figure P6-67

6-68 A torque T of 10 kN · m is applied to the steel ($G = 80$ GPa) shaft shown in Fig. P6-68 without the brass ($G = 40$ GPa) shell. The brass shell is then slipped into place and attached to the steel. After the original torque is released, determine

a. The maximum shearing stress in the brass shell.
b. The maximum shearing stress in the steel shaft.
c. The final rotation of the right end of the steel shaft with respect to its left end.

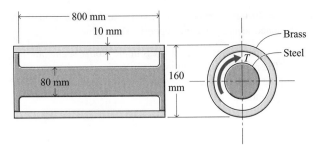

Figure P6-68

Computer Problems

6-69 The hollow circular aluminum alloy ($G = 4000$ ksi) shaft shown in Fig. P6-69 is 5 ft long and is attached to rigid supports at both ends. A 7500-lb · ft torque T is applied at section C, which is located 2 ft from the right end. The outer radius r_o of the shaft must be 2 in.; however, the inner radius r_i can vary (0 in. $\leq r_i \leq 1.9$ in.). Calculate and plot

a. The rotation θ of a section at C with respect to its no-load position as a function of the radius ratio r_i/r_o ($0 \leq r_i/r_o \leq 0.95$).

b. The maximum shearing stress τ_c as a function of the radius ratio r_i/r_o ($0 \leq r_i/r_o \leq 0.95$).

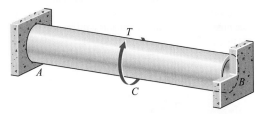

Figure P6-69

6-70 A composite shaft consists of a 2-m-long solid circular steel ($G = 80$ GPa) section securely fastened to a 2-m-long solid circular bronze ($G = 40$ GPa) section, as shown in Fig. P6-70. Both ends of the composite shaft are attached to rigid supports. The maximum shearing stress must not exceed 60 MPa, and the rotation of any cross section in the shaft must not exceed 0.04 rad. The ratio of the diameters d_b/d_s of the two sections can vary ($1/2 \leq d_b/d_s \leq 2$), but the average diameter $(d_b + d_s)/2$ must be 100 mm. Calculate and plot

a. The maximum allowable torque T as a function of the diameter ratio d_b/d_s ($1/2 \leq d_b/d_s \leq 2$).

b. The rotation θ of a section at C as a function of the diameter ratio d_b/d_s ($1/2 \leq d_b/d_s \leq 2$).

c. The maximum shearing stress τ_b in the bronze shaft as a function of the diameter ratio d_b/d_s ($1/2 \leq d_b/d_s \leq 2$).

d. The maximum shearing stress τ_s in the steel shaft as a function of the diameter ratio d_b/d_s ($1/2 \leq d_b/d_s \leq 2$).

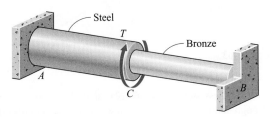

Figure P6-70

6-71 A hollow steel ($G = 12,000$ ksi) shaft is stiffened by filling its center with an aluminum alloy ($G = 4000$ ksi) shaft as shown in Fig. P6-71. If the steel and aluminum parts rotate as a single unit, calculate and plot

a. The rotation of end B with respect to its no-load position $\theta_{B/A}$ as a function of the diameter of the aluminum alloy shaft d_a (0 in. $\leq d_a \leq 3.75$ in.).

b. The shear stress τ_a in the aluminum alloy shaft at the interface between the two shafts as a function of the diameter of the aluminum alloy shaft d_a (0 in. $\leq d_a \leq 3.75$ in.).

c. The shear stress τ_s in the steel shaft at the interface between the two shafts as a function of the diameter of the aluminum alloy shaft d_a (0 in. $\leq d_a \leq 3.75$ in.).

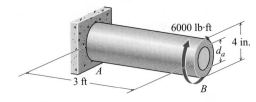

Figure P6-71

6-72 The solid steel ($G = 80$ GPa) shaft shown in Fig. P6-72 is fixed to the wall at C. The flange at A is to be attached to the wall with eight 18-mm-diameter bolts on a 300-mm-diameter circle. However, the bolt holes in the flange have an angular misalignment of 1° with respect to the holes in the wall. If a torque T is applied to the shaft at B, calculate and plot

a. The maximum shearing stresses in both sections of the shaft as a function of the torque T ($0 \leq T \leq 60$ kN · m). (Assume that the bolts are inserted and tightened as soon as the holes align.)
b. The shearing stresses in the bolts as a function of the torque T ($0 \leq T \leq 60$ kN · m).
c. The rotation θ of a section at B as a function of the torque T ($0 \leq T \leq 60$ kN · m).

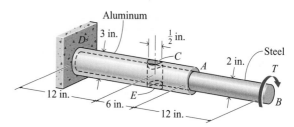

Figure P6-73

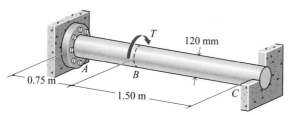

Figure P6-72

6-73 A 2-in.-diameter solid steel ($G = 12{,}000$ ksi and $\tau_{max} = 30$ ksi) shaft and a 3-in.-diameter hollow aluminum alloy ($G = 4000$ ksi and $\tau_{max} = 24$ ksi) shaft are fastened together with a 1/2-in.-diameter brass ($\tau_{max} = 36$ ksi) pin, as shown in Fig. P6-73. Calculate and plot

a. The maximum shear stresses, τ_a in the aluminum shaft and τ_s in the steel shaft, as a function of the torque T applied to the end of the shaft ($0 \leq T \leq 20$ kip · in.).
b. The average shear stress, τ_b on the cross-sectional area of the brass pin at the interface between the shafts as a function of the torque T ($0 \leq T \leq 20$ kip · in.).
c. The rotation of end B with respect to its no-load position $\theta_{B/D}$ as a function of the torque T ($0 \leq T \leq 20$ kip · in.).
d. What is the maximum torque that can be applied to the shaft without exceeding the maximum shear stresses in either the shaft or the pin?

6-74 The torsional assembly shown in Fig. P6-74 consists of a solid steel ($G = 80$ GPa) shaft CD and a hollow aluminum alloy ($G = 28$ GPa) shaft EF that has a bronze ($G = 45$ GPa) core. The ends C and F are fixed to rigid walls, and the bronze core of shaft EF is connected to the flange at E so that the aluminum and bronze and parts act as a single unit. Bolts are used to connect flanges D and E. Bolt clearance in flange E permits flange D to rotate $3°$ before shaft EF carries any of the load. If a torque T_D is applied to flange D, calculate and plot

a. The maximum shearing stresses in each of the materials as a function of the torque T_D ($0 \leq T_D \leq 90$ kN · m).
b. The rotation of flange D with respect to its no-load position $\theta_{D/C}$ as a function of the torque T_D ($0 \leq T_D \leq 90$ kN · m).

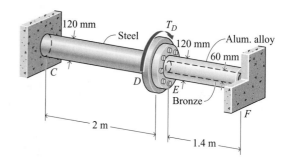

Figure P6-74

6-8 COMBINED LOADING—AXIAL, TORSIONAL, AND PRESSURE VESSEL

In previous sections, formulas were developed for determining normal and shearing stresses on specific planes in axially loaded bars, circular bars being twisted by a torque, and pressure vessels. For example, the normal stress at a point on a transverse cross section of a circular bar subjected only to an axial load P is $\sigma = P/A$. The shearing stress in the same bar being twisted only by a torque T is $\tau = T\rho/J$. If the bar is pulled and twisted at the same time, the combined loading produces both a normal stress σ and a shearing stress τ at a point on the transverse cross section of the bar. As long as the strains are small, these stresses can be computed separately and superimposed on the element. Once these stresses are known, the normal and shearing stresses on other planes through the

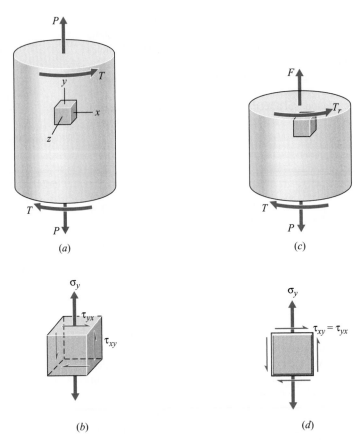

Figure 6-22

point can be determined using the stress transformation equations or Mohr's circle. Furthermore, the maximum normal and shearing stresses at the point can be found using the methods described in Section 2-10.

Consider the circular bar shown in Fig. 6-22a, which is subjected to an axial load P and a torque T. A small element of volume at the outside surface of the bar is shown in Fig. 6-22b. Acting on the top surface of the element is the normal stress σ_y resulting from the axial load P and the shearing stress τ_{yx} resulting from the torque T. The normal stress is tensile because P is a tensile force. The direction of the shearing stress on the top surface of the element is in the direction of the resisting torque shown on the free-body diagram of Fig. 6-22c. Shearing stresses exist on both transverse and longitudinal planes because $\tau_{xy} = \tau_{yx}$. On the surface of the bar a state of plane stress exists, and thus, the volume element of Fig. 6-22b is shown as a plane element in Fig. 6-22d. Once the stresses on the planes shown in Fig. 6-22d are known, the stresses on any other plane through the point as well as the principal stresses and the maximum shearing stress at the point can be found. The following Example Problems illustrate the procedures discussed.

6.15

Example Problem 6-11   The solid 100-mm-diameter shaft shown in Fig. 6-23a is subjected to an axial compressive force $P = 200$ kN and a torque $T = 30$ kN · m. For point A on the outside surface of the shaft, determine

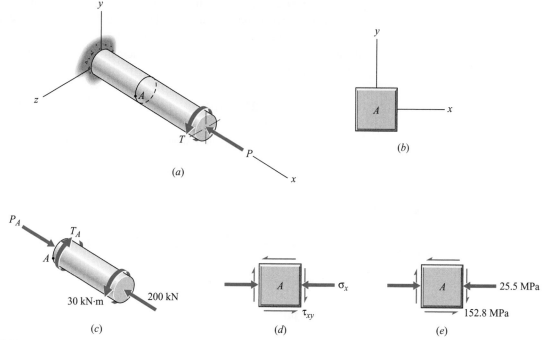

Figure 6-23

(a) The x- and y-components of stress.

(b) The principal stresses and the maximum shearing stress at the point.

SOLUTION

(a) Since point A is on the outside surface (a free surface) of the shaft, a state of plane stress exists at the point. The coordinate system is selected as shown in Fig. 6-23b. This is the same coordinate system for which the equations in Chapter 2 were developed. Passing a transverse plane through point A and isolating the segment of the shaft to the right of point A results in the free-body diagram shown in Fig. 6-23c. The internal forces on the transverse cross section are an axial compressive force P_A of 200 kN and a torque T_A of 30 kN · m. The directions of the stresses on the left face of the element are in accordance with the directions of the forces that produce the stresses, that is, in the directions of the resisting forces. These stresses are shown in Fig. 6-23d. The magnitude of the stresses are determined as follows:

$$\sigma_x = \frac{P_A}{A} = \frac{200(10^3)}{(\pi/4)(0.100)^2} = 25.46(10^6)\,\text{N/m}^2 = 25.46\,\text{MPa}$$

$$\tau_{xy} = \frac{T_A c}{J} = \frac{30(10^3)(0.050)}{(\pi/2)(0.050)^4} = 152.79(10^6)\,\text{N/m}^2 = 152.79\,\text{MPa}$$

▶ There is no normal stress in the y-direction (the circumferential direction) since there is no force to cause such a stress. The shear stresses on the four sides of the element have the same magnitude. The shear stresses that meet at a corner either both point toward the corner or both point away from the corner.

The x- and y-components of stress on the element at A are

$$\sigma_x = 25.46\,\text{MPa} \cong 25.5\,\text{MPa (C)} \qquad \textbf{Ans.}$$

$$\tau_{xy} = 152.79\,\text{MPa} \cong 152.8\,\text{MPa} \qquad \textbf{Ans.}$$

These stresses are shown on the element of Fig. 6-23e. Note that point A could be anywhere on the outside surface along the length of the shaft that is not in the vicinity of the fixed or loaded ends.

(b) Equation 2-15 is used to calculate the principal stresses. The stresses for use in this equation are

$$\sigma_x = -25.46 \text{ MPa} \qquad \sigma_y = 0 \qquad \tau_{xy} = -152.79 \text{ MPa}$$

Substituting these stress components into Eq. 2-15 yields

$$\sigma_{p1,p2} = \frac{\sigma_x + \sigma_y}{2} \pm \sqrt{\left(\frac{\sigma_x - \sigma_y}{2}\right)^2 + \tau_{xy}^2}$$

$$= \frac{-25.46 + 0}{2} \pm \sqrt{\left(\frac{-25.46 - 0}{2}\right)^2 + (-152.79)^2}$$

$$= -12.73 \pm 153.32$$

Thus, the principal stresses are

$$\sigma_{p1} = -12.73 + 153.32 = 140.59 \text{ MPa} \cong 140.6 \text{ MPa (T)} \qquad \textbf{Ans.}$$

$$\sigma_{p2} = -12.73 - 153.32 = -166.05 \text{ MPa} \cong 166.1 \text{ MPa (C)} \qquad \textbf{Ans.}$$

$$\sigma_{p3} = \sigma_z = 0 \qquad \textbf{Ans.}$$

Since the two in-plane principal stresses are of opposite sign and $\sigma_{p3} = \sigma_z = 0$,

$$\tau_{max} = \frac{\sigma_{max} - \sigma_{min}}{2}$$

$$= \frac{140.59 - (-166.05)}{2}$$

$$= 153.32 \text{ MPa} \cong 153.3 \text{ MPa} \qquad \textbf{Ans.}$$

Example Problem 6-12 The thin-walled cylindrical pressure vessel shown in Fig. 6-24a has an inside diameter of 24 in. and a wall thickness of 1/2 in. The vessel is subjected to an internal pressure of 250 psi. In addition, a torque of 150 kip · ft is applied to the vessel through rigid plates on the ends of the vessel. Determine the maximum normal and shearing stresses at a point on the outside surface of the vessel.

SOLUTION

The stresses on an element on the outside surface of the vessel are shown in Fig. 6-24b, where x is in the longitudinal (axial) direction and y is in the hoop direction. The normal stresses are due to the internal pressure and are

$$\sigma_a = \sigma_x = \frac{pr}{2t} = \frac{250(12)}{2(1/2)} = 3000\,\text{psi}$$

$$\sigma_h = \sigma_y = \frac{pr}{t} = \frac{250(12)}{1/2} = 6000\,\text{psi}$$

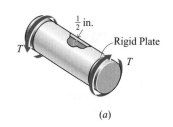

(a)

The magnitude of the shearing stress resulting from the torque is

$$\tau_{xy} = \frac{Tc}{J} = \frac{150(10^3)(12)(12.5)}{(\pi/2)(12.5^4 - 12^4)} = 3894\,\text{psi}$$

and the direction of the shearing stress is the same as the direction of the internal torque.

Equation 2-15 is used to calculate the maximum normal stress (a principal stress). The stresses for use in this equation are

$$\sigma_x = 3000\,\text{psi} \qquad \sigma_y = 6000\,\text{psi} \qquad \tau_{xy} = 3894\,\text{psi}$$

Substituting these stress components into Eq. 2-15 yields

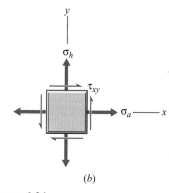

(b)

Figure 6-24

$$\sigma_{p1,p2} = \frac{\sigma_x + \sigma_y}{2} \pm \sqrt{\left(\frac{\sigma_x - \sigma_y}{2}\right)^2 + \tau_{xy}^2}$$

$$= \frac{3000 + 6000}{2} \pm \sqrt{\left(\frac{3000 - 6000}{2}\right)^2 + (3894)^2}$$

$$= 4500 \pm 4173$$

Thus, the principal stresses are

$$\sigma_{p1} = 4500 + 4173 = 8673\,\text{psi}$$

$$\sigma_{p2} = 4500 - 4173 = 327\,\text{psi}$$

$$\sigma_{p3} = \sigma_z = 0$$

The maximum normal stress is

$$\sigma_{\max} = \sigma_{p1} = 8673\,\text{psi} \cong 8670\,\text{psi (T)} \qquad \textbf{Ans.}$$

Since the two in-plane principal stresses have the same sign and $\sigma_{p3} = \sigma_z = 0$,

$$\tau_{\max} = \frac{\sigma_{\max} - \sigma_{\min}}{2}$$

$$= \frac{8673 - 0}{2}$$

$$= 4337\,\text{psi} \cong 4340\,\text{psi} \qquad \textbf{Ans.}$$

PROBLEMS

Introductory Problems

6-75* A 4-in.- diameter shaft is subjected to both a torque of 30 kip · in. and an axial tensile load of 50 kip, as shown in Fig. P6-75. Determine the principal stresses and the maximum shearing stress at point A on the surface of the shaft.

Figure P6-75

6-76* A hollow shaft with an outside diameter of 400 mm and an inside diameter of 300 mm is subjected to both a torque of 350 of kN · m and an axial tensile load of 1500 kN, as shown in Fig. P6-76. Determine

a. The *xyz*-components of stress at a point on the outside surface of the shaft.
b. The principal stresses and the maximum shearing stress at a point on the outside surface of the shaft.

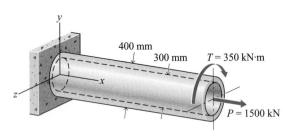

Figure P6-76

6-77 A 2-in.-diameter shaft is used in an aircraft engine to transmit 360 hp at 1500 rpm to a propeller that develops a thrust of 2800 lb. Determine the principal stresses and the maximum shearing stress produced at any point on the outside surface of the shaft.

6-78* A 60-mm-diameter shaft must transmit a torque of unknown magnitude while it is supporting an axial tensile load of 125 kN. Determine the maximum allowable value for the torque if the tensile principal stress at a point on the outside surface of the shaft must not exceed 100 MPa.

6-79 A 4-in.-diameter shaft must support an axial tensile load of unknown magnitude while it is transmitting a torque of 100 kip · in. Determine the maximum allowable value for the axial load if the tensile principal stress at a point on the outside surface of the shaft must not exceed 18,000 psi.

Intermediate Problems

6-80* A steel shaft is loaded and supported as shown in Fig. P6-80. Determine the maximum allowable value for the axial load P if the maximum shearing stress in the shaft is not to exceed 60 MPa and the maximum compressive stress in the shaft is not to exceed 96 MPa.

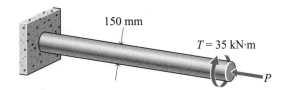

Figure P6-80

6-81* A 6-in.-diameter shaft will be used to support the axial load and torques shown in Fig. P6-81. Determine the principal stresses and the maximum shearing stress at point A on the surface of the shaft.

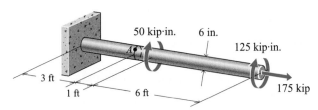

Figure P6-81

6-82 A steel shaft with the left portion solid and the right portion hollow is loaded as shown in Fig. P6-82. If the maximum shearing stress in the shaft must not exceed 80 MPa and the maximum tensile stress in the shaft must not exceed 140 MPa, determine the maximum axial load P that can be applied to the shaft.

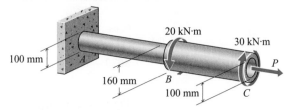

Figure P6-82

6-83* A steel shaft is loaded and supported as shown in Fig. P6-83. If the maximum shearing stress in the shaft must not exceed 10 ksi and the maximum tensile stress in the shaft must not exceed 15 ksi, determine the maximum torque T that can be applied to the shaft.

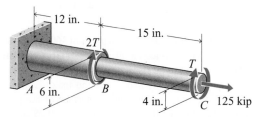

Figure P6-83

6-84 A tube having an inside diameter equal to one-half the outside diameter must transmit a torque of 7.5 kN · m while resisting an axial compressive load of 200 kN. Determine the minimum outside diameter required if the maximum compressive stress in the tube is not to exceed 100 MPa.

6-85 A shaft that is transmitting 240 hp at 1800 rpm must also support an axial tensile load of 20 kip. If the maximum tensile stress in the shaft is not to exceed 15 ksi, determine the minimum diameter required for the shaft.

Challenging Problems

6-86* A 25-mm-diameter steel ($E = 210$ GPa and $v = 0.30$) bar is subjected to a tensile load P and a torque T, as shown in Fig. P6-86. Determine the axial load P and the torque T if the strains indicated by gages a and b on the bar are $\epsilon_a = +1084$ μm/m and $\epsilon_b = -754$ μm/m.

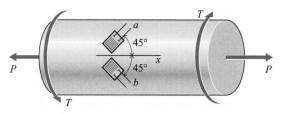

Figure P6-86

6-87* The thin-walled cylindrical pressure vessel shown in Fig. P6-87a has an inside diameter of 20 in. and a wall thick-

ness of 3/8 in. The vessel is subjected to an internal pressure of p and a torque of T. At a point on the outside surface of the steel ($E = 30,000$ ksi and $v = 0.30$) vessel, the strain rosette shown in Fig. P6-87b was used to obtain the following normal strain data: $\epsilon_a = +36$ μin./in., $\epsilon_b = +310$ μin./in., and $\epsilon_c = +150$ μin./in. Gages a and c are oriented in the axial and hoop directions of the vessel, respectively. Determine the pressure p and the torque T.

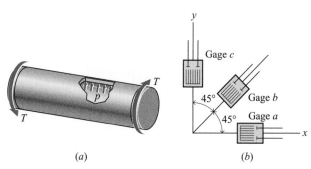

(a)   (b)

Figure P6-87

6-88 A 50-mm-diameter steel ($E = 200$ GPa and $v = 0.30$) bar is subjected to a tensile load P and a torque T, as shown in Fig. P6-88. Determine the axial load P and the torque T if the strains indicated by gages a and b on the bar are $\epsilon_a = +1414$ μm/m and $\epsilon_b = -212$ μm/m.

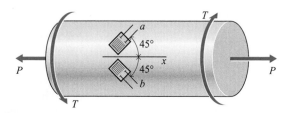

Figure P6-88

6-89 A thin-walled cylindrical pressure vessel with an inside diameter of 4 ft is fabricated by butt-welding 0.6-in.-thick plate with a spiral seam as shown in Fig. P6-89. The pressure in the tank is 360 psi. Additional loads are applied to the cylinder through a rigid end plate as shown in Fig. P6-89. Determine

a. The normal and shearing stresses on the plane of the weld at a point on the outside surface of the tank.
b. The principal stresses and the maximum shearing stress at a point on the inside surface of the tank.

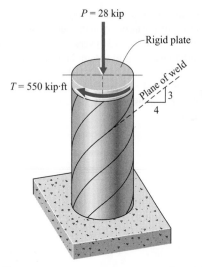

$P = 28$ kip

Rigid plate

$T = 550$ kip·ft

Plane of weld

3
4

Figure P6-89

Computer Problems

6-90 A 120-mm-diameter shaft is used to transmit an axial load of 250 kN and a torque of 20 kN · m. It is proposed to replace the solid shaft with a hollow shaft having the same weight (the same cross-sectional area). Calculate and plot the axial normal stress σ_x, the torsional shear stress τ_{xy}, the principal normal stress σ_{p1}, and the maximum shearing stress τ_{max} at a point on the outside surface of the shaft as a function of the outside diameter d_o (120 mm $\leq d_o \leq$ 300 mm) of the shaft.

6-91 An aircraft engine is designed to transmit 360 hp at 1500 rpm to a propeller that develops a thrust of 2800 lb. Calculate and plot the axial normal stress σ_x, the torsional shear stress τ_{xy}, the principal normal stress σ_{p1}, and the maximum shearing stress τ_{max} at a point on the outside surface of the propeller shaft as a function of the shaft diameter d (0.75 in. $\leq d \leq$ 2.50 in.).

6-9 STRESS CONCENTRATIONS IN CIRCULAR SHAFTS UNDER TORSIONAL LOADINGS

In Section 5-6 it was shown that the introduction of a circular hole or other geometric discontinuity into an axially loaded member can cause a significant increase in the magnitude of the stress (stress concentration) in the immediate vicinity of the discontinuity. This is also the case for circular shafts under torsional loading.

In previous sections of this chapter, it was shown that the magnitude of the maximum shearing stress in a circular shaft of uniform cross section and made of a linearly elastic material is given by Eq. 6-6 as

$$\tau_c = \frac{T_r c}{J} \tag{6-6}$$

Equation 6-6 (torsional loading) can also be used to determine the maximum shearing stress in a tapered shaft if the change in diameter occurs in a gradual manner. For stepped shafts, however, large increases in stress (stress concentrations) occur in the vicinity of the abrupt changes in diameter. These large stresses can be reduced by using a fillet between the parts of the shaft with the different diameters. The magnitude of the maximum shearing stress in the fillet can be expressed in terms of a *stress concentration factor K* as

$$\tau_{max} = K_t \tau_c = K_t \frac{T_r c}{J} \tag{6-12}$$

The stress concentration factor K_t depends on the ratio of the diameters of the two portions of the shaft (D/d) and the ratio of the radius of the fillet to the diameter of the smaller shaft (r/d). Stress concentration factors K_t (based on the net section) for stepped circular shafts and for circular shafts with U-shaped grooves are shown in Fig. 6-25. A careful examination of Fig. 6-25b shows that a generous fillet radius r should be used wherever a change in shaft diameter occurs. Equation 6-12 can be used to determine localized maximum shearing stresses in stepped shafts as long as the value of τ_{max} does not exceed the proportional limit of the material.

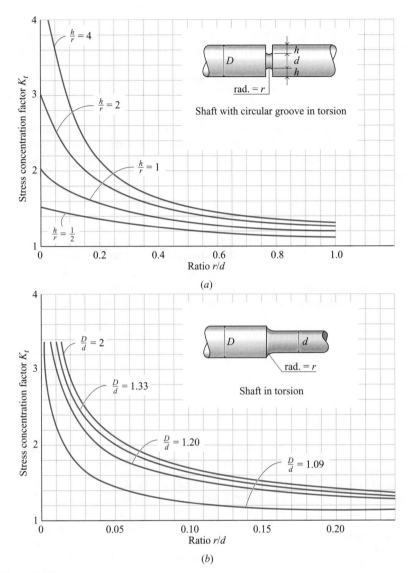

Figure 6-25

Stress concentrations also occur in circular shafts at oil holes and grooves (see Fig. 6-25a), and at keyways used for attaching pulleys and gears to the shaft. Each of these types of discontinuity requires special consideration during the design process.

Example Problem 6-13 A stepped shaft has a 4-in. diameter for one-half of its length and a 2-in. diameter for the other half. If the maximum shearing stress in the shaft must be limited to 8 ksi when the shaft is transmitting a torque of 6280 lb · in., determine the minimum fillet radius needed at the junction between the two portions of the shaft.

SOLUTION

The magnitude of the maximum shearing stress produced by the torque of 6280 lb · in. in the portion of the shaft with the 2-in. diameter is given by Eq. 6-6 as

$$\tau_{max} = \frac{T_r c}{J} = \frac{6280(1)}{(\pi/2)(1^4)} = 3998 \, \text{psi}$$

▶ The actual stress at the change in section is K_t times larger than the nominal stress in the 2-in.-diameter shaft calculated using the shear stress formula Eq. 6-6. The values used for c and J are for the smaller section since the stress concentration graph (Fig. 6-25b) is based on the *net section*. Before using any stress concentration graphs, it is important to determine whether the stress concentration factors are based on the *net section* or the *gross section* properties.

Since the maximum shearing stress in the fillet between the two portions of the shaft must be limited to 8 ksi, the maximum permissible value for the stress concentration factor K_t based on the maximum shearing stress in the small (net) section is

$$K_t = 8/3.998 = 2.001 \cong 2.00$$

The stress concentration factor K_t depends on two ratios (D/d) and (r/d). For the 4-in.-diameter shaft with the 2-in.-diameter turned section, the ratio $D/d = 4/2 = 2$. From the curves presented in Fig. 6-25b, a ratio $D/d = 2$ and a stress concentration factor $K_t = 2.00$ give a ratio $r/d = 0.06$. Thus, the minimum permissible radius for the fillet between the two portions of the shaft is

$$r = 0.06d = 0.06(2) = 0.1200 \, \text{in.} \qquad \textbf{Ans.}$$

PROBLEMS

Introductory Problems

6-92* A fillet with a radius of 12 mm is used at the junction in a stepped shaft where the diameter is reduced from 135 mm to 100 mm. Determine the maximum shearing stress in the fillet when the shaft is transmitting a torque of 10 kN · m.

6-93* A fillet with a radius of 0.15 in. is used at the junction in a stepped shaft where the diameter is reduced from 4.00 in. to 3.00 in. Determine the maximum shearing stress in the fillet when the shaft is transmitting a torque of 4000 lb · ft.

6-94 The small portion of a stepped shaft has a diameter of 50 mm. The radius of the fillet at the junction between the large and small portions is 4.5 mm. If the maximum shearing stress in the fillet must be limited to 40 MPa when the shaft is transmitting a torque of 614 N · m, determine the maximum diameter that can be used for the large portion of the shaft.

6-95 A fillet with a radius of 1/8 in. is used at the junction in a stepped shaft where the diameter is reduced from 8.00 in. to 6.00 in. Determine the maximum torque that the shaft can transmit if the maximum shearing stress in the fillet must be limited to 12 ksi.

Intermediate Problems

6-96* A semicircular groove with a 5-mm radius is required in a 110-mm diameter shaft. If the maximum shearing stress in

the shaft must be limited to 60 MPa, determine the maximum torque that can be transmitted by the shaft.

6-97 A stepped shaft has a 5-in. diameter for one-half of its length and a 4-in. diameter for the other half. If the maximum shearing stress in the fillet between the two portions of the shaft must be limited to 12 ksi when the maximum shearing stress in the 4-in. portion is 8 ksi, determine the minimum radius needed at the junction between the two portions of the shaft.

Challenging Problems

6-98* A shallow crack has been located in a 100-mm-diameter shaft. The crack will be removed by turning down a 200-mm length of the shaft surrounding the crack with a tool bit that has a 5-mm radius. If the maximum shearing stress in the 5-mm fillet must be limited to 60 MPa when the shaft is transmitting a torque of 3.27 kN · m, determine the minimum allowable diameter for the reduced section.

6-99 A 2-in.-diameter shaft contains a 1/2-in.-deep, U-shaped groove that has a 1/4-in. radius at the bottom of the groove. The shaft must transmit a torque of 500 lb · in. If a factor of safety of 3 with respect to failure by yielding is specified, determine the minimum elastic strength in shear required for the shaft material.

6-10 INELASTIC BEHAVIOR OF TORSIONAL MEMBERS

One of the limitations of the torsion formula (Eq. 6-6) is that stresses must be less than the proportional limit of the material. In some design situations, a limited amount of inelastic action can be permitted; therefore, the analysis presented in Section 6-3 will be extended in this section to include some inelastic action in the material. In Section 6-2, the assumption was made that a plane section remains plane and a diameter remains straight for circular sections under pure torsion. This assumption led to the development of Eq. 6-3,

$$\gamma_\rho = \frac{\gamma_c}{c}\rho \tag{6-3}$$

which states that the shearing strain at any point on a transverse cross section of a circular shaft is proportional to the distance of the point from the axis of the shaft. Equation 6-3 is valid for elastic or inelastic action provided the strains are not too large ($\tan \gamma \cong \gamma$). Once the shearing strain distribution is known, the law of variation of the shearing stress τ with radius ρ on the transverse plane can be determined from a shear stress-strain diagram for the material. As an aid to the thought sequence, it is suggested that sketches be drawn of γ vs. ρ, then τ vs. γ, and finally τ vs. ρ, as illustrated for two typical examples in Fig. 6-26. For both examples, Fig. 6-26a shows that the shearing strain γ varies linearly with the radius ρ of the circular shaft or tube. One example, Figs. 6-26b and c, represents a material behavior that strain softens (after the yield point is reached, the shear modulus decreases at the strain increases). The second example, Figs. 6-26d and e, represents a material behavior that is elastoplastic. The resisting torque (or other unknown if the torque is known) can then be evaluated by substituting the

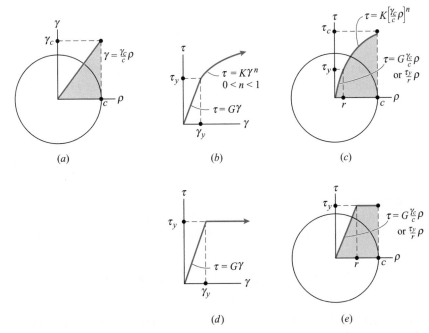

Figure 6-26

expression for τ in terms of ρ into Eq. 6-1:

$$T_r = \int_{\text{area}} \rho \tau_\rho dA \tag{6-1}$$

The procedure is illustrated in the following Example Problems.

Example Problem 6-14   A solid circular steel shaft 4 in. in diameter is subjected to a pure torque of 7π kip · ft. Assume the steel is elastoplastic (see Fig. 6-26d), having a yield point τ_y in shear of 18 ksi and a modulus of rigidity G of 12,000 ksi. Determine the maximum shearing stress in the shaft and the magnitude of the angle of twist in a 10-ft length.

SOLUTION

Assume first that the maximum stress is less than the proportional limit. In this case, if the material is homogeneous and isotropic, the expression $\tau = G\gamma$ applies over the entire cross section (see Fig. 6-26d), and the τ vs. ρ diagram would appear like the γ vs. ρ diagram (Fig. 6-26a), with γ_c replaced by τ_c, making the equation of the curve $\tau = (\tau_c/c)\rho$. Substituting this expression for τ into Eq. 6-1 (dA is shown in Fig. 6-27a) yields

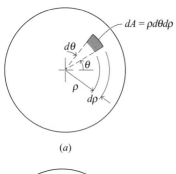

(a)

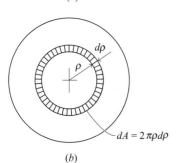

(b)

Figure 6-27

$$7\pi(12) = \int_0^2 \int_0^{2\pi} \rho\left(\frac{\tau_c}{c}\rho\right)(\rho \, d\rho \, d\theta) = \int_0^2 \rho\left(\frac{\tau_c}{2}\rho\right)(2\pi\rho \, d\rho) = \pi\tau_c\left(\frac{2^4}{4}\right)$$

from which τ_c is 21 ksi, which is greater than 18 ksi. Therefore, the assumption is not valid, and the stress distribution curve should be as in Fig. 6-26e with the maximum shearing stress

$$\tau_{\text{max}} = \tau_y = 18 \text{ ksi} \qquad\qquad \textbf{Ans.}$$

Observe that the first integration above replaces the original dA with an annular area of width $d\rho$, as shown in Fig. 6-27b. In the future, dA will be written in this form.

To obtain the angle of twist, Eq. 6-2 will be rewritten as

$$\theta = \frac{\gamma L}{\rho} = \frac{\gamma_y L}{r} \tag{a}$$

where the shearing strain at the yield point γ_y equals $\tau_y/G = 18/12{,}000 = 0.0015$ rad and r is the radius of that part of the cross section that is deforming elastically (Fig. 6-26e). The value of r will be found by means of Eq. 6-1. Thus,

$$T_r = \int_0^r \rho\left(\frac{18}{r}\rho\right)(2\pi\rho \, d\rho) + \int_r^2 \rho(18)(2\pi\rho \, d\rho)$$

The first integral represents that portion of the torque carried by the part of the shaft that is deforming elastically ($0 \le \rho \le r$), and the second integral represents the portion of the torque carried by the part of the shaft where the shearing stress is constant ($r \le \rho \le c$). Thus,

$$7\pi(12) = 2\pi\left(\frac{18}{r}\right)\left(\frac{r^4}{4}\right) + 2\pi(18)\left(\frac{2^3}{3} - \frac{r^3}{3}\right)$$

from which $r^3 = 4.000$ and $r = 1.5873$ in. Now, from Eq. (a), the magnitude of the angle of twist is

$$\theta = \frac{0.0015(10)(12)}{1.5873} = 0.1134 \text{ rad} \qquad\qquad \textbf{Ans.}$$

Example Problem 6-15   A straight shaft with a hollow circular cross section with outside and inside diameters of 3 and 2 in., respectively, is made of a magnesium alloy having the shear stress-strain diagram of Fig. 6-28. Determine the torque required to twist a 5-ft length of the shaft through 0.240 rad.

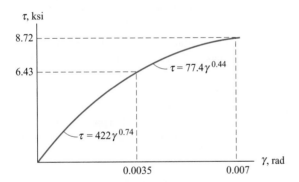

Figure 6-28

SOLUTION
The shear strains at the outside and inside surfaces are obtained from Eq. 6-2. Thus,

$$\gamma_{1.5} = \frac{\rho\theta}{L} = \frac{(1.5)(0.240)}{5(12)} = 0.006 \text{ rad}$$

$$\gamma_{1.0} = \frac{\rho\theta}{L} = \frac{(1)(0.240)}{5(12)} = 0.004 \text{ rad}$$

Since both the minimum and maximum shear strains lie within the right-hand portion of the curve of Fig. 6-28, the shear stress-strain function given for this curve applies for the entire cross section. Thus the γ vs. ρ function is

$$\gamma = \frac{\gamma_c}{c}\rho = \frac{0.006}{1.5}\rho = 0.004\rho$$

The τ vs. ρ function then becomes

$$\tau = 77.4\,\gamma^{0.44} = 77.4(0.004\rho)^{0.44} = 6.818\rho^{0.44}$$

Substituting this expression into Eq. 6-1 yields

$$T_r = \int_{1.0}^{1.5} \rho(6.818\rho^{0.44})(2\pi\rho\,d\rho) = 12.453\rho^{3.44}\bigg]_{1.0}^{1.5}$$

from which

$$T = T_r = 37.8\,\text{kip}\cdot\text{in.} \qquad\qquad \textbf{Ans.}$$

PROBLEMS

Introductory Problems

6-100*   A solid circular 100-mm-diameter shaft is made of an elastoplastic steel ($G = 80$ GPa) that has a shearing yield point of 120 MPa. Determine

a. The torque required to initiate plastic action in the shaft.
b. The percent increase in torque required to produce plastic action over the entire cross section of the shaft. (*Note:* There will always be some elastic action near the axis of the shaft; however, this effect is so small that it can be neglected without producing significant error.)

6-101*   A solid circular 4-in.-diameter shaft is made of an elastoplastic steel ($G = 12,000$ ksi) that has a shearing yield point of 24 ksi. Determine

a. The applied torque when the plastic zone begins at $r = 1.5$ in. (see Fig. 6-26e).
b. The applied torque when the plastic zone begins at $r = 1$ in.

6-102   A hollow shaft has an inside diameter of 50 mm and an outside diameter of 100 mm. The shaft is made of an elastoplastic steel ($G = 80$ GPa) that has a shearing yield point of 140 MPa. Determine

a. The applied torque when the plastic zone begins at $r = 40$ mm (see Fig. 6-26e).
b. The applied torque when the shearing stress at a point on the inside surface of the shaft reaches the yield point.

6-103*   A solid circular 3-in.-diameter shaft is made of an elastoplastic steel ($G = 12,000$ ksi) that has a shearing yield point of 24 ksi. When the plastic zone begins at $r = 3/4$ in. (see Fig. 6-26e), determine

a. The shearing strain at the surface of the shaft.
b. The torque applied to the shaft.

6-104   A hollow shaft with an outside diameter of 100 mm and an inside diameter of 50 mm is twisted through an angle of 8° in a 3-m length. The shaft is made of an elastoplastic steel ($G = 80$ GPa) that has a yield point in shear of 140 MPa. Determine

a. The maximum shearing strain in the shaft.
b. The magnitude of the applied torque.
c. The shearing stress at a point on a cross section at the inside surface of the shaft.

6-105   A solid circular 4-in.-diameter shaft is twisted through an angle of 5° in a 5-ft length. The shaft is made of an elastoplastic steel ($G = 12,000$ ksi) that has a yield point in shear of 18 ksi. Determine

a. The maximum shearing strain in the shaft.
b. The magnitude of the applied torque.
c. The shearing stress at a point on a cross section 1/2 in. from the axis of the shaft.

Intermediate Problems

6-106*   A solid circular 50-mm-diameter shaft is 1 m long. The shaft is made of an aluminum alloy that has a shearing stress-strain diagram that can be approximated by the two straight lines shown in Fig. P6-106. Determine

a. The torque required to develop a maximum shearing stress of 230 MPa in the shaft.
b. The angle of twist when the torque of part a is applied.

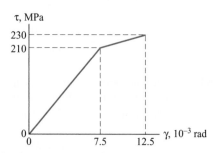

Figure P6-106

6-107*   A hollow shaft has an inside diameter of 2.5 in., an outside diameter of 4 in., and is 3 ft long. The shaft is made of an aluminum alloy that has a shearing stress-strain diagram that can be approximated by the two straight lines shown in Fig. P6-107. Determine

a. The torque required to develop a shearing stress of 30 ksi at a point on a cross section at the inner surface of the shaft.

b. The angle of twist when the torque of part a is applied.

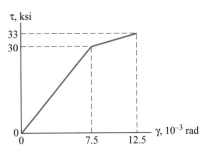

Figure P6-107

6-108   A solid circular 80-mm-diameter shaft is made of a magnesium alloy that has the shearing stress-strain diagram shown in Fig. P6-108. Determine the torque required to twist a 2-m length of the shaft through 0.300 rad.

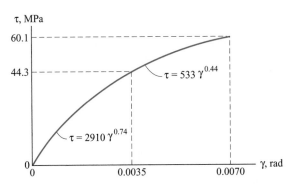

Figure P6-108

6-109   The stress-strain curve for a polymeric material used for small gears and shafts is shown in Fig. P6-109. If the material is used for a circular shaft of radius R that is to transmit a torque T, develop expressions for

a. The angle of twist in terms of T, R, k, and length L.

b. The maximum shearing stress in the shaft in terms of T and R.

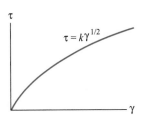

Figure P6-109

Challenging Problems

6-110*   Two 80-mm-diameter steel ($G = 80$ GPa) and bronze ($G = 45$ GPa) shafts are rigidly connected and supported as shown in Fig. P6-110a. The shearing stress-strain diagram for the steel is shown in Fig. P6-110b. The bronze has a proportional limit in shear of 84 MPa. Determine the torque required to produce a maximum shearing stress of 60 MPa in the bronze.

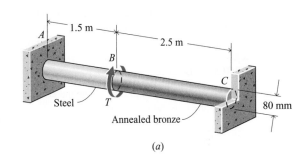

(a)

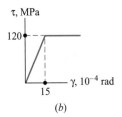

(b)

Figure P6-110

6-111*   The composite shaft shown in Fig. P6-111a consists of a hollow steel cylinder with a 6-in. outside diameter and a 4-in. inside diameter over a solid 2-in.-diameter cold-rolled brass core. The two members are rigidly connected to bar AB at the right end and to the wall at the left end. The shearing stress-strain diagram for the steel is shown in Fig. P6-111b. The brass ($G = 5000$ ksi) has a proportional limit in shear of 30 ksi. Determine the torque required to rotate bar AB through an angle of 0.30 rad.

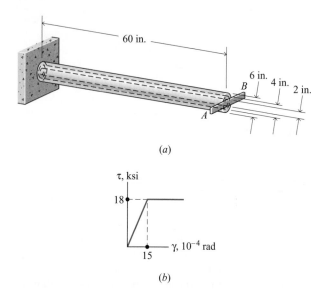

(a)

(b)

Figure P6-111

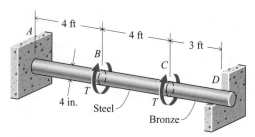

Figure P6-113

6-112 The 160-mm-diameter steel shaft of Fig. P6-112 has a 100-mm-diameter bronze core inserted and securely bonded to the steel ($G = 80$ GPa) in 3 m of the right end. The steel is elastoplastic and has a yield point in shear of 120 MPa. The bronze has a modulus of rigidity G of 40 GPa and a proportional limit in shear of 240 MPa. Determine

a. The maximum shearing stress in each of the materials.
b. The rotation of the free end of the shaft.

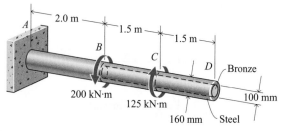

Figure P6-112

6-113 The circular shaft of Fig. P6-113 consists of a steel ($G = 12,000$ ksi) segment ABC securely connected to a bronze ($G = 6000$ ksi) segment CD. Ends A and D of the shaft are securely fastened to rigid supports. The steel is elastoplastic and has a yield point in shear of 18 ksi. The bronze has a proportional limit in shear of 35 ksi. When the two equal torques T are applied, the section at B rotates 0.072 rad in the direction of T. Determine

a. The magnitude of the applied torque T.
b. The maximum shearing stress in each of the materials.

6-114 The 100-mm-diameter shaft shown in Fig. P6-114 is composed of a brass ($G = 40$ GPa) segment and a steel ($G = 80$ GPa) segment. The steel is elastoplastic and has a yield point in shear of 120 MPa. The brass has a proportional limit in shear of 60 MPa. Determine the magnitude of the torque, applied as shown, that is required to produce a maximum shearing stress in the brass of 50 MPa.

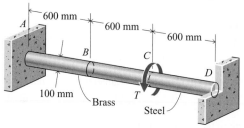

Figure P6-114

Computer Problems

6-115 A 4-in.-diameter solid circular steel shaft is subjected to a torque T. Assume that the steel is elastoplastic (see Fig. 6-26d), having a yield point τ_y in shear of 18 ksi and a modulus of rigidity G of 12,000 ksi. Calculate and plot

a. The location of the yield surface r_y as a function of the torque T ($0 \leq T \leq 25$ kip · ft).
b. The maximum shearing stress τ in the shaft as a function of the torque T ($0 \leq T \leq 25$ kip · ft).
c. The angle of twist θ in a 10-ft length of the shaft as a function of the torque T ($0 \leq T \leq 25$ kip · ft).

6-116 A 120-mm-diameter solid circular shaft is made of a material that has a shear stress-strain diagram similar to Fig. 6-26b. If $G = 16.66$ GPa, $\tau_y = 44.3$ MPa, and $\tau = 602\gamma^{0.44}$ MPa when $\tau > \tau_y$, calculate and plot

a. The torque T required to rotate a 3-m length of the shaft as a function of the angle of twist θ ($0° \leq \theta \leq 25°$).
b. The location of the yield surface r_y as a function of the angle of twist θ ($0° \leq \theta \leq 25°$).
c. The maximum shearing stress τ in the shaft as a function of the angle of twist θ ($0° \leq \theta \leq 25°$).

6-117 A 1-in.-diameter aluminum alloy shaft and a 1/2-in.-diameter steel shaft are rigidly connected and supported as shown in Fig. P6-117a. The shear-strain diagrams for the aluminum alloy and the steel are shown in Figs. P6-117b and c, respectively. Compute and plot the rotation θ for a section at B as a function of the torque T applied at B ($0 \leq T \leq 750$ lb · ft).

6-118 A 20-mm-diameter aluminum alloy shaft and a 25-mm-diameter steel shaft are rigidly connected and supported as shown in Fig. P6-118a. The shear stress-strain diagrams for the aluminum alloy and the steel are shown in Figs. P6-118b and c, respectively. Compute and plot the rotation θ for a section at C as a function of the torque T applied at C ($0 \leq T \leq 475$ N · m).

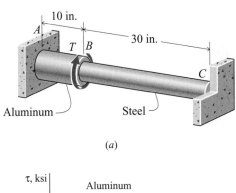

(a)

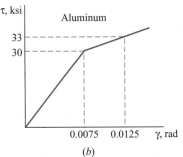

(b)

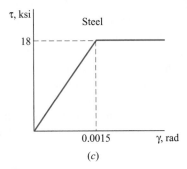

(c)

Figure P6-117

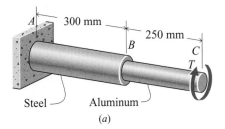

(a)

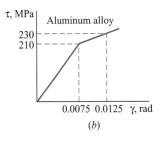

(b)

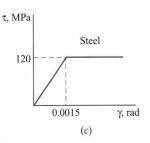

(c)

Figure P6-118

6-11 TORSION OF NONCIRCULAR SECTIONS

Prior to 1820, when A. Duleau published experimental results to the contrary, it was thought that the shearing stresses in any torsionally loaded member were proportional to the distance from its axis. Duleau proved experimentally that this is not true for rectangular cross sections. An examination of Fig. 6-29 will verify Duleau's conclusion. If the stresses in the rectangular bar were proportional to the distance from its axis, the maximum stress would occur at the corners. However,

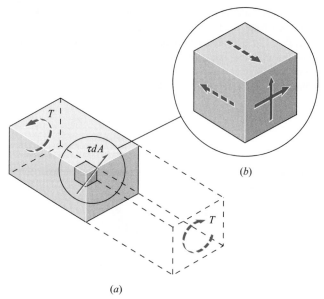

Figure 6-29

if there were a stress of any magnitude at the corner, as indicated in Fig. 6-29*a*, it could be resolved into the components indicated in Fig. 6-29*b*; if these components existed, the two components shown dashed would also exist. These last components cannot exist, because the surfaces on which they are shown are free boundaries. Therefore, the shearing stresses at the corners of the rectangular bar must be zero.

The first correct analysis of the torsion of the prismatic bar of noncircular cross section was published by Saint-Venant in 1855; however, the scope of this analysis is beyond the elementry discussion of this book.[6] The results of Saint-Venant's analysis indicate that, in general, except for members with circular cross sections, every section will warp (not remain plane) when the bar is twisted.

For the case of the rectangular bar shown back in Fig. 6-3*d*, the distortion of the small squares is maximum at the midpoint of a side of the cross section and disappears at the corners. Since this distortion is a measure of shearing strain, Hooke's law requires that the shearing stress be maximum at the midpoint of a side of the cross section and zero at the corners. Equations for the maximum shearing stress and angle of twist for a rectangular section obtained from Saint-Venant's theory are

$$\tau_{\max} = \frac{T}{\alpha a^2 b} \tag{6-13}$$

$$\theta = \frac{TL}{\beta a^3 b G} \tag{6-14}$$

where *a* and *b* are the lengths of the short and long sides of the rectangle, respectively. The numerical factors α and β can be obtained from Fig. 6-30.

---

[6]A complete discussion of this theory is presented in various books, such as *Mathematical Theory of Elasticity*, I. S. Sokolnikoff, 2nd ed., McGraw-Hill, New York, 1956, pp. 109–134.

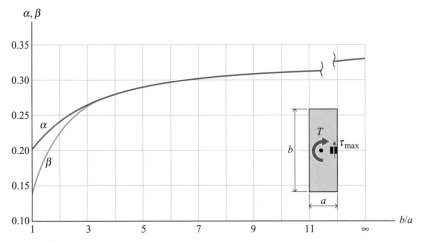

Figure 6-30

6-12 TORSION OF THIN-WALLED TUBES—SHEAR FLOW

Although the elementary torsion theory presented in Sections 6-1, 6-2, and 6-3 is limited to circular sections, one class of noncircular sections that can be readily analyzed by elementary methods is a thin-walled section such as the one illustrated in Fig. 6-31a that represents a noncircular section with a wall of variable thickness t.

A useful concept associated with the analysis of thin-walled sections is shear flow q, defined as the internal shearing force per unit of length of the thin section. Typical units for q are pounds per inch or newtons per meter. In terms of stress, q equals τt, where τ is the average shearing stress across the thickness t. It will be demonstrated that the shear flow on a cross section is constant even though the thickness of the section wall varies. Figure 6-31b shows a block cut from the member of Fig. 6-31a between A and B, and as the member is subjected to pure torsion, the shear forces $V_1 \cdots V_4$ alone (no normal forces) are necessary for equilibrium. Summing forces in the x-direction gives

$$V_1 = V_3$$

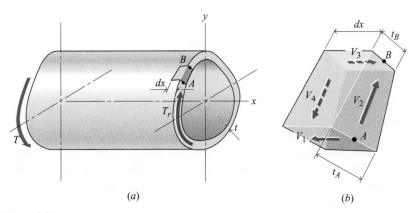

(a)   (b)

Figure 6-31

or

$$q_1 dx = q_3 dx$$

from which

$$q_1 = q_3$$

and, as $q = \tau t$,

$$\tau_1 t_A = \tau_3 t_B \qquad (a)$$

The shearing stresses at point A on the longitudinal and transverse planes have the same magnitude; likewise, the shearing stresses at point B have the same magnitude on the two orthogonal planes; hence, Eq. (a) may be written

$$\tau_A t_A = \tau_B t_B$$

or

$$q_A = q_B$$

which was to be proved.

An expression for the resulting torque on a section can be developed by considering the force dF acting through the center of a differential length of perimeter ds, as shown in Fig. 6-32. The resisting torque T_r is the resultant of the moments of the forces dF; that is,

$$T_r = \int (dF)\rho = \int (q \, ds)\rho = q \int \rho \, ds$$

This integral may be difficult to evaluate by formal calculus; however, the quantity $\rho \, ds$ is twice the area of the triangle shown shaded in Fig. 6-32, which makes the integral equal to twice the area enclosed by the median line A. The resulting expression is

$$T_r = q(2A) \qquad (6\text{-}15)$$

or, in terms of stress,

$$\tau = \frac{T_r}{2At} \qquad (6\text{-}16)$$

where τ is the average shearing stress across the thickness t (and tangent to the perimeter) and is reasonably accurate when t is relatively small. For example, in a round tube with a diameter-to-wall thickness ratio of 20, the stress as given by Eq. 6-16 is 5 percent less than that given by the torsion formula. It must be emphasized that Eq. 6-16 applies only to "closed" sections—that is, sections with a continuous periphery. If the member is slotted longitudinally (see, for example, Fig. 6-33), the resistance to torsion would be diminished considerably from that for the closed section.

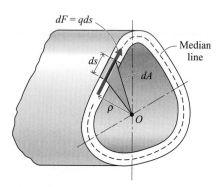

$dF = qds$

ds

Median line

dA

ρ

O

Figure 6-32

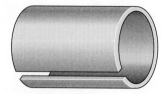

Figure 6-33

■ **Example Problem 6-16**   The aluminum alloy ($G = 4000$ ksi) bar shown in Fig. 6-34 is subjected to a torque $T = 2500$ lb · in. Determine the maximum shearing stress and the angle of twist for the 12-in. length.

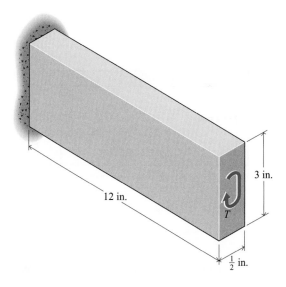

12 in.

3 in.

T

$\frac{1}{2}$ in.

Figure 6-34

SOLUTION

Once α and β are known, Eqs. 6-13 and 6-14 can be used to calculate the maximum shearing stress and the angle of twist. Since $b/a = 3/(1/2) = 6$, the values of α and β from Fig. 6-30 are

$$\alpha = \beta = 0.30$$

and thus,

$$\tau_{\max} = \frac{T}{\alpha a^2 b}$$

$$= \frac{2500}{0.30(1/2)^2(3)} = 11{,}111 \text{ psi} \cong 11{,}110 \text{ psi} \qquad \textbf{Ans.}$$

and

$$\theta = \frac{TL}{\beta a^3 b G}$$

$$= \frac{2500(12)}{0.30(1/2)^3(3)(4000)(10^3)} = 0.06667 \text{ rad} \cong 0.0667 \text{ rad} \qquad \textbf{Ans.}$$

Example Problem 6-17   A rectangular box section of aluminum alloy has outside dimensions 100×50 mm, as shown in Fig. 6-35. The plate thickness is 2 mm for the 50-mm sides and 3 mm for the 100-mm sides. If the maximum shearing stress must be limited to 95 MPa, determine the maximum torque that can be applied to the section. Neglect stress concentrations.

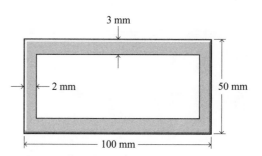

Figure 6-35

▶ Recall that the shear flow $q = \tau t$ is constant. Therefore, as the section thickness decreases, the shear stress must increase, and the maximum shear stress will occur in the thinnest portion of the section.

▶ The area A is the area enclosed by the median line of the section. In this case, A is the area of the rectangle (100-1-1) mm wide and (50-1.5-1.5) mm tall.

SOLUTION

The maximum stress will occur in the thinnest plate; therefore,

$$q = \tau t = 95(10^6)(2)(10^{-3}) = 190(10^3)\,\text{N/m}$$

The torque that can be transmitted by the section is given by Eq. 6-15 as

$$\begin{aligned}
T &= 2q\,A \\
&= 2(190)(10^3)(100-2)(50-3)(10^{-6}) \\
&= 1750\,\text{N}\cdot\text{m}
\end{aligned}$$ **Ans.**

PROBLEMS

Introductory Problems

6-119* The allowable shearing stress for the aluminum alloy ($G = 4000$ ksi) bar shown in Fig. P6-119 is 12 ksi. Determine the maximum permissible torque that may be applied to the bar.

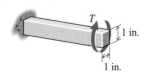

Figure P6-119

6-120* The two bars shown in Fig. P6-120 are made of aluminum ($G = 28$ GPa). The cross-sectional areas and lengths of the two bars are identical. If the maximum shearing stress must be limited to 25 MPa, determine

a. The maximum permissible torque that may be applied to each bar.

b. The angle of twist for each bar when the torque of part a is being applied.

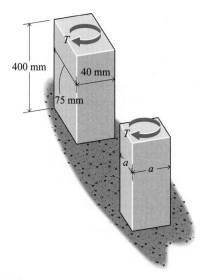

Figure P6-120

6-121 Two aluminum alloy bars ($G = 4000$ ksi) of identical length are rigidly attached to fixed supports at one end. One bar (see Fig. P6-121a) is square. The second bar was machined from a square bar of the same dimensions as the first bar, as shown in Fig. P6-121b. If the maximum shearing stress must be limited to 12 ksi, determine

a. The maximum permissible torque that may be applied to the free end of each bar.
b. The angle of twist for each bar when the torque of part a is being applied if the bars are 3 ft long.

(a)   (b)

Figure P6-121

6-122* A torque of 2.0 kN · m will be applied to the hollow, thin-walled, aluminum section shown in Fig. P6-122. If the maximum shearing stress must be limited to 40 MPa, determine the minimum thickness required for the section.

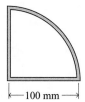

←——100 mm——→

Figure P6-122

6-123 A torque of 125 kip · in. will be applied to the hollow, thin-walled, aluminum alloy section shown in Fig P6-123. If the maximum shearing stress must be limited to 8 ksi, determine the minimum thickness required for the section.

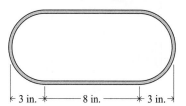

←— 3 in. —*——— 8 in. ———*— 3 in. —→

Figure P6-123

6-124 A 500-mm-wide × 3-mm-thick × 2-m-long aluminum sheet is to be formed into a hollow section by bending through 360° and welding (butt-weld) the long edges together. Assume a median length of 500 mm (no stretching of the sheet due to bending). If the maximum shearing stress must be limited to 75 MPa, determine the maximum torque that can be carried by the hollow section if

a. The shape of the section is a circle.
b. The shape of the section is an equilateral triangle.
c. The shape of the section is a square.
d. The shape of the section is a 150 × 100-mm rectangle.

Intermediate Problems

6-125* A solid rectangular bar of aluminum alloy ($G = 4000$ ksi) is subjected to the torques shown in Fig. P6-125. If $T_1 = 10,000$ lb · in., $T_2 = 30,000$ lb · in., $a = 2$ in., $b = 3$ in., and $L_1 = L_2 = 30$ in., determine

a. The maximum shearing stress in the bar.
b. The rotation of end C with respect to the support at A.

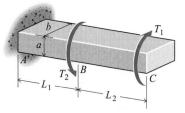

Figure P6-125

6-126* The torsion member shown in Fig. P6-125 has the cross section shown in Fig. P6-126. If $T_2 = T_1 = 2T$, $a = 75$ mm,

$b = 100$ mm, and $L_1 = L_2 = 800$ mm, determine the maximum torque T that may be applied to the bar if the maximum shearing stress in the bar must be limited to 80 MPa.

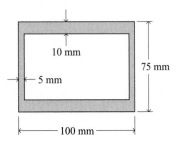

Figure P6-126

6-127 Two torsion members (see Fig. P6-127) made of the same material have the same length and the same weight. For the same allowable shear stress, determine

a. The ratio of the torque that can be carried by the circular bar to the torque that can be carried by the square bar.
b. The ratio of the angles of twist when the torques of part a are being carried by the two sections.

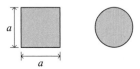

Figure P6-127

6-128 The 50 × 50-mm square torsion member shown in Fig. P6-128 is made of an aluminum alloy ($G = 28$ GPa). If the maximum shearing stress must not exceed 80 MPa and the rotation of end D with respect to end A must not exceed 0.035 rad, determine the maximum torque T that can be carried by the bar.

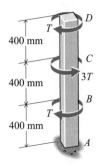

Figure P6-128

Challenging Problems

6-129* A torque box from an airplane wing is shown in Fig. P6-129. Curves AB and CD are 40.2 in. long. The mean depth of the box is 12.5 in. The box is made of an aluminum alloy with an allowable shearing stress of 8 ksi. Determine the maximum torque that can be applied to the box.

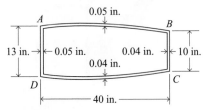

Figure P6-129

6-130 A cross section of an airplane fuselage made of aluminum alloy is shown in Fig. P6-130. For an applied torque of 200 kN · m and an allowable shearing stress of 50 MPa, determine the minimum thickness of sheet (constant for the entire periphery) required to resist the torque.

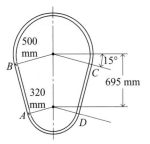

Figure P6-130

6-131 The 1.5 × 1.5-in. square bar of aluminum alloy ($G = 4000$ ksi) shown in Fig. P6-131 is rigidly attached to supports at A and D. Determine the reactions at the supports if $T_1 = 8000$ lb · in., $T_2 = 0$, and $L_1 = L_2 = L_3 = 1.5$ ft.

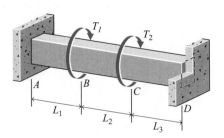

Figure P6-131

6-13 DESIGN PROBLEMS

Design, within a limited context, has been discussed previously in Chapter 5. In that chapter, design was limited to axially loaded members and to pins. This chapter will extend design to solid or hollow circular bars subjected to static torsional loading. Design of these bars will be limited to proportioning a torsionally loaded member to perform a specified function without failure. In this chapter, failure will refer to failure by yielding or failure by fracture.

Furthermore, design will be limited to circular bars made of ductile materials for which the significant failure stress is the shearing stress at yield or fracture. This book will not address the important issue of fatigue loading. That topic is covered in later courses.

For many materials the yield strength in shear is not given in tables such as Tables B-17 and B-18. It will be shown in Chapter 10 that, for ductile materials, the yield strength in shear is either 0.5 or 0.577 times the tensile yield strength, depending on the criterion selected for failure. For the present, the more conservative value of 0.5 will be used. Design for combined loading will be discussed in Chapter 10.

> ### Example Problem 6-18
> A solid circular shaft 4 ft long made of 2014-T4 wrought aluminum is subjected to a torsional load of 10,000 lb · in. If failure is by yielding and a factor of safety (FS) of 2 is specified, select a suitable diameter for the shaft if 2014-T4 wrought aluminum bars are available with diameters in increments of 1/8 in.
>
> **SOLUTION**
> The failure criterion is (refer to Chapter 5)
>
> $$\text{Strength} \geq (\text{Factor of Safety})(\text{Stress})$$
>
> For a torsionally loaded circular shaft, stress refers to the magnitude of the shearing stress $\tau_c = T_r c / J$. For a failure mode of yielding, strength is the yield strength in shear τ_y, which for 2014-T4 wrought aluminum is listed in Table B-17 as 24 ksi. Thus,
>
> $$\tau_y \geq \text{FS}\left(\frac{T_r c}{J}\right) = \text{FS}\left(\frac{Tr}{\pi r^4/2}\right) = \frac{16(\text{FS})(T)}{\pi d^3}$$
>
> Solving for the diameter,
>
> $$d \geq \left[\frac{16(\text{FS})(T)}{\pi \tau_y}\right]^{1/3} = \left[\frac{16(2)(10{,}000)}{\pi(24)(10^3)}\right]^{1/3}$$
> $$d \geq 1.619 \text{ in.}$$
>
> Since bars are available in 1/8-in. increments, the smallest permissible bar is
>
> $$d_{\min} = 1\frac{5}{8} \text{ in.} \qquad \textbf{Ans.}$$

■ **Example Problem 6-19**   A solid circular shaft 2 m long is to transmit 1000 kW at 600 rpm. Failure is by yielding, and the factor of safety is 1.75. If the shaft is made of structural steel, select a suitable diameter for the shaft if bars are available with diameters in increments of 10 mm.

SOLUTION

Since failure is by yielding, the strength is the yield strength in shear. Table B-18 lists the yield strength in tension as 250 MPa, but does not give a value for the yield strength in shear. According to the discussion in Section 6-13, τ_y will be taken as $\sigma_y/2$. Thus $\tau_y = 250/2 = 125$ MPa. Using the results of Example Problem 6-18,

$$d \geq \left[\frac{16(\text{FS})(T)}{\pi \tau_y} \right]^{1/3} \tag{a}$$

The torque may be found using Eq. 6-11

$$\text{Power} = T\omega = T(2\pi N)$$

or

$$T = \frac{\text{Power}}{2\pi N} \tag{b}$$

Combining Eqs. (a) and (b) gives

$$d \geq \left[\frac{8(\text{FS})(\text{Power})}{\pi^2 N \tau_y} \right]^{1/3}$$

$$d \geq \left[\frac{8(1.75)(1000)(10^3)}{\pi^2(600/60)(125)(10^6)} \right]^{1/3}$$

$$d \geq 0.1043 \text{ m} = 104.3 \text{ mm}$$

Since bars are available in 10-mm increments, the smallest permissible bar is

$$d_{\min} = 110 \text{ mm} \qquad\qquad \textbf{Ans.}$$

■ **Example Problem 6-20**   A stepped shaft is subjected to the torques shown in Fig. 6-36. Both segments of the shaft are made of 6061-T6 wrought aluminum. For a factor of safety of 2.0 against failure by yielding, determine the required diameters for the two segments of the shaft.

SOLUTION

For failure by yielding, the significant strength in the shaft is the shearing yield strength. From Table B-17, the shearing yield strength for 6061-T6 wrought

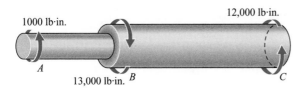

Figure 6-36

aluminum is 26 ksi. Using the results of Example Problem 6-18,

$$d \geq \left[\frac{16(\text{FS})(T)}{\pi \tau_y}\right]^{1/3}$$

For segment AB,

$$d_{AB} \geq \left[\frac{16(2)(1000)}{\pi(26)(10^3)}\right]^{1/3} = 0.7317 \text{ in.}$$

For segment BC,

$$d_{BC} \geq \left[\frac{16(2)(12,000)}{\pi(26)(10^3)}\right]^{1/3} = 1.675 \text{ in.}$$

The minimum diameters are then

$$d_{AB} = 0.732 \text{ in.} \qquad \textbf{Ans.}$$

$$d_{BC} = 1.675 \text{ in.} \qquad \textbf{Ans.}$$

▶ Of course, a fillet will be necessary at the change of diameter to reduce the stress concentration. If the fillet radius r is about 0.7 (0.732 in.) $\cong$ 0.5 in., the stress concentration factor would be about 2, and the actual stress at the change of diameter would be 2 (13 ksi) = 26 ksi = τ_{yield}.

Example Problem 6-21 A steel pipe will be used as a shaft to transmit 100 kW at 120 rpm. Failure is by yielding ($\sigma_y = 250$ MPa), and the factor of safety is 1.5. Determine the lightest-weight standard steel pipe that can be used for the shaft.

SOLUTION
The failure criterion

$$\tau_y \geq (\text{FS})\left(\frac{T_r c}{J}\right)$$

can be solved for (J/c) to yield

$$\frac{J}{c} \geq \frac{\text{FS}(T_r)}{\tau_y} = \frac{\text{FS}(\text{Power})}{\tau_y(2\pi N)}$$

Substituting the given numerical values,

$$\frac{J}{c} \geq \frac{1.5(100)(10^3)}{125(10^6)(2\pi)(120/60)}$$

or

$$\frac{J}{c} \geq 95.49(10^{-6})\,\text{m}^3 = 95.49(10^3)\,\text{mm}^3$$

where

$$\frac{J}{c} = \frac{\pi(r_o^4 - r_i^4)}{2r_0} \geq 95.49(10^3)\,\text{mm}^3 \qquad (a)$$

Equation (a) can be satisfied for an infinite number of hollow pipes having different radius ratios. Table B-14 can be used to select a pipe that satisfies the requirement that $J/c \geq 95.49(10^3)$ mm^3. However, Table B-14 lists the properties I and S, where I is the rectangular second moment of area with respect to a diameter of the pipe, and S is the section modulus, $S = I/c$. Since $J = 2I$ for circular sections (either solid or hollow), the required section modulus for the pipe is

▶ Recall that the polar second moment of area of a cross section is equal to the sum of the second moments relative to the x-and y-axes, $J = I_x + I_y$. Since $I_x = I_y = 1/4\,\pi r^4$ for a circular section, $J = 2I = 2I_x = 2I_y = 1/2\,\pi r^4$.

$$S = \frac{I}{c} = \frac{1}{2}\frac{J}{c} \geq \frac{95.49(10^3)}{2} = 47.75(10^3)\,\text{mm}^3 \qquad (b)$$

The lightest-weight standard pipe in Table B-14, with $S \geq 47.75(10^3)$ mm^3, is a 102-mm-nominal-diameter pipe. Any pipe in Table B-14 that has a section modulus greater than $S = 47.75(10^3)$ mm^3 would satisfy the stress requirement, but the 102-mm pipe is the lightest, since it has the smallest cross-sectional area.

PROBLEMS

Introductory Problems

6-132* A motor is to transmit 150 kW to a piece of mechanical equipment. The power is transmitted through a solid structrual steel shaft. Failure is by yielding, and the factor of safety is 1.25. The designer has the freedom to operate the motor at 60 rpm or at 6000 rpm. For each case, determine the minimum shaft diameter. Shafts are available with diameters in increments of 5 mm. If weight is important, which speed would be used?

6-133* A 3-ft-long steel pipe is subjected to a torque of 1200 lb·ft at each end. The pipe is made of 0.2% C hardened steel, failure is by yielding, and the factor of safety is 1.5. Determine the nominal diameter of the lightest standard-weight steel pipe that can be used for the shaft.

6-134 A standard-weight structural steel pipe must transmit 150 kW at 60 rpm. The failure mode is yielding, and the factor of safety is 1.5.

  a. Select the lightest standard-weight steel pipe that can be used.

  b. If solid structural steel shafts are available with diameters in increments of 10 mm, determine the minimum diameter that can be used.

  c. Compare the weights of the two shafts.

6-135 A shaft is to transmit 100 hp at 200 rpm. The designer has a variety of solid bars and standard steel pipes to select from. Both the bars and the pipes are made of structural steel, the failure mode is yielding, and the factor of safety is 2.

  a. Select the lightest standard-weight steel pipe that can be used.

  b. Select a suitable solid shaft if they are available with diameters in increments of 1/8 in.

  c. If weight is critical, which shaft should be used?

Intermediate Problems

6-136* The motor shown in Fig. P6-136 supplies a torque of 1000 N · m to shaft *ABCDE*. The torques removed at *C*,

D, and E are 500 N · m, 300 N · m, and 200 N · m, respectively. The shaft is the same diameter throughout and is made of 0.4% C hot-rolled steel. For a factor of safety of 3 and failure by yielding, select a suitable diameter for the shaft if shafts are available with diameters in increments of 10 mm.

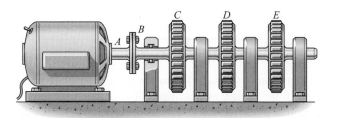

Figure P6-136

6-137 A torque of 30,000 lb · in. is supplied to the factory drive shaft of Fig. P6-137 by a belt that drives pulley A. A torque of 10,000 lb · in. is removed by pulley B and 20,000 lb · in. by pulley C. The shaft is made of structural steel and has a constant diameter over its length. Segment AB of the shaft is 3 ft long, and segment BC is 4 ft long. Failure is by yielding, and the factor of safety is 2.25. Select a suitable diameter for the shaft if shafts are available with diameters in increments of 1/8 in.

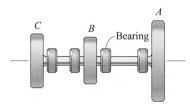

Figure P6-137

Challenging Problems

6-138* The band brake shown in Fig. P6-138 is part of a hoisting machine. The coefficient of friction between the 500-mm-diameter drum and the flat belt is 0.20. The maximum actuating force P that can be applied to the brake arm is 490 N. Rotation of the drum is clockwise. What minimum-size shaft should be used to transmit the resisting torque developed by the brake to the machine if the shaft is to be made of 0.4% C hot-rolled steel? The factor of safety is 3 for failure by yielding. Circular steel bars are available with diameters in increments of 5 mm.

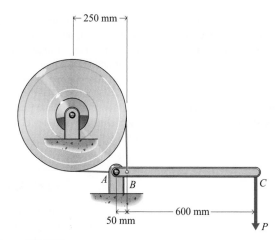

Figure P6-138

6-139 The motor shown in Fig. P6-139 supplies a torque of 380 lb · ft to shaft ABC. The torques removed at gears C and B are 220 lb · ft and 160 lb · ft, respectively. The shaft ABC has a constant diameter and is made of 0.4% C hot-rolled steel, failure is by yielding, and the factor of safety is 2. Determine

a. The minimum allowable diameter of the shaft if shafts are available with diameters in increments of 1/8 in.

b. The minimum allowable diameter of the bolts used in the coupling, if eight bolts are used, the material is structural steel, the mode of failure is yielding, and the factor of safety is 1.5. The diameter of the bolt circle is $d_1 = 3.5$ in., and the bolts are available with diameters in increments of 1/16 in.

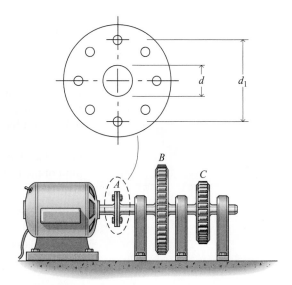

Figure P6-139

6-140 A shaft used to transmit power is constructed by joining two solid segments of shaft with a collar, as shown in Fig. P6-140. The collar has an inside diameter equal to the diameter of the shaft, and both the collar and the shaft are made of the same material. The collar is securely bonded to the shaft segments. Determine the ratio of the diameters of the collar and shaft such that the splice can transmit the same power as the shaft and at the same maximum shearing stress level. Is the solution dependent on the material selected?

Figure P6-140

SUMMARY

The problem of transmitting a torque (a couple) from one plane to a parallel plane is frequently encountered in the design of machinery. The simplest device for accomplishing this function is a circular shaft. The resisting torque is statically equivalent to the sum of the torques produced by the shear stresses

$$T_r = \int_{\text{area}} \rho \, dF = \int_{\text{area}} \rho \tau_\rho \, dA \tag{6-1}$$

where ρ is the distance from the axis of the shaft to the element of area dA. The law of variation of the shearing stress on the transverse plane (τ as a function of radial position ρ) must be known before the integral of Eq. 6-1 can be evaluated. If the assumption is made that a plane transverse cross section before twisting remains plane after twisting and a diameter of the section remains straight, the distortion of the shaft can be expressed as

$$\gamma_c = \frac{c\theta}{L} \quad \text{and} \quad \gamma_\rho = \frac{\rho\theta}{L} \tag{6-2}$$

or

$$\gamma_\rho = \frac{\gamma_c}{c}\rho \tag{6-3}$$

The angle θ is called the angle of twist. Equation 6-3 indicates that the shearing strain is zero at the center of the shaft and increases linearly with respect to the distance ρ from the axis of the shaft. This equation can be combined with Eq. 6-1 once the relationship between the shearing stress τ and the shearing strain γ is known. Since no assumptions have been made about the relationship between the stress and the strain or about the type of material of which the shaft is made, Eq. 6-3 is valid for elastic or inelastic action and for homogeneous or heterogeneous materials, provided the strains are not too large ($\tan \gamma \cong \gamma$). If the assumption is made that Hooke's law ($\tau = G\gamma$) applies (stresses must be below the proportional limit of the material), Eq. 6-3 can be written

$$\tau_\rho = \frac{\tau_c}{c}\rho \tag{6-4}$$

When Eq. 6-4 is substituted into Eq. 6-1, the result is

$$\tau_\rho = \frac{T_r\rho}{J} \quad \text{and} \quad \tau_c = \frac{T_r c}{J} \tag{6-6}$$

where J is the polar second moment of the cross-sectional area of the shaft. Equation 6-6 indicates that the shearing stress τ_ρ, like the shearing strain γ_ρ, is zero at the center of the shaft and increases linearly with respect to the distance ρ from the axis of the shaft. Both the shearing strain γ and the shearing stress τ are maximum when $\rho = c$. Equation 6-6 is known as the elastic torison formula and is valid for both solid and hollow circular shafts.

Frequently, the amount of twist in a shaft is important. Equations 6-2, 6-6, and Hooke's law ($\tau = G\gamma$) can be combined to give

$$\theta = \frac{\gamma_\rho L}{\rho} = \frac{\tau_\rho L}{\rho G} \qquad \text{or} \qquad \theta = \frac{T_r L}{GJ} \qquad \text{(6-7a, b)}$$

The angle of twist determined from the above expressions is for a length of shaft of constant diameter ($J = $ constant), constant material properties ($G = $ constant), and carrying a torque T_r. Ideally, the length of shaft should not include sections too near to (within about one-half shaft diameter of) places where mechanical devices (gears, pulleys, or couplings) are attached. For practical purposes, however, it is customary to neglect distortions at connections and to compute angles as if there were no discontinuities.

If T_r, G, or J is not constant along the length of the shaft, Eq. 6-7b takes the form

$$\theta = \sum_{i=1}^{n} \frac{T_{ri} L_i}{G_i J_i} \qquad \text{(6-7c)}$$

where each term in the summation is for a length L where T_r, G, and J are constant. If T_r, G, or J is a function of x (the distance along the length of the shaft), the angle of twist is found using

$$\theta = \int_0^L \frac{T_r dx}{GJ} \qquad \text{(6-7d)}$$

For a shaft in pure torsion (as in this chapter), the maximum shearing stress occurs on both transverse and longitudinal planes and is given by the elastic torsion formula, Eq. 6-6. Maximum normal stresses occur on planes oriented at 45° with the axis of the shaft and perpendicular to the surface of the shaft. On one of these planes, the normal stress is tension, and on the other the normal stress is compression. Furthermore, the magnitude of the maximum normal stresses is the same as the maximum shearing stress given by Eq. 6-6.

One of the most common uses of a circular shaft is the transmission of power. Power is defined as the time rate of doing work, and the basic relationship for work done by a constant torque T is $W_k = T\phi$, where W_k is work and ϕ is the angular displacement of the shaft in radians. The derivative of W_k with respect to time t gives

$$\text{Power} = \frac{dW_k}{dt} = T\frac{d\phi}{dt} = T\omega \qquad \text{(6-11)}$$

where dW_k/dt is power, T is a constant torque, and ω is the constant angular velocity of the shaft (in radians per second). In the SI system of units, power is given in watts (1 W = 1 N · m/s). In the U.S. customary system of units, power is usually given in units of horsepower (1 hp $= 550$ lb · ft/s $= 33,000$ lb · ft/min). Equation 6-11 shows that for a fixed amount of power, the torque that can be transmitted by a shaft decreases with rotation rate. Conversely, for a fixed torque, the power required to rotate the shaft increases with angular rotation rate.

If a circular shaft is pulled and twisted at the same time, the combined loading produces both a normal stress and a shearing stress at a point on the transverse cross section of the shaft. As long as the strains are small, these stresses can be computed separately (using Eqs. 2-2 and 6-6) and superimposed on the element. Once these stresses are known, the normal and shearing stresses on other planes through the point can be determined using the stress transformation equations or Mohr's circle. The maximum normal stresses and the maximum shearing stresses at the point can be found using the methods described in Section 2-10.

One of the limitations of the elastic torsion formula (Eq. 6-6) is that stresses must be less than the proportional limit of the material. In some design situations, however, a limited amount of inelastic action can be permitted. Equation 6-3, which states that the shearing strain at any point on a transverse cross section of a circular shaft is proportional to the distance of the point from the axis of the shaft

$$\gamma_\rho = \frac{\gamma_c}{c}\rho \tag{6-3}$$

is valid for elastic or inelastic action provided the strains are not too large (tan γ $\cong \gamma$). The variation of the shearing stress must be determined from a shear stress-strain diagram for the material. Finally, the resisting torque can then be evaluated by substituting the expression for τ in terms of ρ into Eq. 6-1

$$T_r = \int_{\text{area}} \rho\tau_\rho \, dA \tag{6-1}$$

and integrating over the cross-sectional area of the shaft.

Although the elementary torsion theory is limited to circular sections, one class of noncircular sections that can be readily analyzed by elementary methods is a thin-walled section. In such sections, the shear flow q (defined as the internal shearing force per unit of length of the thin section) is constant even though the thickness of the section wall varies. The resisting torque T_r is the resultant of the moment of the shear force

$$T_r = 2q \, A \tag{6-15}$$

where $q = \tau t$ is the shear flow, τ is the average shearing stress across the thickness t, and A is the area enclosed by the median line. Solving for the shear stress gives

$$\tau = \frac{T_r}{2At} \tag{6-16}$$

REVIEW PROBLEMS

6-141* A steel ($G = 12,000$ ksi) shaft is loaded and supported as shown in Fig. P6-141. Determine

a. The maximum shearing stress in the shaft.

b. The rotation of a section at the right end with respect to its no-load position.

c. The rotation of a section 7 ft from the left end with respect to its no-load position.

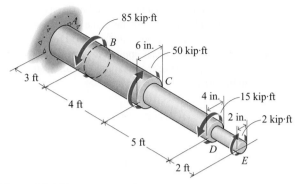

Figure P6-141

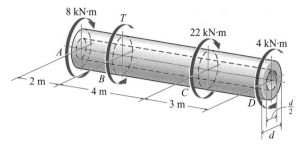

Figure P6-144

6-142* A torque of 12.0 kN · m is supplied to the driving gear *B* of Fig. P6-142 by a motor. Gear *A* takes off 4.0 kN · m of torque, and the remainder is taken off by gear *C*. For an allowable shearing stress of 80 MPa, determine

a. The minimum permissible diameters for the two shafts.
b. The angle of twist of gear *A* with respect to gear *C* if both shafts are made of steel ($G = 80$ GPa) and have diameters of 75 mm.

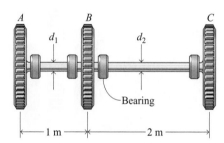

Figure P6-142

6-143 An aluminum alloy ($G = 3800$ ksi) tube will be used to transmit a torque in a control mechanism. The tube has an outside diameter of 1.25 in. and a wall thickness of 0.065 in. Because of the tendency of thin sections to buckle, the maximum compressive stress in the tube must be limited to 8000 psi. Determine

a. The maximum torque that can be applied.
b. The angle of twist in a 3-ft length when a torque of 1000 lb · in. is applied.

6-144* The hollow circular steel ($G = 80$ GPa) shaft of Fig. P6-144 is in equilibrium under the torques indicated. Determine

a. The minimum permissible outside diameter *d* if the maximum shearing stress in the shaft is not to exceed 100 MPa.
b. The rotation of a section at *D* with respect to a section at *A* for the shaft with an outside diameter of 120 mm.

6-145 A solid circular aluminum alloy ($G = 4000$ ksi) shaft is 2.5 in. in diameter and 3 ft long. The maximum permissible angle of twist is 0.052 rad, and the allowable shearing stress is 10 ksi. Determine the maximum horsepower that this shaft can deliver when rotating at 500 rpm.

6-146 A solid circular stepped steel ($G = 80$ GPa) shaft has the dimensions and is subjected to the torques shown in Fig. P6-146. Determine

a. The maximum tensile stress in section *AB* of the shaft.
b. The maximum compressive stress in section *BC* of the shaft.
c. The rotation of a section at *C* with respect to its no-load position.

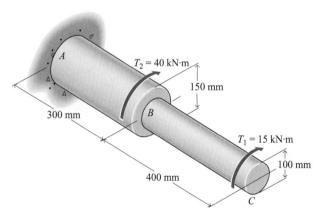

Figure P6-146

6-147* The inner surface of the aluminum alloy ($G = 4000$ ksi) sleeve *A* and the outer surface of the steel ($G = 12,000$ ksi) shaft *B* of Fig. P6-147 are smooth. Both the sleeve and the shaft are rigidly fixed to the wall at *D*. The 0.500-in. diameter pin *C* fills a hole drilled completely through a diameter of the sleeve and shaft. If the average shearing stress on the cross-sectional area of the pin at the interface between the shaft and the sleeve must not exceed 5000 psi, determine

a. The maximum torque T that can be applied to the right end of the steel shaft B.
b. The maximum shearing stress in the aluminum alloy sleeve A when the torque of part a is applied.
c. The rotation of the right end of the shaft when the maximum torque T is applied.

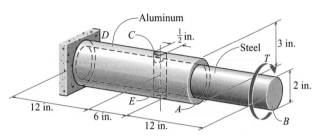

Figure P6-147

6-148 The 160-mm-diameter steel ($G = 80$ GPa) shaft shown in Fig. P6-148 has a 100-mm-diameter bronze ($G = 40$ GPa) core inserted in 3 m of its right end. The bronze is securely bonded to the steel. Determine

a. The maximum shearing stress in each of the materials.
b. The rotation of the free end of the shaft.

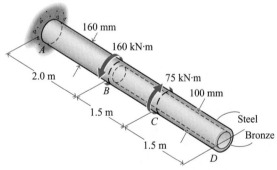

Figure P6-148

6-149 A torque of 1850 lb · ft will be applied to the hollow, thin-walled, aluminum alloy section shown in Fig. P6-149. If the maximum shearing stress must be limited to 8 ksi, determine the minimum thickness required for the section.

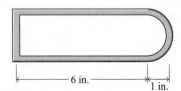

Figure P6-149

6-150* A cross section of the leading edge of an airplane wing is shown in Fig. P6-150. The length along the curve is 660 mm, and the enclosed area is 53,000 mm². Sheet thicknesses are shown on the diagram. For an applied torque of 12 kN · m, determine the magnitude of the maximum shearing stress developed in the section.

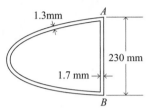

Figure P6-150

6-151* The motor shown in Fig. P6-151 delivers 200 hp at 300 rpm to a piece of equipment at B. The shaft is made from 0.4% C hot-rolled steel ($G = 11,600$ ksi), which is available with diameters in increments of 1/8 in. The length of the shaft between A and B is 3 ft. Determine the shaft diameter required if the maximum shearing stress in the shaft must not exceed 15.9 ksi and the angle of twist between A and B must not exceed 1.5°.

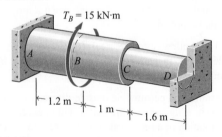

Figure P6-151

6-152 The 100-mm-diameter segment ABC of the shaft shown in Fig. P6-152 is initially not connected to the 60-mm-diameter segment CD. Torque $T_B = 15$ kN · m is applied at section B, and a secure connection between the two segments is then made at C, after which the torque T_B is removed. Determine the resulting maximum shearing stress in segment CD after the torque T_B is removed. The moduli of rigidity are 40 GPa for ABC and 80 GPa for CD.

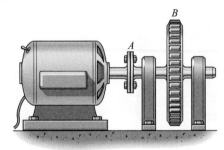

Figure P6-152

Chapter 7
Flexural Loading: Stresses in Beams

7-1 INTRODUCTION

A member subjected to loads applied transverse to the long dimension of the member and which causes the member to bend is a *beam*. The beam, or flexural member, is frequently encountered in structures and machines, and its elementary stress analysis constitutes one of the more interesting facets of mechanics of materials. For example, Fig. 7-1 is a photograph of an I-beam, *AB*, simply supported in a testing machine and loaded at the one-third points. Figure 7-2 depicts the shape (exaggerated) of the beam when loaded.

Before proceeding with a discussion of stress analysis for flexural members, it may be well to classify some of the various types of beams and loadings encountered in practice. Beams are frequently classified on the basis of their supports or reactions. A beam supported by a pin, roller, or smooth surface at the ends and having one span is called a *simple beam* (Fig. 7-3a). A simple support (a pin or roller) will develop a reaction normal to the beam but will not produce a couple. If either or both ends of the beam project beyond the supports, it is called a *simple beam with overhang* (Fig. 7-3b). A beam with more than two simple supports is a

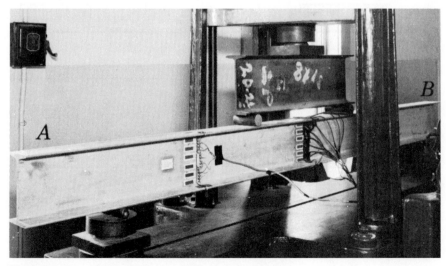

Figure 7-1 Setup for measuring longitudinal strains in a beam.

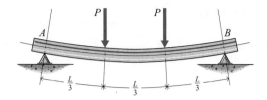

Figure 7-2

continuous beam (Fig. 7-3c). A *cantilever beam* is one in which one end is built into a wall or other support so that the built-in end can neither move transversely nor rotate (Fig. 7-3d). The built-in end is said to be *fixed* if no rotation occurs and *restrained* if a limited amount of rotation occurs. The supports shown in Figs. 7-3d, e, and f represent fixed ends unless otherwise stated. The beams in Figs. 7-3d, e, and f are, in order, a cantilever beam, a beam fixed (or restrained) at the left end and simply supported near the other end (which has an overhang), and a beam fixed (or restrained) at both ends.

   Cantilever beams and simple beams have only two reactions (two forces or one force and a couple), and these reactions can be obtained from a free-body diagram of the beam by applying the equations of equilibrium. Such beams are

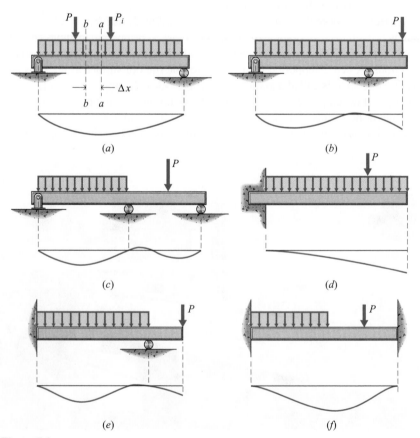

Figure 7-3

said to be statically determinate since the reactions can be obtained from the equations of equilibrium. Beams with more than two reaction components are called *statically indeterminate* because there are not enough equations of equilibrium to determine the reactions. The beams shown in Figs. 7-3a, b, and d are statically determinate, whereas the beams shown in Figs. 7-3c, e, and f are statically indeterminate.

All of the beams shown in Fig. 7-3 are subjected to both concentrated loads and to uniformly distributed loads. Although all of the beams shown in Fig. 7-3 are shown as horizontal, beams may have any orientation. The loads are assumed to act in a plane of symmetry. Distributed loads will be shown on the side of the beam on which they are acting; that is, if drawn on the bottom of the beam, the load is pushing upward and if drawn on the right side of a vertical beam, the load is pushing to the left. Deflection curves (greatly exaggerated) are shown beneath the beams of Fig. 7-3 to assist in visualizing the shapes of the loaded beams.

A free-body diagram of the portion of the beam of Fig. 7-3a between the left end and plane a–a is shown in Fig. 7-4a. A study of this diagram shows that a transverse force V_r and a couple M_r at the section and a force R (a reaction) at the left support are needed to maintain equilibrium. The force V_r is the resultant force due to the shearing stresses acting on the section (on plane a–a) and is called the *resisting shear*. The couple M_r is the resultant moment due to the normal stresses acting on the section (on plane a–a) and is called the *resisting moment*. The magnitudes and senses of V_r and M_r are obtained from the equations of equilibrium $\Sigma F_y = 0$ and $\Sigma M_O = 0$ where O is any axis perpendicular to the xy-plane. The reaction R must be evaluated from a free-body diagram of the entire beam.

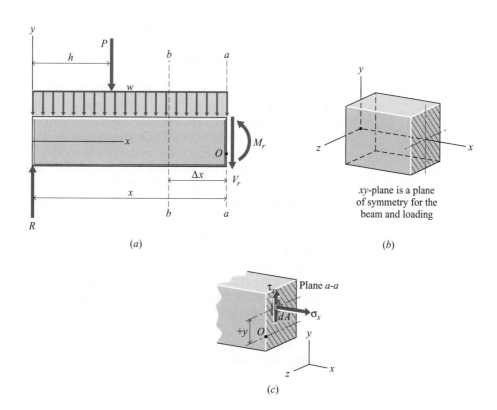

(a)

(b)

xy-plane is a plane of symmetry for the beam and loading

(c)

Figure 7-4

The normal and shearing stresses σ_x and τ_{xy} on plane a–a are related to the resisting moment M_r and the resisting shear V_r by the equations

$$V_r = -\int_{\text{area}} \tau_{xy}\, dA \qquad (7\text{-}1a)$$

$$M_r = -\int_{\text{area}} y\, \sigma_x\, dA \qquad (7\text{-}1b)$$

The resisting shear and moment (V_r and M_r), as shown on Fig. 7-4a, will be defined as positive quantities later in Section 7-5. The normal and shear stresses σ_x and τ_{xy}, as shown on Fig. 7-4c, are defined as positive stresses. The minus signs in Eqs. 7-1a and 7-1b are required to bring these two definitions into agreement. It is obvious from Eqs. 7-1 that the laws of variation of the normal and shearing stresses must be known before the integrals can be evaluated. Thus, the problem is statically indeterminate. For the present, the shearing stresses will be ignored while the normal stresses are studied.

7-2 FLEXURAL STRAINS

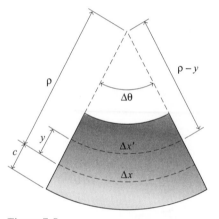

Figure 7-5

A segment of the beam of Fig. 7-4a, between planes a–a and b–b, is shown in Fig. 7-5 with the distortion greatly exaggerated. When Fig. 7-5 was drawn, the assumption was made that a plane section before bending remains a plane after bending. For this to be strictly true, it is necessary that the beam be bent only with couples (no shear on transverse planes). Also, the beam must be so proportioned that it will not buckle and the loads applied so that no twisting occurs (this last limitation will be satisfied if the loads are applied in a plane of symmetry—a sufficient though not a necessary condition). When a beam is bent only with couples, the deformed shape of all longitudinal elements (also referred to as fibers) is an arc of a circle.

Precise experimental measurements indicate that at some distance c above the bottom of the beam, longitudinal elements undergo no change in length. The curved surface formed by these elements (at radius p in Fig. 7-5) is referred to as the *neutral surface of the beam*, and the intersection of this surface with any cross section is called the *neutral axis of the section*. All elements (fibers) on one side of the neutral surface are compressed, and those on the opposite side are elongated. As shown in Fig. 7-5, the fibers above the neutral surface of the beam of Fig. 7-4a (on the same side as the center of curvature) are compressed and the fibers below the neutral surface (on the side opposite the center of curvature) are elongated. The xy-axes of Fig. 7-4c lie in the plane of symmetry; the origin of the coordinate system lies on the neutral surface.

Finally, the assumption is made that all longitudinal elements have the same initial length. This assumption imposes the restriction that the beam be initially straight and of constant cross section; however, in practice, considerable deviation from these last restrictions is often tolerated.

The longitudinal strain ϵ_x experienced by a longitudinal element that is located a distance y from the neutral surface of the beam is determined by using the definition of normal strain as expressed by Eq. 3-1. Thus,

$$\epsilon_x = \frac{\delta}{L} = \frac{L_f - L_i}{L_i}$$

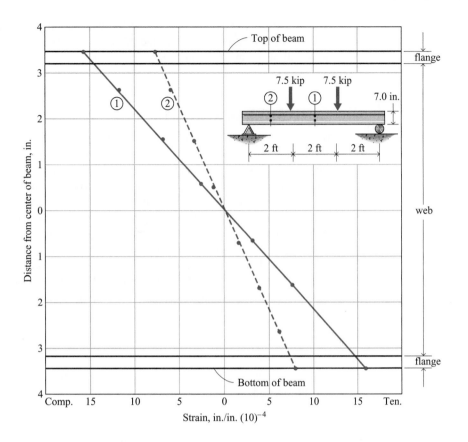

Figure 7-6

where L_f is the final length of the fiber after the beam is loaded and L_i is the initial length of the fiber before the beam is loaded. From the geometry of the beam segment shown in Fig. 7-5,

$$\epsilon_x = \frac{\Delta x' - \Delta x}{\Delta x} = \frac{(\rho - y)(\Delta\theta) - \rho(\Delta\theta)}{\rho(\Delta\theta)} = -\frac{1}{\rho}y \qquad (7\text{-}2)$$

Equation 7-2 indicates that the strain developed in a fiber is directly proportional to the distance of the fiber from the neutral surface of the beam (recall that y is measured from the neutral surface). This variation can be demonstrated experimentally by means of strain gages attached to an I-beam, as shown in Fig. 7-1. The strains, as measured by gages on two different sections, are plotted against the vertical position of the gages on the beam in Fig. 7-6. Curve 1 represents strains on a section at the center of the beam where pure bending occurs (no transverse shear), and curve 2 shows strains at a section near one end of the beam where both flexural (normal) stresses and transverse shearing stresses exist. These curves are both straight lines within the limits of the accuracy of the measuring equipment.[1]

---

[1] A more exact analysis using principles developed in the theory of elasticity indicates that curve 2 should be curved slightly. Note: Other experiments indicate that a plane section of an initially curved beam will also remain plane after bending and that deformations will still be proportional to the distance of the fiber from the neutral surface. The strain, however, will not be proportional to this distance, since each deformation must be divided by a different original length.

Note that Eq. 7-2 is valid for elastic or inelastic action so long as the beam does not twist or buckle and the transverse shearing stresses are small. Problems in this book will be assumed to satisfy these restrictions.

7-3 FLEXURAL STRESSES

With the acceptance of the premise that the longitudinal strain ϵ_x is proportional to the distance of the fiber from the neutral surface of the beam, the law of variation of the normal stress σ_x on the transverse plane can be determined by using a tensile-compressive stress-strain diagram for the material used in fabricating the beam. For many real materials, the tension and compression stress-strain diagrams are identical in the linearly elastic range. Although the diagrams may differ somewhat in the inelastic range, the differences can be neglected for most real problems. For beam problems in this book, the compressive stress-strain diagram will be assumed to be identical to the tensile diagram unless otherwise noted.

For the special case of linearly elastic action, the relationship between stress σ_x and strain ϵ_x is given by Hooke's law, Eq. 4-1a (since the state of stress is uniaxial) as

$$\sigma_x = E\epsilon_x \qquad (a)$$

Substituting Eq. 7-2 into Eq. (a) yields

$$\sigma_x = E\epsilon_x = -\frac{E}{\rho}y \qquad (7\text{-}3)$$

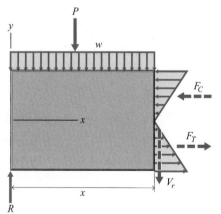

Figure 7-7

Equation 7-3 shows that the normal stress σ_x on the transverse cross section of the beam varies linearly with distance y from the neutral surface. Also, since plane cross sections remain plane, the normal stress σ_x is uniformly distributed in the z-direction (see Fig. 7-4c).

With the law of variation of flexural stress known, Fig. 7-4 can now be redrawn as shown in Fig. 7-7. The forces F_C and F_T are the resultants of the compressive and tensile flexural stresses, respectively. Since the sum of the forces in the x-direction must be zero, F_C is equal to F_T; hence, they form a couple of magnitude M_r.

The resisting moment M_r developed by the normal stresses in a typical beam with loading in a plane of symmetry but of arbitrary cross section, such as the one shown in Fig. 7-8, is given by Eq. 7-1 as

$$M_r = -\int_A y \, dF = -\int_A y \, \sigma_x dA \qquad (b)$$

Since y is measured from the neutral surface, it is first necessary to locate this surface by means of the equilibrium equation $\Sigma F_x = 0$, which gives

$$\Sigma F_x = \int_A dF = \int_A \sigma_x dA = 0 \qquad (c)$$

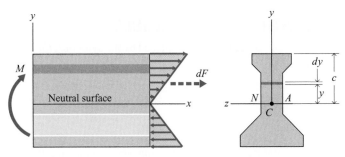

Figure 7-8

Substituting Eq. 7-3 into Eq. (*c*) yields

$$\int_A \sigma_x dA = \int_A \left(-\frac{E}{\rho}y\right)dA$$

$$= -\frac{E}{\rho}\int_A y\, dA = -\frac{E}{\rho}y_C A = 0 \qquad (7\text{-}4)$$

7.1

where y_C is the distance from the neutral axis to the centroidal axis *c–c* of the cross section that is perpendicular to the plane of bending. Since neither (E/ρ) nor *A* is zero, y_C must equal zero. Thus, *for flexural loading and linearly elastic action, the neutral axis passes through the centroid of the cross section.*

Since the normal stress σ_x varies linearly with distance *y* from the neutral surface, the maximum normal stress $\sigma_{\max}$ on the cross section can be written as

$$\sigma_{\max} = -\frac{E}{\rho}c \qquad (7\text{-}5)$$

7.2

where *c* is the distance to the surface of the beam (top or bottom) farthest from the neutral surface. If the quantity (E/ρ) is eliminated between Eqs. 7-3 and 7-5, a useful relationship between the maximum stress $\sigma_{\max}$ on a transverse cross section and the stress σ_x at an arbitrary distance *y* from the neutral surface is obtained. Thus,

$$\sigma_x = \frac{y}{c}\sigma_{\max} = \frac{y}{c}\sigma_c \qquad (7\text{-}6)$$

Substitution of Eq. 7-6 into Eq. (*b*) yields

$$M_r = -\int_A y\,\sigma_x\,dA = -\frac{\sigma_c}{c}\int_A y^2 dA \qquad (7\text{-}7)$$

in which $\sigma_{\max} = \sigma_c$ (equals σ_x evaluated at $y = c$). The integral $\int y^2\, dA$ is called the *second moment of area*. Second moments of area of several common shapes are given in Table A-1, Appendix A. Second moments of more complex areas can usually be derived (without integration) from combinations of these simple shapes, as shown in the next section. A discussion of second moments of area is presented in Appendix A.

7.3

7.4

7-4 THE ELASTIC FLEXURE FORMULA

When the integral $\int_A y^2\, dA$ in Eq. 7-7 is replaced by the symbol I, *the elastic flexure formula* is obtained as

$$\sigma_x = -\frac{M_r y}{I} \tag{7-8}$$

where σ_x is the normal flexural stress on a transverse plane, at a distance y from the neutral surface, M_r is the resisting moment of the section, and I is the second moment of area of the transverse section with respect to the neutral axis. Recall that the neutral axis passes through the centroid of the area.

At any section of the beam, the flexural stress, or normal stress, will be maximum (have the greatest magnitude) at the surface farthest from the neutral axis ($y = c$), and Eq. 7-8 becomes

$$\sigma_{\max} = \frac{M_r c}{I} = \frac{M_r}{S} \tag{7-9}$$

where $S = I/c$ is called the *section modulus of the beam*. Although the section modulus can be readily calculated for a given section, values (magnitudes) of the section modulus are often included in tables to simplify calculations. Observe that for a given area, S becomes larger as the shape is altered to concentrate more of the area as far as possible from the neutral axis. Commercial rolled shapes such as I- and WF-beams and the various built-up sections are intended to optimize the area-section modulus relation.

Thus far, the discussion of flexural behavior has been limited to straight structural members with symmetric cross sections that are loaded in a plane of symmetry. Many other shapes are subjected to flexural loadings, and methods are needed to determine stress distributions in these nonsymmetric shapes. The flexure formula (Eq. 7-8) provides a means for relating the resisting moment M_r at a section of a beam to the normal stress at a point on the transverse cross section. Further insight into the applicability of the flexure formula to nonsymmetric sections can be gained by considering the requirements for equilibrium when the applied moment M does not have a component about the y-axis of the cross section. Thus, from Eq. 7-6 and the equilibrium equation $\Sigma M_y = 0$,

$$\int_A z\, \sigma_x\, dA = \int_A z\frac{\sigma_c}{c} y\, dA = \frac{\sigma_c}{c}\int_A zy\, dA = \frac{\sigma_c}{c} I_{yz} = 0 \tag{a}$$

The quantity I_{yz} is commonly known as the mixed second moment of the cross-sectional area with respect to the centroidal y- and z-axes. Obviously, Eq. (a) can be satisfied only if $I_{yz} = 0$. For symmetric cross sections, $I_{yz} = 0$ (see Appendix A) when the y- and z-axes coincide with the axes of symmetry. For nonsymmetric cross sections, $I_{yz} = 0$ when the y- and z-axes are centroidal principal axes (see Appendix A) for the cross section. Thus, the flexure formula is valid for any cross section, provided y is measured along a principal direction, I is a principal second moment of area, and M is a moment about a principal axis. Problems of this type will be discussed in Section 7-9.

Example Problem 7-1 A timber beam consists of four 2×8-in. planks fastened together to form a box section 8 in. wide $\times$ 12 in. deep, as shown in Fig. 7-9a. If the resisting moment at the section is $M_r = -200(10^3)$ lb $\cdot$ in. determine

7.5

(a) The flexural stress at point A of the cross section.
(b) The flexural stress at point B of the cross section.
(c) The flexural stress at point C of the cross section.
(d) The flexural stress at point D of the cross section.

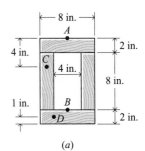

SOLUTION
The resisting moment M_r is assumed to act in the vertical plane of symmetry of the cross section. The neutral axis passes through the centroid of the cross section and is perpendicular to the plane of the resisting moment. As a result of symmetry, the centroid is at the geometric center of the cross section. The second moment of area for the cross section with respect to the neutral axis is found by subtracting the second moment of area for the hollow part of the section (4×8 in.) from the second moment of area for the solid part of the section (8×12 in.). The equation listed in Table A-1 for the second moment of area of a rectangular section is $I = bh^3/12$. Thus, for the cross section shown in Fig. 7-9a,

$$I = \frac{b_s h_s^3}{12} - \frac{b_h h_h^3}{12} = \frac{8(12)^3}{12} - \frac{4(8)^3}{12} = 981.3 \text{ in.}^4$$

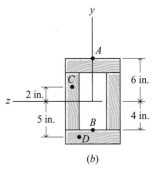

Figure 7-9(a–b)

The distance y, positive upward, from the neutral surface is shown in Fig. 7-9b. Thus,

(a) The flexural (normal) stress at A is

$$\sigma_{xA} = -\frac{M_r y_A}{I} = -\frac{-200(10)^3(+6)}{981.3} = +1222.9 \text{ lb/in.}^2 \cong 1223 \text{ psi (T)}$$
Ans.

▶ Since the resisting moment M_r is negative, the flexural stress σ_x has the same sign as y. When y is positive (the upper portion of the beam), the stress is positive (tension). When y is negative (the lower portion of the beam), the stress is negative (compression).

(b) The flexural stress at B is

$$\sigma_{xB} = -\frac{M_r y_B}{I} = -\frac{-200(10)^3(-4)}{981.3} = -815.2 \text{ lb/in.}^2 \cong 815 \text{ psi (C)}$$
Ans.

(c) The flexural stress at C is

$$\sigma_{xC} = -\frac{M_r y_C}{I} = -\frac{-200(10)^3(+2)}{981.3} = +407.6 \text{ lb/in.}^2 \cong 408 \text{ psi (T)}$$
Ans.

(d) The flexural stress at D is

$$\sigma_{xD} = -\frac{M_r y_D}{I} = -\frac{-200(10)^3(-5)}{981.3} = -1019.1 \text{ lb/in.}^2 \cong 1019 \text{ psi (C)}$$
Ans.

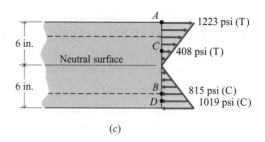

(c)

Figure 7-9(c)

The flexural stress varies linearly with the y-coordinate (see Fig. 7-9b) and is constant with respect to the z-coordinate. The variation of flexural stress over the depth of the beam is shown in Fig. 7-9c.

Alternatively from Eq. 7-8, note on a given cross section that

$$\sigma_x = -\frac{M_r y}{I} \qquad \text{or} \qquad \frac{\sigma_x}{y} = -\frac{M_r}{I} = \text{constant}$$

Therefore, if the stress is known at a point on the cross section, it can be determined at any other point without knowing either the resisting moment M_r or the second moment of area I for the cross section. Thus,

$$\frac{\sigma_{xA}}{y_A} = \frac{\sigma_{xB}}{y_B} = \frac{\sigma_{xC}}{y_C} = \frac{\sigma_{xD}}{y_D}$$

(b) $\quad \sigma_{xB} = \dfrac{y_B}{y_A}\sigma_{xA} = \dfrac{-4}{+6}(+1222.9) = -815.3 \text{ psi} \cong 815 \text{ psi (C)}$ **Ans.**

(c) $\quad \sigma_{xC} = \dfrac{y_C}{y_A}\sigma_{xA} = \dfrac{+2}{+6}(+1222.9) = +407.6 \text{ psi} \cong 408 \text{ psi (T)}$ **Ans.**

(d) $\quad \sigma_{xD} = \dfrac{y_D}{y_A}\sigma_{xA} = \dfrac{-5}{+6}(+1222.9) = -1019.1 \text{ psi} \cong 1019 \text{ psi (C)}$ **Ans.**

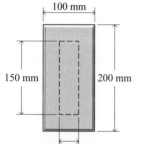

Figure 7-10

■ **Example Problem 7-2** The maximum flexural stress at a certain section (Fig. 7-10) in a beam with a rectangular cross section 100 mm wide × 200 mm deep is 15 MPa. Determine

(a) The resisting moment M_r developed at the section.

(b) The percentage decrease in M_r if the dotted central portion of the cross section shown in Fig. 7-10 is removed.

SOLUTION
The normal stress σ_x at a distance y from the neutral surface on a transverse cross section of a beam is given by Eq. 7-8 as

$$\sigma_x = -\frac{M_r y}{I} \qquad \text{or} \qquad M_r = -\frac{\sigma_x I}{y}$$

(a) For the original cross section,

$$I_{NA} = \frac{100(200)^3}{12} = 66.67(10^6)\,\text{mm}^4 = 66.67(10^{-6})\,\text{m}^4$$

$$|M_r| = \frac{\sigma_x I}{c} = \frac{15(10^6)(66.67)(10^{-6})}{100(10^{-3})}$$

$$= 10.00(10^3)\,\text{N}\cdot\text{m} = 10.00\,\text{kN}\cdot\text{m} \qquad \textbf{Ans.}$$

▶ As a result of symmetry, the centroid is the geometric center of the cross section. The neutral axis is the horizontal axis passing through the centroid.

(b) For the modified cross section,

$$I_{NA} = \frac{100(200)^3}{12} - \frac{50(150)^3}{12} = 52.60(10^6)\,\text{mm}^4 = 52.60(10^{-6})\,\text{m}^4$$

$$|M_r| = \frac{\sigma_x I}{c} = \frac{15(10^6)(52.60)(10^{-6})}{100(10^{-3})}$$

$$= 7.89(10^3)\,\text{N}\cdot\text{m} = 7.89\,\text{kN}\cdot\text{m}$$

$$\text{Percent decrease} = D = \frac{10.00 - 7.89}{10.00}(100) = 21.1\% \qquad \textbf{Ans.}$$

▶ Note that removing 38 percent of the area of the beam (near the neutral axis) resulted in a 21 percent reduction in M_r. If the same amount of area had been removed from the top and bottom of the beam, the reduction in M_r would have been 61 percent!

■ **Example Problem 7-3** A beam has the cross section shown in Fig. 7-11a. On a section where the resisting moment is -75 kN · m, determine

(a) The maximum tensile flexural stress.

(b) The maximum compressive flexural stress.

SOLUTION
The neutral axis is horizontal and passes through the centroid of the cross section. The centroid is located by using the principle of moments as applied to areas. The total area A of the cross section is

$$A = 200(25) + 150(25) + 50(75) = 12,500\,\text{mm}^2$$

The moment of the area M_A about the bottom edge of the cross section is

$$M_A = 200(25)(12.5) + 25(150)(100) + 75(50)(200) = 1,187,500\,\text{mm}^3$$

The distance y_C from the bottom edge of the cross section to the centroid is

$$y_C = \frac{M_A}{A} = \frac{1,187,500}{12,500} = 95\,\text{mm}$$

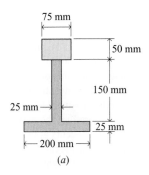

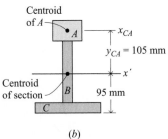

(b)

Figure 7-11(a–b)

as shown in Fig. 7-11b.

Since the centroid of the cross section is not located at the centroid of any of the individual parts (three rectangles), the second moment of the cross section with respect to the horizontal centroidal axis (neutral axis) is found using the parallel axis theorem (see Appendix A). In equation form, the parallel axis theorem for a part of the cross section (one of the rectangles) is

$$I_{x'} = I_{xC} + Ad_y^2$$

where $I_{x'}$ is the second moment of area of the part with respect to the horizontal centroidal axis of the total cross section, I_{xC} is the second moment of area of the part with respect to its horizontal centroidal axis, A is the area of the part, and d_y is the perpendicular distance between the horizontal centroidal axis of the total cross section and the horizontal centroidal axis of the part. In this example, $I_{x'}$ will be found for each of the three parts and added together to find $I_{x'}$ for the complete cross section.

For part A shown in Fig. 7-11b,

$$I_{x'A} = I_{xCA} + A_A d_{yA}^2 = \frac{75(50)^3}{12} + 75(50)(105)^2 = 42.13(10^6)\,\text{mm}^4$$

Similarly, for parts B and C of Fig. 7-11b:

$$I_{x'B} = I_{xCB} + A_B d_{yB}^2 = \frac{25(150)^3}{12} + 25(150)(5)^2 = 7.13(10^6)\,\text{mm}^4$$

$$I_{x'C} = I_{xCC} + A_C d_{yC}^2 = \frac{200(25)^3}{12} + 200(25)(-82.5)^2 = 34.29(10^6)\,\text{mm}^4$$

▶ Since the resisting moment M_r is negative, the top portion of the beam is in tension (σ_x positive) and the bottom portion of the beam is in compression (σ_x negative). The flexural stress varies linearly with distance from the neutral axis. Therefore, the maximum tensile flexural stress occurs at the top surface of the beam and the maximum compressive flexural stress occurs at the bottom surface of the beam.

Adding the second moments of area for the three parts gives the second moment of area for the cross section as

$$I_{x'} = I_{x'A} + I_{x'B} + I_{x'C} = 42.13(10^6) + 7.13(10^6) + 34.29(10^6)$$
$$= 83.55(10^6)\,\text{mm}^4 = 83.55(10^{-6})\,\text{m}^4$$

(a) Since the resisting moment M_r is negative, the maximum tensile flexural stress occurs at the top of the beam and is

$$\sigma_{max} = -\frac{M_r y_t}{I} = -\frac{-75(10^3)(130)(10^{-3})}{83.55(10^{-6})}$$
$$= +116.70(10^6)\,\text{N/m}^2 \cong 116.7\,\text{MPa (T)} \qquad \textbf{Ans.}$$

(b) The maximum compressive stress occurs at the bottom of the beam and is

$$\sigma_{max} = -\frac{M_r y_b}{I} = -\frac{-75(10^3)(-95)(10^{-3})}{83.55(10^{-6})}$$
$$= -85.28(10^6)\,\text{N/m}^2 \cong 85.3\,\text{MPa (C)} \qquad \textbf{Ans.}$$

The linear variation of flexural stress over the depth of the beam is shown in Fig. 7-11c.

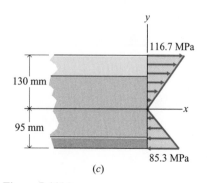

(c)

Figure 7-11(c)

Example Problem 7-4 Determine the largest positive bending moment that can be applied to a WT9 × 38 structural T-beam if the allowable flexural stresses are 20 ksi in tension and 25 ksi in compression.

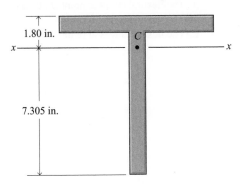

Figure 7-12

SOLUTION

Cross-sectional dimensions of the WT9 × 38 section are given in Table B-11. The X–X axis is the horizontal centroidal axis. Pertinent dimensions are shown in Fig. 7-12. For a WT9 × 38 tee section, $y_c = 1.80$ in. Since the depth of the tee is 9.105 in., the distance from the neutral axis to the bottom of the section is $9.105 - 1.80 = 7.305$ in. The moment acts in the vertical plane of symmetry; thus, the second moment of area is $I_{X-X} = 71.8$ in.[4] Solving Eq. 7-8 for the moment gives

$$M_r = \frac{-\sigma_x I}{y}$$

At the top of the section, $y = +1.80$ in. and σ_x is negative

$$M_r = \frac{-(-25)(71.8)}{1.80} = 997.2 \text{ kip} \cdot \text{in.}$$

At the bottom of the section, $y = -7.305$ in. and σ_x is positive:

$$M_r = \frac{-(+20)(71.8)}{7.305} = 196.6 \text{ kip} \cdot \text{in.}$$

Therefore the largest positive bending moment is

$$M_r = 196.6 \text{ kip} \cdot \text{in.} \qquad \textbf{Ans.}$$

PROBLEMS

MecMovie Activities and Problems

MM7.1 The Centroids Game – Learning the Ropes. Game. Applying the centroid calculation procedure to shapes made up of rectangles.

MM7.2 The Moment of Inertia Game–Starting from Square One. Game. Calculation procedure for moments of inertia applied to shapes comprised of rectangles.

MM7.3 Bending stresses in a flanged shape. Example; Concept checkpoints. Use the flexure formula to determine bending stresses in a flanged shape.

MM7.4 Moments and bending stress. Concept checkpoints. Relating bending moment values to beam tension and compression bending stresses.

Introductory Problems

7-1* The maximum normal stress on a transverse section in a beam with a rectangular cross section 4 in. wide × 6 in. deep is 1000 psi. Determine the resisting moment M_r transmitted by the section.

7-2* Determine the maximum tensile flexural stress if a resisting moment M_r of -10 kN · m is applied to a beam having the cross section shown in Fig. P7-2.

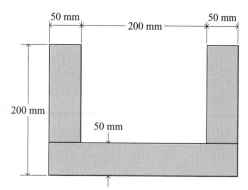

Figure P7-2

7-3 Determine the maximum tensile and compressive flexural stresses if a resisting moment M_r of $+4000$ lb · ft is applied to a beam having the cross section shown in Fig. P7-3.

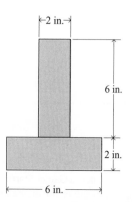

Figure P7-3

7-4* A timber beam consists of three 50 × 200-mm planks fastened together to form an I-beam 200 mm wide × 300 mm deep, as shown in Fig. P7-4. If the flexural stress at point A of the cross section is 7.5 MPa (T), determine

a. The flexural stress at point B of the cross section
b. The flexural stress at point C of the cross section.
c. The flexural stress at point D of the cross section.

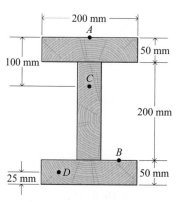

Figure P7-4

7-5 A timber beam is made of three 2 × 6-in. planks fastened together to form an I-beam 6 in. wide × 10 in. deep. If the maximum tensile flexural stress must not exceed 1200 psi, determine the maximum resisting moment M_r that the beam can support.

7-6 The beam of Fig. P7-6 is made of a material that has a tensile and compressive yield strength of 200 MPa. Determine the maximum resisting moment that the beam can support if yielding must be avoided.

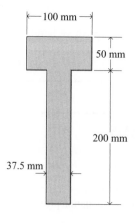

Figure P7-6

7-7* The maximum flexural stress on a transverse cross section of a beam with a 2 × 4-in. rectangular cross section must not exceed 8 ksi. Determine the maximum resisting moment M_r that the beam can support if the neutral axis is

a. Parallel to the 4-in. side.
b. Parallel to the 2-in. side.

7-8 A beam has the cross section shown in Fig. P7-8. On a section where the resisting moment is −10 kN · m, determine

a. The maximum tensile flexural stress.
b. The maximum compressive flexural stress.

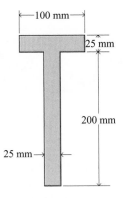

Figure P7-8

Intermediate Problems

7-9* The load-carrying capacity of an S24 × 80 American standard beam (see Appendix B for dimensions) is to be increased by fastening two 8 × 3/4-in. plates to the flanges of the beam, as shown in Fig. P7-9. The maximum flexural stress in both the original and modified beams must be limited to 18 ksi. Determine

a. The maximum resisting moment that the original beam can support.
b. The maximum resisting moment that the modified beam can support.

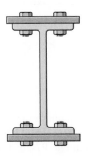

Figure P7-9

7-10* Two L102 × 102 × 12.7-mm structural steel angles (see Appendix B for dimensions) are attached back to back to form a T-section, as shown in Fig. P7-10. Determine the maximum resisting moment M_r that can be supported by the beam if the maximum flexural stress must be limited to 120 MPa.

Figure P7-10

7-11 An I-beam is fabricated by welding two 16 × 2-in. flange plates to a 24 × 1-in. web plate. The beam is loaded in the plane of symmetry parallel to the web. On a section where the resisting moment $M_r = 1000$ kip · ft, determine the maximum flexural stress.

7-12* Determine the maximum tensile and compressive flexural stresses on a section where the resisting moment $M_r = -3$ kN · m if the beam has the cross section shown in Fig. P7-12. The beam is loaded in the plane of symmetry parallel to the web.

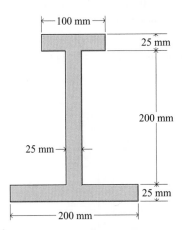

Figure P7-12

7-13 Determine the maximum tensile and compressive flexural stresses on a section where the resisting moment $M_r = -20,000$ lb · in. if the beam has the cross section shown in Fig. P7-13. The beam is loaded in the vertical plane of symmetry.

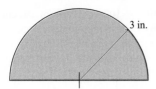

Figure P7-13

7-14 Determine the maximum resisting moment M_r that can be supported by a beam having the cross section shown in Fig. P7-14 if the maximum flexural stress must be limited to 110 MPa.

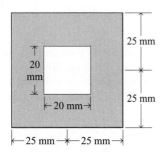

Figure P7-14

Challenging Problems

7-15* A beam has the cross section shown in Fig. P7-15. On a section where the resisting moment is -30 kip · ft, determine

a. The maximum tensile flexural stress.
b. The maximum compressive flexural stress.

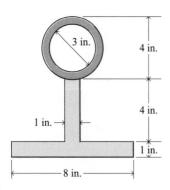

Figure P7-15

7-16* A hardened steel ($E = 210$ GPa) bar with a 50-mm square cross section is subjected to a flexural form of loading that

produces a flexural strain of $+1200$ μm/m at a point on the top surface of the beam. Determine

a. The maximum flexural stress at the point.
b. The resisting moment M_r developed in the beam on a transverse cross section through the point.

7-17 A steel ($E = 29{,}000$ ksi) bar with a rectangular cross section is bent over a rigid mandrel ($R = 12$ in.) as shown in Fig. P7-17. If the maximum flexural stress in the bar is not to exceed the yield strength ($\sigma_y = 36$ ksi) of the steel, determine the maximum allowable thickness h for the bar.

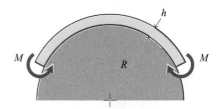

Figure P7-17

7-18* An aluminum alloy ($E = 73$ GPa) bar with a rectangular cross section is bent over a rigid mandrel as shown in Fig. P7-17. The thickness h of the bar is 25 mm. If the maximum flexural stress in the bar must be limited to 100 MPa, determine the minimum allowable radius R for the mandrel.

7-19 The cantilever beam shown in Fig. P7-19 is subjected to a moment $M = 15{,}000$ lb · in at its free end. The beam has a 2×2-in. square cross section. Determine the maximum tensile flexural stress in the beam.

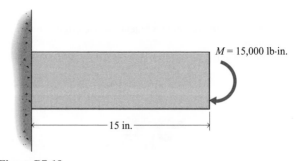

Figure P7-19

7-20 Determine the maximum flexural stress on a section where the resisting moment $M_r = +100$ kN · m if the beam has the cross section shown in Fig. P7-20. The beam is loaded in the vertical plane of symmetry.

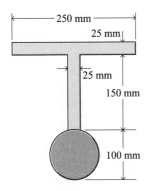

Figure P7-20

250 mm

25 mm

25 mm

150 mm

100 mm

7-21* A steel pipe with an outside diameter of 4 in. and an inside diameter of 3 in. is simply supported at the ends and carries two concentrated loads, as shown in Fig. P7-21. On section A–A, which is 5 ft from the right support, determine

a. The flexural stress at point A on the cross section.
b. The flexural stress at point B on the cross section.

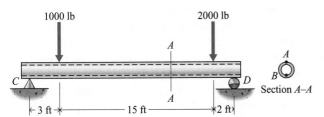

1000 lb

2000 lb

Section A–A

3 ft

15 ft

2 ft

Figure P7-21

7-22 The cantilever beam shown in Fig. P7-22a is subjected to a moment M at its free end. The cross section of the beam is shown in Fig. P7-22b. If the allowable stresses are 90 MPa (T) and 140 MPa (C), determine the maximum moment that can be applied to the beam.

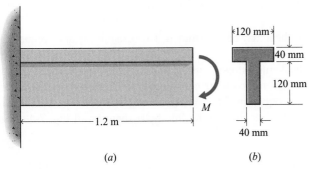

120 mm

40 mm

120 mm

40 mm

M

1.2 m

(a)

(b)

Figure P7-22

Computer Problems

7-23 A beam with a hollow circular cross section (see Fig. P7-23) is being designed to support a maximum moment of 100 kip · in. If the wall thickness is fixed ($t = r_o - r_i = 0.25$ in.),

a. Calculate and plot the maximum flexural stress in the beam as a function of the outside diameter d_o (2.5 in. $\leq d_o \leq$ 6 in.).
b. What is the smallest outside diameter that can be used if the maximum flexural stress in the beam must not exceed 40 ksi?

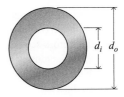

d_i d_o

Figure P7-23

7-24 A beam with a solid rectangular cross section (see Fig. P7-24) is being designed to support a maximum moment of 6 kN · m. If the height of the beam is to be twice the width of the beam ($h = 2b$),

a. Calculate and plot the maximum flexural stress in the beam as a function of the height (50 mm $\leq h \leq$ 400 mm).
b. What is the smallest height that can be used if the maximum flexural stress in the beam must not exceed 20 ksi?

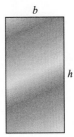

b

h

Figure P7-24

7-25 An S 4 × 9.5 structural steel beam is carrying a moment M that produces a maximum flexural stress of 20 ksi. The moment M must be increased by 75 percent, but the maximum flexural stress must not be increased. In order to strengthen the beam, rectangular steel plates are to be attached to the top and bottom flanges as shown in Fig. P7-25. If the thickness t of the plates must not exceed 1/2 in., prepare a design curve that shows the acceptable values of plate width b as a function of plate thickness t (0 in. $\leq t \leq$ 1/2 in.).

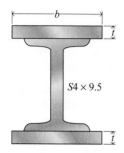

Figure P7-25

7-26   A beam with a solid rectangular cross section (Fig. P7-26a) is being redesigned as an I-beam (Fig. P7-26b) having the same weight per meter (the same cross-sectional area). If the maximum flexural stress in the beam must not exceed 150 MPa and the thickness of the web must not be less than 6 mm, compute

and plot the percent increase in load-carrying capability (M_{max}) of the beam as a function of the thickness (t) of the flanges.

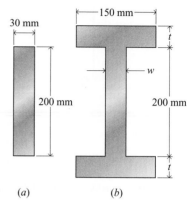

Figure P7-26

7-5   SHEAR FORCES AND BENDING MOMENTS IN BEAMS

The method for determining flexural stresses outlined in Section 7-4 is adequate if one wishes to determine the flexural stresses on any specified transverse cross section of the beam. However, if the maximum flexural stress is required in a beam subjected to a loading that produces a resisting moment that varies with position along the beam, it is desirable to have a method for determining the maximum resisting moment. Similarly, the maximum transverse shearing stress will occur at a section where the resisting shear (V_r of Fig. 7-4a) is maximum and a method for determining such sections is likewise desirable.

In the following section equations for the resisting shear V_r and the resisting moment M_r will be obtained using equilibrium equations. In Section 7-6, relationships will be developed between loads applied to the beam and the resisting shear V_r at a section and between the resisting shear V_r at a section and the resisting moment M_r at that section.

7-5-1   Shear Force and Bending Moment: An Equilibrium Approach
When the equilibrium equation $\Sigma F_y = 0$ is applied to the freebody diagram of Fig. 7-4a, the result can be written as

$$R - wx - P - V_r = 0$$

from which the resisting shear may be determined. The resisting shear is frequently called shear or transverse shear.

The resultant of the flexural stresses on any transverse section has been shown to be a couple (if only transverse loads are considered) and has been designated as M_r. When the equilibrium equation $\Sigma M_O = 0$ (where O is any axis parallel to the neutral axis of the section) is applied to the free-body diagram of Fig. 7-4a, the result can be written as

$$Rx - \frac{wx^2}{2} - P(x - h) - M_r = 0$$

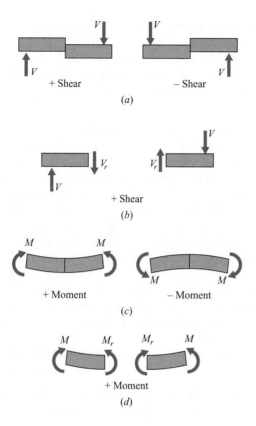

Figure 7-13

from which the resisting moment may be determined. The resisting moment is frequently called bending moment.

The variation of V_r and M_r along the beam can be shown conveniently by means of equations or by means of shear and bending moment diagrams (graphs of V_r and M_r as functions of x).

A sign convention is necessary for the correct interpretation of results obtained from equations or diagrams for shear and moment.[2] The following convention will give consistent results regardless of whether one proceeds from left to right or from right to left along the beam. By definition, the shear at a section is positive when the portion of the beam to the left of the section (for a horizontal beam) tends to move upward with respect to the portion to the right of the section, as shown in Fig. 7-13a. Then for a positive shear force, the transverse shear pushes up on the portion of the beam to the left of the section while the resisting shear pushes downward on the section, as shown in Fig. 7-13b. The converse applies when the portion of the beam is taken to the right of the section. Also, by definition, the bending moment in a horizontal beam is positive at sections for which the top of the beam is in compression and the bottom is in tension, as shown in Fig. 7-13c Then for a positive bending moment, the bending moment acting on the portion of the beam to the left of the section must act clockwise while the resisting moment on the section must act counterclockwise, as shown in Fig. 7-13d. Observe that

---

[2]This sign convention differs from the one presented earlier for resisting forces on a section. The sign convention discussed in this section is the traditional sign convention for beam calculation problems.

the signs of the terms in the preceding equations for V_r and M_r agree with these definitions (sign conventions).

Since M_r and V_r vary with x (Fig. 7-4a), they are functions of x, and equations for M_r and V_r can be obtained from free-body diagrams of portions of the beam. The procedure can be summarized as follows:

- The beam is sectioned at an arbitrary location x.
- A free-body diagram is drawn for the portion of the beam to the left (or for the portion to the right) of the transverse cross section. The shear force and bending moment must be drawn according to the sign convention established earlier.
- Equilibrium equations $\Sigma F_y = 0$ and $\Sigma M_{\text{cut}} = 0$ are written from the free-body diagrams and solved to get the shear force and bending moment at the location x. The equations obtained for $V_r(x)$ and $M_r(x)$ with be valid for a range of x for which the nature of the loading does not change. That is, if the beam is sectioned in a region in which there are no concentrated or distributed loads, the equations will be valid for the entire region $x_L < x < x_R$ where x_L is the location of the nearest load to the left of the section and x_R is the location of the nearest load to the right of the section. No matter where in the region $x_L < x < x_R$ the beam is sectioned, the free-body diagram will look exactly the same and the equations will be exactly the same.
- The process must be repeated for each different segment of the beam.

Example Problem 7-5 will illustrate the procedure to calculate the shear force and the bending moment for any section of a beam.

Example Problem 7-5   A beam is loaded and supported as shown in Fig. 7-14a. Write equations for the shear force V_r and the bending moment M_r for any section of the beam

(a) In the interval AB.
(b) In the interval BC.
(c) In the interval CD.

SOLUTION
A free-body diagram, or load diagram, for the beam is shown in Fig. 7-14b. The reactions R_A and R_D shown on the load diagram are determined from the equations of equilibrium.

$$+\uparrow \Sigma M_D = 0: \qquad R_A(8) - 400(6)(3) - 2000(2) = 0$$
$$+\downarrow \Sigma M_A = 0: \qquad R_D(8) - 400(6)(5) - 2000(6) = 0$$

Solving yields

$$R_A = 1400 \text{ lb} \qquad \text{and} \qquad R_D = 3000 \text{ lb}$$

(a) A free-body diagram of a portion of the beam from the left end to any section between A and B is shown in Fig. 7-14c. The bending moment M_r and the

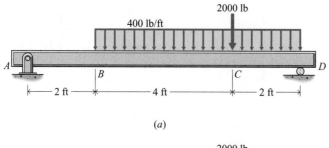

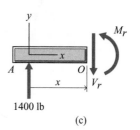

(a)

(c)

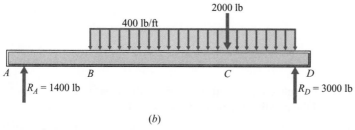

(b)

Figure 7-14(a–c)

transverse shear V_r are shown as positive quantities. From the equilibrium equation $\Sigma F_y = 0$,

$$+\uparrow \Sigma F_y = 0: \qquad\qquad 1400 - V_r = 0$$

Solving yields

$$V_r = 1400 \text{ lb} \qquad (0 < x < 2 \text{ ft}) \qquad\qquad \textbf{Ans.}$$

From the equilibrium equation $\Sigma M_O = 0$,

$$+\downarrow \Sigma M_O = 0: \qquad\qquad -1400(x) + M_r = 0$$

Solving yields

$$M_r = 1400x \text{ lb} \cdot \text{ft} \qquad (0 < x < 2 \text{ ft}) \qquad\qquad \textbf{Ans.}$$

Thus, in the interval AB ($0 < x < 2$ ft), the shear force V_r is constant and the bending moment M_r varies linearly with x.

(b) A free-body diagram of a portion of the beam from the left end to any section between B and C is shown in Fig. 7-14d. The equations of equilibrium for this portion of the beam are

$$+\uparrow \Sigma F_y = 0: \qquad 1400 - 400(x-2) - V_r = 0$$

$$+\downarrow \Sigma M_O = 0: \qquad -1400(x) + 400(x-2)\left(\frac{x-2}{2}\right) + M_r = 0$$

Solving yields

$$V_r = -400x + 2200 \text{ lb} \qquad (2 \text{ ft} < x < 6 \text{ ft}) \qquad \textbf{Ans.}$$

$$M_r = -200x^2 + 2200x - 800 \text{ lb} \cdot \text{ft} \qquad (2 \text{ ft} < x < 6 \text{ ft}) \qquad \textbf{Ans.}$$

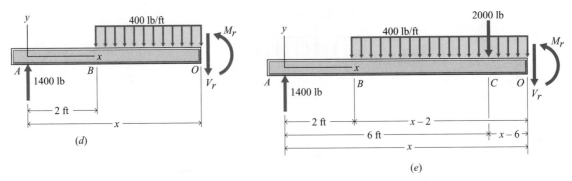

(d)

(e)

Figure 7-14(d–e)

In the interval BC (2 ft $< x <$ 6 ft), the shear force V_r varies linearly with x and the bending moment M_r varies with the square of x.

(c) A free-body diagram of a portion of the beam from the left end to any section between C and D is shown in Fig. 7-14e. The equations of equilibrium for this portion of the beam are

$$+\uparrow \Sigma F_y = 0: \quad 1400 - 400(x - 2) - 2000 - V_r = 0$$

$$+\downarrow \Sigma M_O = 0: -1400(x) + 400(x - 2)\left(\frac{x - 2}{2}\right) + 2000(x - 6) + M_r = 0$$

Solving yields

$$V_r = -400x + 200 \text{ lb} \qquad (6 \text{ ft} < x < 8 \text{ ft}) \qquad \textbf{Ans.}$$

$$M_r = -200x^2 + 200x + 11{,}200 \text{ lb} \cdot \text{ft} \qquad (6 \text{ ft} < x < 8 \text{ ft}) \qquad \textbf{Ans.}$$

In the interval CD (6 ft $< x <$ 8 ft), the shear force V_r varies linearly with x and the bending moment M_r varies with the square of x.

An alternate free-body diagram for interval CD is shown in Fig. 7-14f. The equations of equilibrium for this portion of the beam are

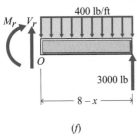

(f)

Figure 7-14(f)

$$+\uparrow \Sigma F_y = 0: \qquad V_r - 400(8 - x) + 3000 = 0$$

$$+\downarrow \Sigma M_O = 0: \qquad -M_r - 400(8 - x)\left(\frac{8 - x}{2}\right) + 3000(8 - x) = 0$$

Solving yields

$$V_r = -400x + 200 \text{ lb} \qquad (6 \text{ ft} < x < 8 \text{ ft}) \qquad \textbf{Ans.}$$

$$M_r = -200x^2 + 200x + 11{,}200 \text{ lb} \cdot \text{ft} \qquad (6 \text{ ft} < x < 8 \text{ ft}) \qquad \textbf{Ans.}$$

▶ Either the free-body diagram of the portion of beam to the left of the section or the free-body diagram of the portion of beam to the right of the section may be used. The shorter section or the section with the fewest forces is most easily solved. Although a section out of the middle of the beam (for example, from B to x, 2 ft $< x <$ 6 ft) could also be used, it is not recommended.

which are identical to the results obtained using the free-body diagram shown in Fig. 7-14e. This is to be expected, since the values of V_r and M_r in Figs. 7-14e and 7-14f are equal in magnitude and opposite in direction, according to Newton's third law.

Example Problem 7-6   An S152 × 19 steel beam is loaded and supported as shown in Fig. 7-15a. On a section 3 m to the right of A, determine

(a) The flexural stress at a point 25 mm below the top of the beam.
(b) The maximum flexural stress on the section.

SOLUTION
The reaction at A is found using the free-body diagram shown in Fig. 7-15b and the equation of equilibrium $\Sigma M_B = 0$. Thus,

$$+\downarrow \Sigma M_B = 0: \qquad 20(2) + 8(6)(7) + 10(12) - R_A(8) = 0$$

from which

$$R_A = 62 \text{ kN}$$

The flexural stresses are required at a specific section of the beam; therefore, bending moment equations for the complete beam are not needed. On a section 3 m to the right of A, the bending moment can be found by using the free-body diagram shown in Fig. 7-15c and the equation of equilibrium $\Sigma M_O = 0$. Thus,

$$+\downarrow \Sigma M_O = 0: \qquad M_r + 8(5)(2.5) - 62(3) + 10(7) = 0$$

from which

$$M_r = +16 \text{ kN} \cdot \text{m}$$

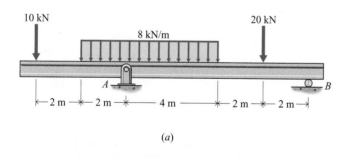

(a)

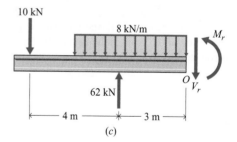

(c)

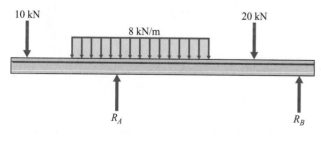

(b)

Figure 7-15(a–c)

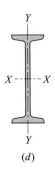

Figure 7-15(d)

(d)

The cross section for the beam is shown in Fig. 7-15d. The X–X–axis is the neutral axis. For an S152 × 19 section (see Appendix B).

$$I = 9.20(10^6)\,\text{mm}^4 = 9.20(10^{-6})\,\text{m}^4$$
$$S = 121(10^3)\,\text{mm}^3 = 121(10^{-6})\,\text{m}^3$$
$$d = 152.4\,\text{mm} = 0.1524\,\text{m}$$

(a) At a point 25 mm below the top of the beam,

$$y = \frac{d}{2} - 25 = \frac{152.4}{2} - 25 = 51.2\,\text{mm} = 0.0512\,\text{m}$$

$$\sigma_x = -\frac{M_r y}{I} = -\frac{16(10^3)(0.0512)}{9.20(10^{-6})}$$

$$= -89.04(10^6)\,\text{N/m}^2 \cong 89.0\,\text{MPa (C)} \qquad \textbf{Ans.}$$

(b) The maximum flexural stress is

$$\sigma_{max} = \frac{M_r}{S} = \frac{16(10^3)}{121(10^{-6})} = 132.23(10^6)\,\text{N/m}^2$$

$$\cong 132.2\,\text{MPa (T) (on the bottom)} \qquad \textbf{Ans.}$$
$$\cong 132.2\,\text{MPa (C) (on the top)} \qquad \textbf{Ans.}$$

PROBLEMS

Introductory Problems

7-27* For the cantilever beam shown in Fig. P7-27, write equations for the shear force V_r and the bending moment M_r for any section of the beam in the interval $0 < x < 4$ ft. Use the coordinate system shown.

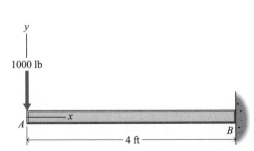

Figure P7-27

any section of the beam in the interval $0 < x < 2$ m. Use the coordinate system shown.

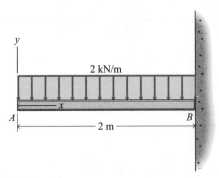

Figure P7-28

7-28* For the cantilever beam shown in Fig. P7-28, write equations for the shear force V_r and the bending moment M_r for

7-29 A beam is loaded and supported as shown in Fig. P7-29. Using the coordinate axes shown, write equations for the shear force V_r and the bending moment M_r for any section of the beam in the interval $0 < x < 10$ ft.

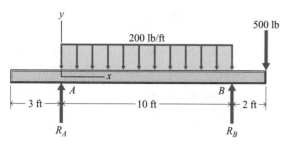

Figure P7-29

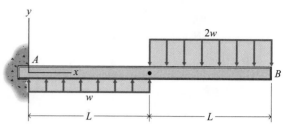

Figure P7-32

7-30* A beam is loaded and supported as shown in Fig. P7-30. Using the coordinate axes shown, write equations for the shear force V_r and the bending moment M_r for any section of the beam in the interval $0 < x < 4$ m.

7-33* The beam shown in Fig. P7-31 has a solid rectangular cross section that is 3 in. wide and 8 in. deep. Determine the maximum tensile flexural stress on a section at $x = 10$ ft.

7-34 The beam shown in Fig. P7-30 is an S254 × 52 steel section. On a section at $x = 5$ m, determine the maximum tensile and compressive flexural stresses.

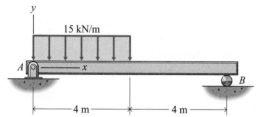

Figure P7-30

Intermediate Problems

7-35* A beam is loaded and supported as shown in Fig. P7-35. Using the coordinate axes shown, write equations for the shear force V_r and the bending moment M_r for any section of the beam

a. In the interval -3 ft $< x < 0$.
b. In the interval $0 < x < 2$ ft.
c. In the interval 2 ft $< x < 8$ ft.
d. In the interval 8 ft $< x < 10$ ft.

7-31 A beam is loaded and supported as shown in Fig. P7-31. Using the coordinate axes shown, write equations for the shear force V_r and the bending moment M_r for any section of the beam in the interval 4 ft $< x < 8$ ft.

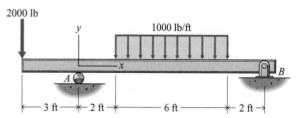

Figure P7-35

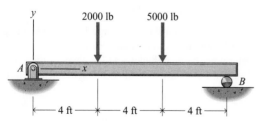

Figure P7-31

7-32 A beam is loaded and supported as shown in Fig. P7-32. Using the coordinate axes shown, write equations for the shear force V_r and the bending moment M_r for any section of the beam in the interval $0 < x < L$.

7-36* A beam is loaded and supported as shown in Fig. P7-36. Using the coordinate axes shown, write equations for the shear force V_r and the bending moment M_r for any section of the beam

a. In the interval -2 m $< x < 0$.
b. In the interval $0 < x < 4$ m.
c. In the interval 4 m $< x < 6$ m.
d. In the interval 6 m $< x < 10$ m.

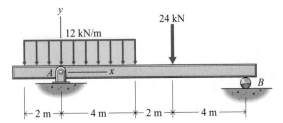

Figure P7-36

7-37 A beam is loaded and supported as shown in Fig. P7-37. Using the coordinate axes shown,

a. Write equations for the shear force V_r and the bending moment M_r for any section of the beam

b. Determine the magnitudes and locations of the maximum shear force and the maximum bending moment in the beam.

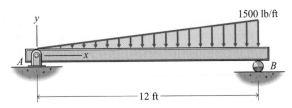

Figure P7-37

7-38* An S178 × 30 steel beam (see Appendix B) is loaded and supported as shown in Fig. P7-38.

a. Using the coordinate axes shown, write equations for the shear force V_r and the bending moment M_r for any section of the beam in the interval 0.5 m < x < 2.5 m.

b. Determine the flexural stress at a point 15 mm above the bottom of the beam on a section at x = 1.5 m.

c. Determine the maximum flexural stress on a section at x = 1.5 m.

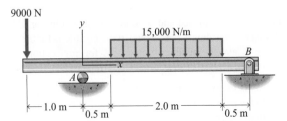

Figure P7-38

7-39 An S8 × 23 steel beam (see Appendix B) is loaded and supported as shown in Fig. P7-39.

a. Using the coordinate axes shown, write equations for the shear force V_r and the bending moment M_r for any section of the beam in the interval 6 ft < x < 10 ft.

b. Determine the flexural stress at a point 1 in. below the top of the beam on a section at x = 3 ft.

c. Determine the maximum flexural stress on a section at x = 3 ft.

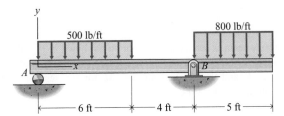

Figure P7-39

7-40 A beam is loaded and supported as shown in Fig. P7-40. On a section 3 m to the right of support A, determine the maximum tensile and compressive flexural stresses if w = 3.0 kN/m.

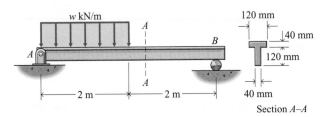

Figure P7-40

Challenging Problems

7-41* A beam is loaded and supported as shown in Fig. P7-41. Using the coordinate axes shown,

a. Write equations for the shear force V_r and the bending moment M_r for any section of the beam.

b. Determine the magnitudes and locations of the maximum shear force and the maximum bending moment in the beam.

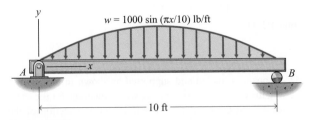

Figure P7-41

7-42* A beam is loaded and supported as shown in Fig. P7-42. Using the coordinate axes shown,

a. Write equations for the shear force V_r and the bending moment M_r for any section of the beam.
b. Determine the magnitudes and locations of the maximum shear force and the maximum bending moment in the beam.

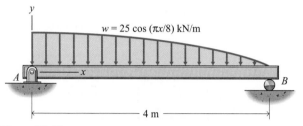

Figure P7-42

7-43 A beam is loaded and supported as shown in Fig. P7-43. Using the coordinate axes shown,

a. Write equations for the shear force V_r and the bending moment M_r for any section of the beam.
b. Determine the magnitudes and locations of the maximum shear force and the maximum bending moment in the beam.

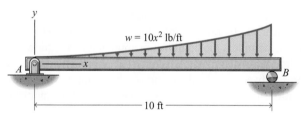

Figure P7-43

7-44 A beam is loaded and supported as shown in Fig. P7-44. Using the coordinate axes shown,

a. Write equations for the shear force V_r and the bending moment M_r for any section of the beam.
b. Determine the magnitudes and locations of the maximum shear force and the maximum bending moment in the beam.

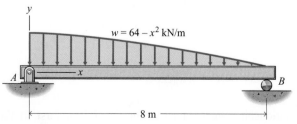

Figure P7-44

7-45 Two C10 × 15.3 steel channels (see Appendix B) are placed back to back to form a 10-in.-deep beam, as shown in Fig. P7-45a. The beam is 10 ft long and is simply supported at its ends. The beam carries a uniformly distributed load of 1000 lb/ft over its entire length and a concentrated load P at the center of the span, as shown in Fig. P7-45b. If the allowable flexural stress on the section at the center of the span is 16,000 psi, determine the maximum permissible value of the concentrated load P.

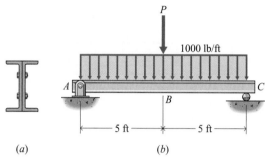

(a)   (b)

Figure P7-45

Computer Problems

7-46 An overhead crane consists of a carriage that moves along a beam as shown in Fig. P7-46. If the carriage moves slowly along the beam, compute and plot

a. The bending moment $M(x)$ in the beam for $b = 1$ m, 2 m, and 2.75 m as a function of x ($0 < x < 5$ m).
b. The bending moments M_B under the left wheel and M_C under the right wheel as a function of b (0.25 m $< b <$ 4.75 m).
c. What is the largest bending moment in the beam? When and where does it occur?

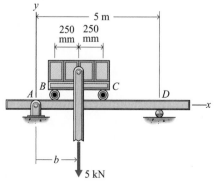

Figure P7-46

7-47   The supports for the beam shown in Fig. P7-47 are symmetrically located. If the distance d from the supports to the ends of the beam is adjustable

a. Compute and plot the bending moment $M(x)$ in the beam for $d = 2$ ft, 3 ft, and 4 ft as a function of x ($0 < x < 15$ ft).

b. Compute and plot the maximum bending moments $(M_{max})_{AB}$ in segment AB and $(M_{max})_{BC}$ in segment BC as functions of d ($0 < d < 6$ ft).

c. What value of d gives the smallest M_{max} in the beam?

7-48   The left support of the beam shown in Fig. P7-48 is fixed at 1 m from the end while the location of the right support is adjustable.

a. Compute and plot the bending moment $M(x)$ in the beam for $d = 1.0$ m, 1.5 m, and 2 m as a function of x ($0 < x < 6$ m).

b. Compute and plot the maximum bending moments $(M_{max})_{AB}$ in segment AB, $(M_{max})_{BC}$ in segment BC, and $(M_{max})_{CD}$ in segment CD as functions of d ($0 < d < 6$ m).

c. What value of d gives the smallest M_{max} in the beam?

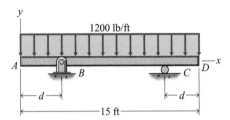

Figure P7-47

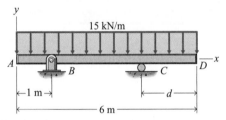

Figure P7-48

7-6   LOAD, SHEAR FORCE, AND BENDING MOMENT RELATIONSHIPS

The equilibrium approach is a fairly simple and straightforward method of obtaining equations for the shear force and bending moment in a beam. However, if the loading on the beam is complex, the equilibrium approach can require several sections and several free-body diagrams. An alternative approach is to derive mathematical relationships between the loads acting on the beam and the shear forces in the beam and relationships between the shear forces and bending moments in the beam.

Consider the beam loaded and supported as shown in Fig. 7-16a. At some location x, the beam is acted on by a distributed load w, a concentrated load P, and a concentrated couple C. A free-body diagram of a segment of the beam centered at the location x is shown in Fig. 7-16b. The upward direction is considered positive for the applied loads w and P, the resisting shears and moments shown are positive according to the sign convention established earlier, and Δx, ΔV, and ΔM may be large or small. The element must be in equilibrium, and force equilibrium $(+\uparrow\Sigma F_y = 0)$ gives

$$V_L + w_{avg}\Delta x + P - (V_L + \Delta V) = 0$$

from which

$$\Delta V = P + w_{avg}\Delta x \qquad (7\text{-}10a)$$

Four important relationships are obtained from Eq. 7-10a. First, if the concentrated force P and the distributed force w are both zero in some region

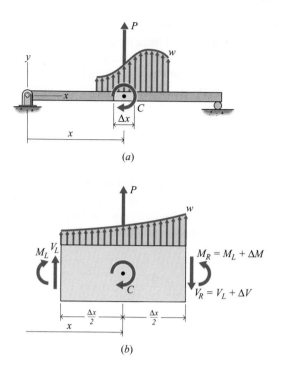

Figure 7-16

of the beam (Δx large or small), then

$$\Delta V = 0 \quad \text{or} \quad V_L = V_R \qquad (7\text{-}10b)$$

That is, in any segment of a beam where there are no loads, the resisting shear force is constant.

Second, if the concentrated load P is not zero, then in the limit as $\Delta x \to 0$,

$$\Delta V = P \quad \text{or} \quad V_R = V_L + P \qquad (7\text{-}10c)$$

That is, across any concentrated load P, the shear force graph (shear force versus x) jumps by the amount of the concentrated load. Furthermore, moving from left to right along the beam, the shear force graph jumps in the direction of the concentrated load.

Third, if the concentrated load P is zero, then in the limit as $\Delta x \to 0$,

$$\Delta V = w_{\text{avg}} \Delta x \to 0$$

and the shear force is a continuous function at x. Dividing through by Δx then gives

$$\lim_{\Delta x \to 0} \frac{\Delta V}{\Delta x} = \frac{dV}{dx} = w \qquad (7\text{-}10d)$$

That is, the slope of the shear force graph at any location (section) x in the beam is equal to the intensity of loading at that section of the beam. Moving from left to right along the beam, if the distributed force is upward, then the slope of the shear force graph ($dV/dx = w$) is positive and the shear force graph is increasing (moving upward). If the distributed force is zero, then the slope of the shear force graph ($dV/dx = 0$) is zero and the shear force is constant.

Finally, in any region of the beam in which Eq. 7-10d is valid (any region in which there are no concentrated loads), the equation can be integrated between definite limits to obtain

$$V_2 - V_1 = \int_{V_1}^{V_2} dV = \int_{x_1}^{x_2} w\, dx \qquad (7\text{-}10e)$$

That is, for any section of the beam acted on by a distributed load w and no concentrated force ($P = 0$), the change in shear between sections at x_1 and x_2 is equal to the area under the load diagram between the two sections.

Similarly, applying moment equilibrium ($+\downarrow \Sigma M_{\text{center}} = 0$) to the free-body diagram of Fig. 7-16b gives

$$(M_L + \Delta M) - M_L - C - V_L \frac{\Delta x}{2} - (V_L + \Delta V)\frac{\Delta x}{2} + a(w_{\text{avg}}\Delta x) = 0$$

or

$$\Delta M = C + V_L \Delta x + \Delta V \frac{\Delta x}{2} - a(w_{\text{avg}}\Delta x) \qquad (7\text{-}11a)$$

in which $\frac{-\Delta x}{2} < a < \frac{\Delta x}{2}$ and in the limit as $\Delta x \to 0$, $a \to 0$ and $w_{\text{avg}} \to w$.

Three important relationships are obtained from Eq. 7-11a. First, if the concentrated couple C is not zero, then in the limit as $\Delta x \to 0$,

$$\Delta M = C \qquad \text{or} \qquad M_R = M_L + C \qquad (7\text{-}11b)$$

That is, across any concentrated couple C, the bending moment graph (bending moment versus x) jumps by the amount of the concentrated couple. Furthermore, moving from left to right along the beam, the bending moment graph jumps upward for a clockwise concentrated couple and jumps downward for a counterclockwise concentrated couple.

Second, if the concentrated couple C and concentrated force P are both zero,[3] then in the limit as $\Delta x \to 0$,

$$\Delta M = V_L \Delta x + \Delta V \frac{\Delta x}{2} - a(w_{\text{avg}}\Delta x) \to 0$$

and the bending moment is a continuous function of x. Dividing through by Δx then gives

$$\lim_{\Delta x \to 0} \frac{\Delta M}{\Delta x} = \frac{dM}{dx} = V \qquad (7\text{-}11c)$$

That is, the slope of the bending moment graph at any location x in the beam is equal to the value of the shear force at that section of the beam. Moving from left

---

[3] If the concentrated force P is not zero, the bending moment will still be continuous at x but it will not be continuously differentiable at x. For a point slightly to the left of P, $dM/dx = V_L$ while for a point slightly to the right of P, $dM/dx = V_R$.

to right along the beam, if the shear force is positive, then $dM/dx = V$ is positive and the bending moment graph is increasing.

Finally, in any region of the beam in which Eq. 7-11c is valid (any region in which there are no concentrated loads or couples), the equation can be integrated between definite limits to obtain

$$M_2 - M_1 = \int_{M_1}^{M_2} dM = \int_{x_1}^{x_2} V\,dx \tag{7-11d}$$

That is, for any section of the beam in which the shear force is continuous ($C = P = 0$), the change in bending moment between sections at x_1 and x_2 is equal to the area under the shear force graph between the two sections.

Note that Eqs. 7-10 through 7-11 were derived with the x-axis positive to the right, the applied loads positive upward, and the resisting shear and moment with signs as indicated in Fig. 7-13. If one or more of these assumptions are changed, the algebraic signs in the equations may need to be altered.

Equations 7-10 through 7-11 can be used to draw shear and bending moment diagrams and to compute values of shear and moment at various sections along a beam.

7-6-1 Shear and Bending Moment Diagrams   Shear and bending moment diagrams provide a convenient method for obtaining maximum values of the resisting shear and bending moment. A shear diagram is a graph in which abscissas represent distances along the beam and ordinates represent the transverse shear at the corresponding sections. A bending moment diagram is a graph in which abscissas represent distances along the beam and ordinates represent the bending moment at the corresponding sections.

Shear and bending moment diagrams can be drawn by calculating values of shear and moment at various sections along the beam and plotting enough points to obtain a smooth curve. Such a procedure is rather time-consuming; therefore, other more rapid methods will be developed using the load, shear force, and bending moment relationships developed in this section.

A convenient arrangement for constructing shear and bending moment diagrams is to draw a free-body diagram of the entire beam and construct shear and bending moment diagrams directly below. Two methods of procedure are used.

The first method consists of writing algebraic equations for the shear force V_r and the bending moment M_r and constructing curves from the equations. This method has the disadvantage that unless the load is uniformly distributed or varies according to a known equation along the entire beam, no single elementary expression can be written for the shear V_r or the bending moment M_r that applies to the entire length of the beam. Instead, it is necessary to divide the beam into intervals bounded by the abrupt changes in the loading. An origin should be selected, positive directions should be shown for the coordinate axes, and the limits of the abscissa (usually x) should be indicated for each interval.

Complete shear and bending moment diagrams should indicate values of shear and moment at each section where the load changes abruptly and at sections where they are maximum or minimum (negative maximum values). Sections where the shear and bending moment are zero should also be located.

The second method consists of drawing the shear diagram from the load diagram and the bending moment diagram from the shear diagram by using Eqs. 7-10 and 7-11. This latter method, though it may not produce a precise curve, is less time-consuming than the first and it does provide the information usually required.

When all loads and reactions are known, the shear and bending moment at the ends of the beam can be determined by inspection. Both shear and bending moment are zero at the free end of a beam unless a force or a couple is applied there, in which case, the shear is the same as the force and the bending moment the same as the couple. At a simply supported or pinned end, the shear must equal the end reaction and the bending moment must be zero. At a built-in or fixed end, the reactions are the shear and the bending moment values.

Once a starting point for the shear diagram is established, the diagram can be sketched by using the definition of shear and the fact that the slope of the shear diagram can be obtained from the load diagram. When positive directions are chosen as upward and to the right, a positive distributed load (acting upward) produces a positive slope on the shear diagram. Similarly, a negative load (acting downward) produces a negative slope on the shear diagram. A concentrated force produces an abrupt change in shear. The change in shear between any two sections is given by the area under the load diagram between the same two sections. The change in shear at a concentrated force is equal to the concentrated force.

A bending moment diagram is drawn from the shear diagram in the same manner. The slope at any point on the bending moment diagram is given by the shear at the corresponding point on the shear diagram, a positive shear produces a positive slope and a negative shear produces a negative slope, when upward and to the right are positive. The change in the bending moment between any two sections is given by the area under the shear diagram between the two sections. A concentrated couple applied to a beam at any section will cause the bending moment at the section to change abruptly by an amount equal to the moment of the couple.

7.6

The choice of which method to use depends on the type of information needed. If only the maximum values of shear force or bending moment are needed, then the second method usually gives these values more easily than the first. If equations of the bending moment are needed (they will be needed in Chapter 8 for finding the deflected shape of the beam), then the equilibrium approach must be used.

Example Problems 7-7 and 7-8 illustrate the two methods for drawing shear and bending moment diagrams.

Example Problem 7-7   A beam is loaded and supported as shown in Fig. 7-17a

(a) Write equations for the shear and the bending moment for any section of the beam in the interval AB.

(b) Write equations for the shear and the bending moment for any section of the beam in the interval BC.

(c) Draw complete shear and bending moment diagrams for the beam.

SOLUTION

A free-body diagram for the beam is shown in Fig. 7-17b. It is not necessary to compute the reactions on a cantilever beam in order to write shear and bending moment equations or to draw shear and bending moment diagrams; however, the

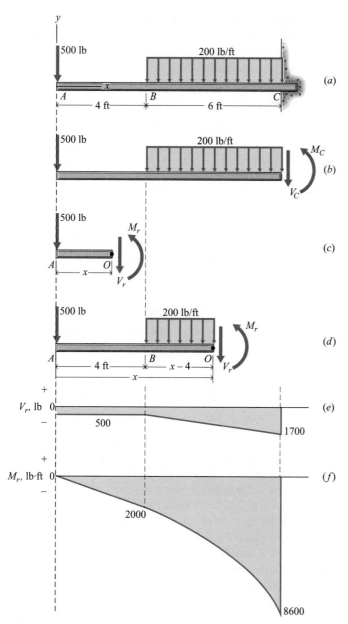

Figure 7-17

reactions provide a convenient check. Thus,

$$+ \uparrow \Sigma F_y = 0: \qquad -500 - 200(6) - V_C = 0$$
$$+ \downarrow \Sigma M_C = 0: \qquad 500(10) + 200(6)(3) + M_C = 0$$

from which

$$V_C = -1700 \text{ lb} \qquad \text{and} \qquad M_C = -8600 \text{ lb} \cdot \text{ft}$$

(a) A free-body diagram of a portion of the beam from the left end to any section between A and B is shown in Fig. 7-17c. The resisting shear V_r and the resisting moment M_r are shown as positive values.
From the equilibrium equation $\Sigma F_y = 0$,

$$+\uparrow \Sigma F_y = 0: \qquad -500 - V_r = 0$$
$$V_r = -500 \text{ lb} \qquad 0 < x < 4 \text{ ft} \qquad \textbf{Ans.}$$

From the equilibrium equation $\Sigma M_O = 0$,

$$+\downarrow \Sigma M_O = 0: \qquad 500(x) + M_r = 0$$
$$M_r = -500x \text{ lb} \cdot \text{ft} \qquad 0 < x < 4 \text{ ft} \qquad \textbf{Ans.}$$

(b) A free-body diagram of a portion of the beam from the left end to any section between B and C is shown in Fig. 7-17d. From the equilibrium equation $\Sigma F_y = 0$,

$$+\uparrow \Sigma F_y = 0: \qquad -500 - 200(x - 4) - V_r = 0$$
$$V_r = 300 - 200x \text{ lb} \qquad 4 < x < 10 \text{ ft} \qquad \textbf{Ans.}$$

From the equilibrium equation $\Sigma M_O = 0$,

$$+\downarrow \Sigma M_O = 0: \quad 500(x) + 200(x-4)(x-4)/2 + M_r = 0$$
$$M_r = -100x^2 + 300x - 1600 \text{ lb} \cdot \text{ft} \qquad 4 < x < 10 \text{ ft} \qquad \textbf{Ans.}$$

(c) The equations for V_r can be plotted in the appropriate intervals to give the shear diagram of Fig. 7-17e. Likewise, the equations for M_r can be plotted to give the bending moment diagram of Fig. 7-17f.

The shear and bending moment diagrams can also be drawn from the load diagram of Fig. 7-17b without writing the shear and moment equations. The shear just to the right of the 500-lb load is -500 lb. The slope of the shear diagram is equal to the distributed load, and since no distributed load is applied between A and B, the slope of the diagram is zero between A and B, From B to C the distributed load is uniform; therefore, the slope of the shear diagram is constant between B and C. The change of shear from B to C is equal to the area under the load diagram between B and C [$\Delta V = 200(6) = 1200$ lb]; therefore, the shear at C is -1700 lb. This shear is the same as the reaction at C, which provides a check.

The bending moment is zero at the free end A. From A to B the shear is constant $(-500$ lb); therefore, the slope of the bending moment diagram is constant, and the bending moment diagram is a straight line between A and B. The shear increases from -500 lb at B to -1700 lb at C; therefore, the slope of the bending moment diagram is negative from B to C and increases uniformly in magnitude. The change in bending moment from A to B is equal to the area under the shear diagram and is $\Delta M = 500(4) = 2000$ lb $\cdot$ ft. The area under the shear diagram from B to C is $\Delta M = (1/2)(500 + 1700)(6) = 6600$ lb $\cdot$ ft. Thus, the bending moment at C is $-2000 + (-6600) = -8600$ lb $\cdot$ ft. This bending moment is the same as the reaction at C.

▶ The shear graph "moves" in the direction of the load. If a concentrated load acts upward, the shear graph jumps upward (when moving from left to right along the beam). The amount of the jump is the magnitude of the concentrated load. If a distributed load acts downward, the shear graph slopes downward (when moving from left to right along the beam). The total change in shear is equal in magnitude to the area under the load diagram.

Example Problem 7-8 A beam is loaded and supported as shown in Fig. 7-18*a*.

(a) Write equations for the shear and the bending moment for any section of the beam in the interval *CD*.

(b) Draw complete shear and bending moment diagrams for the beam.

(c) If the beam is an S457 × 104 American standard steel beam, determine the maximum tensile and compressive flexural stresses in the beam, and state the section(s) where they occur.

SOLUTION

The reactions are determined by using the free-body diagram of the beam shown in Fig. 7-18*b* From the equilibrium equations,

$$+\curvearrowleft \Sigma M_D = 0: \qquad R_A(5) + 4 - 4(7)(1.5) - 8(1.5) = 0$$
$$+\downarrow \Sigma M_A = 0: \qquad R_D(5) - 4 - 4(7)(3.5) - 8(3.5) = 0$$

from which

$$R_A = 10\,\text{kN} \qquad \text{and} \qquad R_D = 26\,\text{kN}$$

(a) A free-body diagram of a portion of the beam from the left end to any section between *C* and *D* is shown in Fig. 7-18*c*.
From the equilibrium equation $\Sigma F_y = 0$,

$$+\uparrow \Sigma F_y = 0: \qquad 10 - 4(x) - 8 - V_r = 0$$
$$V_r = -4x + 2\,\text{kN} \qquad 3.5 < x < 5\,\text{m} \qquad \textbf{Ans.}$$

From the equilibrium equation $\Sigma M_O = 0$,

$$+\downarrow \Sigma M_O = 0: \quad -10(x) + 4(x)(x/2) - 4 + 8(x - 3.5) + M_r = 0$$
$$M_r = -2x^2 + 2x + 32\,\text{kN} \cdot \text{m} \qquad 3.5 < x < 5\,\text{m} \quad \textbf{Ans.}$$

(b) The equations for V_r and M_r in the other intervals can be written in the same manner, and the shear and the bending moment diagrams (Fig. 7-18*d* and *e*) can be obtained by plotting these equations. In this example, the shear diagram will be drawn directly from the load diagram (Fig. 7-18*b*.). The shear just to the right of *A* is 10 kN. From *A* to *C* the shear decreases at a constant rate of 4 kN/m. Thus, the shear just to the left of *C* is

$$V_C = V_A + \Delta V = 10 - 4(3.5) = -4\,\text{kN}$$

The concentrated downward load at *C* causes the shear to change suddenly from −4 kN just to the left of *C* to −12 kN just to the right of *C*. From *C* to *D* the shear continues to decrease at a constant rate of 4 kN/m. Thus, the shear just to the left of *D* is

$$V_D = V_C + \Delta V = -12 - 4(1.5) = -18\,\text{kN}$$

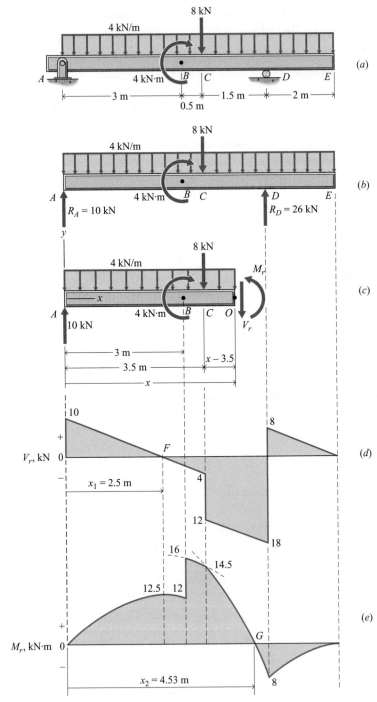

Figure 7-18

The reaction at D causes the shear to change suddenly from -18 kN just to the left of D to $+8$ kN just to the right of D. From D to E the shear continues to decrease at a constant rate of 4 kN/m. Thus, the shear at E is

$$V_E = V_D + \Delta V = +8 - 4(2) = 0$$

Since the distributed load is uniform over the entire beam, the slope of the shear diagram is constant. Points of zero shear, such as point F in Fig. 7-18d, are located from the geometry of the shear diagram. For example, the slope of the shear diagram is 4 kN/m. Therefore,

$$x_1 = 10/4 = 2.5 \text{ m}$$

The moment is zero at A, and the slope of the bending moment diagram (equal to the shear) is 10 kN · m/m. From A to C the shear and, hence, the slope of the bending moment diagram decrease uniformly to zero at F and to -4 at C. The abrupt change of shear at C indicates a sudden change of slope of the bending moment diagram; thus, the two parts of the bending moment diagram at C are not tangent. The slope of the diagram changes from -12 at C to -18 just to the left of D. From D to E the slope changes from $+8$ to zero. The change of moment from A to F is equal to the area under the shear diagram from A to F; therefore,

$$M_F = M_A + \Delta M = 0 + 10(2.5)/2 = 12.5 \text{ kN} \cdot \text{m}$$

Similarly,

$$V_B = V_A + \Delta V = 10 - 4(3) = -2 \text{ kN}$$
$$M_B = M_F + \Delta M = 12.5 - 2(0.5)/2 = 12.0 \text{ kN} \cdot \text{m}$$

The 4-kN · m concentrated couple is applied at B. Since the bending moment just to the left of B is positive and the couple contributes an additional positive moment to sections to the right of B, the bending moment changes abruptly from $+12$ kN · m to $+16$ kN · m. In a similar manner bending moments at C, D, and E are determined from the shear diagram areas as

$$M_C = M_B + \Delta M = 16.0 - (4+2)(0.5)/2 = 14.5 \text{ kN} \cdot \text{m}$$
$$M_D = M_C + \Delta M = 14.5 - (12+18)(1.5)/2 = -8.0 \text{ kN} \cdot \text{m}$$
$$M_E = M_D + \Delta M = -8.0 + 8(2)/2 = 0$$

There is no moment applied at end E of the beam, and if the bending moment at E is not zero, it indicates that an error has occurred. The point G, where the bending moment is zero, can be determined by setting the expression for the bending moment from part (a) equal to zero and solving for x. The result is

$$x_2 = 0.5 \pm \sqrt{16.25} = 4.531 \text{ m} \cong 4.53 \text{ m} \qquad \textbf{Ans.}$$

Note in this example that maximum and minimum bending moments may occur at sections where the shear curve passes through zero. In general, the shear curve may pass through zero at a number of points along the beam, and each such crossing indicates a point of possible maximum bending moment (in engineering, the bending moment with the largest absolute value is the maximum bending moment). It should be emphasized that the shear curve does not indicate the presence of abrupt discontinuities in the bending moment curve; hence, the maximum bending moment may occur where a couple

▶ Point F can also be located using the area under the load diagram. The task is to find how much area is necessary to cause the shear graph to change from 10 kN to 0; that is, 10 kN = (4 kN/m) × (b m), which gives $b = 2.5$ m.

is applied to the beam, rather than where the shear passes through zero. All possibilities should be examined to determine the maximum bending moment.

Since the flexural stress is zero at sections where the bending moment is zero, if a beam must be spliced, the splice should be located at or near such a section.

(c) The depth of an S 457 × 104 American standard steel beam is 457.2 mm (Table B-4). The second moment of area is $358(10^6)$ mm^4, and the section modulus is $1690(10^3)$ mm^3. Since the beam is symmetric with respect to the X–X-axis (NA), the maximum tensile and compressive flexural stresses are equal in magnitude. They occur at a section where the resisting moment has the largest magnitude. From Fig. 7-18e, the largest resisting moment is $+16$ kN · m. Thus,

$$\sigma_x = \frac{-M_r y}{I}$$

which for $y = \pm\left(\dfrac{457.2}{2}\right) = \pm 228.6$ mm gives

$$\sigma_x = \frac{16(10^3)(228.6)(10^{-3})}{358(10^{-6})}$$

$$= 10.217(10^6)\,\text{N/m}^2$$

$$\cong 10.22\,\text{MPa (T or C)} \qquad \textbf{Ans.}$$

These stresses occur on a section at B.

7.7

7.8

7.9

7.10

PROBLEMS

MecMovie Activities and Problems

MM7.5 Rules for constructing shear & moment diagrams. Interactive examples. Exercises to develop skills needed to successfully construct shear force and bending moment diagrams.

MM7.6 Shear force and bending moment diagrams: following the rules. Game. Five-round game emphasizing the six rules needed to construct shear force and bending moment diagrams.

MM7.7 Extruded aluminum beam. Example; Concept checkpoints. Determine maximum bending moments given allowable tension and compression stresses.

MM7.8 Determine maximum bending stress. Example; Concept checkpoints. Determine bending moment diagram and maximum tension and compression bending stresses for a t-shape.

Introductory Problems

7-49* Draw complete shear and bending moment diagrams for the beam shown in Fig. P7-49.

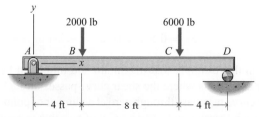

Figure P7-49

7-50* Draw complete shear and bending moment diagrams for the beam shown in Fig. P7-50.

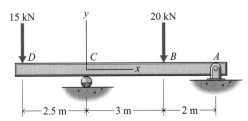

15 kN   20 kN

D   C   B   A

2.5 m — 3 m — 2 m

Figure P7-50

7-51 Draw complete shear and bending moment diagrams for the beam shown in Fig. P7-51.

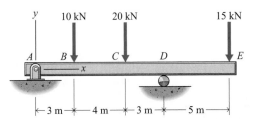

2000 lb/ft   8000 lb

A   B   C   D

12 ft — 4 ft — 8 ft

Figure P7-51

7-52* Draw complete shear and bending moment diagrams for the beam shown in Fig. P7-52.

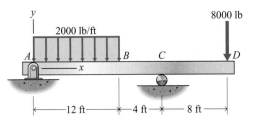

10 kN   20 kN   15 kN

A   B   C   D   E

3 m — 4 m — 3 m — 5 m

Figure P7-52

7-53 Draw complete shear and bending moment diagrams for the beam shown in Fig. P7-53.

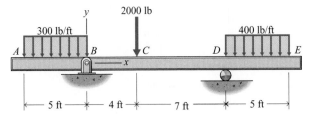

300 lb/ft   2000 lb   400 lb/ft

A   B   C   D   E

5 ft — 4 ft — 7 ft — 5 ft

Figure P7-53

7-54 Draw complete shear and bending moment diagrams for the beam shown in Fig. P7-54.

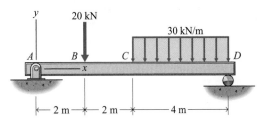

20 kN   30 kN/m

A   B   C   D

2 m — 2 m — 4 m

Figure P7-54

7-55* Draw complete shear and bending moment diagrams for the beam shown in Fig. P7-55.

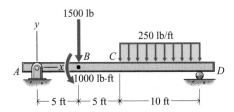

1500 lb   250 lb/ft

A   B   C   D
       1000 lb·ft

5 ft — 5 ft — 10 ft

Figure P7-55

7-56 Draw complete shear and bending moment diagrams for the beam shown in Fig. P7-56.

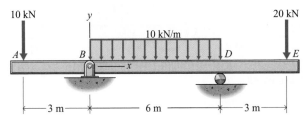

10 kN   10 kN/m   20 kN

A   B   D   E

3 m — 6 m — 3 m

Figure P7-56

Intermediate Problems

7-57* The beam shown in Fig. P7-57a has the cross section shown in Fig. P7-57b. Determine the maximum tensile and compressive flexural stresses in the beam.

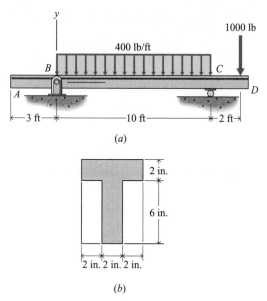

(a)

(b)

Figure P7-57

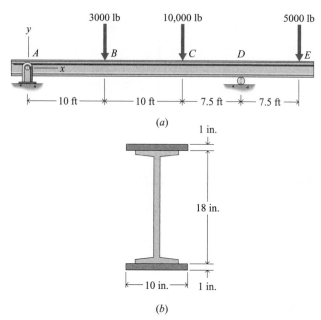

(a)

(b)

Figure P7-59

7-58* A W 102 × 19 wide-flange beam is loaded and supported as shown in Fig. P7-58. Determine the maximum tensile and compressive flexural stresses in the beam.

7-60 A beam is loaded and supported as shown in Fig. P7-60a. Two 250 × 25-mm steel plates and two C 254 × 45 channels are welded together to form the cross section shown in Fig. P7-60b. Determine the maximum tensile and compressive flexural stresses in the beam.

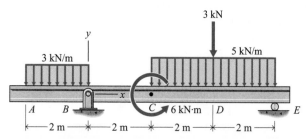

Figure P7-58

7-59 A beam is loaded and supported as shown in Fig. P7-59a. Two 10 × 1-in. steel plates are welded to the flanges of an S 18 × 70 American standard I-beam to form the cross section shown in Fig. P7-59b. Determine the maximum tensile and compressive flexural stresses in the beam.

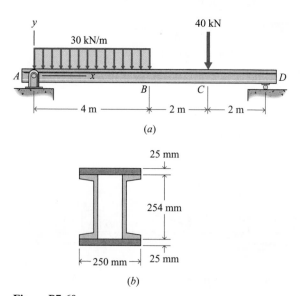

(a)

(b)

Figure P7-60

7-61 A WT8 × 25 structural steel T-section is loaded and supported as shown in Fig. P7-61. Determine the maximum tensile and compressive flexural stresses in the beam.

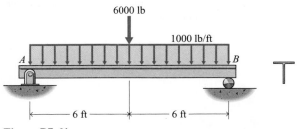

Figure P7-61

7-62* A C254 × 30 structural steel channel is loaded and supported as shown in Fig. P7-62. Determine the maximum tensile and compressive flexural stresses in the beam.

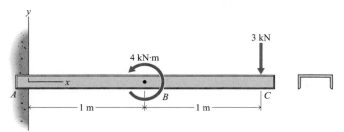

Figure P7-62

7-63 An 8 × 8-in. nominal size structural timber (see Appendix B) is supported by two brick columns, as shown in Fig. P7-63. Assume that the brick columns transmit only vertical forces to the timber beam. The beam supports the roof of a building through three timber columns. Columns *A* and *C* transmit forces of 1.8 kip to the beam; column *B* transmits a force of 2.2 kip.

a. Draw complete shear and bending moment diagrams for the beam.
b. Determine the maximum tensile and compressive flexural stresses in the beam.

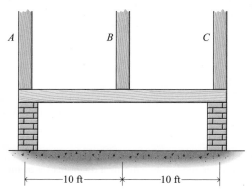

Figure P7-63

Challenging Problems

7-64* Draw complete shear and bending moment diagrams for the beam *ABCD* shown in Fig. P7-64.

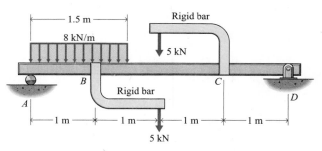

Figure P7-64

7-65* An S 15 × 50 American standard steel beam (see Appendix B) is loaded and supported as shown in Fig. P7-65. The total length of the beam is 15 ft. If the allowable flexural stress is 15,000 psi, determine the maximum permissible value for the distributed load *w*.

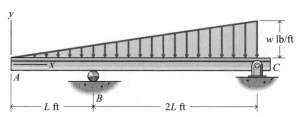

Figure P7-65

7-66 Draw complete shear and bending moment diagrams for segments *AB* and *CD* of the structure shown in Fig. P7-66.

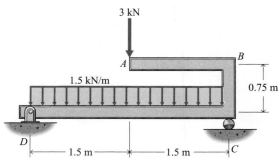

Figure P7-66

7-67* Member AB supports a 55-lb sign, as shown in Fig. P7-67. Determine

a. The maximum tensile flexural stress in the 1/2-in.-nominal-diameter standard steel pipe AB (see Appendix B).
b. The normal stress in the 3/16-in.-diameter wire BC.
c. The shearing stress in the 1/4-in.-diameter pin at A, which is in double shear.

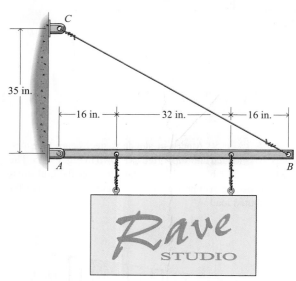

Figure P7-67

7-68 An S457 × 81 American standard steel beam (see Appendix B) is loaded and supported as shown in Fig. P7-68. The segments of the beam are connected with a smooth pin at D.

a. Draw complete shear and bending moment diagrams for the beam.
b. Determine the maximum tensile and compressive flexural stresses in the beam.

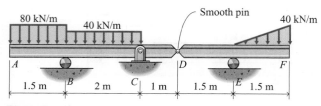

Figure P7-68

7-69 A tractor is moving slowly over a bridge, as shown in Fig. P7-69. The forces exerted on one beam of the bridge by the tractor are 4050 lb by the rear wheels and 1010 lb by the front wheels. Determine the position x of the tractor for which the bending moment in the beam is maximum.

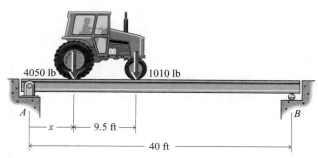

Figure P7-69

7-70* A body with a mass of 1500 kg is supported by a roller on an I-beam, as shown in Fig. P7-70. The roller moves slowly along the beam, thereby causing the shear force V_r and the bending moment M_r to be functions of x.

a. Draw complete shear and bending moment diagrams when the roller is at position x.
b. Determine the position of the roller when the bending moment is maximum.

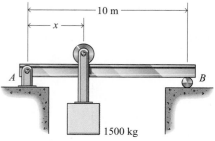

Figure P7-70

7-71 Two beams AD and EH are spliced, as shown in Fig. P7-71. Draw complete shear and bending moment diagrams for beam CF.

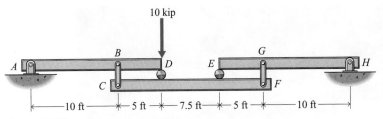

10 kip

Figure P7-71

7-7 SHEARING STRESSES IN BEAMS

The discussion of shearing stresses in beams was delayed while flexural stresses were studied in Section 7-3. This procedure seems to be in keeping with the historical record on the study of beam stresses. From the time of Coulomb's paper, which contained the correct theory of the distribution of flexural stresses, approximately seventy years elapsed before the Russian engineer D. J. Jourawski (1821–1891), while designing timber railroad bridges in 1844–1850, developed the elementary shear stress theory used today. In 1856, Saint-Venant developed a rigorous solution for shearing stresses in beams; however, the elementary solution of Jourawski is the one in general use today by engineers and architects because it yields adequate results and is much easier to apply. The method requires use of the elastic flexure formula in its development; therefore, the formula developed is limited to elastic action. The shearing stress evaluation discussed in this section is used as follows:

1. For timber beams, because of their longitudinal plane of low shear resistance.

2. In design codes for shear stresses in the webs of I-beams.

3. For the evaluation of principal stresses at the junction between the flange and the web in certain wide-flange beams.

If one constructs a beam by stacking flat slabs one on top of another without fastening them together, and then loads this beam in a direction normal to the surface of the slabs, the resulting deformation will appear somewhat like that in Fig. 7-19a. This same type of deformation can be observed by taking a pack of cards and bending them, and noting the relative motion of the ends of the cards with respect to each other. The fact that a solid beam does not exhibit this relative movement of longitudinal elements (see Fig. 7-19b, in which the beam is identical to that of Fig. 7-19a except that the layers are glued together) indicates the presence of shearing stresses on longitudinal planes. Evaluation of these shearing stresses will be determined by means of equilibrium and the free-body diagram of the short portion of a beam with a rectangular cross section shown in Fig. 7-20a.

The normal force dF acting on a differential area $dA = t \, dy$ on a cross section of the beam is equal to $\sigma_x dA$. The resultant of these differential forces is $F = \int \sigma_x dA$ integrated over the area of the cross section, where σ_x is the flexural stress at a distance y from the neutral surface and is given by the expression $\sigma_x = -M_r y/I$. Therefore, the resultant normal force F_1 on the left end of the segment

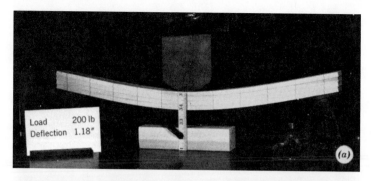

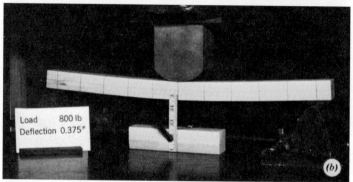

Figure 7-19

from y_1 to the top of the beam is[4]

$$F_1 = -\frac{M}{I} \int_A y \, dA = -\frac{M}{I} \int_{y_1}^{c} y \, (t \, dy)$$

Similarly, the resultant force F_2 on the right side of the element is

$$F_2 = -\frac{(M + \Delta M)}{I} \int_{y_1}^{c} y \, (t \, dy)$$

These forces are shown on the free-body diagram of Fig. 7–20b. Also shown on the free-body diagram of Fig. 7–20b are the resultants of the vertical shear stresses on the left V_L and right V_R sides of the element and the resultant of the horizontal shear stress V_H on the bottom of the element. A summation of forces in the horizontal direction yields

$$V_H = F_2 - F_1 = -\frac{\Delta M}{I} \int_{y_1}^{c} y \, (t \, dy)$$

The average shearing stress τ_{avg} is the horizontal shear force V_H divided by the horizontal shear area $A_s = t \, \Delta x$ between sections A and B. Thus,

$$\tau_{avg} = \frac{V_H}{A_s} = -\frac{\Delta M}{It \, \Delta x} \int_{y_1}^{c} y \, (t \, dy)$$

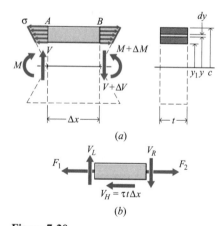

(a)

(b)

Figure 7-20

---

[4]If M is positive, then the normal stress above the neutral axis (where y is positive) will be negative (compression) and the force F_1 will be a compressive force. If M is negative, then the normal stress above the neutral axis will be positive (tension) and the force F_1 will be a tensile force as drawn.

In the limit as $\Delta x \to 0$

$$\tau = \lim_{\Delta x \to 0} \frac{\Delta M}{\Delta x}\left(-\frac{1}{It}\right)\int_{y_1}^{c} t y \, dy = \frac{dM}{dx}\left(-\frac{1}{It}\right)\int_{y_1}^{c} t y \, dy \qquad (a)$$

The shear V_r at the beam section where the stress is to be evaluated is given by Eq. 7-11c as $V_r = dM/dx$. The integral of Eq. (a) is the first moment of the portion of the cross-sectional area between the transverse line where the stress is to be evaluated and the extreme fiber of the beam. This integral is designated Q, and when values of V_r and Q are substituted into Eq. (a), the formula for the horizontal (or longitudinal) shearing stress becomes

$$\tau_H = -\frac{V_r Q}{It} \qquad (b)$$

The minus sign in Eq. (b) is needed to satisfy Eq. 7-1a and is consistent with the sign convention for shearing stresses (Fig. 7-4c). At each point in the beam, the horizontal (longitudinal) and vertical (transverse) shearing stresses have the same magnitude ($\tau_{xy} = \tau_{yx}$); hence, Eq. (b) also gives the vertical shearing stress at a point in a beam (averaged across the width).[5] For the balance of this chapter, magnitudes of V_r and Q will be used to determine the magnitude of the shearing stress τ and Eq. (b) will be written as

$$\tau = \frac{V_r Q}{It} \qquad (7\text{-}12)$$

The sense of the stress τ will be determined from the sense of the shear V_r on transverse planes and from $\tau_{xy} = \tau_{yx}$ on longitudinal planes.

Because the flexure formula was used in the derivation of Eq. 7-12, it is subject to the same assumptions and limitations as the flexure formula. Although the stress given by Eq. 7-12 is associated with a particular point in a beam, it is averaged across the thickness t and hence is accurate only if t is not too great.

The variation of shearing stress on a transverse cross section of a beam will be demonstrated by using the rectangular cross section shown in Fig. 7-21a. The transverse shearing stress at any point of the section at a distance y_1 from the neutral axis is from Fig. 7-21b and Eq. 7-12,

$$\tau = \frac{V_r Q}{It} = \frac{V}{It}\int_A y \, dA = \frac{V}{It}\int_{y_1}^{c} y t \, dy$$

$$= \frac{V}{I}\int_{y_1}^{h/2} y \, dy = \frac{V}{2I}\left[\left(\frac{h}{2}\right)^2 - y_1^2\right] \qquad (c)$$

Equation (c) indicates that the transverse shearing stress on a rectangular cross section has a parabolic distribution, as shown in Fig. 7-21c. The shearing stress acts in the direction of the shear force V that produces the stress. The maximum

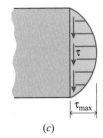

Figure 7-21

---

[5]If the shear force of V is positive (downward on section B), then the horizontal shear stress will be negative (to the right on the bottom of the element) and the vertical shear stress will also be negative (downward on section B—in the same direction as the shear force).

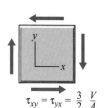

shearing stress occurs when $y_1 = 0$ (at the neutral axis) and has a magnitude

$$\tau_{\max} = \frac{Vh^2}{8I} = \frac{Vh^2}{8(th^3/12)} = \frac{3}{2}\frac{V}{th} = \frac{3}{2}\frac{V}{A} \qquad (7\text{-}13)$$

The maximum shearing stress given by Eq. 7-13 (Fig. 7-22) exists on both the transverse plane and the longitudinal plane along the neutral surface. This equation is useful in the design of timber beams with rectangular cross sections since timber has a low shearing strength parallel to the grain.

Equation 7-13 is valid for a rectangular section and should not be used for other sections. For a rectangular section the maximum shearing stress is 1.5 times the average shearing stress ($\tau_{\mathrm{avg}} = V/A$). For a rectangular section having a depth twice the width, the maximum stress as computed by Saint-Venant's more rigorous method is about 3 percent greater than that given by Eq. 7-13. If the beam is square, the error is about 12 percent. If the width is four times the depth, the error is almost 100 percent, from which one must conclude that, if Eq. 7-12 were applied to a point in the flange of an I-beam or T-section, the result would be worthless. Furthermore, if Eq. 7-12 is applied to sections where the sides of the beam are not parallel, such as a triangular section, the average transverse shearing stress is subject to additional error because the variation of transverse shearing stress is greater when the sides are not parallel.

A second illustration of the variation of shearing stress on a transverse section of a beam will be demonstrated by using the inverted T-shaped beam shown in Fig. 7-23a. For this section,

7.12

$$y_C = \frac{2(10)(7) + 10(2)(1)}{2(10) + 10(2)} = 4 \text{ in.}$$

$$c_1 = 8 \text{ in.} \qquad \text{and} \qquad c_2 = 4 \text{ in.}$$

Note that in Eq. 7-12, V and I are constant for any section, and only Q and t vary for different points in the section. The transverse shearing stress at any point in the stem of the section a distance y_1 from the neutral axis is from Fig. 7-23a

Figure 7-22

$$\tau_{xy} = \tau_{yx} = \frac{3}{2}\frac{V}{A}$$

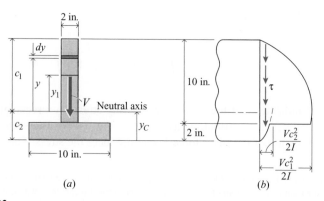

(a)

(b)

Figure 7-23

and Eq. 7-12,

$$\tau = \frac{V_r Q}{It} = \frac{V}{It}\int_{y_1}^{c_1} y\, t\, dy$$

$$= \frac{V}{2I}(c_1^2 - y_1^2) = \frac{V}{2I}(8^2 - y_1^2) \qquad (-2 < y_1 < 8 \text{ in.})$$

An expression for the average shearing stress in the flange can be written in a similar manner and is

$$\tau = \frac{V}{2I}(c_2^2 - y_1^2) = \frac{V}{2I}(4^2 - y_1^2) \qquad (-4 < y_1 < -2 \text{ in.})$$

7.13

These are parabolic equations for the theoretical stress distribution, and the results are shown in Fig. 7-23b. The diagram has a discontinuity at the junction of the flange and stem because the thickness of the section changes abruptly. The distribution in the flange is fictitious because the stress at the top of the flange must be zero (a free surface). From Fig. 7-23b and Eq. 7-12, one may conclude that, in general, the maximum[6] longitudinal and transverse shearing stress occurs at the neutral surface at a section where the transverse shear V_r is maximum. There may be exceptions such as a beam with a cross section in the form of a Greek cross with Q/t at the neutral surface less than the value some distance from the neutral surface.

   Another example of importance is the determination of the shearing stress in an I-beam. Consider the W203 $\times$ 22 section [$I = 20.0(10^6)$ mm^4] shown in Fig. 7-24a, and let the shear V_r on the section be 37.5 kN. Equation 7-12 may be used to calculate the shearing stress at various distances y_1 from the neutral axis of the beam. For example, at the neutral axis ($y_1 = 0$)

$$Q_{NA} = 102(8)(99) + 95(6.2)(47.5) = 108.76(10^3) \text{ mm}^3 = 108.76(10^{-6}) \text{ m}^3$$

$$\tau_{NA} = \frac{V_r Q_{NA}}{It_w} = \frac{37.5(10^3)(108.76)(10^{-6})}{20.0(10^{-6})(6.2)(10^{-3})} = 32.89(10^6) \text{ N/m}^2 \cong 32.9 \text{ MPa}$$

Similarly, at the junction between the web and the flange with $t = 6.2$ mm (in the web),

$$Q_J = 102(8)(99) = 80.78(10^3) \text{ mm}^3 = 80.78(10^{-6}) \text{ m}^3$$

$$\tau_J = \frac{V_r Q_J}{It_w} = \frac{37.5(10^3)(80.78)(10^{-6})}{20.0(10^{-6})(6.2)(10^{-3})} = 24.43(10^6) \text{ N/m}^2 \cong 24.4 \text{ MPa}$$

The shearing stress distribution for the complete cross section of the beam is shown in Fig. 7-24b. Equation 7-12 gives $\tau = 1.485$ MPa at the junction of the web and flange with $t = 102$ mm (in the flange). However, this result is incorrect because the bottom surface of the top flange (or the top surface of the bottom flange) is a free surface and thus $\tau = 0$. A similar result was obtained for the inverted T-section, shown in Fig. 7-23. More advanced methods of the theory of elasticity must be used to derive a correct solution.

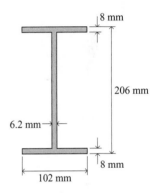

W 203 $\times$ 22 section

(a)

$\tau = 1.485$ MPa
$\tau = 24.4$ MPa
$\tau_{max} = 32.9$ MPa
$\tau_{avg} = \dfrac{V}{A_{web}} = 31.8$ MPa

Shearing stress distribution

(b)

Figure 7-24

---

[6] In this book the term *maximum*, as applied to a longitudinal and transverse shearing stress, will mean the average stress across the thickness t at a point where such average has the maximum value.

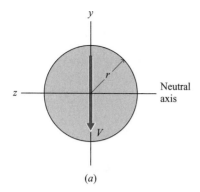

(a)

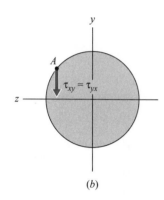

(b)

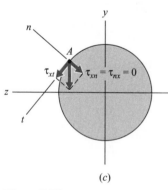

(c)

Figure 7-25

7.14

7.15

The variation of shearing stress over the depth of the web is small, and the shearing stresses in the flanges are small compared to those in the web. As a result, the majority of the shear force V_r is carried by the web. In the design of I-beams, the maximum shearing stress calculated by using Eq. 7-12 is approximated by dividing the shear force V_r by the area of the web, that is,

$$\tau_{\text{avg}} = \frac{V_r}{A_{\text{web}}} \tag{7-14}$$

For the example being considered,

$$\tau_{\text{avg}} = \frac{V_r}{A_{\text{web}}} = \frac{37.5(10^3)}{6.2(10^{-3})(190)(10^{-3})} = 31.83(10^6)\,\text{N/m}^2 \cong 31.8\,\text{MPa}$$

The maximum and average shearing stresses differ by approximately 3 percent. For commercial I-beams, the maximum difference is approximately 10 percent. *Equation 7–14 should only be used to calculate shearing stresses in I-beams; it should not be used for T-, rectangular, or other sections.* Equation 7–14 is specified in design codes for I-beams.

As a final example, consider a beam with a circular cross section. This type of beam is important in the transmission of power, for example, a shaft between a motor and a piece of equipment. Bending loads are induced in shafts by forces at gears, bearings, and pulleys.

Consider a beam with a solid circular cross section subjected to a shear load V, as shown in Fig. 7-25a. According to Eq. 7-12, a shear force V causes a shearing stress τ_{xy} in the direction of V, as shown in Fig. 7-25b. This shearing stress at point A can be resolved into normal (n) and tangential (t) components, as shown in Fig. 7-25c. However, Eq. 2-11 requires that $\tau_{xn} = \tau_{nx}$ (where τ_{nx} is the shearing stress in the x-direction on a plane with outward normal in the n direction). The outside surface of the shaft (beam) is a free surface; therefore, $\tau_{nx} = \tau_{xn} = 0$, which indicates that any shearing stress at point A must be tangent to the surface of the shaft (beam) and not in the direction of the shear force V as required by Eq. 7-12. At the neutral axis, the shearing stress τ_{xt} is in the direction of shear force V and Eq. 7-12 gives

$$\tau_{\text{max}} = \tau_{NA} = \frac{VQ_{NA}}{It_{NA}} = \frac{V(\pi r^2/2)(4r/3\pi)}{(\pi r^4/4)(2r)} = \frac{4V}{3\pi r^2} = \frac{4}{3}\frac{V}{A} \tag{7-15}$$

where A is the cross-sectional area of the beam. The maximum shearing stress given by Eq. 7-15 differs from the maximum shearing stress given by the mathematical theory of elasticity[7] by approximately 5 percent.

Example Problem 7-9 A W254 × 33 wide-flange beam is loaded and supported as shown in Fig. 7-26a. At section A–A of the beam, determine the flexural and shearing stresses and show both stresses on a stress element

(a) At point a on the top surface of the flange.

(b) At point b in the web at the junction of the web and the flange.

(c) At point c on the neutral axis.

---

[7]*The Mathematical Theory of Elasticity.* A. E. H. Love, Dover Publications, 4th ed., New York, 1994.

SOLUTION

A free-body diagram of the beam is shown in Fig. 7-26b. The reactions at supports B and C are determined by using the equilibrium equation $\Sigma M = 0$. Thus,

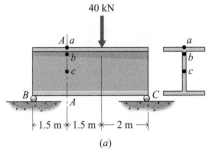

40 kN

(a)

$+\downarrow \Sigma M_C = 0$: $\qquad 40(2) - R_B(5) = 0$

$+\downarrow \Sigma M_B = 0$: $\qquad R_C(5) - 40(3) = 0$

from which

$$R_B = 16 \text{ kN} \qquad \text{and} \qquad R_C = 24 \text{ kN}$$

A free-body diagram of the part of the beam to the left of section A–A is shown in Fig. 7-26c. The shear V_r and moment M_r transmitted by section A–A are determined by using the equilibrium equations $\Sigma F_y = 0$ and $\Sigma M_O = 0$. Thus,

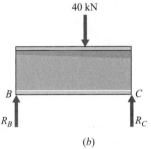

(b)

$+\uparrow \Sigma F_y = 0$: $\qquad 16 - V_r = 0$

$+\downarrow \Sigma M_O = 0$: $\qquad M_r - 16(1.5) = 0$

from which

$$V_r = 16 \text{ kN} \qquad \text{and} \qquad M_r = 24 \text{ kN} \cdot \text{m}$$

(a) For a W254 $\times$ 33 section: $I = 49.1(10^6) \text{ mm}^4$, $d = 258$ mm

$$w_f = 146 \text{ mm}, \quad t_f = 9.1 \text{ mm}, \quad t_w = 6.1 \text{ mm}$$

The flexural stress at point a is given by Eq. 7-8 as

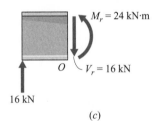

(c)

$$\sigma_{xa} = -\frac{M_r y}{I} = -\frac{24(10^3)(129)(10^{-3})}{49.1(10^{-6})}$$

$$= -63.05(10^6) \text{ N/m}^2 \cong 63.1 \text{ MPa (C)} \qquad \textbf{Ans.}$$

The shearing stress at point a is given by Eq. 7-12. At point a, $Q = 0$, and thus $\tau = 0$. The stresses on an element at a are shown in Fig. 7-26d.

(b) At point b, $y = 129 - 9.1 = 119.9$ mm

$$\sigma_{xb} = -\frac{M_r y}{I} = -\frac{24(10^3)(119.9)(10^{-3})}{49.1(10^{-6})}$$

$$= -58.61(10^6) \text{ N/m}^2 \cong 58.6 \text{ MPa (C)} \qquad \textbf{Ans.}$$

At point b, the first moment Q_b of the area above the point is

$$Q_b = y_C A = 124.45(146)(9.1) = 165.34(10^3) \text{ mm}^3 = 165.34(10^{-6}) \text{ m}^3$$

$$\tau_b = \frac{V_r Q_b}{I_{NA} t_w} = \frac{16(10^3)(165.34)(10^{-6})}{49.1(10^{-6})(6.1)(10^{-3})}$$

$$= 8.832(10^6) \text{ N/m}^2 \cong 8.83 \text{ MPa} \qquad \textbf{Ans.}$$

The stresses on an element at b are shown in Fig. 7-26e.

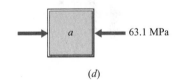

(d)

Figure 7-26(a–d)

▶ At point a the shearing stress is zero. Therefore, $\sigma_{xa} = 63.1$ MPa (C) is a principal stress. The other principal stresses are both zero, and the maximum shear stress is $\tau_{\max} = 31.6$ MPa and occurs on surfaces inclined at 45° to the axis of the beam. It can also be shown that these stresses are larger than the principal stresses and maximum shear stress at any other point on this section of the beam.

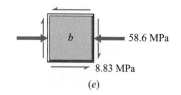

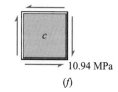

58.6 MPa

8.83 MPa

(e)

10.94 MPa

(f)

Figure 7-26(e–f)

(c) Since point c is on the neutral axis, the flexural stress is zero. The shearing stress is maximum, and an approximate value is given by Eq. 7-14 as

$$\tau_{max} = \frac{V_r}{A_{web}} = \frac{16(10^3)}{6.1(10^{-3})(239.8)(10^{-3})}$$

$$= 10.938(10^6)\,N/m^2 \cong 10.94\,MPa \qquad \textbf{Ans.}$$

The stresses on an element at c are shown in Fig. 7-26f.

The shearing stress at point c can also be obtained using Eq. 7-12. At point c, the first moment Q_c of the area above the point is

$$Q_c = y_C A = 124.45(146)(9.1) + 59.95(119.9)(6.1)$$

$$= 209.2(10^3)\,mm^3 = 209.2(10^{-6})\,m^3$$

$$\tau_c = \frac{V_r Q_c}{I_{NA} t_w} = \frac{16(10^3)(209.2)(10^{-6})}{49.1(10^{-6})(6.1)(10^{-3})} = 11.176(10^6)\,N/m^2 \cong 11.18\,MPa$$

The percent difference between the two values of shearing stress is

$$D = \frac{11.176 - 10.938}{11.176}(100) = 2.13\,\text{percent}$$

Example Problem 7-10   A beam is simply supported and carries a concentrated load of 1800 lb at the center of a 15-ft span, as shown in Fig. 7-27a. If the beam has the T cross section shown in Fig. 7-27b, determine

(a) The average shearing stress on a horizontal plane 4 in. above the bottom of the beam and 6 ft from the left support.

(b) The maximum transverse shearing stress in the beam.

(c) The average shearing stress in the joint between the flange and the stem at a section 6 ft from the left support.

(d) The force transmitted from the flange to the stem by the glue in a 12-in. length of the joint centered 6 ft from the left support.

(e) The maximum tensile flexural stress in the beam.

$P = 1800$ lb

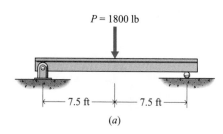

7.5 ft — 7.5 ft

(a)

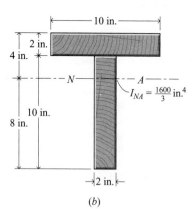

10 in.

2 in.

4 in.

N A

$I_{NA} = \frac{1600}{3}\,in.^4$

10 in.

8 in.

2 in.

(b)

Figure 7-27(a–b)

SOLUTION

The second moment of the cross-sectional area about the neutral axis is

$$I_{NA} = \frac{1}{12}(2)(10)^3 + 2(10)(3)^2 + \frac{1}{12}(10)(2)^3 + 10(2)(3)^2 = 533.3\,in.^4$$

(a) The shear force V_r on a cross section 6 ft from the left support is $+900$ lb, as shown on Fig. 7-27c. The first moment Q_4 for the bottom 4 in. of the stem (see Fig. 7-27e) is

$$Q_4 = y_{C4} A_4 = 6(2)(4) = 48\,in.^3$$

The average shearing stress on a horizontal plane 4 in. above the bottom of the beam is then given by Eq. 7-12 as

$$\tau_4 = \frac{V_r Q_4}{I_{NA} t_4} = \frac{900(48)}{533.3(2)} = 40.50 \text{ psi} \cong 40.5 \text{ psi} \qquad \textbf{Ans.}$$

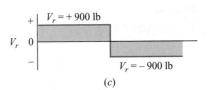

(c)

(b) The maximum transverse shearing stress in the beam will occur at the neutral axis on the cross section supporting the largest shear force V_r. The first moment Q_{NA} for the part of the stem below the neutral axis is

$$Q_{NA} = 4(2)(8) = 64 \text{ in.}^3$$

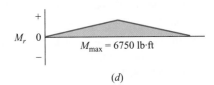

(d)

In the computation of Q, it is immaterial whether one takes the area above or below a transverse line. For example, Q_{NA} for the area above the neutral axis is $Q_{\text{flange}} + Q_{\text{stem}} = 3(10)(2) + 1(2)(2) = 64 \text{ in.}^3$, which is the same as that for the part of the stem below the neutral axis. Since the shear force V_r equals 900 lb on all cross sections of the beam,

$$\tau_{\text{max}} = \frac{V_r Q_{NA}}{I_{NA} t_S} = \frac{900(64)}{533.3(2)} = 54.00 \text{ psi} \cong 54.0 \text{ psi} \qquad \textbf{Ans.}$$

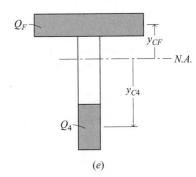

(e)

Figure 7-27(c–e)

(c) The first moment Q_F for the flange of the beam (see Fig. 7-27e) is

$$Q_F = y_{CF} A_F = 3(10)(2) = 60 \text{ in.}^3$$

The average shearing stress on the horizontal plane at the joint between the flange and the stem is then given by Eq. 7-12 as

$$\tau_J = \frac{V_r Q_F}{I_{NA} t_S} = \frac{900(60)}{533.3(2)} = 50.63 \text{ psi} \cong 50.6 \text{ psi} \qquad \textbf{Ans.}$$

▶ The definition of Q is the first moment (relative to the neutral axis) of the portion of the area of the cross section between the transverse line where the stress is to be evaluated and the top of the beam. However, $(y_C A)_{\text{top}} + (y_C A)_{\text{bottom}} = (y_C A)_{\text{total}} = 0$, since y is measured from the centroid (the neutral axis). Therefore, the magnitudes of $(y_C A)_{\text{top}}$ and $(y_C A)_{\text{bottom}}$ are the same, and the area below the section can be used instead if it is more convenient.

(d) The force transmitted from the flange to the stem by the glue is

$$V_g = \tau_J A_J = 50.63(12)(2) = 1215.1 \text{ lb} \cong 1215 \text{ lb} \qquad \textbf{Ans.}$$

7.16

(e) The maximum tensile flexural stress in the beam will occur in a fiber at the bottom of the beam because the resisting moment at all cross sections of the beam is positive. The largest moment occurs at midspan, as shown in Fig. 7-27d. The flexural stress is given by Eq. 7-8 as

7.17

$$\sigma_{\text{max}} = \frac{-M_{\text{max}} c}{I} = \frac{-6750(12)(-8)}{533.3} = 1215.1 \text{ psi} \cong 1215 \text{ psi (T)} \qquad \textbf{Ans.}$$

7.18

PROBLEMS

MecMovie Activities and Problems

MM7.9 Q-tile: The Q Section Property Game. Identify proper area needed to compute Q for transverse shear calculations. Compute Q for various configurations.

MM7.10 Shear stress in a standard steel shape. Example; Try one. Determine the transverse shear stress at specified points in a steel t-shape.

MM7.12 Nailing two box beams. Example; Concept checkpoints. Determine box beam shear capacity for two cross-section configurations.

MM7.13 Determine nail spacing for U-beam. Example; Concept checkpoints. Determine nail spacing for beam made up of three boards.

MM7.14 Bolt spacing in built-up steel beam. Example; Concept checkpoints. Determine maximum bolt spacing for built-up steel beam.

Introductory Problems

7-72* A timber beam is loaded and supported as shown in Fig. P7-72. At section *A–A* of the beam, determine the shearing stresses on horizontal planes that pass through points *a*, *b*, and *c* of the cross section.

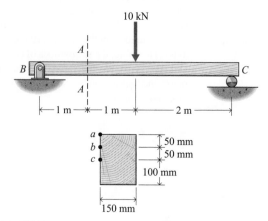

Figure P7-72

7-73* The transverse shear V_r at a certain section of a timber beam is 7000 lb. If the beam has the cross section shown in Fig. P7-73, determine

a. The horizontal shearing stress in the glued joint 2 in. below the top of the beam.
b. The transverse shearing stress at a point 3 in. below the top of the beam.

c. The magnitude and location of the maximum transverse shearing stress on the cross section.

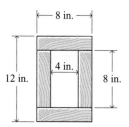

Figure P7-73

7-74 A W 254 × 89 structural steel wide-flange beam (see Appendix B) is loaded and supported as shown in Fig. P7-74. Determine the maximum transverse shearing stress at section *A–A* of the beam

a. Using Eq. 7-12.
b. Using Eq. 7-14.

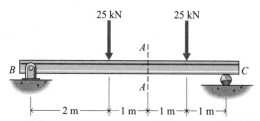

Figure P7-74

7-75* A WT 7 × 34 structural steel beam (see Appendix B) is loaded and supported as shown in Fig. P7-75. Determine the maximum transverse shearing stress at section *A–A* of the beam.

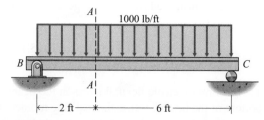

Figure P7-75

7-76 A timber beam 3.5 m long is simply supported at its ends and carries a uniformly distributed load *w* of 6 kN/m over its entire length. If the beam has the cross section shown in Fig. P7-76, determine

a. The maximum horizontal shearing stress in the glued joints between the web and flanges of the beam.
b. The maximum horizontal shearing stress in the beam.

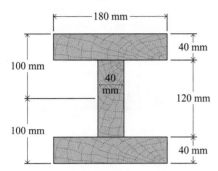

Figure P7-76

7-77 A W10 × 30 structural steel wide-flange beam (see Appendix B) is loaded and supported as shown in Fig. P7-77. At section *A–A* of the beam, determine

a. The maximum transverse shearing stress due to the 4000-lb load.
b. The maximum transverse shearing stress due to the 4000-lb load plus the weight of the beam.

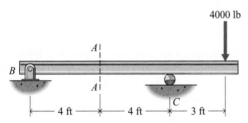

Figure P7-77

Intermediate Problems

7-78* A W 203 × 60 structural steel wide-flange section (see Appendix B) is used for the cantilever beam shown in Fig. P7-78. Determine the maximum flexural and transverse shearing stresses in the beam and state where they occur.

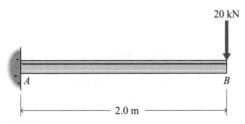

Figure P7-78

7-79* The beam shown in Fig. P7-79*a* is composed of two 1 × 6-in. and two 1 × 3-in. hard maple boards that are glued together as shown in Fig. P7-79*b*. Determine the magnitude and location of

a. The maximum tensile flexural stress in the beam.
b. The maximum horizontal shearing stress in the beam.

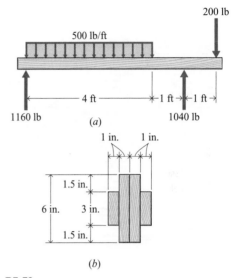

Figure P7-79

7-80 The beam shown in Fig. P7-80*a* is composed of three pieces of timber that are glued together as shown in Fig. P7-80*b*. Determine

a. The maximum horizontal shearing stress in the glued joints.
b. The maximum horizontal shearing stress in the wood.
c. The maximum tensile and compressive flexural stresses in the beam.

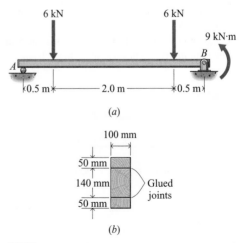

Figure P7-80

7-81 The lintel beam AB shown in Fig. P7-81a has the cross section shown in Fig. P7-81b and is used to support a brick wall over a door opening. The brick wall is assumed to produce a triangular load distribution. The total load carried by the beam is 500 lb. If the beam is simply supported at the ends, determine the maximum flexural and transverse shearing stresses in the beam and state where they occur.

7-83 The timber beam shown in Fig. P7-83a is fabricated by gluing two 1 × 5-in. and two 1 × 4-in. boards together as shown in Fig. P7-83b. Determine

a. The maximum horizontal shearing stress in the glued joints.
b. The maximum horizontal shearing stress in the wood.
c. The maximum tensile and compressive flexural stresses in the beam.

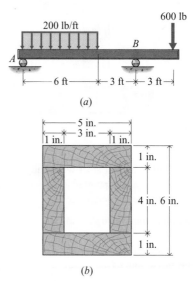

Figure P7-83

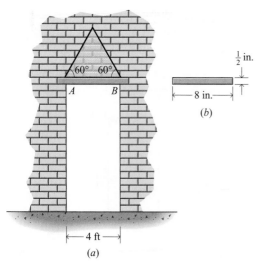

Figure P7-81

7-82* A timber beam is simply supported and carries a uniformly distributed load of 4 kN/m over the full length of the beam. If the beam has the cross section shown in Fig. P7-82 and a span of 6 m, determine

a. The horizontal shearing stress in the glued joint 50 mm below the top of the beam and 1 m from the left support.
b. The horizontal shearing stress in the glued joint 50 mm above the bottom of the beam and 1/2 m from the left support.
c. The maximum horizontal shearing stress in the beam.
d. The maximum tensile flexural stress in the beam.

7-84 A laminated wood beam consists of six 50 × 150-mm planks glued together to form a section 150 mm wide and 300 mm deep, as shown in Fig. P7-84. If the strength of the glue in shear is 825 kPa, determine

a. The maximum uniformly distributed load w that can be applied over the full length of the beam if the beam is simply supported and has a span of 4.5 m.
b. The horizontal shearing stress in the glued joint 50 mm below the top of the beam and 0.6 m from the left support when the load of part a is applied.
c. The maximum tensile flexural stress in the beam when the load of part a is applied.

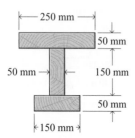

Figure P7-82

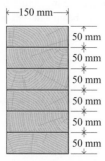

Figure P7-84

Challenging Problems

7-85* A timber beam is fabricated from one 2 × 8-in. and two 2 × 6-in. pieces of lumber to form the cross section shown in Fig. P7-85. The flanges of the beam are fastened to the web with nails that can safely transmit a shear force of 100 lb. If the beam is simply supported and carries a 1000-lb load at the center of 12-ft span, determine

a. The shear force transferred by the nails from the flange to the web in a 12-in. length of the beam.
b. The spacing required for the nails.

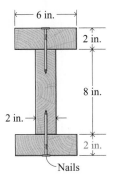

6 in.

2 in.

8 in.

2 in.

2 in.

Nails

Figure P7-85

7-86* A cantilever beam is used to support a concentrated load of 20 kN at the end of the beam. The beam is fabricated by bolting two C457 × 86 steel channels (see Appendix B) back to back to form the H-section shown in Fig. P7-86. If the pairs of bolts are spaced at 300-mm intervals along the beam, determine

a. The shear force carried by each of the bolts.
b. The bolt diameter required if the shear and bearing stresses for the bolts must be limited to 60 MPa and 125 MPa, respectively.

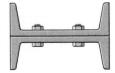

Figure P7-86

7-87 A W 18 × 97 steel beam (see Appendix B) will have 1/2 × 10-in. cover plates welded to the top and bottom flanges as shown in Fig. P7-87. A 20-in. length of the beam for which the shear is constant will be subjected to bending moments of +4600 kip · in. and +2300 kip · in. at the ends. The fillet weld has an allowable load of 2400 lb per in. Determine the number of fillet welds, each 2 in. long, required on each side of the cover plate.

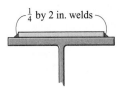
$\frac{1}{4}$ by 2 in. welds

Figure P7-87

7-88 A box beam will be fabricated by bolting two 15 × 260-mm steel plates to two C 305 × 45 steel channels (see Appendix B), as shown in Fig. P7-88. The beam will be simply supported at the ends and will carry a concentrated load of 125 kN at the center of a 5-m span. Determine the bolt spacing required if the bolts have a diameter of 20 mm and an allowable shearing stress of 150 MPa.

Figure P7-88

7-89* A W 21 × 101 steel beam (see Appendix B) is simply supported at the ends and carries a concentrated load at the center of a 20-ft span. The concentrated load must be increased to 125 kip, which requires that the beam be strengthened. It has been decided that two 3/4 × 16-in. steel plates will be bolted to the flanges, as shown in Fig. P7-89. Determine the bolt spacing required if the bolts have a diameter of 3/4 in. and an allowable shearing stress of 17.5 ksi.

Figure P7-89

7-90 A W 356 × 122 steel beam (see Appendix B) has a C 381 × 74 channel bolted to the top flange, as shown in Fig. P7-90. The beam is simply supported at the ends and carries a concentrated load of 96 kN at the center of an 8-m span. If the pairs of bolts are spaced at 500-mm intervals along the beam, determine

a. The shear force carried by each of the bolts.
b. The bolt diameter required if the shear stress for the bolts must be limited to 60 MPa.

Figure P7-90

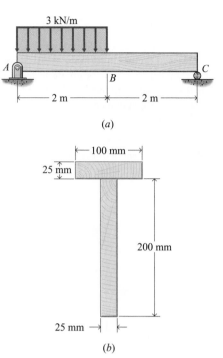

(a)

(b)

Figure P7-92

Computer Problems

7-91 The transverse shear V_r at a certain section of a timber beam is 7500 lb. If the beam has the cross section shown in Fig. P7-91, compute and plot the vertical shearing stress τ as a function of distance y (−6 in. < y < +6 in.) from the neutral axis.

7-93 A timber beam is simply supported and carries a uniformly distributed load w of 360 lb/ft over its entire 18 ft span (Fig. P7-93a). If the beam has the cross section shown in Fig. P7-93b, compute and plot the vertical shearing stress τ as a function of distance y from the neutral axis for a cross section 2 ft from the left end of the beam.

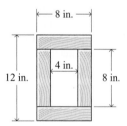

Figure P7-91

7-92 The beam shown in Fig. P7-92a is fabricated by gluing two pieces of timber together to form the cross section shown in Fig. P7-92b. Compute and plot the vertical shearing stress τ as a function of distance y from the neutral axis for a cross section 0.5 m from the left end of the beam.

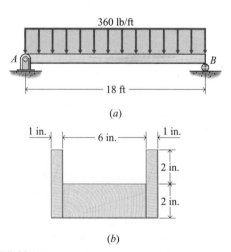

(a)

(b)

Figure P7-93

7-8   PRINCIPAL STRESSES IN FLEXURAL MEMBERS

Methods for finding the flexural stress at any point in a beam were presented in Section 7-4. Procedures for locating the critical sections of a beam (maximum M_r and V_r were developed in Sections 7-5 and 7-6. Methods for determining the transverse and longitudinal shearing stresses at any point in a beam were presented in Section 7-7. However, the discussion of stresses in beams is incomplete without careful consideration of the principal stresses and the maximum shearing stress at carefully selected points on the sections of maximum shear V_r and maximum bending moment M_r.

The flexural stress is maximum at points on the top or bottom edge of a section; the transverse and longitudinal shearing stresses are zero at these points. Consequently, the flexural stresses at points on the top or bottom edge of a given section are also the principal stresses at these points, and the corresponding maximum shearing stress is equal to one-half the flexural stress [$\tau_{max} = (\sigma_p - 0)/2$]. The longitudinal and transverse shearing stresses are normally maximum for a given section at the neutral axis, where the flexural stress is zero. Thus, the evidence presented so far is limited to the extremes in a cross section. One might well wonder what the stress situation might be between these points. Unfortunately, the magnitude of the principal stresses throughout a cross section cannot be expressed for all sections as a simple function of position. However, in order to provide some insight into the nature of the problem, two typical sections will be discussed in the following paragraphs.

For a cantilever beam of rectangular cross section subjected to a concentrated load, the theoretical principal stress variation is indicated for two sections in Fig. 7-28. One observes from this figure that, at a distance from the load of

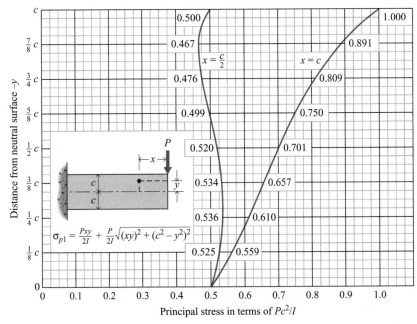

Figure 7-28

one-fourth the depth of the beam, the maximum normal stress does not occur at the surface. However, at a distance of one-half the depth, the maximum flexural stress is the maximum normal stress. For either of these sections, Saint-Venant's principle (see Section 5-6) indicates that the flexure (and transverse shearing stress) formula is inapplicable. Since such a small increase in bending moment is required to overcome the effect of the transverse shearing stress, the conclusion may be drawn that, for a rectangular cross section, in regions where the flexure formula applies, the maximum flexural stress is the maximum normal stress. Although for the rectangular cross section the maximum shearing stress will usually be one-half the maximum normal stress (at a surface of the beam), for materials having a longitudinal plane of weakness (for example, the usual timber beam), the longitudinal shearing stress may frequently be the significant stress—hence, the emphasis on this stress.

The other section to be discussed is the deep, wide-flange section subjected to a combination of large shear and large bending moment. For this combination, the high flexural and transverse shearing stresses occurring simultaneously at the junction of flange and web sometimes yield a principal stress greater than the flexural stress at the surface of the flange (see Example Problem 7-11). In general, at any point in a beam, a combination of large M_r, V_r, Q, and y and a small t should suggest a check on the principal stresses at such a point. Otherwise, the maximum flexural stress will very likely be the maximum normal stress, and the maximum shearing stress will probably occur at the same point.

A knowledge of the directions of the principal stresses may aid in the prediction of the directions of cracks in a brittle material (concrete, for example) and thus may aid in the design of reinforcement to carry the tensile stresses. Curves drawn with their tangent at each point in the directions of the principal stresses are called *stress trajectories*. Since there are, in general, two nonzero principal stresses at each point (plane stress), there are two stress trajectories passing through each point. These curves will be perpendicular since the principal stresses are orthogonal; one set of curves will represent the maximum stresses, whereas the other represents the minimum stresses. The trajectories for a simply supported rectangular beam carrying a concentrated load at the midpoint are shown in Fig. 7-29, with dashed lines representing the directions of the compressive stresses and solid lines showing tensile stress directions. In the vicinities of the load and reactions there are stress concentrations, and the trajectories become much more complicated. Figure 7-29 neglects all stress concentrations.

In order to determine the principal stresses and the maximum shearing stresses at a particular point in a beam, it is necessary to calculate the flexural

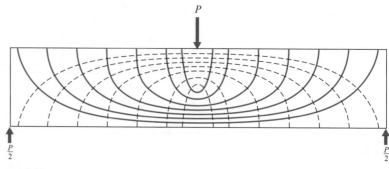

Figure 7-29

stresses and transverse (or longitudinal) shearing stresses at the point. With the stresses on orthogonal planes through the point known, the method of Section 2-10 or 2-11 can be used to calculate the maximum stresses at the point. Example Problem 7-11 illustrates the procedure and provides an example in which the principal stresses at some interior point are greater than the maximum flexural stresses.

■ **Example Problem 7-11**   A W610 × 145 cantilever beam [$I = 1243(10^6)$ mm^4, $S = 4079(10^3)$ mm^3], supported at the left end, carries a uniformly distributed load of 160 kN/m on a span of 2.5 m. Determine the maximum normal and shearing stresses in the beam.

SOLUTION
For a cantilever beam with a uniformly distributed load, both the maximum bending moment and the maximum transverse shear occur on the section at the support. In this case the resisting shear and moment are

$$M_r = -\frac{wL^2}{2} = -\frac{160(2.5)^2}{2} = -500 \text{ kN} \cdot \text{m}$$

$$V_r = wL = 160(2.5) = 400 \text{ kN}$$

The upper half of the cross section of the beam at the wall is shown in Fig. 7-30a. The distribution of flexural stress for this half of the section is shown in Fig. 7-30b, and the distribution of the average transverse shearing stress is as shown in Fig. 7-30c. The vertical stresses σ_y due to the pressure of the load on the top of the beam are considered negligible.

Values will be calculated for three points; namely, at the neutral axis, in the web at the junction of the web and the top flange, and at the top surface. Both the fillets and the stress concentrations at the junction of the web and flange will be neglected. At the neutral axis, the flexural stress is zero; however,

$$Q_{NA} = 304.8(19.7)(295) + 285.1(11.9)(142.6) = 2.255(10^6) \text{ mm}^3$$

$$\tau = \frac{V_r Q}{It} = \frac{400(10^3)(2.255)(10^{-3})}{1243(10^{-6})(0.0119)} = 60.98(10^6) \text{ N/m}^2 \cong 61.0 \text{ MPa}$$

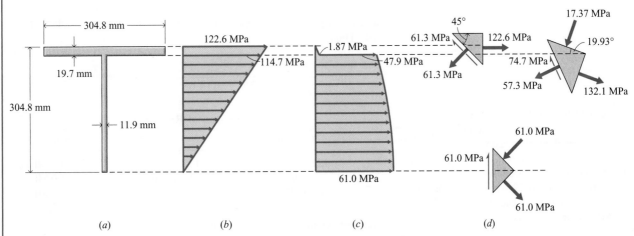

(a)   (b)   (c)   (d)

Figure 7-30

In the web at the junction with the top flange, the flexural stress is

$$\sigma_x = -\frac{M_r y}{I} = -\frac{-500(10^3)(0.2851)}{1243(10^{-6})} = 114.68(10^6)\,\text{N/m}^2 \cong 114.7\,\text{MPa (T)}$$

and the transverse shearing stress is

$$Q_J = 304.8(19.7)(295) = 1.771(10^6)\,\text{mm}^3$$

$$\tau = \frac{V_r Q}{It} = \frac{400(10^3)(1.771)(10^{-3})}{1243(10^{-6})(0.0119)} = 47.89(10^6)\,\text{N/m}^2 \cong 47.9\,\text{MPa}$$

At the top surface, the transverse shearing stress is zero, and the flexural stress is

$$\sigma = -\frac{M_r y}{I} = -\frac{M_r}{S} = -\frac{-500(10^3)}{4079(10^{-6})} = 122.58(10^6)\,\text{N/m}^2 \cong 122.6\,\text{MPa (T)}$$

The principal stresses and maximum shearing stresses for each of the three selected points are shown in Fig. 7-30d. The calculations, from the equations of Section 2-10, for the point at the junction of web and flange are

$$\sigma_{p1,p2} = \frac{\sigma_x + \sigma_y}{2} \pm \sqrt{\left(\frac{\sigma_x - \sigma_y}{2}\right)^2 + \tau_{xy}^2}$$

$$= \frac{114.68 + 0}{2} \pm \sqrt{\left(\frac{114.68 - 0}{2}\right)^2 + (-47.89)^2}$$

$$= 57.34 \pm 74.71$$

$$\sigma_{p1} = 57.34 + 74.71 = +132.05\,\text{MPa} \cong 132.1\,\text{MPa (T)}$$

$$\sigma_{p2} = 57.34 - 74.71 = -17.37\,\text{MPa} \cong 17.37\,\text{MPa (C)}$$

Since σ_{p1} and σ_{p2} have opposite signs, the maximum shearing stress is

$$\tau_{max} = \frac{\sigma_{max} - \sigma_{min}}{2} = \frac{132.05 - (-17.37)}{2} = 74.71\,\text{MPa} \cong 74.7\,\text{MPa (T)}$$

The angle is not particularly important in this case, but it can be obtained from Eq. 2-14. Thus,

$$\theta_p = \frac{1}{2}\tan^{-1}\frac{2\tau_{xy}}{\sigma_x - \sigma_y} = \frac{1}{2}\tan^{-1}\frac{2(-47.89)}{114.68 - 0} = -19.93° = 19.93°\,\downarrow$$

▶ It should be noted that the maximum tensile stress of 132.1 MPa at the junction is 7.75 percent above the maximum tensile flexural stress of 122.6 MPa at the top edge. Also, the maximum shearing stress of 74.7 MPa at the junction is 22.5 percent above the maximum transverse shearing stress of 61.0 MPa at the neutral axis.

Comparison of Results

Stress	Top Edge	Junction	Neutral Axis
σ_{p1}	122.6 MPa (T)	132.1 MPa (T)	61.0 MPa (T)
σ_{p2}	0	17.37 MPa (C)	61.0 MPa (C)
τ_{max}	61.3 MPa	74.7 MPa	61.0 MPa

The other stresses of Fig. 7-30d are obtained by inspection.

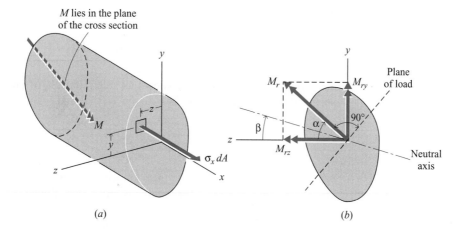

Figure 7-31

cross section, and a plane cross section is assumed to remain plane after bending. Also, the following development is restricted to linearly elastic action.

The resisting moment $\mathbf{M}_r$ is the resultant of the moment of the forces produced by the normal stress σ_x acting on the cross section, as shown in Fig. 7-31a. Since the orientation of the neutral axis is not known, the flexural stress distribution function cannot be expressed in terms of one variable, as in Eq. 7-3 of Section 7-3. However, since a plane section remains plane, the stress variation can be written as

$$\sigma_x = a + k_1 y + k_2 z \qquad (a)$$

where a, k_1, and k_2 are constants.

The net resultant force is zero, and the moments of the force $\sigma_x \, dA$ about the y- and z-axes are equivalent to the components M_{ry} and M_{rz} of the resisting moment $\mathbf{M}_r$. Thus,

$$R = \int_A \sigma_x \, dA = 0$$

$$M_{ry} = \int_A z \, \sigma_x \, dA \qquad (b)$$

$$M_{rz} = -\int_A y \, \sigma_x \, dA$$

Substituting Eq. (a) into Eqs. (b) yields

$$R = a \int_A dA + k_1 \int_A y \, dA + k_2 \int_A z \, dA = 0$$

$$M_{ry} = a \int_A z \, dA + k_1 \int_A yz \, dA + k_2 \int_A z^2 dA \qquad (c)$$

$$M_{rz} = -a \int_A y \, dA - k_1 \int_A y^2 \, dA - k_2 \int_A yz \, dA$$

Since the origin of coordinates is at the centroid of the cross section,

$$\int_A y \, dA = \int_A z \, dA = 0$$

The terms

$$\int_A y^2\, dA \qquad \text{and} \qquad \int_A z^2\, dA$$

are the second moments (also known as area moments of inertia) of the cross-sectional area with respect to the z- and y-axes (I_z and I_y), respectively. The term

$$\int_A yz\, dA$$

is the mixed second moment I_{yz} (also known as the area product of inertia) of the cross-sectional area with respect to the y- and z-axes. Equations (c) reduce to

$$R = aA = 0$$
$$M_{ry} = k_1 I_{yz} + k_2 I_y \qquad\qquad (d)$$
$$M_{rz} = -k_1 I_z - k_2 I_{yz}$$

Solving Eqs. (d) for a, k_1, and k_2 gives

$$a = 0$$
$$k_1 = -\frac{M_{rz} I_y + M_{ry} I_{yz}}{I_y I_z - I_{yz}^2} \qquad\qquad (e)$$
$$k_2 = \frac{M_{ry} I_z + M_{rz} I_{yz}}{I_y I_z - I_{yz}^2}$$

Substituting Eqs. (e) into Eq. (a) yields the elastic flexure formula for unsymmetrical bending. Thus,

$$\sigma_x = -\left[\frac{M_{rz} I_y + M_{ry} I_{yz}}{I_y I_z - I_{yz}^2}\right] y + \left[\frac{M_{ry} I_z + M_{rz} I_{yz}}{I_y I_z - I_{yz}^2}\right] z \qquad\qquad (7\text{-}16)$$

or

$$\sigma_x = \left[\frac{I_z z - I_{yz} y}{I_y I_z - I_{yz}^2}\right] M_{ry} + \left[\frac{-I_y y + I_{yz} z}{I_y I_z - I_{yz}^2}\right] M_{rz} \qquad\qquad (7\text{-}17)$$

Since σ_x is zero at the origin of coordinates ($y = z = 0$), the neutral axis passes through the centroid of the cross section.

The location of points in the cross section where the normal stress is maximum or minimum can be found once the location of the neutral axis is known. Since σ_x is zero at the neutral surface, the orientation of the neutral axis is found by setting Eq. 7-16 equal to zero. Thus,

$$-(M_{rz} I_y + M_{ry} I_{yz})y + (M_{ry} I_z + M_{rz} I_{yz})z = 0$$

or

$$y = \left[\frac{M_{ry} I_z + M_{rz} I_{yz}}{M_{rz} I_y + M_{ry} I_{yz}}\right] z$$

which is the equation of the neutral axis in the yz-plane. The slope of the neutral axis is $dy/dz = \tan \beta$; therefore, the orientation of the neutral axis is given by the expression

$$\tan \beta = \frac{M_{ry}I_z + M_{rz}I_{yz}}{M_{rz}I_y + M_{ry}I_{yz}} \tag{7-18}$$

Equations 7-16, 7-17, and 7-18 have been developed by assuming that the z-component M_{rz} of the resisting moment is positive. If the projection of $\mathbf{M}_r$ is in the negative z-direction, then M_{rz} is negative. The angles α and β (measured from the z-axis) are positive in the clockwise direction.

In some problems, it will be convenient to select the y- and z-axes such that they are principal axes Y and Z. The mixed second moment of area I_{YZ} is then zero and Eqs. 7-16, 7-17, and 7-18 reduce to

$$\sigma_x = -\frac{M_{rZ}Y}{I_Z} + \frac{M_{rY}Z}{I_Y} \tag{7-19}$$

$$\tan \beta = \frac{M_{rY}I_Z}{M_{rZ}I_Y} = \frac{I_Z}{I_Y}\tan \alpha \tag{7-20}$$

Equation 7-20 indicates that the neutral axis is not perpendicular to the plane of loading unless (1) the angle α is zero, in which case the plane of loading is (or is parallel to) a principal plane, or (2) the two principal second moments are equal; this reduces to the special kind of symmetry in which all centroidal second moments are equal (square, circle, and the like). The following Example Problems will illustrate the application of Eqs. 7-16 through 7-20.

■ **Example Problem 7-12** A beam with the T cross section shown in Fig. 7-32a is subjected to a couple **M**, which has a magnitude of 13,600 lb · in. The resisting couple $\mathbf{M}_r$ on the cross section makes an angle $\alpha = 36.87°$ with respect to the z-axis, as shown on Fig. 7-32a. Determine

(a) The orientation of the neutral axis (show its location on a sketch of the cross section).

(b) The maximum tensile and compressive flexural stresses in the beam.

SOLUTION

(a) Because of symmetry, the y-axis is a principal axis (see Appendix A) of the cross section; therefore, the z-axis is the other principal axis. The principal second moments are $I_Z = I_z = 136$ in.^{4} and $I_Y = I_y = 40$ in.^{4} The components of the internal moment are

$$M_{rY} = M_{ry} = -13{,}600 \sin 36.87° = -8160 \text{ lb} \cdot \text{in.}$$
$$M_{rZ} = M_{rz} = +13{,}600 \cos 36.87° = +10{,}880 \text{ lb} \cdot \text{in.}$$

The orientation of the neutral axis is found by using Eq. 7-20. Thus,

$$\tan\beta = \frac{M_{rY}I_Z}{M_{rZ}I_Y} = \frac{-8160(136)}{10{,}880(40)} = -2.550$$

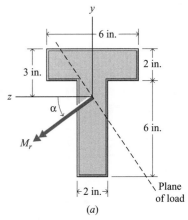

Figure 7-32(a)

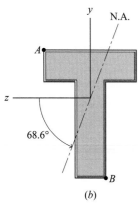

Figure 7-32(b)

from which

$$\beta = -68.59° \cong -68.6°$$ **Ans.**

as shown in Fig. 7-32b.

(b) The maximum tensile and compressive flexural stresses occur at points on the cross section farthest from the neutral axis. These points are labeled A and B on Fig. 7-32b. The flexural stresses at these points are found using Eq. 7-19. Thus,

At point A, $y = 3$ in. and $z = 3$ in.

$$\sigma_{xA} = -\frac{M_{rZ}Y_A}{I_Z} + \frac{M_{rY}Z_A}{I_Y}$$

$$= -\frac{10,880(3)}{136} + \frac{-8160(3)}{40} = -852 \text{ psi} = 852 \text{ psi (C)}$$ **Ans.**

At point B, $y = -5$ in. and $z = -1$ in.

$$\sigma_{xB} = -\frac{M_{rZ}Y_B}{I_Z} + \frac{M_{rY}Z_B}{I_Y}$$

$$= -\frac{10,880(-5)}{136} + \frac{-8160(-1)}{40} = +604 \text{ psi} = 604 \text{ psi (T)}$$ **Ans.**

That part of the cross section to the left of the neutral axis is subjected to compressive flexural stresses, and the part of the cross section to the right of the neutral axis is subjected to tensile flexural stresses.

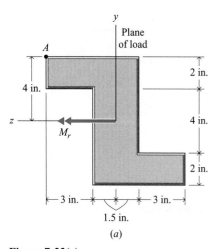

(a)

Figure 7-33(a)

Example Problem 7-13   A beam with the Z cross section shown in Fig. 7-33a is subjected to a moment **M**, which has a magnitude of 300 kip · in. The resisting couple **M**$_r$ on the cross section is in the z-direction, as shown in Fig. 7-33a. The second and mixed moments of area of the cross section with respect to the y- and z-axes are $I_y = 135$ in.^{4}, $I_z = 240$ in.^{4}, and $I_{yz} = 108$ in.^{4} Determine

(a) The flexural stress at point A by using the flexure formula for unsymmetrical bending (Eq. 7-16 or 7-17).

(b) The orientation of the neutral axis.

(c) The maximum tensile and compressive flexural stresses in the beam.

SOLUTION

(a) Since $M_{ry} = 0$, Eqs. 7-16 and 7-17 reduce to

$$\sigma_x = -\left[\frac{I_y y - I_{yz}z}{I_y I_z - I_{yz}^2}\right]M_{rz}$$ (a)

At point A, $y = 4$ in., $z = 4.5$ in., and Eq. (a) yields

$$\sigma_{xA} = -\left[\frac{135(4) - 108(4.5)}{135(240) - (108)^2}\right](300)$$

$$= -0.7813 \text{ ksi} \cong 0.781 \text{ ksi (C)} \qquad \textbf{Ans.}$$

(b) The orientation of the neutral axis is found using Eq. 7-18, which reduces to

$$\tan \beta = \frac{I_{yz}}{I_y}$$

Thus,

$$\tan \beta = \frac{I_{yz}}{I_y} = \frac{108}{135} = 0.800$$

from which

$$\beta = 38.66° \cong 38.7° \qquad \textbf{Ans.}$$

as shown in Fig. 7-33b.

(c) The maximum tensile and compressive flexural stresses occur at points on the cross section farthest from the neutral axis. These points are labeled B and C on Fig. 7-33b. The flexural stresses at these points are found using Eq. (a) of part (a). Thus,

At point B, $y = 4$ in., $z = -1.5$ in., and Eq. (a) yields

$$\sigma_{xB} = -\left[\frac{135(4) - 108(-1.5)}{135(240) - (108)^2}\right](300)$$

$$= -10.156 \text{ ksi} \cong 10.16 \text{ ksi (C)} \qquad \textbf{Ans.}$$

At point C, $y = -4$ in., $z = 1.5$ in., and Eq. (a) yields

$$\sigma_{xC} = -\left[\frac{135(-4) - 108(1.5)}{135(240) - (108)^2}\right](300)$$

$$= +10.156 \text{ ksi} \cong 10.16 \text{ ksi (T)} \qquad \textbf{Ans.}$$

That part of the cross section above the neutral axis is subjected to compressive flexural stresses, and the part of the cross section below the neutral axis is subjected to tensile flexural stresses.

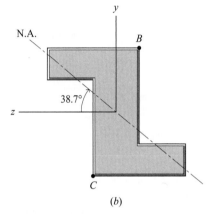

Figure 7-33(b)

 PROBLEMS

Introductory Problems

7-104* A beam with the T cross section shown in Fig. P7-104 is subjected to a moment **M**, which has a magnitude of 20 kN · m. The resisting moment **M**$_r$ on the cross section makes an angle $\alpha = 10°$ with the z-axis, as shown in the figure. Determine

a. The orientation of the neutral axis (show its location on a sketch of the cross section).
b. The maximum tensile and compressive flexural stresses in the beam.

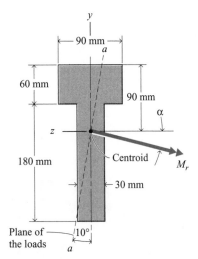

Figure P7-104

7-105* A beam with the I cross section shown in Fig. P7-105 is subjected to a moment **M**, which has a magnitude of 10,000 lb · in. The resisting moment **M**$_r$ on the cross section makes an angle $\alpha = 36.87°$ with the z-axis, as shown in the figure. Determine

a. The orientation of the neutral axis (show its location on a sketch of the cross section).
b. The maximum tensile and compressive flexural stresses in the beam.

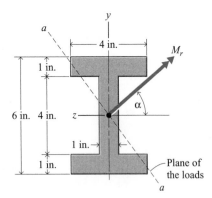

Figure P7-105

7-106 A beam with the Z cross section shown in Fig. P7-106 is subjected to a moment **M**, which has a magnitude of 8000 N · m.

The resisting moment **M**$_r$ on the cross section is in the direction shown in the figure. The second and mixed moments of area for the cross section are $I_y = 4.02(10^6)$ mm^4, $I_z = 16.64(10^6)$ mm^4, and $I_{yz} = -6.05(10^6)$ mm^4. Determine

a. The flexural stress at point A.
b. The orientation of the neutral axis (show its location on a sketch of the cross section).
c. The maximum tensile and compressive flexural stresses in the beam.

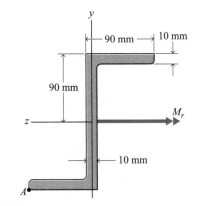

Figure P7-106

7-107 An $8 \times 8 \times 1$-in. angle is used for a beam that is subjected to a moment **M**, which has a magnitude of 7500 lb · ft. The resisting moment **M**$_r$ on the cross section is in the direction shown in Fig. P7-107. The second and mixed moments of area for the cross section are $I_y = I_z = 89.0$ in.^{4} and $I_{yz} = 52.5$ in.^{4} Determine

a. The flexural stress at point A.
b. The orientation of the neutral axis (show its location on a sketch of the cross section).
c. The maximum tensile and compressive flexural stresses in the beam.

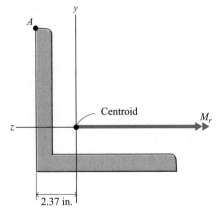

Figure P7-107

Intermediate Problems

7-108* A beam with the cross section shown in Fig. P7-108 is subjected to a moment **M**, which has a magnitude of 2 kN · m. The resisting moment M_r on the cross section is in the direction shown in the figure. Determine

a. The flexural stress at point A.
b. The orientation of the neutral axis (show its location on a sketch of the cross section).
c. The maximum tensile and compressive flexural stresses in the beam.

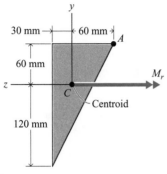

Figure P7-108

7-109* A beam with the cross section shown in Fig. P7-109 is subjected to a moment **M**, which has a magnitude of 50 kip · in. The resisting moment M_r on the cross section is in the direction shown in the figure. Determine

a. The flexural stress at point A.
b. The orientation of the neutral axis (show its location on a sketch of the cross section).
c. The maximum tensile and compressive flexural stresses in the beam.

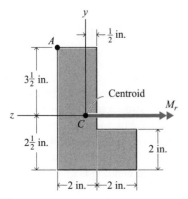

Figure P7-109

7-110 A beam with a rectangular cross section 50 mm wide × 300 mm deep is designed for a system of loads that produces a moment in the xy-plane (the y-axis is parallel to the 300-mm side). In construction, the beam is positioned such that the moment acts in a plane inclined 3° with respect to the xy-plane. Determine the increase in the maximum flexural stress produced by the construction error.

7-111 Determine the minimum moment M_r and its orientation α with respect to the z-axis required to produce a maximum flexural stress of 2000 psi in a beam with a rectangular cross section 6 in. wide × 12 in. deep (the y-axis is parallel to the 12-in. side).

Challenging Problems

7-112* A beam with the cross section shown in Fig. P7-112 is subjected to a moment **M**, which has a magnitude of 20 kN · m. The resisting moment M_r on the cross section is in the direction shown in the figure. Determine

a. The flexural stress at point A.
b. The orientation of the neutral axis (show its location on a sketch of the cross section).
c. The maximum tensile and compressive flexural stresses in the beam.

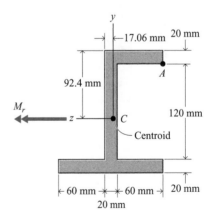

Figure P7-112

7-113* A beam with a triangular cross section is subjected to a moment **M**, which has a magnitude of 10 kip · in. The resisting moment M_r on the cross section makes an angle $\alpha = 26.57°$ with the z-axis, as shown in Fig. P7-113. Determine

a. The flexural stresses at point A, B, and D of the cross section.
b. The orientation of the neutral axis (show its location on a sketch of the cross section).

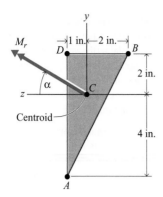

Figure P7-113

7-114 A beam with a rectangular cross section supports an internal moment M_r which is oriented at an angle α with respect to the z-axis, as shown in Fig. P7-114.

a. Develop an expression for the maximum flexural stress in terms of b, h, M, and α.
b. Determine α in terms of b and h to produce the maximum flexural stress for a given moment.
c. Use the expression for α obtained in part b and determine the orientation of the neutral axis when $h = 2b$.

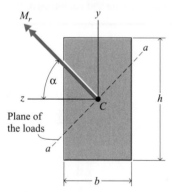

Figure P7-114

7-115 A beam with a Z cross section is subjected to a moment $\mathbf{M}$, which has a magnitude of 20 kip · in. The resisting moment $\mathbf{M}_r$ on the cross section makes an angle $\alpha = 36.87°$ with the z-axis, as shown in Fig. P7-115. The second and mixed moments of area for the cross section are $I_y = 8.83$ in.^{4}, $I_z = 25.4$ in.^{4}, and $I_{yz} = 11.3$ in.^{4} Determine

a. The flexural stress at point A.
b. The orientation of the neutral axis (show its location on a sketch of the cross section).
c. The maximum tensile and compressive flexural stresses in the beam.

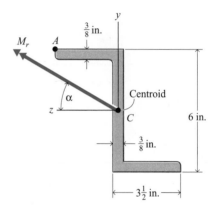

Figure P7-115

7-10 STRESS CONCENTRATIONS UNDER FLEXURAL LOADINGS

In Section 5-6, it was shown that the introduction of a circular hole or other geometric discontinuity into an axially loaded member can cause a significant increase in the magnitude of the stress (stress concentration) in the immediate vicinity of the discontinuity. This is also the case for circular shafts under torsional forms of loading (see Section 6-9) and for flexural members.

In Section 7-4 of this chapter, it was shown that the flexural stress in a beam of uniform cross section in a region of pure bending is given by Eq. 7-8 as

$$\sigma_x = -\frac{M_r y}{I} \tag{7-8}$$

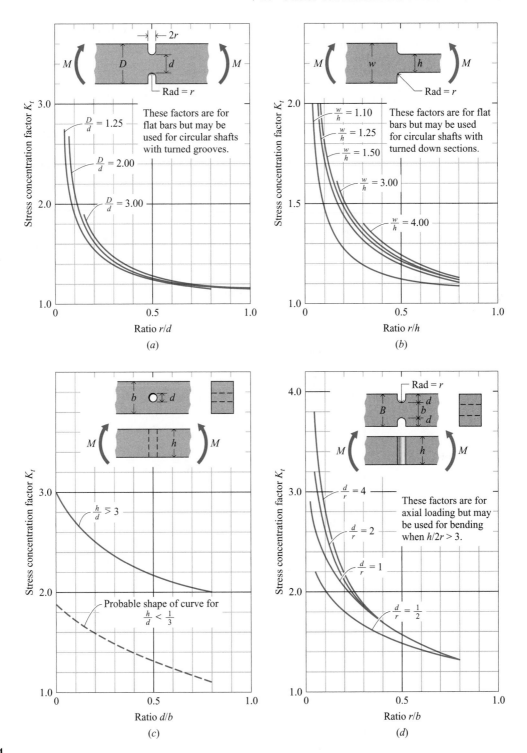

Figure 7-34

If a notch, hole, or other type of discontinuity is introduced into the flexural member, the magnitude of the flexural stress in the vicinity of the discontinuity will increase significantly. Similarly, the magnitude of the flexural stress in regions of abrupt change in a cross section will increase significantly unless adequate fillets are introduced to smooth the transition from one size to the other. In these cases, the magnitude of the flexural stress at the notch, hole, or fillet can be expressed in terms of a *stress concentration factor* K_t as

$$\sigma_x = -K_t\left(\frac{M_r y}{I}\right) \tag{7-21}$$

Since the factor K_t depends only on the geometry of the member, curves can be developed which show the stress concentration factor K_t as a function of the ratios of the parameters involved. Such curves (based on a net section) for fillets, holes, and grooves in flexural members are shown in Fig. 7-34.

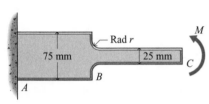

Figure 7-35

Example Problem 7-14 The cantilever spring shown in Fig. 7-35 is made of SAE 4340 heat-treated steel and is 50 mm wide. Determine the maximum safe moment M if a factor of safety FS of 2.5 with respect to failure by fracture is specified and

(a) The radius r is 5 mm.
(b) The radius r is 10 mm.
(c) The radius r is 15 mm.

SOLUTION
The ultimate strength σ_u for heat-treated SAE 4340 steel (see Appendix B) is 1030 MPa. Thus, the allowable stress σ_{all} is

$$\sigma_{all} = \frac{\sigma_u}{FS} = \frac{1030}{2.5} = 412 \text{ MPa}$$

The second moment of area I about the neutral axis (of the small section) for bending of the spring is

$$I = \frac{bh^3}{12} = \frac{50(25^3)}{12} = 65.10(10^3)\,\text{mm}^4 = 65.10(10^{-9})\,\text{m}^4$$

The stress concentration factor K_t for the fillet is obtained from Fig. 7-34b. For $r = 5$ mm,

$$w/h = 75/25 = 3, \ r/h = 5/25 = 0.20, \ \text{which gives } K_t = 1.51.$$

For $r = 10$ mm,

$$w/h = 75/25 = 3, \ r/h = 10/25 = 0.40, \ \text{which gives } K_t = 1.28.$$

For $r = 15$ mm,

$$w/h = 75/25 = 3, \ r/h = 15/25 = 0.60, \ \text{which gives } K_t = 1.18.$$

Using the normal stress σ_x at a point on the bottom surface of segment BC of the beam as the nominal stress, Eq. 7-21 yields

$$M_r = M = -\frac{\sigma_{all} I}{K_t y} = -\frac{412(10^6)(65.10)(10^{-9})}{K_t(-12.5)(10^{-3})} = \frac{2146}{K_t}$$

▶ As the radius of the fillet increases, the stress concentration factor decreases (for a constant w/h).

(a) $M = \dfrac{2146}{K_t} = \dfrac{2146}{1.51} = 1421.2 \text{ N} \cdot \text{m} \cong 1421 \text{ N} \cdot \text{m}$ **Ans.**

(b) $M = \dfrac{2146}{K_t} = \dfrac{2146}{1.28} = 1676.6 \text{ N} \cdot \text{m} \cong 1677 \text{ N} \cdot \text{m}$ **Ans.**

(c) $M = \dfrac{2146}{K_t} = \dfrac{2146}{1.18} = 1818.6 \text{ N} \cdot \text{m} \cong 1819 \text{ N} \cdot \text{m}$ **Ans.**

Stress concentration at the wall has been neglected.

PROBLEMS

Introductory Problems

7-116* An alloy-steel spring, similar to the one shown in Fig. 7-35, has a width of 20 mm and a change in depth at section B from 75 mm to 60 mm. If the radius of the fillet between the two sections is 6 mm, determine the maximum moment that the spring can resist if the maximum flexural stress in the spring must not exceed 80 MPa.

7-117* A stainless-steel spring, similar to the one shown in Fig. 7-35, has a width of 3/4 in. and a change in depth at section B from 3/8 in. to 1/4 in. Determine the minimum acceptable radius for the fillet if the stress concentration factor must not exceed 1.40.

7-118 A 100-mm-diameter cold-rolled stainless-steel (see Appendix B) shaft has a 10-mm-deep groove around the full circumference of the shaft. If the groove has an 8-mm radius at the bottom, determine the percent reduction in strength for flexural-type loadings.

7-119 A stainless-steel bar 3/4 in. wide × 3/8 in. deep has a pair of semicircular grooves cut in the edges of the bar (from top to bottom). If the grooves have a 1/16-in. radius, determine the percent reduction in strength for flexural-type loadings.

Intermediate Problems

7-120* A timber beam 150 mm wide × 200 mm deep has 25-mm-diameter hole drilled from top to bottom of the beam on the centerline of a cross section. Determine the percent reduction in strength produced by the presence of the hole.

7-121 A 3-in.-diameter 0.4% C hot-rolled steel (see Appendix B) shaft has a reduced diameter of 2.73 in. for 12 in. of its length, as shown in Fig. P7-121. If the tool used to turn down the section has a radius of 0.25 in., determine the maximum allowable bending load P that can be applied to the end of the shaft if a factor of safety of 3 with respect to failure by yielding is specified. Neglect stress concentration at the wall.

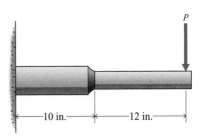

Figure P7-121

Challenging Problems

7-122* A 0.4% C hot-rolled steel (see Appendix B) bar with a rectangular cross section will be loaded as a cantilever beam. The bar has a depth h of 200 mm and has a 25-mm-diameter hole drilled from top to bottom of the beam on the centerline of a cross section where a bending moment of 50 kN · m must be supported. If a factor of safety of 4 with respect to failure by yielding is specified, determine the minimum acceptable width b for the bar.

7-123 A load of 5000 lb is supported by the beams shown in Fig. P7-123. The beams are 2 in. wide × 4 in. deep. The holes for the threaded rods have a 5/8-in. diameter. If the maximum tensile flexural stresses in the beams must not exceed 20 ksi, determine the maximum permissible span for the top beam.

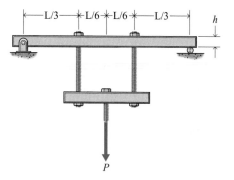

Figure P7-123

Computer Problems

7-124 A steel bar with a rectangular cross section will be loaded as a cantilever beam. The bar has a 10-mm-diameter hole drilled from top to bottom of the beam on the centerline of a cross section where a bending moment of 10 kN · m must be supported without exceeding a stress of 200 MPa. Prepare a curve showing the acceptable combinations of beam width b and beam depth h. Limit the minimum depth of the beam to 30 mm so that the ratio of the beam depth h to the hole diameter d is greater than 3.

7-125 Two beams with rectangular cross sections, similar to the ones shown in Fig. P7-123, support a concentrated load P of 4.5 kip. The top beam has a span of 42 in. If the maximum tensile flexural stress in the top beam must not exceed 15 ksi, prepare a curve showing the acceptable combinations of beam width b and beam depth h if the hole diameters are 3/4 in. Limit the minimum depth of the beam to 2.5 in. so that the ratio of the beam depth h to the hole diameter d is greater than 3.

7-11 INELASTIC BEHAVIOR OF FLEXURAL MEMBERS

A large proportion of structural and machine designs are based on elastic analysis, for which the flexure formula is applicable. However, for some designs, particularly when the weight of the structure is important (aircraft design, for example), the limitation requiring stresses to remain below the proportional limit of the material results in uneconomical or inefficient designs. Therefore, this limitation is sometimes discarded and higher stress levels are tolerated in the design. This section is concerned with the analysis of stresses in instances in which the proportional limit of the material is exceeded or the material does not exhibit a linear stress-strain relationship. In both of these cases, Hooke's law does not apply.

The basic approach to problems involving inelastic action is the same as that outlined in Sections 7-2 and 7-3 for linearly elastic action. A plane section is assumed to remain plane; therefore, a linear distribution of strain exists. If a stress-strain diagram is available for the material, it can be used with the strain distribution to obtain a stress distribution for the beam. The stress and strain distributions shown in Fig. 7-36a and b are for a beam of symmetrical cross section made of a material

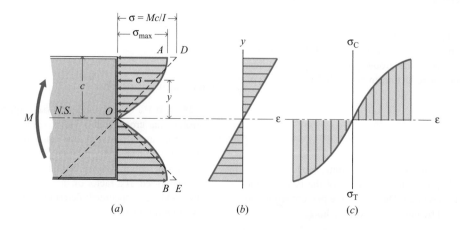

(a)   (b)   (c)

Figure 7-36

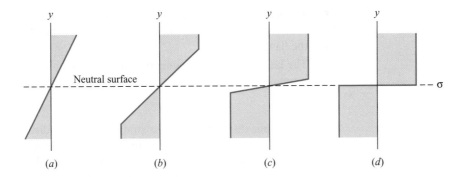

Figure 7-37

having the stress-strain diagram shown in Fig. 7-36c. If an equation can be written for the stress distribution, the resisting moment can be obtained by using Eq. 7-1. If it is not practical to obtain an equation for the stress distribution, the beam depth can be divided into layers of finite thickness and the stress associated with each layer determined. The method of Section 7-3 can then be used to evaluate the resisting moment. As a simplifying technique, the stress-strain diagram can be approximated by a series of straight lines resulting in a simplified stress distribution diagram and a reduction in the amount of work involved.

If the material is elastoplastic (see Section 5–7), the large strains associated with yielding will permit plastic action to take place as indicated in Fig. 7-37. The stress distribution for fully elastic action is shown in Fig. 7-37a, for partially plastic action in Fig. 7-37b, and for essentially complete plastic action in Fig. 7-37c. There must always be a slight amount of elastic action if the strain at the neutral surface is zero. The plastic action shown in Fig. 7-37c is idealized in Fig. 7-37d; this is the assumed distribution used for plastic analysis of structural steel beams.

In members subjected to inelastic action, the neutral axis does not necessarily coincide with a centroidal axis as it does when the action is elastic. If the section is symmetrical and the stress-strain diagrams for tension and compression are identical, the neutral axis for plastic action coincides with the centroidal axis. However, if the section is unsymmetrical (for example, a T-section) or if the stress-strain relations for tension and compression differ appreciably (for example, cast iron), the neutral axis shifts away from the fibers that first experience inelastic action, and it is necessary to locate the axis before the resisting moment can be evaluated. The neutral axis is located by using the equation

$$\Sigma F_x = \int_{\text{area}} \sigma_x \, dA = 0 \qquad (a)$$

and solving for the location of the axis where σ_x is zero. In Eq. (a), x is perpendicular to the cross section of the beam.

The following Example Problems illustrate the concepts of inelastic analysis as applied to the pure bending of beams that are loaded in a plane of summetry.

Example Problem 7-15   A beam having the T cross section of Fig. 7-38 is made of elastoplastic steel ($E = 200$ GPa) with a proportional limit (equal to the yield point) of 240 MPa. Determine

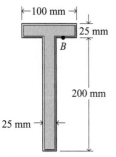

Figure 7-38

(a) The bending moment (applied in the vertical plane of symmetry) that will produce a longitudinal strain of -0.0012 m/m at point B on the lower face of the flange.

(b) The bending moment required to produce completely plastic action in the beam.

SOLUTION

(a) The strain distribution, stress-strain, and stress distribution diagrams are shown in Figs. 7-39a, b, and c, respectively. Because of the inelastic action and unsymmetrical section, the location of the neutral axis must first be determined by using Eq. (a), the stress distribution diagram shown in Fig. 7-39c,

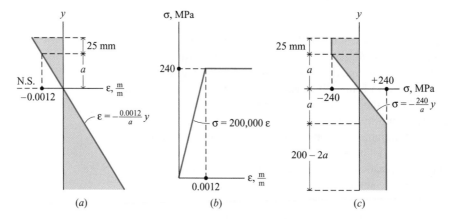

Figure 7-39

▶ The centroid of this section is 87.5 mm below the top of the beam. As the bottom of the beam begins to yield, it carries proportionately less of the moment than it would if it hadn't yielded. Therefore, the neutral axis shifts upward until the top of the beam also begins to yield. When the yield surface has reached the bottom of the flange (Fig. 7-39c), the neutral axis has shifted up to a point only 75 mm below the top of the beam.

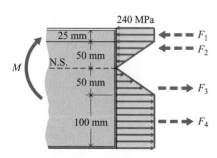

Figure 7-40

and areas from Fig. 7-38. Thus,

$$\int_A \sigma \, dA = -240(10^6)(0.025)(0.100) - \frac{240(10^6)}{2}(a)(0.025)$$
$$+ \frac{240(10^6)}{2}(a)(0.025) + 240(10^6)(0.200 - 2a)(0.025) = 0$$

from which

$$a = 0.050 \text{ m} = 50 \text{ mm}$$

Once the neutral axis is located, the easiest way to find the moment is to use the free-body diagram shown in Fig. 7-40. Moment M must be balanced by the moments of forces F_1, F_2, F_3, and F_4 about the neutral axis. Each force represents a volume (stress times area) from the stress distribution diagram of Fig. 7-39c. Each force passes through the centroid of the part of the area of the stress distribution diagram used to determine the force. For example, F_1 (which results from the stress acting over the flange of the section) is

$$F_1 = \sigma_F A_F = 240(10^6)[0.100(0.025)] = 600(10^3) \text{ N}$$

Force F_1 acts at a distance $d_1 = 50 + 12.5 = 62.5$ mm above the neutral surface. The moment of F_1 with respect to the neutral axis is

$$M_1 = F_1 d_1 = 600(10^3)(62.5)(10^{-3}) = 37.5(10^3) \text{ N} \cdot \text{m}$$

Proceeding in a similar manner for the remaining forces yields

$$M_r = F_1 d_1 + F_2 d_2 + F_3 d_3 + F_4 d_4$$
$$= 240(10^6)(0.100)(0.025)(0.0625)$$
$$+ \frac{1}{2}(240)(10^6)(0.050)(0.025)\left(\frac{2}{3}\right)(0.050)$$
$$+ \frac{1}{2}(240)(10^6)(0.050)(0.025)\left(\frac{2}{3}\right)(0.050)$$
$$+ 240(10^6)(0.100)(0.025)(0.100) = 107.5(10^3) \, \text{N} \cdot \text{m}$$

from which

$$M = M_r = 107.5 \, \text{kN} \cdot \text{m, as shown} \qquad \textbf{Ans.}$$

(b) In this case, the stress distribution diagram would be similar to the diagram shown in Fig. 7-37*d*. If the distance from the neutral surface to the bottom of the stem is designated as distance *a*, Eq. (*a*) yields

$$\int_A \sigma_x \, dA = 240(10^6)(0.025)(a) - 240(10^6)(0.025)(0.200 - a)$$
$$- 240(10^6)(0.100)(0.025) = 0$$

From which $a = 150$ mm; therefore, the neutral surface is 75 mm below the top of the flange. The plastic moment obtained by using the free-body diagram method is

$$M_r = 240(10^6)(0.100)(0.025)(0.0625)$$
$$+ 240(10^6)(0.050)(0.025)(0.025)$$
$$+ 240(10^6)(0.150)(0.025)(0.075) = 112.5(10^3) \, \text{N} \cdot \text{m}$$

Therefore,

$$M_p = M_r = 112.5 \, \text{kN} \cdot \text{m} \qquad \textbf{Ans.}$$

Example Problem 7-16 A beam having the cross section of Fig. 7-41*a* is made of magnesium alloy having the approximate stress-strain diagram of Fig. 7-41*b*. Determine the magnitude of the bending moment applied in the vertical plane of symmetry necessary to produce a flexural stress of magnitude 14.4 ksi at point *A*, which is at the lower surface of the top flange.

SOLUTION
Because of symmetry, the neutral axis coincides with the centroidal *z*-axis; therefore, the strain and stress distribution diagrams are as shown in Figs. 7-41*c* and *d*. Since the specified stress is 14.4 ksi at the lower surface of the top flange ($y = 2$ in.), the strains (from Fig. 7-41*b*) at $y = 2$ in. and $y = 3$ in. are 0.0035 in./in. and 0.00525 in./in., respectively. Thus, the first stress-strain function (see

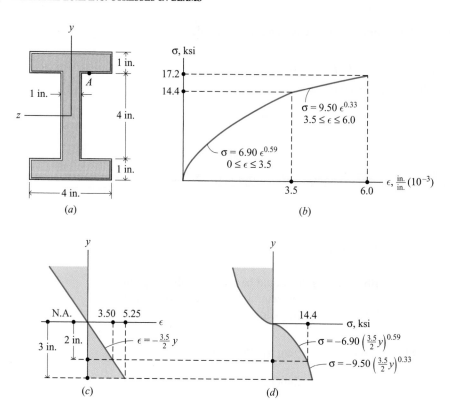

Figure 7-41

Fig. 7-41b) is valid for the entire web and the second function is valid for the entire flange. Once the stress distributions are known, Eq. 7-1 can be used to calculate the bending moment. Thus,

▶ For clarity on Fig. 7-41, a factor of 10^{-3} was left off of the strain ϵ in the equations on the figures. When the 10^{-3} is included, the equations on the figure become $\epsilon = \dfrac{0.0035}{2}y$; $\sigma = 6.90(1000\epsilon)^{0.59} = 405\epsilon^{0.59} = 405\left(\dfrac{0.0035}{2}y\right)^{0.59}$, where $0 \leq \epsilon \leq 0.0035$ in./in.; and $\sigma = 9.50(1000\epsilon)^{0.33} = 93.1\epsilon^{0.33} = 93.1\left(\dfrac{0.0035}{2}y\right)^{0.33}$, where $0.0035 \leq \epsilon \leq 0.006$ in./in.

$$M_r = \int_A \sigma_x \, y \, dA$$

$$= 2\left[\int_0^2 405\left(\frac{0.0035}{2}y\right)^{0.59}(y)(1\,dy) + \int_2^3 93.1\left(\frac{0.0035}{2}y\right)^{0.33}(y)(4\,dy)\right]$$

$$= 2\left[\int_0^2 9.569y^{1.59}\,dy + \int_2^3 45.837y^{1.33}\,dy\right]$$

$$= 2\left[\frac{9.569y^{2.59}}{2.59}\right]_0^2 + 2\left[\frac{45.837y^{2.33}}{2.33}\right]_2^3$$

$$= 44.49 + 311.00 = 355.49 \text{ kip} \cdot \text{in.} \cong 355 \text{ kip} \cdot \text{in.}$$ **Ans.**

Example Problem 7-17 A beam is to be fabricated from three 6 × 1-in. plates that will be welded together to form a symmetrical I-section, half of which is shown in Fig. 7-42b. Determine the maximum elastic and plastic bending moments (M_e and M_p) that the beam can support if the material is

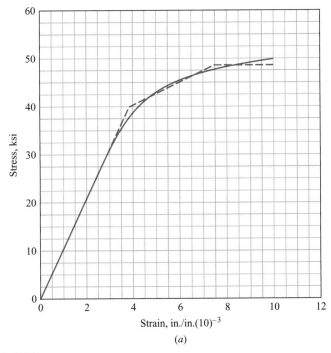

Figure 7-42(a)

(a) Elastoplastic steel with a proportional limit (equal to the yield point) of 40 ksi.

(b) An aluminum alloy with the stress-strain diagram shown in Fig. 7-42a. Assume the diagram is the same in tension and compression and limit the strain to 0.010.

SOLUTION

(a) For the elastic moment, the flexure formula $\sigma_x = M_r c/I$ applies. The second moment of area I about the neutral axis is 166 in.[4] Therefore,

$$M_e = \frac{\sigma_x I}{c} = \frac{40(166)}{4} = 1660 \text{ kip} \cdot \text{in.} \qquad \textbf{Ans.}$$

For this material, the ideal stress distribution of Fig. 7-37d may be assumed when computing the plastic moment, which is

$$M_p = [40(6)(1)(3.5) + 40(3)(1)(1.5)](2) = 2040 \text{ kip} \cdot \text{in.} \qquad \textbf{Ans.}$$

To check the degree of approximation involved in assuming the ideal stress distribution for this section, the maximum strain before strain hardening takes place will be assumed to be 16 times the elastic strain (a reasonable value for structural steel). When this strain is reached in the outer fibers, the maximum elastic strain occurs at $(1/16)(4)$ or 1/4 in. above and below the neutral surface; this means that the actual stress distribution diagram is like that of Fig. 7-37c, the distance from the neutral surface to the point where the stress is constant being 1/4 in. This distribution makes the moment of 2040 kip · in. computed

▶ Since the stress is essentially constant over the entire section, the moment of the stress on the top flange is just the stress times the area times the distance to the centroid of the area. Similarly, the moment of the stress on the upper half of the web is the stress times the area times the distance to the centroid of the area. The moment of the stress on the bottom flange and on the bottom half of the web are identical to that of the top.

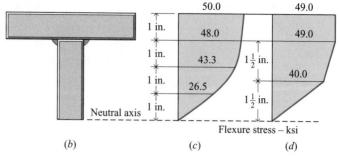

Figure 7-42(b–d)

above too large by the amount

$$\Delta M_r = \frac{40}{2}\left(\frac{1}{4}\right)(1)\left(\frac{1}{3}\right)\left(\frac{1}{4}\right)(2) = 0.833 \text{ kip} \cdot \text{in.}$$

or 0.04 percent.

(b) The elastic moment is given by the flexure formula, where σ is the proportional limit that, from Fig. 7-42a, is 35 ksi. Then,

$$M_e = \frac{\sigma_x I}{c} = \frac{35(166)}{4} = 1453 \text{ kip} \cdot \text{in.} \qquad \textbf{Ans.}$$

The plastic moment will first be determined with the aid of the stress distribution diagram of Fig. 7-42c that was obtained from Fig. 7-42a by assuming the maximum strain in the beam to be 0.010, and since the strains are proportional to the distance from the neutral surface of the beam, the stress for any point in the cross section may be read from the diagram of Fig. 7-42a. With the assumption that the curve in Fig. 7-42c is a series of chords connecting the points at 1-in. vertical intervals, the following solution is obtained:

▶ Connecting the points at 1-in. vertical intervals on the stress diagram of Fig. 7-42c with straight lines results in a series of trapezoids. Each trapezoid can be replaced with a rectangular distributed load and a triangular distributed load. The moments of each of these pieces are then calculated by replacing the distributed loads with a concentrated force equal in magnitude to the area under the load and acting through the centroid of the load distribution.

$$M_p = 2\left[6(1)(48)\left(\frac{7}{2}\right) + (50 - 48)\left(\frac{1}{2}\right)(6)(1)\left(3 + \frac{2}{3}\right)\right.$$
$$+ 43.4(1)(1)\left(\frac{5}{2}\right) + (48 - 43.3)\left(\frac{1}{2}\right)\left(2 + \frac{2}{3}\right) + 26.5(1)(1)\left(\frac{3}{2}\right)$$
$$\left. + (43.3 - 26.5)\left(\frac{1}{2}\right)\left(1 + \frac{2}{3}\right) + \frac{26.5}{2}\left(\frac{2}{3}\right)\right]$$
$$= 2414 \text{ kip} \cdot \text{in.} \cong 2410 \text{ kip} \cdot \text{in.} \qquad \textbf{Ans.}$$

This result will be checked by assuming the curve below the flange to be a parabola. The flange moment will be the same as above and equals 2060 kip · in. Then,

$$M_p = 2060 + \frac{2}{3}(48)(3)(1)\left(\frac{5}{8}\right)(3)(2) = 2420 \text{ kip} \cdot \text{in.} \qquad \textbf{Ans.}$$

One more technique will be applied to this problem. The stress-strain diagram of Fig. 7-42a will be approximated by the dotted lines shown, and the resulting stress distribution diagram for the beam will be as in Fig. 7-42*d*. With the use of this diagram, the plastic moment becomes

$$M_p = 2\left[49(6)(1)\left(\frac{7}{2}\right) + 40\left(\frac{3}{2}\right)(1)\left(\frac{3}{2} + \frac{3}{4}\right)\right.$$

$$\left. + (49 - 40)\left(\frac{1}{2}\right)\left(\frac{3}{2}\right)(1)\left(\frac{3}{2} + 1\right) + 40\left(\frac{1}{2}\right)\left(\frac{3}{2}\right)(1)\left(\frac{2}{3}\right)\left(\frac{3}{2}\right)\right]$$

$$= 2422 \text{ kip} \cdot \text{in.} \cong 2420 \text{ kip} \cdot \text{in.} \qquad \textbf{Ans.}$$

The last three results indicate that any of the three techniques is adequate for this problem.

From the preceding discussion and examples, one may observe that, if the load capacity of a beam is based on the plastic moment rather than the elastic moment, a considerable savings in material may be realized (with the same factor of safety). Furthermore, in the design of statically indeterminate beams and frames, the plastic method of analysis is considerably less time-consuming than the elastic analysis. This is not to say that the elastic method of analysis should be discarded as outmoded, for there are many designs that must be based on elastic action, and for some situations, particularly when repeated loading is involved, the application of plastic analysis may be dangerous.

Further discussion of plastic design may be found in *Fundamentals of Structural Steel Design*.[8]

PROBLEMS

Introductory Problems

7-126* Determine the maximum elastic and plastic bending moments for a W203 × 60 steel beam having a proportional limit (equal to the yield point) of 250 MPa.

7-127* Determine the maximum elastic and plastic bending moments for a W33 × 201 steel beam having a proportional limit (equal to the yield point) of 36 ksi.

7-128 Determine the maximum elastic and plastic bending moments for a W762 × 196 steel beam having a proportional limit (equal to the yield point) of 250 MPa.

7-129* Determine the ratio of the plastic moment to the maximum elastic moment for a beam of elastoplastic material with a rectangular cross section.

7-130 Determine the ratio of the plastic moment to the maximum elastic moment for a beam of elastoplastic material with a circular cross section.

Intermediate Problems

7-131* Determine the ratio of the plastic moment to the maximum elastic moment for a beam of elastoplastic material with the cross section shown in Fig. P7-131.

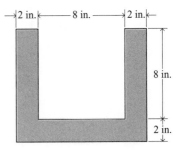

Figure P7-131

---

[8] *Fundamentals of Structural Steel Design*, William T. Segui, PWS–Kent Publishing Boston: Co., 1989.

7-132* Determine the ratio of the plastic moment to the maximum elastic moment for a beam of elastoplastic material with a square cross section if

a. The neutral axis is located as shown in Fig. P7-132a.
b. The neutral axis is located as shown in Fig. P7-132b.

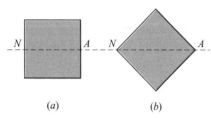

(a)   (b)

Figure P7-132

7-133 A beam of elastoplastic material (yield point of 36 ksi) has the cross section shown in Fig. P7-133. Determine

a. The location of the neutral axis when the stress in the outer fibers of the top flange reaches the yield point.
b. The moment required to produce the condition of part a.
c. The ratio of the plastic moment to the maximum elastic moment for this cross section.

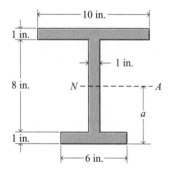

Figure P7-133

7-134 A beam of elastoplastic material (yield point of 240 MPa) has the cross section shown in Fig. P7-134. Determine

a. The location of the neutral axis when the stress in the outer fibers of the top flange reaches the yield point.
b. The moment required to produce the condition of part a.
c. The ratio of the plastic moment to the maximum elastic moment for this cross section.

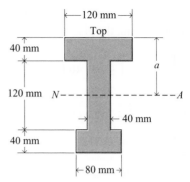

Figure P7-134

7-135* A beam of elastoplastic material (yield point of 36 ksi) has the cross section shown in Fig. P7-135. Determine

a. The location of the neutral axis when the stress in the outer fibers of the top flange reaches the yield point.
b. The moment required to produce the condition of part a.
c. The ratio of the plastic moment to the maximum elastic moment for this cross section.

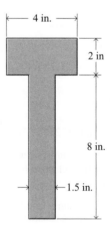

Figure P7-135

7-136 A WT305 × 70 steel beam of elastoplastic material has a proportional limit (equal to the yield point) of 250 MPa. Determine

a. The location of the neutral axis when the stress in the outer fibers of the top flange reaches the yield point.
b. The moment required to produce the condition of part a.
c. The ratio of the plastic moment to the maximum elastic moment for this cross section.

Challenging Problems

7-137* A beam of rectangular cross section is made of a material for which the stress-strain diagram in tension can be represented by the expression $\sigma = K\epsilon^{1/2}$. The shape of the diagram is the same in tension and compression. Develop an expression similar to the flexure formula for relating flexural stress and applied moment.

7-138 A beam having the T cross section shown in Fig. P7-138a is made of a magnesium alloy that has the stress-strain diagram shown in Fig. P7-138b. The beam is subjected to a bending moment that produces a maximum flexural stress of 99.3 MPa (T). When this moment is applied, the neutral surface is located at the junction of the flange and stem. Determine

a. The dimension c.

b. The bending moment applied to the beam.

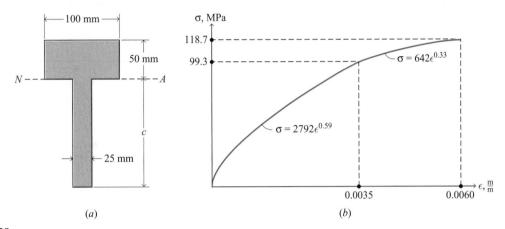

(a)

(b)

Figure P7-138

7-12 SHEARING STRESSES IN THIN-WALLED OPEN SECTIONS—SHEAR CENTER

One of the assumptions made in the development of the elastic flexure formula was that the loads were applied in a plane of symmetry. When this assumption is not satisfied, the beam will, in general, twist about a longitudinal axis. It is possible, however, to place the loads in such a plane that the beam will not twist. When any load is applied in such a plane, the line of action of the load will pass through the *shear center* (also known as the *center of flexure* or *center of twist*). The shear center is defined as the point in the plane of the cross section through which the resultant of the transverse shearing stresses due to flexure (no torsion) will pass for any orientation of transverse loads. For cross-sectional areas having two axes of symmetry, the shear center coincides with the centroid of the cross-sectional area. For those sections with one axis of symmetry, the shear center is always located on the axis of symmetry. If a beam having a cross section with one axis of symmetry is positioned such that the plane of symmetry is the neutral axis for flexural stresses, the plane of the loads must be perpendicular to the neutral axis but cannot pass through the centroid if bending is to occur without twisting. The channel is a common structural shape with one axis of symmetry. It is normally used with the axis of symmetry as the neutral axis since the section modulus is relatively large in this position.

Thin-walled open sections such as channels, angles, wide-flange sections, and the like develop significant shearing stresses during bending. Since the shearing stresses can be assumed to be uniform through the thickness, procedures based on longitudinal equilibrium similar to those of Section 7-7 can be used

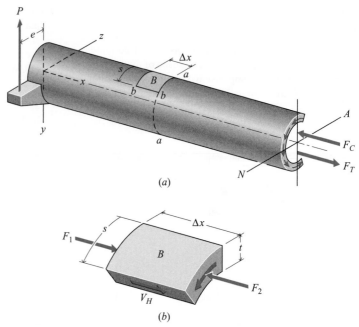

Figure 7-43

to establish shearing stress distributions and to locate the shear centers for these sections.

Consider a cantilever beam of arbitrary cross section but constant thickness (see Fig. 7-43*a*) loaded in a plane parallel to one of its principal planes. The normal stress σ_x in any longitudinal fiber is then given by the elastic flexure formula as $\sigma_x = -M_r y/I$. The shearing stresses on section *a–a* can be determined by considering longitudinal equilibrium of the small block *B* shown in Fig. 7-43*a*. An enlarged free-body diagram of the block is shown in Fig. 7-43*b*. Since the moments on sections at the ends of the block are M and $M + \Delta M$, the resultant forces F_1 and F_2 are

$$F_1 = \int_{A_S} \sigma_x \, dA = -\int_{A_S} \frac{M_r y}{I} \, dA = -\int_0^s \frac{My}{I}(t \, ds)$$

and

$$F_2 = \int_{A_S} \sigma_x \, dA = -\int_{A_S} \frac{(M + \Delta M)y}{I} \, dA = -\int_0^s \frac{(M + \Delta M)y}{I}(t \, ds)$$

where A_S is the area of the cross section between the free edge of the block and the longitudinal plane *b–b* at a distance *s* from the free edge. A summation of forces in the longitudinal direction on the block of Fig. 7-43*b* yields

$$V_H = F_2 - F_1 = -\int_{A_S} \frac{(\Delta M)y}{I}(dA) = -\frac{\Delta M}{I}\int_0^s y(t \, ds)$$

The average shearing stress on the longitudinal plane b–b is V_H divided by the area. Thus,

$$\tau = \lim_{\Delta x \to 0} \frac{\Delta M}{\Delta x}\left(-\frac{1}{It}\right)\int_{A_S} y\, dA = \frac{dM}{dx}\left(-\frac{1}{It}\right)\int_{A_S} y\, dA$$
$$= \frac{dM}{dx}\left(-\frac{1}{It}\right)\int_0^s y\,(t\, ds) \tag{a}$$

where dM/dx equals V_r, the shear at the section where the shearing stresses are to be evaluated.

Equation (a) is identical in form to Eq. (a) of Section 7-7, and the terms have identical meanings. The integral is the first or static moment with respect to the neutral axis of the cross-sectional area to one side of the longitudinal plane b–b. This integral is usually referred to as Q_s. Thus,

$$\tau = -\frac{V_r Q_s}{It} \tag{b}$$

which is identical in form to Eq. (b) of Section 7-7. The minus sign in Eq. (b) is needed to satisfy Eq. 7-1a and is consistent with the sign convention for shearing stresses (see Fig. 7-4c). At each point in the beam, the longitudinal and transverse shearing stresses have the same magnitude (since $\tau_{xy} = \tau_{yx}$); hence, Eq. (b) also gives the transverse shearing stress at a point in a beam (averaged across the width). As was done in Section 7-7, magnitudes of V_r and Q_s will be used to determine the magnitude of the shearing stress τ, and Eq. (b) will be written as

$$\tau = \frac{V_r Q_s}{It} \tag{7-22}$$

The sense of τ will be determined from the sense of the shear V_r on transverse planes and from $\tau_{xy} = \tau_{yx}$ on longitudinal planes. The shearing stresses are uniform through the thickness and act tangent to the surface of the beam. This same shearing stress acts on the transverse cross section at a distance s from the free edge of the section. The stresses on the transverse cross section "flow" in a continuous direction, as shown in Fig. 7-43a, and at the neutral axis they have the same sense as the shear V_r. The shearing force per unit length of cross section is frequently referred to as the shear flow q. Shear flow was previously discussed in Section 6-12.

The procedure for locating a shear center will be illustrated by considering a channel section loaded as a cantilever beam (see Fig. 7-44). Since all fibers in a flange can be considered to be located at a distance $h/2$ from the neutral axis, the shearing stress distribution obtained from Eq. (a) is

$$\tau = \frac{V_r}{It}\int_0^s y\,(t\, ds) = \frac{V_r}{It}\int_0^s \left(\frac{h}{2}\right)(t\, ds) = \frac{V_r hs}{2I} \tag{c}$$

This result shows that the horizontal shearing stress in the flange varies linearly from zero at the outer edge to $V_r hb/(2I)$ at the web. The resultant shearing force

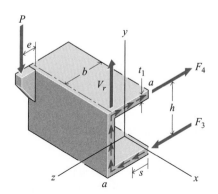

Figure 7-44

on the bottom flange is

$$F_3 = (\tau_{avg})(area) = \left(\frac{V_r h b}{4I} \right)(bt_1) = \frac{V_r h b^2 t_1}{4I} \qquad (d)$$

as shown in Fig. 7-44. A similar analysis indicates that the force F_4 on the upper flange is equal in magnitude and opposite in direction to F_3. These two forces constitute a clockwise twisting moment that must be balanced by the resultant force in the web and the applied load if the beam is not to twist.

The resultant of the shearing forces in the web is V_r and must be equal to the applied load P for equilibrium. It is assumed that the vertical shearing force transmitted by the flanges is small enough to neglect (see Section 7-7). The forces V_r and P will provide a counterclockwise couple to balance the couple due to F_3 and F_4 if P is located a distance e from the center of the web, as indicated in Fig. 7-44. When the moments of these two couples are equated, the result is

$$Pe = V_r e = h F_3 = \frac{V_r h^2 b^2 t_1}{4I} = \frac{P h^2 b^2 t_1}{4I}$$

from which

$$e = \frac{h^2 b^2 t_1}{4I} \qquad (e)$$

If the load P of Fig. 7-44 were applied in the z-direction, the forces corresponding to F_3 and F_4 (entirely different magnitudes) would be equal, and both would be directed to oppose the load P. There would also be two shearing forces in the web (parallel to the web) of equal magnitude and oppositely directed; hence, a moment equation would establish the fact that the shear center lies on the z-axis of symmetry. If there is to be no twisting of the beam (buckling is assumed not to occur), any load must pass through the intersection of this axis of symmetry and the line of action of force P of Fig. 7-44. If the load is inclined, it may be resolved into components parallel to these axes, and the stresses may be determined using the principle of superposition.

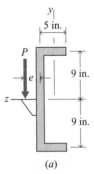

Figure 7-45(a)

▶ Figure 7-45a is similar to Fig. 7-44; that is, Fig. 7-45a is a two-dimensional representation of Fig. 7-44.

Example Problem 7-18
A channel section is being used for a cantilever beam that supports a load P at the free end. Centerline dimensions for the channel are shown in Fig. 7-45a. All metal in the channel section is 1/2 in. thick. Assume that all of the section is effective in resisting flexural stresses and that only the web resists vertical shearing stresses.

(a) Locate the shear center of the section with respect to the center of the web.

(b) Prepare a sketch showing the distribution of shearing stress on the cross section when the vertical shear at the section is 12 kip.

SOLUTION

(a) The location of the shear center for a thin-walled channel section is given by Eq. (e) as

$$e = \frac{h^2 b^2 t}{4I}$$

For the given channel section,

$$I = \frac{1}{12}(1/2)(18)^3 + 2\left[\frac{1}{12}(5)(1/2)^3 + 5(1/2)(9)^2\right] = 648 \text{ in.}^4$$

with $h = 18$ in., $b = 5$ in., and $t = 1/2$ in.

$$e = \frac{h^2 b^2 t}{4I} = \frac{(18)^2(5)^2(1/2)}{4(648)} = 1.5625 \text{ in.} \cong 1.563 \text{ in.} \qquad \textbf{Ans.}$$

▶ For thin sections it is customary to neglect the first term in brackets for the flanges; that is, the term $(5)(1/2)^3/12$. Then, the second moment of area $I = 648$ in.^{4} This will be done in subsequent problems. All calculations are done using centerline dimensions.

(b) The shearing stress at any point in a flange of the channel section at a distance s from the outer edge of the flange (see Fig. 7-45b) is given by Eq. 7-22 as

$$\tau = \frac{V_r Q_s}{It} = \frac{V_r(st)(h/2)}{It} = \frac{V_r h s}{2I}$$

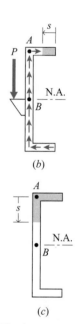

(b)

(c)

Figure 7-45(b–c)

which indicates that the stress distribution in the flange is linear. At the outer edge of the flange, $s = 0$; therefore, $\tau = 0$. At point A (the intersection of the centerlines of the flange and web), $s = 5$ in.; therefore,

$$\tau_A = \frac{V_r h s}{2I} = \frac{12(18)(5)}{2(648)} = 0.8333 \text{ ksi} \cong 833 \text{ psi}$$

Thus, the shearing stress in the top flange varies in a linear manner from 0 at the outer edge to 833 psi at the centerline of the web.

The first moment Q_s for a point in the web at a distance s (see Fig. 7-45c) from the centerline of the top flange is

$$Q_s = Q_f + Q_w = 5(1/2)(9) + (1/2)(s)(9 - 0.5s)$$
$$= 22.5 + 4.5s - 0.25s^2$$

Since V_r, I, and t are constants, the shear stress distribution in the web is quadratic (parabolic) with the maximum shearing stress occurring at point B (the neutral axis). Thus,

At point A at the top of the web ($s = 0$):

$$Q_s = 22.5 \text{ in.}^3$$
$$\tau_A = \frac{V_r Q_s}{It} = \frac{12(22.5)}{648(1/2)} = 0.8333 \text{ ksi} \cong 833 \text{ psi}$$

At point B at the neutral axis ($s = 9$ in.):

$$Q_s = 22.5 + 4.5(9) - 0.25(9)^2 = 42.75 \text{ in.}^3$$
$$\tau_B = \frac{V_r Q_s}{It_w} = \frac{12(42.75)}{648(1/2)} = 1.5833 \text{ ksi} \cong 1583 \text{ psi}$$

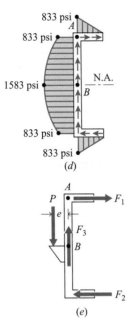

(d)

(e)

Figure 7-45(d–e)

▶ Since the distribution of shear stress in the flanges is linear, the average stress is just half the maximum stress $\tau_{avg} = 0.833/2$. Multiplying this average stress times the area of the flange gives the resultant shear force in the flange F_1. The distribution of shear stress in the web is parabolic, and the resultant shear force F_3 is calculated as the area under the stress diagram. The resultant of the rectangular portion of the load in the web is the stress $(833\ \text{lb/in.}^2)$ times the area it acts on $(18\ \text{in.} \times 1/2\ \text{in.})$. The resultant of the parabolic portion of the load in the web is two-thirds of the stress $(1583 - 833\ \text{lb/in.}^2)$ times the area of the web $(18\ \text{in.} \times 1/2\ \text{in.})$.

The distribution of shearing stresses on the cross section is shown in Fig. 7-45d. The stresses "flow" in a manner to oppose the applied load P as indicated in Fig. 7-45d.

The resultant shear force F_1 in the top flange (see Fig. 7-45e) is

$$F_1 = \tau_{avg} A_f = \frac{1}{2}(0.8333)(5)(1/2) = 1.0416\ \text{kip}$$

The resultant shear force F_2 in the lower flange is equal in magnitude and opposite in direction to F_1, as shown in Fig. 7-45e. The resultant shear force F_3 in the web (since the distribution is parabolic) is

$$F_3 = 833(18)(1/2) + \frac{2}{3}(1583 - 833)(18)(1/2) = 11,997 \cong 12\ \text{kip}$$

which shows that the force in the web is approximately equal to the applied force.

Applying the equilibrium equations $\Sigma F_y = 0$ and $\Sigma M_B = 0$ to the part of the beam between the free end and the section shown in Fig. 7-45a yields

$$+\uparrow \Sigma F_y = 0: \qquad F_3 - P = 11.997 - 12 = -0.003 \cong 0$$
$$+\downarrow \Sigma M_B = 0: \qquad Pe - F_1 h = 12(1.5625) - 1.0416(18) = 0.0012 \cong 0$$

which verifies the accuracy of the above results and the assumption that only the web resists vertical shearing stresses.

Example Problem 7-19 All metal in the cross section shown in Fig. 7-46a is 5 mm thick. The dimensions shown are centerline dimensions for the flanges and the webs. Assume that all of the section is effective in resisting flexural stresses and that only the web resists vertical shearing stresses.

(a) Locate the shear center of the cross section with respect to the center of the right web for $P = 15\ \text{kN}$.

(b) Determine the maximum shearing stress produced on the cross section by a vertical force P of 15 kN.

SOLUTION

(a) The shearing stress at point A (see Fig. 7-46b) in the top flange of the cross section is given by Eq. 7-22 as

$$\tau = \frac{V_r Q_s}{It}$$

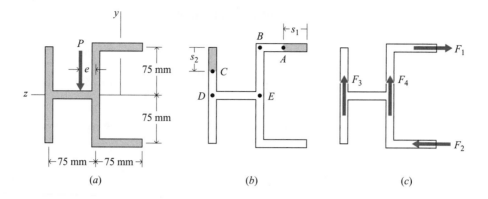

Figure 7-46

For the total cross section,

$$I = \frac{1}{12}(5)(150)^3 + \frac{1}{12}(5)(150)^3$$

$$+ 2\left[\frac{1}{12}(75)(5)^3 + 5(75)(75)^2\right] + \frac{1}{12}(75)(5)^3$$

$$= 7.034(10^6)\,\text{mm}^4 = 7.034(10^{-6})\,\text{m}^4$$

▶ Neglecting the third and fifth terms of the expression for I gives $I = 7.034(10^6)\,\text{mm}^4$.

For point A in the top flange,

$$Q_s = 5(s_1)(75) = 375s_1\,\text{mm}^3 = 375s_1(10^{-9})\,\text{m}^3$$

$$\tau_A = \frac{V_r Q_s}{It} = \frac{15(10^3)(375s_1)(10^{-9})}{7.034(10^{-6})(5)(10^{-3})} = 0.15994s_1(10^6)\,\text{N/m}^2$$

At point B in the top flange ($s_1 = 75$ mm),

$$\tau_B = 0.15994(75)(10^6) = 12.00(10^6)\,\text{N/m}^2 = 12.00\,\text{MPa}$$

Thus, the shearing stress in the top flange varies in a linear manner from 0 at the outer edge to 12.00 MPa at the centerline of the web. The resultant shearing force on the top flange (see Fig. 7-46c) is

$$F_1 = (\tau_{\text{avg}})(\text{area}) = \frac{1}{2}(12.00)(10^6)(75)(5)(10^{-6}) = 2250\,\text{N}$$

A similar analysis indicates that the force F_2 on the lower flange (see Fig. 7-46c) is equal in magnitude and opposite in direction to F_1.
    For point C in the left web (see Fig. 7-46b),

$$Q_s = 5(s_2)(75 - 0.5s_2)$$
$$= (375s_2 - 2.5s_2^2)\,\text{mm}^3 = (375s_2 - 2.5s_2^2)(10^{-9})\,\text{m}^3$$

Since V_r, I, and t are constants, the shear stress distribution in the web is parabolic with the maximum stress occurring at the neutral axis. At the top of the left web $s_2 = 0$; therefore, $Q_s = 0$ and $\tau = 0$.

At point D at the neutral axis ($s_2 = 75$ mm),

$$Q_s = [375(75) - 2.5(75)^2] = 14.063(10^3)\,\text{mm}^3 = 14.063(10^{-6})\,\text{m}^3$$

$$\tau_D = \frac{V_r Q_s}{It_w} = \frac{15(10^3)(14.063)(10^{-6})}{7.034(10^{-6})(5)(10^{-3})} = 5.998(10^6)\,\text{N/m}^2 \cong 6.00\,\text{MPa}$$

Since the stress distribution in the left web is parabolic, the resultant shearing force in the left web is

$$F_3 = \frac{2}{3}(5.998)(10^6)(5)(150)(10^{-6}) = 2999\text{N} \cong 3.00\,\text{kN}$$

▶ The resultant of the stress distribution in the left web F_3 is the "area" under the parabola, which is two-thirds of the "base" (5 mm × 150 mm) times the "height" (5.998 MPa).

Summing moments about point E in the right web yields

$$+\!\downarrow \Sigma M_E = 0: \qquad\qquad Pe - F_1(150) - F_3(75) = 0$$

from which

$$e = \frac{F_1(150) + F_3(75)}{P} = \frac{2250(150) + 2999(75)}{15,000} = 37.5\,\text{mm} \qquad \textbf{Ans.}$$

(b) At point B in the right web,

$$Q_s = 5(75)(75) = 28.13(10^3)\,\text{mm}^3 = 28.13(10^{-6})\,\text{m}^3$$

$$\tau_B = \frac{V_r Q_s}{It} = \frac{15(10^3)(28.13)(10^{-6})}{7.034(10^{-6})(5)(10^{-3})} = 11.997(10^6)\,\text{N/m}^2 \cong 12.00\,\text{MPa}$$

At point E in the right web,

$$Q_s = 5(75)(75) \quad\; + 5(75)(37.5) = 42.19(10^3)\,\text{mm}^3 = 42.19(10^{-6})\,\text{m}^3$$

$$\tau_E = \tau_{\text{max}} = \frac{V_r Q_s}{It} = \frac{15(10^3)(42.19)(10^{-6})}{7.034(10^{-6})(5)(10^{-3})}$$

$$= 17.994(10^6)\,\text{N/m}^2 \cong 17.99\,\text{MPa} \qquad \textbf{Ans.}$$

As a check on the above results,

$$F_4 = 12.00(10^6)(5)(150)(10^{-6}) + \frac{2}{3}(17.99 - 12.00)(10^6)(150)(5)(10^{-6})$$

$$= 11,995\,\text{N} \cong 12.00\,\text{kN}$$

▶ The resultant of the stress distribution in the right web F_4 is the "area" under the parabola, which is two-thirds of the "base" (5 mm × 150 mm) times the "height" (17.99–12.00 MPa) plus the "area" under the rectangle, which is equal to the "base" (5 mm × 150 mm) times the "height" (12.00 MPa).

The resultant vertical shearing force V_r from the two webs is

$$V_r = F_3 + F_4 = 2.999 + 11.995 = 14.99\,\text{kN}$$

which verifies the accuracy of the above results and the assumption that only the web resists vertical shearing stresses.

PROBLEMS

Introductory Problems

7-139* A channel is to be fabricated with a depth of 7 in., a web thickness of 1/4 in., and a flange width of 4 in. Structural considerations require the shear center of the channel to be located 2 in. from the center of the web. Determine the required thickness of the flanges.

7-140* All metal in the cross section shown in Fig. P7-140 has a thickness of 2.5 mm. The dimensions shown are centerline dimensions for the flanges and the web.

   a. Locate the shear center of the section with respect to the center of the web.

   b. Determine the shearing stress at point O when the vertical shear at the section is 2.5 kN.

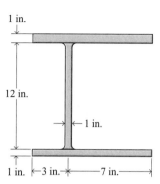

Figure P7-141

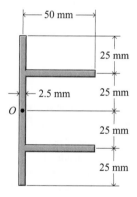

Figure P7-140

7-141 An eccentric H-section is made by welding two 1 × 10-in. steel flanges to a 1 × 12-in. steel web, as shown in Fig. P7-141. The section is used as a 5-ft cantilever beam that carries a concentrated load of 100 kip at the free end. Assume that all of the section is effective in resisting flexural stresses and that only the web resists vertical shearing stresses.

   a. Locate the shear center of the section with respect to the center of the web.

   b. Prepare a sketch showing the distribution of shearing stress throughout the cross section.

7-142 Two 14-mm-thick plates are to be welded to the flanges of a channel (see Fig. P7-142) that is 460 mm deep overall, has flanges that are 100 mm wide × 16 mm thick, and has a web that is 14 mm thick. Determine

   a. The width b of the plates if the shear center of the section must be located at the center of the web.

   b. The maximum vertical shearing stress at the section when the vertical shear at the section is 40 kN.

Figure P7-142

Intermediate Problems

7-143* All metal in the cross section shown in Fig. P7-143 has a thickness of 1/4 in. The dimensions shown are centerline dimensions for the flanges and the web.

   a. Locate the shear center of the section with respect to the center of the web.

   b. Determine the shearing stress at point O when the vertical shear at the section is 1500 lb.

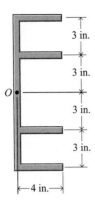

3 in.

3 in.

O

3 in.

3 in.

← 4 in. →

Figure P7-143

7-144* Locate the shear center for the cross section shown in Fig. P7-144 and determine the maximum shearing stress produced on the cross section by a vertical shear of 6 kN.

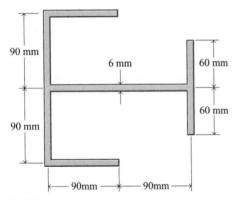

90 mm

6 mm

60 mm

60 mm

90 mm

← 90mm → 90mm →

Figure P7-144

7-145 A thin-walled box section (see Fig. P7-145) is used as a cantilever beam. Locate the shear center of the section with respect to the center of the web.

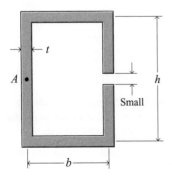

t

A

h

Small

← b →

Figure P7-145

Challenging Problems

7-146* A thin-walled cylindrical tube cut longitudinally to make a semicylinder is used as a cantilever beam. The load acts parallel to the cut section as shown in Fig. P7-146.

a. Locate the shear center of the section with respect to the center of the tube (dimension e).
b. Determine the shearing stress at point A if the radius R of the section is 25 mm, the thickness t is 2.5 mm, and the load P is 440 N.

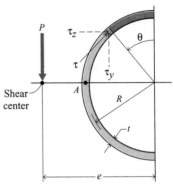

P

τ_z

θ

τ

τ_y

Shear center

A

R

t

e

Figure P7-146

7-147* A thin-walled slotted tube is used as a cantilever beam. The beam is loaded as shown in Fig. P7-147.

a. Locate the shear center of the section with respect to the center of the tube (dimension e).
b. Determine the shearing stress at point A if the radius R of the section is 2 in., the thickness t is 0.10 in., and the load P is 110 lb.

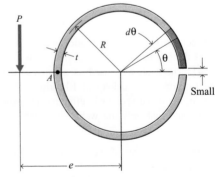

P

$d\theta$

R

t

θ

A

Small

← e →

Figure P7-147

7-148 A cantilever beam with the cross section shown in Fig. P7-148 is subjected to a vertical concentrated load of 4 kN at the free end. The section has a constant thickness of 4 mm.

a. Locate the shear center of the section with respect to the center of the web.

b. Prepare a sketch showing the distribution of shearing stress throughout the cross section. The dimensions shown in Fig. P7-148 are centerline dimensions for the flanges, web, and extensions. Assume that all of the section is effective in resisting flexural stresses but that the vertical shearing force resisted by the flanges is negligible. The vertical shearing force in the 20-mm extensions is not negligible.

7-149 Locate the shear center for the cross section shown in Fig. P7-149 and determine the maximum shearing stress produced on the cross section by a vertical shear force of 300 lb.

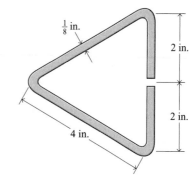

Figure P7-149

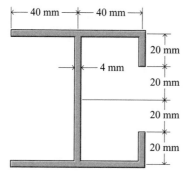

Figure P7-148

7-13 FLEXURAL STRESSES IN BEAMS OF TWO MATERIALS

The method of flexural stress computation covered in Section 7-4 is sufficiently general to cover symmetrical beams composed of longitudinal elements (layers) of different materials. However, for many real beams of two materials (often referred to as reinforced beams), a method can be developed to allow the use of the elastic flexure formula, thus reducing the computational labor involved. The method is applicable to elastic design only.

The assumption of a plane section remaining plane is still valid, provided the different materials are securely bonded together to provide the necessary resistance to longitudinal shearing stresses. Therefore, the usual linear transverse distribution of longitudinal strains ($\epsilon_x = -y/\rho$) is valid.

The beam of Fig. 7-47, composed of a central portion of material A and two outer layers of material B, will serve as a model for the development of the stress distribution. The section is assumed to be symmetrical with respect to the xy- and xz-planes, and the moment is applied in the xy-plane. The strains at point a in material A and point b in material B are related using the linear strain relationships; $\epsilon_a = -a/\rho$ and $\epsilon_b = -b/\rho$ or

$$\epsilon_b = \frac{b}{a}\epsilon_a$$

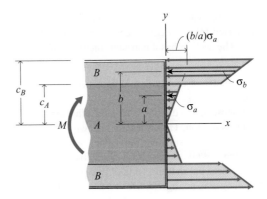

Figure 7-47

As long as neither material is subjected to stresses above the proportional limit, Hooke's law applies and the linear strain relationship gives

$$\frac{\sigma_b}{E_B} = \frac{b}{a}\left(\frac{\sigma_a}{E_A}\right)$$

Thus,

$$\sigma_b = \frac{b}{a}\left(\frac{E_B}{E_A}\right)\sigma_a \qquad (a)$$

From this relation it is evident that, at the junction between the two materials where distance a and b are equal, there is an abrupt change in stress determined by the ratio $n = E_B/E_A$ of the two moduli.

If Eq. (a) is used, the normal force on a differential end area of an element of material B is given by the expression

$$dF_B = \sigma_b \, dA = n\left(\frac{b}{a}\right)(\sigma_a)dA = \left(\frac{b}{a}\sigma_a\right)(nt)\, dy \qquad c_A \le y \le c_B$$

where t is the width of the beam at a distance b from the neutral surface. The first factor in parentheses represents the linear stress distribution in a homogeneous material A. The second factor in parentheses may be interpreted as the extended width of the beam from $y = c_A$ to $y = c_B$ if material B were replaced by material A, thus resulting in an equivalent or transformed cross section for a beam of homogeneous material. The transformed section is obtained by replacing either material by an equivalent amount of the other material as determined by the ratio n of their elastic moduli. The location of the neutral axis (through the centroid) and the second moment of area for the transformed section (one material) can be found in the usual way. In addition, the flexure formula can be applied to determine the flexural stresses in the transformed section. The stress in the material that was transformed is found using Eq. (a).

The method is not limited to two materials; however, the use of more than two materials in one beam would be unusual. The method is illustrated by the following example problem.

7.19

7.20

Example Problem 7-20   A timber beam 4 in. wide × 8 in. deep has a 3.5-in.-wide × 1/2-in.-deep aluminum alloy plate securely fastened to its bottom face, as shown in Fig. 7-48a. The moduli of elasticity for the timber and aluminum alloy are 1250 ksi and 10,000 ksi, respectively. Determine the maximum flexural stress in each material if the applied moment is 75 kip·in.

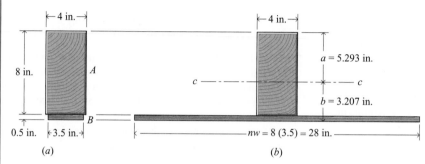

Figure 7-48

SOLUTION

The ratio of moduli for the aluminum and timber is

$$n = \frac{E_B}{E_A} = \frac{E_a}{E_t} = \frac{10,000}{1250} = 8$$

The actual cross section (timber A and aluminum B) and the transformed timber cross section are shown in Figs. 7-48a and b, respectively. The neutral axis of the transformed section is located by the principle of moments as

$$y_C = \frac{\Sigma M}{\Sigma A} = \frac{28(1/2)(1/4) + 4(8)(4.5)}{28(1/2) + 4(8)} = 3.207 \text{ in.}$$

above the bottom of the section. The second moment of area of the transformed timber cross section with respect to the neutral axis is

$$I = \frac{1}{12}(28)(1/2)^3 + 28(1/2)(2.957)^2$$

$$+ \frac{1}{12}(4)(8)^3 + 4(8)(1.293)^2 = 346.9 \text{ in.}^4$$

The maximum flexural stress in the timber is

$$\sigma_{tmax} = \frac{Mc}{I} = \frac{75(10^3)(5.293)}{346.9} = 1.1444(10^3) \text{ psi}$$

$$\cong 1144 \text{ psi} \qquad\qquad\qquad\qquad\qquad\text{**Ans.**}$$

▶ If the aluminum alloy plate were replaced with a piece of wood 8 × 3.5 in. = 28 in. wide, the force and moment on the wooden flange would be the same as the force and moment on the aluminum plate even though the stresses at points on the wooden flange would be only one-eighth the stress at an equivalent point on the aluminum plate.

and for the aluminum is

$$\sigma_{a\max} = \frac{b}{a}\left(\frac{E_a}{E_t}\right)\sigma_{t\max} = \frac{3.207}{5.293}(8)(1144.4)$$
$$\cong 5550 \text{ psi} \qquad\qquad \textbf{Ans.}$$

PROBLEMS

MecMovie Activities and Problems

MM7.18 Introducing the transformed area method. Example; Try one. Determine bending stresses in a composite beam using the transformed area method.

MM7.19 Aluminum and brass composite beam. Example; Try one. Given allowable stresses for two materials, determine the largest allowable moment that can be applied to the beam cross section.

Introductory Problems

7-150* A timber beam 150 mm wide × 350 mm deep has a 150-mm-wide × 15-mm-thick steel plate fastened securely to its top face. The moduli of elasticity for the timber and steel are 10 GPa and 200 GPa, respectively. Determine the maximum flexural stress in the timber when the maximum flexural stress in the steel is 75 MPa (T).

7-151* A timber beam 6 in. wide × 12 in. deep has a 6-in.-wide × 1/2-in.-thick steel plate fastened securely to its bottom face. The moduli of elasticity for the timber and steel are 1500 ksi and 30,000 ksi, respectively. Determine the maximum flexural stress in the steel when the maximum flexural stress in the timber is 1250 psi (C).

7-152 A composite beam 225 mm wide × 300 mm deep × 4 m long is made by bolting two 100-mm-wide × 300-mm-deep timber planks to the sides of a 25 × 300-mm structural aluminum plate, as shown in Fig. P7-152. The moduli of elasticity for the timber and aluminum are 8 GPa and 73 GPa, respectively. Determine the maximum tensile flexural stress in each

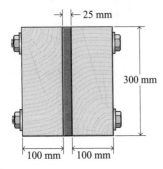

Figure P7-152

of the materials if the composite beam is simply supported and carries a concentrated load 30 kN in the center of the beam.

7-153 A 4-in-wide × 6-in-deep timber cantilever beam 6 ft long is reinforced by bolting two 1/2 × 6-in. structural steel plates to the sides of the timber beam, as shown in Fig. P7-153. The moduli of elasticity for the timber and steel are 1600 ksi and 29,000 ksi, respectively. Determine the maximum tensile flexural stress in each of the materials when a couple $M = -10$ kip · ft is applied to the free end of the beam.

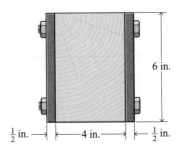

Figure P7-153

Intermediate Problems

7-154* A 50-mm-wide × 80-mm-deep wood ($E = 10$ GPa) beam will be reinforced with 3-mm-thick structural aluminum ($E = 70$ GPa) plates on its top and bottom faces. A maximum bending moment of 3 kN·m must be resisted by the composite beam. If the allowable flexural stresses are 15 MPa in the wood and 135 MPa in the aluminum, determine the minimum width required for the aluminum plates.

7-155 A cantilever beam 6 ft long carries a concentrated load of 4000 lb at the free end. The beam consists of a 4-in.-wide × 10-in.-deep timber section reinforced with 4-in.-wide × 3/4-in.-thick steel plates on the top and bottom surfaces. The moduli of elasticity for the wood and steel are 1600 ksi and 30,000 ksi, respectively. Determine the maximum flexural stresses in the wood and in the steel.

7-156 A 50-mm-wide × 125-mm-deep polymer ($E = 1.40$ GPa) beam will be reinforced with a 6-mm-thick brass ($E = 100$ GPa) plate on its bottom face. If the allowable flexural stresses are 6 MPa in the polymer and 60 MPa in the brass, determine the width of brass plate required to have the allowable stresses in the two materials occur simultaneously.

Challenging Problems

7-157* A timber beam 8 in. wide × 15 in. deep has an 8-in.-wide × 1/2-in.-thick steel plate securely fastened to its bottom face. The beam will be simply supported, will have a span of 16 ft, and will carry a uniformly distributed load over its entire length. The moduli of elasticity for the wood and steel are 1600 ksi and 30,000 ksi, respectively. If the allowable flexural stresses are 1600 psi in the wood and 18,000 psi in the steel, determine the maximum allowable magnitude for the distributed load.

7-158* A 150-mm wide × 300-mm-deep timber beam 5 m long is reinforced with 150-mm-wide × 15-mm-thick steel plates on the top and bottom faces. The beam is simply supported and carries a uniformly distributed load of 20 kN/m over its entire length. The moduli of elasticity for the timber and steel are 13 GPa and 200 GPa, respectively. Determine the maximum tensile flexural stresses in the timber and in the steel.

7-159 A 6-in-wide × 12-in.-deep timber ($E = 1500$ ksi) beam will be reinforced with steel ($E = 30,000$ ksi) plates on its top and bottom faces. The beam will be simply supported, will have a span of 20 ft, and will carry a concentrated load of 5000 lb at the center of the span. If the allowable flexural stresses are 1 ksi in the wood and 10 ksi in the steel, prepare a curve showing the acceptable combinations of plate width and plate thickness. Limit the plate thicknesses to values less than 3/4 in.

7-160 A timber beam 200 mm wide × 350 mm deep has a 200-mm-wide × 16-mm-thick steel plate securely fastened to its bottom face. The moduli of elasticity for the wood and steel

are 12 GPa and 200 GPa, respectively. If the allowable flexural stresses are 10 MPa in the wood and 75 MPa in the steel, determine the maximum load P that can be applied at the center of a simply supported beam having a span of 4 m.

Computer Problems

7-161 A timber beam 8 in. wide × 15 in. deep is to be strengthened by adding 8-in.-wide × t-in.-thick steel plates to its top and bottom faces. The moduli of elasticity for the wood and steel are 1600 ksi and 30,000 ksi, respectively. If the allowable flexural stresses are 2.4 ksi in the wood and 18 ksi in the steel,

a. Determine the maximum moment that can be carried by the beam without the steel plates.
b. Compute and plot the percent increase in moment-carrying capacity of the beam gained by adding the steel plates, for $0 \leq t \leq 2$ in.
c. Compute and plot the maximum stresses in the wood and in the steel when the beam is loaded to capacity for $0 \leq t \leq 2$ in.

7-162 A timber beam 150 mm wide × 300 mm deep is to be strengthened by fastening 50-mm-thick × w-mm-wide aluminum alloy plates to its top and bottom faces. The moduli of elasticity for the wood and aluminum alloy are 13 GPa and 73 GPa, respectively.

a. If a maximum bending moment of 75 kN · m must be resisted by the composite beam, compute and plot the maximum flexural stresses in the wood and in the aluminum for $0 \leq w \leq 150$ mm.
b. If the allowable flexural stresses are 15 MPa in the wood and 135 MPa in the aluminum alloy, determine the minimum width w for the aluminum alloy plates.

7-14 FLEXURAL STRESSES IN REINFORCED CONCRETE BEAMS

Concrete is widely used in beam construction because it is economical, readily available, fireproof, and exhibits a reasonable compressive strength. However, concrete has relatively little tensile strength; therefore, concrete beams must be reinforced with another material, usually steel, that can resist the tensile forces.

The design of reinforced concrete beams is beyond the scope of this book. According to MacGregor,[9] reinforced concrete beams are designed when the structure reaches a limit state (when the structure or an element of the structure becomes unfit for its intended use). The limit state is divided into two groups: one leads to collapse of the structure; the second does not cause collapse of the structure. The second limit state is referred to as the serviceability limit state and is based on a

---

[9]*Reinforced Concrete Mechanics and Design*, 3rd ed., J. G. McGregor, Upper Saddle River, N. J. Prentice Hall, 1997.

linear distribution of strain, as in Section 7-13. Thus, at the service load the beam acts elastically.

The transformed section method of Section 7-13 provides a satisfactory procedure for analyzing reinforced concrete beam problems for the serviceability state. The transformed section used for these problems consists of the actual concrete on the compression side of the neutral axis plus the equivalent amount of hypothetical concrete (which is able to develop tensile stresses) on the tension side of the neutral axis required to replace the steel reinforcing rods. The actual concrete on the tension side of the beam is assumed to crack to the neutral surface; therefore, it has no tensile load-carrying ability and is neglected.

The solution for maximum stresses in a given beam or for the maximum bending moment with given allowable stresses consists of three steps.

1. Locate the neutral axis for the transformed section.
2. Determine the second moment of area of the transformed section with respect to the neutral axis.
3. Use the flexure formula to determine the required stresses or moment.

In the design of a reinforced concrete beam to carry a specified moment with a balanced design (a balanced design means that the allowable stresses in the two materials are reached simultaneously), the following four equations can be written in terms of four unknown properties of the cross section and solved simultaneously.

1. The flexure formula for the allowable stress in the concrete.
2. The flexure formula for the allowable stress in the steel.
3. The moment equation for the location of the neutral axis.
4. The equation for the second moment of area of the cross section with respect to the neutral axis.

The design of a reinforced concrete beam also requires consideration of other factors such as the bond (shearing) stresses between the concrete and reinforcing steel, the diagonal tensile stresses that may be developed, and the amount of concrete that is needed beyond the reinforcing bars. Discussion of such topics can be found in textbooks devoted to reinforced concrete design.

The procedure for analyzing or designing reinforced concrete beams using the transformed section method is illustrated in the following examples.

Example Problem 7-21   A simple reinforced concrete beam carries a uniformly distributed load of 1500 lb/ft on a span of 16 ft. The beam has a rectangular cross section 12 in. wide $\times$ 21 in. deep, and 2 in.^{2} of steel reinforcing rods are placed with their centers 3 in. from the bottom of the beam, as shown in Fig. 7-49a. The moduli of elasticity for the concrete and steel are 2500 ksi and 30,000 ksi, respectively. Determine the maximum flexural stress in the concrete and the average normal stress in the steel.

SOLUTION
The ratio of the moduli of elasticity is

$$ n = \frac{E_s}{E_c} = \frac{30{,}000}{2500} = 12 $$

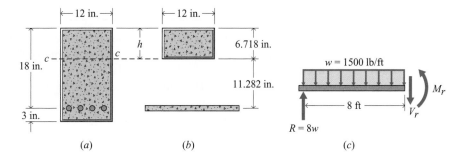

Figure 7-49

which means that $A_{hc} = nA_s = 12(2) = 24$ in.^{2} of hypothetical concrete that resists tension is required for the transformed cross section. The actual and transformed cross sections are shown in Figs. 7-49a and b. The transformed steel area is at the same point in the cross section as the actual steel area. The principle of moments with respect to the centroidal axis for the transformed cross section gives

$$12h(h/2) = 24(18 - h)$$

from which

$$h = 6.718 \text{ in.} \quad \text{and} \quad 18 - h = 11.282 \text{ in.}$$

The second moment of area of the transformed cross section with respect to the neutral axis is

$$I = \frac{1}{12}b_1 h_1^3 + 24(11.282)^2 + \frac{1}{12}(12)(6.718)^3$$
$$+ 12(6.718)(6.718/2)^2$$

where b_1 and h_1 are the base and height, respectively, of the transformed steel. It is usually assumed that the term $b_1 h_1^3/12$ is negligible compared to the other terms in the expression for I. Thus

$$I = 4268 \text{ in.}^4$$

For a simple beam with a uniformly distributed load, the maximum moment occurs at the center of the span and, from the free-body diagram of Fig. 7-49c, is

$$M_r = 8w(8) - 8w(4) = +32w \text{ lb·ft}$$

The flexure formula applied to the transformed cross section gives the maximum stress in the concrete as

$$\sigma_c = -\frac{M_r h}{I} = -\frac{32(1500)(12)(6.718)}{4268} = -906.6 \text{ psi} \cong 907 \text{ psi (C)} \qquad \textbf{Ans.}$$

Since the stresses in the transformed cross section of the beam have a linear distribution, the average stress in the steel is [Eq. (a) of Section 7-13]

$$\sigma_s = \frac{18 - h}{h}(n)\sigma_c = \frac{11.282}{6.718}(12)(906.6) = 18{,}270 \text{ psi (T)} \qquad \textbf{Ans.}$$

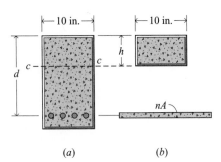

(a)   (b)

Figure 7-50

Example Problem 7-22   Determine the distance from the top of the beam to the center of the steel and the area of steel required for balanced design of a reinforced concrete beam that is to be 10 in. wide and that must resist a bending moment of 600 kip·in. The allowable stresses in the concrete and steel are 1000 psi and 18,000 psi, respectively. The moduli of elasticity for the concrete and steel are 2500 ksi and 30,000 ksi, respectively.

SOLUTION
The ratio of the moduli of elasticity is

$$n = \frac{E_s}{E_c} = \frac{30,000}{2500} = 12$$

The actual and transformed cross sections for the beam are shown in Figs. 7-50a and b. The four unknown quantities are d, h, A, and I for the transformed section where A is the area of the steel. The four available equations involving these four unknowns are as follows:

1. The flexure formula for the maximum stress in the concrete is

$$\sigma_c = -\frac{M_r h}{I} = -\frac{600(10^3)h}{I} = -1000 \text{ psi}$$

2. The flexure formula modified for the average stress in the steel is

$$\sigma_s = \frac{d-h}{h}(n)\sigma_c = \frac{600(10^3)(12)(d-h)}{I} = 18,000 \text{ psi}$$

3. The principle of moments with respect to the neutral axis gives

$$10h(h/2) = 12A(d-h)$$

4. The second moment of area with respect to the neutral axis is

$$I = \frac{1}{3}(10)h^3 + 12A(d-h)^2$$

Simultaneous solution of these four equations yields

$$h = 7.44 \text{ in.} \qquad d = 18.61 \text{ in.} \qquad A = 2.07 \text{ in.}^2 \qquad \textbf{Ans.}$$

PROBLEMS

Introductory Problems

7-163*   A reinforced concrete beam (see Fig. P7-163) is 10 in. wide with a depth to the center of the steel of 18 in. The tension reinforcement consists of three 1-in.-diameter steel bars. The ratio of the modulus of elasticity of the steel to that of the concrete is 12. If the allowable stresses are 1000 psi in the concrete and 18,000 psi in the steel, determine the maximum moment that the beam can support.

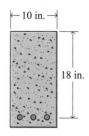

Figure P7-163

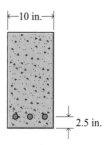

Figure P7-165

7-164* A reinforced concrete beam (see Fig. P7-164) has a 200-mm-wide × 350-mm-deep cross section with four 15-mm-diameter steel bars placed 75 mm from the bottom of the beam. The maximum moment supported by the beam is 15 kN·m. The moduli of elasticity of the concrete and steel are 15 GPa and 200 GPa, respectively. Determine the maximum average tensile stress in the steel and the maximum compressive stress in the concrete at the section of maximum moment.

Intermediate Problems

7-166* A simply supported reinforced concrete beam 200 mm wide with a depth to the center of steel of 300 mm has a span of 4 m. The tension reinforcement consists of three 16-mm-diameter steel bars. The ratio of the moduli of elasticity is 12. If the allowable stresses are 6.5 MPa in the concrete and 120 MPa in the steel, determine the maximum load per meter of length that can be uniformly distributed over the middle half of the beam.

7-167 A simply supported reinforced concrete beam is 8 in. wide and has a span of 12 ft. The tension reinforcement, which is located 16 in. below the top surface of the beam, consists of three 7/8-in.-diameter steel bars. The moduli of elasticity of the concrete and steel are 2400 ksi and 30,000 ksi, respectively. If the allowable stresses are 1000 psi in the concrete and 16,000 psi in the steel, determine the maximum load per foot of length that can be uniformly distributed over the full length of the beam.

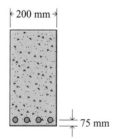

Figure P7-164

Challenging Problems

7-168* A steel reinforced concrete beam of balanced design has a width of 300 mm and a depth to center of steel of 500 mm. Use 16.5 GPa and 198 GPa for the moduli of elasticity of the concrete and steel, respectively. If the allowable stresses are 7 MPa in the concrete and 125 MPa in the steel, determine

a. The required cross-sectional area for the steel rods.
b. The maximum moment that can be resisted by the beam.

7-169 A steel reinforced concrete beam of balanced design is needed to support a uniformly distributed load of 1000 lb/ft on a simply supported span of 16 ft. The width of the beam must be 10 in., and the allowable stresses are 800 psi in the concrete and 16,000 psi in the steel. The moduli of elasticity of the concrete and steel are 2400 ksi and 30,000 ksi, respectively. Determine

a. The required cross-sectional area for the steel rods.
b. The depth from the top surface of the beam to the center of the steel rods.

7-165 A simply supported reinforced concrete beam carries a uniformly distributed load of 820 lb/ft on a span of 13 ft. The beam has a rectangular cross section 10 in. wide × 18 in. deep. The tension reinforcement consists of three 3/4-in.-diameter steel bars placed 2.5 in. from the bottom of the beam, as shown in Fig. P7-165. The moduli of elasticity of the concrete and steel are 2200 ksi and 30,000 ksi, respectively. Determine the average tensile stress in the steel and the maximum compressive stress in the concrete at the section of maximum moment.

7-15 FLEXURAL STRESSES IN CURVED BEAMS

One of the assumptions made in the development of the flexural stress theory in Section 7-2 was that all longitudinal elements of a beam have the same length, thus restricting the theory to initially straight beams of constant cross section. Although considerable deviation from this restriction can be tolerated in real problems, when the initial curvature of the beam becomes significant, the linear variation of strain over the cross section is no longer valid, even though the assumption of the plane cross section remaining plane is valid. A theory will now be developed for a beam, subjected to pure bending, having a constant cross section and a constant or slowly varying initial radius of curvature in the plane of bending. The development will be limited to linearly elastic action.

Figure 7-51a is the elevation of part of such a beam. The xy-plane is the plane of bending and a plane of symmetry. The radius of curvature (distance to the center of curvature) of the neutral surface is R and the radius of curvature to some other surface in the beam (located at a distance y from the neutral surface) is ρ. Since a plane section before bending remains plane after bending, the longitudinal deformation of any element will be proportional to the distance of the element from the neutral surface, as indicated in Fig. 7-51a, from which

$$\delta = \frac{\delta_i}{b}y \qquad (a)$$

The unstrained length of any longitudinal element is $\rho\theta$; therefore, Eq. (a) in terms of strain becomes

$$\rho\theta\epsilon_x = \frac{r_i\theta\epsilon_i}{b}y$$

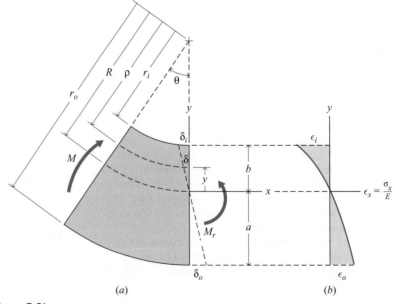

(a) $\qquad\qquad\qquad\qquad\qquad$ (b)

Figure 7-51

from which the expression for the longitudinal strain distribution becomes

$$\epsilon_x = \frac{r_i \epsilon_i}{b} \frac{y}{\rho} = \frac{r_i \epsilon_i}{b} \frac{y}{R - y} \qquad (b)$$

which shows that the strain does not vary linearly with y (Fig. 7-51b), as was the case for the initially straight beam. Since the action is elastic, Hooke's law applies; therefore, $\epsilon_x = \sigma_x / E$ (for $\sigma_y = \sigma_z = 0$), which when substituted into Eq. (b) yields

$$\sigma_x = \frac{r_i \sigma_i}{b} \frac{y}{\rho} = \frac{r_i \sigma_i}{b} \frac{y}{R - y} \qquad (c)$$

which indicates that the flexural stress distribution as well as the strain distribution is not linear with y.

The location of the neutral surface is obtained from the equation $\Sigma F_x = 0$; thus,

$$\int_A \sigma_x \, dA = \frac{r_i \sigma_i}{b} \int_A \frac{y}{\rho} \, dA = 0$$

Since $y = R - \rho$, where R is the radius of the neutral surface,

$$\frac{r_i \sigma_i}{b} \int_A \frac{R - \rho}{\rho} \, dA = 0$$

Since r_i, σ_i, and b are not zero, it follows that

$$\int_A \frac{R - \rho}{\rho} \, dA = 0 \qquad (7\text{-}23)$$

Equation 7-23 may be written

$$\int_A \frac{R - \rho}{\rho} \, dA = R \int_A \frac{dA}{\rho} - \int_A dA = 0$$

or

$$R = \frac{A}{\displaystyle\int_A \frac{dA}{\rho}} \qquad (7\text{-}24)$$

In general, Eq. 7-23 or 7-24 should be solved for R for each specific problem; however, the general solution for a rectangular cross section of width t is easily obtained as follows:

$$\int_{r_i}^{r_o} \frac{R - \rho}{\rho} t \, d\rho = \int_{r_i}^{r_o} t \frac{R d\rho}{\rho} - \int_{r_i}^{r_o} t \, d\rho = 0$$

Since R is a constant,

$$R \ln(r_o / r_i) - (r_o - r_i) = 0$$

from which the radius of the neutral surface is

$$R = \frac{r_o - r_i}{\ln(r_o/r_i)} \qquad (d)$$

The term $\int_A \dfrac{dA}{\rho}$ in Eq. 7-24 is tabulated and shown in Table B-20 for several cross sections.

The resisting moment in terms of the flexural stress is obtained from the equilibrium equation $\Sigma M = 0$; thus,

$$M_r = -\int_A (\sigma_x \, dA)y = -\frac{r_i\sigma_i}{b}\int_A \frac{y^2}{\rho}dA = -\frac{r_i\sigma_i}{b}\int_A \frac{(R-\rho)^2}{\rho}dA \qquad (e)$$

The value of R for a given problem, obtained from Eq. 7-24, can be substituted in Eq. (e) and a solution for M_r thus obtained. However, it will, in general, be found more convenient to write Eq. (e) in the following form:

$$M_r = \frac{r_i\sigma_i}{b}\left[\int_A (R-\rho)dA - R\int_A \frac{R-\rho}{\rho}dA\right]$$

From Eq. 7-23, the second integral in the brackets is zero, and when $R-\rho$ is replaced by y in the first integral, the resisting moment is given by the expression

$$M_r = \frac{r_i\sigma_i}{b}\int_A y\,dA = \frac{r_i\sigma_i}{b}Ay_C \qquad (f)$$

where y_C is the y-coordinate of the centroid of the cross-sectional area A measured from the neutral axis NA. For a positive moment, y_C must always be negative, indicating that the neutral axis of the cross section is always displaced from the centroid toward the center of curvature. Replacing $(r_i\sigma_i)/b$ in Eq. (f) by $(\rho\sigma_x)/y$ from Eq. (c) and solving for σ_x gives

$$\sigma_x = \frac{My}{\rho Ay_C} = \frac{My}{(R-y)Ay_C} \qquad (7\text{-}25)$$

which is the expression for the elastic flexural stress at any point in an initially curved beam.

The preceding development is for pure bending and neglects radial compressive stresses that occur within the material. These compressive stresses are usually very small. If the beam is loaded with forces (instead of couples), additional stresses will occur on the radial planes. Because the action is elastic, the principle of superposition applies and the additional normal stresses can be added to the flexural stresses obtained from Eq. 7-25.

The application of Eqs. 7-24 and 7-25 is illustrated in the following examples.

Example Problem 7-23 A segment of a curved beam (see Fig. 7-52a) of high-strength steel which has a proportional limit of 95 ksi has the trapezoidal cross section shown in Fig. 7-52b. The beam is subjected to a moment M of -100 kip · in.

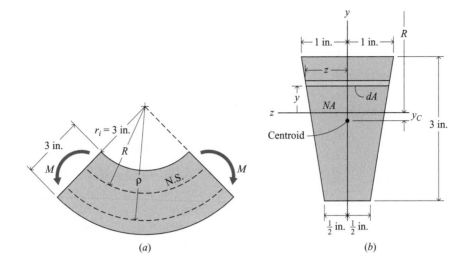

Figure 7-52

(a) Determine the flexural stresses at the top and bottom surfaces.
(b) Sketch the flexural stress distribution in the beam.
(c) Determine the percentages of error if the flexure formula for a straight beam (Eq. 7-8) were used for part (a).

SOLUTION

The first step will be to locate the neutral surface by using Table B-20 of Appendix B. The cross-sectional area is

$$A = \frac{2+1}{2}(6-3) = 4.5 \text{ in.}^2$$

and

$$\int_A \frac{dA}{\rho} = \frac{2(6)-1(3)}{6-3}\ln\frac{6}{3} - 2 + 1 = 1.0794 \text{ in.}$$

The value of R is found using Eq. 7-24:

$$R = \frac{A}{\displaystyle\int_A \frac{dA}{\rho}} = \frac{4.5}{1.0794} = 4.1688 \text{ in.}$$

The radial distance from the center of curvature to the centroid of the cross-sectional area is

$$r_C = \frac{3(4+1)+6(2+2)}{3(2+1)} = 4.3333 \text{ in.}$$

Therefore, $y_C = R - r_C = 4.1688 - 4.3333 = -0.1645$ in.

(a) The required stresses from Eq. 7-25 are

At the bottom of the beam ($y = -1.8312$ in. and $\rho = r_o = 6$ in.):

$$\sigma_{xB} = \frac{M_r y}{\rho A y_C} = \frac{-100(-1.8312)}{6(4.5)(-0.1645)} = -41.23 \text{ ksi} \cong 41.2 \text{ ksi (C)} \qquad \textbf{Ans.}$$

At the top of the beam ($y = +1.688$ in. and $\rho = r_i = 3$ in.):

$$\sigma_{xT} = \frac{M_r y}{\rho A y_C} = \frac{-100(+1.1688)}{3(4.5)(-0.1645)} = +52.63 \text{ ksi} \cong 52.6 \text{ ksi (T)} \qquad \textbf{Ans.}$$

These stresses are well below the proportional limit of the material.

(b) Plotting the two stresses from part (a) and zero stress at the neutral surface will indicate that the curve must be shaped as in Fig. 7-53.

(c) The cross-sectional second moment of area with respect to the centroidal axis parallel to the neutral axis is

$$I = \frac{1}{12}(1)(3)^3 + 1(3)(0.1667)^2 + \frac{1}{36}(1)(3)^3$$
$$+ \frac{1}{2}(1)(3)(0.3333)^2 = 3.25 \text{ in.}^4$$

At the bottom of the beam ($y = -1.6667$ in.):

$$\sigma_{xB} = -\frac{M_r y}{I} = -\frac{-100(-1.6667)}{3.25} = -51.28 \text{ ksi} \cong 51.3 \text{ ksi (C)}$$

At the top of the beam ($y = +1.3333$ in.):

$$\sigma_{xT} = -\frac{M_r y}{I} = -\frac{-100(+1.3333)}{3.25} = +41.02 \text{ ksi} \cong 41.0 \text{ ksi (T)}$$

Therefore, the errors are

$$\frac{51.28 - 41.23}{41.23}(100) = +24.37 \cong 24.4\% \text{ high at bottom}$$

$$\frac{41.02 - 52.63}{52.63}(100) = -22.06 \cong 22.1\% \text{ low at top} \qquad \textbf{Ans.}$$

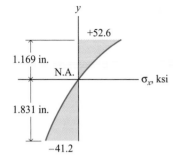

Figure 7-53

Example Problem 7-24 The elevation and cross section of a segment of a punch press frame are shown in Fig. 7-54. The frame is made of a gray cast iron that has a proportional limit in tension of 100 MPa. Determine the maximum tensile and compressive flexural stresses produced by a moment M of -40 kN·m.

SOLUTION

Assume that the neutral axis of the cross section is to the right of the flange and apply Eq. 7-24; thus, using Table B-20 results in

$$A = \sum [b(r_o - r_i)] = 240(60) + 40(240)$$
$$= 24,000 \text{ mm}^2$$

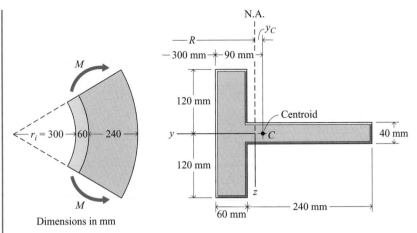

Dimensions in mm

Figure 7-54

and

$$\int_A \frac{dA}{\rho} = \sum \left[b \ln \frac{r_o}{r_i} \right] = 240 \ln \frac{360}{300} + 40 \ln \frac{600}{360}$$
$$= 64.190 \text{ mm}$$

Therefore,

$$R = \frac{A}{\displaystyle\int_A \frac{dA}{\rho}} = \frac{24{,}000}{64.190} = 373.89 \text{ mm}$$

Therefore, the assumption about the location of the neutral axis is correct.

The location of the centroid of the cross section with respect to the center of curvature is found using the principle of moments for areas. Thus,

$$r_C = \frac{60(240)(330) + 40(240)(480)}{60(240) + 40(240)} = 390 \text{ mm}$$

The distance from the neutral axis to the centroid is

$$y_C = R - r_C = 373.89 - 390 = -16.11 \text{ mm}$$

From Eq. 7-25 the stresses are:

At the outside of the stem ($y = -226.11$ mm and $\rho = 600$ mm):

$$\sigma_{xo} = \frac{M_r y}{\rho A y_C} = \frac{-40(10^3)(-226.11)(10^{-3})}{600(10^{-3})(24{,}000)(10^{-6})(-16.11)(10^{-3})}$$
$$= -38.99(10^6) \text{ N/m}^2 \cong 39.0 \text{ MPa (C)} \qquad\qquad \textbf{Ans.}$$

At the inside of the flange ($y = +73.89$ mm and $\rho = 300$ mm):

$$\sigma_{xi} = \frac{M_r y}{\rho A y_C} = \frac{-40(10^3)(+73.89)(10^{-3})}{300(10^{-3})(24{,}000)(10^{-6})(-16.11)(10^{-3})}$$

$$= 25.48(10^6) \text{ N/m}^2 \cong 25.5 \text{ MPa (T)} \qquad \textbf{Ans.}$$

The low stresses indicate elastic action.

PROBLEMS

Introductory Problems

7-170 A curved rectangular beam with a width t, a depth d (in the radial direction), and an inside radius r_i of $10d$ is subjected to a moment M in the plane of curvature. Show that the error in computing the maximum flexural stress by Eq. 7-8 (the flexure formula) instead of Eq. 7-25 (the curved beam formula) is approximately 3.7 percent low.

7-171* A curved beam having an inside radius of 6 in. has the cross section shown in Fig. P7-171. The beam is subjected to a moment M in the plane of curvature. Determine the dimension b needed to make the flexural stress at the inside curved surface equal in magnitude to the flexural stress at the outside curved surface of the beam.

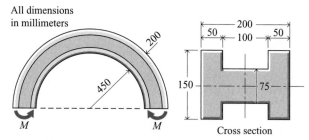

All dimensions in millimeters

Figure P7-172

7-173* Determine the maximum tensile and compressive flexural stresses in the curved beam of Fig. P7-173 if the magnitude of the moment M shown in the figure is 30 kip · ft.

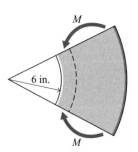

Figure P7-171

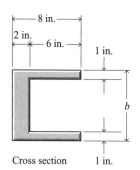

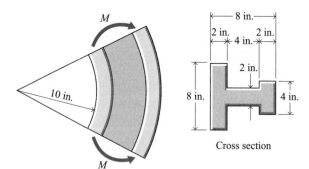

Figure P7-173

Intermediate Problems

7-172 Determine the maximum tensile and compressive flexural stresses in the curved beam of Fig. P7-172 if the magnitude of the moment M shown in the figure is 20 kN · m.

Challenging Problems

7-174* The curved beam of Fig. P7-174 is subjected to a bending moment M in the plane of curvature, as shown in the figure. Determine the maximum permissible magnitude for M if the flexural stress is not to exceed 35 MPa (T) or 140 MPa (C).

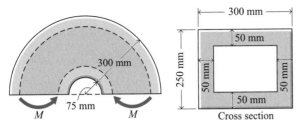

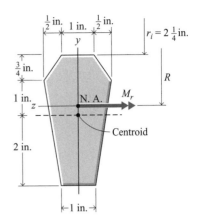

Figure P7-174

Cross section

7-175 The cross section of a segment of a crane hook is shown in Fig. P7-175. The curvature and loading are in the xy-plane. Determine the flexural stresses at the inside curved surface and at the outside curved surface if the magnitude of the resisting moment M_r shown on the cross section is 70 kip · in.

Figure P7-175

7-16 COMBINED LOADING: AXIAL, PRESSURE, FLEXURAL, AND TORSIONAL

The stresses and strains produced by the fundamental types of loads (axial, pressure of thin-walled vessels, torsional, and flexural) have now been analyzed. Many machine and structural elements are subjected to a combination of any two or all four of these types of loads, and a procedure for calculating the normal and shearing stresses resulting from such loads at a point on a given plane is required. The procedure used to solve such problems is the same as that previously developed for axial and torsional loads.

First, the member is sectioned to expose a cross section normal to the axis of the member, and a free-body diagram of the part of the member to one side of the section is drawn. Then, the internal forces at the section are found using the equations of equilibrium. The internal forces may be a combination of an axial force, a shear force, a torque, or a bending moment. Using the formulas previously developed, the normal and shear stresses are then calculated separately at the point of interest on the cross section for each of the internal forces. Normal stresses due to the axial force, the internal pressure, and the bending moment are then added (or subtracted) to obtain the normal stress at the point. Similarly, the shearing stresses at the point due to the shear force and torque are added (or subtracted) to obtain the shearing stress at the point. Once the normal and shearing stresses at a point are known, the stress transformation equations (Eqs. 2-12 and 2-13) can be used to determine normal and shearing stresses on other planes through the point. Principal stresses and maximum shearing stresses at the point and the planes on which they act can be determined by using Mohr's circle or Eqs. 2-14 through 2-18.

The combined normal and shearing stresses at a point on a cross section can be determined using the above procedure (method of superposition), provided the combined stresses do not exceed the proportional limit of the material. The following example problems illustrate the procedure.

7.21

Example Problem 7-25 The solid 100-mm-diameter shaft shown in Fig. 7-55a is subjected to an axial compressive force $P = 200$ kN and a vertical force $V = 100$ kN. For point A on the outside surface of the shaft, determine

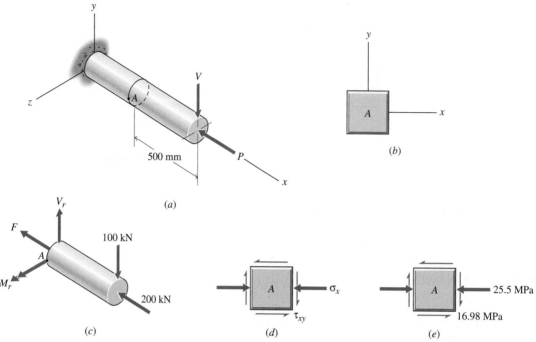

Figure 7-55

(a) The x- and y-components of stress.

(b) The principal stresses and the maximum shearing stress at the point.

SOLUTION

(a) Since point A is on the outside surface (a free surface) of the shaft, a state of plane stress exists at the point. The coordinate system is selected as shown in Fig. 7-55b. This is the same coordinate system for which the equations in this chapter were developed. Passing a transverse plane through point A and isolating the segment of the shaft to the right of point A results in the free-body diagram shown in Fig. 7-55c. The internal forces on the transverse cross section are an axial compressive force $F = 200$ kN, a shear force $V_r = 100$ kN, and a moment $M_r = 50$ kN · m. The directions of the stresses on the left face of the element are in accordance with the directions of the forces that produce the stresses, that is, in the directions of the internal forces. These stresses are shown in Fig. 7-55d. The magnitudes of the stresses are determined as follows.

$$\sigma_x = \frac{F}{A} = \frac{200(10^3)}{(\pi/4)(0.100)^2} = 25.46(10^6)\,\text{N/m}^2 = 25.46\,\text{MPa}$$

$$\tau = \frac{V_r Q}{It} = \frac{V_r(y_C A)}{It} = \frac{V_r\left[\left(\dfrac{4r}{3\pi}\right)\left(\dfrac{\pi r^2}{2}\right)\right]}{\left(\dfrac{\pi r^4}{4}\right)(2r)} = \frac{4V_r}{3\pi r^2}$$

$$= \frac{4(100)(10^3)}{3\pi(0.050)^2} = 16.977(10^6)\,\text{N/m}^2 = 16.977\,\text{MPa}$$

There is no normal stress due to M_r since point A is on the neutral axis.

The x- and y-components of stress on the element at A are

$$\sigma_x = 25.46 \text{ MPa} \cong 25.5 \text{ MPa (C)} \qquad \textbf{Ans.}$$

$$\tau_{xy} = -16.977 \text{ MPa} \cong -16.98 \text{ MPa} \qquad \textbf{Ans.}$$

There is no normal stress in the y-direction (the circumferential direction), since there is no force to cause such a stress. The shear stresses on the four sides of the element have the same magnitude. The shear stresses that meet at a corner either both point toward the corner or both point away from the corner. These stresses are shown on the element of Fig. 7-55e.

(b) Equation (2-15) is used to calculate the principal stresses. The stresses for use in this equation are

$$\sigma_x = -25.46 \text{ MPa} \qquad \sigma_y = 0 \qquad \tau_{xy} = -16.977 \text{ MPa}$$

Substituting these stress components into Eq. 2-15 yields

$$\sigma_{p1,p2} = \frac{\sigma_x + \sigma_y}{2} \pm \sqrt{\left(\frac{\sigma_x - \sigma_y}{2}\right)^2 + \tau_{xy}^2}$$

$$= \frac{-25.46 + 0}{2} \pm \sqrt{\left(\frac{-25.46 - 0}{2}\right)^2 + (-16.977)^2}$$

$$= -12.73 \pm 21.22$$

Thus, the principal stresses are

$$\sigma_{p1} = -12.73 + 21.22 = +8.490 \text{ MPa} \cong 8.49 \text{ MPa (T)} \qquad \textbf{Ans.}$$

$$\sigma_{p2} = -12.73 - 21.22 = -33.95 \text{ MPa} \cong 34.0 \text{ MPa (C)} \qquad \textbf{Ans.}$$

$$\sigma_{p3} = \sigma_z = 0 \qquad \textbf{Ans.}$$

Since the two in-plane principal stresses are of opposite sign and $\sigma_{p3} = \sigma_z = 0$,

$$\tau_{max} = \frac{\sigma_{max} - \sigma_{min}}{2}$$

$$= \frac{8.490 - (-33.95)}{2}$$

$$= 21.2 \text{ MPa} \qquad \textbf{Ans.}$$

Example Problem 7-26   The cast iron frame of a small press is shaped as shown in Fig. 7-56a. The cross section a–a of the frame is shown in Fig. 7-56b. Axis c–c passes through the centroid of the cross section. For a load Q of 16 kip, and assuming linearly elastic action, determine

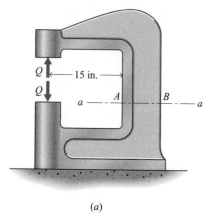

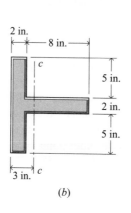

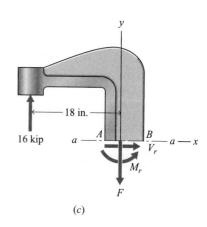

(a)   (b)   (c)

Figure 7-56(a–c)

(a) The normal stress distribution on section a–a.

(b) The location of the neutral axis (line of zero stress).

(c) The principal stresses for the critical points on section a–a.

SOLUTION

(a) The frame is sectioned at a–a, and a free-body diagram for the part of the frame above the section is shown in Fig. 7-56c. Since the load acts in a plane of symmetry, there are three independent equations of equilibrium. The internal forces at the section are found using the equations of equilibrium as follows:

$$+ \rightarrow \Sigma F_x = 0: \qquad V_r = 0 \qquad\qquad V_r = 0$$

$$+ \uparrow \Sigma F_y = 0: \qquad 16 - F = 0 \qquad\quad F = 16 \text{ kip}$$

$$+\downarrow\Sigma M_{c-c} = 0: \qquad M_r - 16(18) = 0 \quad M_r = 288 \text{ kip} \cdot \text{in.}$$

Thus, there are two internal forces on the section, an axial force F which produces a constant normal stress $\sigma_{y1} = F/A$ over the section, and a bending moment M_r, which produces a linear variation of normal stress $\sigma_{y2} = -M_r x/I$ over the section. The cross-sectional area A and the second moment I of the cross-sectional area with respect to the centroidal axis c–c are

$$A = 12(2) + 8(2) = 40 \text{ in.}^2$$

$$I = \frac{1}{12}(12)(2)^3 + 12(2)(2)^2 + \frac{1}{12}(2)(8)^3 + 2(8)(3)^2 = 333.3 \text{ in.}^4$$

The stresses σ_{y1} and σ_{y2} due to the internal forces are

$$\sigma_{y1} = \frac{F}{A} = \frac{16{,}000}{40} = +400 \text{ psi} = 400 \text{ psi (T)}$$

which is constant over the cross-sectional area. The maximum tensile flexural stress occurs at the left edge of section a–a (point A) and is

$$\sigma_{y2A} = -\frac{M_r x_A}{I} = -\frac{288(10^3)(-3)}{333.3} = +2592 \text{ psi} = 2592 \text{ psi (T)}$$

The maximum compressive flexural stress occurs at the right edge of section a–a (point B) and is

$$\sigma_{y2B} = -\frac{M_r x_B}{I} = -\frac{288(10^3)(+7)}{333.3} = -6049 \text{ psi} = 6049 \text{ psi (C)}$$

The distributions of stresses σ_{y1} and σ_{y2} are shown in Figs. 7-56d and e, respectively. Superimposing the normal stresses at points A and B gives

$$\sigma_{yA} = \sigma_{y1A} + \sigma_{y2A} = 400 + 2592 = +2992 \text{ psi} \cong 2990 \text{ psi (T)}$$
$$\sigma_{yB} = \sigma_{y1B} + \sigma_{y2B} = 400 - 6049 = -5649 \text{ psi} \cong 5650 \text{ psi (C)}$$

The distribution of normal stress on section a–a is shown in Fig. 7-56f.

(b) The location of the neutral axis (line of zero stress) can be determined as the place where the compressive flexural stress is 400 psi because this stress will just balance the axial tensile stress of 400 psi. Thus,

$$\sigma_y = -\frac{M_r x}{I} = -\frac{288(10^3)(x)}{333.3} = -400 \text{ psi}$$

from which

$$x = \frac{333.3(400)}{288(10^3)} = 0.4629 \text{ in.} \cong 0.463 \text{ in.} \qquad \textbf{Ans.}$$

▶ The normal stress due to bending varies from zero at the centroidal axis (c–c on Fig. 7-56b) to a maximum at the edges of the "beam." The neutral axis (the line of zero stress) lies 0.463 in. to the right of the centroidal axis.

to the right of the centroidal axis of the cross section.

(c) With no shearing stresses on section a–a, the normal stresses are principal stresses, and the critical points, as observed from the stress distribution of Fig. 7-56f, are the left and right edges of the section.
The principal stresses for the right edge are

$$\sigma_p = 5650 \text{ psi (C)} \qquad \text{and} \qquad 0 \qquad \textbf{Ans.}$$

The principal stresses for the left edge are

$$\sigma_p = 2990 \text{ psi (T)} \qquad \text{and} \qquad 0 \qquad \textbf{Ans.}$$

The principal stresses are shown on elements A and B of Fig. 7-56g.

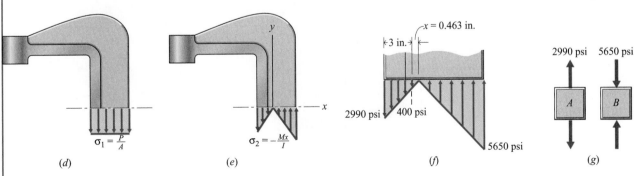

(d) (e) (f) (g)

Figure 7-56(d–g)

Example Problem 7-27   A gray cast iron compression member is subjected to a vertical load $Q = 100$ kN, as shown in Fig. 7-57a. Determine the normal stresses at each corner of section $ABCD$.

SOLUTION
Since stresses are required at points A, B, C, and D, the member is sectioned through these points, and a free-body diagram of the portion of the member above the section is drawn, as shown in Fig. 7-57b. In general, the internal forces acting on a section of a member consist of shear forces V_x and V_z, an axial force F, a torque $T_r = M_y$, and bending moments M_x and M_z. The six equations of equilibrium used to determine these internal forces are

$$\Sigma F_x = 0: \qquad V_x = 0 \qquad\qquad\qquad V_x = 0$$
$$\Sigma F_y = 0: \qquad F - 100 = 0 \qquad\qquad F = -100 \text{ kN} = 100 \text{ kN (C)}$$
$$\Sigma F_z = 0: \qquad V_z = 0 \qquad\qquad\qquad V_z = 0$$
$$\Sigma M_x = 0: \qquad M_x - 100(0.060) = 0 \qquad M_x = 6 \text{ kN} \cdot \text{m}$$
$$\Sigma M_y = 0: \qquad T_r = 0 \qquad\qquad\qquad T_r = 0$$
$$\Sigma M_z = 0: \qquad M_z - 100(0.075) = 0 \qquad M_z = 7.5 \text{ kN} \cdot \text{m}$$

Thus, there are three internal forces on the section: (1) an axial force F which produces a constant normal stress $\sigma_1 = F/A$ over the section; and (2) two bending moments M_x and M_z each of which produces a linear variation of normal stress $\sigma = Mc/I$ over the section. The cross-sectional area A and the second moments I_x and I_z of the area with respect to the centroidal x and z axes are

$$A = 150(120) = 18.00(10^3) \text{ mm}^2 = 18.00(10^{-3}) \text{ m}^2$$

$$I_x = \frac{1}{12}(150)(120)^3 = 21.60(10^6) \text{ mm}^4 = 21.60(10^{-6}) \text{ m}^4$$

$$I_z = \frac{1}{12}(120)(150)^3 = 33.75(10^6) \text{ mm}^4 = 33.75(10^{-6}) \text{ m}^4$$

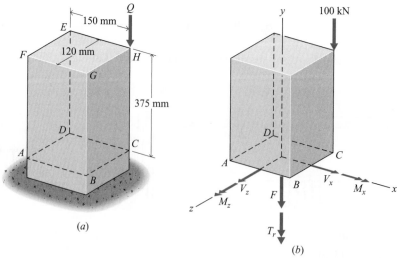

(a)

(b)

Figure 7-57(a–b)

The axial force F produces a uniformly distributed compressive stress on plane $ABCD$. The magnitude of the stress is

$$\sigma_1 = \frac{F}{A} = \frac{100(10^3)}{18.00(10^{-3})} = 5.556(10^6)\,\text{N/m}^2 = 5.556\,\text{MPa}$$

This stress is shown in Fig. 7-57c.

The bending moment M_x produces a normal stress that varies linearly with respect to z and is zero on the neutral axis (the x-axis). Along line AB, the normal stress is tensile; and, along line CD, the normal stress is compressive. The magnitude of these stresses is

$$\sigma_2 = \frac{M_x c_2}{I_x} = \frac{6(10^3)(60)(10^{-3})}{21.60(10^{-6})} = 16.667(10^6)\,\text{N/m}^2 = 16.667\,\text{MPa}$$

These stresses are shown in Fig. 7-57d.

Similarly, bending moment M_z produces a compressive stress along edge BC of plane $ABCD$ and an equal tensile stress along edge AD. The magnitude of these stresses is

$$\sigma_3 = \frac{M_z c_3}{I_z} = \frac{7.5(10^3)(75)(10^{-3})}{33.75(10^{-6})} = 16.667(10^6)\,\text{N/m}^2 = 16.667\,\text{MPa}$$

These stresses are shown in Fig. 7-57e.

The normal stresses at the corners of section $ABCD$ are found by superimposing the component stresses shown in Figs. 7-57c, d, and e. The maximum compressive stress occurs at C and is

$$\sigma_{yC} = -5.556 - 16.667 - 16.667 = -38.89\,\text{MPa} \cong 38.9\,\text{MPa (C)} \qquad \textbf{Ans.}$$

The maximum tensile stress occurs at A and is

$$\sigma_{yA} = -5.556 + 16.667 + 16.667 = +27.78\,\text{MPa} \cong 27.8\,\text{MPa (T)} \qquad \textbf{Ans.}$$

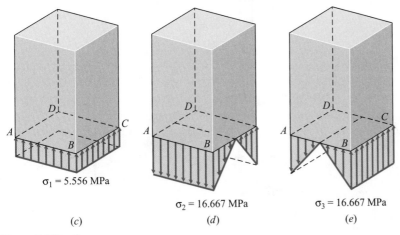

$\sigma_1 = 5.556\,\text{MPa}$

$\sigma_2 = 16.667\,\text{MPa}$

$\sigma_3 = 16.667\,\text{MPa}$

(c)

(d)

(e)

Figure 7-57(c–e)

The stresses at the remaining two corners are

$$\sigma_{yB} = -5.556 + 16.667 - 16.667 = -5.556 \text{ MPa} \cong 5.56 \text{ MPa (C)} \qquad \textbf{Ans.}$$

$$\sigma_{yD} = -5.556 - 16.667 + 16.667 = -5.556 \text{ MPa} \cong 5.56 \text{ MPa (C)} \qquad \textbf{Ans.}$$

Example Problem 7-28   A 100-mm-diameter shaft is loaded and supported, as shown in Fig. 7-58a. Determine

(a) The normal and shearing stresses at points A, B, C, and D on a section at the wall. Neglect stress concentrations.

(b) The principal stresses and maximum shearing stresses at points A, B, C, and D of section $ABCD$.

SOLUTION

A free-body diagram of the part of the shaft to the left of section $ABCD$ is shown in Fig. 7-58b. The internal forces acting on section $ABCD$ are shear forces V_y and V_z, an axial force F, a torque $T_r = M_x$, and bending moments M_y and M_z. The six equations of equilibrium used to determine these internal forces are

$$
\begin{array}{lll}
\Sigma F_x = 0: & F - 150 = 0 & F = 150 \text{ kN} \\
\Sigma F_y = 0: & V_y - 5 = 0 & V_y = 5 \text{ kN} \\
\Sigma F_z = 0: & V_z = 0 & V_z = 0 \\
\Sigma M_x = 0: & T_r + 5(0.600) = 0 & T_r = -3 \text{ kN} \cdot \text{m} \\
\Sigma M_y = 0: & M_y = 0 & M_y = 0 \\
\Sigma M_z = 0: & M_z + 5(0.750) = 0 & M_z = -3.75 \text{ kN} \cdot \text{m}
\end{array}
$$

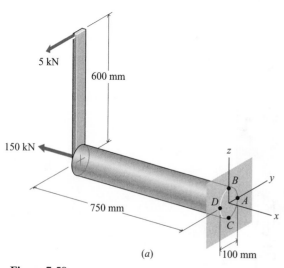

(a)

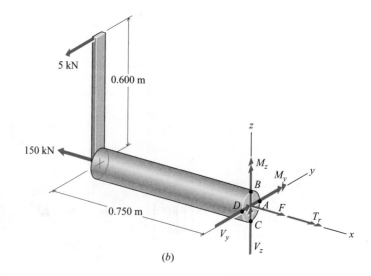

(b)

Figure 7-58

Thus, there are four internal forces on the section: (1) an axial force F that produces a constant tensile stress over the section, (2) a shear force V_y that produces shearing stresses at points B and C of the section but zero shearing stresses at points A and D, (3) a torque T_r that produces the same shearing stress at all surface points of the shaft, and (4) a bending moment M_z that produces a tensile stress at point A and an equal compressive stress at point D. The cross-sectional area A, the first moment Q, the second moment I, and the polar second moment J are

$$A = \frac{\pi}{4}(d)^2 = \frac{\pi}{4}(100)^2 = 7854 \text{ mm}^2 = 7854(10^{-6}) \text{ m}^2$$

$$Q = \frac{\pi}{2}(r)^2\left(\frac{4r}{3\pi}\right) = \frac{2}{3}(r)^3 = \frac{2}{3}(50)^3 = 83.33(10^3) \text{ mm}^3 = 83.33(10^{-6}) \text{ m}^3$$

$$I = \frac{\pi}{4}(r)^4 = \frac{\pi}{4}(50)^4 = 4.909(10^6) \text{ mm}^4 = 4.909(10^{-6}) \text{ m}^4$$

$$J = \frac{\pi}{2}(d)^4 = \frac{\pi}{2}(50)^4 = 9.817(10^6) \text{ mm}^4 = 9.817(10^{-6}) \text{ m}^4$$

(a) The magnitude of the normal stress produced by axial force F is

$$\sigma_1 = \frac{F}{A} = \frac{150(10^3)}{7854(10^{-6})} = 19.099(10^6) \text{ N/m}^2 = 19.099 \text{ MPa}$$

The magnitude of the normal stress produced by moment M_z is

$$\sigma_2 = \frac{M_z c}{I} = \frac{3.75(10^3)(50)(10^{-3})}{4.909(10^{-6})} = 38.195(10^6) \text{ N/m}^2 = 38.195 \text{ MPa}$$

The magnitude of the shearing stress produced by torque T_r is

$$\tau_1 = \frac{T_r c}{J} = \frac{3(10^3)(50)(10^{-3})}{9.817(10^{-6})} = 15.280(10^6) \text{ N/m}^2 = 15.280 \text{ MPa}$$

The magnitude of the shearing stress produced by shear force V_y is

$$\tau_2 = \frac{V_y Q}{It} = \frac{5(10^3)(83.33)(10^{-6})}{4.909(10^{-6})(100)(10^{-3})} = 0.8487(10^6) \text{ N/m}^2 = 0.8487 \text{ MPa}$$

These component stresses are shown in Fig. 7-59 for points A, B, C, and D. The state of stress at each point is plane stress. Therefore, the superimposed

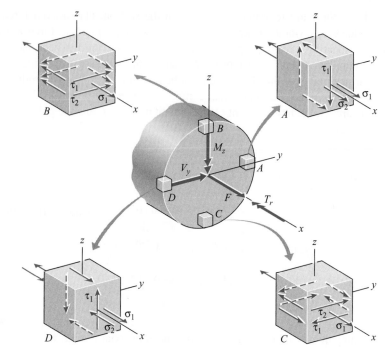

Figure 7-59

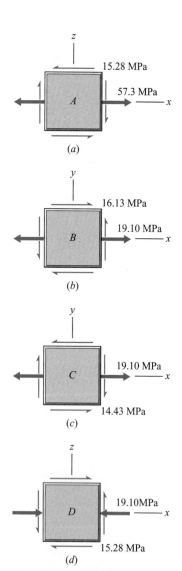

states of stress for points A, B, C, and D can be represented on two-dimensional stress elements, as shown in Figs. 7-60a, b, c, and d, respectively. **Ans.**

(b) The principal stresses at each of the points is obtained by using Eq. 2-15. Thus, for point A shown in Fig. 7-60a,

$$\sigma_{p1,p2} = \frac{\sigma_x + \sigma_z}{2} \pm \sqrt{\left(\frac{\sigma_x - \sigma_z}{2}\right)^2 + \tau_{xz}^2}$$

$$= \frac{57.3 + 0}{2} \pm \sqrt{\left(\frac{57.3 - 0}{2}\right)^2 + (-15.28)^2}$$

$$= 28.65 \pm 32.47$$

$$\sigma_{p1} = 28.65 + 32.47 = +61.12 \text{ MPa} \cong 61.1 \text{ MPa (T)}$$

$$\sigma_{p2} = 28.65 - 32.47 = -3.820 \text{ MPa} \cong 3.82 \text{ MPa (C)}$$

$$\sigma_{p3} = \sigma_y = 0$$

Since σ_{p1} and σ_{p2} have opposite signs, the maximum shearing stress is given by Eq. 2-18 as

$$\tau_{max} = \frac{\sigma_{max} - \sigma_{min}}{2} = \frac{61.12 - (-3.820)}{2} = 32.47 \text{ MPa} \cong 32.5 \text{ MPa}$$

Proceeding in a similar fashion for the remaining points yields the principal stresses and the maximum shearing stress as

Point	σ_{p1}	σ_{p2}	σ_{p3}	τ_{max}	
A	61.6 MPa (T)	3.82 MPa (C)	0	32.5 MPa	**Ans.**
B	28.3 MPa (T)	9.20 MPa (C)	0	18.75 MPa	**Ans.**
C	26.9 MPa (T)	7.75 MPa (C)	0	17.30 MPa	**Ans.**
D	8.47 MPa (T)	27.6 MPa (C)	0	18.02 MPa	**Ans.**

This Example Problem illustrates that it may be necessary to determine stresses at a number of points in order to locate the most severely stressed point.

 PROBLEMS

MecMovie Activities and Problems

MM7.20 C-clamp normal stresses. Example; Try one. Combined normal stresses due to axial force and bending moment.

MM7.21 Precast concrete beam and corbel. Example; Try one. Combined normal stresses due to axial force and bending moment.

MM7.22 The tree–combined axial and bending. Concept checkpoints. Determine axial force, bending moment, moment of inertia, axial stress, bending stress, and combined stress values for a simple tree structure.

MM7.23 Jib crane boom. Example. Try one. Combined normal stresses due to axial force and bending moment.

MM7.24 Beams bending about two axes. Two examples; Concept checkpoints. Combined normal stresses due to bending moments about two axes plus transverse shear stress.

Introductory Problems

7-176* A hollow shaft 1 m long with an outside diameter of 400 mm and an inside diameter of 300 mm is subjected to both a horizontal load $V = 500$ kN and an axial tensile load $P = 1500$ kN, as shown in Fig. P7-176. Determine the principal stresses and the maximum shearing stress at point A, which is on the outside surface of the shaft next to the wall.

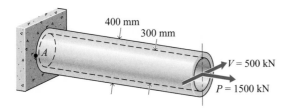

Figure P7-176

7-177* The T-section shown in Fig. P7-177 is used as a short post to support a compressive load $P = 150$ kip. The load is applied on the centerline of the stem at a distance $e = 2$ in. from the centroid of the cross section. Determine the normal stresses at points C and D on section AB.

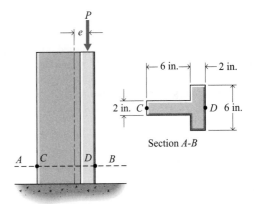

Figure P7-177

7-178 A human femur is modeled as shown in Fig. P7-178. The abductor muscle force is $M = 4060$ N, and the femoral load is $J = 5210$ N.

a. Determine the maximum tensile and compressive stresses at section a–a if the section is modeled as a solid circular section 27 mm in diameter.

b. Determine the maximum tensile and compressive stresses at section a–a if the section is modeled as a hollow cylinder of outside diameter 27 mm and inside diameter 16 mm.

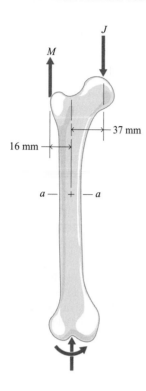

Figure P7-178

7-179* A 4-in.-diameter shaft is subjected to a torque of 30 kip · in., an axial tensile load of 50 kip, and a vertical load of 5 kip, as shown in Fig. P7-179. Determine the principal stresses and the maximum shearing stress at point A on the surface of the shaft. The transverse section through A is 24 in. from the right end of the shaft.

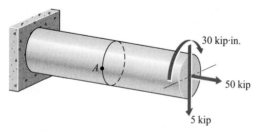

Figure P7-179

7-180 The cross section of the straight vertical portion of the coil-loading hook shown in Fig. P7-180a is shown in Fig. P7-180b. The horizontal distance from the line of action of the applied load to the inside face CD of the cross section is 600 mm. Determine the maximum tensile and compressive stresses on section $CDEF$ for a 40 kN load.

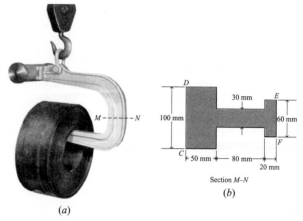

(a)

Section M–N

(b)

Figure P7-180

7-181 A structural member with a rectangular cross section supports a 15-kip load, as shown in Fig. P7-181. Determine the distribution of normal stress on section AB of the member.

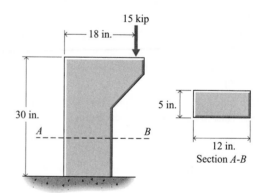

Figure P7-181

7-182* The cross section of the steel member shown in Fig. P7-182 is a rectangle 100 mm wide × 150 mm deep. Determine the maximum normal stress on a vertical section at the wall. Neglect stress concentrations.

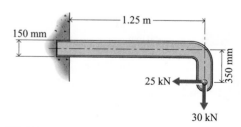

Figure P7-182

7-183* The estimated maximum total force P to be exerted on the C-clamp shown in Fig. P7-183 is 450 lb. If the normal stress on section A–A is not to exceed 16,000 psi, determine the minimum allowable value for the dimension h of the cross section.

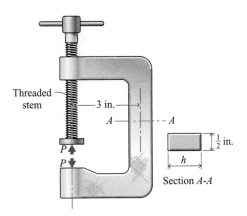

Section A-A

Figure P7-183

7-184 Determine the magnitudes of the maximum tensile and compressive normal stresses on the transverse plane B–B in the straight portion of the structure shown in Fig. P7-184. The member is braced perpendicular to the plane of symmetry.

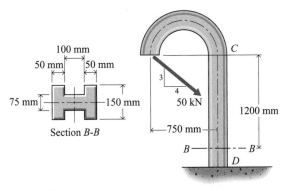

Figure P7-184

7-185 The clamp of Fig. P7-185 is used to hold two boards. If the clamping force is 80 lb, determine the maximum tensile and compressive stresses at section a–a. The clamp has a 1/2 × 3/16-in. rectangular cross section.

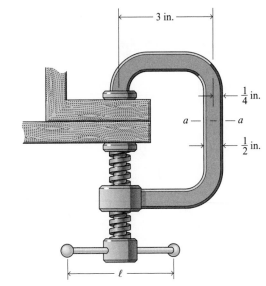

Figure P7-185

Intermediate Problems

7-186* A 150-mm-diameter shaft will be used to support the axial load, the torsional loads, and the shear load shown in Fig. P7-186. Determine the principal stresses and the maximum shearing stress at point A on the top surface of the shaft.

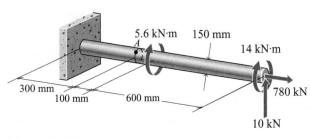

Figure P7-186

7-187* A 30-lb force P is applied to the brake pedal of an automobile as shown in Fig P7-187. Force Q is applied to the brake cylinder. Determine the maximum tensile and compressive normal stresses on section a–a, which is midway between points A and B. Section a–a may be modeled as a 3/16 × 1-in. rectangle.

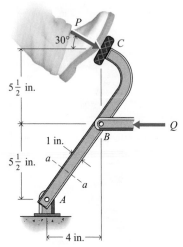

Figure P7-187

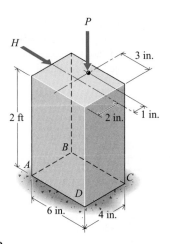

Figure P7-189

7-188   An automobile engine with a mass of 360 kg is supported by an engine hoist, as shown in Fig. P7-188. Determine the maximum tensile and compressive normal stresses on section a–a if member ABC is a hollow square 100×100-mm section with a wall thickness of 20 mm.

7-190   A cylindrical pressure vessel is 6 m long and is simply supported at the ends, as shown in Fig. P7-190. The inside diameter of the vessel is 1200 mm, and the wall thickness is 4 mm. The vessel and contents weigh 10 kN/m, and the contents exert a uniform internal pressure of 200 kPa. Determine the principal stresses and maximum shear stress at points A and B on the outside surface of the vessel.

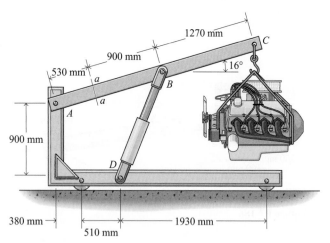

Figure P7-188

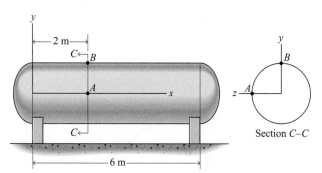

Figure P7-190

7-189*   A short post supports a vertical force $P = 9600$ lb and a horizontal force $H = 800$ lb, as shown in Fig. P7-189. Determine the vertical normal stresses at corners A, B, C, and D of the post. Neglect stress concentrations.

7-191   The output from a strain gage located on the bottom surface of the hat section shown in Fig. P7-191 will be used to indicate the magnitude of the load P applied to the section. The hat section is made of aluminum alloy ($E = 10{,}600$ ksi and $v = 1/3$) and is 1 in. wide. When the maximum load $P = 110$ lb is applied to the section, the strain gage should read $\epsilon = +1000$ μm/m. Plot a curve showing the combinations of thickness t and height h that will satisfy the specification. Limit the range of h from 0 to 2 in.

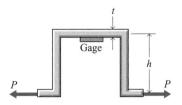

Figure P7-191

7-192* Determine the normal stresses on a transverse section at points A, B, C, and D of the rectangular post shown in Fig. P7-192. Neglect stress concentrations.

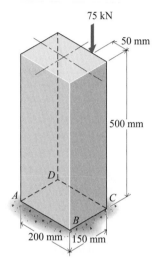

Figure P7-192

7-193 A short post supports a vertical force $P = 4000$ lb, a horizontal force $H = 500$ lb, and a horizontal force $Q = 400$ lb as shown in Fig. P7-193. Determine the normal stresses on a transverse section at corners A, B, C, and D of the post. Neglect stress concentrations.

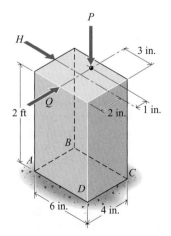

Figure P7-193

7-194* Determine the normal stresses on a transverse section at points A, B, C, and D of the rectangular post shown in Fig. P7-194. Neglect stress concentrations. Also, locate the neutral axis for the section.

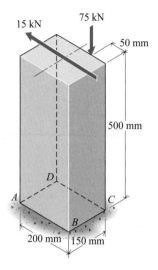

Figure P7-194

7-195 The homogeneous sign shown in Fig. P7-195 weighs 55 lb; the weight of other members is negligible. The cross section of member AB is a rectangle 1 in. deep and 2 in. wide. Determine the principal stresses and maximum shear stress

a. At the top of member AB at section a–a, which is midway between the points of attachment of the sign.
b. At the bottom of member AB at section a–a.
c. Halfway between the top and bottom of member AB at section a–a.

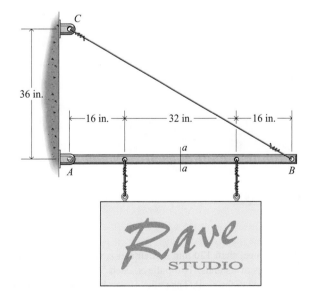

Figure P7-195

Challenging Problems

7-196* Four strain gages are mounted at 90° intervals around the circumference of a 100-mm-diameter steel ($E = 210$ GPa and $v = 0.30$) bar, as shown in Fig. P7-196. As a result of axial and flexural loading, the four gages indicate longitudinal strains of $\epsilon_1 = -200$ μm/m, $\epsilon_2 = +820$ μm/m, $\epsilon_3 = +600$ μm/m, and $\epsilon_4 = -420$ μm/m. Determine the axial load P and the two moments M_y and M_z.

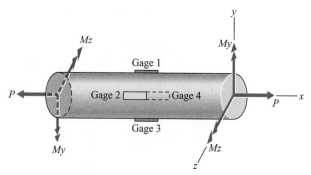

Figure P7-196

7-197* Determine the principal stresses and the maximum shearing stress at points on the top and bottom of section a–a of the pipe system shown in Fig. P7-197. The pipe has an outside diameter of 1 in. and a wall thickness of 1/8 in.

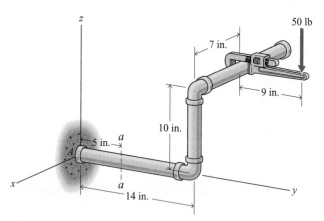

Figure P7-197

7-198 A steel shaft 120 mm in diameter is supported in flexible bearings at its ends. Two pulleys, each 500 mm in diameter, are keyed to the shaft. The pulleys carry belts that produce the forces shown in Fig. P7-198. Determine the principal stresses

and the maximum shearing stress at point A on the top surface of the shaft.

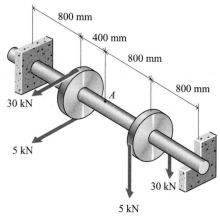

Figure P7-198

7-199* Three strain gages are mounted on a 1-in.-diameter aluminum alloy ($E = 10,000$ ksi and $v = 1/3$) rod, as shown in Fig. P7-199. When loads P and Q are applied to the rod, they produce longitudinal strains $\epsilon_A = +550$ μin./in., $\epsilon_B = +400$ μin./in., and $\epsilon_c = -300$ μin./in. Determine the magnitudes of loads P and Q and the location x of load P.

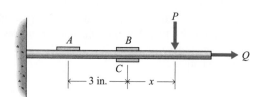

Figure P7-199

7-200 A thin-walled cylindrical pressure vessel with an inside diameter of 500 mm is fabricated by butt-welding 15-mm-thick plate with a spiral seam as shown in Fig. P7-200. The pressure in the tank is 2500 kPa. Additional loads are applied to the cylinder through a rigid end plate as shown in Fig. P7-200. Determine

a. The normal and shearing stresses on the plane of the weld at point B on the outside surface of the tank.

b. The principal stresses and the maximum shearing stress at point A on the outside surface of the tank.

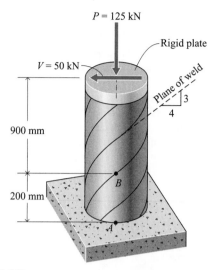

$P = 125$ kN

Rigid plate

$V = 50$ kN

Plane of weld

3
4

900 mm

B

200 mm

A

Figure P7-200

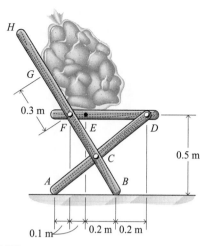

H

G

0.3 m

F   E   D

0.5 m

C

A   B

0.1 m   0.2 m   0.2 m

Figure P7-202

7-201   A solid shaft 4 in. in diameter is acted on by forces P and Q, as shown in Fig. P7-201. Determine the principal stresses and the maximum shearing stress at points A and B on the surface of the shaft.

7-203*   Four strain gages are mounted at 90° intervals around the circumference of a 4-in.-diameter steel ($E = 29{,}000$ ksi and $v = 0.30$) shaft, as shown in Fig. P7-203. All of the gages are oriented at an angle of 45° with respect to a line on the surface of the shaft that is parallel to the axis of the shaft. Determine the torque T, shear V, and moment M at the cross section where the gages are located if $\epsilon_A = +450\ \mu\text{in./in.}$, $\epsilon_B = +325\ \mu\text{in./in.}$, $\epsilon_C = +550\ \mu\text{in./in.}$, and $\epsilon_D = +675\ \mu\text{in./in.}$

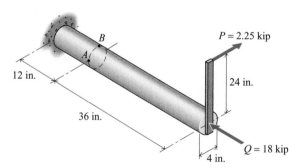

B

A

12 in.

$P = 2.25$ kip

24 in.

36 in.

$Q = 18$ kip

4 in.

Figure P7-201

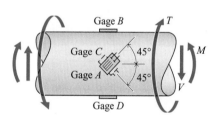

Gage B

T

Gage C

45°

M

Gage A

45°

V

Gage D

Figure P7-203

7-202*   A bag of potatoes is resting on a chair, as shown in Fig. P7-202. The force exerted by the potatoes on the frame at one side of the chair is equivalent to horizontal and vertical forces of 24 N and 84 N, respectively, at E and a force of 28 N perpendicular to member BH at G. Determine the maximum tensile and compressive normal stresses on a section midway between pins C and F. Member BH has a 10×30-mm cross section; the 30-mm dimension lies in the plane of the page.

7-204   The homogeneous advertising billboard shown in Fig. P7-204 has a mass of 250 kg and is subjected to a wind load of 1.5 kPa (assumed uniformly distributed over the area of the billboard). The billboard is supported by a pipe of outside diameter 900 mm and wall thickness 10 mm. Determine the principal stresses and maximum shear stresses at points A and B on the outside surface of the pipe. Neglect the weight of the pipe and stress concentrations.

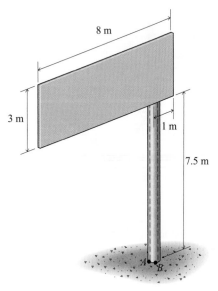

Figure P7-204

7-205 The frame shown in Fig. P7-205 is constructed of 4 × 4-in. timbers. Determine and show on sketches the principal and maximum shearing stresses at points G and H.

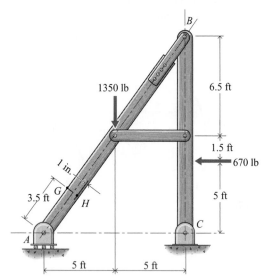

Figure P7-205

Computer Problems

7-206 The bracket shown in Fig. P7-206 is loaded in a plane of symmetry. At section a–a the cross section of the bracket is

that of an I-beam 240 mm tall by 120 mm wide, as shown in the figure. For the transverse section a–a

a. Calculate and plot the normal stress σ_x and the shear stress τ_{xy} as functions of the position y (-120 mm $< y <$ 120 mm).

b. Where is the normal stress equal to zero (measured relative to the neutral axis of bending)?

c. Calculate and plot the principal normal stresses σ_{p1} and σ_{p2} and the maximum shear stress τ_{max} as functions of the position y (-120 mm $< y <$ 120 mm).

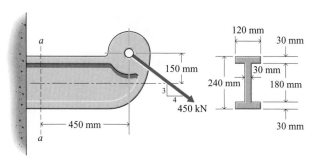

Figure P7-206

7-207 As the C-clamp shown in Fig. P7-207 is tightened, the web of the C-section is subjected to flexural stresses as well as axial tensile stresses. If the web of the section has a width of $h = 1$ in. and a thickness of $b = 1/4$ in., and the clamp has a capacity of $d = 3$ in.

a. Calculate and plot the maximum tensile and compressive stresses on section a–a as a function of the clamp force P (0 lb $< P <$ 500 lb).

b. Determine s, the distance from the inside edge of the web to the point where the normal stress is zero. Is s a function of P?

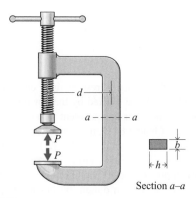

Figure P7-207

7-17   DESIGN PROBLEMS

Again in this chapter, design will be limited to proportioning a member (in this case, a straight beam) to perform a specified function without exceeding specified levels of stress. Failure by excessive deformation (excessive elastic deflection) will be discussed in Chapter 8. Design problems involving combined loading will be presented in Chapter 10. Failure by yielding or failure by fracture, which results from excessive normal (flexural) stresses or shearing stresses, must be considered when designing beams. The elastic flexure formula and the shearing stress formula developed in Sections 7-4 and 7-7 and used to calculate flexural stresses and transverse (or longitudinal) shearing stresses in beams are

$$\sigma_{\max} = \frac{M_r c}{I} = \frac{M_r}{S} \qquad \text{and} \qquad \tau = \frac{V_r Q}{It}$$

To determine maximum normal and shearing stresses in beams, sections must be located where M_r and V_r are maximum (critical sections). A shear force diagram is used to locate the critical section of the beam where V_r is maximum. The critical section for flexure, the section where M_r is maximum, is found from a bending moment diagram. In general, the absolute maximum value of M_r is used for design purposes. However, care must be exercised for beams with cross sections that are nonsymmetrical with respect to the neutral axis (such as T-beams) and beams made of materials with different properties in tension and compression. In these cases, both the largest positive and the largest negative values of M_r must be considered.

Beam design consists of finding a cross-sectional shape so that flexural and shearing stresses do not exceed permissible values, called allowable values. For a safe design

$$\text{Strength} \geq (\text{Factor of safety}) (\text{Stress}) \qquad (a)$$

where strength is a material property and stress is computed using either Eq. 7-9 or 7-12. Equation (a) may be written

$$\frac{\text{Strength}}{\text{Factor of safety}} \geq \text{Stress} \qquad (b)$$

where (Strength/Factor of safety) is the allowable stress. If the symbol σ_{all} is used for allowable stress, Eq. (b) may be written for flexure as

$$\sigma_{\text{all}} \geq \frac{M_r c}{I} = \frac{M_r}{S} \qquad (7\text{-}26)$$

and

$$\tau_{\text{all}} \geq \frac{V_r Q}{It} \qquad (7\text{-}27)$$

for shear.

Experience indicates that beam design is usually governed by flexural stresses. Thus, a beam is usually designed for flexure using Eq. 7-26, and then checked for shearing stress using Eq. 7-27. If the shearing stress is less than the allowable shearing stress, this procedure is adequate. If the allowable shearing stress has been exceeded, the beam is redesigned and the process is repeated. However, the shearing stress (longitudinal) may be the controlling factor for beams made of timber. The Example Problems that follow illustrate the procedures for designing straight beams loaded in a plane of symmetry.

Other design factors, such as local web yielding, web crippling, and side-sway web buckling for steel beams may be found in *Steel Structures Design and Behavior*.[10] For wood beams, additional design factors such as load duration, moisture content, temperature, and beam stability are discussed in *Design of Wood Structures*.[11]

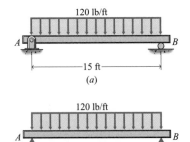

120 lb/ft

A ⟶ *B*

|← 15 ft →|

(*a*)

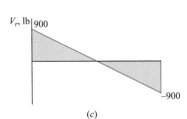

120 lb/ft

A ⟶ *B*

900 lb   900 lb

(*b*)

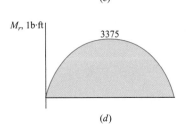

V_r, lb | 900

−900

(*c*)

M_r, lb·ft |

3375

(*d*)

Figure 7-61

Example Problem 7-29   An air-dried Douglas fir beam of rectangular cross section is to support the load shown in Fig. 7-61*a*. If the allowable flexural stresses in tension and compression are 1200 psi and the allowable shearing stress is 100 psi, determine the lightest weight standard structural timber that can be used.

SOLUTION
First, load (free-body), shear force, and bending moment diagrams are drawn, as shown in Figs. 7-61*b*, *c*, and *d*, respectively. From these diagrams it is determined that the maximum shear force is 900 lb and the maximum bending moment is 3375 lb · ft. Using the maximum bending moment in Eq. 7-26 yields the required minimum section modulus for the beam as

$$S \geq \frac{M_{max}}{\sigma_{all}} = \frac{3375(12)}{1200} = 33.75 \text{ in.}^3$$

Table B-15 in Appendix B contains a listing of standard structural timbers together with values for their section modulus S. Note that the properties listed are for dressed (actual size) timbers. Since the lightest-weight beam is wanted, a beam with $S \geq 33.75$ in.^{3} and the smallest weight per unit length is a 2 × 12-in. nominal size timber. For this beam, the actual values of S and I are 35.8 in.^{3} and 206 in.^{4}, respectively. Equation 7-27 can now be used to see whether or not the allowable shearing stress requirement is met. The maximum shearing stress for a beam with a rectangular cross section occurs at the neutral axis and is given by

$$\tau_{max} = \frac{V_r Q}{It} = 1.5\frac{V_r}{A}$$

where V_r is the absolute value of the maximum shear force and A is the cross-sectional area of the beam. Thus,

$$\tau_{max} = 1.5\frac{V_r}{A} = 1.5\frac{900}{(1.625)(11.5)} = 72.24 \text{ psi} < 100 \text{ psi}$$

Equation 7-27 is satisfied, since $\tau_{all} \geq \tau_{max}$ or 100 psi > 72.24 psi. The maximum shearing stress is within the allowable limit; therefore, the beam selected is satisfactory. However, the analysis assumed that the beam was weightless, whereas

---

[10] *Steel Structures Design and Behavior*, 4th ed., C. G. Salmon and J. F. Johnson, Harper Collins, New York, 1996.
[11] *Design of Wood Structures*, 3rd ed., D. E. Breyer, McGraw-Hill, New York, 1993.

the beam selected weighs 5.19 lb/ft. The bending moment diagram for the uniformly distributed weight of the beam is similar to Fig. 7-61d, and M_{max} for the weight is $M_{max} = 146$ lb · ft. Adding the maximum bending moments for the applied loading and the beam weight gives

$$M_{max} = 3375 + 146 = 3521 \text{ lb} \cdot \text{ft}$$

The required section modulus then becomes

$$S \geq \frac{M_{max}}{\sigma_{all}} = \frac{3521(12)}{1200} = 35.21 \text{ in.}^3$$

If the value of S (35.21 in.³ for this example) had been greater than S for the original beam selected (35.8 in.³ for this example), the entire procedure would be repeated based on the section modulus for the beam with the applied loading plus the weight. Thus, the procedure is a trial-and-error process.

Similarly,

$$V_{r\,max} = 900 + \frac{1}{2}(5.19)(15) = 939 \text{ lb}$$

Thus,

$$\tau_{max} = 1.5\frac{V_r}{A} = 1.5\frac{939}{(1.625)(11.5)} = 75.37 \text{ psi} < 100 \text{ psi}$$

The original beam selected had a section modulus of $S = 35.8$ in.³ Thus, the 2×12-in. nominal beam is satisfactory. **Ans.**

Example Problem 7-30   Select the lightest wide-flange beam that can be used to support the load shown in Fig. 7-62a. The allowable flexural stresses in tension and compression are 160 MPa and the allowable shearing stress is 82 MPa.

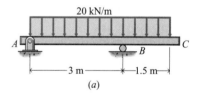

(a)

Figure 7-62(a)

SOLUTION
Load (free-body), shear force, and bending moment diagrams for the beam are shown in Figs. 7-62b, c, and d, respectively. From these diagrams it is determined that the maximum shear force is 37.5 kN and the maximum bending moment is 22.50 kN · m. Using the maximum bending moment in Eq. 7-26 yields the required section modulus for the beam as

$$S \geq \frac{M_{max}}{\sigma_{all}} = \frac{22.50(10^3)}{160(10^6)} = 140.63(10^{-6}) \text{ m}^3 = 140.63(10^3) \text{ mm}^3$$

Table B-2 in Appendix B contains a listing of the properties of wide-flange sections and from this listing it is determined that the lightest section with $S \geq 140.63(10^3)$ mm³ (with respect to the x–x-axis), is a W203 $\times$ 22 section.

When designing wide-flange beams or American standard beams, it is often assumed that the entire shear load is carried by the web of the beam and that it is

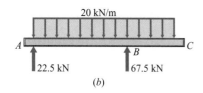

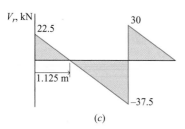

(b)

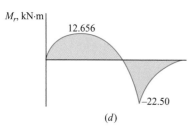

(c)

M_r, kN·m

12.656

−22.50

(d)

Figure 7-62(b–d)

uniformly distributed (see Section 7-7). Thus,

$$\tau_{avg} = \frac{V_r}{A_{web}} = \frac{37.5(10^3)}{6.2(10^{-3})[206 - 2(8)](10^{-3})}$$
$$= 31.8(10^6)\,N/m^2 = 31.8\,MPa$$

Since τ_{avg} is less than $\tau_{all} = 82$ MPa, the W203 × 22 section is satisfactory.

As in the previous example, the weight of the beam should be considered. The mass per unit length for this beam is 22 kg/m, and thus the weight per unit length is $W = mg = 22(9.81) = 215.8$ N/m, or 0.2158 kN/m. The weight per unit length of the beam is about 1.1 percent of the applied force per unit length. The actual section modulus $S = 193(10^3)$ mm³ is about 27 percent higher than the required minimum value of $S_{min} = 140.63\,(10^3)$ mm³. Thus, the maximum flexural stress is less than the allowable value even when the weight of the beam is considered and the W203 × 22 wide-flange section is acceptable.

Since the cross section is symmetric, the maximum tensile stress and the maximum compressive stress both occur on the section where the magnitude of the bending moment is a maximum. For a negative bending moment, the top of the beam will be in tension and the bottom of the beam will be in compression. The sign of the shear force is completely arbitrary, and the largest (absolute value) shear force is used in the design calculations.

As discussed in Section 7-8, the principal stresses (maximum normal stress) at the junction of the flange and web may exceed the maximum flexural (normal) stresses at the top and bottom of the beam. Also, the maximum shear stress at the junction may exceed the maximum shearing stress at the neutral axis. Design codes set the allowable flexural stress and the allowable shear stress at a level such that the methods used in the Example Problems are acceptable.

PROBLEMS

Introductory Problems

7-208* A 5-m-long simply supported timber beam is loaded with 6700-N concentrated loads applied 2 m from each support. If the allowable flexural stress is 9 MPa and the allowable horizontal shearing stress is 0.6 MPa, select the lightest standard structural timber that can be used to support the loading.

7-209* A 20-ft-long simply supported timber beam is loaded with 1800-lb concentrated loads applied 6 ft from each support. If the allowable flexural stress is 1900 psi and the allowable horizontal shearing stress is 90 psi, select the lightest standard structural timber that can be used to support the loading.

7-210 The lever shown in Fig. P7-210 is used to lift a 275-kg rock. Select a standard steel pipe to perform the task. The allowable flexural stress is 135 MPa. Neglect the effects of shear.

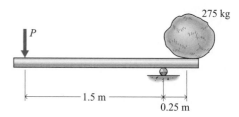

Figure P7-210

Intermediate Problems

7-211* A 16-ft-long simply supported beam is loaded with a uniform load of 4000 lb/ft over its entire length. If the allowable flexural stress is 22 ksi and the allowable vertical shearing stress is 14.5 ksi, select the lightest structural steel wide-flange beam that can be used to support the loading.

7-212* A structural steel beam is subjected to the loading shown in Fig. P7-212. If the allowable flexural stress is 152 MPa and the allowable vertical shearing stress is 100 MPa, select the lightest American standard beam that can be used to support the loading.

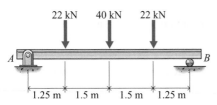

Figure P7-212

7-213 Select the lightest wide-flange beam that can be used to support the loading shown in Fig. P7-213. The allowable flexural stress is 22 ksi, and the allowable vertical shearing stress is 14.5 ksi.

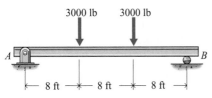

Figure P7-213

Challenging Problems

7-214* A 15-kN load is supported by a roller on an I-beam, as shown in Fig. P7-214. The roller moves slowly along the beam, thereby causing the shear force and bending moment to be functions of x. Select the lightest permissible American standard beam to support the loading. The allowable flexural stress is 152 MPa, and the allowable vertical shearing stress is 100 MPa. Neglect the weight of the beam.

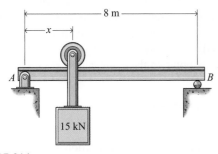

Figure P7-214

7-215 The floor framing plan for a residential dwelling is shown in Fig. P7-215. The floor decking is to be supported by 2-in. nominal width joists spaced 16 in. apart. Each joist is to span 12 ft and is simply supported at the ends. The floor decking is subjected to a uniform loading of 60 lb/ft^2, which includes the live load plus an allowance for the dead load of the flooring system. The joists are made of construction-grade Douglas fir with an allowable flexural stress of 1200 psi and an allowable horizontal shearing stress of 120 psi. Determine the required nominal depth of the joists.

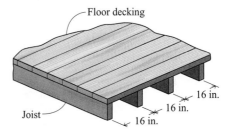

Figure P7-215

7-216 A carriage moves slowly along a simply supported I-beam, as shown in Fig. P7-216. Select the lightest permissible wide-flange beam to support the loading if the allowable flexural stress is 165 MPa and the allowable vertical shearing stress is 100 MPa. Note that the shear force and bending moment are functions of b, the position of the left-hand wheel.

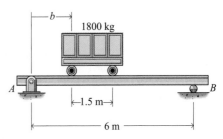

Figure P7-216

SUMMARY

A member subjected to loads applied transverse to the long dimension of the member and which cause the member to bend is known as a beam. A beam supported by pins, rollers, or smooth surfaces at the ends is called a simple beam. A simple support will develop a reaction normal to the beam but will not produce a couple. A cantilever beam has one end built into a wall or other support. The built-in end is said to be fixed if no rotation occurs and restrained if a limited amount of rotation occurs.

Cantilever beams and simple beams have only two reactions (two forces or one force and a couple), and these reactions can be obtained from a free-body diagram of the beam by applying the equations of equilibrium. Such beams are said to be statically determinate. Beams with more than two reaction components are called statically indeterminate since there are not enough equations of equilibrium to determine the reactions.

A free-body diagram of a portion of a beam with a cross section exposed by an imaginary cut shows that a transverse force V_r and a couple M_r at the cut section are needed to maintain equilibrium. The force V_r is the resultant force due to the shearing stresses. The couple M_r is the resultant moment due to the normal stresses. The magnitudes and senses of V_r and M_r are obtained from the equations of equilibrium $\Sigma F_y = 0$ and $\Sigma M_O = 0$, where O is any axis perpendicular to the xy-plane.

The normal and shearing stresses σ and τ on a transverse plane of a beam are related to the resisting moment M_r and the shear V_r by the equations

$$V_r = -\int_{\text{area}} \tau_{xy}\, dA \tag{7-1a}$$

$$M_r = -\int_{\text{area}} y\, \sigma_x\, dA \tag{7-1b}$$

It is obvious from Eqs. 7-1 that the laws of variation of the normal and shearing stresses must be known before the integrals can be evaluated.

The variation of normal stress on a plane is obtained by assuming that a plane section before bending remains a plane after bending. For this to be strictly true, it is necessary that the beam be bent only with couples. When a beam is bent with couples, the deformed shape of all longitudinal elements (also referred to as fibers) is an arc of a circle. Precise experimental measurements indicate that at some distance c above the bottom of the beam, longitudinal elements undergo no change in length. The curved surface formed by these elements is referred to as the neutral surface of the beam, and the intersection of this surface with any cross section is called the neutral axis of the section. All elements (fibers) on one side of the neutral surface are compressed, while those on the opposite side are elongated. As a result, the normal strain at any point on the plane can be expressed as

$$\epsilon_x = -\frac{1}{\rho} y \tag{7-2}$$

Equation 7-2 indicates that the strain in a fiber is proportional to the distance of the fiber from the neutral surface of the beam. Equation 7-2 is valid for elastic or inelastic action so long as the transverse shearing stresses are small.

Since the longitudinal strain ϵ_x is proportional to the distance of the fiber from the neutral surface of the beam, the normal stress σ_x on the plane (for linearly elastic action) is given by Hooke's law as

$$\sigma_x = E\epsilon_x = -\frac{E}{\rho}y \qquad (7\text{-}3)$$

Substituting Eq. 7-3 into Eq. 7-1b yields

$$M_r = -\int_A y\sigma_x \, dA = -\frac{E}{\rho}\int_A y^2 \, dA$$

The integral $\int_A y^2 \, dA$ is called the second moment of area. When the integral $\int_A y^2 dA$ is replaced by the symbol I, the elastic flexure formula is obtained as

$$\sigma_x = -\frac{M_r y}{I} \qquad (7\text{-}8)$$

where σ_x is the flexural stress at a distance y from the neutral surface and on a transverse plane, M_r is the resisting moment of the section, and I is the second moment of area of the transverse section with respect to the neutral axis.

At any section of the beam, the flexural stress will be maximum (have the greatest magnitude) at the surface farthest from the neutral axis ($y = c$), and Eq. 7-8 becomes

$$\sigma_{\max} = \frac{M_r c}{I} = \frac{M_r}{S} \qquad (7\text{-}9)$$

where $S = I/c$ is called the section modulus of the beam. Although the section modulus can be readily calculated for a given section, values (magnitudes) of the modulus are often included in tables to simplify calculations. For a given area, S becomes larger as the shape is altered to concentrate more of the area as far as possible from the neutral axis. Commercial rolled shapes such as I-beams and the various built-up sections are intended to optimize the area-section modulus relation.

If the maximum flexural stress is required in a beam subjected to a loading that produces a bending moment that varies with position along the beam, it is desirable to have a method for determining the maximum moment. Similarly, the maximum transverse shearing stress will occur at a section where the resisting shear is maximum. Shear and bending moment diagrams provide a method for obtaining maximum values of shear and moment. A shear diagram is a graph in which abscissas represent distance along the beam and ordinates represent the transverse shear at the corresponding sections. By definition, the shear at a section is positive when the portion of the beam to the left of the section (for a horizontal beam) tends to move upward with respect to the portion to the right of the sections. A moment diagram is a graph in which abscissas represent distances along the beam and ordinates represent the bending moment at the corresponding sections. Also by definition, the bending moment in a horizontal beam is positive at sections for which the top of the beam is in compression and the bottom is in tension.

The equilibrium approach is a fairly simple and straightforward method of getting equations for the shear force and bending moment in a beam. However, if the loading on the beam is complex, the equilibrium approach can require several cuts and several free-body diagrams. Alternatively, four simple relationships (which were developed by using equilibrium considerations) that are used to construct shear and moment diagrams are

$$\frac{dV}{dx} = w \tag{7-10d}$$

That is, the slope of the shear diagram at any location x in the beam is equal to the intensity of loading at that section of the beam.

$$V_2 - V_1 = \int_{x_1}^{x_2} w \, dx \tag{7-10e}$$

That is, for any section of the beam acted on by a distributed load w and no concentrated force ($P = 0$), the change in shear between sections at x_1 and x_2 is equal to the area under the load diagram between the two sections.

$$\frac{dM}{dx} = V \tag{7-11c}$$

That is, the slope of the bending moment diagram at any location x in the beam is equal to the value of the shear force at that section of the beam.

$$M_2 - M_1 = \int_{x_1}^{x_2} V \, dx \tag{7-11d}$$

That is, for any section of the beam in which the shear force is continuous ($C = P = 0$), the change in bending moment between sections at x_1 and x_2 is equal to the area under the shear diagram between the two sections.

At each point in a beam, the horizontal (longitudinal) and vertical (transverse) shearing stresses have the same magnitude and are given by the expression

$$\tau = \frac{V_r Q}{It} \tag{7-12}$$

where Q is the first moment (relative to the neutral axis) of the portion of the area of the cross section between the transverse line where the shear stress is to be evaluated and the top of the beam. The sense of the stress τ is the same as the sense of the shear V_r on the transverse plane and is determined from $\tau_{xy} = \tau_{yx}$ on a longitudinal plane. Because the flexure formula was used in the derivation of Eq. 7-12, it is subject to the same limitations as the flexure formula. Although the stress given by Eq. 7-12 is associated with a particular point in a beam, it is averaged across the thickness t and hence is accurate only if t is not too great.

The shearing stress given by Eq. 7-12 is zero at the top and bottom of the beam and varies quadratically between the top and bottom of the beam. Usually, the maximum shear stress will occur at the neutral axis. However, if there is a sudden change in the width of the beam, as at the junction between the web and the flange of an I-beam, there will be a corresponding jump in the shearing stress. Therefore, if the neutral axis is not the thinnest section of the beam, the maximum shear stress may occur at the point closest to the neutral axis where the thickness of the beam changes.

Equations 7-8, 7-9, and 7-12 apply only to linearly elastic action in a beam of a single material. The flexural stress was assumed to vary linearly from zero at the neutral axis to a maximum at the edge of the beam. In addition, the flexural stress was assumed to be constant across the width of the beam. Before these formulas can be applied to a beam of two materials, the section must first be transformed into an equivalent cross section for a beam of a homogeneous material. The transformed section is obtained by replacing either material by an equivalent amount of the other material as determined by the ratio n of their elastic moduli. For example, the force on an element of area at a distance y from the neutral axis is unchanged if the "stronger" material is replaced with a piece of the "weaker" material that is $n = E_s/E_w$ times as wide as the "stronger" material. The actual stress in the "stronger" material is then n times larger than the stress calculated at the same point in the transformed section.

For some designs, the limitation requiring stresses to remain below the proportional limit of the material results in uneconomical or inefficient designs. Therefore, this limitation is sometimes discarded and higher stress levels are tolerated in the design. The basic approach to solving problems involving inelastic action is the same as that outlined in Sections 7-2 and 7-3 for linearly elastic action. The neutral axis is located by using the equation

$$\Sigma F_x = \int_{area} \sigma_x \, dA = 0$$

and solving for the location of the axis where σ_x is zero. Once the neutral axis is located, the resisting moment is found using Eq. 7-1b.

One of the assumptions made in the development of the elastic flexure formula was that the loads were applied in a plane of symmetry. When this assumption is not satisfied, the beam will, in general, twist about a longitudinal axis. It is possible, however, to place the loads in such a plane that the beam will not twist. When any load is applied in such a plane, the line of action of the load will pass through the shear center (also known as the center of flexure or center of twist). For cross sections having an axis of symmetry, the shear center is always located on the axis of symmetry. For cross-sectional areas having two axes of symmetry, the shear center coincides with the centroid of the cross-sectional area. If a beam having a cross section with only one axis of symmetry is positioned such that the plane of symmetry is the neutral axis for flexural stresses, the plane of the loads must be perpendicular to the neutral axis but cannot pass through the centroid if bending is to occur without twisting.

Many machine and structural elements are subjected to a combination of axial, torsional, internal pressure, and flexural loads. Then, the internal forces at a section may be a combination of an axial force, a shear force, a torque, and/or a bending moment. If the strains are not too large, the stresses at a point of interest on the cross section can be calculated separately for each of the internal forces. Normal stresses due to the axial force, the internal pressure, and the bending moment are then added (or subtracted) to obtain the normal stress at the point. Similarly, the shearing stresses at the point due to the shear force and the torque are added (or subtracted) to obtain the shearing stress at the point. Once the total normal and shearing stresses at a point are known, the stress transformation equations can be used to determine normal and shearing stresses on other planes through the point. Principal stresses and maximum shearing stresses at the point and the planes on which they act can be determined by using Mohr's circle or Eqs. 2-14 through 2-18.

REVIEW PROBLEMS

7-217* A beam has the cross section shown in Fig. P7-217. If the flexural stress at point A is 2000 psi (T), determine

a. The maximum flexural stress on the section.
b. The resisting moment, M, at the section.

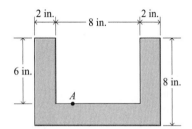

Figure P7-217

7-218* A T-beam has the cross section shown in Fig. P7-218. Determine the maximum tensile and compressive flexural stresses on a cross section of the beam where the resisting moment being transmitted is 100 kN · m.

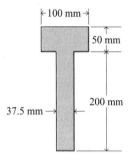

Figure P7-218

7-219 A beam is loaded and supported as shown in Fig. P7-219.

a. Draw complete shear force and bending moment diagrams for the beam.
b. Using the coordinate axes shown, write equations for the shear force and bending moment for any section of the beam in the interval $0 < x < 10$ ft.

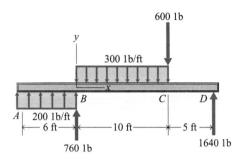

Figure P7-219

7-220* A beam is loaded and supported as shown in Fig. P7-220.

a. Draw complete shear force and bending moment diagrams for the beam.
b. Using the coordinate axes shown, write equations for the shear force and bending moment for any section of the beam in the interval $0 < x < 4$ m.

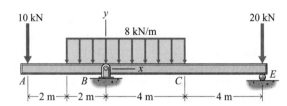

Figure P7-220

7-221 The beam shown in Fig. P7-221a has the cross section shown in Fig. P7-221b. Determine the maximum tensile and compressive flexural stresses in the beam.

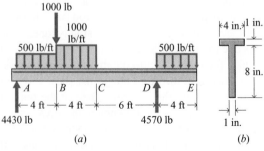

Figure P7-221

7-222 A WT 305 × 70 structural tee (see Appendix B) is loaded and supported as a beam (with the flange on top) as shown in Fig. P7-222. Determine

a. The maximum tensile flexural stress in the beam.
b. The maximum compressive flexural stress in the beam.
c. The maximum vertical shearing stress in the beam.
d. The vertical shearing stress at a point in the stem just below the flange on the cross section where the maximum vertical shearing stress occurs.

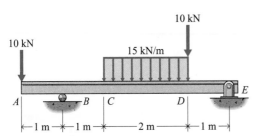

Figure P7-222

7-223* An 8 × 8 × 1-in. angle is used for a beam that is subjected to a moment of magnitude 6000 lb · ft. The resisting moment $\mathbf{M}_r$ on the cross section is in the direction shown on Fig. P7-223. The second and mixed moments of area for the cross section are $I_y = I_z = 89.0$ in.^{4} and $I_{yz} = 52.5$ in.^{4} Determine the maximum tensile and compressive flexural stresses and state where on the cross section they occur.

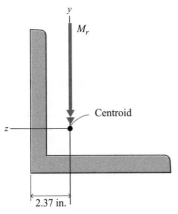

Figure P7-223

7-224 A steel bar, similar to the bar shown in Fig. 7-34c, is subjected to a bending moment M of 1400 N · m. If $h = 75$ mm, $b = 50$ mm, and $d = 20$ mm, determine the maximum flexural stress in the bar.

7-225* Determine the maximum elastic and plastic bending moments for a W14 × 120 steel beam having a proportional limit (equal to the yield point) of 36 ksi.

7-226* Locate the shear center for the cross section shown in Fig. P7-226 and determine the maximum shearing stress produced on the cross section by a vertical shear of 2.5 kN.

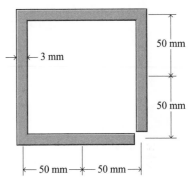

Figure P7-226

7-227 A 2-in.-wide × 3-in.-deep polymer ($E = 300$ ksi) beam will be reinforced with 1/8-in.-thick structural aluminum ($E = 10,000$ ksi) plates on its top and bottom faces. A maximum bending moment of 10 kip · in. must be resisted by the composite beam. If the allowable flexural stresses are 1 ksi in the polymer and 20 ksi in the aluminum, determine the minimum width required for the aluminum plates.

7-228 A simply supported reinforced concrete beam carries a uniformly distributed load of 10 kN/m on a span of 5 m. The beam has a rectangular cross section 250 mm wide × 450 mm deep. Four 20-mm-diameter steel reinforcing rods are placed 50 mm from the bottom of the beam, as shown in Fig. P7-228. The modulii of elasticity of the concrete and steel are 17 MPa and 200 MPa, respectively. Determine the average tensile stress in the steel and the maximum compressive stress in the concrete at the section of maximum moment.

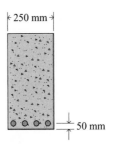

Figure P7-228

7-229* A crane hook has the dimensions and cross section shown in Fig. P7-229. The allowable stresses on plane $a\text{–}a$ are 12,000 psi (T) and 16,000 psi (C). Determine the capacity P of the hook.

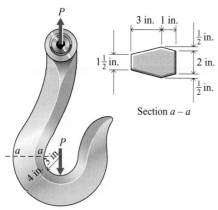

Figure P7-229

7-230 The beam shown in Fig. P7-230 has a 75×200-mm rectangular cross section and is loaded in a plane of symmetry. Determine the principal and maximum shearing stresses at point A.

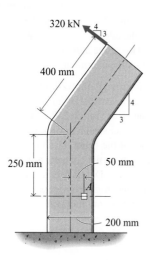

Figure P7-230

7-231 A steel shaft 4 in. in diameter is supported in flexible bearings at its ends. Two pulleys, each 2 ft in diameter, are keyed to the shaft. The pulleys carry belts that produce the forces shown in Fig. P7-231. Determine the principal stress and the maximum shearing stress at point A on the top surface of the shaft.

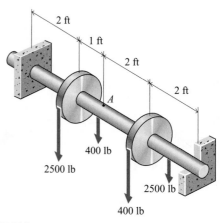

Figure P7-231

7-232* Select the lightest pair of structural steel angles (see Appendix B) that may be used for the beam of Fig. P7-232 if the maximum flexural stress must be limited to 60 MPa. The angles will be fastened back to back to form a T-section.

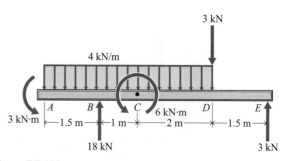

Figure P7-232

Chapter 8
Flexural Loading: Beam Deflections

8-1 INTRODUCTION

Important relations between applied load and stress (flexural and shear) in a beam were presented in Chapter 7. A beam design, however, is frequently not complete until the amount of deflection has been determined for the specified load. Failure to control beam deflections within proper limits in building construction is frequently reflected by the development of cracks in plastered walls and ceilings. Beams in many machines must deflect just the right amount for gears or other parts to make proper contact. In innumerable instances the requirements for a beam involve a given load-carrying capacity with a specified maximum deflection.

The deflection of a beam depends on the stiffness of the material and the dimensions of the beam as well as on the applied loads and supports. Four common methods for calculating beam deflections owing to flexural stresses are presented here: (1) the integration method, (2) the singularity function method, (3) the superposition method, and (4) an energy method.

8-2 THE DIFFERENTIAL EQUATION OF THE ELASTIC CURVE

When a straight beam is loaded and the action is elastic, the centroidal axis of the beam is a curve defined as the elastic curve. An example of an elastic curve is shown in Fig. 8-1. The deflection (displacement) of the elastic curve (measured from the undeformed centroidal axis) is given the symbol v and is positive when measured in the upward direction (positive y-direction). An enlarged view of a portion of the elastic curve is shown in Fig. 8-2. To find the displacement (deflection) v at point A in the elastic curve, consider the slope of the elastic curve at A, and the angle θ:

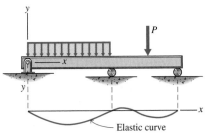

Figure 8-1

$$\text{Slope} = \frac{dv}{dx} = \tan \theta \qquad (a)$$

For a small slope, $\tan \theta \approx \theta$ (measured in radians), and

$$\theta = \frac{dv}{dx} \qquad (b)$$

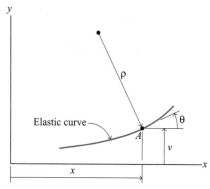

Figure 8-2

The curvature, $1/\rho$, from the calculus is

$$\frac{1}{\rho} = \frac{d^2v/dx^2}{[1 + (dv/dx)^2]^{3/2}} \qquad (c)$$

Since the slope is small, the denominator of Eq. (c) may be written as

$$[1 + (dv/dx)^2]^{3/2} \cong 1$$

and Eq. (c) becomes

$$\frac{1}{\rho} = \frac{d^2v}{dx^2} = \frac{d\theta}{dx} \qquad (d)$$

When Eq. 7-3

$$\sigma_x = E\varepsilon_x = E\left(\frac{-y}{\rho}\right) \qquad (7\text{-}3)$$

and Eq. 7-8

$$\sigma_x = \frac{-M_r y}{I} \qquad (7\text{-}8)$$

are combined with Eq. (d), the result is

$$\frac{1}{\rho} = \frac{M_r}{EI} = \frac{d^2v}{dx^2} \qquad (e)$$

Thus,

$$EI\frac{d^2v}{dx^2} = M_r \qquad (8\text{-}1)$$

where d^2v/dx^2 is the curvature ($1/\rho$) of the elastic curve, E is the modulus of elasticity of the beam material, I is the second moment of area of the cross-sectional area with respect to the neutral axis, and M_r is the resisting moment. Note that E, I, and M_r may be functions of the coordinate x.

The sign convention for bending moments established in Section 7-5 will be used for Eq. 8-1. Both E and I are always positive; therefore, the signs of the bending moment and the second derivative must be consistent. With the coordinate axes shown in Fig. 8-3, the slope changes from positive to negative in the interval from A to B; therefore, the second derivative is negative, which agrees with the sign convention for the moment established in Section 7-5. For the interval BC, both d^2v/dx^2 and M are positive.

Figure 8-3 also reveals that the signs of the bending moment and the second derivative are also consistent when the origin of the coordinate system is selected at the right end of the beam with x positive to the left and v positive upward.

Equations 7-10d, 7-11c, and 8-1 provide a means for correlating the successive derivatives of the deflection v of the elastic curve with the physical quantities

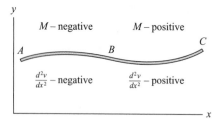

Figure 8-3

that they represent in beam action. They are

$$\text{Deflection} = v$$

$$\text{Slope} = \frac{dv}{dx}$$

$$\text{Moment} = EI\,\frac{d^2v}{dx^2} \quad \text{(from Eq. 8-1)}$$

$$\text{Shear} = \frac{dM}{dx} \text{ (from Eq. 7-11}c\text{)} = EI\,\frac{d^3v}{dx^3} \quad \text{(for } EI \text{ constant)}$$

$$\text{Load} = \frac{dV}{dx} \text{ (from Eq. 7-10}d\text{)} = EI\,\frac{d^4v}{dx^4} \quad \text{(for } EI \text{ constant)}$$

where the signs are as given in Section 7-5.

Before developing specific methods for calculating beam deflections, it is advisable to consider the assumptions used in the development of the basic relation, Eq. 8-1. All of the limitations that apply to the flexure formula apply to the calculation of deflections because the flexure formula was used in the derivation of Eq. 8-1. It is further assumed that

1. The square of the slope of the beam is negligible compared to unity.
2. The beam deflection due to shearing stresses is negligible (a plane section is assumed to remain plane).
3. The values of E and I remain constant for any interval along the beam. In case either of them varies and can be expressed as a function of the distance x along the beam, a solution of Eq. 8-1 that takes this variation into account may be possible.

8-3 DEFLECTION BY INTEGRATION

Whenever the assumptions of the previous section are essentially correct and the bending moment can be readily expressed as an integrable function of x, Eq. 8-1 can be solved for the deflection v of the elastic curve of a beam at any point x along the beam. The constants of integration can be evaluated from the applicable boundary or matching conditions.

A *boundary condition* is defined as a known set of values for x and v, or x and dv/dx, at a specific location along the beam. One boundary condition can be used to determine one and only one constant of integration. For example, a roller or pin at any point in a beam (Figs. 8-4a and b) represents a simple support at which the beam cannot deflect (unless otherwise stated in the problem) but can rotate. At a fixed end, as represented by Figs. 8-4c and d, the beam can neither deflect nor rotate unless otherwise stated. Thus, a boundary condition is $v = 0$ at the pin support of Fig. 8-4a, and $v = 0$ at the roller support of Fig. 8-4b. At the fixed support shown in Fig. 8-4c (or d), the boundary conditions are $v = 0$ and $dv/dx = 0$.

Many beams are subjected to abrupt changes in loading along the beam, such as concentrated loads, reactions, or even distinct changes in the amount of uniformly distributed load. Because the expressions for the bending moment on the left and right of any abrupt change in load are different functions of x, it is impossible to write a single equation for the bending moment in terms of ordinary algebraic functions that is valid for the entire length of the beam. This can be resolved by writing separate bending moment equations for each interval of the

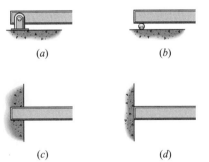

(a)   (b)

(c)   (d)

Figure 8-4

beam. Although the intervals are bounded by abrupt changes in load, the beam is continuous at such locations; therefore, the slope and the deflection at the junction of adjacent intervals must match. A *matching condition* is defined as the equality of slope or deflection, as determined at the junction of two intervals from the elastic curve equations for both intervals. One matching condition (for example, at x equals $L/3$, v from the left equation equals v from the right equation) can be used to determine one and only one constant of integration.

The procedure for obtaining beam deflections when matching conditions are required is lengthy and tedious. A method is presented in Section 8-5 in which singularity functions are used to write a single equation for the bending moment that is valid for the entire length of the beam; this eliminates the need for matching conditions and, accordingly, reduces the labor involved.

Calculating the deflection of a beam by the double integration method, that is, integrating Eq. 8-1 twice, involves four definite steps, and the following sequence for these steps is strongly recommended.

1. Select the interval or intervals of the beam to be used; next, place a set of co-ordinate axes on the beam with the origin at one end of an interval and then indicate the range of values of x in each interval. For example, two adjacent intervals might be

$$0 \le x \le L/3 \qquad \text{and} \qquad L/3 \le x \le L$$

2. List the available boundary and matching conditions (where two or more adja-cent intervals are used) for each interval selected. Remember that two conditions are required to evaluate the two constants of integration for each interval used.

3. Express the bending moment as a function of x for each interval selected and equate it to $EI(d^2v/dx^2)$.

4. Solve the differential equation or equations from step 3 and evaluate all constants of integration. Check the resulting equations for dimensional homogeneity. Calculate the deflection at specific points when required.

The following examples illustrate the use of the double integration method for calculating beam deflections.

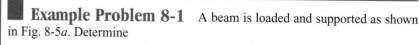

8.1

Example Problem 8-1   A beam is loaded and supported as shown in Fig. 8-5*a*. Determine

(a) The equation of the elastic curve.

(b) The deflection at the left end of the beam.

(c) The slope at the left end of the beam.

SOLUTION

(a) The beam is sectioned at position x, and a free-body diagram is drawn for the segment of the beam to the left of the section, as shown in Fig. 8-5*b*. The notations $V_r(x)$ and $M_r(x)$ indicate that the shear and bending moment are functions of x. The resisting shear force and bending moment are both shown in the positive directions defined in Chapter 7. Summing moments about a horizontal axis in the plane of the section eliminates the shear and yields

$$+\downarrow\ \Sigma M_O = 0: \qquad\qquad\qquad M_r(x) + \frac{wx^2}{2} = 0$$

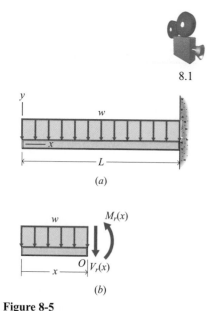

Figure 8-5

from which

$$M_r(x) = -\frac{wx^2}{2} \qquad 0 \le x \le L$$

Equation 8-1 then gives

$$EI\frac{d^2v}{dx^2} = -\frac{wx^2}{2} \qquad 0 \le x \le L$$

where E and I are constant. Successive integration gives

$$EI\frac{dv}{dx} = -\frac{wx^3}{6} + C_1 \qquad (a)$$

and

$$EIv = -\frac{wx^4}{24} + C_1x + C_2 \qquad (b)$$

where C_1 and C_2 are constants of integration to be determined using the boundary conditions. The available boundary conditions are

$$\frac{dv}{dx} = 0 \qquad \text{when} \qquad x = L \qquad (c)$$

$$v = 0 \qquad \text{when} \qquad x = L \qquad (d)$$

▶ Boundary conditions are simply conditions that must be satisfied by a function [in this case $v(x)$] at the boundaries or ends of the interval. The cantilevered connection at the wall provides whatever force is necessary to prevent vertical motion ($v = 0$ at $x = L$) and whatever moment is necessary to prevent rotation ($dv/dx = \theta = 0$ at $x = L$).

Substituting Eq. (c) into Eq. (a) and Eq. (d) into Eq. (b) gives

$$0 = -\frac{wL^3}{6} + C_1 \qquad C_1 = \frac{wL^3}{6}$$

and

$$0 = -\frac{wL^4}{24} + \frac{wL^3}{6}(L) + C_2 \qquad C_2 = -\frac{wL^4}{8}$$

Substituting values of C_1 and C_2 into Eq. (b) gives the equation of the elastic curve for the beam. Thus,

$$v = -\frac{w}{24EI}(x^4 - 4L^3x + 3L^4) \qquad 0 \le x \le L \qquad \textbf{Ans.}$$

Note that the elastic curve equation is dimensionally homogeneous.
(b) At the left end of the beam, $x = 0$; therefore,

$$v = -\frac{w}{24EI}(0 - 0 + 3L^4) = -\frac{wL^4}{8EI} = \frac{wL^4}{8EI} \downarrow \qquad \textbf{Ans.}$$

The negative sign indicates that the deflection is downward.
(c) The slope of the elastic curve is given by Eq. (a) as

$$\frac{dv}{dx} = -\frac{w}{6EI}(x^3 - L^3) \qquad 0 \le x \le L$$

from which the slope at the left end of the beam, where $x = 0$, is

$$\frac{dv}{dx} = -\frac{w}{6EI}\,(0 - L^3) = +\frac{wL^3}{6EI} = \frac{wL^3}{6EI} \nearrow \qquad \textbf{Ans.}$$

The plus sign indicates that the slope of the beam at the left end is positive (upward and to the right).

Example Problem 8-2   For the beam loaded and supported as shown in Fig. 8-6a, determine

(a) The equation of the elastic curve for the interval between the supports.
(b) The deflection midway between the supports.
(c) The point of maximum deflection between the supports.
(d) The maximum deflection in the interval between the supports.

SOLUTION

(a) From a free-body diagram of the beam and the equation $\Sigma M_B = 0$,

$$+\uparrow \Sigma M_B = 0: \qquad R_A(L) - \frac{wL^2}{12} - wL\left(\frac{L}{2}\right) = 0$$

$$R_A = +\frac{7wL}{12} = \frac{7wL}{12} \uparrow$$

▶ The origin of coordinates is often placed at the left support to simplify the determination of the constants of integration. Since the deflection is zero at a support, this choice makes $v = 0$ at $x = 0$, which then makes $C_2 = 0$.

As indicated in Fig. 8-6b, the origin of coordinates is selected at the left support, and the interval to be used is $0 \leq x \leq L$. The two required boundary conditions are $v = 0$ when $x = 0$ and $v = 0$ when $x = L$. From the free-body diagram of the portion of the beam shown in Fig. 8-6b, Eq. 8-1 yields

$$EI\,\frac{d^2v}{dx^2} = M_r(x) = \frac{7wL}{12}x - \frac{wL^2}{12} - wx\left(\frac{x}{2}\right) \qquad 0 \leq x \leq L$$

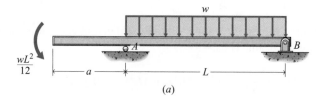

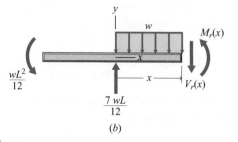

(a)

(b)

Figure 8-6

Successive integration gives

$$EI\frac{dv}{dx} = \frac{7wL}{24}x^2 - \frac{wL^2}{12}x - \frac{w}{6}x^3 + C_1$$

and

$$EIv = \frac{7wL}{72}x^3 - \frac{wL^2}{24}x^2 - \frac{w}{24}x^4 + C_1x + C_2$$

where C_1 and C_2 are constants of integration to be determined using the boundary conditions. Substitution of the boundary condition $v = 0$ when $x = 0$ gives

$$C_2 = 0$$

Substitution of the remaining boundary condition $v = 0$ when $x = L$ gives

$$C_1 = -\frac{wL^3}{72}$$

Therefore, the elastic curve equation is

$$v = -\frac{w}{72EI}(3x^4 - 7Lx^3 + 3L^2x^2 + L^3x) \qquad 0 \le x \le L \qquad \textbf{Ans.}$$

(b) The deflection midway between the supports is obtained by substituting $x = L/2$ into the elastic curve equation. Thus,

$$v = -\frac{w}{72EI}\left[3\left(\frac{L}{2}\right)^4 - 7L\left(\frac{L}{2}\right)^3 + 3L^2\left(\frac{L}{2}\right)^2 + L^3\left(\frac{L}{2}\right)\right]$$

from which

$$v = -\frac{wL^4}{128EI} = -\frac{7.81(10^{-3})wL^4}{EI} = \frac{7.81(10^{-3})wL^4}{EI} \downarrow \qquad \textbf{Ans.}$$

(c) The maximum deflection occurs where the slope dv/dx is zero

$$\frac{dv}{dx} = -\frac{w}{72EI}(12x^3 - 21Lx^2 + 6L^2x + L^3) = 0 \qquad 0 \le x \le L$$

from which

$$12x^3 - 21Lx^2 + 6L^2x + L^3 = 0$$

The solution of this cubic equation in x gives the point of maximum deflection at $x = -0.1162L$; $x = 0.541L$; and $x = 1.325L$. Thus,

$$x = 0.541L \text{ to the right of the left support} \qquad \textbf{Ans.}$$

▶ The cubic equation $12x^3 - 21Lx^2 + 6L^2x + L^3 = 0$ has three roots; $x = -0.1162L$, $x = +0.541L$, and $x = +1.325L$. Only the middle root has physical significance in this problem.

(d) The maximum deflection can readily be obtained by substituting $0.541L$ for x in the elastic curve equation. The result is

$$v_{max} = -\frac{7.88(10^{-3})wL^4}{EI} = \frac{7.88(10^{-3})wL^4}{EI} \downarrow \qquad \textbf{Ans.}$$

Example Problem 8-3   For the beam loaded and supported as shown in Fig. 8-7a, determine the deflection of the right end.

SOLUTION

From the free-body diagram of Fig. 8-7b, the equations of equilibrium give

$$+\uparrow \Sigma F_y = 0: \qquad V_A - w\,(L/3) = 0 \qquad\qquad V_A = wL/3 \uparrow$$
$$+\downarrow \Sigma M_A = 0: \qquad M_A - w\,(L/3)(5L/6) = 0 \qquad M_A = 5wL^2/18 \downarrow$$

In Fig. 8-7b there are two intervals to be considered, namely, the loaded portion of the beam and the unloaded portion of the beam. The loaded portion of the beam must be used because it contains the point where the deflection is required. However, a quick check reveals the absence of boundary conditions in this interval. It therefore becomes necessary to use both intervals as well as matching and boundary conditions; hence, the origin of coordinates is selected at the left end of the beam, as shown. The intervals are

$$0 \le x \le 2L/3 \qquad \text{and} \qquad 2L/3 \le x \le L$$

The available boundary conditions are

$$\frac{dv}{dx} = 0 \qquad \text{when} \qquad x = 0$$
$$v = 0 \qquad \text{when} \qquad x = 0$$

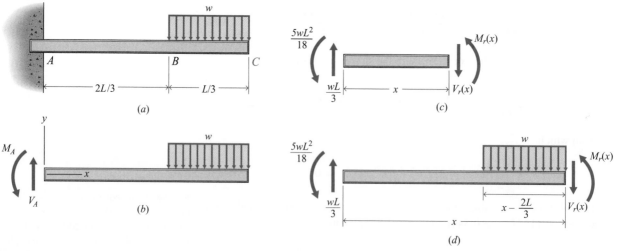

Figure 8-7

The available matching conditions when $x = 2L/3$ are

$$\frac{dv}{dx} \text{ from the left equation} = \frac{dv}{dx} \text{ from the right equation}$$
$$v \text{ from the left equation} = v \text{ from the right equation}$$

where the left equation is for the interval $0 \leq x \leq 2L/3$ and the right equation is for the interval $2L/3 \leq x \leq L$. Four conditions (two boundary and two matching) are sufficient for the evaluation of the four constants of integration (two in each of the two elastic curve differential equations); therefore, the problem can be solved in the following manner.

From the free-body diagram of Fig. 8-7c where the beam is sectioned in the unloaded interval, Eq. 8-1 yields for $0 \leq x \leq 2L/3$

$$EI\frac{d^2v}{dx^2} = M_r(x) = \frac{wL}{3}x - \frac{5wL^2}{18} \qquad (a)$$

From the free-body diagram of Fig. 8-7d where the beam is sectioned in the loaded interval, Eq. 8-1 yields for $2L/3 \leq x \leq L$

$$EI\frac{d^2v}{dx^2} = M_r(x) = \frac{wL}{3}x - \frac{5wL^2}{18} - w\left(x - \frac{2L}{3}\right)\left(\frac{x - 2L/3}{2}\right) \qquad (b)$$

Integration of Eqs. (a) and (b) gives

$$EI\frac{dv}{dx} = \frac{wL}{6}x^2 - \frac{5wL^2}{18}x + C_1 \qquad 0 \leq x \leq 2L/3 \qquad (c)$$

$$EI\frac{dv}{dx} = \frac{wL}{6}x^2 - \frac{5wL^2}{18}x - \frac{w}{6}\left(x - \frac{2L}{3}\right)^3 + C_3 \qquad 2L/3 \leq x \leq L \qquad (d)$$

Substitution of the boundary condition $\frac{dv}{dx} = 0$ when $x = 0$ into Eq. (c) gives

$$C_1 = 0$$

Since the beam has a continuous slope at $x = 2L/3$

$$\frac{dv}{dx} \text{[from Eq. (c) at } x = 2L/3\text{]} = \frac{dv}{dx} \text{[from Eq. (d) at } x = 2L/3\text{]}$$

which gives

$$C_3 = 0$$

Integration of the resulting differential equations gives

$$EIv = \frac{wL}{18}x^3 - \frac{5wL^2}{36}x^2 + C_2 \qquad 0 \leq x \leq 2L/3$$

$$EIv = \frac{wL}{18}x^3 - \frac{5wL^2}{36}x^2 - \frac{w}{24}\left(x - \frac{2L}{3}\right)^4 + C_4 \qquad 2L/3 \leq x \leq L$$

Use of the remaining boundary condition yields $C_2 = 0$. Use of the final matching condition yields $C_4 = 0$.

The deflection of the right end of the beam can now be obtained from the elastic curve equation for the right interval by replacing x by its value L. The result is

$$EIv = \frac{wL}{18}(L^3) - \frac{5wL^2}{36}(L^2) - \frac{w}{24}\left(L - \frac{2L}{3}\right)^4$$

from which

$$v = -\frac{163}{1944}\frac{wL^4}{EI} = \frac{163}{1944}\frac{wL^4}{EI} \downarrow \qquad \textbf{Ans.}$$

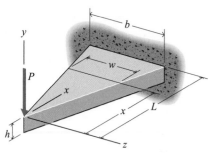

Figure 8-8

Example Problem 8-4   Determine the deflection at the left end of the cantilever beam with variable width shown in Fig. 8-8.

SOLUTION
Since the width w of the beam varies linearly with respect to position x along the length of the beam, the second moment of area I of the cross section also varies linearly with respect to position x. Thus, at position x,

$$I = \frac{wh^3}{12} = \frac{(bx/L)\,h^3}{12} = \left(\frac{bh^3}{12}\right)\left(\frac{x}{L}\right) = I_L\left(\frac{x}{L}\right)$$

where $I_L = bh^3/12$ is the second moment of area of the cross section of the beam at the support. This variation in second moment of area I must be included in the integration process used to determine the equation of the elastic curve for the beam. The equation of the elastic curve for the beam is obtained by successive integration from Eq. 8-1. Thus,

$$E\frac{d^2v}{dx^2} = \frac{M_r(x)}{I} = -\frac{Px}{I} = -\frac{Px}{I_L(x/L)} = -\frac{PL}{I_L}$$

Successive integration gives

$$EI_L\frac{dv}{dx} = -PLx + C_1$$

$$EI_Lv = -\frac{PLx^2}{2} + C_1x + C_2$$

From the boundary conditions,

$$\frac{dv}{dx} = 0 \quad \text{when} \quad x = L, \quad C_1 = PL^2$$

$$v = 0 \quad \text{when} \quad x = L, \quad C_2 = -\frac{PL^3}{2}$$

Thus, the equation of the elastic curve is

$$v = -\frac{PL}{EI_L}\left(\frac{x^2}{2} - Lx + \frac{L^2}{2}\right)$$

At the left end of the beam where $x = 0$, the deflection is

$$v = -\frac{PL^3}{2EI_L} = \frac{6PL^3}{Ebh^2} \downarrow \qquad\qquad \textbf{Ans.}$$

It is also interesting to note how the maximum flexural stress varies along the length of the beam. From Eq. 7-9,

$$\sigma_{max} = \frac{Mc}{I} = \frac{Px\left(\dfrac{h}{2}\right)}{\left(\dfrac{bh^3}{12}\right)\left(\dfrac{x}{L}\right)} = \frac{6PL}{bh^2} \qquad (a)$$

▶ Equation (a) indicates that the maximum flexural stress does not depend on position x but is constant along the entire length of the beam. This type of beam is frequently referred to as a *constant stress beam*.

▍PROBLEMS

MecMovie Activities and Problems

MM8.1 Beam boundary condition game. Determine appropriate boundary conditions necessary to determine beam deflections using the double integration method.

Introductory Problems

8-1* A beam is loaded and supported as shown in Fig. P8-1. Determine

   a. The equation of the elastic curve. Use the designated axes.
   b. The deflection at the left end of the beam.
   c. The slope at the left end of the beam.

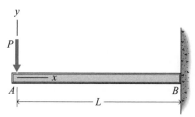

Figure P8-1

8-2* A beam is loaded and supported as shown in Fig. P8-2. Determine

   a. The equation of the elastic curve. Use the designated axes.
   b. The deflection at the right end of the beam.
   c. The slope at the right end of the beam.

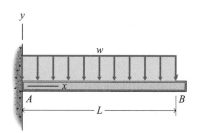

Figure P8-2

8-3 A beam is loaded and supported as shown in Fig. P8-3. Determine

   a. The equation of the elastic curve. Use the designated axes.
   b. The deflection midway between the supports.
   c. The slope at the left end of the beam.

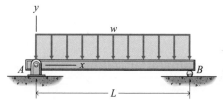

Figure P8-3

8-4* A beam is loaded and supported as shown in Fig. P8-4. Determine

   a. The equation of the elastic curve. Use the designated axes.
   b. The deflection at the left end of the beam.
   c. The slope at the left end of the beam.

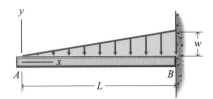

Figure P8-4

8-5 A beam is loaded and supported as shown in Fig. P8-5. Determine

a. The equation of the elastic curve. Use the designated axes.
b. The deflection midway between the supports.
c. The slope at the right end of the beam.

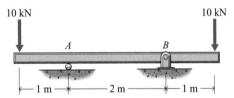

Figure P8-5

8-6 For the steel beam [$E = 200$ GPa and $I = 32.0(10^6)$ mm^4] shown in Fig. P8-6, determine the deflection at a section midway between the supports.

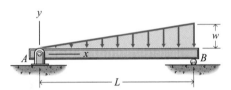

Figure P8-6

8-7* For the steel beam ($E = 30,000$ ksi and $I = 32.1$ in.^{4}) shown in Fig. P8-7, determine the deflection at a section midway between the supports.

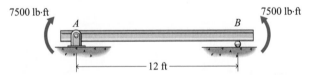

Figure P8-7

8-8 The cantilever beam shown in Fig. P8-8a is fabricated from three 30×120-mm aluminum ($E = 70$ GPa) bars, as shown in Fig. P8-8b. Determine the deflection at the left end of the beam.

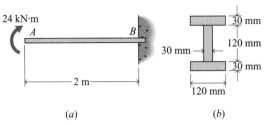

(a)   (b)

Figure P8-8

8-9 The beam shown in Fig. P8-9 is a W10 $\times$ 30 structural steel ($E = 29,000$ ksi) wide-flange section (see Appendix B). Determine

a. The equation of the elastic curve. Use the designated axes.
b. The deflection at the left end of the beam if $w = 2000$ lb/ft and $L = 10$ ft.

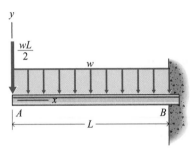

Figure P8-9

8-10* A cantilever beam is fixed at the left end and carries a uniformly distributed load w over the full length of the beam. In addition, the right end is subjected to a moment of $3wL^2/8$, as shown in Fig. P8-10. Determine

a. The equation of the elastic curve. Use the designated axes.
b. The maximum deflection in the beam if $I = 2.5 (10^6)$ mm^4, $E = 210$ GPa, $L = 3$ m, and $w = 1500$ N/m.

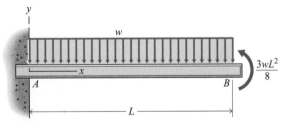

Figure P8-10

8-11* A beam is loaded and supported as shown in Fig. P8-11. Determine

a. The equation of the elastic curve. Use the designated axes.
b. The deflection midway between the supports.

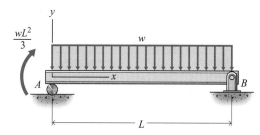

Figure P8-11

8-12 A beam is loaded and supported as shown in Fig. P8-12. Determine

a. The equation of the elastic curve for the interval between the supports. Use the designated axes.
b. The deflection midway between the supports.

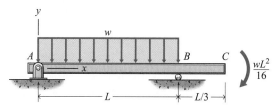

Figure P8-12

Intermediate Problems

8-13* A beam is loaded and supported as shown in Fig. P8-13. Determine

a. The equation of the elastic curve. Use the designated axes.
b. The slope at the left and right ends of the beam.
c. The deflection midway between the supports.

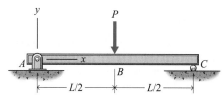

Figure P8-13

8-14* A 100 × 300-mm timber having a modulus of elasticity of 8 GPa is loaded and supported as shown in Fig. P8-14. Determine

a. The deflection at the 7-kN load.
b. The deflection at the free end of the beam.

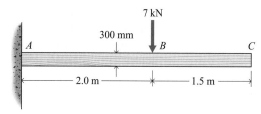

Figure P8-14

8-15 Determine the maximum deflection for the beam shown in Fig. P8-15.

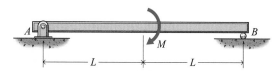

Figure P8-15

8-16* A boy with a mass of 60 kg is standing on a 40 × 300-mm wood ($E = 10$ GPa) diving board, as shown in Fig. P8-16. If the length AB is 0.6 m and the length BC is 1.5 m, determine the maximum deflection in the diving board when the boy is standing on the end of the board.

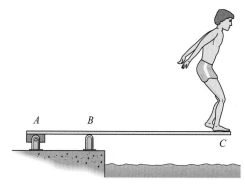

Figure P8-16

8-17 Determine the maximum deflection for the beam shown in Fig. P8-17.

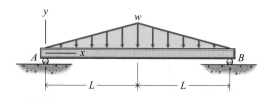

Figure P8-17

8-18 A timber beam 150 mm wide × 300 mm deep is loaded and supported as shown in Fig. P8-18. The modulus of elasticity of the timber is 10 GPa. A pointer is attached to the right end of the beam. The load acts at the midpoint of the span. Determine

a. The deflection of the right end of the pointer.
b. The maximum deflection of the beam.

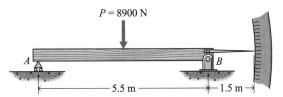

Figure P8-18

8-19* The cantilever beam shown in Fig. P8-19 has a second moment of area of I in the interval AB and a second moment of area of $2I$ in the interval BC. Determine the deflection at end A of the beam.

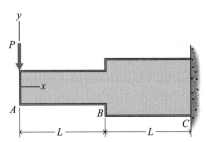

Figure P8-19

8-20 A beam is loaded and supported as shown in Fig. P8-20. Determine

a. The equation of the elastic curve of member AB. Use the designated axes.
b. The slope at the right end of the beam AB.
c. The deflection midway between the supports.

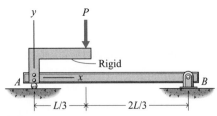

Figure P8-20

8-21 A beam is loaded and supported as shown in Fig. P8-21. Determine

a. The maximum deflection between the supports.
b. The deflection at the right end of the beam.

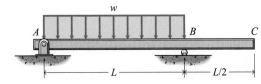

Figure P8-21

8-22* A beam is loaded and supported as shown in Fig. P8-22. Determine

a. The equation of the elastic curve. Use the designated axes.
b. The deflection at the left end of the beam.

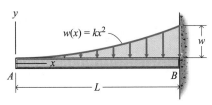

Figure P8-22

8-23* Determine the deflection midway between the supports for beam AB of Fig. P8-23. Segment BC of the beam is rigid.

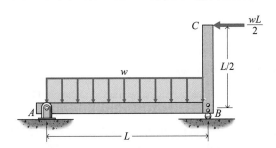

Figure P8-23

8-24 The beam shown in Fig. P8-24 is a WT 203 × 37 structural steel ($E = 200$ GPa) T-section (see Appendix B). Determine

a. The equation of the elastic curve for the region of the beam between the supports. Use the designated axes.
b. The deflection midway between the supports if $w = 5.5$ kN/m and $L = 3.5$ m.

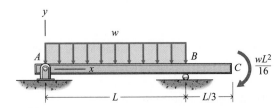

Figure P8-24

8-25 The beam shown in Fig. P8-25 is a W8 × 40 structural steel ($E = 29{,}000$ ksi) wide-flange section (see Appendix B). Determine

a. The equation of the elastic curve for the region of the beam between the supports. Use the designated axes.
b. The deflection midway between the supports if $w = 240$ lb/ft and $L = 16$ ft.

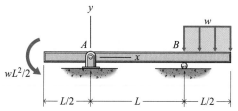

Figure P8-25

Challenging Problems

8-26* The cantilever beam ABC shown in Fig. P8-26 has a second moment of area of $2I$ in the interval AB and a second moment of area of I in the interval BC. Determine

a. The deflection at section B.
b. The deflection at section C.

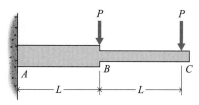

Figure P8-26

8-27* A beam AB is loaded and supported as shown in Fig. P8-27. The load P is applied through a collar that can be positioned on the load bar DE at any location in the interval $L/4 < a < 3L/4$. Determine

a. The equation of the elastic curve for beam AB.
b. The location of the load P for maximum deflection at end B.
c. The location of load P for zero deflection at end B.

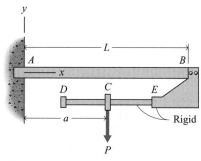

Figure P8-27

8-28 The simply supported beam $ABCD$ shown in Fig. P8-28 has a second moment of area of $2I$ in the center section BC and a second moment of area of I in the other two sections near the supports. Determine

a. The deflection at section B.
b. The maximum deflection in the beam.

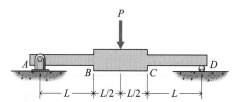

Figure P8-28

8-29* Determine the maximum deflection for the beam shown in Fig. P8-29.

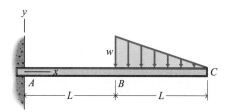

Figure P8-29

8-30 A beam is loaded and supported as shown in Fig. P8-30. Determine

a. The deflection at the left end of the beam.
b. The deflection midway between the supports.

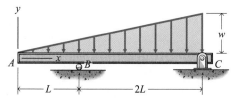

Figure P8-30

8-31 The cantilever beam ABC shown in Fig. P8-31 has a second moment of area of $4I$ in the interval AB and a second moment of area of I in the interval BC. Determine

a. The deflection at section B.
b. The deflection at section C.

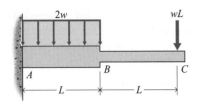

Figure P8-31

8-32* A beam is loaded and supported as shown in Fig. P8-32. Determine the deflection midway between the supports.

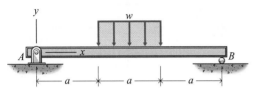

Figure P8-32

8-33* A beam is loaded and supported as shown in Fig. P8-33. Determine

a. The slope at the left end of the beam.
b. The maximum deflection between the supports.

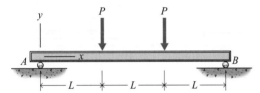

Figure P8-33

8-34 A beam is loaded and supported as shown in Fig. P8-34. Determine the deflection at the left end of the distributed load.

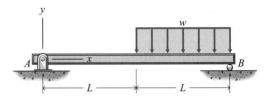

Figure P8-34

8-35 The cantilever beam ABC shown in Fig. P8-35 has a second moment of area of $4I$ in the interval AB and a second moment of area of I in the interval BC. Determine

a. The deflection at section B.
b. The deflection at section C.

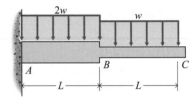

Figure P8-35

8-4 DEFLECTIONS BY INTEGRATION OF SHEAR FORCE OR LOAD EQUATIONS

In Section 8-3, the equation of the elastic curve was obtained by integrating Eq. 8-1 and applying the appropriate boundary conditions to evaluate the two constants of integration. In a similar manner, the equation of the elastic curve can be obtained from load and shear force equations. The differential equations that relate deflection v to load $w(x)$ or deflection v to shear force $V(x)$ are obtained by substituting Eq. 7-11c or Eq. 7-11d, respectively, into Eq. 8-1. Thus,

$$EI \frac{d^2v}{dx^2} = M(x) \tag{8-1}$$

$$EI \frac{d^3v}{dx^3} = V(x) \tag{8-2}$$

$$EI \frac{d^4v}{dx^4} = w(x) \tag{8-3}$$

When Eq. 8-2 or 8-3 is used to obtain the equation of the elastic curve, either three or four integrations will be required instead of the two integrations required with

Eq. 8-1. These additional integrations will introduce additional constants of integration. The boundary conditions, however, now include conditions on the shear forces and bending moments, in addition to the conditions on slopes and deflections. Use of a particular differential equation is usually based on mathematical convenience or personal preference. In those instances when the expression for the load is easier to write than the expression for the moment, Eq. 8-3 would be preferred over Eq. 8-1. The following examples illustrate the use of Eq. 8-3 for calculating beam deflections.

Example Problem 8-5 A beam is loaded and supported as shown in Fig. 8-9. Determine

(a) The equation of the elastic curve.

(b) The maximum deflection of the beam.

SOLUTION

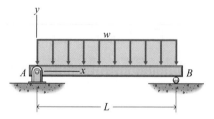

Figure 8-9

(a) Since the equation for the load distribution $[w(x) = w = \text{constant}]$ is given, Eq. 8-3 will be used to determine the equation of the elastic curve. In Section 7-6 (see Fig. 7-16), the upward direction was considered positive for a distributed load w; therefore, Eq. 8-3 is written as

$$EI\frac{d^4v}{dx^4} = w(x) = -w$$

Successive integration gives

$$EI\frac{d^3v}{dx^3} = V(x) = -wx + C_1$$

$$EI\frac{d^2v}{dx^2} = M(x) = -\frac{wx^2}{2} + C_1x + C_2$$

$$EI\frac{dv}{dx} = -\frac{wx^3}{6} + C_1\frac{x^2}{2} + C_2x + C_3$$

$$EIv = -\frac{wx^4}{24} + C_1\frac{x^3}{6} + C_2\frac{x^2}{2} + C_3x + C_4$$

The four constants of integration are determined by applying the boundary conditions. Thus,

At $x = 0$, $v = 0$; therefore, $C_4 = 0$

At $x = 0$, $M = 0$; therefore, $C_2 = 0$

At $x = L$, $M = 0$; therefore, $C_1 = \dfrac{wL}{2}$

At $x = L$, $v = 0$; therefore, $C_3 = -\dfrac{wL^3}{24}$

▶ The constant C_1 could also have been determined from a boundary condition involving the shear force V. For example, the shear force jumps upward by $wL/2$ across the left support. Therefore, at $x = 0$ the shear force equation gives $V = w(0) + C_1 = wL/2$ and the first constant of integration is $C_1 = wL/2$.

Thus,

$$v = -\frac{w}{24EI}[x^4 - 2Lx^3 + L^3x] \qquad \textbf{Ans.}$$

(b) The maximum deflection occurs at $x = L/2$, which when substituted into the equation of the elastic curve gives

$$v_{max} = -\frac{5wL^4}{384EI} = \frac{5wL^4}{384EI} \downarrow \qquad \textbf{Ans.}$$

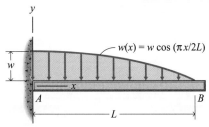

Figure 8-10

Example Problem 8-6 A beam is loaded and supported as shown in Fig. 8-10. Determine

(a) The equation of the elastic curve.
(b) The deflection at the right end of the beam.
(c) The support reactions V_A and M_A at the left end of the beam.

SOLUTION

(a) Since the equation for the load distribution is given and the moment equation is not easy to write, Eq. 8-3 will be used to determine the deflections. In Section 7-6 (see Fig. 7-16), the upward direction was considered positive for a distributed load w; therefore, Eq. 8-3 is written as

$$EI\frac{d^4v}{dx^4} = w(x) = -w \cos\frac{\pi x}{2L}$$

Successive integration gives

$$EI\frac{d^3v}{dx^3} = V(x) = -\frac{2wL}{\pi}\sin\frac{\pi x}{2L} + C_1$$

$$EI\frac{d^2v}{dx^2} = M(x) = \frac{4wL^2}{\pi^2}\cos\frac{\pi x}{2L} + C_1 x + C_2$$

$$EI\frac{dv}{dx} = \frac{8wL^3}{\pi^3}\sin\frac{\pi x}{2L} + C_1\frac{x^2}{2} + C_2 x + C_3$$

$$EIv = -\frac{16wL^4}{\pi^4}\cos\frac{\pi x}{2L} + C_1\frac{x^3}{6} + C_2\frac{x^2}{2} + C_3 x + C_4$$

The four constants of integration are determined by applying the boundary conditions. Thus,

$$\text{At } x = 0, v = 0; \qquad \text{therefore, } C_4 = \frac{16wL^4}{\pi^4}$$

$$\text{At } x = 0, \frac{dv}{dx} = 0; \qquad \text{therefore, } C_3 = 0$$

$$\text{At } x = L, V = 0; \qquad \text{therefore, } C_1 = \frac{2wL}{\pi}$$

$$\text{At } x = L, M = 0; \qquad \text{therefore, } C_2 = -\frac{2wL^2}{\pi}$$

Thus,

$$v = -\frac{w}{3\pi^4 EI}\left[48L^4\cos\frac{\pi x}{2L} - \pi^3 Lx^3 + 3\pi^3 L^2 x^2 - 48L^4\right] \quad \textbf{Ans.}$$

(b) The deflection at the right end of the beam is

$$v_B = v_{x=L} = -\frac{w}{3\pi^4 EI}(-\pi^3 L^4 + 3\pi^3 L^4 - 48L^4)$$

$$= -\frac{(2\pi^3 - 48)wL^4}{(3\pi^4 EI)} = -0.04795wL^4/EI \quad \textbf{Ans.}$$

(c) The shear force $V(x)$ and bending moment $M(x)$ at any distance x from the support are

$$V(x) = \frac{2wL}{\pi}\left[1 - \sin\frac{\pi x}{2L}\right]$$

$$M(x) = \frac{2wL}{\pi^2}\left[2L\cos\frac{\pi x}{2L} + \pi x - \pi L\right]$$

Thus, the support reactions at the left end of the beam are

$$V_A = V_{x=0} = \frac{2wL}{\pi} \quad \textbf{Ans.}$$

$$M_A = M_{x=0} = -\frac{2(\pi - 2)wL^2}{\pi^2} \quad \textbf{Ans.}$$

PROBLEMS

Introductory Problems

8-36* For the beam and loading shown in Fig. P8-36, determine

  a. The equation of the elastic curve.
  b. The maximum deflection for the beam.

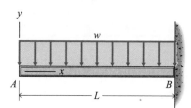

Figure P8-36

8-37* For the beam and loading shown in Fig. P8-37, determine

  a. The equation of the elastic curve.
  b. The maximum deflection for the beam.

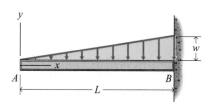

Figure P8-37

8-38 For the beam and loading shown in Fig. P8-38, determine

  a. The equation of the elastic curve.
  b. The deflection midway between the supports.

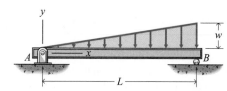

Figure P8-38

Intermediate Problems

8-39* A beam is loaded and supported as shown in Fig. P8-39. Determine

a. The equation of the elastic curve.
b. The deflection at the left end of the beam.
c. The support reactions V_B and M_B.

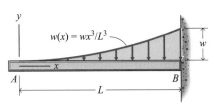

Figure P8-39

8-40* A beam is loaded and supported as shown in Fig. P8-40. Determine

a. The equation of the elastic curve.
b. The maximum deflection of the beam.

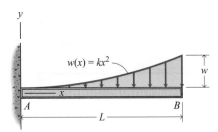

Figure P8-40

8-41 A beam is loaded and supported as shown in Fig. P8-41. Determine

a. The equation of the elastic curve.
b. The deflection midway between the supports.
c. The maximum deflection of the beam.
d. The support reactions R_A and R_B.

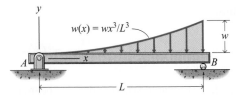

Figure P8-41

Challenging Problems

8-42* A beam is loaded and supported as shown in Fig. P8-42. Determine

a. The equation of the elastic curve.
b. The deflection at the left end of the beam.
c. The support reactions V_B and M_B.

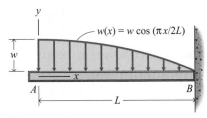

Figure P8-42

8-43* A beam is loaded and supported as shown in Fig. P8-43. Determine

a. The equation of the elastic curve.
b. The deflection midway between the supports.
c. The maximum deflection of the beam.
d. The slope at the left end of the beam.
e. The support reactions R_A and R_B.

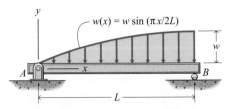

Figure P8-43

8-44 A beam is loaded and supported as shown in Fig. P8-44. Determine

a. The equation of the elastic curve.
b. The deflection midway between the supports.
c. The slope at the left end of the beam.
d. The support reactions R_A and R_B.

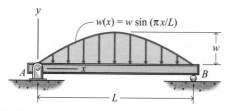

Figure P8-44

8-45 A beam is loaded and supported as shown in Fig. P8-45. Determine

a. The equation of the elastic curve.
b. The deflection at the left end of the beam.

c. The slope at the left end of the beam.
d. The support reactions V_B and M_B.

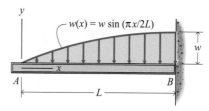

Figure P8-45

8-5 SINGULARITY FUNCTIONS

The double integration method of Section 8-3 becomes tedious and time-consuming when several intervals and several sets of matching conditions are required. The labor involved in solving problems of this type, however, can be diminished by making use of singularity functions following the method developed in 1862 by the German mathematician A. Clebsch (1833–1872).[1]

Singularity functions are closely related to the unit step function used by the British physicist O. Heaviside (1850–1925) to analyze the transient response of electrical circuits. Singularity functions will be used here for writing one bending moment equation that applies in all intervals along a beam, thus eliminating the need for matching conditions.

To illustrate the use of singularity functions, consider the beam loaded as shown in Fig. 8-11. The terms R_L and R_R represent support reactions at the left and right supports, respectively. The moment equations at the four designated sections are

$$M_1 = R_L x \qquad\qquad\qquad\qquad\qquad\qquad 0 < x < x_1$$
$$M_2 = R_L x - P(x - x_1) \qquad\qquad\qquad x_1 < x < x_2$$
$$M_3 = R_L x - P(x - x_1) + M_A \qquad\qquad x_2 < x < x_3$$
$$M_4 = R_L x - P(x - x_1) + M_A - \frac{w}{2}(x - x_3)^2 \qquad x_3 < x < L$$

Note that the origin of the coordinate system is at the left end of the beam, with positive x to the right. Each time a section "jumps over" a discontinuity in load (at x_1, x_2, and x_3), an additional term appears in the moment equation. For example, in section 2 the moment equation involves all loads to the left of the section; the same is true for section 3 and section 4.

First, consider the moment equation

$$M_2 = R_L x - P(x - x_1) \qquad x_1 < x < x_2$$

The moment equation for both M_1 and M_2 can be represented by a single moment equation

$$M = R_L x - P\langle x - x_1 \rangle^1 \qquad 0 < x < x_2$$

[1]For a rather complete history of the Clebsch method and the numerous extensions thereof, see "Clebsch's Method for Beam Deflections," Walter D. Pilkey, *Journal of Engineering Education*, January 1964, p. 170.

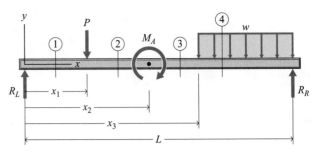

Figure 8-11

where $\langle x - x_1 \rangle^1$ is ignored when $x < x_1$ and replaced by parentheses $(x - x_1)^1$ when $x \geq x_1$. That is, the bracket $\langle x - x_1 \rangle^1 = 0$ when $x < x_1$ and $\langle x - x_1 \rangle^1 = (x - x_1)^1$ when $x \geq x_1$. Therefore,

$$M = M_1 = R_L x \qquad\qquad\qquad 0 < x < x_1$$
$$M = M_2 = R_L x - P(x - x_1) \qquad x_1 < x < x_2$$

The exponent 1 in $\langle x - x_1 \rangle^1$ has the same meaning as the exponent 1 (which is not usually written) in $(x - x_1)$.

For a section at 3, the moment equation can be written

$$M = M_3 = R_L x - P\langle x - x_1 \rangle^1 + M_A \langle x - x_2 \rangle^0 \qquad 0 < x < x_3$$

For $x < x_2$, the term $\langle x - x_2 \rangle^0$ is zero, and for $x \geq x_2$ the term $\langle x - x_2 \rangle^0 = (x - x_2)^0 = 1$; that is, the zero power of the $\langle x - x_2 \rangle^0$ term is unity for $x \geq x_2$.

Proceeding in a similar fashion, the moment equation for the entire beam can be written using a single expression:

$$M = M_4 = R_L x - P\langle x - x_1 \rangle^1 + M_A \langle x - x_2 \rangle^0 - \frac{w}{2} \langle x - x_3 \rangle^2 \qquad 0 < x < L$$

The terms $\langle x - x_1 \rangle^1$, $\langle x - x_2 \rangle^0$, and $\langle x - x_3 \rangle^2$ are called *singularity functions*.

A singularity function of x is written as $\langle x - x_0 \rangle^n$, where n is any integer (positive or negative) including zero, and x_0 is a constant equal to the value of x at the initial boundary of a specific interval along a beam. The brackets $\langle\,\rangle$ are replaced by parentheses $(\,)$ when $x \geq x_0$ and by zero when $x < x_0$. Selected properties of singularity functions required for beam-deflection problems are listed here for emphasis and ready reference.

$$\langle x - x_0 \rangle^n = \begin{cases} (x - x_0)^n & \text{when } n > 0 \text{ and } x \geq x_0 \\ 0 & \text{when } n > 0 \text{ and } x < x_0 \end{cases}$$

$$\langle x - x_0 \rangle^0 = \begin{cases} 1 & \text{when } x \geq x_0 \\ 0 & \text{when } x < x_0 \end{cases}$$

$$\int \langle x - x_0 \rangle^n dx = \frac{1}{n+1} \langle x - x_0 \rangle^{n+1} + C \qquad \text{when } n \geq 0$$

$$\frac{d}{dx} \langle x - x_0 \rangle^n = n \langle x - x_0 \rangle^{n-1} \qquad\qquad \text{when } n \geq 1$$

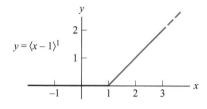

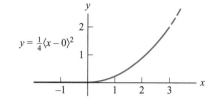

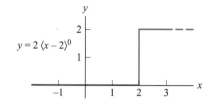

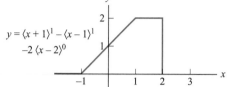

Figure 8-12

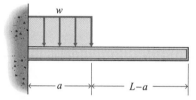

Figure 8-13

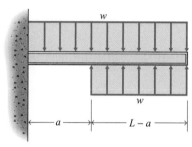

Figure 8-14

Several examples of singularity functions are shown in Fig. 8-12.

Distributed loadings that are sectionally continuous (the distributed load cannot be represented by a single function of x for all values of x) are readily obtained by superposition, as illustrated in the following examples.

A special word of caution is warranted for distributed loadings. For the beam shown in Fig. 8-11 the distributed load is extended to the right end of the beam. If the distributed load does not extend to the right end of the beam (as in Fig. 8-13), the distributed load should be replaced with an equivalent loading in which each load extends to the right end of the beam (as in Fig. 8-14). Example Problems 8-9 and 8-10 illustrate how to handle such distributed loadings.

Example Problem 8-7   A cantilever beam is loaded and supported as shown in Fig. 8-15a. Use singularity functions to determine the deflection

(a) At a distance $x = L$ from the support.

(b) At the right end of the beam.

SOLUTION

A free-body diagram of the entire beam is shown in Fig. 8-15b. The reactions at the fixed support are found using the equations of equilibrium

$$+ \uparrow \Sigma F_y = 0: \qquad V_A + P - P = 0 \qquad V_A = 0$$
$$+ \downarrow \Sigma M_A = 0: \qquad -M_A + PL - P(2L) = 0 \qquad M_A = -PL$$

The expression for the resisting moment for the entire beam, using the appropriate singularity functions, is

$$M_r(x) = -PL + P\langle x - L \rangle^1 \qquad 0 \le x \le 2L$$

The resisting moment is substituted into Eq. 8-1

$$EI\frac{d^2v}{dx^2} = -PL + P\langle x - L \rangle \qquad 0 \le x \le 2L \qquad (a)$$

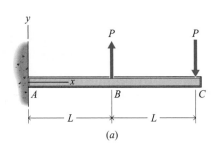

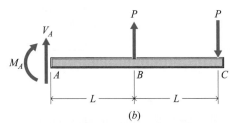

Figure 8-15

If the flexural rigidity EI is constant, Eq. (a) can be integrated twice to get

$$EI\frac{dv}{dx} = -PLx + \frac{P\langle x - L\rangle^2}{2} + C_1 \qquad 0 \leq x \leq 2L \qquad (b)$$

$$EIv = \frac{-PLx^2}{2} + \frac{P\langle x - L\rangle^3}{6} + C_1x + C_2 \qquad 0 \leq x \leq 2L \qquad (c)$$

where C_1 and C_2 are constants of integration to be determined using the boundary conditions. The available boundary conditions are

$$\text{when } x = 0, \qquad \frac{dv}{dx} = 0 \qquad (d)$$

$$\text{when } x = 0, \qquad v = 0 \qquad (e)$$

For the range of x that contains $x = 0$, $\langle x - L\rangle^n = 0$, and Eqs. ($b$) and ($c$) become

$$EI\frac{dv}{dx} = -PLx + C_1 \qquad (f)$$

$$EIv = \frac{-PLx^2}{2} + C_1x + C_2 \qquad (g)$$

Using Eq. (f) and boundary condition (d) gives $C_1 = 0$; Eq. (g) and boundary condition (e) gives $C_2 = 0$. Equations (b) and (c) are then written as

$$EI\frac{dv}{dx} = -PLx + \frac{P\langle x - L\rangle^2}{2} \qquad 0 \leq x \leq 2L \qquad (b')$$

$$EIv = \frac{-PLx^2}{2} + \frac{P\langle x - L\rangle^3}{6} \qquad 0 \leq x \leq 2L \qquad (c')$$

(a) For $x = L$, Eq. (c') is written

$$EIv = \frac{-PL^3}{2} + \frac{P(L-L)^3}{6} = \frac{-PL^3}{2}$$

which gives

$$v_{x=L} = \frac{-PL^3}{2EI} = \frac{PL^3}{2EI} \downarrow \qquad \text{Ans.}$$

(b) At the right end of the beam, $x = 2L$, and Eq. (c') is

$$EIv = \frac{-PL(2L)^2}{2} + \frac{P(2L-L)^3}{6} = \frac{-11PL^3}{6}$$

which gives

$$v_{x=2L} = \frac{-11PL^3}{6EI} = \frac{11PL^3}{6EI} \downarrow \qquad \textbf{Ans.}$$

■ **Example Problem 8-8**   For the beam loaded and supported as shown in Fig. 8-16a, determine the deflection of the right end.

SOLUTION

The free-body diagram of Fig. 8-16b and the equations of equilibrium $\Sigma F_y = 0$ and $\Sigma M_O = 0$ yield

$+\uparrow \Sigma F_y = 0$: $V_A - w(L/3) = 0$ $V_A = wL/3 \uparrow$

$+\downarrow \Sigma M_O = 0$: $M_A - w(L/3)(5L/6) = 0$ $M_A = 5wL^2/18 \downarrow$

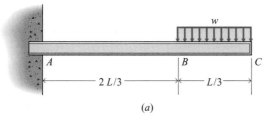

(a)

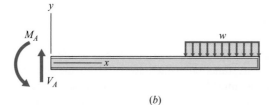

(b)

Figure 8-16

The internal resisting moment (obtained from another free-body diagram and equilibrium)

$$M_r = \frac{-5wL^2}{18} + \frac{wLx}{3} - \frac{w}{2}\left\langle x - \frac{2L}{3}\right\rangle^2$$

is substituted into Eq. 8-1 to get

$$EI\frac{d^2v}{dx^2} = -\frac{5wL^2}{18} + \frac{wLx}{3} - \frac{w}{2}\left\langle x - \frac{2L}{3}\right\rangle^2$$

The first integration gives

$$EI\frac{dv}{dx} = -\frac{5wL^2x}{18} + \frac{wLx^2}{6} - \frac{w}{6}\left\langle x - \frac{2L}{3}\right\rangle^3 + C_1$$

The boundary condition $dv/dx = 0$ when $x = 0$ gives $C_1 = 0$ because the term in the brackets is zero when $x \leq 2L/3$. Integrating again gives

$$EIv = -\frac{5wL^2x^2}{36} + \frac{wLx^3}{18} - \frac{w}{24}\left\langle x - \frac{2L}{3}\right\rangle^4 + C_2$$

The boundary condition $v = 0$ when $x = 0$ gives $C_2 = 0$ because the term in the brackets is zero when $x \leq 2L/3$.

The deflection at the right end of the beam is obtained by substituting $x = L$ in the elastic curve equation. The result is

$$v = -\frac{163}{1944}\frac{wL^4}{EI} = \frac{163}{1944}\frac{wL^4}{EI} \downarrow \qquad \textbf{Ans.}$$

This result agrees with that of Example Problem 8-3.

▊ **Example Problem 8-9**   Determine the deflection at the left end of the beam of Fig. 8-17a.

SOLUTION
The free-body diagram of Fig. 8-17b and the equilibrium equation $\Sigma M_B = 0$ yield

$$+\uparrow \Sigma M_B = 0: \quad R_L(L) - w(L/2)(7L/4) + wL^2/2 = 0 \qquad R_L = 3wL/8 \uparrow$$

To express the moment of the distributed load at the left end of the beam in terms of singularity functions which are valid for the full length of the beam, the distributed load must be represented on the free-body diagram by equivalent distributed loads on the top and bottom of the beam as shown in Fig. 8-17c. When the expression for the bending moment obtained from the free-body diagram of

▶ Because singularity functions do not allow distributed loads to terminate in the middle of the beam, the actual load is replaced with an equivalent load that is expressible in terms of singularity functions. The original distributed load is extended all the way to the end of the beam, and a canceling distributed load is applied to the beam from $x = -L/2$ to $x = L$.

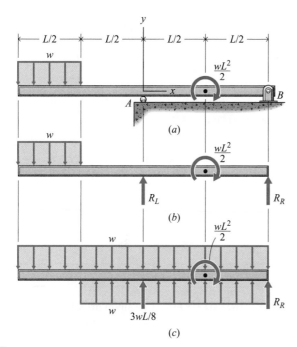

Figure 8-17

Fig. 8-17c is substituted in Eq. 8-1, the result is

$$EI \frac{d^2v}{dx^2} = -\frac{w}{2}(x+L)^2 + \frac{w}{2}\left\langle x+\frac{L}{2}\right\rangle^2 + \frac{3wL}{8}\langle x-0\rangle^1 + \frac{wL^2}{2}\left\langle x-\frac{L}{2}\right\rangle^0$$

where the first term after the equals sign represents the distributed load on the top of the beam and the second term represents the distributed load on the bottom of the beam. The effect of the two terms is to terminate the distributed load at $x = -L/2$. The boundary conditions are $v = 0$ when $x = 0$, and $v = 0$ when $x = L$. Two integrations of the moment equation give

$$EI \frac{dv}{dx} = -\frac{w}{6}(x+L)^3 + \frac{w}{6}\left\langle x+\frac{L}{2}\right\rangle^3 + \frac{3wL}{16}\langle x-0\rangle^2 + \frac{wL^2}{2}\left\langle x-\frac{L}{2}\right\rangle^1 + C_1$$

and

$$EIv = -\frac{w}{24}(x+L)^4 + \frac{w}{24}\left\langle x+\frac{L}{2}\right\rangle^4 + \frac{wL}{16}\langle x-0\rangle^3$$
$$+ \frac{wL^2}{4}\left\langle x-\frac{L}{2}\right\rangle^2 + C_1 x + C_2$$

The first boundary condition, $v = 0$ when $x = 0$, gives

$$0 = -\frac{wL^4}{24} + \frac{wL^4}{384} + 0 + 0 + 0 + C_2 \qquad C_2 = +\frac{5}{128}wL^4$$

▶ Note that even though the origin of coordinates is at the left support (so $v = 0$ at $x = 0$), the constant of integration C_2 is not zero. At $x = 0$, $(x+L)^4 = L^4$; $\left\langle x+\frac{L}{2}\right\rangle^4 = \frac{L^4}{16}$; $\langle x-0\rangle^3 = 0^3 = 0$; and $\left\langle x-\frac{L}{2}\right\rangle^2 = \left\langle -\frac{L}{2}\right\rangle^2 = 0$ (because a singularity function is zero whenever its argument is negative).

The second boundary condition, $v = 0$ when $x = L$, gives

$$0 = -\frac{16wL^4}{24} + \frac{81wL^4}{384} + \frac{wL^4}{16} + \frac{wL^4}{16} + C_1 L + \frac{5wL^4}{128} \qquad C_1 = +\frac{7}{24}wL^3$$

The deflection at the left end is obtained by substituting $x = -L$ in the elastic curve equation. The result is

$$EIv = 0 + 0 + 0 + 0 + \frac{7}{24}wL^3(-L) + \frac{5}{128}wL^4 = -\frac{97}{384}wL^4$$

Thus:

$$v = -\frac{97}{384}\frac{wL^4}{EI} = \frac{97}{384}\frac{wL^4}{EI} \downarrow \qquad \textbf{Ans.}$$

Example Problem 8-10 Use singularity functions to write a single equation for the bending moment at any section of the beam shown in Fig. 8-18a.

SOLUTION

The loading on the beam of Fig. 8-18a can be considered a combination of the loadings shown in Fig. 8-18b, c, d, and e, where downward-acting loads are shown

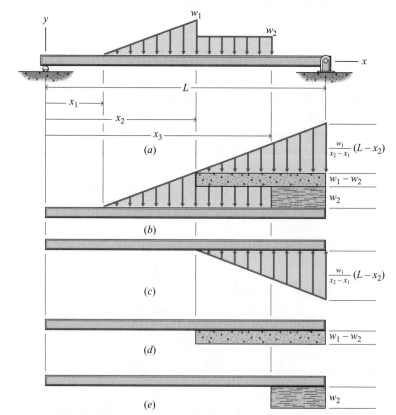

Figure 8-18

on top of the beam and upward-acting loads on the bottom. The magnitude of the linearly varying load at any point $x \geq x_1$ is

$$w = \frac{w_1(x - x_1)}{x_2 - x_1}$$

The moment of the linearly varying load at any point $x \geq x_1$ is

$$M(x) = -\frac{1}{2}\left[\frac{w_1}{x_2 - x_1}(x - x_1)\right](x - x_1)(x - x_1)/3$$

$$= -\frac{w_1}{6(x_2 - x_1)}(x - x_1)^3$$

Once the moment term for the linearly varying load is introduced in the moment equation at $x = x_1$, its effect continues to the end of the beam. As a result, terms must be introduced at the appropriate locations to terminate the effects. The effect is reduced to that of a constant distributed load of magnitude w_1 by introducing the linearly distributed load of Fig. 8-18c at $x = x_2$. The magnitude of this constant distributed load is reduced from w_1 to w_2 by introducing the constant distributed load of magnitude $w_1 - w_2$ at $x = x_2$, as shown in Fig. 8-18d. Finally, the constant distributed load of magnitude w_2 is terminated at $x = x_3$ by introducing the constant distributed load w_2, as shown in Fig. 8-18e. The moment equation for the beam is then written in terms of singularity functions as

▶ The initial distributed load is a linear function of x, $w = ax + b$. The constants a and b must be chosen so that $w = 0$ when $x = x_1$, $(0 = ax_1 + b)$, and $w = w_1$ when $x = x_2$, $(w_1 = ax_2 + b)$. Together, these two conditions give $a = \dfrac{w_1}{x_2 - x_1}$ and $b = \dfrac{w_1 x_1}{x_2 - x_1}$, so that $w = -\dfrac{w_1}{x_2 - x_1}(x - x_1)$. The slope of the second triangular load has to be the same as the first—it just starts later; $w = -\dfrac{w_1}{x_2 - x_1}(x - x_2)$.

$$M(x) = R_L x - \frac{w_1}{6(x_2 - x_1)}\langle x - x_1 \rangle^3 + \frac{w_1}{6(x_2 - x_1)}\langle x - x_2 \rangle^3$$

$$+ \frac{w_1 - w_2}{2}\langle x - x_2 \rangle^2 + \frac{w_2}{2}\langle x - x_3 \rangle^2 \qquad \textbf{Ans.}$$

▌ PROBLEMS

Introductory Problems

8-46* A cantilever beam is loaded and supported as shown in Fig. P8-46. Use singularity functions to determine the deflection

a. At a distance $x = L$ from the support.
b. At the right end of the beam.

8-47* The gangplank between a fishing boat and a dock consists of a wood ($E = 1700$ ksi) plank 10 ft long, 12 in. wide, and 2 in. thick. If the plank is modeled as the simply supported beam shown in Fig. P8-47, determine the deflection of the gangplank directly under the 165-lb man when he is at B ($x = 6$ ft).

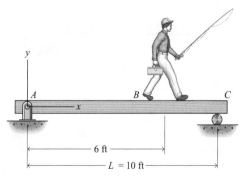

Figure P8-47

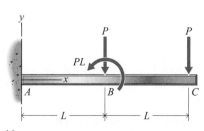

Figure P8-46

8-48 A beam is loaded and supported as shown in Fig. P8-48. Use singularity functions to determine the deflection

a. At a distance $x = L$ from the left support.
b. At the middle of the span.
c. At a distance $x = 2L$ from the left support.

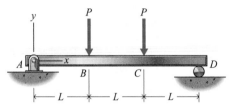

Figure P8-48

8-49* A beam is loaded and supported as shown in Fig. P8-49. Use singularity functions to determine the deflection

a. At the right end of the beam.
b. At a section midway between the supports.

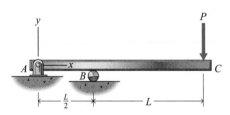

Figure P8-49

8-50* A beam is loaded and supported as shown in Fig. P8-50. Use singularity functions to determine the deflection at the middle of the span.

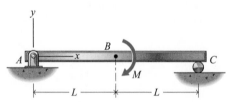

Figure P8-50

8-51 A beam is loaded and supported as shown in Fig. P8-51. Use singularity functions to determine the deflection

a. At the point of application of the concentrated load $3P$.
b. At the right end of the beam.

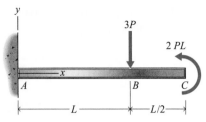

Figure P8-51

8-52 A beam is loaded and supported as shown in Fig. P8-52. Use singularity functions to determine the deflection

a. At the left end of the beam.
b. At a point midway between the supports.

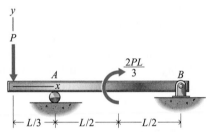

Figure P8-52

Intermediate Problems

8-53* Use singularity functions to determine the deflection at the left end of the cantilever beam shown in Fig. P8-53.

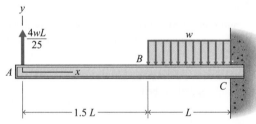

Figure P8-53

8-54* A cantilever beam is loaded and supported as shown in Fig. P8-54. Use singularity functions to determine the deflection

a. At a distance $x = L$ from the support.
b. At the right end of the beam.

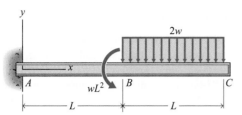

Figure P8-54

8-55 A beam is loaded and supported as shown in Fig. P8-55. Use singularity functions to determine the deflection

a. At $x = L$.
b. At a section midway between the supports.

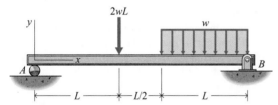

Figure P8-55

8-56* A beam is loaded and supported as shown in Fig. P8-56. Use singularity functions to determine the deflection

a. At $x = L$.
b. At $x = 3L/2$.

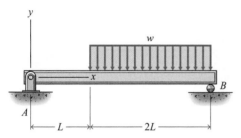

Figure P8-56

8-57 A cantilever beam is loaded and supported as shown in Fig. P8-57. Use singularity functions to determine the deflection at the left end of the beam.

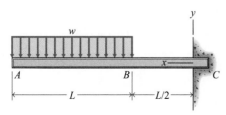

Figure P8-57

8-58 A beam is loaded and supported as shown in Fig. P8-58. Use singularity functions to determine the deflection

a. At the right end of the beam.
b. At a section midway between the supports.

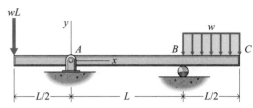

Figure P8-58

Challenging Problems

8-59* Use singularity functions to determine the deflection at the right end of the cantilever beam shown in Fig. P8-59.

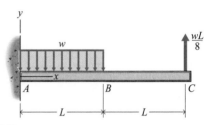

Figure P8-59

8-60* A beam is loaded and supported as shown in Fig. P8-60. Use singularity functions to determine the deflection

a. At the left end of the distributed load.
b. At a section midway between the supports.
c. At the right end of the distributed load.

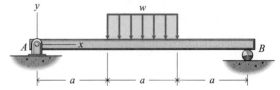

Figure P8-60

8-61 A cantilever beam is loaded and supported as shown in Fig. P8-61. Use singularity functions to determine the deflection

a. At a distance $x = L$ from the support.
b. At the right end of the beam.

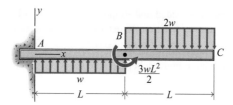

Figure P8-61

8-62 A beam is loaded and supported as shown in Fig. P8-62. Use singularity functions to determine

a. The deflection midway between the supports.
b. The maximum deflection of the beam.

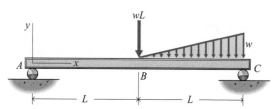

Figure P8-62

8-63* A beam is loaded and supported as shown in Fig. P8-63. Use singularity functions to determine the deflection at the middle of the span.

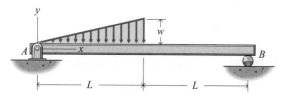

Figure P8-63

8-64 A beam is loaded and supported as shown in Fig. P8-64. Use singularity functions to determine

a. The deflection at the middle of the span.
b. The maximum deflection in the beam.

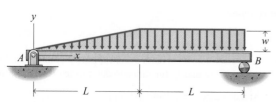

Figure P8-64

Computer Problems

8-65 A 160-lb diver walks slowly onto a diving board. The diving board is a wood ($E = 1800$ ksi) plank 10 ft long, 18 in. wide, and 2 in. thick, and it is modeled as the cantilever beam shown in Fig. P8-65. For the diver at positions $a = nL/5$ ($n = 1, 2, 3, 4, 5$), compute and plot the deflection curve for the diving board (plot v as a function of x for $0 \leq x \leq 10$ ft).

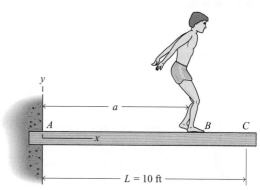

Figure P8-65

8-66 A diving board consists of a wood ($E = 12$ GPa) plank that is pinned at the left end and rests on a movable support, as shown in Fig. P8-66. The board is 3 m long, 500 mm wide, and 80 mm thick. If a 70-kg diver stands at the end of the board,

a. Compute and plot the deflection curve for the beam (plot v as a function of x for $0 \leq x \leq 3$ m) for the right support at positions $b = 0.5$ m, 1.0 m, and 1.5 m.
b. If the stiffness of the board is defined as the ratio of the diver's weight to the deflection at the end of the board ($k = W/v$), compute and plot the stiffness of the board as a function of b for $0 \leq b \leq 1.5$ m. Does the stiffness depend on the weight of the diver?

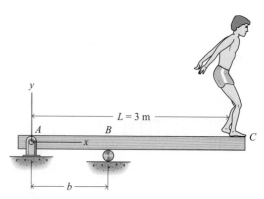

Figure P8-66

8-67 The gangplank between a fishing boat and a dock consists of a wood ($E = 1800$ ksi) plank 10 ft long, 12 in. wide, and 2 in. thick. If the plank is modeled as the simply supported beam shown in Fig. P8-67, compute and plot the deflection curve for the gangplank (plot v as a function of x for $0 \leq x \leq$ 10 ft) as a 170-lb man walks across the plank. Plot curves for the man at positions $a = nL/5$ ($n = 1, 2, 3, 4$).

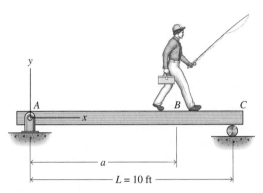

Figure P8-67

8-68 A bridge over a small stream on a golf course consists of a wood deck (weighing 1000 N/m) laid over two wood ($E =$ 12 GPa) beams. Each of the beams is 5 m long, 100 mm wide, and h mm high; each beam carries half of the loading on the bridge. It is desired that the bridge support the weight of a loaded golf cart (total weight of 2400 N) with a maximum deflection of no more than 50 mm. Model the beams and loading as shown in Fig. P8-68 and

a. Determine the minimum depth h_{min} of the beams that will support the given weight.
b. If $h = 125$ mm, compute and plot the deflection curve for the beam (plot v as a function of x for $0 \leq x \leq L$) for the cart at positions $a = 1$ m, 2 m, and 3 m.

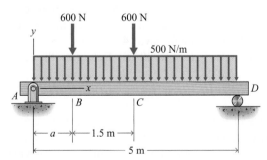

Figure P8-68

8-69 A Cub Scout troop is marching across a small footbridge consisting of a wood ($E = 1800$ ksi) plank 8 ft long, 12 in. wide, and 1.5 in. thick. The Scouts are separated by 2 ft and are modeled as static, concentrated loads (75 lb each) on a simply supported beam, as shown in Fig. P8-69. Compute and plot the deflection curve of the bridge (plot v as a function of x

for $0 \leq x \leq 8$ ft) as the troop marches across the bridge. (Initially, only one Scout is on the bridge—at position B. Then two Scouts are on the bridge—at positions B and C. The troop continues to march across until finally only a single Scout is on the bridge—at position D.)

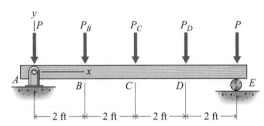

Figure P8-69

8-70 A small weight $W = 100$ N is suspended from the end of a meter stick that rests between two smooth pegs, as shown in Fig. P8-70. The wide (30-mm) side of the wood ($E = 12$ GPa) stick is horizontal, and the thin (10-mm) side is vertical. The approximation used in simple beam theory

$$EI \frac{d^2v}{dx^2} = M(x) \qquad (a)$$

is not very good here since neither the deflections nor slopes are small. Instead, the exact differential equation

$$EI \frac{d^2v/dx^2}{[1 + (dv/dx)^2]^{3/2}} = M(x) \qquad (b)$$

should be solved.

a. Compute and plot the deflection curve (plot v as a function of x for $0 \leq x \leq 1$ m) for the beam using simple beam theory [Eq. (a)].
b. On the same graph plot the deflection curve (plot v as a function of x for $0 \leq x \leq 1$ m) for the beam using the exact differential equation [Eq. (b)]. (*Note:* This curve will not go all the way to $x = 1$ m. It is the length of the stick that is 1 m, not the horizontal reach of the stick.)

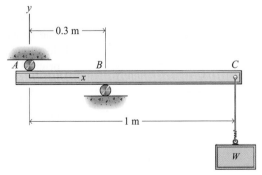

Figure P8-70

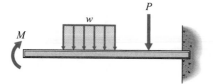

8.2

Figure 8-19

8-6 DEFLECTIONS BY SUPERPOSITION

The method of superposition is based on the fact that the resultant effect of several loads acting on a member simultaneously is the sum of the contributions from each of the loads applied individually. The results for the separate loads are frequently available from previous work or easily determined by previous methods. In such instances, the superposition method becomes a powerful concept or tool for finding stresses, deflections, and the like. The method is applicable in all cases in which a linear relation exists between the stresses or deflections and the applied loads.

To show that beam deflections can be accurately determined by the method of superposition, consider the cantilever beam of Fig. 8-19 with loads M, w, and P. To determine the deflection at any point of this beam by the double integration method, it is necessary to express the bending moment in terms of the applied loads. For each interval along the beam, the value of M_r is the algebraic sum of the moments due to the separate loads. After two successive integrations, the solution for the deflection at any point will still be the algebraic sum of the contributions for each applied load. Furthermore, for any given value of x, the relation between applied load and resulting deflection will be linear. It is evident, therefore, that the deflection of a beam is the sum of the deflections produced by the individual loads. Once the deflections produced by a few typical individual loads have been determined by one of the methods already presented, the superposition method provides a means of rapidly solving a wide variety of more complicated problems by various combinations of known results. As more data become available, a wider range of problems can be solved by superposition.

The data in Appendix B (Table B-19) are provided for use in mastering the superposition method. No attempt is made to give a large number of results because such data are readily available in various handbooks. The data given and the illustrative examples are for the purpose of making the concept and methods clear.

> **Example Problem 8-11** A 16-ft-long, simply supported beam carries a uniformly distributed load of 500 lb/ft and a concentrated load of 1000 lb, as shown in Fig. 8-20a. The beam is rough sawn (4 in. wide × 8 in. tall, $I = 170.67$ in.^{4}) out of air-dried Douglas fir ($E = 1900$ ksi). Determine the deflection v_c at the center of the beam.

SOLUTION

The deflection at the center of the beam consists of two parts, v_1 due to the distributed load and v_2 due to the concentrated load. As shown in Fig. 8-20b, the original beam with two loads can be replaced by two beams, each carrying one of the two loads.

From case 7 of Table B-19 (Appendix B), the deflection at the center of the beam due to the distributed load, with $L = 16$ ft $= 192$ in., is given by

$$v_1 = -\frac{5wL^4}{384EI} = -\frac{5\,(500/12)\,(192)^4}{384\,(1.9)\,(10^6)\,(170.67)} = -2.2736 \text{ in.}$$

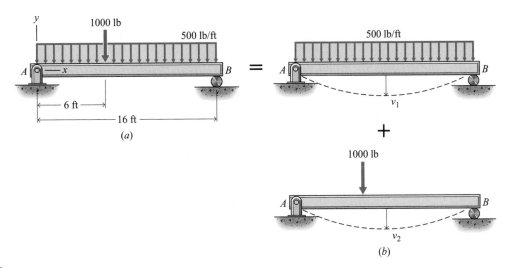

Figure 8-20

Similarly, from case 5 of Table B-19 (in which $b = 6$ ft $= 72$ in. is the shorter distance between the concentrated load and the end of the beam), the deflection at the center of the beam due to the concentrated load is given by

$$v_2 = -\frac{Pb(3L^2 - 4b^2)}{48EI} = -\frac{1000\,(72)\,[3(192)^2 - 4(72)^2]}{48\,(1.9)\,(10^6)\,(170.67)} = -0.4156 \text{ in.}$$

8.3

Consequently, the total deflection of the center of the beam is

$$v_c = v_1 + v_2 = (-2.2736) + (-0.4156)$$
$$= -2.6892 \text{ in.} \cong 2.69 \text{ in.} \downarrow \qquad \textbf{Ans.}$$

■ Example Problem 8-12 A 5-m long cantilever beam carries a uniformly distributed load of 7.5 kN/m and a concentrated load of 25 kN, as shown in Fig. 8-21a. The steel ($E = 28$ GPa) beam is a wide-flange section [$I = 500(10^{-6})$ m⁴]. Determine the deflection at the right end of the beam.

8.4

SOLUTION
As shown in Fig. 8-21b, the cantilever beam with two loads can be replaced by two beams, each carrying one of the two loads. The elastic curve for the concentrated load is shown greatly exaggerated in Fig. 8-21b. The deflection at the right end is given as $v_1 + v_2$ where v_1 is the deflection at the location of the concentrated load and v_2 is the additional deflection of the unloaded 2 m. From case 1 of Table B-19 (Appendix B),

8.5

$$v_1 = -\frac{PL^3}{3EI} = -\frac{25(10^3)(3)^3}{3(28)(10^9)(500)(10^{-6})} = -0.016071 \text{ m}$$

8.6

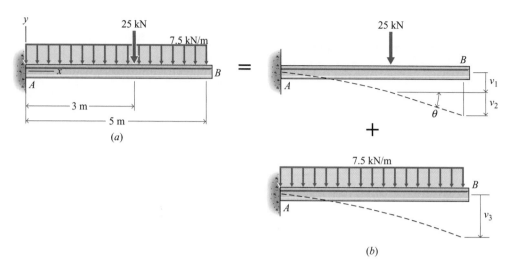

Figure 8-21

8.7

8.8

▶ The correct expression for v_2 should be $\tan\theta = \dfrac{v_2}{L}$. However, since the deflection (and angle of deflection) is small, $\tan\theta \cong \theta$ and $v_2 \cong \theta L$.

8.9

From the concentrated load to the end of the beam, the slope is constant. Again from case 1 of Table B-19, the constant slope of the beam is

$$\theta = -\frac{PL^2}{2EI} = -\frac{25(10^3)(3)^2}{2(28)(10^9)(500)(10^{-6})} = -0.008036 \text{ rad}$$

and the deflection v_2 is given by

$$v_2 = \theta L = -0.008036(2) = -0.016071 \text{ m}$$

Consequently, the total deflection of the right end of the beam due to the concentrated load is

$$v_1 + v_2 = (-0.016071) + (-0.016071) = -0.03214 \text{ m} = 32.14 \text{ mm} \downarrow$$

The elastic curve for the distributed load is also shown (greatly exaggerated) in Fig. 8-21b. From case 2 of Table B-19 (Appendix B), the deflection of the right end of the beam due to the distributed load is

$$v_3 = -\frac{wL^4}{8EI} = -\frac{7.5(10^3)(5)^4}{8(28)(10^9)(500)(10^{-6})} = -0.04185 \text{ m} = 41.85 \text{ mm} \downarrow$$

Finally, the total deflection of the right end of the beam due to both loads is

$$v = (v_1 + v_2) + v_3 = (-32.14) + (-41.85)$$
$$= -73.99 \text{ mm} \cong 74.0 \text{ mm} \downarrow \qquad \textbf{Ans.}$$

Example Problem 8-13   A simply supported beam carries a uniformly distributed load of $w = 1000$ lb/ft over its entire length and a concentrated load of $P = 2000$ lb, as shown in Fig. 8-22a. The steel ($E = 29,000$ ksi) beam is a wide-flange section ($I = 518$ in.^{4}). Determine the deflection directly under the 2000-lb load when $L_1 = 6$ ft and $L_2 = 4$ ft.

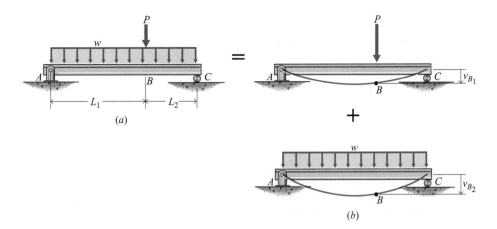

Figure 8-22

SOLUTION

The loading of Fig. 8-22a can be replaced by the two loads shown in Fig. 8-22b. Directly beneath the separate loadings of Fig. 8-22b are shown elastic curves (greatly exaggerated). These elastic curves are like the ones shown in cases 5 and 7, respectively, of Table B-19 (Appendix B). However, the equations of the elastic curves (in the last column of the table) must be used to determine the deflections at $x = L_1 = 6$ ft.

For the concentrated load $P = 2000$ lb,

$$v_1 = \frac{-Pbx}{6EIL}(L^2 - b^2 - x^2)$$

Using $b = 4$ ft ($= 48$ in.), $x = 6$ ft ($= 72$ in.), $L = 10$ ft ($= 120$ in.), and the other data given in the problem statement,

$$v_{B1} = \frac{(-2000)(48)(72)}{6(29)(10^6)(518)(120)}[(120)^2 - (48)^2 - (72)^2] = -0.004417 \text{ in.}$$

For the uniformly distributed load $w = 1000$ lb/ft $\left(= \dfrac{1000}{12}\dfrac{\text{lb}}{\text{in.}}\right)$,

$$v_2 = \frac{-wx}{24EI}(x^3 - 2Lx^2 + L^3)$$

$$v_{B2} = \frac{-(1000/12)(72)}{24(29)(10^6)(518)}[(72)^3 - 2(120)(72)^2 + (120)^3] = -0.014264 \text{ in.}$$

The total deflection at B is then

$$v_B = v_{B1} + v_{B2} = (-0.004417) + (-0.014264) = -0.018681 \text{ in.}$$
$$\cong -0.01868 \text{ in.} = 0.01868 \text{ in.} \downarrow \qquad\qquad \textbf{Ans.}$$

Example Problem 8-14   For the beam in Fig. 8-23a, determine the maximum deflection when E is $12(10^6)$ psi and I is 81 in.[4]

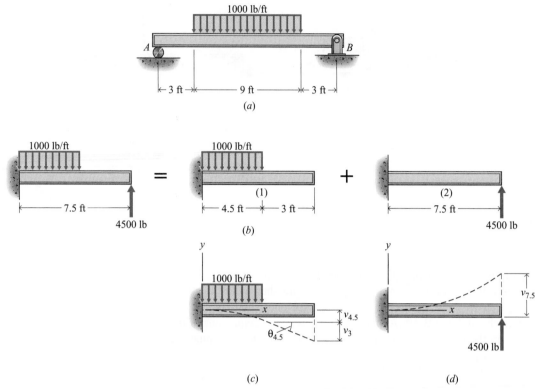

Figure 8-23

▶ Because the slope at the middle of the beam is zero, the left half of the beam could be replaced with a cantilevered support without changing any of the stresses, forces, or moments in the right half of the beam.

SOLUTION

From symmetry and the equilibrium equation $\Sigma F_y = 0$, the reactions are equal and 4500 lb upward. Because of the symmetrical loading, the slope of the beam is zero at the center of the span, and the right (or left) half of the beam can be considered a cantilever beam with two loads. As shown in Fig. 8-23b, for the right half, the cantilever with two loads can be replaced by two beams (designated 1 and 2), each carrying one of the two loads.

The elastic curve (exaggerated) for part 1, as shown in Fig. 8-23c, gives the deflection at the right end as $v_{4.5} + v_3$, where $v_{4.5}$ is the deflection at the end of the uniformly distributed load and v_3 is the additional deflection of the unloaded 3 ft. From case 2 of Table B-19 of Appendix B,

$$v_{4.5} = -\frac{wL^4}{8EI} = -\frac{(1000/12)[4.5(12)]^4}{8(12)(10^6)(81)} = -0.09113 \text{ in.}$$

Similarly,

$$\theta_{4.5} = -\frac{wL^3}{6EI} = -\frac{(1000/12)[4.5(12)]^3}{6(12)(10^6)(81)} = -0.002250 \text{ rad}$$

from which

$$v_3 = \theta L = (-0.002250)(3)(12) = -0.0810 \text{ in.}$$

Consequently, the total deflection of the right end for part 1 is

$$v_{7.5} = v_{4.5} + v_3 = (-0.09113) + (-0.0810)$$
$$= -0.17213 \text{ in.} = 0.17213 \text{ in.} \downarrow$$

The elastic curve (exaggerated) for part 2 is shown in Fig. 8-23d. From case 1 of Table B-19 of Appendix B,

$$v_{7.5} = +\frac{PL^3}{3EI} = +\frac{4500[7.5(12)]^3}{3(12)(10^6)(81)} = +1.1250 \text{ in.} = 1.1250 \text{ in.} \uparrow$$

The algebraic sum of the deflections for parts 1 and 2 is

$$v_R = v_1 + v_2 = (-0.17213) + (1.1250)$$
$$= +0.9529 \text{ in.} \cong 0.953 \text{ in.} \uparrow$$

which means that the right end of the beam is 0.953 in. above the center. Obviously, the right end does not move, and the maximum deflection is at the center and is

$$y_{\max} = 0.953 \text{ in.} \downarrow \qquad \qquad \textbf{Ans.}$$

Example Problem 8-15 A beam is loaded and supported as shown in Fig. 8-24a. Use the method of superposition to determine the deflection

(a) At a point midway between the supports.

(b) At the right end of the beam.

SOLUTION

(a) The deflection at a point midway between the supports is determined by using the beam shown in Fig. 8-24b. The effects of the loaded overhang on span AC of the beam can be represented by a shear force $V = wL$ and a moment $M = wL^2/2$. Since the shear force V does not contribute to the deflection at any point in span AC of the beam, the deflection at the middle of the span, as shown in Figs. 8-24c and d, can be expressed as

$$v_B = v_w + v_M$$

The deflections v_w and v_M are listed in cases 7 and 8 of Table B-19 of Appendix B, respectively. Thus,

$$v_B = -\frac{5w(2L)^4}{384EI} + \frac{(wL^2/2)(2L)^2}{16EI} = -\frac{wL^4}{12EI} = \frac{wL^4}{12EI} \downarrow \qquad \textbf{Ans.}$$

(b) The deflection at the right end of the beam is produced by the combined effects of the distributed load on the overhang and the rotation of the cross section of the beam at support C, as shown in Figs. 8-24c, d, and e. Thus,

$$v_D = \theta_w L + \theta_M L + v_w$$

▶ The support reactions at A and C are unchanged if the distributed load on the overhang CD is replaced with an equivalent force-couple at C. Also, the shear force and bending moment at points between the supports are unchanged if the distributed load on the overhang CD is replaced with an equivalent force-couple at C. Therefore, the deflection and slope at points between the supports will be unchanged if the distributed load on the overhang CD is replaced with an equivalent force-couple at C.

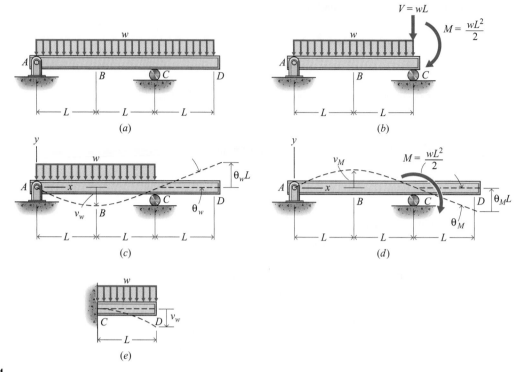

Figure 8-24

The angles θ_w and θ_M and the deflection y_w are listed in cases 7, 8, and 2 of Table B-19 of Appendix B, respectively. Thus,

$$v_D = \frac{w(2L)^3(L)}{24EI} - \frac{(wL^2/2)(2L)(L)}{3EI} - \frac{wL^4}{8EI} = -\frac{wL^4}{8EI} = \frac{wL^4}{8EI} \downarrow \qquad \textbf{Ans.}$$

PROBLEMS

MecMovie Activities and Problems

MM8.2 8 Skills: Part I. Theory; Concept checkpoints. Series of skills necessary to solve beam deflection problems using the superposition method.

MM8.3 8 Skills: Part II. Theory; Concept checkpoints. Series of skills necessary to solve beam deflection problems using the superposition method.

MM8.4 Superposition warm-up. Example; Concept checkpoints. Examples and concept checkpoints pertaining to four basic superposition skills.

MM8.5 One simple beam, one load, three cases. Concept checkpoints. Determine beam deflections at various points in a simply supported beam with two overhangs.

Use the method of superposition to determine deflections in the following problems.

Introductory Problems

8-71* For the cantilever beam shown in Fig. P8-71, determine

a. The slope and deflection at section B.
b. The slope and deflection at section C.

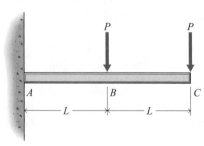

Figure P8-71

8-72* Determine the deflection midway between the supports of the beam shown in Fig. P8-72 when $P = 13.5$ kN, $L = 3$ m, $I = 80 (10^6)$ mm^4, and $E = 200$ GPa.

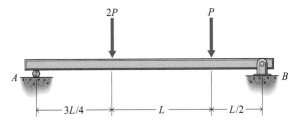

Figure P8-72

8-73 Determine the deflection at the right end of the cantilever beam shown in Fig. P8-73.

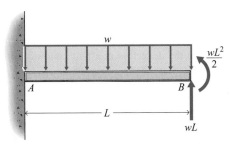

Figure P8-73

8-74* Determine the deflection at the right end of the cantilever beam shown in Fig. P8-74.

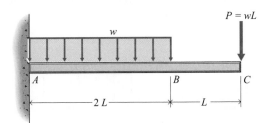

Figure P8-74

8-75 Determine the deflection at the right end of the cantilever beam shown in Fig. P8-75.

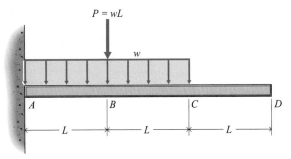

Figure P8-75

8-76 For the beam shown in Fig. P8-76, determine

a. The deflection midway between the supports.
b. The deflection at the right end of the beam.

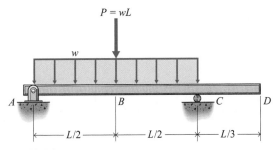

Figure P8-76

8-77* For the beam shown in Fig. P8-77, determine

a. The deflection at B.
b. The deflection at end C.

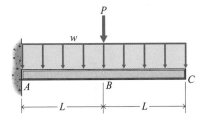

Figure P8-77

8-78 Determine the deflection midway between the supports of the beam shown in Fig. P8-78.

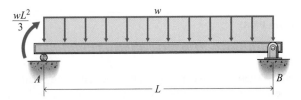

Figure P8-78

Intermediate Problems

8-79* Determine the deflection at the right end of the beam shown in Fig. P8-79.

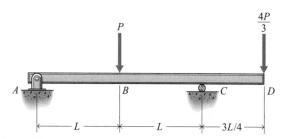

Figure P8-79

8-80* Determine the deflection at the right end of the beam shown in Fig. P8-80.

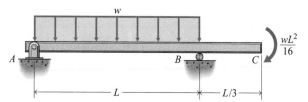

Figure P8-80

8-81 Member AB of Fig. P8-81 is the flexural member of a scale that is used to weigh food in a microwave oven. Determine the deflection of point C when $W = 5$ lb, $L = 2$ in., and $EI = 100$ lb · in.^{2}

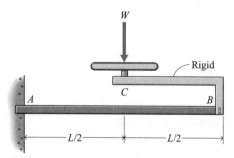

Figure P8-81

8-82* Determine the deflection at the free end of the cantilever beam shown in Fig. P8-82 when $w = 7$ kN/m, $L = 1.8$ m, $I = 130(10^{-6})$ m^4, and $E = 200$ GPa.

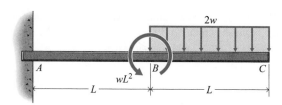

Figure P8-82

8-83 Determine the deflection at the right end of the cantilever beam shown in Fig. P8-83.

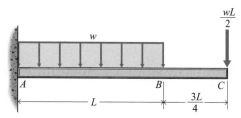

Figure P8-83

8-84 Determine the deflection at a point midway between the supports of the beam shown in Fig. P8-84.

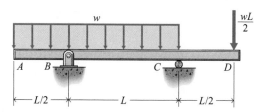

Figure P8-84

8-85* Determine the deflection at the right end of the cantilever beam shown in Fig. P8-85.

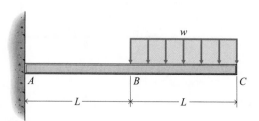

Figure P8-85

8-86 Determine the deflection at the right end of the cantilever beam shown in Fig. P8-86.

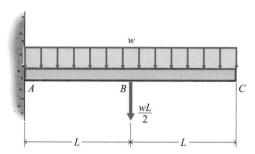

Figure P8-86

Challenging Problems

8-87* Determine the midspan deflection of beam AC of Fig. P8-87. Both beams have the same flexural rigidity.

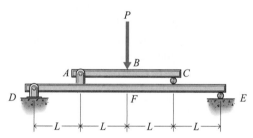

Figure P8-87

8-88* Determine the deflection at the right end of the beam shown in Fig. P8-88.

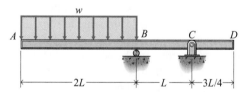

Figure P8-88

8-89 For the cantilever beam shown in Fig. P8-89, determine the deflection of the beam

a. At point B.
b. At point C.
c. At the free end, D.

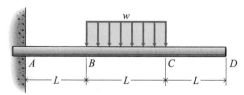

Figure P8-89

8-90* For the cantilever beam shown in Fig. P8-90, $w = 7.5$ kN/m, $L = 3$ m, $I = 180\ (10^6)$ mm^4, and $E = 200$ GPa. Determine the deflection of the beam

a. At point B.
b. At the free end, C.

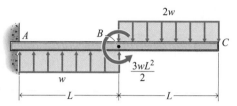

Figure P8-90

8-91 A simply supported beam is loaded as shown in Fig. P8-91. The inverted L-shaped bracket at C is rigid. Determine the deflection of the beam ABC at the middle of its span.

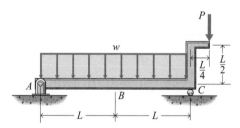

Figure P8-91

8-92 For the cantilever beam shown in Fig. P8-92, determine the deflection of the beam

a. At point B.
b. At the free end, C.

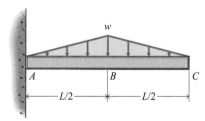

Figure P8-92

8-93* A simply supported beam carries a uniformly distributed load of 400 lb/ft and a concentrated load of 2000 lb, as shown in Fig. P8-93. If the flexural rigidity of the beam is $EI = 350\ (10^6)$ lb · in.^{2}, determine the deflection of the beam

a. Midway between the supports.
b. At the right end of the beam.
c. At B, directly under the concentrated load.

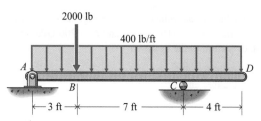

Figure P8-93

8-94 A simply supported beam is loaded and supported as shown in Fig. P8-94. If $E = 200$ GPa and $I = 90\,(10^6)$ mm^4, determine the deflection of the beam

a. Midway between the supports.
b. At the left end of the beam.
c. At a point 2 m to the right of the support at B.

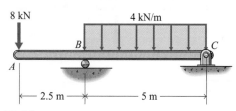

Figure P8-94

8-7 DEFLECTIONS DUE TO SHEARING STRESS

As mentioned in Section 8-2, the beam deflections calculated so far neglect the deflection produced by the shearing stresses in the beam. For short, heavily loaded beams this deflection can be significant, and an approximate method for evaluating such deflections will now be developed. The deflection of the neutral surface dv due to shearing stresses in the interval dx along the beam of Fig. 8-25 is

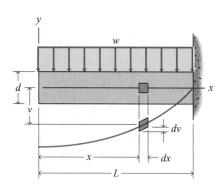

Figure 8-25

$$dv = \gamma\, dx = \frac{\tau}{G}\, dx = \frac{V_r Q}{GIt}\, dx$$

from which, since the shear in Fig. 8-25 is negative,

$$\frac{GIt}{Q}\frac{dv}{dx} = -V_r \qquad (8\text{-}4)$$

Since the vertical shearing stress varies from top to bottom of a beam, the deflection due to shear is not uniform. This nonuniform deflection due to shear is reflected in a slight warping of the cross sections of the beam. Equation 8-4 gives values too high because the maximum shearing stress (at the neutral surface) is used and also because the rotation of the differential shear element is ignored.

In order to obtain an idea of the relative amount of beam deflections due to shearing stress, consider a rectangular cross section for the beam of Fig. 8-25 and use the maximum stress for which

$$\tau_{\text{max}} = \frac{V_r Q}{It} = \frac{3}{2}\frac{V_r}{A}$$

or

$$\frac{Q}{It} = \frac{3}{2A}$$

where A is the cross-sectional area of the beam. The expression for dv becomes

$$dv = \frac{3wx\, dx}{2AG}$$

where V_r is replaced by its value $-wx$. Integration along the entire beam gives the change in v due to shear as[2]

$$v_s = \frac{3w}{2AG} \int_0^L x\,dx = \frac{3wL^2}{4AG}$$

which equals the deflection at the left end. For this same beam the magnitude of the deflection at the left end due to flexural stresses is

$$v_f = \frac{wL^4}{8EI} = \frac{3wL^4}{2EAd^2}$$

and the magnitude of the total deflection at the left end becomes

$$v = v_f + v_s = \frac{3wL^4}{2EAd^2} + \frac{3wL^2}{4AG} = \frac{3wL^2}{4AG}\left(\frac{2L^2G}{d^2E} + 1\right) \qquad (a)$$

Equation (a) indicates that the ratio of the two deflections increases as the square of L/d, which means that the deflection due to shear is of importance only in the case of very short, deep beams.

■ **Example Problem 8-16**   A structural steel ($E = 29{,}000$ ksi and $G = 11{,}000$ ksi) cantilever beam with a rectangular cross section 2 in. wide × 4 in. deep supports a concentrated load of 1000 lb at the end of a 3-ft span. Determine the percent increase in deflection at the free end of the beam resulting from the shearing stresses.

SOLUTION
The deflection dv of the neutral surface due to shearing stresses in an interval dx along the beam is

$$dv = \gamma\,dx = \frac{\tau}{G}dx = \frac{V_r Q}{ItG}dx$$

For a beam with a rectangular cross section, $I = bh^3/12$ and $Q = bh^2/8$. Also, $V_r = -P$ for a cantilever beam that supports a concentrated load P at the free end. Thus,

$$dv = -\frac{3P}{2bhG}dx$$

The deflection at v_s at the free end of the beam due to shearing stresses is

$$v_s = \int_0^L dv = -\frac{3PL}{2bhG} = -\frac{3\,(1000)\,(3)\,(12)}{2\,(2)\,(4)\,(11)\,(10^6)} = -0.0006136 \text{ in.}$$

---

[2] As stated, this result is too high because the maximum shear stress was used at every point. Using an energy method (such as Castigliano's theorem, discussed in the next section) which averages the shear stress across the cross section would yield $v_s = 3wL^2/5AG$ and Eq. (a) would become

$$v = \frac{3wL^4}{2EAd^2} + \frac{3wL^2}{5AG} = \frac{3wL^2}{5AG}\left(\frac{5L^2G}{2d^2E} + 1\right).$$

The deflection v_f at the free end of the beam due to flexure is

$$v_f = -\frac{PL^3}{3EI} = -\frac{4PL^3}{Ebh^3} = -\frac{4(1000)[3(12)]^3}{29(10^6)(2)(4)^3} = -0.05028 \text{ in.}$$

Therefore,

$$\text{Increase} = \frac{0.0006136}{0.05028}(100) = 1.220\% \qquad \textbf{Ans.}$$

PROBLEMS

8-95* A structural steel ($E = 29{,}000$ ksi and $G = 11{,}000$ ksi) cantilever beam with a 4-in.-diameter circular cross section supports a concentrated load of 1200 lb at the end of a 4-ft span. Determine the percent increase in deflection at the free end of the beam resulting from the shearing stresses.

8-96* An aluminum alloy ($E = 73$ GPa and $G = 28$ GPa) cantilever beam with a rectangular cross section 50 mm wide × 100 mm deep supports a uniformly distributed load of 5 kN/m over a 1.5-m span. Determine the percent increase in deflection at the free end of the beam resulting from the shearing stresses.

8-97 A structural steel ($E = 29{,}000$ ksi and $G = 11{,}000$ ksi) beam with a hollow rectangular cross section 3 in. wide × 5 in. deep is made from 1/2-in.-thick plate. The beam is simply supported and carries a concentrated load of 4000 lb at the center of an 8-ft span. Determine the percent increase in deflection at the center of the span resulting from the shearing stresses.

8-98 A W203 × 60 structural steel ($E = 200$ GPa and $G = 76$ GPa) wide-flange section is used as a simply supported beam to support a distributed load of 20 kN/m over a 4-m span. Determine the percent increase in deflection at the center of the span resulting from the shearing stresses.

8-8 DEFLECTIONS BY ENERGY METHODS—CASTIGLIANO'S THEOREM

Strain energy techniques are frequently used to analyze the deflections of beams and structures. Of the many available methods, the application of Castigliano's theorem, to be developed here, is one of the most widely used. It was presented in 1873 by the Italian engineer Alberto Castigliano (1847–1884). Although the theorem will be derived by considering the strain energy stored in beams, it is applicable to any structure for which the force-deformation relations are linear.

The concept of strain energy is illustrated in Fig. 8-26, in which Fig. 8-26a represents a bar of uniform cross section subjected to a slowly applied axial load P and held at the upper end by a support assumed to be rigid. From the load-deformation diagram (Fig. 8-26b) the work W_k done in elongating the bar an amount δ_2 is

$$W_k = \int_0^{\delta_2} P\,d\delta \qquad (a)$$

where P is some function of δ. The work done on the bar must equal the change in energy of the material,[3] and this energy change, because it involves the strained configuration of the material, is termed *strain energy U*. If δ is expressed in terms

---

[3]Known as Clapeyron's theorem, after the French engineer B. P. E. Clapeyron (1799–1864).

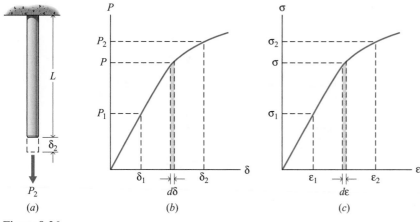

Figure 8-26

of axial strain ($\delta = L\epsilon$) and P in terms of axial stress ($P = A\sigma$), Eq. (a) becomes

$$W_k = U = \int_0^{\epsilon_2} (\sigma)(A)(L)\,d\epsilon = AL\int_0^{\epsilon_2} \sigma\,d\epsilon \qquad (b)$$

where σ is a function of ϵ (See Fig. 8-26c).

If Hooke's law applies,

$$\epsilon = \frac{\sigma}{E} \qquad \text{and} \qquad d\epsilon = \frac{1}{E}d\sigma$$

Eq. (b) becomes

$$U = \left(\frac{AL}{E}\right)\int_0^{\sigma_1} \sigma\,d\sigma$$

or

$$U = AL\left(\frac{\sigma_1^2}{2E}\right) \qquad (c)$$

Equation (c) gives the elastic strain energy (which is, in general, recoverable[4]) for axial loading of a material obeying Hooke's law. The quantity in parentheses, $\sigma_1^2/(2E)$, is the elastic strain energy u in tension or compression per unit volume, or *strain energy intensity*, for a particular value of σ. For shear loading the expression would be identical except that σ would be replaced by τ and E by G.

If the beam shown in Fig. 8-27 is slowly and simultaneously loaded by the two forces P_1 and P_2 with resulting deflections v_1 and v_2, the strain energy U of the beam is, by Clapeyron's theorem, equal to the work done by the forces. Therefore,

$$U = \frac{1}{2}P_1 v_1 + \frac{1}{2}P_2 v_2 \qquad (d)$$

---

[4]Elastic hysteresis is neglected here as an unnecessary complication.

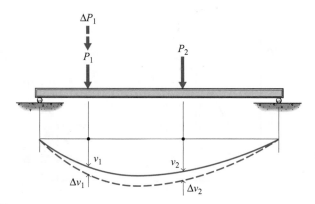

Figure 8-27

Let the force P_1 be increased by a small amount ΔP_1 (while P_1 and P_2 remain constant), and let Δv_1 and Δv_2 be the changes in deflection due to this incremental load. Because the forces P_1 and P_2 are already present, the increase in the strain energy is

$$\Delta U = \frac{1}{2}\Delta P_1 \Delta v_1 + P_1 \Delta v_1 + P_2 \Delta v_2 \qquad (e)$$

If the order of loading is reversed so that the incremental force, ΔP_1, is applied first, followed by P_1 and P_2, the resulting strain energy is

$$U + \Delta U = \frac{1}{2}\Delta P_1 \Delta v_1 + \Delta P_1 v_1 + \frac{1}{2}P_1 v_1 + \frac{1}{2}P_2 v_2 \qquad (f)$$

The resulting strain energy must be independent of the order of loading; hence, by combining Eqs. (d), (e), and (f), one obtains

$$\Delta P_1 v_1 = P_1 \Delta v_1 + P_2 \Delta v_2 \qquad (g)$$

Equations (e) and (g) can be combined to give

$$\frac{\Delta U}{\Delta P_1} = v_1 + \frac{1}{2}\Delta v_1$$

or upon taking the limit as ΔP_1 approaches zero[5]

$$\frac{\partial U}{\partial P_1} = v_1 \qquad (h)$$

For general cases in which there are many loads involved, Eq. (h) is written as

$$\frac{\partial U}{\partial P_i} = v_i \qquad (8\text{-}5a)$$

---

[5]The partial derivative is used because the strain energy is a function of both P_1 and P_2.

The following is a statement of Castigliano's theorem:

If the strain energy of a linearly elastic structure is expressed in terms of the system of external loads, the partial derivative of the strain energy with respect to a concentrated external load is the deflection of the structure at the point of application and in the direction of that load.

By a similar development, Castigliano's theorem can also be shown to be valid for applied moments and the resulting rotations (or changes in slope) of the structure. Thus,

$$\frac{\partial U}{\partial M_i} = \theta_i \tag{8-5b}$$

If the deflection is required either at a point where there is no unique point load or in a direction not aligned with the applied load, a dummy load is introduced at the desired point acting in the proper direction. The deflection is obtained by first differentiating the strain energy with respect to the dummy load and then taking the limit as the dummy load approaches zero. Also, for the application of Eq. 8-5b, either a unique point moment or a dummy moment must be applied at point i. The moment will be in the direction of rotation at the point. Note that if the loading consists of a number of point loads, all expressed in terms of a single parameter (e.g., P, $2P$, $3P$, wL, $2wL$), and if the deflection is wanted at one of the applied loads, one must either write the moment equation with this load as a separate identifiable term or add a dummy load (say Q) at the point so that the partial derivative can be taken with respect to this load only.

It was previously shown that the strain energy per unit volume for uniaxial stress is $\sigma^2/(2E)$; hence, the total strain energy under uniaxial stress is

$$U = \int_{\text{vol}} \frac{\sigma^2}{2E} dV$$

For a beam of constant (or slowly varying) cross section subjected to pure bending, the principal stresses are parallel to the axis of the beam; therefore, using Eq. 7-3, the strain energy becomes

$$U = \frac{1}{2E} \int_{\text{vol}} \left(-\frac{M_r y}{I} \right)^2 dV \tag{i}$$

By writing dV as $dA\, dx$ (where x is measured along the axis of the beam), Eq. (i) becomes

$$U = \frac{1}{2E} \int_0^L \left(\frac{M_r^2}{I^2} \int_{\text{area}} y^2 dA \right) dx = \frac{1}{2E} \int_0^L \frac{M_r^2}{I} dx \tag{8-6}$$

Equation 8-6 was developed for a beam loaded in pure bending. However, most real beams will be subjected to transverse loads that induce shearing stresses and, in the case of distributed loading, transverse normal stresses. It is assumed that the transverse normal stress is small enough to neglect, and in Section 8-7, it was shown that, except for short, deep beams, the deflection due to shearing stresses is also small enough to neglect. Hence, Eq. 8-6 which neglects strain energy due to these stresses, is applicable to the usual real beams. In applying Eq. 8-5 or Eq. 8-6, it is usually much simpler to apply Leibnitz's rule[6] to differentiate under the integral sign so that

$$v_i = \frac{\partial U}{\partial P_i} = \frac{1}{E} \int_0^L \frac{M_r}{I} \frac{\partial M_r}{\partial P_i} dx \qquad (8\text{-}7)$$

Example Problem 8-17
The cantilever beam shown in Fig. 8-28a is subjected to a concentrated load P at the left end. Determine the deflection and slope at the left end of the beam using Castigliano's theorem. The flexural rigidity is constant. Neglect the deflection due to shear.

SOLUTION
The deflection is found using Eqs. 8-5a and 8-7, and the slope is found using Eq. 8-5b. For the deflection

$$v_i = \frac{\partial U}{\partial P_i} = \frac{1}{EI} \int M_r \frac{\partial M_r}{\partial P_i} dx \qquad (a)$$

and for the slope

$$\theta_i = \frac{\partial U}{\partial M_i} = \frac{1}{EI} \int M_r \frac{\partial M_r}{\partial M_i} dx \qquad (b)$$

To apply Eq. (a) one needs a force in the direction of the deflection at the point i where the deflection is to be found. The force P at point A satisfies these conditions. In a similar manner, to find θ_A one needs a moment M_A at the point i where the slope is to be found. Such a moment does not exist; thus a dummy moment is placed at A. In Fig. 8-28b the dummy moment is shown dashed to indicate that it is fictitious. The moment M_A is set equal to zero after the partial derivatives in Eqs. (a) and (b) are performed.

Using the free-body diagram of Fig. 8-28c and the moment equation of equilibrium,

$$+\downarrow \Sigma M_o = 0: \qquad\qquad M_r + Px + M_A = 0$$

gives

$$M_r = -Px - M_A \qquad 0 \le x \le L \qquad (c)$$

Combining Eqs. (a) and (c) gives

$$v_A = \frac{1}{EI} \int_0^L M_r \frac{\partial M_r}{\partial P} dx = \frac{1}{EI} \int_0^L (-Px - M_A)(-x) dx$$

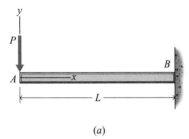

(a)

(b)

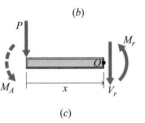

(c)

Figure 8-28

---

[6]*Advanced Calculus*, W. Kaplan, Addison-Wesley, Reading, Mass., 1953, p. 219.

Setting $M_A = 0$ and carrying out the integration gives

$$v_A = \frac{PL^3}{3EI}$$ **Ans.**

The positive sign for v_A indicates that the deflection is in the direction of the load P. This result agrees with case 1 of Table B-19 (Appendix B).

Combining Eqs. (b) and (c) gives

$$\theta_A = \frac{1}{EI} \int_0^L M_r \frac{\partial M_r}{\partial M_A} dx = \frac{1}{EI} \int_0^L (-Px - M_A)(-1)\, dx$$

Setting $M_A = 0$ and carrying out the integration gives

$$\theta_A = \frac{PL^2}{2EI}$$ **Ans.**

The positive sign for θ_A indicates a slope to the left and down, to correspond to the direction of M_A. This expression for θ_A also agrees with case 1 of Table B-19 (Appendix B).

■ **Example Problem 8-18** Determine the deflection at the free end of a cantilever beam of constant cross section and length L that is loaded with a force P at the free end and a distributed load that varies linearly from zero at the free end to w_o at the support, as shown in Fig. 8-29a.

SOLUTION

From the free-body diagram of Fig. 8-29b, the moment equation is

$$M_r = -Px - \frac{w_o x^3}{6L}$$

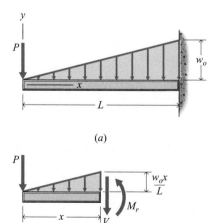

(a)

(b)

Figure 8-29

The partial derivative of M_r with respect to P is $-x$, and Eq. 8-7 becomes

$$EIv = \int_0^L M_r \frac{\partial M_r}{\partial P} dx$$

$$= \int_0^L \left(-Px - \frac{w_o x^3}{6L} \right) (-x)\, dx$$

$$= \int_0^L \left(Px^2 + \frac{w_o x^4}{6L} \right) dx = +\frac{PL^3}{3} + \frac{w_o L^4}{30}$$

The positive signs indicate deflection in the direction of the force P; hence,

$$v = \frac{PL^3}{3EI} + \frac{w_o L^4}{30EI} \downarrow \qquad\qquad \textbf{Ans.}$$

The result can be verified by referring to Table B-19 (cases 1 and 3) in Appendix B.

■ **Example Problem 8-19**   Determine the deflection at the center of a simply supported beam of constant cross section and span L carrying a uniformly distributed load w over its entire length.

SOLUTION

The deflection is required at a point where there is no unique point load. Thus, a dummy load P is introduced at the center of the beam in the direction of the desired deflection. In the free-body diagram of Fig. 8-30, the dashed force P represents the dummy load. The moment equation is

$$M_r = M_w + M_p = \frac{wLx}{2} - \frac{wx^2}{2} + \frac{Px}{2} - P\left\langle x - \frac{L}{2} \right\rangle^1$$

where the quantity $\langle x - L/2 \rangle^1$ is zero for all $x \le L/2$ (see Section 8-5). The partial derivative of M_r with respect to P is

$$\frac{\partial M_r}{\partial P} = \frac{x}{2} - \left\langle x - \frac{L}{2} \right\rangle^1$$

The dummy force P is equated to zero after the partial derivative is taken, and the deflection is given by

$$EIv = \int_0^L M_r \frac{\partial M_r}{\partial P} dx$$

$$= \int_0^L \left(\frac{wLx}{2} - \frac{wx^2}{2} \right)\left(\frac{x}{2} - \left\langle x - \frac{L}{2} \right\rangle^1 \right) dx$$

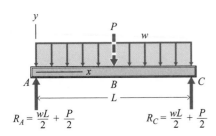

Figure 8-30

$R_A = \dfrac{wL}{2} + \dfrac{P}{2}$ $R_C = \dfrac{wL}{2} + \dfrac{P}{2}$

which, for ease of integration, can be written as

$$EIv = \frac{w}{4} \int_0^L (Lx^2 - x^3)dx + \frac{w}{4} \int_{L/2}^L (L^2x - 3Lx^2 + 2x^3)dx = \frac{5wL^4}{384}$$

Since the deflection is positive, it is in the direction of the force,

$$v = \frac{5wL^4}{384EI} \downarrow \qquad \qquad \textbf{Ans.}$$

This result corresponds to case 7 of Table B-19 (Appendix B).

ALTERNATE SOLUTION
The moment equations may be written as

$$M_r = \frac{wLx}{2} + \frac{Px}{2} - \frac{wx^2}{2} \qquad 0 \le x \le \frac{L}{2}$$

and

$$M_r = \frac{wLx}{2} + \frac{Px}{2} - \frac{wx^2}{2} - P\left(x - \frac{L}{2}\right) \qquad \frac{L}{2} \le x \le L$$

The partial derivatives of M_r with respect to P are

$$\frac{\partial M_r}{\partial P} = \frac{x}{2} \qquad 0 \le x \le \frac{L}{2}$$

and

$$\frac{\partial M_r}{\partial P} = \frac{x}{2} - x + \frac{L}{2} = \frac{L}{2} - \frac{x}{2} \qquad \frac{L}{2} \le x \le L$$

Setting the dummy force P equal to zero and using Eq. 8-7 gives

$$EIv = \int_0^{L/2} \left(\frac{wLx}{2} - \frac{wx^2}{2}\right)\left(\frac{x}{2}\right)dx + \int_{L/2}^L \left(\frac{wLx}{2} - \frac{wx^2}{2}\right)\left(\frac{L}{2} - \frac{x}{2}\right)dx = \frac{5wL^4}{384}$$

which is the same as the result obtained previously using singularity functions. Note further, however, that the strain energies in the two segments of the beam are equal due to symmetry. Therefore, the integration need only extend over half of the beam if the strain energy is doubled. Thus, from Eq. 8-7,

$$EIv = 2\int_0^{L/2} \left(\frac{wLx}{2} - \frac{wx^2}{2}\right)\left(\frac{x}{2}\right)dx = \frac{5wL^4}{384}$$

PROBLEMS

Introductory Problems

8-99* Determine the deflection and slope at the left end of the cantilever beam shown in Fig. P8-99.

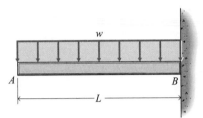

Figure P8-99

8-100* A beam is loaded and supported as shown in Fig. P8-100. Determine the deflection at the concentrated load P.

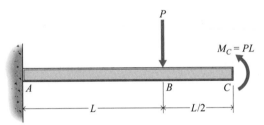

Figure P8-100

8-101 Determine the deflection and slope at the left end of the cantilever beam shown in Fig. P8-101.

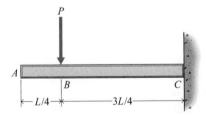

Figure P8-101

8-102* A beam is loaded and supported as shown in Fig. P8-102. Determine the deflection at the concentrated load P.

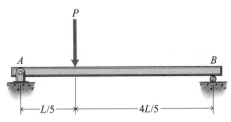

Figure P8-102

8-103 A beam is loaded and supported as shown in Fig. P8-103. Determine the deflection

a. At the concentrated load P.
b. At a point midway between the supports.

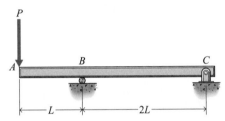

Figure P8-103

Intermediate Problems

8-104* A beam is loaded and supported as shown in Fig. P8-104. Determine the deflection at the midpoint of the distributed load.

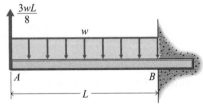

Figure P8-104

8-105* Determine the deflection at the left end of the cantilever beam shown in Fig. P8-105.

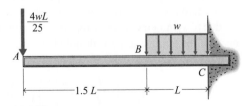

Figure P8-105

8-106 For the beam shown in Fig. P8-106, determine the slope and deflection at the section in the beam where the couple M is applied.

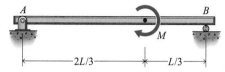

Figure P8-106

8-107* Determine the deflection at point B of the beam shown in Fig. P8-107.

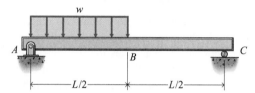

Figure P8-107

8-108 Determine the deflections at points A and B of the beam shown in Fig. P8-108.

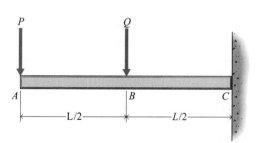

Figure P8-108

8-109 A beam is loaded and supported as shown in Fig. P8-109. Determine the deflection

a. At the left end of the beam.
b. At a point midway between the supports.

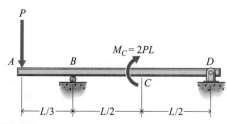

Figure P8-109

Challenging Problems

8-110* Determine the deflection at a section midway between the supports when the beam is loaded and supported as shown in Fig. P8-110.

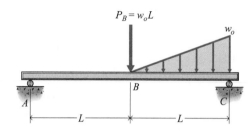

Figure P8-110

8-111* A beam is loaded and supported as shown in Fig. P8-111. Determine the deflection at the left end of the beam.

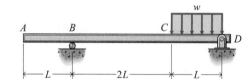

Figure P8-111

8-112 The cantilever beam shown in Fig. P8-112 has a second moment of area of $2I$ in the interval AB and I in the interval BC. Determine the deflection at point C.

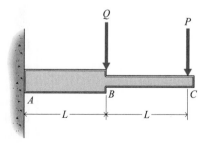

Figure P8-112

8-113 Determine the deflection at point B of the beam shown in Fig. P8-113. The second moment of area is $2I$ in the center section BC and I in the sections near the supports.

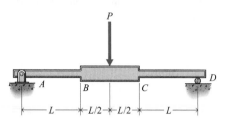

Figure P8-113

8-114 Determine the deflection midway between the supports of the beam shown in Fig. P8-114.

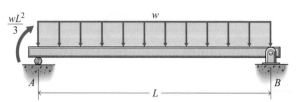

Figure P8-114

8-9 STATICALLY INDETERMINATE BEAMS

A beam, subjected only to transverse loads, with more than two reaction components, is statically indeterminate because the equations of equilibrium are not sufficient to determine all the reactions. In such cases the geometry of the deformation of the loaded beam is used to obtain the additional relations needed for an evaluation of the reactions (or other unknown forces). For problems involving elastic action, each additional constraint on a beam provides additional information concerning slopes or deflections. Such information, when used with the appropriate slope or deflection equations, yields expressions that supplement the independent equations of equilibrium.

8-9-1 The Integration Method
For statically determinate beams, known slopes and deflections were used to obtain boundary and matching conditions, from which the constants of integration in the elastic curve equation could be evaluated. For statically indeterminate beams, the procedures are identical. However, the moment equations will contain reactions or loads that cannot be evaluated from the available equations of equilibrium, and one additional boundary condition is needed for the evaluation of each such unknown. For example, if a beam is subjected to a force system for which there are two independent equilibrium equations and if there are four unknown reactions or loads on the beam, two boundary or matching conditions are needed in addition to those necessary for the determination of the constants of integration. These extra boundary conditions, when substituted in the appropriate elastic curve equations (slope or deflection), will yield the necessary additional equations. The following examples illustrate the method.

Example Problem 8-20
A beam is loaded and supported as shown in Fig. 8-31a. Determine the reactions at A and B.

SOLUTION
From the free-body diagram of Fig. 8-31b it is seen that there are three unknown reaction components (M_A, V_A, R_B) and that only two independent equations of equilibrium are available. The additional unknown requires the use of the elastic curve equation, for which one extra boundary condition is required in addition to the two required for the constants of integration. Because three boundary

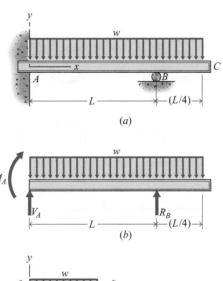

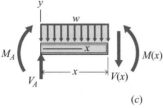

Figure 8-31

conditions are available in the interval between the supports, only one elastic curve equation needs to be written. The origin of coordinates is arbitrarily placed at the wall and for the interval $0 \le x \le L$, the boundary conditions are: when $x = 0$, $dv/dx = 0$; when $x = 0$, $v = 0$; and when $x = L$, $v = 0$. From Fig. 8-31c and Eq. 8-1,

$$EI \frac{d^2v}{dx^2} = M(x) = V_A x + M_A - \frac{wx^2}{2} \qquad 0 \le x \le L$$

Integration gives

$$EI \frac{dv}{dx} = \frac{V_A x^2}{2} + M_A x - \frac{wx^3}{6} + C_1$$

The first boundary condition $dv/dx = 0$ when $x = 0$ gives $C_1 = 0$. A second integration yields

$$EIv = \frac{V_A x^3}{6} + \frac{M_A x^2}{2} - \frac{wx^4}{24} + C_2$$

The second boundary condition $v = 0$ when $x = 0$ gives $C_2 = 0$, and the last boundary condition $v = 0$ when $x = L$ gives

$$0 = \frac{V_A L^3}{6} + \frac{M_A L^2}{2} - \frac{wL^4}{24}$$

which reduces to

$$4V_A L + 12M_A = wL^2 \tag{a}$$

The equation of equilibrium $\Sigma M_B = 0$ for the free-body diagram of Fig. 8-31b yields

$$+\uparrow \Sigma M_B = 0: \qquad V_A L + M_A - w\left(\frac{5L}{4}\right)\left(\frac{3L}{8}\right) = 0$$

which reduces to

$$32V_A L + 32M_A = 15wL^2 \tag{b}$$

Simultaneous solution of Eqs. (a) and (b) gives

$$M_A = -\frac{7wL^2}{64} = \frac{7wL^2}{64}\downarrow \qquad\qquad \textbf{Ans.}$$

and

$$V_A = +\frac{37wL}{64} = \frac{37wL}{64}\uparrow \qquad\qquad \textbf{Ans.}$$

▶ An alternate solution would be to place the origin of coordinates at the right support and write the moment equation for the interval $0 \le x \le L$. This equation would involve only one unknown, the reaction R_B. Upon integration and evaluation of the constants, the third boundary condition would directly yield the value of R_B. The two independent equilibrium equations could then be used to evaluate M_A and V_A.

Finally, the equation $\Sigma F_y = 0$ for Fig. 8-31b gives

$$+\uparrow \Sigma F_y = 0: \qquad R_B + V_A - w(5L/4) = 0$$

from which

$$R_B = +\frac{43wL}{64} = \frac{43wL}{64}\uparrow \qquad\qquad \textbf{Ans.}$$

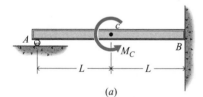

(a)

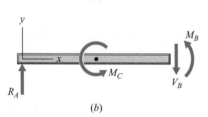

(b)

Figure 8-32

Example Problem 8-21   A beam is loaded and supported as shown in Fig. 8-32a. Determine

(a) The reactions at supports A and B.
(b) The deflection at the middle of the span.

SOLUTION

(a) There are three unknown reaction components (M_B, V_B, and R_A) on the free-body diagram of Fig. 8-32b. Because there are three unknown reaction components and only two independent equations of equilibrium, the problem is statically indeterminate and a deformation equation will be needed in order to solve for the three unknowns. Additional complications arise because two moment equations are needed, one in the interval $0 \le x \le L$ and another in the interval $L \le x \le 2L$. Integrating each moment equation twice results in four constants of integration, along with the unknown reaction R_A. To solve the resulting equations requires three boundary conditions (one at A and two at B) and two matching conditions at the point where the moment is applied. The lengthy algebraic computations can be simplified if one uses singularity functions. The moment equation is

$$EI \frac{d^2v}{dx^2} = M(x) = R_A x - M_C \langle x - L \rangle^0 \qquad 0 \le x \le 2L$$

which upon integration yields

$$EI \frac{dv}{dx} = \frac{R_A x^2}{2} - M_C \langle x - L \rangle^1 + C_1$$

$$EIv = \frac{R_A x^3}{6} - \frac{M_C}{2} \langle x - L \rangle^2 + C_1 x + C_2$$

Using the boundary condition $v = 0$ when $x = 0$ gives $C_2 = 0$. The boundary condition $dv/dx = 0$ when $x = 2L$ gives

$$2R_A L^2 - M_C L + C_1 = 0 \qquad\qquad (a)$$

and the boundary condition $v = 0$ when $x = 2L$ gives

$$\frac{4R_A L^3}{3} - \frac{M_C L^2}{2} + 2C_1 L = 0 \qquad\qquad (b)$$

Solving Eqs. (a) and (b) simultaneously yields

$$C_1 = -\frac{M_C L}{8} \qquad \text{and} \qquad R_A = +\frac{9M_C}{16L} = \frac{9M_C}{16L} \uparrow \qquad \textbf{Ans.}$$

The remaining two unknowns are found using the equations of equilibrium and the free-body diagram shown in Fig. 8-32b.

$$+\uparrow \Sigma F_y = 0: \qquad R_A - V_B = \frac{9M_C}{16L} - V_B = 0$$

$$+\downarrow \Sigma M_B = 0: \qquad -R_A(2L) + M_C + M_B$$

$$= -\frac{9M_C}{16L}(2L) + M_C + M_B = 0$$

The reactions at B obtained from the above two equations are

$$V_B = +\frac{9M_C}{16L} = \frac{9M_C}{16L} \downarrow \qquad \text{and} \qquad M_B = +\frac{M_C}{8} = \frac{M_C}{8} \uparrow \qquad \textbf{Ans.}$$

(b) Setting $x = L$ in the elastic curve equation

$$EIv = \frac{R_A x^3}{6} - \frac{M_C}{2} \langle x - L \rangle^2 + C_1 x + C_2$$

yields

$$EIv = \frac{1}{6} \left(\frac{9M_C}{16L} \right)(L)^3 - \frac{1}{2}(M_C)(0)^2 + \left(-\frac{M_C L}{8} \right)(L) + 0$$

from which the deflection at the middle of the span is

$$v = -\frac{M_C L^2}{32EI} = \frac{M_C L^2}{32EI} \downarrow \qquad \textbf{Ans.}$$

PROBLEMS

MecMovie Activities and Problems

MM8.6 Propped cantilevers. Example; Concept checkpoints. Determine the roller reaction for a propped cantilever.

MM8.7 Beam on three supports. Concept checkpoints. Determine one roller reaction for a simply supported beam on three supports using superposition.

Introductory Problems

8-115* A beam is loaded and supported as shown in Fig. P8-115. Determine the reactions at the supports A and B.

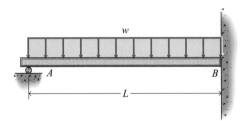

Figure P8-115

8-116* When the moment M is applied to the beam shown in Fig. P8-116, the slope at the left end of the beam is zero. Determine the magnitude of the moment M.

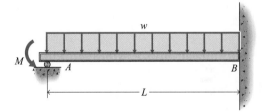

Figure P8-116

8-117 A beam is loaded and supported as shown in Fig. P8-117. Determine the magnitude of the moment M required to make the slope at the left end of the beam zero.

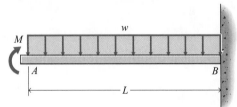

Figure P8-117

8-118* When the moment M is applied to the left end of the cantilever beam shown in Fig. P8-118, the slope at the left end of the beam is zero. Determine the magnitude of the moment M.

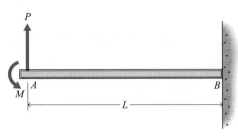

Figure P8-118

8-119 When the load P is applied to the right end of the cantilever beam shown in Fig. P8-119, the deflection at the right end of the beam is zero. Determine the magnitude of the load P.

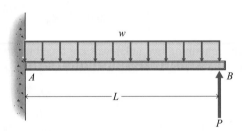

Figure P8-119

8-120 A beam is loaded and supported as shown in Fig. P8-120. Determine the reactions at the supports A and B.

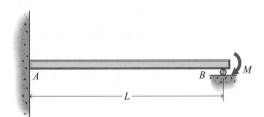

Figure P8-120

Intermediate Problems

8-121* A beam is loaded and supported as shown in Fig. P8-121. Determine the reactions at the supports A and B.

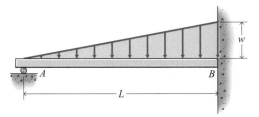

Figure P8-121

8-122* A beam is loaded and supported as shown in Fig. P8-122. Determine the magnitude of the load P required to make the slope of the beam zero at the right end.

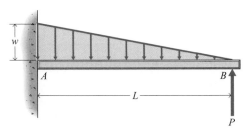

Figure P8-122

8-123 Determine the support reactions for the beam loaded and supported as shown in Fig. P8-123.

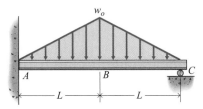

Figure P8-123

8-124* A beam is loaded and supported as shown in Fig. P8-124. Determine the reactions at the supports A and B.

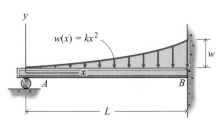

Figure P8-124

8-125 A beam is loaded and supported as shown in Fig. P8-125. Determine

a. The reactions at the supports A and B.
b. The deflection at the middle of the span.

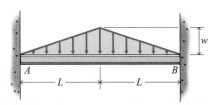

Figure P8-125

8-126 A beam is loaded and supported as shown in Fig. P8-126. Determine the reactions at the supports A and B.

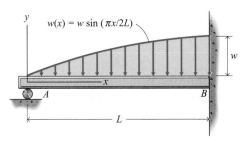

Figure P8-126

8-127* A beam is loaded and supported as shown in Fig. P8-127. Determine the reactions at the supports B, C, and D.

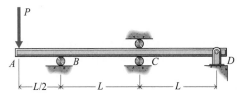

Figure P8-127

8-128* A beam is loaded and supported as shown in Fig. P8-128. Determine

a. The reactions at supports A and B.
b. The deflection at the middle of the span.

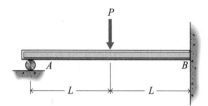

Figure P8-128

8-129 A beam is loaded and supported as shown in Fig. P8-129. Determine

a. The reactions at the supports A and B.
b. The deflection at the middle of the span.

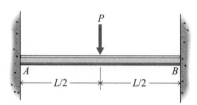

Figure P8-129

Challenging Problems

8-130* A beam is loaded and supported as shown in Fig. P8-130. Determine

a. The reactions at the supports A, B, and C.
b. The moment over the middle support.

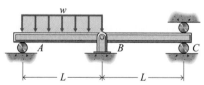

Figure P8-130

8-131* A beam is loaded and supported as shown in Fig. P8-131. Determine the reactions at the supports A, C, and D.

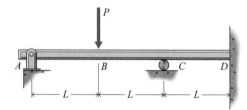

Figure P8-131

8-132 A beam is loaded and supported as shown in Fig. P8-132. Determine the reactions at the supports A, B, and C.

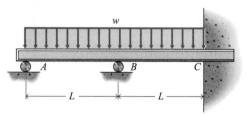

Figure P8-132

8-133* A beam is loaded and supported as shown in Fig. P8-133. Determine the reactions at the supports A and C.

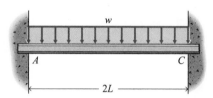

Figure P8-133

8-134 A beam is loaded and supported as shown in Fig. P8-134. Determine

a. The reactions at the supports A and D.
b. The deflection at B if $E = 200$ GPa and $I = 350 (10^6)$ mm^4.

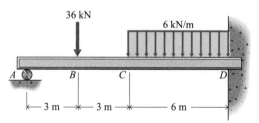

Figure P8-134

8-135 A beam is loaded and supported as shown in Fig. P8-135. Determine the reactions at the supports A, B, and D.

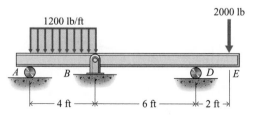

Figure P8-135

Computer Problems

8-136 A 10-m-long beam carries a uniformly distributed load and is supported as shown in Fig. P8-136. Horizontal reactions at the supports may be neglected. If the beam is constructed of wood ($EI = 1500$ kN $\cdot$ m^2), compute and plot

a. The deflection curves for the beam (plot v as a function of x for $0 \le x \le 10$ m) for center support locations of $b = 2$ m, 4 m, and 7 m.

b. The bending moment distribution along the beam (plot M_r as a function of x for $0 \le x \le 10$ m) for center support locations of $b = 2$ m, 4 m, and 7 m.

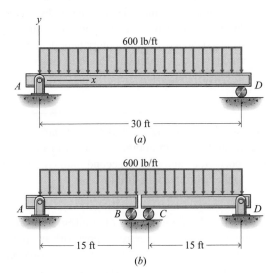

Figure P8-136

8-137 A 30-ft-long beam carries a uniformly distributed load and is simply supported as shown in Fig. P8-137a. Modified designs of the beam consist of replacing the single long beam with a pair of shorter beams (Fig. P8-137b) or adding a center support to the long beam (Fig. P8-137c). If the beams are all constructed of wood ($EI = 12,000$ kip $\cdot$ ft^2), compute and plot the deflection curves for the three cases (plot v as a function of x for $0 \le x \le 30$ ft). Plot all three cases on the same graph. (Neglect horizontal reactions at the supports.)

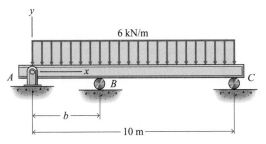

(a)

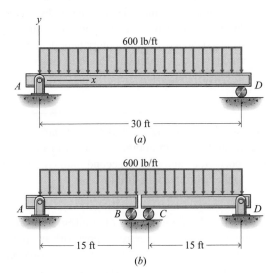

(b)

Figure P8-137(a–b)

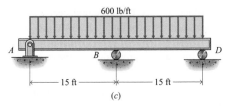

(c)

Figure P8-137(c)

8-138 A 6-m-long WT178 × 51 structural steel section is used for the cantilever beam shown in Fig. P8-138. The flange is at the top of the beam. If a roller support is added to the beam at B, compute and plot

a. The deflection curves for the beam (plot v as a function of x for $0 \le x \le 6$ m) for support locations of $b = 3$ m, 4 m, and 5 m.

b. The bending moment distribution along the beam (plot M_r as a function of x for $0 \le x \le 6$ m) for support locations of $b = 3$ m, 4 m, and 5 m.

c. The maximum tensile and compressive flexural stresses in the beam as a function of b for 1 m $\le b \le 6$ m. What range of b would be acceptable for this beam?

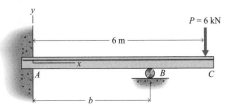

Figure P8-138

8-139 A concentrated load $P = 3900$ lb moves slowly across the beam shown in Fig. P8-139. Horizontal reactions at the supports may be neglected. The 16-ft-long beam is a W 4 × 13 structural steel section. Compute and plot

a. The deflection curves for the beam (plot v as a function of x for $0 \le x \le 16$ ft) for load locations of $b = 3$ ft, 7 ft, and 11 ft.

b. The bending moment distribution along the beam (plot M_r as a function of x for $0 \le x \le 16$ ft) for load locations of $b = 3$ ft, 7 ft, and 11 ft.

c. The maximum flexural stresses in the beam as a function of b for $1 \le b \le 8$ ft. What range of b would be acceptable for this beam?

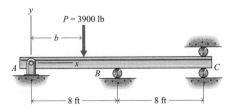

Figure P8-139

8-140 A 6-m-long S457 × 104 structural steel section is used for the two-span beam shown in Fig. P8-140. The beam supports a uniformly distributed load w of 100 kN/m. The support at B settles with time until it provides no resistance to deflection of the beam.

a. Determine b_{max}, the maximum amount that the support B will settle.
b. Compute and plot the deflection curves for the beam (plot v as a function of x for $0 \le x \le 6$ m) for support B initially ($b = 0$) and settled ($b = b_{max}/3$ and $b = 2b_{max}/3$).
c. Compute and plot the bending moment distribution along the beam (plot M_r as a function of x for $0 \le x \le 6$ m) for support B initially ($b = 0$) and settled ($b = b_{max}/3$ and $b = 2b_{max}/3$).
d. Compute and plot the maximum flexural stresses in the beam as a function of b ($0 \le b \le b_{max}$). What range of b would be acceptable for this beam?

8-141 A 20-ft-long S12 × 35 structural steel section is used for the two-span beam shown in Fig. P8-141. The beam supports a uniformly distributed load w of 1800 lb/ft. The support at A settles with time until it provides no resistance to deflection of the beam.

a. Determine a_{max}, the maximum amount that the support A will settle.
b. Compute and plot the deflection curves for the beam (plot v as a function of x for $0 \le x \le 20$ ft) for support A initially ($a = 0$) and settled ($a = a_{max}/3$ and $a = 2a_{max}/3$).
c. Compute and plot the bending moment distribution along the beam (plot M_r as a function of x for $0 \le x \le 20$ ft) for support A initially ($a = 0$) and settled ($a = a_{max}/3$ and $a = 2a_{max}/3$).
d. Compute and plot the maximum flexural stresses in the beam as a function of a ($0 \le a \le a_{max}$). What range of a would be acceptable for this beam?

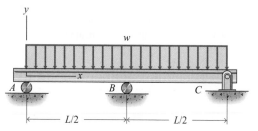

Figure P8-140

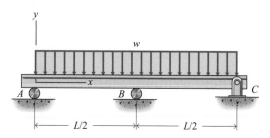

Figure P8-141

8-9-2 The Superposition Method

The concept (discussed in Section 8-6) that a slope or deflection due to several loads in the algebraic sum of the slopes or deflections due to each of the loads acting individually is frequently used to provide the deformation equations needed to supplement the equilibrium equations in the solution of statically indeterminate beam problems. To provide the necessary deformation equations, selected restraints are removed and replaced by unknown loads (forces and couples); the deformation diagrams corresponding to individual loads (both known and unknown) are sketched; and the component deflections or slopes are summed to produce the known configuration. The following examples illustrate the use of superposition for this purpose.

Example Problem 8-22   A steel ($E = 30,000$ ksi) beam 20 ft long is simply supported at the ends and at the midpoint, as shown in Fig. 8-33a. Determine the reactions at supports A, B, and C. The second moment of area of the cross section with respect to the neutral axis is 100 in.[4]

▶ The center support exerts whatever force is necessary to prevent the center of the beam from settling. Therefore, the problem has been reformulated: "Determine the upward force at B that will make the deflection at B equal to zero."

SOLUTION
With three unknown reactions and only two equations of equilibrium available, the beam is statically indeterminate. Replace the center support with an unknown applied load. The resulting simply supported beam is equivalent to two beams with individual loads, as shown in Fig. 8-33b. The resulting deflection at the midpoint of the beam is the deflection v_R due to R_B plus the deflection v_w due to

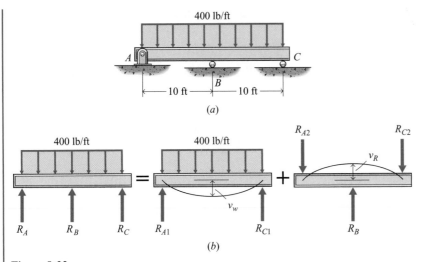

Figure 8-33

the uniform load; that is,

$$v = v_R + v_w = 0 \qquad (a)$$

The deflection v_w can be obtained from case 7 of Table B-19 in Appendix B, and is

$$v_w = -\frac{5wL^4}{384EI} = -\frac{5(400/12)[20(12)]^4}{384(30)(10^6)(100)} = -0.4800 \text{ in.}$$

The deflection v_R can be obtained in terms R_B from case 6 of Table B-19 in Appendix B, and is

$$v_R = \frac{R_B L^3}{48EI} = \frac{R_B[20(12)]^3}{48(30)(10^6)(100)} = 96(10^{-6})R_B \text{ in.}$$

When these values are substituted in Eq. (a), the result is

$$v = 96(10^{-6})R_B - 0.4800 = 0$$

from which

$$R_B = +5000 \text{ lb} = 5000 \text{ lb} \uparrow \qquad \textbf{Ans.}$$

The equilibrium equation $\Sigma F_y = 0$ and symmetry give

$$R_A = R_C = \frac{1}{2}[400(20) - 5000] = +1500 \text{ lb} = 1500 \text{ lb} \uparrow \qquad \textbf{Ans.}$$

The arithmetic will frequently be simplified in beam-deflection problems if expressions for the deflections are substituted in the deflection equation in symbol form. In this example, Eq. (*a*) becomes

$$\frac{R_B L^3}{48EI} - \frac{5wL^4}{384EI} = 0$$

which reduces to

$$R_B = \frac{5wL}{8}$$

or

$$R_B = \frac{5\,(400)\,(20)}{8} = 5000 \text{ lb} \uparrow$$

▶ The built-in support at the right end exerts whatever force is necessary to prevent
the right end of the beam from settling or
rotating. Therefore, the problem has been
reformulated: "Determine the force and moment at *B* that will make the deflection and
slope of the beam at *B* equal to zero."

Example Problem 8-23   A beam is loaded and supported as shown
in Fig. 8-34*a*. Determine the reactions at supports *A* and *B*.

SOLUTION

There are four unknown reactions (a shear and moment at each end), and only
two equations of equilibrium are available; therefore, the beam is statically indeterminate, and two deformation equations are necessary. The constraint at
the right end can be replaced with an unknown force and couple. The resulting
cantilever beam is equivalent to three beams with individual loads, as shown in
Fig. 8-34*b*. Note that the unknown shear and moment at the right end are both
shown as positive values so that the algebraic sign of the result will be correct.
From the geometry of the constrained beam, the resultant slope and the resultant
deflection at the right end are both zero. The slope and deflection at the end of
each of the three replacement beams can be obtained from the expressions in
Table B-19 of Appendix B. Thus, the first beam with load *P* (see case 1 of Table
B-19) has a constant slope from *P* to the end of the beam, which is

$$\theta_P = -\frac{Pa^2}{2EI}$$

The deflection v_P at the end is made up of two parts v_1 for a beam of length *a*,
and v_2 the added deflection of the tangent segment (straight line) from *P* to the

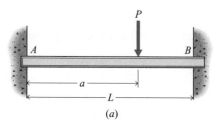

(*a*)

Figure 8-34(a)

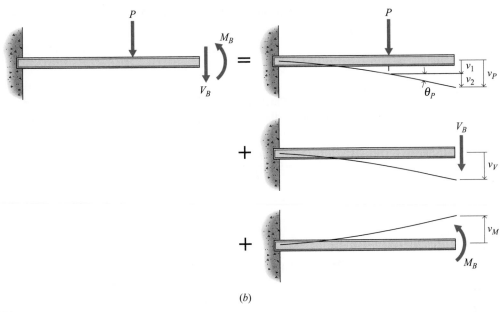

(b)

Figure 8-34(b)

end of the beam. This deflection is

$$v_p = v_1 + v_2 = -\frac{Pa^3}{3EI} + (L - a)\theta_P$$

$$= -\frac{Pa^3}{3EI} + (L - a)\left(-\frac{Pa^2}{2EI}\right) = \frac{Pa^3}{6EI} - \frac{Pa^2L}{2EI}$$

The slope and deflection at the end of the beam due to the shear V_B (also from case 1 of Table B-19) are

$$\theta_V = -\frac{V_B L^2}{2EI} \quad \text{and} \quad v_V = -\frac{V_B L^3}{3EI}$$

Finally, the slope and deflection at the right end of the beam due to M_B (see case 4 of Table B-19) are

$$\theta_M = \frac{M_B L}{EI} \quad \text{and} \quad v_M = \frac{M_B L^2}{2EI}$$

Since the resultant slope is zero,

$$\theta_P + \theta_V + \theta_M = -\frac{Pa^2}{2EI} - \frac{V_B L^2}{2EI} + \frac{M_B L}{EI} = 0$$

Similarly,

$$v_P + v_V + v_M = \frac{Pa^3}{6EI} - \frac{Pa^2L}{2EI} - \frac{V_B L^3}{3EI} + \frac{M_B L^2}{2EI} = 0$$

The simultaneous solution of these two equations yields

$$M_B = -\frac{Pa^2(L-a)}{L^2} \quad \text{and} \quad V_B = -\frac{Pa^2(3L-2a)}{L^3} \qquad \textbf{Ans.}$$

When the equations of equilibrium are applied to a free-body diagram of the entire beam, the shear and moment at the left end are found to be

$$M_A = -\frac{Pa(L-a)^2}{L^2} \quad \text{and} \quad V_A = +\frac{P(L^3 - 3a^2L + 2a^3)}{L^3} \qquad \textbf{Ans.}$$

Example Problem 8-24 The 4-in.-wide × 6-in.-deep timber ($E = 1200$ ksi) beam shown in Fig. 8-35a is fixed at the left end and supported at the right end with an aluminum alloy ($E = 10,000$ ksi) tie rod that has a cross-sectional area of 0.125 in.^{2} Determine the tension in the tie rod if it is unstretched before the load is applied to the timber beam.

SOLUTION
There are three unknown support reactions acting on this beam (a force and a moment at A and a force at B) and only two equations of equilibrium. Thus, the problem is statically indeterminate. The constraint at the right end of the beam (the aluminum rod) can be replaced with a concentrated force at B (equal to the tension force in the tie rod), as shown in Fig. 8-35b. At point B, the deflection of point B for the tie rod must be equal to the deflection of point B for the beam,

$$v_{rod} = v_{beam} = v_w + v_F \qquad (a)$$

▶ Note that since the rod is subjected to a tensile force, its length will increase. Point B on the rod then moves downward, which gives the minus sign for the deformation of point B for the rod, v_{rod}.

The deflection of point B for the tie rod is found using Eq. 5-2

$$v_{rod} = -\left(\frac{FL}{EA}\right)_{rod} = \frac{-F(30)(12)}{10(10^6)(0.125)} = -288F(10^{-6}) \qquad (b)$$

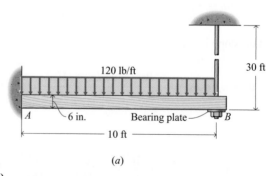

(a)

Figure 8-35(a)

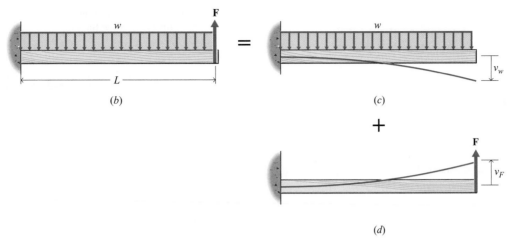

(b) = *(c)*

+

(d)

Figure 8-35(b–d)

The deflection of point B for the beam is found using cases 2 and 1 of Table B-19 (Appendix B),

$$
\begin{aligned}
v_{beam} = v_w + v_F &= \frac{-wL^4}{8EI} + \frac{PL^3}{3EI} \\
&= \frac{-(120/12)[(10)(12)]^4}{8(1.2)(10^6)[(4)(6)^3/12]} + \frac{F[(10)(12)]^3}{3(1.2)(10^6)[(4)(6)^3/12]}
\end{aligned}
\tag{c}
$$

Substituting Eqs. (b) and (c) into Eq. (a) and solving yields

$$
F = 431 \text{ lb} \qquad \textbf{Ans.}
$$

PROBLEMS

Use the method of superposition to solve the following problems.

Introductory Problems

8-142* A beam is loaded and supported as shown in Fig. P8-142. Determine the reactions at the supports A and B.

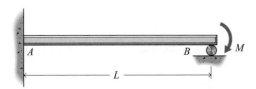

Figure P8-142

8-143* A beam is loaded and supported as shown in Fig. P8-143. Determine the reactions at the supports A and B.

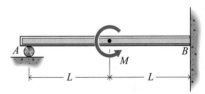

Figure P8-143

8-144 A beam is loaded and supported as shown in Fig. P8-144. Determine the magnitude of the load P required to make the slope at the left end of the beam zero.

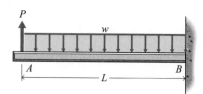

Figure P8-144

8-145* A beam is loaded and supported as shown in Fig. P8-145. Determine the magnitude of the moment M required to make the deflection at the left end of the beam zero.

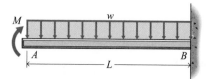

Figure P8-145

8-146 A beam is loaded and supported as shown in Fig. P8-146. Determine the reactions at the supports A and B.

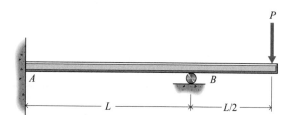

Figure P8-146

8-147 A beam is loaded and supported as shown in Fig. P8-147. Determine the reactions at the supports A and B.

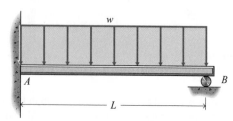

Figure P8-147

8-148* A beam is loaded and supported as shown in Fig. P8-148. Determine the reactions at the supports A and B.

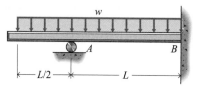

Figure P8-148

8-149 A beam is loaded and supported as shown in Fig. P8-149. Determine the reactions at the supports A, B, and C.

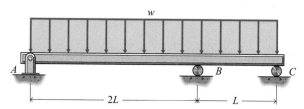

Figure P8-149

Intermediate Problems

8-150* A beam is loaded and supported as shown in Fig. P8-150. When the load P is applied, the slope at the right end of the beam is zero. Determine

a. The magnitude of the load P.
b. The reactions at the supports A and B.

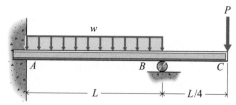

Figure P8-150

8-151* A beam is loaded and supported as shown in Fig. P8-151. Determine

a. The reactions at the supports A, B, and C.
b. The deflection at the middle of span AB.

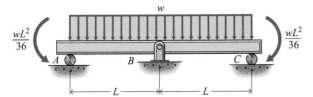

Figure P8-151

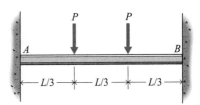

Figure P8-154

8-152 A beam is loaded and supported as shown in Fig. P8-152. Determine

a. The reactions at the supports A and C.
b. The deflection at the right end of the distributed load.

8-155 A beam is loaded and supported as shown in Fig. P8-155. Determine the magnitude of the moment M required to make

a. The slope at the right end of the beam zero.
b. The deflection at the right end of the beam zero.

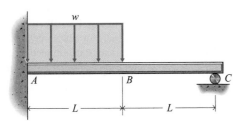

Figure P8-152

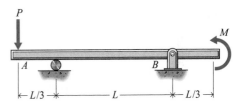

Figure P8-155

8-153* A beam is loaded and supported as shown in Fig. P8-153. Determine

a. The reactions at the supports A and C.
b. The deflection at the right end of the distributed load.

8-156* Draw complete shear force and bending moment diagrams for the beam shown in Fig. P8-156.

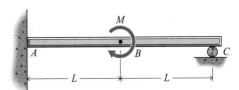

Figure P8-156

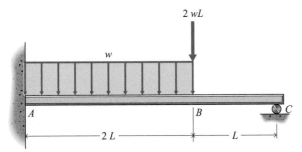

Figure P8-153

8-157 Draw complete shear force and bending moment diagrams for the beam shown in Fig. P8-157.

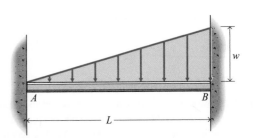

8-154 A beam is loaded and supported as shown in Fig. P8-154. Determine

a. The reactions at the supports A and B.
b. The deflection at the middle of the span.

Figure P8-157

Challenging Problems

8-158* Two beams are loaded and supported as shown in Fig. P8-158. Determine the reactions at the supports A, B, and D. Both beams have the same flexural rigidity.

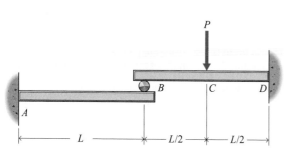

Figure P8-158

8-159* A steel ($E = 29,000$ ksi and $I = 120$ in.^{4}) beam is loaded and supported as shown in Fig. P8-159. The post BD is a 6 × 6-in. timber ($E = 1500$ ksi) that is braced to prevent buckling. Determine the load carried by the post if it is unstressed before the 530-lb/ft distributed load is applied.

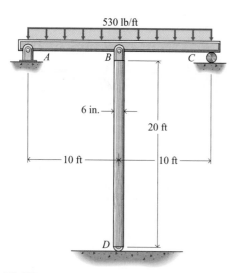

Figure P8-159

8-160 An aluminum ($E = 70$ GPa) beam is loaded and supported as shown in Fig. P8-160. When the beam is not loaded, it rests lightly on the spring of modulus 900 kN/m. The second moment of area of the beam with respect to its neutral axis is $30(10^6)$ mm^4. Determine the deformation of the spring after the 80-kN/m distributed load is applied.

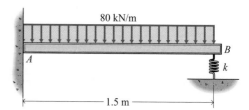

Figure P8-160

8-161* Two steel ($E = 29,000$ ksi) beams support a 1600-lb concentrated load, as shown in Fig. P8-161. In the unloaded condition, beam AB touches but exerts no force on beam CD. Beam AB is an S4 × 9.5 American standard section, and beam CD is an S5 × 14.75 section (see Appendix B). Determine

a. The maximum flexural stress in each beam.
b. The maximum transverse shearing stress in each beam.

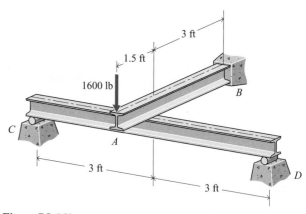

Figure P8-161

8-162 A uniformly distributed load of 7 kN/m is supported by two 100 × 100-mm timber ($E = 8.5$ GPa) beams arranged as shown in Fig. P8-162. Beam AB is fixed at the wall, and beam CD is simply supported. Before the load is applied, the beams are in contact at B, but the reaction at B is zero. After the 7-kN/m distributed load is applied, determine

a. The maximum flexural stress in each beam.
b. The maximum longitudinal shearing stress in each beam.

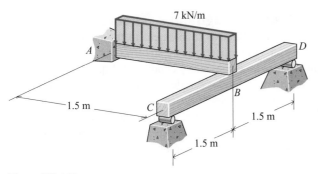

Figure P8-162

8-163 A beam is loaded and supported as shown in Fig. P8-163.

a. Determine the reactions at the supports A and C.
b. Draw complete shear force and bending moment diagrams for the beam.

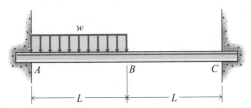

Figure P8-163

8-164* A beam is loaded and supported as shown in Fig. P8-164.

a. Determine the reactions at the supports A and C.
b. Draw complete shear force and bending moment diagrams for the beam.

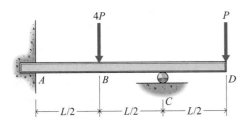

Figure P8-164

8-165* A timber ($E = 1800$ ksi) beam is loaded and supported as shown in Fig. P8-165. The tension in the 1/2-in.-diameter rod BD is zero before the load is applied to the beam. Determine the axial stress in the tie rod BD if

a. The tie rod BD is made of steel ($E = 30{,}000$ ksi).
b. The tie rod BD is made of an aluminum alloy ($E = 10{,}000$ ksi).

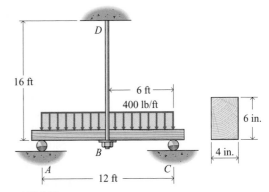

Figure P8-165

8-166 In Fig. P8-166, the aluminum alloy tie rod passes through a hole in the aluminum alloy cantilever beam and through the coil spring positioned on the end of the tie rod. Before loading, there is a clearance of 2.5 mm between the bottom of the beam and the top of the spring. The cross-sectional area of the tie rod is 100 mm², the second moment of area of the cross section of the beam with respect to its neutral axis is $40(10^6)$ mm⁴, the modulus of elasticity of the aluminum alloy is 70 GPa, and the spring modulus is 1000 kN/m. Determine the axial stress in the tie rod when $M = 9$ kN · m and $w = 90$ kN/m.

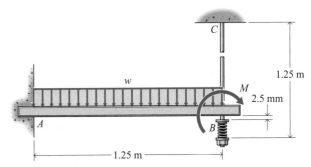

Figure P8-166

8-167 A beam is loaded and supported as shown in Fig. P8-167.

a. Determine the reactions at the supports A, B, and C.
b. Draw complete shear force and bending moment diagrams for the beam.

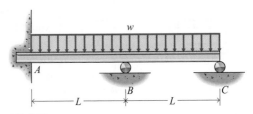

Figure P8-167

8-9-3 Energy Method—Castigliano's Theorem
Castigliano's theorem (see Section 8-8) is an effective supplement to the equations of equilibrium in the solution of statically indeterminate structures. If the deflection at some point in a structure is known (a boundary condition), Eq. 8-5a, may be applied at this point and set equal to the known deflection. The resulting equation will contain one or more unknown forces. If the equation contains more than one unknown, additional equations must be written, such as equilibrium equations or the application of Eq. 8-5b. However, there are certain restrictions necessary to ensure that the equations are independent; also, some remarks regarding procedure may be in order. The following outline may be found helpful.

1. For any system of loads and reactions, as many independent energy equations (Eq. 8-5a or 8-5b) can be written as there are redundant unknowns—defined as unknowns not necessary to maintain equilibrium of the structure. This restriction is equivalent to the restriction requiring one boundary condition for each extra unknown.

2. If only one independent energy equation can be written and the moment equation contains two unknowns (forces or couples, including dummy loads), one of the unknowns must be expressed in terms of the other by means of an equilibrium equation. This case is illustrated in the alternate solution of Example Problem 8-25, which follows. If two independent energy equations can be written, these can contain two unknowns that can be obtained from a solution of the two energy equations, as illustrated in Example Problem 8-26, which follows.

3. If the loading changes along the beam, the moment equation can be written for the entire span using the singularity notation of Section 8-5 and, after multiplication by $\partial M/\partial P_i$ or $\partial M/M_i$ (also using singularity notation), integration can be performed over the entire span. However, this procedure requires integration by parts of such expressions as $x\langle x - L\rangle^1\, dx$. Sometimes it may be more convenient to write an ordinary algebraic moment equation for each interval, multiply by $\partial M/\partial P_i$ or $\partial M/\partial M_i$ (different for each interval), and integrate over each interval; then add the results of all the integrations.

Example Problem 8-25   A beam is loaded and supported as shown in Fig. 8-36a. Use Castigliano's theorem to determine the reactions at supports A and B.

SOLUTION
The free-body diagram of Fig. 8-36b indicates that the problem is statically indeterminate, because there are three unknown reaction components and only two independent equations of equilibrium available.

If the portion of the beam to the right of the reaction R_B was removed and replaced by an equivalent shearing force and bending moment at the transverse section above R_B, neither R_B nor the elastic curve in the interval $0 \le x \le L$ would be changed. Hence, it is necessary to deal with only the strain energy of the beam in the interval $0 \le x \le L$. With the coordinate system placed as shown, the resulting moment equation obtained from the free-body diagram in Fig. 8-36c is

$$M_r = R_B x - \frac{w}{2}\left(x + \frac{L}{4}\right)^2 = R_B x - \frac{wx^2}{2} - \frac{wLx}{4} - \frac{wL^2}{32}$$

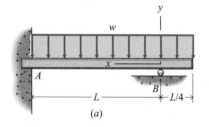

(a)

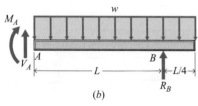

(b)

Figure 8-36(a–b)

From Eq. 8-5a, the deflection at the right support is given by

$$v_B = \frac{\partial U}{\partial R_B} = \frac{1}{EI} \int_0^L M_r \frac{\partial M_r}{\partial R_B} dx$$

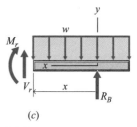

(c)

Figure 8-36(c)

Since the partial derivative of M_r with respect to R_B is x, the expression for the deflection is

$$v_B = \frac{1}{EI} \int_0^L \left(R_B x - \frac{wx^2}{2} - \frac{wLx}{4} - \frac{wL^2}{32} \right) (x) dx$$

$$= \frac{1}{EI} \int_0^L \left(R_B x^2 - \frac{wx^3}{2} - \frac{wLx^2}{4} - \frac{wL^2 x}{32} \right) dx$$

which upon integration and substitution of limits becomes

$$v_B = \frac{1}{EI} \left(\frac{R_B L^3}{3} - \frac{wL^4}{8} - \frac{wL^4}{12} - \frac{wL^4}{64} \right)$$

Since the support is unyielding ($v_B = 0$ when $x = 0$), the above expression can be equated to zero and solved for R_B, yielding

$$R_B = +\frac{43wL}{64} = \frac{43wL}{64} \uparrow \qquad \textbf{Ans.}$$

The other reaction components (M_A and V_A) are obtained from the free-body diagram shown in Fig. 8-36b and the equilibrium equations $\Sigma M_A = 0$ and $\Sigma F_y = 0$. Thus,

$$+\uparrow \Sigma M_A = 0: \qquad M_A + w\,(5L/4)\,(5L/8) - R_B\,(L) = 0$$

$$M_A + \frac{25wL^2}{32} - \frac{43wL}{64}\,(L) = 0$$

from which

$$M_A = -\frac{7wL^2}{64} = \frac{7wL^2}{64} \downarrow \qquad \textbf{Ans.}$$

and

$$+\uparrow \Sigma F_y = 0: \qquad V_A + R_B - w\,(5L/4) = V_A + \frac{43wL}{64} - \frac{5wL}{4} = 0$$

from which

$$V_A = +\frac{37wL}{64} = \frac{37wL}{64} \uparrow \qquad \textbf{Ans.}$$

ALTERNATE SOLUTION

This solution will make use of the free-body diagram of Fig. 8-36d, from which the moment equation is

$$M_r = M_A + V_A x - \frac{wx^2}{2} \qquad 0 \le x \le L$$

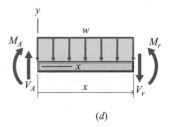

(d)

Figure 8-36(d)

Since there are two unknowns in the moment equation and only one independent energy equation can be written, the equilibrium equation $\Sigma M_B = 0$ (Fig. 8-36b) will be used to obtain a relation between the two unknowns. Thus,

$$+\uparrow \Sigma M_B = 0: \qquad\qquad M_A + V_A L - \frac{5wL}{4}\left(\frac{3L}{8}\right) = 0$$

from which

$$M_A = \frac{15wL^2}{32} - V_A L \qquad \text{or} \qquad V_A = \frac{15wL}{32} - \frac{M_A}{L}$$

Eliminating M_A from the moment equation for M_r gives

$$M_r = V_A(x - L) + \frac{15wL^2}{32} - \frac{wx^2}{2}$$

and then

$$\frac{\partial M_r}{\partial V_A} = x - L$$

From Eq. 8-5a, the deflection at the left support is given by

$$v_A = \frac{\partial U}{\partial V_A} = \frac{1}{EI}\int_0^L M_r \frac{\partial M_r}{\partial V_A}dx$$

$$= \frac{1}{EI}\int_0^L \left[V_A(x - L)^2 - \frac{wx^2}{2}(x - L) + \frac{15wL^2}{32}(x - L)\right]dx = 0$$

from which

$$V_A = +\frac{37wL}{64} = \frac{37wL}{64}\uparrow \qquad\qquad\qquad \textbf{Ans.}$$

The other reaction components (M_A and R_B) can be obtained from equilibrium equations.

■ **Example Problem 8-26** Determine the reactions at the left end of the beam of Fig. 8-37a.

SOLUTION

The free-body diagram of Fig. 8-37b indicates four unknown reaction components, and since there are only two independent equations of equilibrium, two supplementary equations are necessary. These are obtained by applying Eqs. 8-5a and 8-5b and solving the resulting simultaneous equations for M_0 and V_0. With the origin at the left support, the boundary conditions are $v = 0$ and $\theta = 0$ at $x = 0$. Since the moment equation is

$$M_r = M_0 + V_0 x - P\langle x - a\rangle^1$$

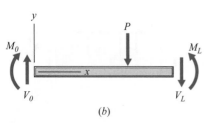

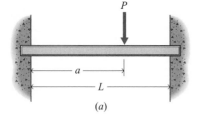

(a)

(b)

Figure 8-37

the deflection at the left end is given by Eq. 8-5a as

$$v_0 = \frac{\partial U}{\partial V_0} = \frac{1}{EI} \int_0^L M_r \frac{\partial M_r}{\partial V_0} dx$$

$$= \frac{1}{EI} \int_0^L \left[M_0 + V_0 x - P\langle x - a \rangle^1 \right] (x) \, dx$$

$$= \frac{1}{EI} \left[\frac{M_0 x^2}{2} + \frac{V_0 x^3}{3} - P\left(\frac{x}{2} \langle x - a \rangle^2 - \frac{1}{6} \langle x - a \rangle^3 \right) \right]_0^L$$

where the last term was integrated by parts. Substitution of the limits gives

$$EI v_0 = \frac{M_0 L^2}{2} + \frac{V_0 L^3}{3} - \frac{PL}{2} (L - a)^2 + \frac{P}{6} (L - a)^3$$

Application of the boundary condition $v_0 = 0$ gives

$$\frac{M_0 L^2}{2} + \frac{V_0 L^3}{3} - \frac{PL^3}{3} + \frac{PL^2 a}{2} - \frac{P a^3}{6} = 0 \qquad (a)$$

When Eq. 8-5b is applied, the rotation at the left end is given by the expression

$$\theta_0 = \frac{\partial U}{\partial M_0} = \frac{1}{EI} \int_0^L M_r \frac{\partial M_r}{\partial M_0} dx$$

$$= \frac{1}{EI} \int_0^L (M_0 + V_0 x - P\langle x - a \rangle^1) dx$$

$$= \frac{1}{EI} \left[M_0 x + \frac{V_0 x^2}{2} - \frac{P}{2} \langle x - a \rangle^2 \right]_0^L$$

Substitution of the limits gives

$$EI \theta_0 = M_0 L + \frac{V_0 L^2}{2} - \frac{P}{2} (L - a)^2$$

Application of the boundary condition $\theta_0 = 0$ gives

$$M_0 L + \frac{V_0 L^2}{2} - \frac{PL^2}{2} + PaL - \frac{P a^2}{2} = 0 \qquad (b)$$

Equations (a) and (b) are solved to obtain the following results:

$$V_0 = P - \frac{3 P a^2}{L^2} + \frac{2 P a^3}{L^3} \qquad \textbf{Ans.}$$

and

$$M_0 = -Pa + \frac{2 P a^2}{L} + \frac{P a^3}{L^2} \qquad \textbf{Ans.}$$

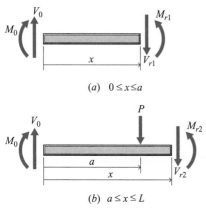

(a) $0 \leq x \leq a$

(b) $a \leq x \leq L$

Figure 8-38

ALTERNATE SOLUTION

Integration by parts can be eliminated by considering the resisting moments in the two segments of the beam separately. Consider first the segment shown in Fig. 8-38a, where the resisting moment is

$$M_{r1} = V_0 x + M_0 \qquad 0 \leq x \leq a$$

Next, consider the segment shown in Fig. 8-38b, where the resisting moment is

$$M_{r2} = V_0 x + M_0 - P(x - a)$$
$$= V_0 x + M_0 - Px + Pa \qquad a \leq x \leq L$$

The strain energy of the beam is

$$U = U_1 + U_2$$

and the deflection becomes

$$v_0 = \frac{1}{EI} \int_0^a M_{r1} \frac{\partial M_{r1}}{\partial V_0} dx + \frac{1}{EI} \int_a^L M_{r2} \frac{\partial M_{r2}}{\partial V_0} dx \qquad (c)$$

and the slope is

$$\theta_0 = \frac{1}{EI} \int_0^a M_{r1} \frac{\partial M_{r1}}{\partial M_0} dx + \frac{1}{EI} \int_a^L M_{r2} \frac{\partial M_{r2}}{\partial M_0} dx \qquad (d)$$

Setting both $v_0 = 0$ and $\theta_0 = 0$ and carrying out the mathematics indicated in Eqs. (c) and (d) results in Eqs. (a) and (b) again, and the rest of the solution is the same as before.

PROBLEMS

Use Castigliano's theorem to solve the following problems.

Introductory Problems

8-168* A beam is loaded and supported as shown in Fig. P8-168. Determine the reaction at the support B.

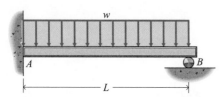

Figure P8-168

8-169* A beam is loaded and supported as shown in Fig. P8-169. When the couple Q is applied, the slope at the right end of the beam is zero. Determine the magnitude of the couple Q.

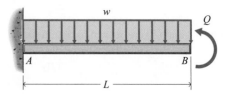

Figure P8-169

8-170 A beam is loaded and supported as shown in Fig. P8-170. Determine the reactions at the supports A and B.

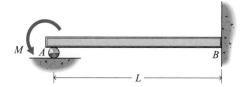

Figure P8-170

8-171* A beam is loaded and supported as shown in Fig. P8-171. Determine the reactions at the supports A and B.

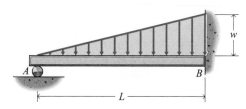

Figure P8-171

8-172 A beam is loaded and supported as shown in Fig. P8-172. Determine the reactions at the supports A, B, and C.

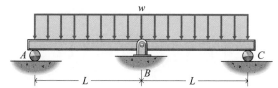

Figure P8-172

8-173 A beam is loaded and supported as shown in Fig. P8-173. Determine the reactions at the supports A and B.

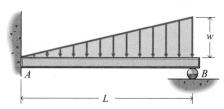

Figure P8-173

Intermediate Problems

8-174* A beam is loaded and supported as shown in Fig. P8-174. Determine the reactions at the supports A and B.

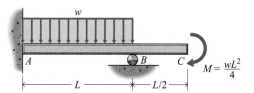

Figure P8-174

8-175* A beam is loaded and supported as shown in Fig. P8-175. Determine

    a. The reactions at the supports, A, B, and D.
    b. The deflection under the concentrated load P.

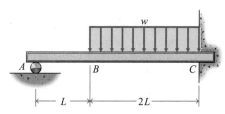

Figure P8-175

8-176 A beam is loaded and supported as shown in Fig. P8-176. Determine the reactions at the supports A and C.

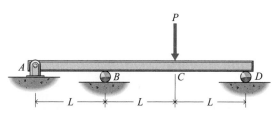

Figure P8-176

8-177 A beam is loaded and supported as shown in Fig. P8-177. Determine the reactions at the supports A and B.

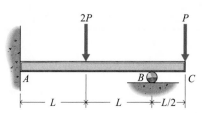

Figure P8-177

8-178* A beam is loaded and supported as shown in Fig. P8-178. Determine the reactions at the supports A and B.

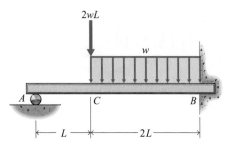

Figure P8-178

8-179 A beam is loaded and supported as shown in Fig. P8-179. Determine the reactions at the supports A and D.

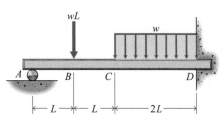

Figure P8-179

Challenging Problems

8-180* The beam shown in Fig. P8-180 is simply supported at the left end and framed into a column at the right end. When the load is applied, the ends of the beam remain at the same level, but the right end rotates (due to loading the adjacent span) clockwise until the slope of the elastic curve is $wL^3/(24EI)$ downward to the right. Determine the reaction at the support A.

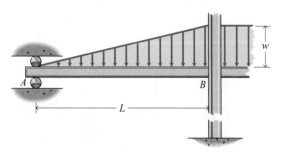

Figure P8-180

8-181* Draw complete shear force and bending moment diagrams for the beam shown in Fig. P8-181.

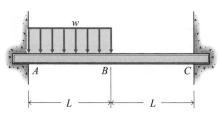

Figure P8-181

8-182 Draw complete shear force and bending moment diagrams for the beam shown in Fig. P8-182.

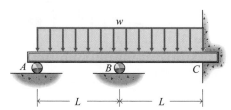

Figure P8-182

8-183* Draw complete shear force and bending moment diagrams for the beam shown in Fig. P8-183.

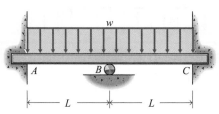

Figure P8-183

8-184 An S178 × 30 American standard section is used for a beam that is loaded and supported, as shown in Fig. P8-184. Determine

a. The reactions at the supports A, B, and C.
b. The maximum flexural and transverse shearing stresses in the beam if $w = 10$ kN/m and $L = 2$ m.

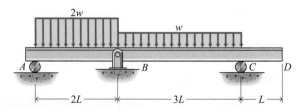

Figure P8-184

8-185 An S6 × 17.25 American standard section is used for a beam that is loaded and supported, as shown in Fig. P8-185. Determine

a. The reactions at the supports A, B, and C.
b. The maximum flexural and transverse shearing stresses in the beam if $w = 500$ lb/ft and $L = 6$ ft.

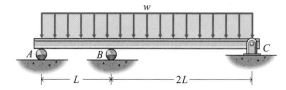

Figure P8-185

8-10 DESIGN PROBLEMS

In Chapter 7 when the design of beams was discussed, either flexural strength or shear strength was the controlling parameter. In this chapter an additional parameter, deflection, will be introduced. Thus, the design of a beam may be based on flexural stress, shearing stress, or deflection. The design procedure is similar to that presented in Chapter 7. The beam is first designed based on flexural stress, and then checked for shearing stress and deflection. If the shearing stress and deflection are within allowable limits, the design is satisfactory. If either the shearing stress or the deflection is greater than the allowable value, the beam must be redesigned until all of the allowable limits are satisfied. Clearly, this is a trial-and-error process. The following examples illustrate the procedures for designing beams where allowable limits are given for flexural stress, shearing stress, and deflection.

Example Problem 8-27   An air-dried Douglas fir timber ($E = 13$ GPa) beam is loaded as shown in Fig. 8-39a. If the allowable flexural stress is 8 MPa, the allowable shearing stress is 0.7 MPa, and the allowable deflection is 14 mm, determine the lightest-weight standard structural timber that can be used for the beam.

SOLUTION
Load (free-body), shear force, and bending moment diagrams for the beam of Fig. 8-39a are shown in Figs. 8-39b, c, and d, respectively. Since the load is uniformly distributed over the full length of the beam,

$$R_A = R_B = \frac{1}{2}(850)(5) = 2125\,\text{N}$$

$$V_{max} = 2125\,\text{N} \qquad M_{max} = \frac{1}{2}(2125)(2.5) = 2656\,\text{N}\cdot\text{m}$$

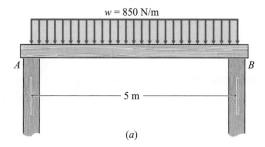

$w = 850$ N/m

A B

5 m

(a)

Figure 8-39(a)

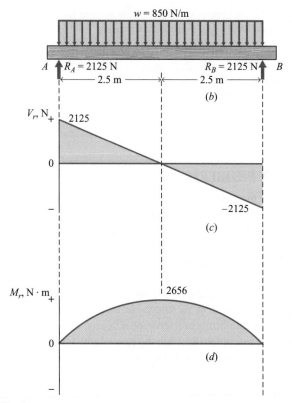

Figure 8-39(b–d)

The minimum section modulus needed to satisfy the allowable value of the flexural stress is given by Eq. 7-9 as

$$S \geq \frac{M_{max}}{\sigma_{all}} = \frac{2656}{8(10^6)} = 332.0(10^{-6})\,\text{m}^3 = 332.0(10^3)\,\text{mm}^3$$

The lightest-weight standard structural timber listed in Table B-16 (Appendix B) with $S \geq 332.0(10^3)\,\text{mm}^3$ is a timber with nominal dimensions of 51×254 mm. Some properties of this timber that will be needed later are

$$\text{Mass/unit length} = 6.38\,\text{kg/m}$$
$$\text{Area} = 9.88(10^3)\,\text{mm}^2 = 9.88(10^{-3})\,\text{m}^2$$
$$I = 48.3(10^6)\,\text{mm}^4 = 48.3(10^{-6})\,\text{m}^4$$
$$S = 400(10^3)\,\text{mm}^3 = 400(10^{-6})\,\text{m}^3$$

For a rectangular section, the area is $A = bh$, the second moment of area is $I = bh^3/12$, and the maximum shearing stress occurs at the neutral axis. Then,

$$Q = \frac{h}{4}\left(b\,\frac{h}{2}\right) = \frac{bh^2}{8}$$

and the maximum shearing stress is

$$\tau = \frac{V_r Q}{It} = \frac{V_r(bh^2/8)}{(bh^3/12)b} = 1.5\frac{V_r}{bh} = 1.5\frac{V_r}{A} = 1.5\frac{2125}{9.88(10^{-3})}$$

$$= 0.3226(10^6)\,\text{N/m}^2 = 0.3226\,\text{MPa} < 0.7\,\text{MPa}$$

Thus, the shearing stress requirement is satisfied.

For a simply supported beam with a uniformly distributed load, case 7 of Table B-19 (Appendix B) gives the maximum deflection as

$$|v_{max}| = \frac{5wL^4}{384EI} = \frac{5(850)(5)^4}{384(13)(10^9)(48.3)(10^{-6})}$$

$$= 11.017(10^{-3})\,\text{m} = 11.017\,\text{mm} < 14\,\text{mm}$$

Therefore, the deflection requirement is satisfied. The 51 × 254-mm standard structural timber satisfies the requirements for flexural stress, shearing stress, and deflection; however, the analysis thus far neglected the weight of the beam. The timber beam weighs $(6.38)(9.81) = 62.59$ N/m. Adding this uniformly distributed load to the applied load gives a uniformly distributed load $w = 850 + 62.6 = 912.6$ N/m. For this loading, the maximum shear force is 2282 N and the maximum bending moment is 2852 N · m. The section modulus needed to satisfy the flexural stress requirement is

$$S \geq \frac{M_{max}}{\sigma_{all}} = \frac{2852}{8(10^6)} = 356.5(10^{-6})\,\text{m}^3 = 356.5(10^3)\,\text{mm}^3 < 400(10^3)\,\text{mm}^3$$

where $S = 400\,(10^3)$ mm^3 is the actual section modulus of the cross section originally selected (51 × 254 mm nominal). Thus, the 51 × 254-mm timber satisfies the flexural stress requirement. The maximum shearing stress and the maximum deflection with the beam weight included are

$$\tau_{max} = 1.5\frac{V_{max}}{A} = 1.5\frac{2282}{9.88(10^{-3})}$$

$$= 0.3465(10^6)\text{N/m}^2 = 0.3465\,\text{MPa} < 0.7\,\text{MPa}$$

$$|v_{max}| = \frac{5wL^4}{384EI} = \frac{5(912.6)(5)^4}{384(13)(10^9)(48.3)(10^{-6})}$$

$$= 11.828(10^{-3})\,\text{m} = 11.828\,\text{mm} < 14\,\text{mm}$$

Thus, the 51 × 254-mm standard structural timber satisfies all requirements with the weight of the beam included.

Example Problem 8-28 A structural steel ($E = 29,000$ ksi) beam is loaded as shown in Fig. 8-40a. If the allowable flexural stress is 24,000 psi, the allowable shearing stress is 14,000 psi, and the allowable deflection midway between supports A and B is 0.5 in., determine the lightest American standard section (S-shape) that can be used for the beam.

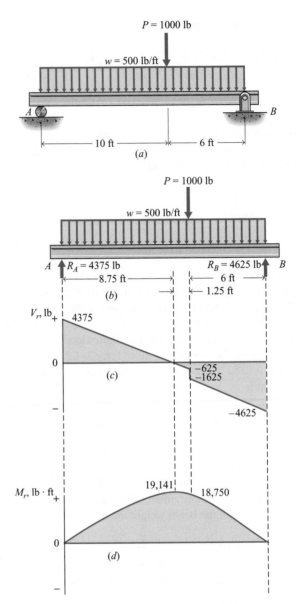

Figure 8-40

SOLUTION
Load (free-body), shear force, and bending moment diagrams for the beam of
Fig. 8-40a are shown in Figs. 8-40b, c, and d, respectively. The equilibrium
equations yield

$$+\circlearrowleft \ \Sigma M_B = 0: \quad R_A(16) - 500(16)(8) - 1000(6) = 0 \quad R_A = 4375 \text{ lb}$$
$$+\uparrow \ \Sigma F_y = 0: \quad 4375 - 500(16) - 1000 + R_B = 0 \quad R_B = 4625 \text{ lb}$$

From the shear force and bending moment diagrams,

$$V_{\max} = 4625 \text{ lb}$$
$$M_{\max} = \frac{1}{2}(4375)(8.75) = 19{,}141 \text{ lb} \cdot \text{ft}$$

The minimum section modulus needed to satisfy the allowable value of the flexural stress is given by Eq. (7-9) as

$$S \geq \frac{M_{max}}{\sigma_{all}} = \frac{19{,}141\,(12)}{24{,}000} = 9.571 \text{ in.}^3$$

The lightest weight American Standard beam (S-shape) listed in Table B-3 (Appendix B) with $S \geq 9.571$ in.^{3} is an S7 × 15.3 section. For this section,

$$S = 10.5 \text{ in.}^3 \qquad t_{web} = 0.252 \text{ in.}$$
$$d = 7.00 \text{ in.} \qquad I = 36.7 \text{ in.}^4$$
$$t_f = 0.392 \text{ in.}$$

The average value of the shearing stress in the web is

$$\tau_{avg} = \frac{V_{max}}{A_{web}} = \frac{4625}{0.252\,[7.00 - 2\,(0.392)]} = 2953 \text{ psi} \ll 14{,}000 \text{ psi}$$

Thus, the shearing stress requirement is satisfied, because the maximum shearing stress in the web of an American standard beam is only slightly larger than the average shearing stress.

The deflection at midspan is found by using the method of superposition and Table B-19 (Appendix B). The given loading (Fig. 8-41a) is equivalent to the two loads, parts (1) and (2), as shown in Fig. 8-41b. For part (1), the midspan deflection is given as case 7 of Table B-19 (Appendix B). It is

$$v_1 = -\frac{5wL^4}{384EI} = -\frac{5(500/12)[16(12)]^4}{384(29)(10^6)(36.7)} = -0.6927 \text{ in.}$$

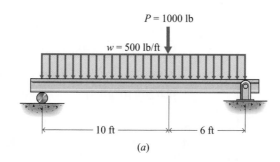

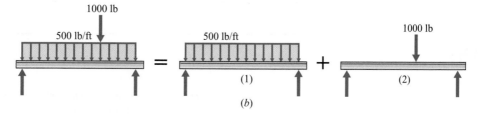

Figure 8-41

For part (2), the midspan deflection is given as case 5 of Table B-19 (Appendix B). It is

$$
v_2 = -\frac{Pb(3L^2 - 4b^2)}{48EI} = -\frac{1000(72)[3(192)^2 - 4(72)^2]}{48(29)(10^6)(36.7)} = -0.12664 \text{ in.}
$$

The midspan deflection is the algebraic sum of v_1 and v_2 and is

$$
|v_{\text{mid}}| = 0.6927 + 0.12664 = 0.8193 \text{ in.} > 0.5 \text{ in.}
$$

Since $|v_{\text{mid}}|$ is greater than v_{all}, a new section must be selected with sufficient I to satisfy the deflection requirement, Thus,

$$
|v_{\text{mid}}| \geq \frac{5wL^4}{384EI} + \frac{Pb(3L^2 - 4b^2)}{48EI}
$$

Solving for I yields

$$
I \geq \frac{5wL^4}{384E|v_{\text{mid}}|} + \frac{Pb(3L^2 - 4b^2)}{48E|v_{\text{mid}}|}
$$

$$
= \frac{5(500/12)(192)^4}{384(29)(10^6)(0.5)} + \frac{1000(72)[3(192)^2 - 4(72)^2]}{48(29)(10^6)(0.5)} = 60.14 \text{ in.}^4
$$

The lightest S-shape in Table B-3 (Appendix B) with $I \geq 60.14$ in.^{4} is an S8 × 23 beam with $I = 64.9$ in.^{4} and $S = 16.2$ in.^{3}. The S8 × 23 beam satisfies the requirements for flexural stress, shearing stress, and deflection.

Consider now the effect of the beam's weight on the deflection. The S8 × 23 beam weighs 23 lb/ft. Adding the weight of the beam to the 500 lb/ft applied distributed load gives a uniformly distributed load $w = 500 + 23 = 523$ lb/ft, resulting in maximum values of shear force and bending moment of 4809 lb and 19,870 lb · ft, respectively. The value of I required as a result of this increase in load is

$$
I \geq \frac{5wL^4}{384E|v_{\text{mid}}|} + \frac{Pb(3L^2 - 4b^2)}{48E|v_{\text{mid}}|}
$$

$$
= \frac{5(523/12)(192)^4}{384(29)(10^6)(0.5)} + \frac{1000(72)[3(192)^2 - 4(72)^2]}{48(29)(10^6)(0.5)} = 62.49 \text{ in.}^4
$$

For the S8 × 23 beam, $I = 64.9$ in.^{4} > 62.49 in.^{4}; therefore, the addition of the weight of the beam does not change the selection of the beam. An S8 × 23 beam satisfies all requirements of flexural stress, shearing stress, and deflection.

PROBLEMS

Introductory Problems

8-186* A 3-m-long simply supported beam is loaded with a uniformly distributed load of 2.6 kN/m over its entire length. The

beam is made of air-dried Douglas fir ($E = 13$ GPa) with an allowable flexural stress of 8 MPa and an allowable shearing stress of 0.7 MPa. The maximum deflection at the center of the span must not exceed 10 mm. Select the lightest standard structural timber that can be used for the beam.

8-187* An air-dried Douglas fir ($E = 1900$ ksi) beam is simply supported and has a span of 16 ft. The beam is subjected to a uniformly distributed load of 800 lb/ft over its entire length. If the allowable flexural stress is 1200 psi, the allowable shearing stress is 90 psi, and the allowable deflection at the middle of the span is $1/2$ in., select the lightest standard structural timber that can be used to support the load.

8-188 A standard structural steel ($E = 200$ GPa) pipe is to support the load shown in Fig. P8-188. The allowable flexural stress and deflection are 150 MPa and 5 mm, respectively. Select the lightest permissible standard steel pipe that can be used to support the load. Neglect the effects of shear.

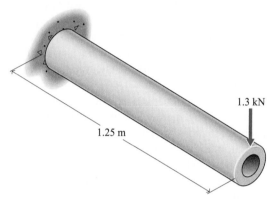

1.3 kN

1.25 m

Figure P8-188

8-189 A portion of a pedestrian walkway along the side of a bridge is shown in Fig. P8-189. Cantilever beams support the loading, one of which is shown. The beams are select structural eastern hemlock ($E = 1200$ ksi) with allowable flexural and shearing stresses of 1300 psi and 80 psi, respectively. The allowable deflection is 0.2 in. Select the lightest standard structural timber that can be used for the beams.

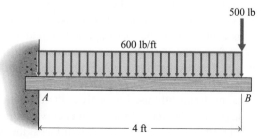

500 lb

600 lb/ft

A

B

4 ft

Figure P8-189

Intermediate Problems

8-190* A solid circular shaft made of ASTM A36 steel ($E = 200$ GPa) is supported by bearings spaced 1.5 m apart. The shaft is to support a 4-kN load perpendicular to the shaft; the load may be placed at any point between the bearings. The allowable flexural stress is 152 MPa, the allowable shearing stress is 100 MPa, and the allowable deflection is 5 mm. If shafts are available with diameters in increments of 5 mm, determine the smallest-diameter shaft that can be used to support the load. Neglect the weight of the shaft.

8-191 A simply supported structural steel ($E = 29,000$ ksi) beam has a span of 24 ft and carries a uniformly distributed load of 1200 lb/ft. The beam has an allowable flexural stress of 24 ksi, an allowable shearing stress of 14 ksi, and an allowable deflection of $1/360$ of the span. Select the lightest American standard (S-shape) beam that can be used to support the loading.

Challenging Problems

8-192* The simply supported structural steel ($E = 200$ GPa) beam shown in Fig. P8-192 has an allowable flexural stress of 165 MPa, an allowable shearing stress of 100 MPa, and an allowable deflection of $1/360$ of the span. Select the lightest wide-flange beam that can be used to support the loading shown in the figure.

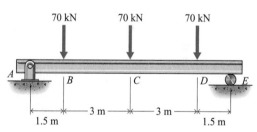

70 kN   70 kN   70 kN

A

B

C

D

E

3 m

3 m

1.5 m

1.5 m

Figure P8-192

8-193 The simply supported beam shown in Fig. P8-193 is made of air-dried Douglas fir ($E = 1900$ ksi) with an allowable flexural stress of 1900 psi, an allowable shearing stress of 85 psi, and an allowable deflection of $1/360$ of the span. Select the lightest standard structural timber that can be used to support the loads shown in the figure.

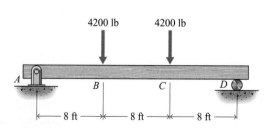

4200 lb   4200 lb

A

B

C

D

8 ft

8 ft

8 ft

Figure P8-193

SUMMARY

A beam design is frequently not complete until the amount of deflection has been determined for the specified load. Failure to control beam deflections within proper limits in building construction is frequently reflected by the development of cracks in plastered walls and ceilings. The deflection of a beam depends on the stiffness of the material and the dimensions of the beam as well as on the applied loads and type of supports.

When a straight beam is loaded and the action is elastic, the centroidal axis of the beam is a curve defined as the elastic curve. The relationship between the modulus of elasticity, second moment of area, curvature, and resisting moment is

$$EI\, \frac{d^2v}{dx^2} = M_r(x) \tag{8-1}$$

which is the differential equation for the elastic curve of a beam where the resisting moment M_r is a function of x.

Whenever the bending moment can be readily expressed as an integrable function of x, Eq. 8-1 can be solved for the deflection v of the elastic curve of a beam at any point x along the beam. The constants of integration are evaluated from the applicable boundary conditions.

Many beams are subjected to abrupt changes in loading along the beam. Because the expressions for the bending moment on the left and right of any abrupt change in load are different functions of x, it is impossible to write a single equation for the bending moment in terms of ordinary algebraic functions that is valid for the entire length of the beam. This can be resolved by writing separate bending moment equations for each interval of the beam. Although the intervals are bounded by abrupt changes in load, the beam is continuous at such locations; therefore, the slope and the deflection at the junction of adjacent intervals must match. A matching condition is defined as the equality of slope or deflection, as determined at the junction of two intervals from the elastic curve equations for both intervals. One matching condition (for example, at x equals $L/3$, v from the left equation equals v from the right equation) can be used to determine one and only one constant of integration.

The double integration method for determining beam deflections becomes tedious and time-consuming when several intervals and several sets of matching conditions are required. The labor involved in solving problems of this type, however, can be diminished by making use of singularity functions. Singularity functions are used to write one bending moment equation that applies in all intervals along a beam, thus eliminating the need for matching conditions.

The method of superposition for determining beam deflections is based on the fact that the resultant effect of several loads acting simultaneously on a member is the sum of the contributions from each of the loads applied individually. The results for the separate loads are frequently available from previous work or easily determined by previous methods. The results for several common loads are listed in Table B-19 of Appendix B.

Strain energy techniques are frequently used to analyze the deflections of beams and structures. By Castigliano's theorem, if the strain energy of a linearly elastic structure is expressed in terms of the system of external loads, the partial derivative of the strain energy with respect to a concentrated external load is the deflection of the structure at the point of application and in the direction of that load

$$\frac{\partial U}{\partial P_i} = v_i \tag{8-5a}$$

$$\frac{\partial U}{\partial M_i} = \theta_i \tag{8-5b}$$

If the deflection is required either at a point where there is no unique point load or in a direction not aligned with the applied load, a dummy load is introduced at the desired point acting in the proper direction. The deflection is obtained by first differentiating the strain energy with respect to the dummy load and then taking the limit as the magnitude of the dummy load approaches zero.

A beam, subjected only to transverse loads, with more than two reaction components is statically indeterminate because the equations of equilibrium are not sufficient to determine all the reactions. The additional relations needed for an evaluation of reactions (or other unknown forces) are obtained from deformation (slope or deflection) equations.

REVIEW PROBLEMS

8-194* The boards for a concrete form are to be bent to a circular curve of 5-m radius. What maximum thickness can be used if the stress is not to exceed 15 MPa? The modulus of elasticity for the wood is 10 GPa.

8-195* The cantilever beam shown in Fig. P8-195a is fabricated from two 1 × 3-in. steel ($E = 30{,}000$ ksi) bars, as shown in Fig. P8-195b. Determine

a. The radius of curvature of the beam.
b. The deflection at the right end of the beam.
c. The deflection 3 ft from the support.

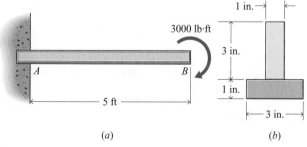

(a)   *(b)*

Figure P8-195

8-196 A beam is loaded and supported as shown in Fig. P8-196. Determine

a. The equation of the elastic curve. Use the designated axes.

b. The slope at the right end of the beam.
c. The deflection at the right end of the beam.

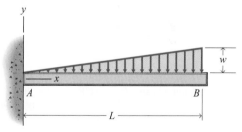

Figure P8-196

8-197* A cantilever beam is loaded and supported as shown in Fig. P8-197. Determine the deflection at the free end of the beam.

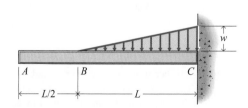

Figure P8-197

8-198 A cantilever beam is fabricated by bolting two structural steel ($E = 200$ GPa) C127 × 10 channel sections together, as shown in Fig. P8-198. Determine the deflection at the right end of the beam.

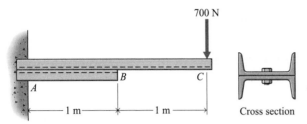

700 N

B   C

A

1 m   1 m   Cross section

Figure P8-198

8-199 Select the lightest structural steel ($E = 29,000$ ksi) wide-flange or American standard beam (Appendix B) that can be used for the beam shown in Fig. P8-199 if the maximum flexural stress must not exceed 10 ksi and if the maximum deflection must not exceed 0.200 in. when $L = 8$ ft and $w = 2000$ lb/ft.

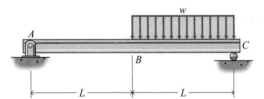

w

A

B   C

L   L

Figure P8-199

8-200* An aluminum beam [$E = 70$ GPa and $I = 20(10^6)$ mm^4] is loaded and supported as shown in Fig. P8-200. Determine the deflection at the middle of span BD.

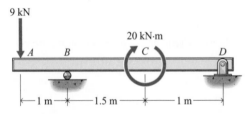

9 kN

20 kN·m

A   B   C   D

1 m   1.5 m   1 m

Figure P8-200

8-201 In Fig. P8-201, beam AC is made of brass and beam BC is made of steel. The second moment of the cross-sectional area of beam BC with respect to its neutral axis is twice that of beam AC, and the modulus of elasticity of the steel is twice

that of the brass. The reaction at C is zero before the load w is applied. Determine

a. The reaction at C on beam BC.
b. The reactions at the supports A and B.

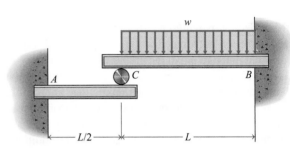

w

A   C   B

L/2   L

Figure P8-201

8-202* The steel ($E = 200$ GPa) beam AB of Fig. P8-202 is fixed at ends A and B and supported at the center by the pin-connected timber ($E = 10$ GPa) struts CD and CE. The cross-sectional area of each strut is 6400 mm^2, and the second moment of the cross-sectional area of the beam with respect to its neutral axis is $25(10^6)$ mm^4. Determine the force in each strut after the 6-kN/m distributed load is applied to the beam.

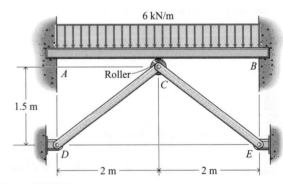

6 kN/m

A   Roller   B

C

1.5 m

D   E

2 m   2 m

Figure P8-202

8-203* A beam is loaded and supported as shown in Fig. P8-203. When the loads are applied, the center support settles an amount equal to $wL^4/(12EI)$. Determine the reactions at the supports A, B, and C.

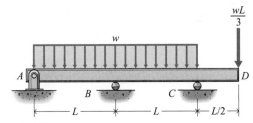

Figure P8-203

8-204 A beam is loaded and supported as shown in Fig. P8-204. Determine

a. The reactions at the supports A and D.
b. The deflection at B if $E = 200$ GPa and $I = 350(10^6)$ mm^4.

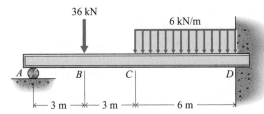

Figure P8-204

8-205 A beam is loaded and supported as shown in Fig. P8-205. Determine the reactions at the supports B, C, and D.

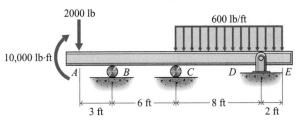

Figure P8-205

8-206* A beam is loaded and supported as shown in Fig. P8-206. Determine

a. The reactions at the supports A, B, and C.
b. The deflection midway between the supports B and C.

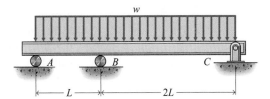

Figure P8-206

Chapter 9
Columns

9-1 INTRODUCTION

In their simplest form, columns are long, straight, prismatic bars subjected to compressive, axial loads. As long as a column remains straight, it can be analyzed by the methods of Chapters 1–5; however, if a column begins to deform laterally, the deflection may become large and lead to catastrophic failure. This situation, called *buckling*, can be defined as the sudden large deformation of a structure due to a slight increase of an existing compressive load under which the structure had exhibited little, if any, deformation before the load was increased. For example, a yardstick will support a compressive load of several pounds without discernible lateral deformation, but once the load becomes large enough to cause the yardstick to "bow out" a slight amount, any further increase of load produces large lateral deflections.

Buckling of such a column is caused not by failure of the material of which the column is composed but by deterioration of what was a stable state of equilibrium to an unstable one. The three states of equilibrium can be illustrated with a ball at rest on a surface, as shown in Figure 9-1. The ball in Fig. 9-1a is in a stable equilibrium position at the bottom of the pit because gravity will cause it to return to its equilibrium position if perturbed. The ball in Fig. 9-1b is in a neutral equilibrium position on the horizontal plane because it will remain at any new position to which it is displaced, tending neither to return to nor move farther from its original position. The ball in Fig. 9-1c, however, is in an unstable equilibrium position at the top of a hill because, if it is perturbed, gravity will cause it to move even farther from its original location until it eventually finds a stable equilibrium position at the bottom of another pit.

As the compressive load on a column is gradually increased from zero, the column is at first in a state of stable equilibrium. During this state, if the column

(a) $\qquad$ (b) $\qquad$ (c)

Figure 9-1

is perturbed by inducing small lateral deflections, it will return to its straight configuration when the lateral loads are removed corresponding to Fig. 9-1a. As the load is increased further, a critical value is reached at which the column is on the verge of experiencing a lateral deflection so, if it is perturbed, it will not return to its straight configuration (this is similar to Fig. 9-1b). The load cannot be increased beyond this value unless the column is restrained laterally; should the lateral restraints be removed, the slightest perturbation will trigger large lateral deflections (like Fig. 9-1c). For long, slender columns, the critical buckling load (the maximum load for which the column is in stable equilibrium) occurs at stress levels much less than the proportional limit for the material. This indicates that this type of buckling is an elastic phenomenon.

9-2 BUCKLING OF LONG, STRAIGHT COLUMNS

The first solution for buckling of long, slender columns was published in 1757 by the Swiss mathematician Leonhard Euler (1707–1783). Although the results of this section can be used only for long, slender columns, the analysis, similar to that used by Euler, is mathematically revealing and helps explain the behavior of columns.

The purpose of this analysis is to determine the minimum axial compressive load for which a column will experience lateral deflections. A straight, slender, pin-ended column of length L centrically loaded by axial compressive forces P at each end is shown in Fig. 9-2a. A *pin-ended column* is supported such that the bending moment and lateral movement are zero at the ends. In Fig. 9-2b, the load P has been increased sufficiently to cause a lateral deflection. Axes are selected with the origin at end A of the column. Summing moments about end A of the free-body diagram of Fig. 9-2b shows that the lateral force at B must be zero, and thus there can be no lateral force at A either. If the column is sectioned at an arbitrary position x, a free-body diagram of the portion below the section will appear as shown in

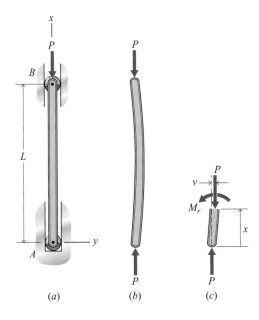

(a)   (b)   (c)

Figure 9-2

Fig. 9-2c. The two forces constitute a couple of magnitude Pv that must equal the resisting moment M_r; thus, $M_r = -Pv$. The differential equation for the elastic curve, as given by Eq. 8-1, becomes

$$EI\frac{d^2v}{dx^2} = M_r = -Pv$$

or

$$\frac{d^2v}{dx^2} + \frac{P}{EI}v = 0 \tag{a}$$

Equation (a) is a homogeneous second-order linear differential equation with constant coefficients. Established methods for the solution of such equations show the solution to be of the form

$$v = A \sin px + B \cos px \tag{b}$$

where A, B, and p are constants. By differentiating Eq. (b), substituting the results into Eq. (a), and collecting like terms, the following expression is obtained for evaluating p and C:

$$\left(-p^2 + \frac{P}{EI}\right)(A \sin px + B \cos px) = 0$$

from which it follows that

$$p^2 = \frac{P}{EI}$$

The constants A and B can be obtained from the two boundary conditions that the deflection of the elastic curve is zero at the ends; that is, at $x = 0$, $v = 0$ and at $x = L$, $v = 0$. Using the boundary condition that $v = 0$ at $x = 0$ and Eq. (b) yields

$$0 = A \sin 0 + B \cos 0$$
$$= 0 + B$$

and thus $B = 0$. Equation (b) then becomes

$$v = A \sin px \tag{b'}$$

Using the boundary condition that $v = 0$ at $x = L$ and Eq. (b') then yields

$$0 = A \sin pL \tag{c}$$

Equation (c) is satisfied if $A = 0$, but then $v = 0$ for all values of load, and the member does not buckle. Thus, for buckling to occur, $\sin pL = 0$ and

$$pL = n\pi \qquad n = 1,2,3,4, \ldots \tag{d}$$

or

$$\sqrt{\frac{P}{EI}}\,L = n\pi$$

which may be written

$$P = \frac{n^2\pi^2 EI}{L^2} \qquad n = 1,2,3,4,\ldots \qquad\qquad (e)$$

The least value of P (called P critical or P_{cr}) occurs when $n = 1$, for which Eq. (e) becomes

$$P_{cr} = \frac{\pi^2 EI}{L^2} \qquad\qquad (9\text{-}1)$$

The *critical buckling load* given by Eq. 9-1 is called the *Euler buckling load*.[1]

The second moment of the cross-sectional area (I) is relative to the axis about which bending occurs. For a pin-ended, centrically loaded column with no intermediate bracing to restrain lateral motion, bending occurs about the "weak" axis—the axis of minimum second moment of area. When I is replaced by Ar^2, where A is the cross-sectional area and r is the radius of gyration about the axis of bending, Eq. 9-1 becomes

$$\frac{P_{cr}}{A} = \frac{\pi^2 E}{(L/r)^2} = \sigma_{cr} \qquad\qquad (9\text{-}2)$$

The dimensionless quantity L/r is called the *slenderness ratio* and is determined for the axis about which bending tends to occur. Equation 9-2 is a particularly useful form if the critical stress σ_{cr} is of concern in the design. Either Eq. 9-1 or 9-2 can be used if the critical force P_{cr} is the primary concern in the design.

Since the analysis above is based on simple beam bending theory, it is valid only as long as the stresses remain in the linearly elastic range. Since the proportional limit is difficult to measure, the yield strength is often used instead of the proportional limit as the limiting stress, and the smallest slenderness ratio for which the Euler buckling load equation is valid occurs when $\sigma_{cr} = \sigma_y$ (the yield strength).

The Euler buckling load as given by Eq. 9-1 or 9-2 agrees well with experimental data if the slenderness ratio is large ($L/r > 140$ for steel columns). Short compression members ($L/r < 40$ for steel columns) can be treated as compression blocks where yielding occurs before buckling. Many columns lie between these extremes where neither solution is applicable. The intermediate length columns are analyzed by using empirical formulas described in Section 9-4.

Example Problem 9-1 An 8-ft-long pin-ended, timber [$E = 1.9(10^6)$ psi and $\sigma_e = 6400$ psi] column has a 2 × 4-in. rectangular cross section. Determine

---

[1] While the analysis predicts the buckling load, it does not determine the corresponding lateral deflection δ. This deflection can assume any nonzero value small enough that the nonlinear factor $[1 + (dv/dx)^2]^{3/2}$ in the curvature expression is approximately unity.

(a) The slenderness ratio.

(b) The Euler buckling load.

(c) The ratio of the axial stress under the action of the buckling load to the elastic strength σ_e of the material.

SOLUTION

(a) To determine the slenderness ratio, the minimum radius of gyration must be calculated. The second moment of area of a rectangular cross section is $bh^3/12$ and the area is bh; therefore, the radius of gyration is $\sqrt{I/A}$ or $h/2\sqrt{3}$. The minimum radius of gyration is found by using the centroidal axis parallel to the longer side of the rectangle. Thus, $b = 4$ in. and $h = 2$ in. so that

$$r = h/2\sqrt{3} = 2/2\sqrt{3} = 0.5774 \text{ in.}$$

The slenderness ratio is then found to be

$$\frac{L}{r} = \frac{8(12)}{0.5774} = 166.3 \qquad \textbf{Ans.}$$

▶ The Euler buckling load could also have been found using Eq. 9-1; $P_{cr} = \dfrac{\pi^2 EI}{L^2} = \dfrac{\pi^2(1.9)(10^6)[(4)(2^3)/(12)]}{[(8)(12)]^2} = $ 5424 lb ≅ 5420 lb

(b) The Euler buckling load is found by using Eq. 9-2. Thus,

$$P_{cr} = \frac{\pi^2 EA}{(L/r)^2} = \frac{\pi^2(1.9)(10^6)(2)(4)}{(166.3)^2} = 5424 \text{ lb} \cong 5420 \text{ lb} \qquad \textbf{Ans.}$$

(c) The axial stress under the action of the buckling load is

$$\sigma_{cr} = \frac{P_{cr}}{A} = \frac{5424}{2(4)} = 678 \text{ psi}$$

and therefore,

▶ This demonstrates that buckling can occur at stresses well below the elastic limit of a material for sufficiently slender columns!

$$\frac{\sigma_{cr}}{\sigma_e} = \frac{678}{6400} = 0.1059 \qquad \textbf{Ans.}$$

Example Problem 9-2   A 12-ft-long pin-ended column is made of 6061-T6 aluminum alloy. The column has a hollow circular cross section with an outside diameter of 5 in. and an inside diameter of 4 in. Determine

(a) The smallest slenderness ratio for which the Euler buckling load equation is valid.

(b) The critical buckling load.

SOLUTION

(a) The smallest slenderness ratio for which the Euler buckling load equation is valid occurs when $\sigma_{cr} = \sigma_y$ (yield strength). Rewriting Eq. 9-2 as

$$\frac{L}{r} = \sqrt{\frac{\pi^2 E}{\sigma_{cr}}} = \sqrt{\frac{\pi^2 E}{\sigma_y}}$$

and substituting $E = 10(10^6)$ psi and $\sigma_y = 40,000$ psi (see Appendix B)

$$\left(\frac{L}{r}\right)_{min} = \sqrt{\frac{\pi^2 E}{\sigma_y}} = \sqrt{\frac{\pi^2(10)(10^6)}{40(10^3)}} = 49.67 \cong 49.7$$ **Ans.**

▶ While the theoretical limit for validity of Eq. 9-1 or 9-2 is $\sigma_{cr} = \sigma_y$, experimental results indicate a limiting slenderness ratio at least 50 percent greater than the minimum given by $\sigma_{cr} = \sigma_y$. In this problem, the actual slenderness ratio (89.96) is 1.8 times the minimum ratio (49.7), so the Euler buckling load is expected to be valid.

(b) The critical buckling load is found using Eq. 9-1.

$$P_{cr} = \frac{\pi^2 EI}{L^2}$$

First, check that the slenderness ratio of the column exceeds 49.7.

$$A = \frac{\pi}{4}(5^2 - 4^2) = 7.069 \text{ in.}^2$$

$$I = \frac{\pi}{64}(5^4 - 4^4) = 18.113 \text{ in.}^4$$

$$r = \sqrt{\frac{I}{A}} = \sqrt{\frac{18.113}{7.069}} = 1.6007$$

$$\frac{L}{r} = \frac{12(12)}{1.6007} = 89.96 > 49.7$$

Thus, the critical buckling load is

$$P_{cr} = \frac{\pi^2 EI}{L^2} = \frac{\pi^2(10)(10^6)(18.113)}{144^2} = 86,210 \text{ lb} \cong 86.2 \text{ kip}$$ **Ans.**

Example Problem 9-3   Two 51 × 51 × 3.2-mm structural steel angles 3 m long will be used as a pin-ended column. Determine the slenderness ratio and the Euler buckling load if

(a) The two angles are not connected, and each acts as an independent member.
(b) The two angles are fastened together, as shown in Fig. 9-3, to act as a unit.

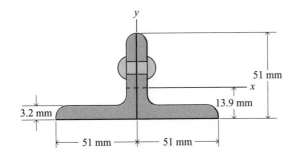

Figure 9-3

SOLUTION

(a) If the angles are not connected and each acts independently, the slenderness ratio is determined by using the minimum radius of gyration of the individual cross sections. From Appendix B, the minimum radius of gyration $r_{\min}$ for this angle is 10.1 mm about the ZZ-axis. Thus, the slenderness ratio is

$$\frac{L}{r} = \frac{L}{r_{\min}} = \frac{3000}{10.1} = 297 \qquad \text{Ans.}$$

▶ The critical stress in each portion of the column is $\sigma_{cr} = P_{cr}/A = \pi^2 E/(L/r)^2$. Therefore, the total load that the column can carry is P_{cr}, which equals σ_{cr} times the total area of 624 mm^2.

The area for each angle is 312 mm^2; therefore, the cross-sectional area for the column is 624 mm^2. The modulus of elasticity for structural steel is 200 GPa; therefore, the buckling load is

$$P_{cr} = \frac{\pi^2 EA}{(L/r)^2} = \frac{\pi^2(200)(10^9)(624)(10^{-6})}{(297)^2}$$

$$= 13.964(10^3)\,\text{N} \cong 13.96\,\text{kN} \qquad \text{Ans.}$$

(b) With the two angles connected as shown in Fig. 9-3, both I_x and I_y or r_x and r_y must be known in order to determine the minimum radius of gyration. Cross-sectional properties for the angles are given in Appendix B. The value of I_y for the two angles is obtained by using the parallel axis theorem; thus,

$$I_y = 2(I_C + Ad^2) = 2\left(Ar_C^2 + Ad^2\right)$$

and

$$r_y = \sqrt{I_y/2A} = \sqrt{\frac{2A(r_C^2 + d^2)}{2A}} = \sqrt{r_C^2 + d^2}$$

The above expression indicates that the radius of gyration for the two angles is the same as that for one angle, a fact obtained directly from the definition of radius of gyration. Then,

$$r_y = \sqrt{(15.9)^2 + (13.9)^2} = 21.12\,\text{mm}$$

In a similar fashion, the radius of gyration about the x-axis is the radius of gyration $r_x = 15.9$ mm for a single angle, since $d = 0$ for this axis. This means that the column tends to buckle about the x-axis, where the slenderness ratio is

$$\frac{L}{r} = \frac{L}{r_{\min}} = \frac{3000}{15.9} = 188.68 \cong 188.7 \qquad \text{Ans.}$$

The corresponding Euler buckling load is

$$P_{cr} = \frac{\pi^2 EA}{(L/r)^2} = \frac{\pi^2(200)(10^9)(624)(10^{-6})}{(188.68)^2}$$

$$= 34.599(10^3)\,\text{N} \cong 34.6\,\text{kN} \qquad \text{Ans.}$$

PROBLEMS

Introductory Problems

9-1* A structural steel rod 1 in. in diameter and 40 in. long will be used to support an axial compressive load P. Determine

a. The slenderness ratio.
b. The smallest slenderness ratio for which the Euler buckling load equation is valid.
c. The Euler buckling load.

9-2* A hollow, circular structural steel column 6 m long has an outside diameter of 125 mm and an inside diameter of 100 mm. Determine

a. The slenderness ratio.
b. The smallest slenderness ratio for which the Euler buckling load equation is valid.
c. The Euler buckling load.

9-3 A wood column of air-dried Douglas fir is 10 ft long and has a 4 × 4-in. rectangular cross section. Determine

a. The slenderness ratio.
b. The smallest slenderness ratio for which the Euler buckling load equation is valid.
c. The Euler buckling load.

9-4* A 5-m-long column with the cross section shown in Fig. P9-4 is constructed from four pieces of timber. The timbers are nailed together so that they act as a unit. Determine

a. The slenderness ratio.
b. The Euler buckling load. Use $E = 14$ GPa for the timber.
c. The axial stress in the column when the Euler load is applied.

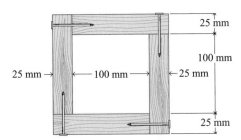

Figure P9-4

9-5 A yardstick has a rectangular cross section 1 1/8 in. wide × 5/32 in. thick. Determine

a. The slenderness ratio.
b. The Euler buckling load. Use $E = 1900$ ksi for the wood.

9-6 Determine the allowable compressive load that an 89-mm-nominal-diameter standard structural steel pipe can support if it is 5 m long and a factor of safety of 2 is specified.

9-7 Determine the allowable compressive load that a 5 × 5-in. air-dried red oak timber can support if it is 18 ft long and a factor of safety of 3 is specified.

Intermediate Problems

9-8* A 2.5-m-long column with the cross section shown in Fig. P9-8 is constructed from two pieces of air-dried Douglas fir timber. The timbers are nailed together so that they act as a unit. Determine

a. The slenderness ratio.
b. The smallest slenderness ratio for which the Euler buckling load equation is valid.
c. The Euler buckling load.

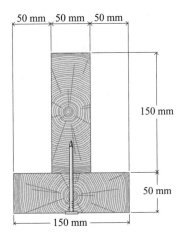

Figure P9-8

9-9* A WT6 × 36 structural steel section (see Appendix B for cross-sectional properties) is used for a 15-ft-long column. Determine

a. The slenderness ratio.
b. The smallest slenderness ratio for which the Euler buckling load equation is valid.
c. The Euler buckling load.

9-10 A WT152 × 89 structural steel section (see Appendix B for cross-sectional properties) is used for a 6-m-long column. Determine

a. The slenderness ratio.
b. The Euler buckling load. Use $E = 200$ GPa for the steel.
c. The axial stress in the column when the Euler load is applied.

9-11 Determine the maximum allowable compressive load for a 10-ft-long aluminum ($E = 10,000$ ksi) column having the cross section shown in Fig. P9-11 if a factor of safety of 2.25 is specified.

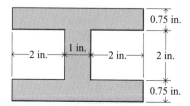

Figure P9-11

9-12 Two L 127 × 76 × 12.7-mm structural steel angles (see Appendix B for cross-sectional properties) are used for a column that is 4.5 m long. Determine the total compressive load required to buckle the two members if

a. They act independently of each other.
b. They are riveted together as shown in Fig. P9-12.

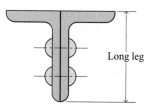

Figure P9-12

9-13* Determine the maximum allowable compressive load for a 10-ft-long aluminum alloy ($E = 10,000$ ksi) column having the cross section shown in Fig. P9-13 if a factor of safety of 2.50 is specified.

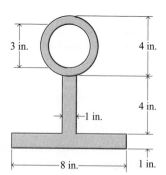

Figure P9-13

9-14 Determine the maximum allowable compressive load for a 6.5-m-long steel ($E = 200$ GPa) column having the cross section shown in Fig. P9-14 if a factor of safety of 1.92 is specified.

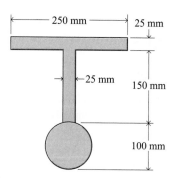

Figure P9-14

Challenging Problems

9-15* The 2-in.-diameter solid circular column shown in Fig. P9-15 is made of an aluminum alloy [$E = 10,000$ ksi and $\alpha = 12.5(10^{-6})/°$F]. If the column is initially stress free, determine the temperature increase that will cause the column to buckle.

Figure P9-15

9-16* A 60-kN load is supported by a tie rod AB and a pipe strut BC, as shown in Fig. P9-16. The tie rod has a diameter of 30 mm and is made of steel with a modulus of elasticity of 210 GPa and a yield strength of 360 MPa. The pipe strut has an inside diameter of 50 mm and a wall thickness of 15 mm and is made of an aluminum alloy with a modulus of elasticity of 73 GPa and a yield strength of 280 MPa. Determine the factor of safety with respect to failure by yielding or buckling for the structure.

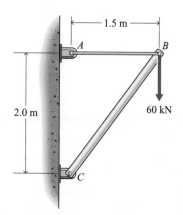

Figure P9-16

9-17 A column 20 ft long is made by riveting three S10 × 25.4 structural steel sections (see Appendix B for cross-sectional properties) together as shown in Fig. P9-17. Determine the maximum compressive load that this column can support. Use $E = 29,000$ ksi.

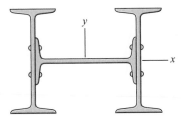

Figure P9-17

9-18* Two C229 × 30 structural steel channels (see Appendix B for cross-sectional properties) are used for a column that is 12 m long. Determine the total compressive load required to buckle the two members if they are laced together back to back 150 mm apart, as shown in Fig. P9-18.

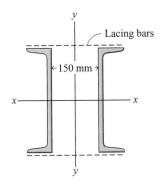

Figure P9-18

9-19 A simple pin-connected truss is loaded and supported as shown in Fig. P9-19. The members of the truss were fabricated by bolting two C10 × 30 channel sections (see Appendix B for cross-sectional properties) back to back to form an H-section.

The channels are made of structural steel with a modulus of elasticity of 29,000 ksi and a yield strength of 36 ksi. Determine the maximum load P that can be applied to the truss if a factor of safety of 1.75 with respect to failure by yielding and a factor of safety of 4 with respect to failure by buckling are specified.

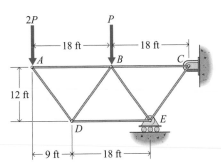

Figure P9-19

9-20 A simple pin-connected truss is loaded and supported, as shown in Fig. P9-20. All members of the truss are WT102 × 43 sections (see Appendix B for cross-sectional properties) made of structural steel with a modulus of elasticity of 200 GPa and a yield strength of 250 MPa. Determine

a. The factor of safety with respect to failure by yielding.
b. The factor of safety with respect to failure by buckling.

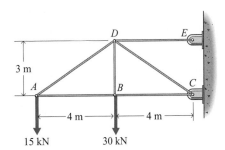

Figure P9-20

9-3 EFFECTS OF DIFFERENT IDEALIZED END CONDITIONS

The Euler buckling formula, as expressed by either Eq. 9-1 or Eq. 9-2, was derived for a column with pinned ends. The Euler equation changes for columns with different end conditions such as the four common ones shown in Fig. 9-4.

While it is possible to set up the differential equation with the appropriate boundary conditions to determine the Euler equation for each new case, a more common approach makes use of the concept of an effective length. The pin-ended column, by definition, has zero bending moments at each end. The length L in

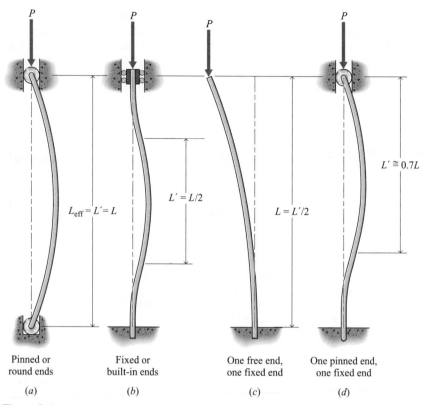

Pinned or round ends
(a)

Fixed or built-in ends
(b)

One free end, one fixed end
(c)

One pinned end, one fixed end
(d)

Figure 9-4

the Euler equation, therefore, is the distance between successive points of zero bending moment. All that is needed to modify the Euler column formula for use with other end conditions is to replace L by L', where L' is defined as the *effective length* of the column (the distance between two successive inflection points or points of zero moment).

The ends of the column in Fig. 9-4b are built in or fixed. Since the deflection curve is symmetrical, the distance between successive points of zero moment (inflection points) is half the length of the column. Thus, the effective length L' of a fixed-end column for use in the Euler column formula is half the true length ($L' = 0.5L$). The column in Fig. 9-4c, being fixed at one end and free at the other end, has zero moment only at the free end. If a mirror image of this column is visualized below the fixed end, however, the effective length between points of zero moment is seen to be twice the actual length of the column ($L' = 2L$). The column in Fig. 9-4d is fixed at one end and pinned at the other end. The effective length of this column cannot be determined by inspection, as could be done in the previous two cases; therefore, it is necessary to solve the differential equation to determine the effective length. This procedure yields $L' = 0.7L$.

A pin-ended column is usually loaded through a pin that, as a result of friction, is not completely free to rotate; hence, there will always be an indeterminate moment at the ends of a pin-connected column that will reduce the distance between the inflection points to a value less than L. Also, it is impossible to support a column so that all rotation is eliminated; therefore, the effective length of the column in Fig. 9-4b will be somewhat greater than $L/2$. As a result, it is usually necessary to modify the effective column lengths indicated by the ideal end conditions. The amount of the corrections will depend on the individual application. In summary, the term L/r in all column formulas in this book is interpreted to mean the effective slenderness ratio L'/r. In the problems in this book, the length given for a member is assumed to be the effective length unless otherwise noted.

Example Problem 9-4 A structural steel ($E = 29{,}000$ ksi) column 10 ft long must support an axial compressive load P, as shown in Fig. 9-5. The column has a 1×2-in. rectangular cross section. The left end of the column is fixed; the pin and bracket arrangement at the right end allows rotation about the pin but prevents rotation about a vertical axis. Determine the maximum safe load for the column if a factor of safety of 2 with respect to failure by buckling is specified.

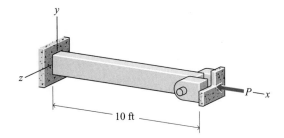

Figure 9-5

SOLUTION

The relationship between factor of safety and loads is

$$FS = \frac{P_u}{P_{all}} \qquad (a)$$

For failure by buckling, the ultimate load P_u is the Euler buckling load P_{cr}, which is

$$P_{cr} = \frac{\pi^2 EI}{(L')^2} \qquad (b)$$

Substituting Eq. (b) into Eq. (a) and rewriting yields

$$P_{all} = \frac{\pi^2 EI}{(L')^2 (FS)} \qquad (c)$$

The second moment of area I and the effective length L' depend on the plane in which the column buckles; either the xy- or xz-plane.

If buckling occurs in the xy-plane, the left end of the column is fixed and the right end is pinned. Thus, the effective length L' is

$$L' = 0.7L = 0.7(10)(12) = 84 \text{ in.}$$

and the second moment of area I is

$$I = \frac{1}{12}bh^3 = \frac{1}{12}(1)(2)^3 = 0.6667 \text{ in.}^4$$

Therefore, for buckling in the xy-plane,

$$P_{all} = \frac{\pi^2 EI}{(L')^2(FS)} = \frac{\pi^2(29)(10^6)(0.6667)}{(84)^2(2)} = 13{,}522 \text{ lb}$$

For buckling in the xz-plane, both the left end of the column and the right end of the column are fixed. Thus, the effective length L' is

$$L' = 0.5L = 0.5(10)(12) = 60 \text{ in.}$$

and the second moment of area I is

$$I = \frac{1}{12}hb^3 = \frac{1}{12}(2)(1)^3 = 0.16667 \text{ in.}^4$$

Therefore, for buckling in the xz-plane:

$$P_{all} = \frac{\pi^2 EI}{(L')^2(FS)} = \frac{\pi^2(29)(10^6)(0.16667)}{(60)^2(2)} = 6626 \text{ lb}$$

For a column with this cross section and end conditions, buckling will occur in the xz-plane and the safe load that may be applied is

$$P_{all} = 6626 \text{ lb} \cong 6.63 \text{ kip} \qquad \textbf{Ans.}$$

▌ PROBLEMS

Introductory Problems

9-21* A finished 2 × 4 timber (actual size 1 5/8 in. × 3 1/2 in. × 10 ft long) is used as a fixed-end, fixed-end column. If the modulus of elasticity for the timber is 1600 ksi and a factor of safety of 3 with respect to failure by buckling is specified, determine the maximum safe load for the column.

9-22* An L102 × 76 × 6.4-mm aluminum alloy ($E = 70$ GPa) angle is used for a fixed-end, pinned-end column having an actual length of 3 m. Determine the maximum safe load for the column if a factor of safety of 1.75 with respect to failure by buckling is specified. See Appendix B for cross-sectional properties; they are the same as those for a steel angle of the same size.

9-23 A W8 × 15 structural steel section (see Appendix B for cross-sectional properties) is used for a fixed-end, free-end column having an actual length of 10 ft. Determine the maximum

safe load for the column if a factor of safety of 2 with respect to failure by buckling is specified. Use $E = 29{,}000$ ksi.

9-24 Determine the maximum load that a 50-mm × 75-mm × 2.5-m-long aluminum alloy ($E = 73$ GPa) bar can support with a factor of safety of 3 with respect to failure by buckling if it is used as a fixed-end, pinned-end column.

9-25* A 6-in. × 6-in. × 20-ft-long timber ($E = 1900$ ksi) is used as a fixed-end, pinned-end column to support a 40,000-lb load. Determine the factor of safety based on the Euler buckling load.

9-26* A W254 × 33 structural steel ($E = 200$ GPa) section is used for a column with an actual length of 6 m. The column can be considered fixed at both ends for bending about the axis of the cross section with the smallest second moment of area and pinned at both ends for bending about the axis with the largest second moment of area. Determine the maximum axial compressive load P that can be supported by the column if a factor of safety of 1.9 with respect to failure by buckling is specified.

9-27 A W10 × 22 structural steel ($E = 29,000$ ksi) section is used for a column with an actual length of 20 ft. The column can be considered pinned at one end and fixed at the other end for bending about the axis of the cross section with the largest second moment of area and fixed at both ends for bending about the axis with the smallest second moment of area. Determine the maximum axial compressive load P that can be supported by the column if a factor of safety of 3 with respect to failure by buckling is specified.

9-28 A S127 × 15 structural steel ($E = 200$ GPa) section (see Appendix B for cross-sectional properties) will be used for a 12-m-long pinned-end, pinned-end column to support a 60-kN load. Equally spaced lateral braces will be installed to prevent buckling about the weak axis. If the braces offer no restraint to bending of the column and no restraint to buckling about the strong axis, determine

a. The spacing required for the lateral braces.
b. The maximum load that the column can support once the lateral braces are installed.

Intermediate Problems

9-29* A WT7 × 24 structural steel section (see Appendix B for cross-sectional properties) is used for a column with an actual length of 20 ft. If the modulus of elasticity for the steel is 29,000 ksi and a factor of safety of 2 with respect to failure by buckling is specified, determine the maximum safe load for the column under the following support conditions.

a. Pinned-pinned.
b. Fixed-free.
c. Pinned-fixed.
d. Fixed-fixed.

9-30* A solid circular rod with diameter D, length L, and modulus of elasticity E will be used to support an axial compressive load P. The support system for the rod is shown in Fig. P9-30. Determine the critical buckling load in terms of D, L, and E if buckling occurs in the plane of the page.

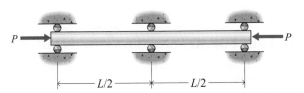

Figure P9-30

9-31 The structural steel ($E = 29,000$ ksi) bar shown in Fig. P9-31 has a 1.0-in. diameter and is 6 ft long. The support at the top

of the bar permits free vertical movement but no lateral movement or rotation. The bottom of the bar is free to move laterally but cannot rotate. Determine the maximum load P that can be applied if a factor of safety of 2.5 with respect to failure by buckling is specified.

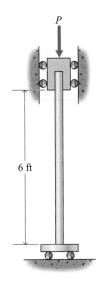

Figure P9-31

9-32* A structural steel ($E = 200$ GPa) bar has a diameter of 50 mm, is 5 m long, and supports an axial compressive load P, as shown in Fig. P9-32. End A is fixed. The support at end B permits free movement in the x- and z-directions but no rotation about the z-axis. Determine the maximum load P that can be applied if a factor of safety of 2 with respect to failure by buckling is specified.

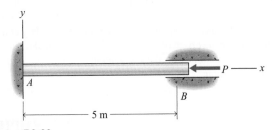

Figure P9-32

9-33 A 2-in.-diameter by 24-ft-long solid, circular aluminum alloy ($E = 10,000$ ksi) bar is used to transmit a 4000-lb force, as shown in Fig. P9-33. If the supports permit buckling only in the plane of the page, determine the factor of safety with respect to failure by buckling.

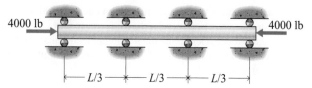

Figure P9-33

9-34 Two 25-mm-diameter structural steel (200 GPa) columns support a 400-kg mass, as shown in Fig. P9-34. The two short struts at the sides of the mass prevent horizontal movement of the mass but offer no constraint to vertical motion. Determine the factor of safety with respect to failure by buckling.

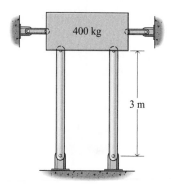

Figure P9-34

Challenging Problems

9-35 Verify the effective length for the fixed-end, fixed-end column shown in Fig. 9-4b by solving the differential equation of the elastic curve and applying the appropriate boundary conditions. Place the origin of the xy-coordinate system at the lower end of the column with the x-axis along the axis of the undeformed column.

9-36 Verify the effective length for the fixed-end, free-end column shown in Fig. 9-4c by solving the differential equation of the elastic curve and applying the appropriate boundary conditions. Place the origin of the xy-coordinate system at the lower end of the column with the x-axis along the axis of the undeformed column.

9-37 A free-body diagram for the fixed-end, pinned-end column shown in Fig. 9-4d is shown in Fig. P9-37. Use the free-body diagram to develop the differential equation of the elastic curve. Verify that the effective length for the fixed-end, pinned-end column is $L' = 0.7L$ by solving the differential equation of the elastic curve and applying the appropriate boundary conditions.

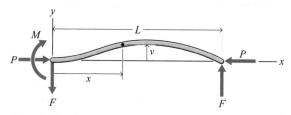

Figure P9-37

9-38 A rigid block is supported by two fixed-end, fixed-end columns, as shown in Fig. P9-38a. Determine the effective lengths of the columns by solving the differential equation of the elastic curve and applying the appropriate boundary conditions. Assume that buckling occurs as shown in Fig. P9-38b.

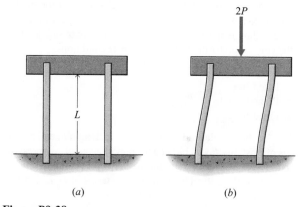

Figure P9-38

9-4 EMPIRICAL COLUMN FORMULAS—CENTRIC LOADING

Euler's formula, Eq. 9-1 or 9-2, is valid when the axial compressive stress for a column is less than the yield strength. The range of usefulness of the Euler formula is seen in Figure 9-6, where the critical stress ($\sigma_{cr} = P_{cr}/A$) is plotted versus the slenderness ratio (L/r). The plot is for structural steel for which $E = 29,000$ ksi (200 GPa) and $\sigma_{\text{yield}} = 36$ ksi (250 MPa). The Euler curve is truncated at 36 ksi because the critical stress cannot exceed the yield strength. For structural steel, this occurs when $L/r = 89$. The dashed portion of the Euler curve indicates that the calculation of the critical stress is no longer useful because the stress exceeds the

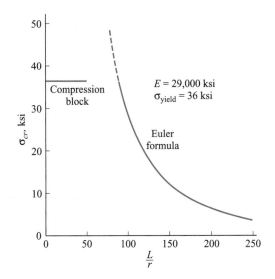

Figure 9-6

yield strength. For practical purposes, columns with large, slenderness ratios are useless because they support only small loads. At the low end of the slenderness scale, the column would behave essentially as a short compression block, and the critical stress would be the compressive strength of the material (for metals, usually the yield strength). The extent of this range, the *compression-block* range, is a matter of judgment or is dictated by specifications. The range between the compression block and the slender ranges is known as the *intermediate* range. Neither compression-block theory nor the Euler formula gives results in the intermediate range that agree with test results.

Experimental results of many tests on axially loaded columns are shown in Fig. 9-7. The Euler curve agrees with experimental data for slenderness ratios in the slender range. Experimental data and compression-block theory agree in the

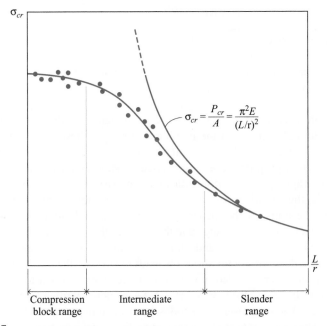

Figure 9-7

Table 9-1 Some Representative Column Codes for Centric Loading

Code No.	Source	Material	Compression-Block and/or Intermediate-Range Formulas and Limitations (L/r is the effective ratio L'/r)	Slender Range
1	a	Structural steel with a yield point σ_y	$0 \leq \dfrac{L}{r} \leq C_c \quad \sigma_{all} = \dfrac{\sigma_y}{FS}\left[1 - \dfrac{1}{2}\left(\dfrac{L/r}{C_c}\right)^2\right]$ $C_c^2 = \dfrac{2\pi^2 E}{\sigma_y}$ $FS = \dfrac{5}{3} + \dfrac{3}{8}\left(\dfrac{L/r}{C_c}\right) - \dfrac{1}{8}\left(\dfrac{L/r}{C_c}\right)^3$	$\dfrac{L}{r} \geq C_c$ $\sigma_{all} = \dfrac{\pi^2 E}{1.92(L/r)^2}$
2	b	2014-T6 (Alclad) Aluminum alloy	$\dfrac{L}{r} \leq 12 \quad \sigma_{all} = 28$ ksi $= 193$ MPa $12 \leq \dfrac{L}{r} \leq 55 \quad \sigma_{all} = \left[30.7 - 0.23\left(\dfrac{L}{r}\right)\right]$ ksi $= \left[212 - 1.585\left(\dfrac{L}{r}\right)\right]$ MPa	$\dfrac{L}{r} \geq 55$ $\sigma_{all} = \dfrac{54,000}{(L/r)^2}$ ksi $= \dfrac{372(10^3)}{(L/r)^2}$ MPa
3	b	6061-T6 Aluminum alloy	$\dfrac{L}{r} \leq 9.5 \quad \sigma_{all} = 19$ ksi $= 131$ MPa $9.5 \leq \dfrac{L}{r} \leq 66 \quad \sigma_{all} = \left[20.2 - 0.126\left(\dfrac{L}{r}\right)\right]$ ksi $= \left[139 - 0.868\left(\dfrac{L}{r}\right)\right]$ MPa	$\dfrac{L}{r} \geq 66$ $\sigma_{all} = \dfrac{51,000}{(L/r)^2}$ ksi $= \dfrac{351(10^3)}{(L/r)^2}$ MPa
4	c	Timber with a rectangular cross section $b \times d$ where $d < b$	$\dfrac{L}{d} \leq 11 \quad \sigma_{all} = F_c^*$ $11 \leq \dfrac{L}{d} \leq k \quad \sigma_{all} = F_c\left[1 - \dfrac{1}{3}\left(\dfrac{L/d}{k}\right)^4\right]$ $k = 0.671\sqrt{E/F_c}$	$k \leq \dfrac{L}{d} \leq 50$ $\sigma_{all} = \dfrac{0.30E}{(L/d)^2}$

a. *Manual of Steel Construction*, 9th ed., American Institute of Steel Construction, New York, 1989.

b. *Specifications for Aluminum Structures*, Aluminum Association, Inc., Washington, D.C., 1986.

c. *Timber Construction Manual*, 3rd ed., American Institute of Timber Construction, John Wiley & Sons, Inc., New York, 1985.

*F_c is the allowable stress for a short block in compression parallel to the grain.

compression-block range. Test results are not in agreement with compression-block theory or the Euler theory in the intermediate range. These intermediate-length columns may be analyzed by empirical formulas.

For design purposes, the entire range of stresses for a given material is covered by an appropriate set of specifications known as a column code. Depending on the code, the empirical formula may be specified for the intermediate range along with the limits of the intermediate range; the code may also specify the Euler formula in the slender range and the limits of the slender range; or, the code may specify the critical stress and limits of the range for the compression-block range. If the code is written for allowable loads or stresses, the factor of safety will either be specified or included in the constants for the empirical formula. A few representative codes are listed in Table 9-1. Each code is written for allowable (or safe) stresses. A few representative codes for steel, aluminum, and timber will be discussed. The codes are taken from references listed in the footnotes of Table 9-1.

   The formulas in Table 9-1 represent column equations that have been in-corporated in various design codes. Slenderness ratios in this table are always the effective slenderness ratio L'/r. If no end conditions are specified in the problems presented later, the stated length is the effective length. Note that the use of high-strength materials will increase the allowable load ($P_{all} = \sigma_{all}A$) for short columns but will have little effect on the load-carrying capacity of long columns, because the critical load (Euler load) depends on Young's modulus, not on the elastic strength of the material. Note also that the use of fixed or restrained ends, which has the effect of reducing the length of the column, materially increases the load-carrying capacity of slender columns but has much less influence on short compression members.

   The discussion so far has been concerned with primary instability, in which the column deflects as a whole into a smooth curve. No discussion of compression loading is complete without reference to local instability in which the member fails locally by crippling of thin sections. Thin open sections such as angles, channels, and H-sections are particularly sensitive to crippling failure. The design of such members to avoid crippling failure is usually governed by specifications controlling the width–thickness ratios of outstanding flanges. Closed-section members of thin material (thin-walled tubes, for example) must also be examined for crippling failure when the members are short. A discussion of these and other design issues may be found in references such as a and b listed in Table 9-1.

Example Problem 9-5   Two structural steel C10 × 25 channels are latticed 5 in. back to back, as shown in Figs. 9-8a and b, to form a column. Determine the maximum allowable axial load for effective lengths of 25 ft and 40 ft. Use Code 1 for structural steel (see Appendix B for properties).

SOLUTION
Both I_x and I_y (see Fig. 9-8b) or r_x and r_y must be known to determine the minimum radius of gyration. Properties of the channel section are given in Appendix B. The value of I_y for two channels is obtained by using the parallel axis theorem; thus,

$$I_y = 2(I_C + Ad^2) = 2(Ar_C^2 + Ad^2)$$

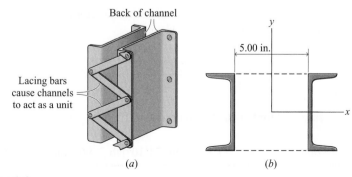

Back of channel

Lacing bars
cause channels
to act as a unit

5.00 in.

y

x

(a)   (b)

Figure 9-8

and

$$r_y = \sqrt{I_y/(2A)} = \sqrt{\frac{2A(r_C^2 + d^2)}{2A}} = \sqrt{r_C^2 + d^2}$$

The above expression indicates that the radius of gyration for the two channels is the same as that for one channel, a fact obtained directly from the definition of radius of gyration. Then,

$$r_y = \sqrt{(0.676)^2 + (2.50 + 0.617)^2} = 3.19 \text{ in.}$$

which is less than the tabular value of 3.52 in. for r_x. This means that the column tends to buckle with respect to the y-axis, and the slenderness ratios are

$$\frac{L}{r} = \frac{12(25)}{3.19} = 94.0 \quad \text{and} \quad \frac{12(40)}{3.19} = 150.5$$

From Code 1,

$$C_c^2 = \frac{2\pi^2 E}{\sigma_y} = \frac{2\pi^2(29,000)}{36} = 15,901 \qquad C_c = 126.1$$

The slenderness ratios above indicate that the 25-ft column is in the intermediate range because $L/r < C_c$ or $94.0 < 126.1$; hence, the factor of safety is

$$\text{FS} = \frac{5}{3} + \frac{3}{8}\left(\frac{94.0}{126.1}\right) - \frac{1}{8}\left(\frac{94.0}{126.1}\right)^3 = 1.894$$

The allowable stress is

$$\sigma_{\text{all}} = \frac{\sigma_y}{\text{FS}}\left[1 - \frac{1}{2}\left(\frac{L/r}{C_c}\right)^2\right] = \frac{36(10^3)}{1.894}\left[1 - \frac{1}{2}\left(\frac{94.0}{126.1}\right)^2\right] = 13,726 \text{ psi}$$

Hence, the safe load for the 25-ft column is

$$P = \sigma_{\text{all}}(A) = 13,726(2)(7.35) = 201.8(10^3) \text{ lb} \cong 202 \text{ kip} \qquad \textbf{Ans.}$$

The 40-ft column has a slenderness ratio of 150.5; therefore, it is in the slender range since $L/r > C_c$ or $150.5 > 126.1$. Hence,

$$\sigma_{\text{all}} = \frac{\pi^2 E}{1.92(L/r)^2} = \frac{\pi^2(29)(10^6)}{1.92(150.5)^2} = 6581 \text{ psi}$$

Hence, the safe load for the 40-ft column is

$$P = \sigma_{\text{all}}(A) = 6581(2)(7.35) = 96.74(10^3) \text{ lb} \cong 96.7 \text{ kip} \qquad \textbf{Ans.}$$

▶ The Euler buckling load is inversely proportional to the square of the length L. Therefore, it might be expected that reducing the length of a column by a factor n (in this case $n = 40/25 = 1.6$) would increase the buckling load by a factor of n^2 [in this case $n^2 = (1.6)^2 = 2.56$]. In this case, however, the safe load for the 25-ft-long column is only about 2.1 times that for the 40-ft-long column.

Example Problem 9-6 A Douglas fir ($E = 11$ GPa and $F_c = 7.6$ MPa) timber column with an effective length of 3.5 m has a 150×200-mm rectangular cross section. Determine the maximum compressive load permitted by Code 4 of Table 9-1.

SOLUTION

Code 4 of Table 9-1 indicates that the allowable stress (and hence the allowable load) depends on the value of L/d, the ratio of the effective length of the column to the smallest cross-sectional dimension and a factor k specified by the code. These values are

$$\frac{L}{d} = \frac{3.5(10^3)}{150} = 23.33$$

and

$$k = 0.671\sqrt{\frac{E}{F_c}} = 0.671\sqrt{\frac{11(10^9)}{7.6(10^6)}} = 25.53$$

Since $11 \leq L/d \leq k$ or $11 < 23.33 < 25.53$, the column is in the intermediate range and the allowable stress is given by the code as

$$\sigma_{all} = F_c\left[1 - \frac{1}{3}\left(\frac{L/d}{k}\right)^4\right] = 7.6\left[1 - \frac{1}{3}\left(\frac{23.33}{25.53}\right)^4\right] = 5.833 \text{ MPa}$$

Thus, the allowable load is

$$P = \sigma_{all}(A) = 5.833(10^6)(150)(200)(10^{-6})$$
$$= 174.99(10^3) \text{ N} \cong 175.0 \text{ kN} \qquad \textbf{Ans.}$$

▶ Note that $L/d = 23.33$ is the ratio of the effective length of the column to the smallest cross-sectional dimension and not the slenderness ratio $L/r = \dfrac{L}{\sqrt{I/A}} = \dfrac{3500}{\sqrt{56.25 \times 10^6/30 \times 10^3}} = 80.83$. The corresponding Euler buckling load for this column would be $P_{cr} = \dfrac{\pi^2 EA}{(L/r)^2} = 498 \times 10^3$ N, which is 2.8 times larger than the allowable load permitted by Code 4 of Table 9-1.

PROBLEMS

Introductory Problems

9-39* An air-dried red oak ($E = 1800$ ksi and $F_c = 4.6$ ksi) timber column with an effective length of 5 ft has a 3×3-in. rectangular cross section. Determine the maximum compressive load permitted by Code 4.

9-40* Douglas fir ($E = 11$ GPa and $F_c = 7.6$ MPa) timber columns with 200×300-mm rectangular cross sections will be used to support axial compressive loads. Determine the maximum compressive load permitted by Code 4 if

a. The effective length of the column is 2 m.
b. The effective length of the column is 4 m.
c. The effective length of the column is 6 m.

9-41 A 2.5-in.-diameter standard-weight steel pipe column is 8 ft long, is pinned at both ends, and supports an axial compressive load P. If $E = 29,000$ ksi and $\sigma_y = 36$ ksi, determine the maximum load permitted by Code 1.

9-42* A W254 × 89 structural steel ($E = 200$ GPa and $\sigma_y = 250$ MPa) column is pinned at both ends, is 3 m long, and supports an axial compressive load P. Determine the maximum load permitted by Code 1.

9-43 A 3.0-in.-diameter solid circular 6061-T6 aluminum alloy bar is to be used as a column with an effective length of 30 in. Determine the maximum axial compressive load P permitted by Code 3.

9-44 A 2014-T6 aluminum alloy tube with an outside diameter of 100 mm and an inside diameter of 80 mm is used for a column with an effective length of 1.0 m. Determine the maximum axial compressive load P permitted by Code 2.

Intermediate Problems

9-45* Three structural steel bars with a 1 × 4-in. rectangular cross section will be used for an 8-ft-long fixed-ended column. Determine the maximum compressive load permitted by Code 1 if

a. The three bars act as independent axially loaded members.
b. The three bars are welded together to form an H-column.

9-46* Three hollow circular structural steel tubes with inside diameters of 50 mm and outside diameters of 80 mm will be used for a 3.5-m-long pin-ended column. Determine the maximum compressive load permitted by Code 1 if

a. The three tubes act as independent axially loaded members.
b. The three tubes are welded together as shown in Fig. P9-46.

Figure P9-46

9-47 Two C10 × 15.3 structural steel channels 12 ft long are used as a fixed-ended, pin-ended column. Determine the maximum load permitted by Code 1 if the channels are welded together to form a 10 × 5.2-in. box section, as shown in Fig. P9-47.

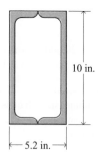

Figure P9-47

9-48 Four L76 × 76 × 12.7-mm structural steel angles 5 m long are used as a fixed-ended column. Determine the maximum load permitted by Code 1 if the angles are fastened together as shown in Fig. P9-48.

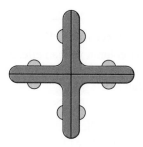

Figure P9-48

9-49* Four L4 × 3 × 3/8-in. structural steel angles 11 ft long are used as a pin-ended column. Determine the maximum load permitted by Code 1 if the angles are welded together to form a 6 × 8-in. box section as shown in Fig. P9-49.

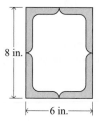

Figure P9-49

9-50 Two L102 × 76 × 9.5-mm structural steel angles 7 m long are used as a pin-ended column. Determine the maximum load permitted by Code 1 if the angles are fastened together to form a 102 × 76-mm box section as shown in Fig. P9-50.

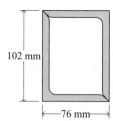

Figure P9-50

9-51 An L5 × 5 × 3/4-in. aluminum alloy 2014-T6 angle will be used as a fixed-ended, pin-ended column to support a load of 120 kip. The cross-sectional properties of steel and aluminum angles are the same (see Appendix B). Determine the maximum permissible length permitted by Code 2.

9-52* A column of aluminum alloy 2014-T6 is composed of two L127 × 127 × 19.1-mm angles riveted together as shown in Fig. P9-52. The length between end connections is 3 m, and the end connections are such that there is no restraint to bending about the y-axis; but restraint to bending about the x-axis reduces the effective length to 2.1 m. Determine the maximum axial compressive load permitted by Code 2. The cross-sectional properties of aluminum and steel angles are the same (see Appendix B).

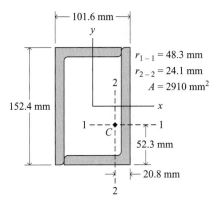

$r_{1-1} = 48.3$ mm
$r_{2-2} = 24.1$ mm
$A = 2910$ mm²

Figure P9-54

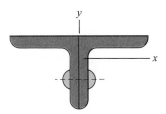

Figure P9-52

9-55 Three 2 × 4-in. timber studs are nailed together to form an 8-ft-long column with the cross section shown in Fig. P9-55. The column is fixed at the bottom and pinned at the top. If $E = 1600$ ksi and $F_c = 1100$ psi, use Code 4 to determine the maximum permissible axial compressive load that may be applied.

9-53* A strut of aluminum alloy 6061-T6 having the cross section shown in Fig. P9-53 is to carry an axial compressive load of 165 kip. The strut is 4 ft long and is fixed at the bottom and pinned at the top. Determine the dimension d of the cross section using Code 3.

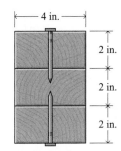

Figure P9-55

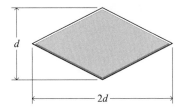

Figure P9-53

9-56 A sand bin is supported by four 200 × 250-mm fixed-ended rectangular timber columns 4.5 m long. Assume that the load is equally divided among the four columns and that the column loads are axial. If the columns are made of timber with $E = 13$ GPa and $F_c = 9$ MPa, determine the maximum load permitted by Code 4.

Challenging Problems

9-57* A 25-ft-long plate and angle column consists of four L 5 × 3 1/2 × 1/2-in. structural steel angles riveted to a 10 × 1/2-in. structural steel plate as shown in Fig. P9-57. Determine the maximum safe load permitted by Code 1 if

a. The column is pinned at both ends.
b. The column is fixed at the base and pinned at the top.

9-54 Two L152 × 89 × 12.7-mm angles of aluminum alloy 2014-T6 are welded together as shown in Fig. P9-54 to form a 4.75-m-long pin-ended column. The pins provide no restraint to bending about the x-axis but reduce the effective length to 3.25 m for bending about the y-axis. The cross-sectional properties of one angle are given on the figure, where C is the centroid of one angle. Determine the maximum axial compressive load permitted by Code 2.

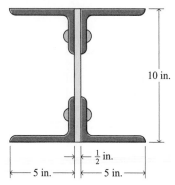

Figure P9-57

9-58* Four C178 × 22 structural steel channels 12 m long are used to fabricate a column with the cross section shown in Fig. P9-58. The column is fixed at the base and pinned at the top. The pin at the top offers no restraint to bending about the *y*-axis; but for bending about the *x*-axis, the pin provides restraint sufficient to reduce the effective length to 7.5 m. Determine the maximum axial compressive load permitted by Code 1.

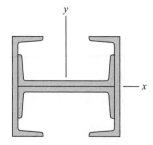

Figure P9-58

9-59 The machine part shown in Fig. P9-59 is made of SAE 4340 heat-treated steel and carries an axial compressive load. The pins at the ends offer no restraint to bending about the axis of the pin, but restraint about the perpendicular axis reduces

the effective length to 36 in. Determine the maximum load permitted by Code 1.

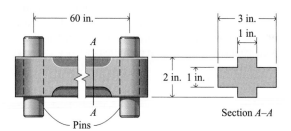

Figure P9-59

9-60 A connecting rod made of SAE 4340 heat-treated steel has the cross section shown in Fig. P9-60. The pins at the ends of the rod are parallel to the *x*-axis and are 1250 mm apart. Assume that the pins offer no restraint to bending about the *x*-axis but provide essentially complete fixity for bending about the *y*-axis. Determine the maximum axial compressive load permitted by Code 1.

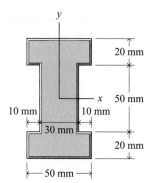

Figure P9-60

9-5 ECCENTRICALLY LOADED COLUMNS

Although a given column will support the maximum load when the load is applied centrically, it is sometimes necessary to apply an eccentric load to a column. For example, a beam supporting a floor load in a building may in turn be supported by an angle riveted or welded to the side of a column, as shown in Fig. 9-9. Frequently, the floorbeam is framed into the column with a stiff connection, and the column is then subjected to a bending moment due to continuity of the floorbeam and column. The eccentricity of the load, or the bending moment, will increase the stress in the column or reduce its load-carrying capacity. Two methods will be presented for computing the allowable load on a column subjected to an eccentric load (or an axial load combined with a bending moment).

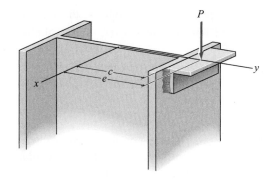

Figure 9-9

9-5-1 Allowable Stress Method

This method is based on the specification that the sum of the direct and bending stresses ($P/A + Mc/I$) shall not exceed the allowable stress prescribed for a centric load by the appropriate column formula,

$$\frac{P}{A} + \frac{Mc}{I} \leq \sigma_{\text{all}} \tag{9-3}$$

In Eq. 9-3, σ_{all} is the allowable stress for a centric load and is calculated using the equations of Table 9-1 and using the largest value of the slenderness ratio for the cross section irrespective of the axis about which bending occurs. Values for c and I used in calculating the bending stress, however, depend on the axis about which the bending occurs. Equation 9-3, which is prescribed by most modern codes, usually produces a conservative design.

9-5-2 Interaction Method

One of the modern expressions for treating combined loads is known as the *interaction formula*, of which several types are in use. The analysis for compression members subjected to bending and direct stress may be derived as follows. It is assumed that the stress in the column can be written as

$$\frac{P}{A} + \frac{Mc}{I} \leq \sigma_{\text{all}}$$

and, when this expression is divided by σ_{all}, it becomes

$$\frac{P/A}{\sigma_{\text{all}}} + \frac{Mc/I}{\sigma_{\text{all}}} \leq 1 \tag{9-4}$$

When considering eccentrically loaded columns, the value of σ_{all} will, in general, be different for the two terms. In the first term P/A represents an axial stress on a column; therefore, the value of σ_{all} should be the average allowable stress on an axially loaded column as obtained by an empirical formula such as those presented in Table 9-1. In the second term, Mc/I represents the flexural stress induced in the member as a result of the eccentricity of the load or an applied bending moment;

therefore, the corresponding value of σ_{all} should be the allowable flexural stress. Since the two values of σ_{all} are different, one recommended form of Eq. 9-4 is the following interaction formula:

$$\frac{P/A}{\sigma_a} + \frac{Mc/I}{\sigma_b} \leq 1 \qquad (9\text{-}5)$$

in which P/A is the average axial stress on the eccentrically loaded column, σ_a is the allowable average axial stress for an axially loaded column (note that the greatest value of L/r should be used to calculate σ_a), Mc/I is the flexural stress in the column, and σ_b is the allowable flexural stress.

When Eq. 9-5 is used to select the most economical section, it will usually not be possible to obtain a section that will exactly satisfy the equation. Any section that makes the sum of the terms on the left side of the equation less than unity is considered safe, and the safe section giving the largest sum (less than unity) is the most efficient section.

Example Problem 9-7

A W457 × 144 wide-flange section is used for the column shown in Fig. 9-10. The column is made of steel ($E = 200$ GPa, $\sigma_y = 290$ MPa, $\sigma_b = 190$ MPa) and has an effective length of 6 m. An eccentric load P ($e = 125$ mm) is applied on the centerline of the web as shown in the figure. Determine the maximum safe load according to

(a) The allowable stress method.
(b) The interaction method.

SOLUTION

The cross-sectional properties for a W457 X 144 wide-flange section are

$$A = 18{,}365 \text{ mm}^2 \qquad c = 472/2 = 236 \text{ mm}$$
$$r_{min} = 67.3 \text{ mm} \qquad I_{min} = 83.7(10^6) \text{ mm}^4$$
$$r_{max} = 199 \text{ mm} \qquad I_{max} = 728(10^6) \text{ mm}^4$$

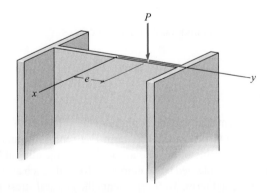

Figure 9-10

Since the column is made of steel, Code 1 of Table 9-1 will be used to calculate σ_{all}. The equation for determining σ_{all} depends on the slenderness range.

$$C_c = \sqrt{\frac{2\pi^2 E}{\sigma_y}} = \sqrt{\frac{2\pi^2(200)(10^9)}{290(10^6)}} = 116.68$$

$$\frac{L}{r_{min}} = \frac{6(10^3)}{67.3} = 89.15 < C_c = 116.68$$

Since $L/r_{min} < C_c$, the intermediate column formula is applicable. The factor of safety FS for use in the intermediate column formula is

$$FS = \frac{5}{3} + \frac{3}{8}\left(\frac{L/r}{C_c}\right) - \frac{1}{8}\left(\frac{L/r}{C_c}\right)^3 = \frac{5}{3} + \frac{3}{8}\left(\frac{89.15}{116.68}\right) - \frac{1}{8}\left(\frac{89.15}{116.68}\right)^3 = 1.90$$

$$\sigma_{all} = \frac{\sigma_y}{FS}\left[1 - \frac{1}{2}\left(\frac{L/r}{C_c}\right)^2\right] = \frac{290(10^6)}{1.90}\left[1 - \frac{1}{2}\left(\frac{89.15}{116.68}\right)^2\right]$$

$$= 108.08(10^6)\,\text{N/m}^2 = 108.08\,\text{MPa} = \sigma_a$$

(a) Using the allowable stress method, Eq. 9-3 gives

$$\frac{P}{A} + \frac{Mc}{I} = \frac{P}{A} + \frac{Pec}{I} \leq \sigma_{all}$$

$$\frac{P}{18,365(10^{-6})} + \frac{P(125)(236)(10^{-6})}{728(10^{-6})} \leq 108.08(10^6)$$

$$P \leq 1138(10^3)\,\text{N}$$

Therefore, the maximum safe load that can be applied according to the allowable stress method is

$$P_{max} = 1138(10^3)\,\text{N} = 1138\,\text{kN} \qquad \textbf{Ans.}$$

(b) Using the interaction method, Eq. 9-5 gives

$$\frac{P/A}{\sigma_a} + \frac{Mc/I}{\sigma_b} = \frac{P/A}{\sigma_a} + \frac{Pec/I}{\sigma_b} \leq 1$$

$$\frac{P/[18,365(10^{-6})]}{108.08(10^6)} + \frac{P(125)(236)(10^{-6})/[728(10^{-6})]}{190(10^6)} \leq 1$$

$$P \leq 1395(10^3)\,\text{N}$$

Therefore, the maximum safe load that can be applied according to the interaction method is

$$P_{max} = 1195(10^3)\,\text{N} = 1395\,\text{kN} \qquad \textbf{Ans.}$$

PROBLEMS

Introductory Problems

9-61* A hollow square steel member with outside and inside dimensions of 5-in. and 4-in. (the walls are 1/2 in. thick) functions as a pin-ended column 10 ft long. Determine the maximum load that the column can carry if the load is applied with a known eccentricity of 9/16 in., as shown in Fig. P9-61. Use the interaction method and let $E = 29,000$ ksi, $\sigma_y = 36$ ksi, and $\sigma_b = 24$ ksi.

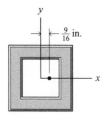

Figure P9-61

9-62* A hollow circular steel member with outside and inside diameters of 150 mm and 120 mm (the walls are 15 mm thick) functions as a pin-ended column 4 m long. Determine the maximum load that the column can carry if the load is applied 20 mm from the axis of the member, as shown in Fig. P9-62. Use the allowable stress method and let $E = 200$ GPa and $\sigma_y = 250$ MPa.

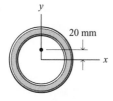

Figure P9-62

9-63 A 2-in.-diameter steel strut is subjected to an eccentric compressive load, as shown in Fig. P9-63. The effective length for bending about the x-axis is 75 in.; but for bending about the y-axis, the end conditions reduce the effective length to 50 in. Determine the maximum load that the strut can carry. Use the interaction method and let $E = 29,000$ ksi, $\sigma_y = 36$ ksi, and $\sigma_b = 24$ ksi.

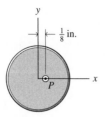

Figure P9-63

9-64 Determine the maximum load P that can be applied to the timber column shown in Fig. P9-64 if $E = 12$ GPa and the allowable stress for compression parallel to the grain is 9 MPa. Use the allowable stress method.

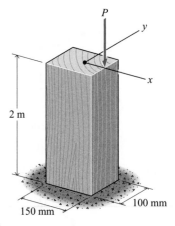

Figure P9-64

Intermediate Problems

9-65* A W14 × 82 structural steel section is used for a 20-ft-long pin-ended column. The load is applied at a point on the centerline of the web 5 in. from the axis of the column. If $E = 29,000$ ksi, $\sigma_y = 36$ ksi, and $\sigma_b = 24$ ksi, determine the maximum safe load according to

a. The allowable stress method.
b. The interaction method.

9-66* A W 356 × 64 structural steel section is used for a 7-m-long fixed-ended column. The load is applied at a point on the centerline of the web 150 mm from the axis of the column. If $E = 200$ GPa, $\sigma_y = 250$ MPa, and $\sigma_b = 160$ MPa, determine the maximum safe load according to

a. The allowable stress method.
b. The interaction method.

9-67 Two C8 × 18.75 structural steel sections 25 ft long are laced back to back, as shown in Fig. P9-67, to form a pin-ended column. Determine the maximum load permitted by the interaction method if the load is applied at point A of the cross section. Let $E = 29,000$ ksi, $\sigma_y = 36$ ksi, and $\sigma_b = 24$ ksi.

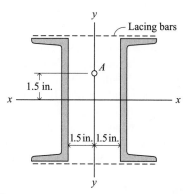

Figure P9-67

9-68 A hollow square steel member with outside and inside dimensions of 100 mm and 70 mm (the walls are 15 mm thick) functions as a pin-ended column 4 m long. Determine the maximum load that the column can carry if the load is applied with a known eccentricity of 15 mm along a diagonal of the square as shown in Fig. P9-68. Use the allowable stress method and let $E = 200$ GPa and $\sigma_y = 250$ MPa.

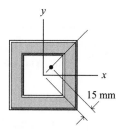

Figure P9-68

Challenging Problems

9-69* A WT8 × 25 structural steel T-section is to be used as a compression member to transmit an eccentric load of 100 kip. The effective length of the member is 10 ft. Determine the maximum eccentricity e permitted if the point of load application is on the centerline of the stem between the outer edge of the flange and the centroidal axis of the column. Use the interaction method with $E = 29{,}000$ ksi, $\sigma_y = 36$ ksi, and $\sigma_b = 24$ ksi.

9-70* Two 50 × 150-mm structural steel bars 5 m long will be welded together, as shown in Fig. P9-70, and used for a pin-ended column. Determine the maximum load permitted if the load is applied at the point on the cross section indicated in Fig. P9-70. Use the allowable stress method and let $E = 200$ GPa and $\sigma_y = 250$ MPa.

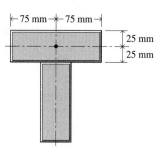

Figure P9-70

9-71 Two L6 × 3 1/2 × 1/2-in, structural steel ($E = 29{,}000$ ksi and $\sigma_y = 36$ ksi) angles are welded together as shown in Fig. P9-71 to form a column with an effective length of 12 ft. Use the allowable stress method to determine

a. The maximum axial compressive load P that can be supported by the column.
b. The maximum and minimum values for the distance d when a 50-kip load is applied.

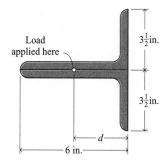

Figure P9-71

9-72 A 2014-T6 aluminum alloy compression member with an effective length of 1.25 m has the T cross section shown in Fig. P9-72.

a. Determine the maximum axial compressive load P permitted by Code 2.
b. Use the allowable stress method to determine the maximum bending moment M in the yz-plane that can be applied as shown when the column is supporting a 175-kN axial load.

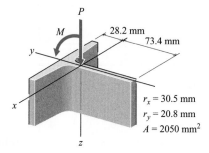

Figure P9-72

9-6 DESIGN PROBLEMS

In previous chapters design usually involved strength as a controlling parameter. Since buckling is an elastic phenomenon, the modulus of elasticity (stiffness) is a more significant parameter than is strength (such as yield strength if failure is by yielding) if the column length is in the slender range. If the column is in the intermediate range, both yield strength and stiffness may be important parameters. When designing columns using the representative codes listed in Table 9-1, a designer must be aware of several factors. The codes are for specific materials, that is, materials with a specific value of yield strength and modulus of elasticity. In addition, some codes include a factor of safety, while others require that a factor of safety be introduced. Each of the codes has a range of applicability for the slenderness ratio. Finally, all codes in Table 9-1 are for axially loaded members.

For example, Code 2 is limited to a specific material, 2014-T6 aluminum alloy. The factor of safety (*FS*) is included in the code. If the slenderness ratio lies between 12 and 55, the column is in the intermediate range and the empirical column formula is valid. If the slenderness ratio exceeds 55, a form of the Euler formula is used with a factor of safety of approximately 1.94 included. For a slenderness ratio less than 12, the axially loaded member is in the compression-block range where buckling does not occur.

In this text, design will be limited to columns subjected to axial loads. Factors such as residual stress, out-of-straightness, and local buckling may be found in design codes established by professional organizations.

The following examples illustrate the use of the codes in Table 9-1.

Example Problem 9-8   Select the lightest structural steel wide-flange section listed in Appendix B to support an axial compressive load of 150 kip as a 15-ft column. Use Code 1.

SOLUTION
When a rolled section is to be selected to support a specified load, it is usually necessary to make several trial solutions since there is no direct relationship between areas and radii of gyration for different structural shapes. The best section is usually the section with the least area (smallest mass) that will support the load. A minimum area can be obtained by assuming $L/r = 0$. The load-carrying capacity of various sections with areas larger than this minimum can then be calculated, using the proper column formula, to determine the lightest one that will carry the specified load.

If L/r is small,

$$FS \cong \frac{5}{3} = 1.667$$

$$\sigma_{all} = \frac{\sigma_y}{FS} = \frac{36,000}{1.667} = 21,596 \text{ psi}$$

$$A_{min} = \frac{P}{\sigma_{all}} = \frac{150,000}{21,596} = 6.946 \text{ in.}^2$$

A column should be selected from Appendix B with an area greater than 6.95 in.² for the first trial. In this case, try a W8 × 24 section for which A is 7.08 in.² and

$r_{\min}$ is 1.61 in. The value of L/r for this column is

$$\frac{L}{r} = \frac{15(12)}{1.61} = 111.8$$

To determine which of the equations for the allowable stress is applicable, first determine the value of C_c.

$$C_c = \sqrt{\frac{2\pi^2 E}{\sigma_y}} = \sqrt{\frac{2\pi^2(29)(10^6)}{36,000}} = 126.1$$

Since $L/r = 111.8 < 126.1$, the column is in the intermediate range where the factor of safety is

$$FS = \frac{5}{3} + \frac{3}{8}\left(\frac{L/r}{C_c}\right) - \frac{1}{8}\left(\frac{L/r}{C_c}\right)^3 = \frac{5}{3} + \frac{3}{8}\left(\frac{111.8}{126.1}\right) - \frac{1}{8}\left(\frac{111.8}{126.1}\right)^3 = 1.912$$

and the allowable stress is

$$\sigma_{\text{all}} = \frac{\sigma_y}{FS}\left[1 - \frac{1}{2}\left(\frac{L/r}{C_c}\right)^2\right] = \frac{36,000}{1.912}\left[1 - \frac{1}{2}\left(\frac{111.8}{126.1}\right)^2\right] = 11,428 \text{ psi}$$

The allowable load is

$$P_{\text{all}} = \sigma_{\text{all}}(A) = 11,428(7.08) = 80,910 \text{ lb} \cong 80.9 \text{ kip}$$

This load is less than the design load; therefore, a column with either a larger area, a larger radius of gyration, or both must be investigated. As a second trial value, use a W12 × 30 section for which A is 8.79 in.^{2} and r is 1.52 in. For this section L/r is 118.4 (intermediate range), the factor of safety is 1.92, the allowable stress is 10,485 psi, and the load this column can support is

$$P_{\text{all}} = \sigma_{\text{all}}(A) = 10,485(8.79) = 92,200 \text{ lb} = 92.2 \text{ kip}$$

This load is also less than the design load of 150 kip. For the third trial, use a W8 × 40 section for which A is 11.7 in.^{2} and r is 2.04 in. For this section L/r is 88.2 (intermediate range), the factor of safety is 1.89, the allowable stress is 14,542 psi, and the load this column can support is

$$P_{\text{all}} = \sigma_{\text{all}}(A) = 14,542(11.7) = 170,140 \text{ lb} \cong 170.1 \text{ kip}$$

Since the 40-lb/ft column (W8 × 40) is stronger than necessary and the 30-lb/ft column (W12 × 30) is not strong enough, any other section investigated should weigh between 30 and 40 lb/ft. The only other wide-flange section in Appendix B that might satisfy the requirements is a W8 × 31 section for which A is 9.13 in.^{2} and r is 2.02 in. For this section L/r is 89.1 (intermediate range), the factor of safety is 1.89, the allowable stress is 14,293 psi, and the load this column can support is 130,500 lb, which is less than the design load of 150 kip.

Thus, a

W8 × 40 section should be used. **Ans.**

The above is not necessarily the best procedure. Different designers have different approaches to the trial-and-error procedure, and for certain problems one approach may be better than another. The important point to make here is that the problem of design involving rolled shapes (other than simple geometric shapes) is, in general, solved by trial and error.

Example Problem 9-9   Determine the dimensions necessary for a 500-mm rectangular strut to carry an axial load of 6.75 kN. The material is aluminum alloy 2014-T6, and the width of the strut is to be twice the thickness. Use Code 2.

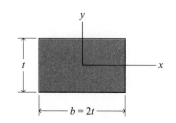

Figure 9-11

SOLUTION

The code is represented by three different equations that depend on the value of L/r, which in turn will depend on the equation used. Thus, it will be necessary to assume that one of the equations applies and use it to obtain the dimension of the column, after which the value of L/r must be calculated and used to check the validity of the equation used. Assume L/r is less than 55 but greater than 12, in which case the straight-line equation is valid. The cross section is shown in Fig. 9-11, and the least second moment of area I_x is equal to $bt^3/12$, the area A is equal to bt, and the least radius of gyration is

$$r = \sqrt{\frac{bt^3/12}{bt}} = 0.2887t$$

The slenderness ratio is

$$\frac{L}{r} = \frac{0.500}{0.2887t} = \frac{1.7319}{t}$$

and, when this value and the expression for the area are substituted in the straight-line formula of Code 2, it becomes

$$\sigma_{\text{all}} = \frac{P_{\text{all}}}{A} = [212 - 1.585(L/r)](10^6)$$

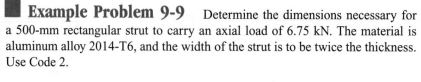

from which

$$t = 14.08(10^{-3}) \, \text{m} = 14.08 \, \text{mm}$$

The value of L/r for this thickness is

$$\frac{L}{r} = \frac{1.7319}{0.01408} = 123.0$$

which is greater than 55 and indicates that the straight-line formula is not valid. The problem must be solved again using the Euler equation. Thus,

$$\frac{P}{A} = \frac{6.75(10^3)}{2t^2} = \frac{372(10^9)}{(1.7319/t)^2}$$

from which

$$t^4 = 27.22(10^{-9}) \text{ m}^4 \quad \text{and} \quad t = 12.84(10^{-3}) \text{ m} = 12.84 \text{ mm}$$

The value of L/r is 134.9 for this thickness, which confirms the use of the Euler formula. The dimensions of the cross section are

$$t = 12.84 \text{ mm} \quad \text{and} \quad b = 25.7 \text{ mm} \qquad \textbf{Ans.}$$

PROBLEMS

Introductory Problems

9-73* A column 10 ft long must support an axial compressive load of 70,000 lb. Select the lightest standard-weight structural steel pipe that can be used. Use Code 1.

9-74* Select the lightest standard-weight structural steel pipe that can be used to support an axial compressive load of 200 kN as a 4-m-long column. Use Code 1.

9-75 Select the lightest structural steel wide-flange section that can be used to support an axial compressive load of 200 kip as a 12-ft-long column. Use Code 1.

9-76 A 7-m-long structural steel column will be used to support an axial compressive load of 400 KN. Select the lightest wide-flange section that can be used. Use Code 1.

Intermediate Problems

9-77* A square 2014-T6 aluminum alloy member must support a 20,000-lb axial compressive load as a 12-ft-long column. Use Code 2 to determine the minimum cross-sectional area required.

9-78* A 2014-T6 aluminum alloy strut with a length of 4 m will be used to support an axial compressive load of 15 kN. Determine the minimum dimensions required if the width of the strut is to be twice the thickness. Use Code 2.

9-79 A Douglas fir ($E = 1800$ ksi and $F_c = 1350$ psi) timber column 14 ft long will be used to support an axial compressive load of 60 kip. Use Code 4 to determine the lightest structural timber that can be used.

9-80 A Douglas fir ($E = 12$ GPa and $F_c = 9.3$ MPa) timber column 4 m long will be used to support an axial compressive load of 100 kN. Use Code 4 to determine the lightest structural timber that can be used.

Challenging Problems

9-81* The structure shown in Fig. P9-81 consists of a solid-steel tie rod BC and a standard-weight structural steel pipe AB. The tie rod has been adequately designed. Using Code 1, determine the lightest pipe that can be used to support the load. The effective length of pipe AB is 9 ft. Neglect the weight of the structure.

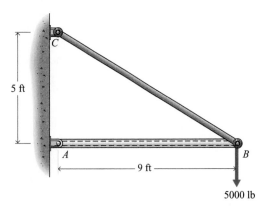

Figure P9-81

9-82*   Member *ABC* of the structure shown in Fig. P9-82 supports a uniformly distributed load of 30 kN/m. Select the lightest standard-weight structural steel pipe that can be used for member *BD*. Use Code 1 and consider *BD* to be a pin-ended member. Neglect the weight of the structure.

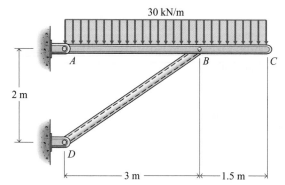

Figure P9-82

9-83   Select the lightest structural steel wide-flange section that can be used for the compression members of the truss shown in Fig. P9-83. Assume that buckling is limited to the plane of the structure and that the tension members have been adequately designed. Use Code 1.

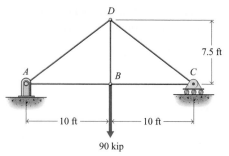

Figure P9-83

9-84   A structural steel standard-weight pipe is used for a spreader bar, as shown in Fig. P9-84. If the structure is to support a load *P* = 40 kN and the cables have been adequately designed, determine the minimum size of pipe needed to support the load using Code 1. The effective length of the spreader bar is 1.5 m. Neglect the weight of the structure.

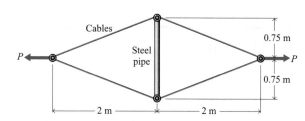

Figure P9-84

SUMMARY

Columns are long, straight, prismatic bars subjected to compressive loads. As long as a column remains straight, it can be analyzed as an axially loaded member, however, if a column begins to deform laterally, the deflection may become large and lead to catastrophic failure called buckling. Buckling of a column is caused by deterioration of what was a stable state of equilibrium to an unstable one, not by failure of the material of which the column is composed. For long, slender columns, the maximum load for which the column is in stable equilibrium (the critical buckling load) occurs at stress levels much less than the proportional limit for the material.

For a straight, slender, pin-ended column that is centrically loaded by axial compressive forces *P* at the ends and that has experienced a small lateral deflection, the differential equation for the elastic curve is

$$EI\frac{d^2v}{dx^2} = M_r = -Pv$$

which has the solution

$$v = A \sin px + B \cos px$$

The minimum value of load *P* for a nontrivial solution is

$$P_{cr} = \frac{\pi^2 EI}{L^2} \tag{9-1}$$

The value given by Eq. 9-1 is called the critical buckling load or the Euler load. The second moment of the cross-sectional area I in Eq. 9-1 refers to the axis about which bending occurs. When I is replaced by Ar^2, where r is the radius of gyration about the axis of bending, Eq. 9-1 becomes

$$\frac{P_{cr}}{A} = \frac{\pi^2 E}{(L/r)^2} = \sigma_{cr} \qquad (9\text{-}2)$$

The quantity L/r is called the slenderness ratio and is determined for the axis about which bending tends to occur. For a pin-ended, centrically loaded column, bending occurs about the axis of minimum second moment of area (minimum radius of gyration).

Equation 9-1 agrees well with experiment if the slenderness ratio is large ($L/r > 140$ for steel columns). Short compression members ($L/r < 40$ for steel columns) can be treated as compression blocks where yielding occurs before buckling. Columns that lie between these extremes are analyzed by using empirical formulas (column design codes).

REVIEW PROBLEMS

9-85* A 20-ft-long timber ($E = 1200$ ksi and $\sigma_e = 2.4$ ksi) column has the cross section shown in Fig. P9-85. The timbers are nailed together so that they act as a unit. Determine

a. The slenderness ratio.
b. The smallest slenderness ratio for which the Euler buckling load equation is valid.
c. The Euler buckling load.
d. The axial stress in the column when the Euler load is applied.

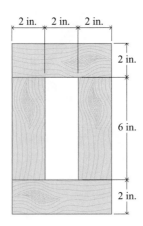

Figure P9-85

9-86* Determine the maximum compressive load that a WT178 × 51 structural steel column (see Appendix B for cross-sectional properties) can support if it is 8 m long and a factor of safety of 1.92 is specified.

9-87 Determine the maximum compressive load that a W36 × 160 structural steel column (see Appendix B for cross-sectional properties) can support if it is 30 ft long and a factor of safety of 2.24 is specified.

9-88* A 3-m-long column with the cross section shown in Fig. P9-88 is fabricated from three pieces of timber ($E = 13$ GPa and $\sigma_e = 35$ MPa). The timbers are nailed together so that they act as a unit. Determine

a. The slenderness ratio.
b. The smallest slenderness ratio for which the Euler buckling load equation is valid.
c. The Euler buckling load.
d. The axial stress in the column when the Euler load is applied.

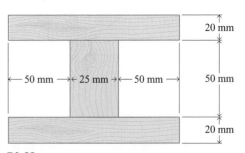

Figure P9-88

9-89 Determine the maximum allowable compressive load for a 12-ft-long aluminum alloy ($E = 10,600$ ksi) column having the cross section shown in Fig. P9-89 if a factor of safety of 2.25 is specified.

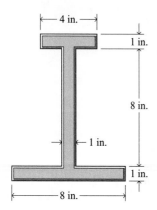

Figure P9-89

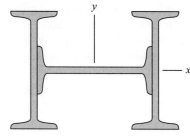

Figure P9-91

9-90 A 25-mm-diameter tie rod and a pipe strut with an inside diameter of 100 mm and a wall thickness of 25 mm are used to support a 100-kN load as shown in Fig. P9-90. Both the tie rod and the pipe strut are made of structural steel with a modulus of elasticity of 200 GPa and a yield strength of 250 MPa. Determine

a. The factor of safety with respect to failure by yielding.
b. The factor of safety with respect to failure by buckling.

9-92* A structural steel W356×122 wide-flange section will be used for a 9-m-long pin-ended column. Determine the maximum axial compressive load permitted by Code 1 if

a. The column is unbraced throughout is total length.
b. The column is braced such that the effective length for bending about the y-axis is reduced to 6 m.

9-93 A cold-rolled steel tension bar AB and a structural steel ($E = 29{,}000$ ksi and $\sigma_y = 36$ ksi) compression strut BC are used to support a load $P = 100$ kip, as shown in Fig. P9-93. Assume that the pins at B and C offer no restraint to bending about the x-axis but provide end conditions which are essentially fixed at C and free at B for bending about the y-axis. Select the lightest structural steel tee section listed in Appendix B that can be used for the strut BC. Use Code 1.

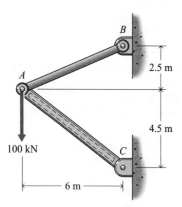

Figure P9-90

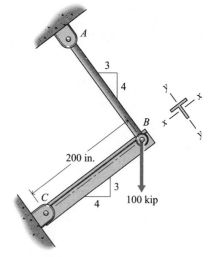

Figure P9-93

9-91* Three S10 × 35 structural steel sections 30 ft long are used to fabricate a column with the cross section shown in Fig. P9-91. The column is fixed at the base and pinned at the top. The pin at the top offers no restraint to bending about the y-axis; but for bending about the x-axis, the pin provides restraint sufficient to reduce the effective length to 20 ft. Determine the maximum axial compressive load permitted by Code 1.

9-94 The compression member AB of the truss shown in Fig. P9-94 is a structural steel W254 × 67 wide-flange section with the xx-axis lying in the plane of the truss. The member is continuous from A to B. Consider all connections to be the equivalent of pin ends. If Code 1 applies, determine

a. The maximum safe load for member AB.

b. The maximum safe load for member AB if the bracing member CD is removed.

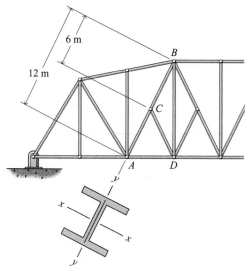

Figure P9-94

9-95* A W8 × 40 structural steel compression member has an effective length of 25 ft. The member is subjected to an axial load P and a bending moment of 10 kip · ft about the x-axis of the cross section (see Appendix B). Use the interaction formula with an allowable flexural stress of $0.66\sigma_y$ to determine the maximum safe load P.

9-96 Two C229 × 30 structural steel channels are laced back to back with a separation of 120 mm, as shown in Fig. P9-96, to form a column with an effective length of 10 m. The load P will be applied at a point 50 mm from the axis of the column on the axis of symmetry parallel to the backs of the channels. If $E = 200$ GPa and $\sigma_y = 250$ MPa, use the allowable stress method to determine the maximum load P that can be supported by the column.

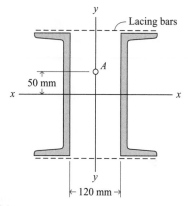

Figure P9-96

9-97 A structural steel ($E = 29{,}000$ ksi) column 20 ft long must support an axial compressive load of 110 kip. The column can be considered pinned at both ends for bending about one axis and fixed at both ends for bending about the other axis. Select the lightest wide-flange or American standard section (see Appendix B) that can be used for the column. Use Code 1.

10-1 INTRODUCTION

This chapter consists of two parts. Part A discusses the concept of strain energy (previously introduced in Chapter 8 on beam deflections) and its application to stress and deformation determinations in members subjected to impact loading. Part B discusses theories of failure for isotropic materials and application of the theories for predicting failure of members subjected to combined static loading.

Part A

Energy Methods

10-2 STRAIN ENERGY

The concept of strain energy was introduced in Section 8-8 by considering the work done by a slowly applied axial load P in elongating a bar of uniform cross section A by an amount δ, as shown in Fig. 10-1a. From the load-deformation diagram (Fig. 10-1b), the work W_k done in elongating the bar is

$$W_k = \int_0^\delta P\, d\delta \qquad (a)$$

Since the work done on the bar must equal the strain energy U stored in the bar, the expression for strain energy in terms of axial stress and axial strain (Fig. 10-1c) is

$$W_k = U = \int_0^\epsilon \sigma A L\, d\epsilon = AL \int_0^\epsilon \sigma\, d\epsilon \qquad (b)$$

When the stress remains below the elastic limit of the material, Hooke's law applies and $d\epsilon$ may be expressed as $d\sigma/E$. Equation (b) then becomes

$$U = \frac{AL}{E} \int_0^\sigma \sigma\, d\sigma$$

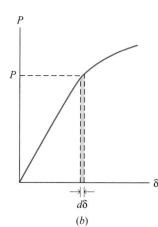
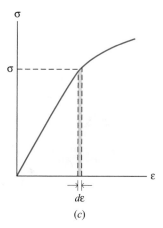

(a) (b) (c)

Figure 10-1

615

or

$$U = AL\left(\frac{\sigma^2}{2E}\right) \tag{10-1}$$

Equation 10-1 gives the elastic strain energy (which is, in general, recoverable) for axial loading of a material obeying Hooke's law. The quantity in parentheses is the elastic strain energy u in tension or compression per unit volume for a uniaxial state of stress,

$$u = \frac{\sigma^2}{2E} \tag{10-2}$$

which is called the *strain energy intensity*.

The integral $\int \sigma\, d\epsilon$ of Eq. (b) represents the area under the stress-strain curve (Fig. 10-1c) and, if evaluated from zero to the elastic limit (for practical purposes, the proportional limit), yields a property known as the *modulus of resilience*. The modulus of resilience is defined as the maximum strain energy per unit volume that a material will absorb without inelastic deformation and is the area under the straight-line portion of the stress-strain diagram, as shown in Fig. 10-2. Customary units are inch-pounds per cubic inch or newton-meters per cubic meter. For practical purposes the yield strength and proportional limit are the same, and thus the modulus of resilience u_R (Fig. 10-2) is

$$u_R = \frac{\sigma_y^2}{2E} \tag{10-3}$$

The area under the entire stress-strain curve from zero to rupture (Fig. 10-3) gives the property known as the *modulus of toughness* u_T and denotes the energy per unit volume necessary to rupture the material. The modulus of resilience and the modulus of toughness have the same units.

For the shear loading shown in Fig. 10-4a, the strain energy intensity is the area under the shearing stress–shearing strain diagram shown in Fig. 10-4b and is

$$u = \int_0^\gamma \tau\, d\gamma \tag{c}$$

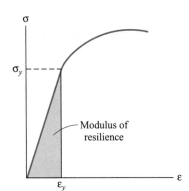

Figure 10-2

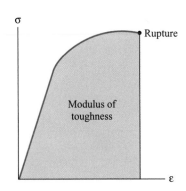

Figure 10-3

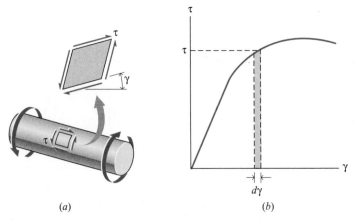

(a)　　　　(b)

Figure 10-4

When the shearing stress remains below the elastic limit of the material, $\tau = G\gamma$, and Eq. (c) may be expressed as

$$u = \int_0^\tau \tau \frac{d\tau}{G} = \frac{\tau^2}{2G} \tag{10-4}$$

Thus, for shear loading, the expression for strain energy intensity is identical to that for axial loading if σ is replaced by τ and E by G.

The strain energy may be obtained by integrating the strain energy intensity over the volume of the material being stressed. When the state of stress in a body can be represented by a nonzero normal stress, the strain energy is

$$U = \int_V \frac{\sigma^2}{2E}\,dV \tag{10-5}$$

where V is the volume of the body. A similar expression may be written for shearing stress and is

$$U = \int_V \frac{\tau^2}{2G}\,dV \tag{10-6}$$

Equations (10-5) and (10-6) may be used to determine the strain energy stored in a body that is subjected to an axial load, a torsional load, transverse shear, or a bending moment. Note that Eqs. 10-5 and 10-6 are scalar equations.

10-3 ELASTIC STRAIN ENERGY FOR VARIOUS LOADS

Equations for elastic strain energy for members subjected to axial loads, torsional loads, bending moments, and transverse shear will be developed in this section.

10-3-1 Strain Energy for Axial Loading
Consider the bar shown in Fig. 10-5, which has a uniform cross-sectional area A, a length L, and is subjected to an axial force P. The strain energy in the bar may be found using Eq. 10-5 and is

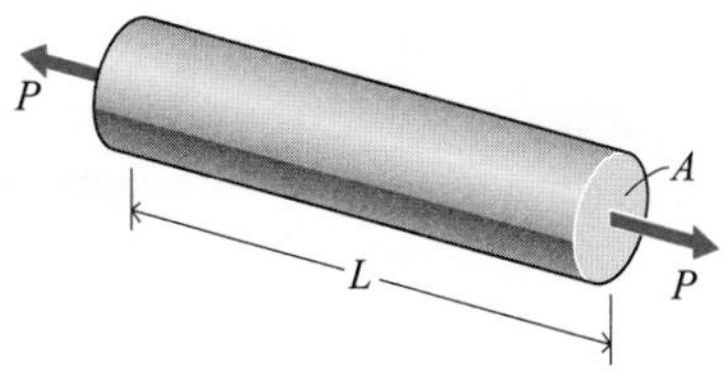

Figure 10-5

$$U = \int_V \frac{\sigma^2}{2E}\,dV = \int_0^L \frac{(P/A)^2}{2E}\,(A\,dx) \tag{a}$$

where $dV = A\,dx$, and x is measured along the length of the bar. Simplifying Eq. (a) yields an expression for the strain energy for axial loading. Thus,

$$U = \frac{P^2 L}{2AE} \tag{10-7}$$

If the bar is slightly tapered or if P or E changes along the length of the bar, the expression for strain energy is

$$U = \int_0^L \frac{P^2}{2AE}\,dx \tag{10-8}$$

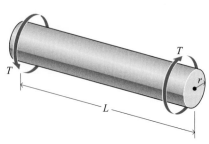

Figure 10-6

10-3-2 Strain Energy for Torsional Loading

The bar shown in Fig. 10-6, has a uniform cross section of radius r, a length L, and is subjected to a torsional load (torque) T. The strain energy in the bar may be found using Eq. 10-6 and is

$$U = \int_V \frac{\tau^2}{2G}\, dV = \int_V \frac{(T\rho/J)^2}{2G}\, dV \qquad (b)$$

where $\tau = T\rho/J$. Since $dV = dA\, dx$ and ρ is the radius to a generic point in the cross section, Eq. (b) becomes

$$U = \int_0^L \frac{T^2}{2GJ^2}\left(\int_A \rho^2\, dA\right) dx$$

The integral in parentheses is J (the polar second moment of area for the cross section of the bar). Thus, the expression for the strain energy for torsional loading is

$$U = \frac{T^2 L}{2GJ} \qquad (10\text{-}9)$$

If the bar is slightly tapered or if T or G changes along the length of the bar, the expression for strain energy is

$$U = \int_0^L \frac{T^2}{2GJ}\, dx \qquad (10\text{-}10)$$

10-3-3 Strain Energy for Bending Loads

Consider the beam shown in Fig. 10-7a, which has a uniform cross section A and a length L. A free-body diagram of a portion of the beam is shown in Fig. 10-7b. On a cross section at a position x along the beam, a bending moment M_r and a transverse shear V_r are required for equilibrium. Both M_r and V_r are functions of position x. Consider first the effect of the bending moment M_r. The strain energy in the beam resulting from the bending moment M_r is given by Eq. 10-5 as

$$U = \int_V \frac{\sigma^2}{2E}\, dV = \int_V \frac{(-M_r y/I)^2}{2E}\, dV \qquad (c)$$

where $\sigma = -M_r y/I$. Since $dV = dA\, dx$, Eq. (c) becomes

$$U = \int_0^L \frac{M_r^2}{2EI^2}\left(\int_A y^2\, dA\right) dx$$

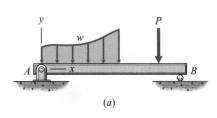

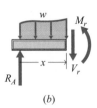

Figure 10-7

The integral in parentheses is I (the second moment of area for the cross section of the beam). Thus, the expression for the strain energy for bending is

$$U = \int_0^L \frac{M_r^2}{2EI}\, dx \qquad (10\text{-}11)$$

Since M_r is a function of x, the relationship between M_r and x must be known before the integral of Eq. 10-11 can be evaluated.

10-3-4 Strain Energy for Transverse Shear
The strain energy in the beam resulting from the transverse shear V_r shown in Fig. 10-7b can be found using Eq. 10-6. Thus,

$$U = \int_V \frac{\tau^2}{2G} dV = \int_V \frac{(V_r Q/It)^2}{2G} dV \qquad (d)$$

where $\tau = V_r Q/It$. Since $dV = dA\, dx$, Eq. (d) becomes

$$U = \int_0^L \frac{V_r^2}{2GI^2} \left(\int_A \frac{Q^2}{t^2} dA \right) dx \qquad (10\text{-}12)$$

The integral in parentheses in Eq. 10-12 depends on the shape of the cross section of the beam. Evaluation of this integral will be illustrated in the Example Problems.

Example Problem 10-1
The two solid circular bars shown in Fig. 10-8 are securely attached at section B. The modulus of elasticity E for both bars is 200 GPa. If the axial load P is 50 kN, determine the strain energy for the assembly.

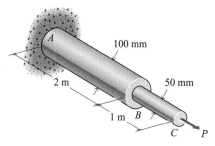

Figure 10-8

SOLUTION
Each segment of the assembly, AB and BC, has a uniform cross section and is subjected to the axial load $P = 50$ kN. The strain energy of the assembly can be found by applying Eq. 10-7 to each of the segments and adding the results. Thus,

$$U = U_{AB} + U_{BC}$$

$$= \left(\frac{P^2 L}{2AE} \right)_{AB} + \left(\frac{P^2 L}{2AE} \right)_{BC}$$

$$= \frac{P^2}{2E} \left[\left(\frac{L}{A} \right)_{AB} + \left(\frac{L}{A} \right)_{BC} \right]$$

$$= \frac{[50(10^3)]^2}{2(200)(10^9)} \left[\frac{2}{(\pi/4)(0.100)^2} + \frac{1}{(\pi/4)(0.050)^2} \right]$$

$$= 4.7746\ \text{N} \cdot \text{m} \cong 4.77\ \text{N} \cdot \text{m} \qquad \textbf{Ans.}$$

Of the strain energy of the assembly, 33.3% exists in segment AB and 66.7% exists in segment BC.

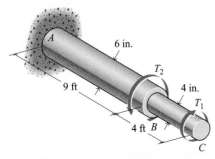

Figure 10-9

Example Problem 10-2   Segments AB and BC of the assembly shown in Fig. 10-9 are made of the same material ($G = 11,000$ ksi) and are securely attached at section B. If $T_1 = 50$ kip $\cdot$ in. and $T_2 = 70$ kip $\cdot$ in., determine the strain energy of the assembly.

SOLUTION

Segment BC is subjected to a constant internal torque of 50 kip $\cdot$ in., and segment AB is subjected to a constant internal torque of 20 kip $\cdot$ in. Because both segments have a uniform cross section and are subjected to a constant internal torque, the strain energy of the assembly can be found by applying Eq. 10-9 to each of the segments and adding the results. Thus,

$$
\begin{aligned}
U &= U_{AB} + U_{BC} \\
&= \left(\frac{T^2 L}{2GJ}\right)_{AB} + \left(\frac{T^2 L}{2GJ}\right)_{BC} \\
&= \frac{1}{2G}\left[\left(\frac{T^2 L}{J}\right)_{AB} + \left(\frac{T^2 L}{J}\right)_{BC}\right] \\
&= \frac{1}{2(11)(10^3)}\left[\frac{(20)^2(9)(12)}{(\pi/2)(3)^4} + \frac{(50)^2(4)(12)}{(\pi/2)(2)^4}\right] \\
&= 0.23246 \text{ kip} \cdot \text{in.} \cong 232 \text{ lb} \cdot \text{in.} \qquad \textbf{Ans.}
\end{aligned}
$$

Of the strain energy of the assembly, 6.6% exists in segment AB and 93.4% exists in segment BC.

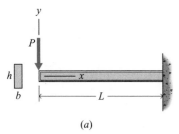

(a)

Figure 10-10(a)

Example Problem 10-3   The cantilever beam shown in Fig. 10-10a has a constant cross section and is subjected to a concentrated force P at its free end.

(a) Determine the strain energy in the beam due to bending.
(b) Determine the strain energy in the beam due to transverse shear.
(c) Compare the results of parts (a) and (b).

SOLUTION

(a) The strain energy due to bending (with $M_r = -Px$) is given by Eq. 10-11 as

$$
U_M = \int_0^L \frac{M_r^2}{2EI}\,dx = \int_0^L \frac{(-Px)^2}{2EI}\,dx = \frac{P^2}{2EI}\int_0^L x^2\,dx = \frac{P^2 L^3}{6EI} \qquad \textbf{Ans.}
$$

(b) The strain energy due to transverse shear (with $V_r = -P$) is given by Eq. 10-12 as

$$
U_V = \int_0^L \frac{V_r^2}{2GI^2}\left(\int_A \frac{Q^2}{t^2}\,dA\right)dx = \int_0^L \frac{P^2}{2GI^2}\left(\int_A \frac{Q^2}{t^2}\,dA\right)dx
$$

However, from Fig. 10-10*b*,

$$Q = A'y_C = b\left(\frac{h}{2} - y\right)\left(y + \frac{h/2 - y}{2}\right) = \frac{b}{2}\left(\frac{h^2}{4} - y^2\right)$$

$$t = b$$

$$dA = b\,dy$$

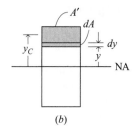

(b)

Figure 10-10(b)

Thus,

$$U_V = \frac{P^2 L}{2GI^2}\int_{-h/2}^{h/2} \frac{1}{b^2}\left(\frac{b}{2}\right)^2\left(\frac{h^2}{4} - y^2\right)^2 b\,dy = \frac{P^2 L b h^5}{240 G I^2}$$

For a rectangular cross section $I = bh^3/12$; therefore,

$$U_V = \frac{3P^2 L}{5Gbh} = \frac{3P^2 L}{5GA} \qquad\qquad \textbf{Ans.}$$

(c) The total strain energy for the beam is

$$U = U_M + U_V = \frac{P^2 L^3}{6EI} + \frac{3P^2 L}{5GA} = \frac{P^2 L^3}{6EI}\left[1 + \frac{3Eh^2}{10GL^2}\right] \qquad (a)$$

For a material such as steel, the ratio $E/G \cong 2.6$ and Eq. (*a*) becomes

$$U = \frac{P^2 L^3}{6EI}\left[1 + 0.78\frac{h^2}{L^2}\right] = \frac{P^2 L^3}{6EI}D$$

where $D = 1 + 0.78(h/L)^2$ is a measure of the effect on the strain energy of the transverse shear. The following table illustrates the percentage of the strain energy resulting from transverse shear for rectangular beams with different h/L ratios.

h/L	$0.78(h/L)^2$	%D
1/5	0.0312	3.03
1/6	0.0217	2.12
1/8	0.0122	1.21
1/10	0.0078	0.77

Thus, for a beam with $h/L = 1/5$ (a very short deep beam), the error in neglecting the strain energy due to transverse shear is approximately 3 percent, whereas for a beam with $h/L = 1/10$ (a long, slender beam), the error is less than 1 percent. Equation (*a*) indicates that the contribution to the total strain energy of the transverse shear increases as the square of h/L, which means that the strain energy due to transverse shear is of importance only in the case of very short deep beams.

PROBLEMS

Introductory Problems

10-1* The tension member shown in Fig. P10-1 consists of a steel pipe A, which has an outside diameter of 6 in. and an inside diameter of 4.5 in., and a solid aluminum alloy bar B, which has an outside diameter of 4 in. The moduli of elasticity for the steel and aluminum alloy are 29,000 ksi and 10,600 ksi, respectively. Determine

a. The elastic strain energy for the steel pipe.
b. The elastic strain energy for the aluminum bar.
c. The elastic strain energy for the entire member.

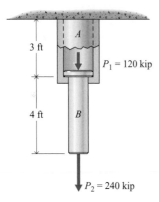

Figure P10-1

10-2* The compression member shown in Fig. P10-2 consists of a solid aluminum alloy bar A, which has an outside diameter of 100 mm; a brass tube B, which has an outside diameter of 150 mm and an inside diameter of 100 mm; and a steel pipe C, which has an outside diameter of 200 mm and an inside diameter of 125 mm. The moduli of elasticity for the aluminum alloy, brass, and steel are 73 GPa, 100 GPa, and 210 GPa, respectively. Determine

a. The elastic strain energy for each segment of the member.
b. The elastic strain energy for the entire member.

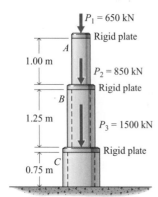

Figure P10-2

10-3 The 4-in.-diameter shaft shown in Fig. P10-3 is composed of brass ($G = 5000$ ksi) and steel ($G = 12,000$ ksi) sections that are rigidly connected. A 100-kip · in. torque is applied at section C. Determine

a. The elastic strain energy for each segment of the shaft.
b. The elastic strain energy for the entire shaft.

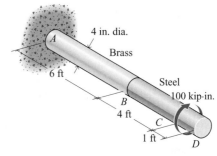

Figure P10-3

10-4* A stepped steel ($G = 80$ GPa) shaft has the dimensions and is subjected to the torques shown in Fig. P10-4. Determine

a. The elastic strain energy for each segment of the shaft.
b. The elastic strain energy for the entire shaft.

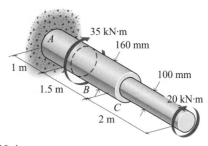

Figure P10-4

10-5 The cantilever beam shown in Fig. P10-5 is subjected to a uniformly distributed load w over its entire length L. Determine the elastic strain energy in the beam due to bending. Neglect the effects of transverse shear.

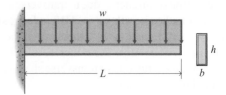

Figure P10-5

10-6 Determine the elastic strain energy in the beam loaded and supported as shown in Fig. P10-6. Neglect the effects of transverse shear.

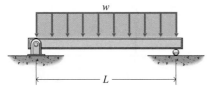

Figure P10-6

Intermediate Problems

10-7* A cantilever beam with a rectangular cross section has a width b, a depth $2c$, and a span L. The beam has a couple M applied at the free end. Compare the elastic strain energy stored in the beam as a result of bending with the elastic strain energy stored in a bar of the same size that is axially loaded to the same maximum tensile stress level.

10-8* Determine the elastic strain energy due to transverse shear for the cantilever beam shown in Fig. P10-5.

10-9 Determine the elastic strain energy due to transverse shear for the simply supported beam shown in Fig. P10-6 if the beam has a rectangular cross section with width b and depth w.

10-10 A simply supported beam with a rectangular cross section has a width b, a depth $2c$, and a span L. The beam carries a uniformly distributed load w over its entire length. Compare the elastic strain energy stored in the beam as a result of bending with the elastic strain energy stored in a bar of the same size that is axially loaded to the same maximum tensile stress level. Neglect the energy resulting from the shearing stresses.

10-11* A torque of 740 lb · ft is applied to the right end of the shaft AB of Fig. P10-11. Both shafts are made of steel ($G = 11,000$ ksi) and have a diameter of 2 in. The mean diameter of bevel gear C is twice that of bevel gear B. Determine the elastic strain energy due to torsion of each shaft.

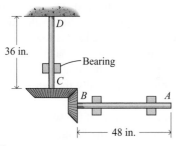

Figure P10-11

Challenging Problems

10-12* A square bar with a cross-sectional area A and a length L is made from a homogeneous material which has a specific weight γ and a modulus of elasticity E. Determine the elastic strain energy stored in the bar (as a result of its own weight) if it hangs vertically while suspended form one end. Express the result in terms of γ, L, A, and E.

10-13* A hollow circular shaft with outside diameter d, inside diameter $d/2$, and length L is subjected to a constant torque of magnitude T. Compare the elastic strain energy stored in this shaft as a result of the torsional loading with the elastic strain energy stored in a bar of the same size that is axially loaded to the same maximum tensile stress level.

10-14 The solid cylindrical shaft of Fig. P10-14 is subjected to a uniformly distributed torque q. Determine the elastic strain energy in the shaft in terms of q, L, G, and c.

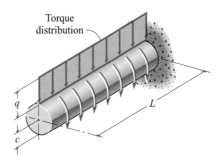

Figure P10-14

10-15* The solid cylindrical shaft of Fig. P10-15 is subjected to a distributed torque that varies linearly from 0 at the left end to q at the right end. Determine the elastic strain energy in the shaft in terms of q, L, G, and c.

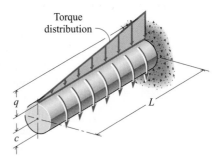

Figure P10-15

10-16 The hollow circular tapered shaft of Fig. P10-16 is subjected to a constant torque T. Determine the elastic strain energy in the shaft in terms of T, L, G, and r.

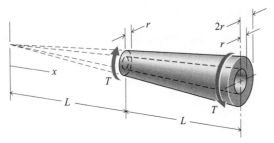

Figure P10-16

10-17 A simply supported beam with a rectangular cross section has a width b, a depth $2c$, and a span L. The beam carries a concentrated load P at midspan, as shown in Fig. P10-17a.

Determine the elastic strain energy stored in the beam as a result of bending. Neglect the energy resulting from the shearing stresses.

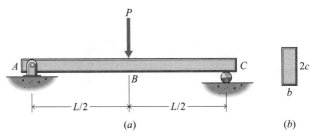

Figure P10-17

10-4 IMPACT LOADING

When the motion of a body is changed (accelerated), the force necessary to produce this acceleration is called a *dynamic force* or *load*. For example, the force an elbow in a pipeline exerts on the fluid in the pipe to change its direction of flow, the pressure on the wings of an airplane pulling out of a dive, the collision of two automobiles, and a man jumping on a diving board are all examples of dynamic loading. A suddenly applied load is called an *impact load*. The last two of the preceding examples are considered impact loads. Under impact loading, if there is elastic action, the loaded system will vibrate until equilibrium is established. A dynamic load may be expressed in terms of mass times the acceleration of the mass center, in terms of the rate of change of the momentum, or in terms of the change of the kinetic energy of the body.

Energy methods can be used to obtain solutions for many problems in mechanics of materials. When an energy approach is used, the magnitude of the load is expressed in terms of the kinetic energy delivered to the loaded system; hence, it is often referred to as an *energy load*. For example, a particle of mass m moving with a speed v possesses a kinetic energy $mv^2/2$; if this particle is stopped by a body, the energy absorbed by the body (the loaded system) is some fractional part of $mv^2/2$, the balance of the energy being converted into sound, heat, and permanent deformation of the striking particle.

In the loaded system, dynamic loading produces stresses and strains, the magnitude and distribution of which will depend not only on the usual parameters encountered previously but also on the velocity of propagation of the strain waves through the material of which the system is composed. This latter consideration, although very important when loads are applied with high velocities, may often be neglected when the velocity of application of the load is low. The velocities are considered to be low when the loading time permits the material to act in the same manner as it does under static loading; that is, the relations between stress and strain and between load and deflection are essentially the same as those already developed for static loading.

Consider the member shown in Fig. 10-11a. If the weight W is slowly lowered (static load) until it comes in contact with the flange at B, the member will elongate by an amount δ_{st}, the static deflection, as shown in Fig. 10-11b. When the

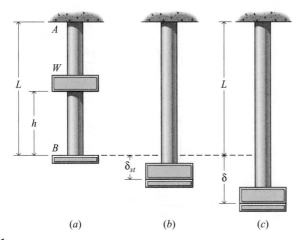

Figure 10-11

weight is dropped from height h, tensile impact stresses are developed in member AB. The stresses are maximum when the deflection of the member is δ (the velocity of W is zero), as shown in Fig. 10-11c, after which the weight moves in an upward direction. Of interest are the maximum stress and deflection of member AB (Fig. 10-11c).

Two methods will be presented to determine maximum values for the stress and for the deflection: the strain energy method and the work–kinetic energy method.

10-4-1 Strain Energy Method In the discussion that follows it is assumed that the stresses are in the elastic range, the material behavior is the same as that for slowly applied loads, and energy losses during impact are negligible. In addition it is assumed that the weights of the member and flange (Fig. 10-11a) are negligible when compared to the weight W.

The work done by the falling weight is equal to the elastic strain energy stored in the member AB. The work done by W is $W(h + \delta)$. The elastic strain energy stored in member AB is given by Eq. 10-7 as $U = P^2L/2AE$ where P is the axial force in the member when it has deformed by an amount δ. For axial loading, $\sigma = P/A$, and thus $U = \sigma^2 AL/2E$. Equating the work and strain energy expressions gives

$$W(h + \delta) = \frac{\sigma^2 AL}{2E} \qquad (a)$$

The deflection δ is

$$\delta = \frac{PL}{AE} = \frac{\sigma L}{E} \qquad (b)$$

Substituting Eq. (b) into Eq. (a) and simplifying yields

$$AL\sigma^2 - 2WL\sigma - 2WhE = 0 \qquad (c)$$

which has the solution

$$\sigma = \frac{W}{A} \pm \sqrt{\left(\frac{W}{A}\right)^2 + \frac{2WhE}{AL}} \tag{10-13}$$

or

$$\sigma = \frac{W}{A}\left[1 \pm \sqrt{1 + \frac{2hAE}{WL}}\right] \tag{10-14}$$

The plus sign preceding the radical in Eq. 10-14 is used, as the negative sign has no physical significance. The maximum deflection is found by substituting Eq. 10-13 or Eq. 10-14 into Eq. (b). The results are

$$\delta = \frac{WL}{AE} + \sqrt{\left(\frac{WL}{AE}\right)^2 + \frac{2WhL}{AE}} \tag{10-15}$$

or

$$\delta = \frac{WL}{AE}\left[1 + \sqrt{1 + \frac{2hAE}{WL}}\right] \tag{10-16}$$

The static deflection δ_{st} may be written

$$\delta_{st} = \frac{FL}{AE} = \frac{WL}{AE} \tag{d}$$

When Eq. (d) is substituted into Eq. 10-16, the deflection due to impact becomes

$$\delta = \delta_{st}\left[1 + \sqrt{1 + \frac{2h}{\delta_{st}}}\right] \tag{10-17}$$

The axial stress due to a static load, $\sigma_{st} = F/A = W/A$, may be substituted into Eq. 10-14 to yield

$$\sigma = \sigma_{st}\left[1 + \sqrt{1 + \frac{2h}{\delta_{st}}}\right] \tag{10-18}$$

The term in the brackets in Eqs. 10-17 and 10-18 is called the *impact factor*.

Of particular interest are the values of δ and σ when $h = 0$ and when h is much greater than δ_{st}. The maximum deflection and stress when the weight is dropped from $h = 0$ are given by Eqs. 10-17 and 10-18 as

$$\delta = 2\delta_{st} \tag{10-19}$$

$$\sigma = 2\sigma_{st} \tag{10-20}$$

where δ_{st} is the deflection when the weight W is slowly lowered onto the flange and σ_{st} is the corresponding axial stress. Thus, the maximum deflection and the maximum stress for a suddenly applied force are double the values when the force is slowly applied.

For the case where h is large compared to δ_{st}, Eqs. 10-17 and 10-18 become

$$\delta = \sqrt{2\delta_{st}h} \qquad (10\text{-}21)$$

$$\sigma = \sqrt{\frac{2E}{AL}Wh} \qquad (10\text{-}22)$$

Thus, using Eq. 10-22, it is seen that the stress may be decreased by increasing the area A or the length L of the member, or by decreasing the modulus of elasticity E. If the member were statically loaded, then $\sigma = F/A = W/A$, and the stress is independent of L or E.

Using work-energy principles from physics or dynamics,

$$Wh = mgh = \frac{1}{2}mV^2$$

where V is the speed of the weight as it makes contact with the flange. Thus, from Eq. 10-22

$$\sigma = \sqrt{\left(\frac{2E}{AL}\right)\left(\frac{mV^2}{2}\right)} = \sqrt{\frac{mEV^2}{AL}} \qquad (10\text{-}23)$$

10-4-2 Work–Kinetic Energy Method

The work–kinetic energy method uses a concept learned in physics and/or dynamics courses. The principle of work and energy states that

$$U_{1\to2} = T_2 - T_1 \qquad (e)$$

where $U_{1\to2}$ is the total work done on a body, and T_1 and T_2 are the kinetic energies of the body before and after the work is done.

Consider the cantilever beam loaded with a falling weight W, as shown in Fig. 10-12. The variation of the end deflection v of the beam with respect to time t is shown in Fig. 10-12b, where δ is the maximum deflection resulting from the falling weight transferring kinetic energy $(mV^2/2)$ to the beam. At the instant δ occurs, the kinetic energy is zero, and the beam begins to vibrate. Only maximum deflections and stresses are of interest in this book; analysis of the vibration characteristics are studied in other courses. In the analysis that follows, it is assumed that the weight W stays in contact with the beam during the deflection δ, that the mass of the beam is negligible, and that the stress-strain behavior of the beam is the same for both static and dynamic loading. Furthermore, no energy is dissipated at the beam supports or contact area between the weight W and the beam.

The quantity δ_{st} in Fig. 10-12a is the static deflection when the weight W is slowly lowered onto the beam. In Fig. 10-12c, P is the largest dynamic force exerted on the beam, and it occurs where the tip of the beam has deflected an amount δ. Since the stress-strain behavior of the beam material is the same for a static or dynamic load, and the load-deflection is assumed to be linear, the relationship

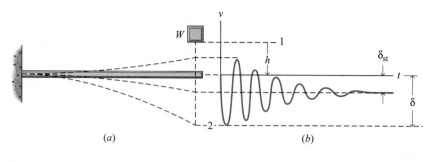

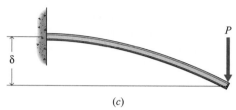

Figure 10-12

between load and deflection (refer to Fig. 10-13) is

$$\frac{P}{\delta} = \frac{W}{\delta_{st}} \tag{f}$$

Applying the work-energy principle to the weight W using Eq. (e) gives

$$U_{1\rightarrow2} = T_2 - T_1 = 0$$

since the kinetic energies are zero at positions 1 (W is released from rest) and 2 (W reaches maximum deflection). Thus,

$$U_{1\rightarrow2} = W(h + \delta) - \frac{1}{2}P\delta = 0 \tag{g}$$

Substituting Eq. (f) into Eq. (g) and rearranging give

$$P^2 - 2WP - \frac{2W^2h}{\delta_{st}} = 0 \tag{h}$$

Solving Eq. (h) for P gives

$$P = W\left[1 + \sqrt{1 + \frac{2h}{\delta_{st}}}\right] \tag{10-24}$$

which may be written, using Eq. (f), as

$$\delta = \delta_{st}\left[1 + \sqrt{1 + \frac{2h}{\delta_{st}}}\right] \tag{10-25}$$

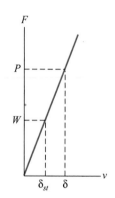

Figure 10-13

Equation 10-25, for a weight dropped on a beam, is the same as Eq. 10-17 for a weight dropped on a bar. As before, the term in brackets is the impact factor.

The static deflection in Eq. 10-25 may be computed using the results of Table B-19 in Appendix B. The impact factor is then found using the term in the brackets of Eq. 10-24 or 10-25. Equation 10-24 is used to determine P, which is used as a static force to compute stresses. The method is illustrated in Example Problem 10-8.

Example Problem 10-4   A 30-mm-diameter aluminum alloy ($E = 70$ GPa) rod 1 m long is fitted with a flange at the bottom, as shown in Fig. 10-14. A collar W, which slides freely on the rod, is dropped from a height of 800 mm. Determine the maximum mass that the collar may have if the yield strength ($\sigma_y = 270$ MPa) of the rod must not be exceeded.

SOLUTION
The value of weight W can be determined by using Eq. 10-14 with $\sigma = 270$ MPa (the yield strength of the rod). Thus,

$$\sigma = \frac{W}{A}\left[1 + \sqrt{1 + \frac{2hAE}{WL}}\right]$$

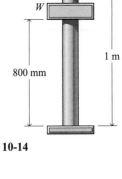

Figure 10-14

Solving for W yields

$$W = \frac{\sigma^2 AL}{2\sigma L + 2hE}$$

$$= \frac{(270)^2(10^{12})(\pi/4)(0.030)^2(1)}{2(270)(10^6)(1) + 2(0.800)(70)(10^9)} = 457.9 \text{ N}$$

Thus, the maximum mass for the collar is

$$m = \frac{W}{g} = \frac{457.9}{9.81} = 46.68 \text{ kg} \cong 46.7 \text{ kg} \qquad \textbf{Ans.}$$

Example Problem 10-5   A weight of 30 lb is dropped from a height of 4 ft onto the center of a small rigid platform, as shown in Fig. 10-15. The two steel ($E = 30,000$ ksi) rods supporting the platform have 1 × 2-in. rectangular cross sections and are 8 ft long. Determine

(a) The maximum tensile stress developed in the rods.
(b) The maximum deflection of the platform.

SOLUTION
(a) The maximum tensile stress developed in the rods can be found by using either Eq. 10-13 or Eq. 10-14. Using Eq. 10-13, and noting that

$$A = 2(1)(2) = 4 \text{ in.}^2$$

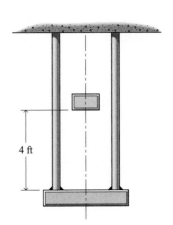

Figure 10-15

give

$$\sigma = \frac{W}{A} + \sqrt{\left(\frac{W}{A}\right)^2 + \frac{2WhE}{AL}}$$

$$= \frac{30}{4} + \sqrt{\left(\frac{30}{4}\right)^2 + \frac{2(30)(4)(12)(30)(10^6)}{4(8)(12)}}$$

$$= 15{,}008 \text{ psi} \cong 15.01 \text{ ksi} \qquad \text{Ans.}$$

Note that the static stress produced by the 30-lb weight would be

$$\sigma = \frac{W}{A} = \frac{30}{4} = 7.50 \text{ psi}$$

(b) The maximum deflection experienced by the platform would be

$$\delta = \frac{\sigma L}{E} = \frac{15{,}008(8)(12)}{(30)(10^6)} = 0.0480 \text{ in.} \qquad \text{Ans.}$$

Note that the static deflection produced by the 30-lb weight would be

$$\delta = \frac{\sigma L}{E} = \frac{7.50(8)(12)}{(30)(10^6)} = 0.0000240 \text{ in.}$$

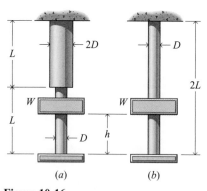

(a)   (b)

Figure 10-16

Example Problem 10-6   The two rods shown in Fig. 10-16 are made of the same material, have the same length $2L$, and have circular cross sections with the same minimum diameter D. Compare the energy-absorbing capacities of the two rods for axial impact loads if the maximum axial stresses in the rods must not exceed the yield strength σ_y of the material.

SOLUTION
For a rod of uniform cross section and length L, the elastic strain energy in the rod is given by Eq. 10-1 as

$$U = \frac{\sigma^2 AL}{2E} = \frac{\sigma^2 V}{2E}$$

For the stepped rod of Fig. 10-16a, the elastic strain energy U_a is the sum of the strain energies of the two parts. Note that the lower segment of the stepped rod will be subjected to a stress σ_y whereas the upper segment will be subjected (as a result of its larger cross-sectional area) to a stress $\sigma_y/4$. If the volume of the lower segment is V, the volume of the upper segment will be $4V$. Therefore,

$$U_a = \frac{\sigma_y^2 V}{2E} + \frac{(\sigma_y/4)^2(4V)}{2E} = \frac{5\sigma_y^2 V}{8E}$$

For the rod with the uniform cross section shown in Fig. 10-16b, the elastic strain energy is

$$U_b = \frac{\sigma_y^2(2V)}{2E} = \frac{\sigma_y^2 V}{E}$$

Comparing U_a and U_b shows that $U_a = \dfrac{5}{8}\ U_b$. Thus, even though the volume of the stepped rod is 2.5 times the volume of the uniform rod, the stepped rod can absorb only 5/8 as much energy as the uniform rod.

Example Problem 10-7   The cantilever beam shown in Fig. 10-17a is made of steel ($E = 30{,}000$ ksi and $\sigma_y = 36$ ksi). Determine the height h from which the 10-lb weight W can be dropped if the yield stress is not to be exceeded.

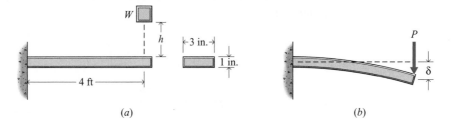

(a)   (b)

Figure 10-17

SOLUTION
The static deflection is found using case 1 of Table B-19 (Appendix B):

$$\delta_{st} = \frac{WL^3}{3EI} = \frac{10[(4)(12)]^3}{3[30(10^6)][3(1^3)/12]} = 49.15(10^{-3})\ \text{in.}$$

The impact factor IF is

$$IF = 1 + \sqrt{1 + \frac{2h}{\delta_{st}}} = 1 + \sqrt{1 + \frac{2h}{49.15(10^{-3})}} \qquad (a)$$

and the load P to cause the maximum beam deflection is computed using Eq. 10-24:

$$P = W\left[1 + \sqrt{1 + \frac{2h}{\delta_{st}}}\right] = 10(IF)$$

The maximum flexural stress in the beam occurs at the support and is (neglecting stress concentration)

$$\sigma = \frac{Mc}{I} = \frac{(PL)c}{I} = \frac{10(IF)Lc}{I}$$

where $\sigma = 36{,}000$ psi, the yield stress. Therefore,

$$36{,}000 = \frac{10(IF)(4)(12)(0.5)}{(3)(1)^3/12} = 960(IF)$$

$$IF = 37.5$$

which, when substituted into Eq. (a), yields

$$h = 32.7 \text{ in.} \qquad\qquad \textbf{Ans.}$$

Example Problem 10-8   The beam of Fig. 10-18 is 75 mm wide and 25 mm deep. Each of the supporting coil springs has a modulus of 9 kN/m. Determine the maximum stress in the beam when the block of weight $W = 45$ N is dropped from a height $h = 15$ mm. The modulus of elasticity of the beam is 200 GPa.

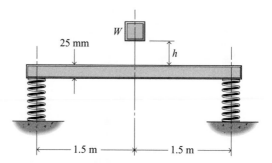

Figure 10-18

SOLUTION
The static deflection of the beam is

$$\delta_{st} = (\delta_{st})_{beam} + (\delta_{st})_{springs}$$
$$= \frac{WL^3}{48EI} + \frac{F_{spring}}{k} \qquad\qquad (a)$$

where the static deflection of the beam is found from case 6 of Table B-19 (Appendix B). Substituting the given data into Eq. (a) gives

$$\delta_{st} = \frac{45(3)^3}{48(200)(10^9)[(0.075)(0.025)^3/12]} + \frac{45/2}{9000}$$
$$= 1.2960(10^{-3}) + 2.5(10^{-3}) = 3.796(10^{-3}) \text{ m}$$

The impact factor is

$$IF = 1 + \sqrt{1 + \frac{2h}{\delta_{st}}} = 1 + \sqrt{1 + \frac{2(0.015)}{3.796(10^{-3})}} = 3.984$$

Using Eq. 10-24, the impact load is determined to be

$$P = W(IF) = 45(3.984) = 179.28 \text{ N}$$

The magnitude of the maximum stress in the beam is

$$\sigma = \frac{M_r c}{I} = \frac{(P/2)(L/2)c}{I} = \frac{PLc}{4I}$$

$$= \frac{179.28(3)(0.025/2)}{4[(0.075)(0.025)^3/12]} = 17.211(10^6) \text{ N/m}^2$$

$$\cong 17.21 \text{ MPa} \qquad\qquad \textbf{Ans.}$$

If the weight W were slowly lowered onto the beam, the maximum stress would be

$$\sigma = \frac{M_r c}{I} = \frac{(W/2)(L/2)c}{I} = \frac{WLc}{4I}$$

$$= \frac{45(3)(0.025/2)}{4[(0.075)(0.025)^3/12]} = 4.32(10^6) \text{ N/m}^2$$

$$= 4.32 \text{ MPa}$$

Note that $\sigma = \sigma_{st}\ (IF)$; that is,

$$\sigma_{st} = \frac{\sigma}{IF} = \frac{17.211(10^6)}{3.984} = 4.320(10^6) \text{ N/m}^2$$

$$\cong 4.32 \text{ MPa}$$

PROBLEMS

Introductory Problems

10-18* A cold-rolled bronze ($E = 100$ GPa) bar with a cross-sectional area of 2500 mm² is to be used as a tension member subjected to axial energy loads, as shown in Fig. P10-18. The allowable axial tensile stress is 70 MPa. Determine the minimum length of bar required if a 20-kg mass is dropped from a height of $h = 1$ m.

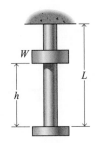

Figure P10-18

10-19* A solid circular steel ($E = 30,000$ ksi) bar with $L = 50$ in. is to be used as a tension member subjected to axial energy loads, as shown in Fig. P10-18. The allowable axial tensile stress is 25 ksi. Determine the minimum diameter required if the weight $W = 30$ lb is dropped from a height of $h = 40$ in.

10-20 A 25-mm-diameter steel ($E = 200$ GPa) rod 900 mm long is supported at the top end and fitted with a loading flange at the bottom end, as shown in Fig. P10-20. A collar W slides freely on the rod. Determine the maximum mass that the collar may have without exceeding the proportional limit (250 MPa) of the rod if

a. The collar is dropped from a height of 600 mm.
b. The collar is suddenly applied to the stop (dropped from a very small height).
c. The collar is slowly applied to the flange (static load).

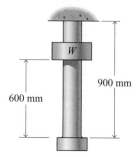

Figure P10-20

10-21 A weight of 10 lb is dropped from a height of 8 in. onto the free end of a steel ($E = 30,000$ ksi) cantilever beam. The beam is 2 in. wide, 3 in. deep, and 30 in. long. Determine

a. The maximum deflection produced by the falling weight.
b. The static load needed to produce the same deflection.
c. The maximum flexural stress caused by the falling weight.

10-22* A timber ($E = 8.2$ GPa) beam 150 mm wide, 100 mm deep, and 3 m long is simply supported at the ends. From what maximum height can a 14-kg mass be dropped onto the center of the beam without causing the maximum flexural stress to exceed 10 MPa?

10-23 A timber ($E = 1200$ ksi) beam 6 in. wide, 4 in. deep, and 10 ft long is simply supported at the ends, as shown in Fig. P10-23. A 30-lb weight is dropped onto the center of the beam from a height of 10 in. Determine

a. The maximum deflection due to the falling weight.
b. The static load needed to produce the same deflection.
c. The maximum flexural stress produced by the falling weight.

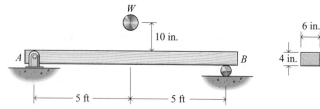

Figure P10-23

Intermediate Problems

10-24* A flat piece of steel ($E = 200$ GPa) 10 mm thick and 2 m long has a width of 25 mm for 0.5 m of its length and a width of 50 mm for the remaining 1.5 m. A 2-kg collar W is dropped onto a loading flange at the end of the bar, as shown in Fig. P10-24. Determine the maximum height h from which the collar can be dropped if the axial tensile stress in the rod is not to exceed 200 MPa. Neglect the effect of stress concentrations.

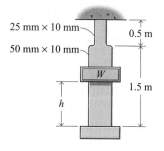

Figure P10-24

10-25* A collar W is dropped from a height of 10 in. onto the loading flange at the end of the steel ($E = 30,000$ ksi) bar shown in Fig. P10-25. If the maximum stress is not to exceed 18 ksi, determine the maximum allowable weight of the collar if

a. The bar has the cross section shown in Fig. P10-25.
b. The bar has a uniform cross section of 2 in.² throughout its entire length.

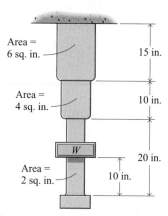

Figure P10-25

10-26 The beam of Fig. P10-26 is made of an aluminum alloy ($E = 70$ GPa). The 240-N block B is dropped from a height of 12 mm onto the top of the coil spring C, which has a modulus of 36 kN/m. Determine the maximum stress developed in the beam.

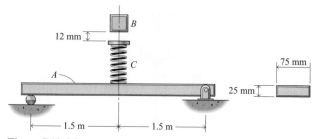

Figure P10-26

10-27* A weight of 50 lb is dropped (see Fig. P10-27) from a height of 4 in. onto a helical spring that has a modulus of 100 lb/in. Determine

a. The maximum deflection of the spring.
b. The static load that would produce the same deflection.

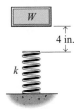

Figure P10-27

10-28 The simply supported 6061-T6 aluminum alloy ($E = 70$ GPa) beam A of Fig. P10-28 is 75 mm wide × 25 mm deep. The center support is a helical spring with a modulus of 18 kN/m. The spring is initially unstressed and in contact with the beam. The 22-kg block drops 50 mm onto the top of the beam. Determine

a. The impact factor.
b. The static deflection.
c. The maximum flexural stress in the beam when the block drops onto the beam.

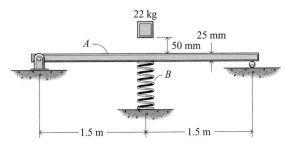

Figure P10-28

10-29 Using Fig. P10-29 and Eqs. 10-14 and 10-16, show that

$$\sigma = \frac{W}{A}\sqrt{\frac{2h}{\delta_{st}}}$$

and

$$\delta = \delta_{st}\sqrt{\frac{2h}{\delta_{st}}}$$

when W is dropped from a height h that is large compared to the static deflection

$$\delta_{st} = \frac{WL}{AE}$$

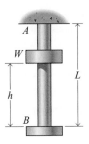

Figure P10-29

Challenging Problems

10-30* A 10-kg mass m is dropped from a height $h = 100$ mm onto the steel ($E = 200$ GPa) beam shown in Fig. P10-30. Determine

a. The maximum deflection at B.
b. The maximum flexural stress produced by the falling mass.

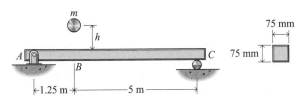

Figure P10-30

10-31* The diver shown in Fig. P10-31 weighs 145 lb and jumps from a height of 2 ft onto the end of a diving board. The board is made of wood ($E = 1800$ ksi) and is 2 ft wide and 2 in. thick. Determine the maximum flexural stress in the diving board.

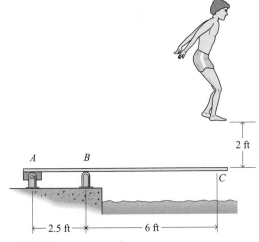

Figure P10-31

10-32 The aluminum alloy ($E = 70$ GPa) beams shown in Fig. P10-32 have cross sections that are 25 mm deep × 100 mm wide. The helical spring C between the beams A and B has a modulus of 20 kN/m. The 5-kg block is dropped onto beam A with an impact factor of 4. Determine

a. The height h from which the block is dropped.
b. The maximum stress developed in each beam.

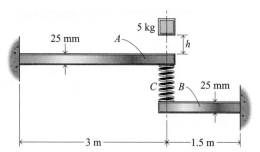

Figure P10-32

10-33 The beam AB shown in Fig. P10-33 has a flexural rigidity EI and is securely fastened to the rigid arm C. The modulus of the helical spring is $6EI/L^3$. The structure rotates about a shaft at B. Determine the height h, in terms of $W, L, E,$ and I, from which W must be dropped to have an impact factor of 4.

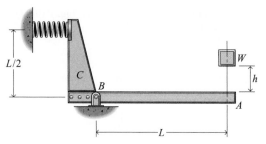

Figure P10-33

10-34 Beams A and B of Fig. P10-34 are made of wood ($E = 8$ GPa) and have cross sections that are 50 mm deep × 150 mm wide. If the maximum flexural stress developed in beam B is 8 MPa, determine

a. The height h from which the 9-kg mass m is dropped.
b. The impact factor.

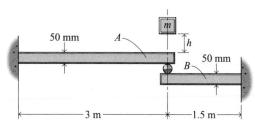

Figure P10-34

Part B
Theories of Failure for Static Loading

10-5 INTRODUCTION

A tension test of an axially loaded member is easy to conduct, and the results, for many types of materials, are well known. When such a member fails, the failure occurs at a specific principal (axial) stress, a definite axial strain, a maximum shearing stress of one-half the axial stress, and a specific amount of strain energy per unit volume of stressed material. Since all of these limits are reached simultaneously for an axial load, it makes no difference which criterion (stress, strain, or energy) is used for predicting failure in another axially loaded member of the same material.

For an element subjected to biaxial or triaxial loading, however, the situation is more complicated because the limits of normal stress, normal strain, shearing stress, and strain energy existing at failure for an axial load are not all reached simultaneously. In other words, the cause of failure, in general, is unknown. In such cases, it becomes important to determine the best criterion for predicting failure, because test results are difficult to obtain and the combinations of loads are endless. Several theories have been proposed for predicting failure of various types of material subjected to many combinations of loads. Unfortunately, none of the theories agrees with test data for all types of materials and combinations of loads. Several of the more common theories of failure are presented and briefly explained in the following sections.

10-6 FAILURE THEORIES FOR DUCTILE MATERIALS

10-6-1 Maximum-Normal-Stress Theory[1]
The maximum-normal-stress theory predicts failure of a specimen subjected to any combination of loads when the maximum normal stress at any point in the specimen reaches the axial failure stress as determined by an axial tensile or compressive test of the same material.

The maximum-normal-stress theory is presented graphically in Fig. 10-19*b* for an element subjected to biaxial principal stresses in the *p*1 and *p*2 directions, as shown in Fig. 10-19*a*. The limiting stress σ_f is the failure stress for this material

---

[1]Often called Rankine's theory after W. J. M. Rankine (1820–1872), an eminent engineering educator in England.

637

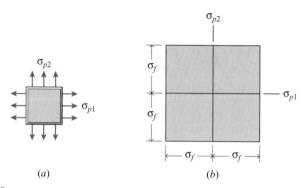

Figure 10-19

when loaded axially and is assumed equal in tension and compression. Any combination of biaxial principal stresses σ_{p1} and σ_{p2} represented by a point inside the square of Fig. 10-19b is safe according to this theory, whereas any combination of stresses represented by a point outside of the square will cause failure of the element on the basis of this theory.

   Once the principal stresses have been found, they may be ordered $\sigma_{p1} > \sigma_{p2} > \sigma_{p3}$. For a state of biaxial stress (plane stress), one of the principal stresses is zero (assumed to be σ_{p3} in Fig. 10-19a). If the principal stress with the largest magnitude is tension, the maximum-normal-stress theory predicts failure when

$$\sigma_{p1} = \sigma_f$$

where σ_f is the failure stress for uniaxial tensile loading. When the principal stress with the largest magnitude is compression, the maximum-normal-stress theory predicts failure when

$$\sigma_{p3} = \sigma_f$$

where σ_f is the failure stress for uniaxial compressive loading. This failure theory will be compared to experimental data for several materials later in this section.

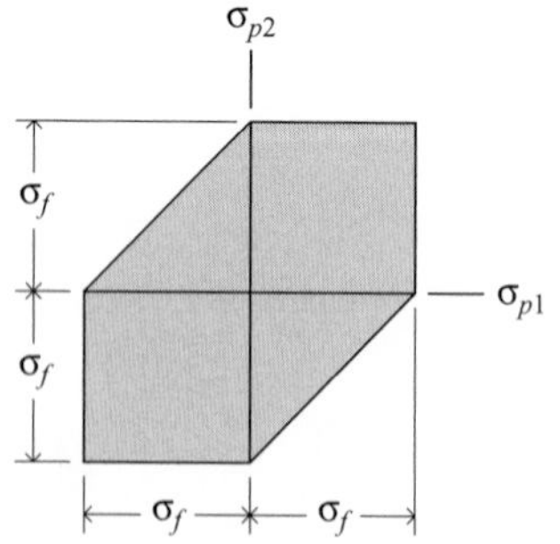

10-6-2 Maximum-Shear-Stress Theory[2]   The maximum-shear-stress theory predicts failure of a specimen subjected to any combination of loads when the maximum shear stress at any point reaches the failure stress τ_f equal to $\sigma_f/2$, as determined by an axial tension or axial compression test of the same material. For ductile materials the shearing elastic limit, as determined from a torsion test (pure shear), is greater than one-half the tensile elastic limit (with an average value of τ_f about 0.57 σ_f). This means that the maximum-shear-stress theory errs on the conservative side by being based on the limit obtained from an axial test.

   The maximum-shear-stress theory is presented graphically in Fig. 10-20 for an element subjected to biaxial (σ_{p3} is equal to zero) principal stresses, as shown in Fig. 10-19a. In the first and third quadrants, σ_{p1} and σ_{p2} have the same sign, and the maximum shearing stress is half the numerically larger principal stress σ_{p1} or σ_{p3}, as explained in Section 2-10. According to the maximum-shear-stress theory,

Figure 10-20

---

[2]Sometimes called Coulomb's theory because it was originally stated by him in 1773. More frequently called Guest's theory or law because of the work of J. J. Guest in England in 1900.

failure occurs when

$$\tau_{max} = \tau_f$$

where τ_f is the failure stress determined from a uniaxial tension or compression test, $\tau_f = \sigma_f/2$. In quadrant one or three

$$\tau_{max} = \frac{\sigma_{max} - \sigma_{min}}{2}$$

$$= \frac{\sigma_{p1}}{2} \qquad \text{(First quadrant)}$$

$$= \frac{\sigma_{p3}}{2} \qquad \text{(Third quadrant)}$$

where $\sigma_{p1} > \sigma_{p2} > \sigma_{p3}$. The maximum-shear-stress theory predicts failure in the first and third quadrants when

$$\sigma_{max} = \sigma_f$$

and is the same as the maximum-normal-stress theory of failure.

In the second and fourth quadrants, where σ_{p1} and σ_{p2} are of an opposite sign, the maximum shearing stress is half the arithmetical sum of the two principal stresses. In the fourth quadrant, the equation of the boundary, or limit stress, line is

$$\sigma_{p1} - \sigma_{p2} = \sigma_f$$

and in the second quadrant the relation is

$$\sigma_{p1} - \sigma_{p2} = -\sigma_f$$

A comparison of this theory with experimental data will be presented later in this section.

10-6-3 Maximum-Distortion-Energy Theory[3] The maximum-distortion-energy theory predicts failure of a specimen subjected to any combination of loads when the distortion component of the strain energy intensity of any portion of the stressed member reaches the failure value of the distortion component of the strain energy intensity as determined from an axial tension or compression test of the same material. This theory assumes that the portion of the strain energy producing volume change is ineffective in causing failure by yielding. Supporting evidence comes from experiments showing that homogeneous materials can withstand very high hydrostatic stresses without yielding. The portion of the strain energy producing the element's change of shape is assumed to be completely responsible for the failure of the material by inelastic action.

The strain energy of distortion is most readily computed by determining the total strain energy of the stressed material and subtracting the strain energy corresponding to the volume change. In Section 10-2, the quantity $\sigma^2/2E$ was

---

[3]Frequently called the Huber-Hencky-von Mises theory, because it was proposed by M. T. Huber of Poland in 1904 and independently by R. von Mises of Germany in 1913. The theory was further developed by H. Hencky and von Mises in Germany and the United States.

defined as the strain energy per unit volume for a member subjected to a slowly applied axial load. This expression can also be written as

$$u = \frac{\sigma^2}{2E} = \frac{\sigma\epsilon}{2}$$

where u is the strain energy per unit volume (strain energy intensity) and σ and ϵ are the slowly applied axial stress and strain. This equation assumes that the stress does not exceed the proportional limit.

When an elastic element is subjected to triaxial loading, the stresses can be resolved into three principal stresses such as σ_{p1}, σ_{p2}, and σ_{p3}, where $p1$, $p2$, and $p3$ are the principal axes. These stresses will be accompanied by three principal strains related to the stresses by Eq. 4-5 of Section 4-3. If it is assumed that the loads are applied simultaneously and gradually, the stresses and strains will increase in the same manner. The total strain energy per unit volume is the sum of the energies produced by each of the stresses (energy is a scalar quantity and can be added algebraically regardless of the directions of the individual stresses); thus,

$$u = \frac{1}{2}(\sigma_{p1}\epsilon_{p1} + \sigma_{p2}\epsilon_{p2} + \sigma_{p3}\epsilon_{p3})$$

When the strains are expressed in terms of the stresses, this equation becomes

$$u = \frac{1}{2E}\left[\sigma_{p1}^2 + \sigma_{p2}^2 + \sigma_{p3}^2 - 2v\left(\sigma_{p1}\sigma_{p2} + \sigma_{p2}\sigma_{p3} + \sigma_{p3}\sigma_{p1}\right)\right]$$

The strain energy can be resolved into two components u_v and u_d, resulting from a volume change and a distortion, respectively, by considering the principal stresses to be made up of two sets of stresses, as indicated in Figs. 10-21a, b, and c. The state of stress in Fig. 10-21c will result in distortion only (no volume change) if the sum of the three normal strains is zero. That is,

$$\begin{aligned} E[\epsilon_{p1} + \epsilon_{p2} + \epsilon_{p3}]_d &= [(\sigma_{p1} - p) - v(\sigma_{p2} + \sigma_{p3} - 2p)] \\ &\quad + [(\sigma_{p2} - p) - v(\sigma_{p3} + \sigma_{p1} - 2p)] \\ &\quad + [(\sigma_{p3} - p) - v(\sigma_{p1} + \sigma_{p2} - 2p)] = 0 \end{aligned}$$

which reduces to

$$(1 - 2v)(\sigma_{p1} + \sigma_{p2} + \sigma_{p3} - 3p) = 0$$

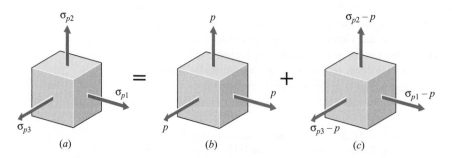

(a)   (b)   (c)

Figure 10-21

Therefore,

$$p = \frac{1}{3}(\sigma_{p1} + \sigma_{p2} + \sigma_{p3})$$

The three normal strains due to p are, from Eq. 4-5,

$$\epsilon_v = (1 - 2v)\frac{p}{E}$$

and the energy resulting from the hydrostatic stress (the volume change) is

$$u_v = 3\left(\frac{p\epsilon_v}{2}\right) = \frac{3}{2}\frac{1 - 2v}{E}p^2$$

$$= \frac{1 - 2v}{6E}[\sigma_{p1} + \sigma_{p2} + \sigma_{p3}]^2$$

The energy resulting from the distortion (change of shape) is

$$u_d = u - u_v$$
$$= \frac{1}{6E}[3(\sigma_{p1}^2 + \sigma_{p2}^2 + \sigma_{p3}^2) - 6v(\sigma_{p1}\sigma_{p2} + \sigma_{p2}\sigma_{p3} + \sigma_{p3}\sigma_{p1})$$
$$- (1 - 2v)(\sigma_{p1} + \sigma_{p2} + \sigma_{p3})^2]$$

When the third term in the brackets is expanded, the expression can be rearranged to give

$$u_d = \frac{1 + v}{6E}[(\sigma_{p1}^2 - 2\sigma_{p1}\sigma_{p2} + \sigma_{p2}^2) + (\sigma_{p2}^2 - 2\sigma_{p2}\sigma_{p3} + \sigma_{p3}^2)$$
$$+ (\sigma_{p3}^2 - 2\sigma_{p3}\sigma_{p1} + \sigma_{p1}^2)]$$
$$= \frac{1 + v}{6E}[(\sigma_{p1} - \sigma_{p2})^2 + (\sigma_{p2} - \sigma_{p3})^2 + (\sigma_{p3} - \sigma_{p1})^2] \qquad (a)$$

The maximum-distortion-energy theory of failure assumes that inelastic action will occur whenever the energy given by Eq. (a) exceeds the limiting value obtained from a tensile test. For this test, only one of the principal stresses will be nonzero. If this stress is called σ_f, the value of u_d becomes

$$(u_d)_f = \frac{1 + v}{3E}\sigma_f^2$$

and when this value is substituted in Eq. (a), it becomes

$$2\sigma_f^2 = (\sigma_{p1} - \sigma_{p2})^2 + (\sigma_{p2} - \sigma_{p3})^2 + (\sigma_{p3} - \sigma_{p1})^2 \qquad (b)$$

for failure by yielding.

When a state of plane stress exists, assuming σ_{p3} equals zero, Eq. (b) becomes

$$\sigma_{p1}^2 - \sigma_{p1}\sigma_{p2} + \sigma_{p2}^2 = \sigma_f^2$$

This last expression is the equation of an ellipse with its major axis along the line σ_{p1} equals σ_{p2}, as shown in Fig. 10-22. A comparison of this theory with experimental data will be presented later in this section.

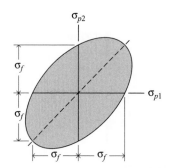

Figure 10-22

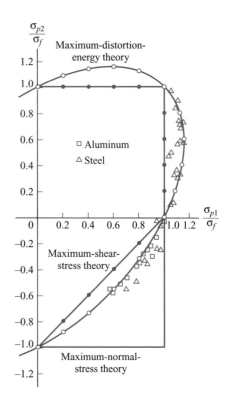

Figure 10-24

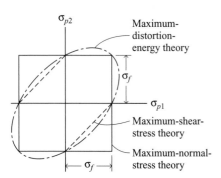

Figure 10-23

The graphic representations of Figs. 10-19, 10-20, and 10-22 are superimposed in Fig. 10-23 for convenient comparison of the different theories. The failure theories are compared to experimental data in Fig. 10-24.[4] In the first quadrant, failure is predicted most closely by the maximum-distortion-energy theory. The maximum-normal-stress and maximum-shear-stress theories are identical in the first quadrant; both theories are conservative because the experimental data lie outside the prediction envelope.

In the fourth quadrant, the principal stresses have opposite signs. The best correlation between theory and experimental data is for the maximum-distortion-energy theory. In this quadrant, the maximum-shear-stress theory is conservative because the experimental data lie outside the prediction envelope. The prediction envelope for the maximum-normal-stress theory lies outside the experimental data in the fourth quadrant; therefore, the theory overpredicts failure stresses and is unsafe.

■ **Example Problem 10-9**   At a point in a structural member subjected to plane stress, the state of stress is as shown in Fig. 10-25. Determine which, if any, of the theories of failure will predict failure by yielding for this state of stress. The yield strength of the material in tension and compression is 36 ksi.

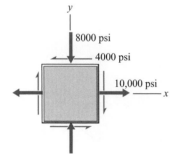

Figure 10-25

[4]*Handbook of Mechanics, Materials, and Structures*, Alexander Blake, Ed., Wiley-Interscience, New York, 1985.

SOLUTION

In order to apply the theories of failure, the principal stresses must first be determined. The given stresses for use in Eq. 2-15 are

$$\sigma_x = +10{,}000 \text{ psi} \qquad \sigma_y = -8000 \text{ psi} \qquad \tau_{xy} = -4000 \text{ psi}$$

When these values are substituted in Eq. 2-15, the principal stresses are found to be

$$\sigma_{p1,p3} = \frac{\sigma_x + \sigma_y}{2} \pm \sqrt{\left(\frac{\sigma_x - \sigma_y}{2}\right)^2 + \tau_{xy}^2}$$

$$= \frac{10{,}000 + (-8000)}{2} \pm \sqrt{\left(\frac{10{,}000 - (-8000)}{2}\right)^2 + (-4000)^2}$$

$$= 1000 \pm 9849$$

$$\sigma_{p1} = 1000 + 9849 = 10{,}849 \text{ psi} \cong 10{,}850 \text{ psi (T)}$$

$$\sigma_{p2} = \sigma_z = 0$$

$$\sigma_{p3} = 1000 - 9849 = -8849 \text{ psi} \cong 8850 \text{ psi (C)}$$

where the principal stresses have been ordered such that $\sigma_{p1} > \sigma_{p2} > \sigma_{p3}$. Thus, the largest principal stress is 10,850 psi tension and the smallest principal stress is 8850 psi compression.

The maximum-normal-stress theory predicts that failure will not occur if

$$\sigma_{\max} < \sigma_f$$

Since

$$\sigma_{\max} = 10{,}850 \text{ psi} < \sigma_f = \sigma_y = 36{,}000 \text{ psi}$$

failure will not occur according to the maximum-normal-stress theory.

The maximum-shear-stress theory of failure predicts that the state of stress shown in Fig. 10-25 is safe if

$$\tau_{\max} < \tau_f$$

Since the principal stresses have opposite signs,

$$\tau_{\max} = \frac{\sigma_{\max} - \sigma_{\min}}{2} = \frac{10{,}850 - (-8850)}{2} = 9850 \text{ psi}$$

The failure stress (yield strength) is 36,000 psi; therefore,

$$\tau_f = \frac{\sigma_f}{2} = \frac{36{,}000}{2} = 18{,}000 \text{ psi}$$

Since

$$\tau_{\max} = 9850 \text{ psi} < \tau_f = 18{,}000 \text{ psi}$$

failure will not occur according to the maximum-shear-stress theory.

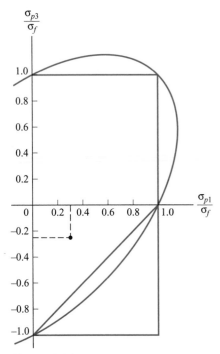

Figure 10-26

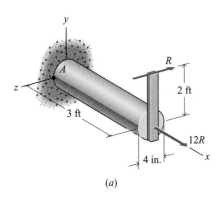

(a)

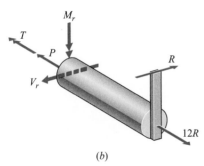

(b)

Figure 10-27

The maximum-distortion-energy theory of failure states that the state of stress shown in Fig. 10-25 is safe if

$$\sigma_{p1}^2 - \sigma_{p1}\sigma_{p3} + \sigma_{p3}^2 < \sigma_f^2$$
$$(10{,}850)^2 - 10{,}850(-8850) + (-8850)^2 = (17{,}090)^2 < (36{,}000)^2$$

Thus, failure will not occur according to the maximum-distortion-energy theory of failure.

Alternatively, the ratios $\sigma_{p1}/\sigma_f = 10{,}850/36{,}000 = 0.301$ and $\sigma_{p3}/\sigma_f = -8850/36{,}000 = -0.246$ can be plotted on Fig. 10-24 and compared to the failure envelopes. The result is shown in Fig. 10-26. Clearly, the state of stress shown in Fig. 10-25 is within all of the failure envelopes; therefore, yielding of the structural member will not occur.

Example Problem 10-10   The solid circular shaft of Fig. 10-27a has a proportional limit of 64 ksi. Determine the value of the load R for failure by yielding as predicted by each of the theories of failure. Assume that point A is the most severely stressed point.

SOLUTION

Passing a section through point A, drawing a free-body diagram of the portion of the body to the right of the section (see Fig. 10-27b), and applying the equations of equilibrium yields the following internal reactions if R is expressed in kip

$$P = 12R \text{ kip} \qquad T_r = -24R \text{ kip} \cdot \text{in.}$$
$$V_r = R \text{ kip} \qquad M_r = 36R \text{ kip} \cdot \text{in.}$$

The shearing stress at A due to the shear force V_r is zero because $Q = 0$. The magnitude of the flexural stress at A is

$$\sigma_1 = \frac{M_r c}{I} = \frac{36R(2)}{\pi(4)^4/64} = \frac{18R}{\pi} \text{ ksi (T)}$$

The axial stress at A is

$$\sigma_2 = \frac{P}{A} = \frac{12R}{\pi(4)^2/4} = \frac{3R}{\pi} \text{ ksi (T)}$$

The sum of the two tensile normal stresses at A becomes

$$\sigma = \sigma_1 + \sigma_2 = \frac{21R}{\pi} \text{ ksi (T)}$$

The magnitude of the torsional shearing stress at A is

$$\tau = \frac{T_r c}{J} = \frac{24R(2)}{\pi(4)^4/32} = \frac{6R}{\pi} \text{ksi}$$

These stresses on an element at A are shown in Fig. 10-28. Substituting the stresses $\sigma_x = 21R/\pi$ ksi, $\sigma_y = 0$, and $\tau_{xy} = 6R/\pi$ ksi into the principal stress-equation (Eq. 2-15) gives

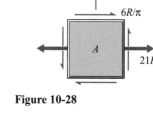

Figure 10-28

$$\sigma_{p1,p3} = \frac{\sigma_x + \sigma_y}{2} \pm \sqrt{\left(\frac{\sigma_x - \sigma_y}{2}\right)^2 + \tau_{xy}^2}$$

$$= \frac{21R/\pi + 0}{2} \pm \sqrt{\left(\frac{21R/\pi - 0}{2}\right)^2 + (6R/\pi)^2}$$

$$= \frac{10.5R}{\pi} \pm \frac{12.09R}{\pi}$$

Ordering the principal stresses such that $\sigma_{p1} > \sigma_{p2} > \sigma_{p3}$ gives

$$\sigma_{p1} = \frac{22.59R}{\pi} \text{ ksi} \qquad \sigma_{p2} = 0 \qquad \sigma_{p3} = -\frac{1.59R}{\pi} \text{ ksi}$$

Since the principal stresses have opposite signs,

$$\tau_{max} = \frac{\sigma_{max} - \sigma_{min}}{2} = \frac{22.59R/\pi - (-1.59R/\pi)}{2} = \frac{12.09R}{\pi} \text{ ksi}$$

According to the maximum-normal-stress theory

$$\sigma_{p1} = \frac{22.59R}{\pi} = \sigma_f = 64 \text{ ksi}$$

from which

$$R = \frac{64\pi}{22.59} = 8.90 \text{ kip} \qquad\qquad \textbf{Ans.}$$

According to the maximum-shear-stress theory

$$\tau_{max} = \frac{12.09R}{\pi} = \tau_f = \frac{\sigma_f}{2} = \frac{64}{2}\text{ksi}$$

from which

$$R = \frac{32\pi}{12.09} = 8.315 \text{ kip} \cong 8.32 \text{ kip} \qquad\qquad \textbf{Ans.}$$

According to the maximum-distortion-energy theory

$$\sigma_{p1}^2 - \sigma_{p1}\sigma_{p3} + \sigma_{p3}^2 = \sigma_f^2$$

$$\left(\frac{22.59R}{\pi}\right)^2 - \frac{22.59R}{\pi}\left(-\frac{1.59R}{\pi}\right) + \left(-\frac{1.59R}{\pi}\right)^2 = (64)^2$$

which gives

$$R = \frac{64\pi}{23.43} = 8.581 \text{ kip} \cong 8.58 \text{ kip} \qquad \textbf{Ans.}$$

In this example, the maximum-shear-stress theory is seen to be the most conservative, and the maximum-normal-stress theory gives the least conservative result. It would not be wise to use the result from the maximum-normal-stress theory since it is not safe when the principal stresses have opposite signs (see Fig. 10-24).

Example Problem 10-11   A thin-walled cylindrical pressure vessel is subjected to an internal pressure of 6 MPa. The inside diameter of the vessel is 500 mm. The yield strength of the material from which the vessel is made is 250 MPa. If a factor of safety of 2.5 is desired, determine the minimum wall thickness according to the maximum-distortion-energy theory.

SOLUTION

The hoop and axial stresses are given by Eqs. 5-8a and b, respectively. These stresses are

$$\sigma_h = \frac{pr}{t} = \frac{6(10^6)(0.250)}{t} = \frac{1.5000(10^6)}{t}$$

$$\sigma_a = \frac{pr}{2t} = \frac{0.7500(10^6)}{t}$$

Since there is no shear stress on the hoop and axial planes, σ_h and σ_a are principal stresses. At a point on the outside surface of the vessel,

$$\sigma_{p1} = \sigma_h = \frac{1.5000(10^6)}{t} \qquad \sigma_{p2} = \sigma_a = \frac{0.7500(10^6)}{t} \qquad \sigma_{p3} = 0$$

According to the maximum-distortion-energy theory, failure occurs (for a state of plane stress) whenever

$$\sigma_{p1}^2 - \sigma_{p1}\sigma_{p2} + \sigma_{p2}^2 = \sigma_f^2$$

where $\sigma_f = \dfrac{\sigma_y}{FS} = \dfrac{250}{2.5} = 100 \text{ MPa}$. Thus

$$\left[\frac{1.5000(10^6)}{t}\right]^2 - \left[\frac{1.5000(10^6)}{t}\right]\left[\frac{0.7500(10^6)}{t}\right] + \left[\frac{0.7500(10^6)}{t}\right]^2$$
$$= \left[100(10^6)\right]^2$$

which yields

$$t = 12.99(10^{-3}) \text{ m} = 12.99 \text{ mm}$$

If a point on the inside surface of the pressure vessel is selected, then

$$\sigma_{p1} = \frac{1.5000(10^6)}{t} \qquad \sigma_{p2} = \frac{0.7500(10^6)}{t} \qquad \sigma_{p3} = -p = -6\,\text{MPa}$$

For this three-dimensional state of stress, the maximum-distortion-energy theory of failure is

$$(\sigma_{p1} - \sigma_{p2})^2 + (\sigma_{p2} - \sigma_{p3})^2 + (\sigma_{p3} - \sigma_{p1})^2 = 2\sigma_f^2$$

which upon substitution of $\sigma_f = 100$ MPa and the principal stresses yields

$$t = 13.71\,\text{mm} \qquad\qquad\qquad\qquad \textbf{Ans.}$$

PROBLEMS

Introductory Problems

10-35* A machine component fabricated from a material with a proportional limit in tension and compression of 60 ksi is subjected to a biaxial state of stress. The principal stresses are 30 ksi (T) and 50 ksi (C). Determine which, if any, of the theories will predict failure by yielding for this state of stress.

10-36* A machine component fabricated from a material with a proportional limit in tension and compression of 380 MPa is subjected to a biaxial state of stress. The principal stresses are 180 MPa (T) and 270 MPa (C). Determine which, if any, of the theories will predict failure by yielding for this state of stress.

10-37 At a point on the free surface of an alloy steel machine component the principal stresses are 45 ksi (T) and 25 ksi (C). What minimum proportional limit is required according to each of the theories if failure by yielding is to be avoided?

10-38* The state of stress at a point on the surface of a machine component is shown in Fig. P10-38. If the yield strength of the material is 250 MPa, determine which, if any, of the theories will predict failure by yielding for this state of stress.

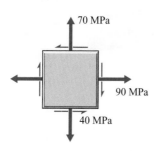

Figure P10-38

10-39 A material with a proportional limit of 36 ksi in tension and compression is subjected to a biaxial state of stress. The principal stresses are 18 ksi (T), 16 ksi (T), and $\sigma_z = 0$. Determine the factor of safety with respect to failure by yielding according to each of the theories of failure.

10-40 At a point on the free surface of an aluminum alloy machine component the principal stresses are 120 MPa (T) and 180 MPa (C). What minimum proportional limit is required according to each of the theories if failure by yielding is to be avoided?

10-41* A material with a yield strength in tension and compression of 60 ksi is subjected to the biaxial state of stress shown in Fig. P10-41. Determine the factor of safety with respect to failure by yielding according to each of the theories of failure.

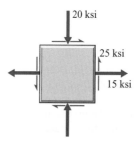

Figure P10-41

10-42 A point on the free surface of a machine component is subjected to the state of stress shown in Fig. P10-42. If the yield strength of the material is 250 MPa, determine the factor of safety with respect to failure by yielding according to each of the theories of failure.

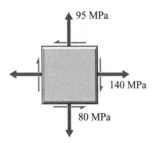

Figure P10-42

Intermediate Problems

10-43* A thin-walled cylindrical pressure vessel is capped at the ends and is subjected to an internal pressure. The inside diameter of the vessel is 5 ft, and the wall thickness is 1.5 in. The vessel is made of steel with a tensile and compressive yield strength of 36 ksi. Determine the internal pressure required to initiate yielding according to

a. The maximum-shear-stress theory.
b. The maximum-distortion-energy theory.

10-44* The solid circular shaft shown in Fig. P10-44 is subjected to a torque T. The yield strength of the material in tension and compression is 400 MPa. Determine the largest permissible value of the torque T according to

a. The maximum-shear-stress theory.
b. The maximum-distortion-energy theory.

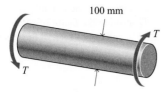

Figure P10-44

10-45 The shaft shown in Fig. P10-45 is made of steel having a proportional limit of 60 ksi in tension or compression. If a factor of safety of 3.0 with respect to failure by yielding is specified, determine the maximum permissible value for the axial load P according to the maximum-shear-stress theory of failure.

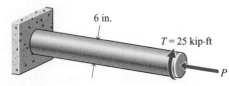

Figure P10-45

10-46 A shaft, similar to the one shown in Fig. P10-45, has a 150-mm diameter and is made of steel having a yield strength in tension and compression of 360 MPa. The applied loads are $P = 2200$ kN and $T = 38$ kN $\cdot$ m. If failure is by yielding, determine the factor of safety according to

a. The maximum-shear-stress theory of failure.
b. The maximum-distortion-energy theory of failure.

10-47 The yield strength σ_y of a ductile material may be determined from a tension test (Fig. P10-47a). Using the three theories of failure discussed in Section 10-6, show that the yield strength τ_y of a member determined from a torsion test (Fig. P10-47b) is predicted to be

a. $\tau_y = \sigma_y$; maximum-normal-stress theory.
b. $\tau_y = 0.5\sigma_y$; maximum-shear-stress theory.
c. $\tau_y = 0.577\sigma_y$; maximum-distortion-energy theory.

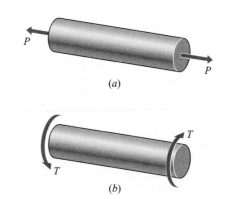

Figure P10-47

10-48* The hollow steel ($\sigma_y = 250$ MPa) shaft shown in Fig. P10-48 is subjected to a torque $T = 40$ kN $\cdot$ m. The factor of safety with respect to failure by yielding is 1.5. Determine the maximum permissible inside diameter for the shaft according to the maximum-shear-stress theory of failure.

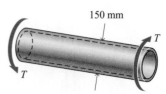

Figure P10-48

10-49 A solid circular steel ($\sigma_y = 36$ ksi) shaft is subjected to an axial tensile load $P = 10$ kip and a bending moment

$M = 5$ kip · in. If a factor of safety of 2.0 for failure by yielding is desired, determine the minimum permissible diameter for the shaft according to the maximum-shear-stress theory.

Challenging Problems

10-50 A solid circular shaft has a diameter d and is subjected to a bending moment M and a torque T. The shaft is made from a ductile material with a yield strength σ_y. Show that the minimum diameter d of the shaft may be found according to the maximum-shear-stress theory from the equation

$$d = \left[\frac{32FS}{\pi \sigma_y} \sqrt{(M)^2 + (T)^2} \right]^{1/3}$$

where FS is the factor of safety.

10-51 The solid circular shaft shown in Fig. P10-51 is subjected to an axial tensile force P, a bending moment M, and a torque T. The shaft is made from a ductile material with a yield strength σ_y. Show that the minimum diameter d of the shaft may be found according to the maximum-shear-stress theory from the equation

$$\frac{\sigma_y}{2FS} = \frac{2}{\pi d^3} \sqrt{(Pd + 8M)^2 + (8T)^2}$$

where FS is the factor of safety.

Figure P10-51

10-52* The shaft shown in Fig. P10-52 is made of steel having a proportional limit of 360 MPa in tension or compression. If a factor of safety of 2.0 with respect to failure by yielding is specified, determine the minimum permissible diameter D according to

a. The maximum-shear-stress theory of failure.
b. The maximum-distortion-energy theory of failure.

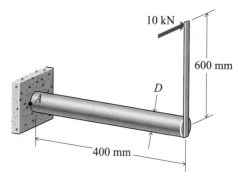

Figure P10-52

10-53* The shaft shown in Fig. P10-53 is made of an aluminum alloy that has a proportional limit of 48 ksi in tension or compression. If a factor of safety of 2.0 with respect to failure by yielding is specified, determine the maximum allowable value for the load R according to

a. The maximum-shear-stress theory of failure.
b. The maximum-distortion-energy theory of failure.

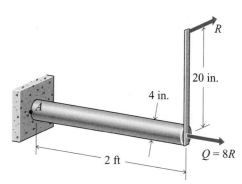

Figure P10-53

10-54 A structural steel pipe used to transmit steam has an inside diameter of 300 mm. If the steam pressure is 5.5 MPa and a factor of safety of 4 with respect to failure by yielding is specified, determine the minimum permissible wall thickness of the pipe according to the maximum-shear-stress theory. The wall of the pipe does not carry any axial load.

10-55* A steel shaft 4 in. in diameter is supported in flexible bearings at its ends. Two pulleys, each 24 in. in diameter, are keyed to the shaft. The pulleys carry belts that are loaded as shown in Fig. P10-55. The steel has a proportional limit of 40 ksi in tension and compression and 23 ksi in shear. If a factor of safety of 2.5 with respect to failure by yielding is specified, determine the maximum allowable belt tension P according to

a. The maximum-shear-stress theory of failure.
b. The maximum-distortion-energy theory of failure.

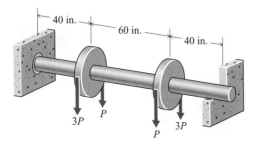

Figure P10-55

10-56* A thick-walled hydraulic cylinder with an inside diameter of 150 mm is required to operate under a maximum internal pressure of 50 MPa. The cylinder is to be made of steel with a proportional limit of 275 MPa and a Poisson's ratio of 0.30. Determine the minimum outside diameter required if a factor of safety of 2 with respect to failure by yielding according to the maximum-shear-stress theory of failure is specified. The wall of the cylinder is not required to carry axial load.

10-57 Determine the maximum allowable internal pressure to which a closed-end, thick-walled cylinder with an inside diameter of 6 in. and an outside diameter of 10 in. may be subjected

if the cylinder is made of an alloy steel with a proportional limit of 80 ksi and a Poisson's ratio of 0.30. A factor of safety of 2.5 with respect to failure by yielding according to the maximum-distortion-energy theory of failure is specified.

10-58 A solid circular shaft acting as a cantilever beam is subjected to the loading shown in Fig. P10-58. The material is 2024-T4 wrought aluminum with a yield strength in tension and compression of 330 MPa. If the factor of safety for failure by yielding is 2.5, determine the minimum permissible diameter of the shaft according to the maximum-shear-stress theory of failure. Neglect the effects of transverse shear.

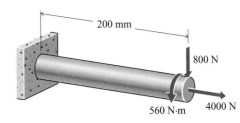

Figure P10-58

10-7 FAILURE THEORIES FOR BRITTLE MATERIALS

Brittle materials, unlike ductile materials, do not yield; therefore, failure is by fracture, and the critical stress is the fracture stress (or the ultimate strength). The ultimate strength in compression is greater than the ultimate strength in tension. The ultimate shear strength of a brittle material is approximately equal to the ultimate tensile strength; this is not the case for ductile materials. Of the theories proposed to predict fracture of brittle materials, only two will be presented in this text: the Coulomb-Mohr theory and the maximum-normal-stress theory.

10-7-1 Coulomb-Mohr and Maximum-Normal-Stress Theories
The tensile test and the compression test are the basis for the Coulomb-Mohr theory. A Mohr's circle for each of these tests is shown in Fig. 10-29a, where σ_{ut} is the ultimate tensile strength and σ_{uc} is the ultimate compressive strength. According to the Coulomb-Mohr theory, failure (fracture) occurs for any state of stress whose Mohr's circle is tangent to the envelope of the two circles shown in Fig. 10-29a (the envelope is the line that is tangent to the two circles, as shown in Fig. 10-29b). Mohr's circle for an arbitrary state of stress is shown dashed in Fig. 10-29b, where σ_{p1} and σ_{p3} are the principal stresses (the principal stresses have been ordered $\sigma_{p1} > \sigma_{p2} > \sigma_{p3}$). When Mohr's circle (dashed) for any state of stress is tangent to the envelope of Fig. 10-29b, failure occurs. The principal stresses σ_{p1} and σ_{p3}, the ultimate tensile strength σ_{ut}, and the ultimate compressive strength σ_{uc} are related by the equation

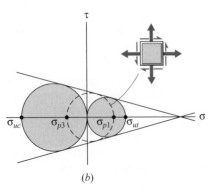

(a)

(b)

Figure 10-29

$$\frac{\sigma_{p1}}{\sigma_{ut}} - \frac{\sigma_{p3}}{\sigma_{uc}} = 1 \qquad (10\text{-}26)$$

where σ_{uc} is the magnitude of the compressive strength, and σ_{p1} and σ_{p3} carry algebraic signs. A state of stress is safe when

$$\frac{\sigma_{p1}}{\sigma_{ut}} - \frac{\sigma_{p3}}{\sigma_{uc}} \leq 1 \qquad (10\text{-}27)$$

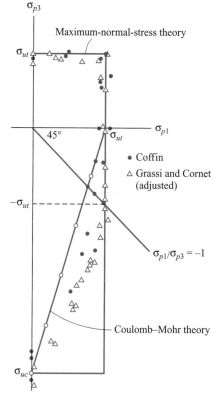

Comparison between theory and experimental data (for gray cast iron) is shown in Fig. 10-30, which is adapted from Blake.[5] In the first quadrant the principal stresses have the same sign, and the maximum-normal-stress theory and the Coulomb-Mohr theory are identical. Note that the failure envelope for the maximum-normal-stress theory is not square as was the case for ductile materials, where the tensile and compressive yield strengths were equal (see Fig. 10-19*b*). The two theories give different predictions of failure in the fourth quadrant, where the principal stresses have opposite signs. In the fourth quadrant, the maximum-normal-stress theory is unsafe because the data lie inside the failure envelope. The Coulomb-Mohr theory is conservative because the data lie outside the failure envelope.

The line for pure torsion, $\sigma_{p1} = -\sigma_{p3}$, is also shown in Fig. 10-30. This line gives two predictions for the ultimate shear strength, that is, where the line intersects the envelopes for the two failure theories. As previously mentioned, one of the characteristics of a brittle material is that the ultimate shear strength is approximately equal to the ultimate tensile strength. This is predicted by the maximum-normal-stress theory, whereas the Coulomb-Mohr theory predicts a value somewhat less.

Figure 10-30

Example Problem 10-12
At the critical point in a machine component, the principal stresses (plane stress) are as shown in Fig. 10-31. The failure strengths for the material are $\sigma_{ut} = 2.5$ ksi and $\sigma_{uc} = 16$ ksi. Use the Coulomb-Mohr theory to determine if this state of stress is safe.

SOLUTION
The state of stress is safe if the stresses satisfy Eq. 10-27.

$$\frac{\sigma_{p1}}{\sigma_{ut}} - \frac{\sigma_{p3}}{\sigma_{uc}} \leq 1$$

Substituting the principal stresses from Fig. 10-31 (after noting that σ_{p3} is compressive and that σ_{uc} is the magnitude of the ultimate compressive strength) into Eq. 10-26 yields

$$\frac{2}{2.5} - \frac{(-4)}{16} = 1.05 > 1$$

Thus, the state of stress shown in Fig. 10-31 is unsafe. **Ans.**

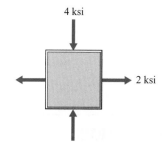

Figure 10-31

[5]*Handbook of Mechanics, Materials, and Structures*, Alexander Blake, Ed., Wiley-Interscience, New York, 1985.

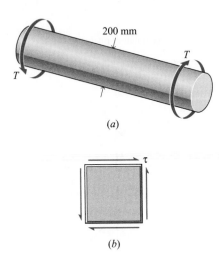

(a)

(b)

Figure 10-32

■ **Example Problem 10-13**   A 200-mm-diameter solid circular shaft is subjected to a torque T, as shown in Fig. 10-32a. The shaft is made of a material with an ultimate tensile strength of 620 MPa and an ultimate compressive strength of 820 MPa. Determine the maximum permissible value for the torque T according to the Coulomb-Mohr theory of failure.

SOLUTION
The state of stress for a shaft subjected to a torsional load T is shown in Fig. 10-32b. The shearing stress τ is given by Eq. 6-6 as

$$\tau = \frac{Tc}{J} = \frac{T(0.100)}{\pi(0.200)^4/32} = 636.6T$$

The principal stresses are

$$\sigma_{p1} = -\sigma_{p3} = \tau = 636.6T \qquad \sigma_{p2} = 0$$

The maximum permissible value of T according to the Coulomb-Mohr theory is given by Eq. 10-26 as

$$\frac{\sigma_{p1}}{\sigma_{ut}} - \frac{\sigma_{p3}}{\sigma_{uc}} = 1$$

or

$$\frac{636.6T}{620(10^6)} - \frac{(-636.6T)}{820(10^6)} = 1$$

Solving for T yields

$$T = 0.5546(10^6)\,\text{N} \cdot \text{m} \cong 555\,\text{kN} \cdot \text{m} \qquad \textbf{Ans.}$$

■ **PROBLEMS**

Introductory Problems

10-59* The state of plane stress at the critical point in a machine component is shown in Fig. P10-59. The failure strengths for the material are 26 ksi in tension and 97 ksi in compression. Use the Coulomb-Mohr theory to determine if this state of stress is safe.

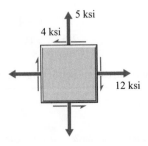

Figure P10-59

10-60* Two states of plane stress are shown in Fig. P10-60. The failure strengths for the material are 152 MPa in tension and 572 MPa in compression. Use the Coulomb-Mohr theory to determine if these states of stress are safe.

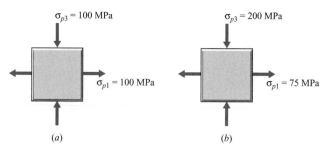

$\sigma_{p3} = 100$ MPa $\sigma_{p3} = 200$ MPa

$\sigma_{p1} = 100$ MPa $\sigma_{p1} = 75$ MPa

(a)   (b)

Figure P10-60

10-61 The state of plane stress at the critical point in a machine component made of cast iron is shown in Fig. P10-61. The failure strengths for this material are 30 ksi in tension and 108 ksi in compression. Use the Coulomb-Mohr theory to determine if this state of stress is safe.

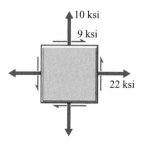

10 ksi

9 ksi

22 ksi

Figure P10-61

Intermediate Problems

10-62* A solid circular cast iron shaft is subjected to the loads shown in Fig. P10-62. The failure strengths for this material are 214 MPa in tension and 750 MPa in compression. Use the Coulomb-Mohr theory to determine the minimum permissible diameter for the shaft. Neglect the effects of transverse shear.

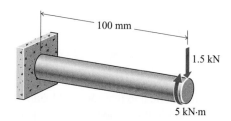

100 mm

1.5 kN

5 kN·m

Figure P10-62

10-63 The solid circular cast iron shaft shown in Fig. P10-63 is subjected to a torque $T = 240$ kip · in. and a bending moment $M = 110$ kip · in. The failure strengths for this material are 43 ksi in tension and 140 ksi in compression. Use the Coulomb-Mohr theory to determine the minimum permissible diameter for this shaft.

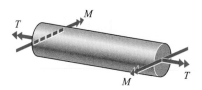

T M

M T

Figure P10-63

10-64 A thin-walled cylindrical pressure vessel has an inside diameter of 300 mm and a wall thickness of 5 mm. The vessel is made of a material with $\sigma_{ut} = 276$ MPa and $\sigma_{uc} = 340$ MPa. Use the Coulomb-Mohr theory to determine the maximum internal pressure that the vessel can safely support.

Challenging Problems

10-65* A solid circular gray cast iron shaft is loaded as shown in Fig. P10-65. The failure strengths for this type of cast iron are 36.5 ksi in tension and 124 ksi in compression. Use the Coulomb-Mohr theory to determine the minimum permissible diameter for the shaft. Neglect the effects of transverse shear.

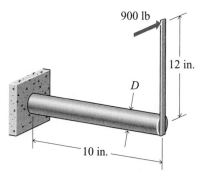

900 lb

12 in.

D

10 in.

Figure P10-65

10-66 The C-clamp shown in Fig. P10-66 is made of a gray cast iron that has an ultimate tensile strength of 180 MPa and an ultimate compressive strength of 670 MPa. Use the Coulomb-Mohr theory to determine the maximum load P that can be applied safely to the clamp.

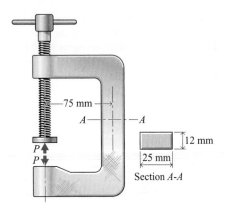

Figure P10-66

10-67 A 4-in.-diameter solid circular shaft is loaded as shown in Fig. P10-67. The shaft is made of a gray cast iron that has an ultimate tensile strength of 22 ksi and an ultimate compressive strength of 82 ksi. Use the Coulomb-Mohr theory to determine if the shaft can safely support the loading shown.

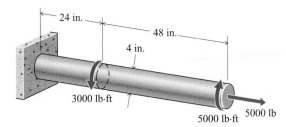

Figure P10-67

SUMMARY

The strain energy stored in a material is equal to the work done in deforming the material. For an axial load, the strain energy intensity (strain energy per unit volume) u is given by

$$u = \int_0^\epsilon \sigma \, d\epsilon$$

The modulus of resilience is defined as the maximum strain energy per unit volume that a material will absorb without inelastic deformation and is the area under the straight-line portion of the stress-strain diagram. For practical purposes, the yield strength and proportional limit are the same, and thus the modulus of resilience is

$$u_R = \frac{\sigma_y^2}{2E} \qquad (10\text{-}3)$$

The area under the entire stress-strain curve from zero to rupture gives the property known as the modulus of toughness u_T and denotes the energy per unit volume necessary to rupture the material.

The total strain energy stored in a material may be obtained by integrating the strain energy intensity over the volume of the material being stressed. When the state of stress in a body can be represented by a nonzero normal stress, the strain energy is

$$U = \int_V \frac{\sigma^2}{2E} dV \qquad (10\text{-}5)$$

where V is the volume of the body. The expressions for strain energy intensity and strain energy for shear loading are identical to that for axial loading if normal stress σ is replaced by shear stress τ and Young's modulus E is replaced by the shear modulus G.

The elastic strain energy for axial loading of a bar which has a uniform cross-sectional area A and is subjected to an axial force P over its entire length L is given by

$$U = \frac{P^2 L}{2 AE} \tag{10-7}$$

For a bar which has a uniform cross section of radius r and is subjected to a torsional load T_r over its entire length L, the strain energy is given by

$$U = \frac{T_r^2 L}{2 GJ} \tag{10-9}$$

The strain energies in a beam resulting from a bending moment M_r and from a shear force V_r are given by

$$U = \int_0^L \frac{M_r^2}{2 EI} dx \tag{10-11}$$

$$U = \int_0^L \frac{V_r^2}{2 GI^2} \left(\int_A \frac{Q^2}{t^2} dA \right) dx \tag{10-12}$$

The integral in the parentheses of Eq. 10-12 depends on the shape of the cross section of the beam. Also, since M_r and V_r are functions of x, the relationships between M_r and x and between V_r and x must be known before the integrals of Eqs. 10-11 and 10-12 can be evaluated. The strain energy due to transverse shear is of importance only in the case of very short, deep beams. For a short, deep beam (depth/length = 1/5), the error in neglecting the strain energy due to transverse shear is approximately 3 percent, whereas for a long, slender beam (depth/length = 1/10) the error is less than 1 percent.

When dynamic loads are applied to a structural member, the magnitude and distribution of stresses and strains depend on the velocity of propagation of the strain waves through the material of which the system is composed as well as on the usual parameters encountered previously. The maximum stresses and maximum deflection occur when all of the kinetic and potential energy of the load are converted to strain energy stored in the material. Assuming that the stresses are in the elastic range, the material behavior is the same as that for slowly applied loads, and energy losses during impact are negligible, the maximum normal stress and the maximum deflection for a suddenly applied axial force are double the values obtained when the force is slowly applied.

A tension test of an axially loaded member is easy to conduct, and the results, for many types of materials, are well known. When such a member fails, the failure occurs at a specific principal (axial) stress σ_f. At this point, the principal (axial) strain is $\epsilon_f = \sigma_f / E$, the maximum shearing stress is $\tau_f = \sigma_f / 2$, and the strain energy per unit volume of stressed material is $u = \sigma_f^2 / 2E$. For an element subjected to biaxial or triaxial loading, however, the situation is more complicated because the limits of normal stress, normal strain, shearing stress, and strain energy existing at failure for an axial load are not all reached simultaneously. In such cases, it becomes important to determine the best criterion for predicting failure. Several theories have been proposed for predicting failure of various types of material subjected to many combinations of loads. Unfortunately, none of the theories agrees with test data for all types of materials and combinations of loads.

The maximum-normal-stress theory (also called Rankine's theory) predicts failure of a specimen subjected to any combination of loads when the maximum normal stress at any point in the specimen reaches the axial failure stress as determined by an axial tensile or compressive test of the same material, $\sigma_{max} = \sigma_f$. The maximum-shear-stress theory (also called Coulomb's theory or Guest's theory) predicts failure of a specimen subjected to any combination of loads when the maximum shear stress at any point reaches the failure stress as determined by an axial tension or axial compression test of the same material, $\tau_{max} = \tau_f = \sigma_f/2$. For ductile materials, the shearing elastic limit, as determined from a torsion test (pure shear), is about $0.57\sigma_f$. Therefore, the maximum-shear-stress theory errs on the conservative side by being based on the limit obtained from an axial test.

The maximum-distortion-energy theory (also called the Huber-Hencky-von Mises theory) predicts failure of a specimen subjected to any combination of loads when the distortion component of the strain energy intensity of any portion of the stressed member reaches the failure value of the distortion component of the strain energy intensity as determined from an axial tension or compression test of the same material. This theory assumes that the portion of the strain energy producing volume change is ineffective in causing failure by yielding. The strain energy of distortion is most readily computed by determining the total strain energy of the stressed material and subtracting the strain energy corresponding to the volume change.

Failure of a ductile material is predicted most closely by the maximum-distortion-energy theory. When the two largest principal normal stresses have the same sign, the maximum-normal-stress and the maximum-shear-stress theories are identical and both theories are conservative. When the two largest principal stresses have opposite signs, the maximum-shear-stress theory is conservative while the maximum-normal-stress theory overpredicts failure stresses and is unsafe.

Failure of a brittle material is closely predicted by the Coulomb-Mohr theory. This theory gives conservative results.

REVIEW PROBLEMS

10-68* The 100-mm-diameter shaft shown in Fig. P10-68 is composed of aluminum alloy ($G = 28$ GPa) and steel ($G = 80$ GPa) sections that are rigidly connected. Determine

a. The elastic strain energy for each segment of the shaft.
b. The elastic strain energy for the complete shaft.

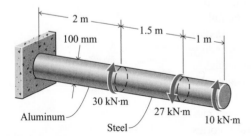

Figure P10-68

10-69* Axial loads are applied at sections A, B, C, and D of the steel ($E = 30,000$ ksi) bar shown in Fig. P10-69. If the bar has a cross-sectional area of 3 in.², determine

a. The elastic strain energy for each segment of the shaft.
b. The elastic strain energy for the complete shaft.

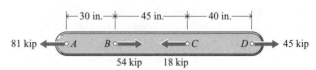

Figure P10-69

10-70 A solid circular steel shaft of diameter d and length L is subjected to a constant torque T. Compare the total elastic strain energy stored in this shaft with the total elastic strain energy stored in an axially loaded bar of the same size and at the same maximum tensile stress level.

10-71* A weight of 40 lb is dropped from a height of 3 ft onto the center of a small rigid platform as shown in Fig. P10-71. The two steel ($E = 30,000$ ksi) rods supporting the platform each have cross-sectional areas of 2.5 in.^{2} and are 8 ft long. Determine

a. The impact factor.
b. The maximum tensile stress developed in the rods.
c. The maximum deflection of the platform.

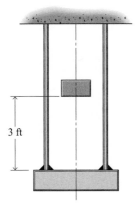

Figure P10-71

10-72 An S 305 × 74 structural steel section (see Appendix B) is used as a simply supported beam 3 m long. What weight falling on the center of the span from a height of 0.75 m will produce a maximum flexural stress of 120 MPa in the beam? The web of the beam is vertical.

10-73 When the 90-lb block B of Fig. P10-73 was dropped from a point 2 in. above the top of the coil spring C (modulus $= 1700$ lb/ft), point D on the steel ($E = 29,000$ ksi) cantilever beam A was observed to deflect 2.4 in. downward. Determine the impact factor.

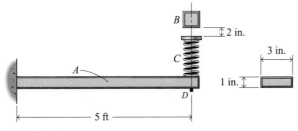

Figure P10-73

10-74* The bronze ($E = 80$ GPa) beam of Fig. P10-74 is 75 mm wide × 25 mm deep. Each of the supporting coil springs has a modulus of 10 kN/m. From what height should the block W, with a mass of 5 kg, be dropped in order to produce a total deflection at the center of the beam equal to four times the deflection produced by the same mass when it is slowly applied to the beam?

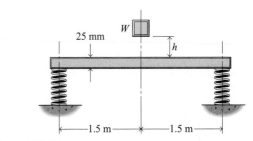

Figure P10-74

10-75* A vertical energy load is applied to the coil spring (modulus $= 200$ lb/in.) of the beam-spring system shown in Fig. P10-75. If the spring itself absorbs 720 lb · in. of energy, determine the energy absorbed by the steel ($E = 30,000$ ksi) beam.

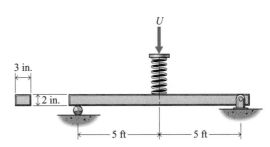

Figure P10-75

10-76 The frame shown in Fig. P10-76 is subjected to a load P of 100 kN. The material is ductile and has a proportional limit of 220 MPa in tension and compression. Determine the factor of safety with respect to failure by yielding according to

a. The maximum-shearing-stress theory of failure.
b. The maximum-distortion-energy theory of failure.

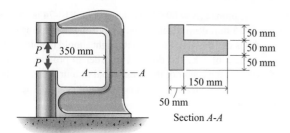

50 mm
50 mm
50 mm

150 mm

50 mm

Section *A-A*

Figure P10-76

10-77*   A material with a proportional limit of 36 ksi in tension and compression is subjected to a biaxial state of stress. Two of the principal stresses are $\sigma_{p1} = 20$ ksi (T) and $\sigma_{p3} = \sigma_z = 0$. Determine the factor of safety with respect to failure by yielding according to each of the theories of failure if

a. The principal stress $\sigma_{p2} = 12$ ksi tension.
b. The principal stress $\sigma_{p2} = 12$ ksi compression.

10-78   A state of plane stress at the critical point in a machine component is shown in Fig. P10-78. The material is brittle with $\sigma_{ut} = 68$ MPa and $\sigma_{uc} = 206$ MPa. Use the Coulomb-Mohr theory of failure to determine if the state of stress is safe.

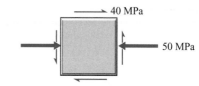

40 MPa

50 MPa

Figure P10-78

Appendix A
Second Moments of Area

A-1 INTRODUCTION

The centroid of an area is located by considering the first moment of the area about an axis. This computation requires evaluation of an integral of the form $\int_A x\, dA$. In the analysis of stresses and deflections in beams and shafts, an expression of the form $\int_A x^2\, dA$ is frequently encountered in which dA represents an element of area and x represents the distance from the element to some axis in, or perpendicular to, the plane of the area. An expression of the form $\int_A x^2\, dA$ is known as the second moment of the area. In the analysis of the angular motion of rigid bodies, an expression of the form $\int_m r^2\, dm$ is encountered, in which dm represents an element of mass and r represents the distance from the element to some axis. Euler[1] gave the name "moment of inertia" to expressions of the form $\int_m r^2\, dm$. Because of the similarity between the two types of integrals, both have become widely known as moments of inertia. In this text, the integrals involving areas will be referred to as "second moments of area." Methods used to determine second moments of area are discussed in this appendix.

A-2 SECOND MOMENT OF PLANE AREAS

The *second moment of an area* with respect to an axis will be denoted by the symbol I for an axis in the plane of the area and by the symbol J for an axis perpendicular to the plane of the area. The particular axis about which the second moment is taken will be denoted by subscripts. Thus, the second moments of the area A shown in Fig. A-1 with respect to x- and y-axes in the plane of the area are

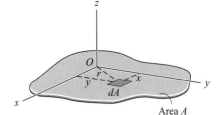

Figure A-1

$$I_x = \int_A y^2 dA \qquad \text{and} \qquad I_y = \int_A x^2 dA \qquad \text{(A-1)}$$

The quantities I_x and I_y are sometimes referred to as the rectangular second moments of the area A.

   Similarly, the second moment of the area A shown in Fig. A-1 with respect to a z-axis, which is perpendicular to the plane of the area at the origin O of the

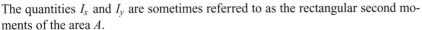

[1]Leonhard Euler (1707–1783), a noted Swiss mathematician and physicist.

xy-coordinate system, is

$$J_z = \int_A r^2 dA = \int_A (x^2 + y^2)\, dA \tag{A-2}$$

Thus,

$$J_z = \int_A x^2 dA + \int_A y^2 dA = I_y + I_x \tag{A-3}$$

The quantity J_z is known as the polar second moment of the area A.

The second moment of an area can be visualized as the sum of a number of terms, each consisting of an area multiplied by a distance squared. Thus, the dimensions of a second moment are a length raised to the fourth power (L^4). Common units are mm^4 and in.^{4} Also, the sign of each term summed to obtain the second moment is positive, because either a positive or negative distance squared is positive. Therefore, the second moment of an area is always positive.

A-2-1 Parallel Axis Theorem for Second Moments of Area

When the second moment of an area has been determined with respect to a given axis, the second moment with respect to a parallel axis can be obtained by means of the parallel axis theorem. If one of the axes (say the x-axis) passes through the centroid of the area, as shown in Fig. A-2, the second moment of the area about a parallel x'-axis located a distance d_x from the x-axis is

$$I_{x'} = \int_A (y + d_x)^2 dA = \int_A y^2 dA + 2d_x \int_A y\, dA + d_x^2 \int_A dA$$

since d_x is the same for every element of area dA. The first integral is the second moment I_x of the area with respect to the x-axis and the last integral is the total area A. Therefore,

$$I_{x'} = I_x + 2d_x \int_A y\, dA + d_x^2 A \tag{a}$$

The integral $\int_A y\, dA$ is the first moment of the area with respect to the x-axis. Since the x-axis passes through the centroid C of the area, the first moment is zero and Eq. (a) becomes

$$I_{x'} = I_{xC} + d_x^2 A \tag{A-4}$$

where I_{xC} is the second moment of the area with respect to the x-axis through the centroid and d_x is the distance between the x- and x'-axes. In a similar manner it can be shown that

$$J_{z'} = J_{zC} + \left(d_x^2 + d_y^2\right) A = J_{zC} + d_z^2 A \tag{A-5}$$

where J_{zC} is the polar second moment of the area with respect to the z-axis through the centroid and d_z is the distance between the z- and z'-axes.

The *parallel axis theorem* states that the second moment of an area with respect to any axis in the plane of the area is equal to the second moment of the area with respect to a parallel axis through the centroid of the area added to the

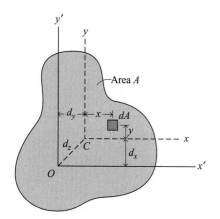

Figure A-2

product of the area and the square of the distance between the two axes (refer to Eqs. A-4 and A-5). The theorem also indicates that the second moment of an area with respect to an axis through the centroid of the area is less than that for any parallel axis because

$$I_{xC} = I_{x'} - d_x^2 A \tag{A-6}$$

As a point of caution, note that the parallel axis theorem (Eq. A-4) is valid only for transfers to or from a centroidal axis. That is, if x' and x'' are two parallel axes (neither of which passes through the centroid of an area), then

$$
\begin{aligned}
I_{x''} &= I_{xC} + d_{x2}^2 A = \left(I_{x'} - d_{x1}^2 A\right) + d_{x2}^2 A \\
&= I_{x'} + \left(d_{x2}^2 - d_{x1}^2\right) A \neq I_{x'} + (d_{x2} - d_{x1})^2 A
\end{aligned}
$$

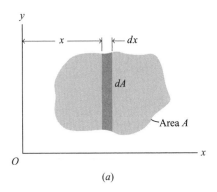

(a)

A-2-2 Second Moments of Areas by Integration

Rectangular and polar second moments of area were defined in Section A-2. When the second moment of a plane area with respect to a line is determined by using Eq. A-1 or A-2, it is possible to select the element of area dA in various ways and to express the area of the element in terms of either polar or Cartesian coordinates. In some cases, an element of area with dimensions $dA = dy\, dx$, as shown in Fig. A-1, may be required. This type of element has the slight advantage that it can be used for calculating both I_x and I_y, but it has the greater disadvantage of requiring double integration. Most problems can be solved with less work by choosing elements of the type shown in Figs. A-3a and A-3b. The following should be considered when selecting an element of area dA for a specific problem.

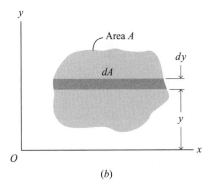

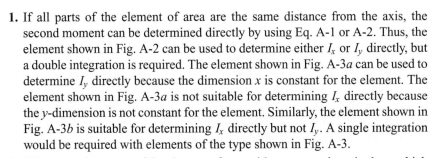

(b)

Figure A-3

1. If all parts of the element of area are the same distance from the axis, the second moment can be determined directly by using Eq. A-1 or A-2. Thus, the element shown in Fig. A-2 can be used to determine either I_x or I_y directly, but a double integration is required. The element shown in Fig. A-3a can be used to determine I_y directly because the dimension x is constant for the element. The element shown in Fig. A-3a is not suitable for determining I_x directly because the y-dimension is not constant for the element. Similarly, the element shown in Fig. A-3b is suitable for determining I_x directly but not I_y. A single integration would be required with elements of the type shown in Fig. A-3.

2. If the second moment of the element of area with respect to the axis about which the second moment of the area is to be found is known, the second moment of the area can be found by summing the second moments of the individual elements that make up the area. For example, if the second moment dI_x for the rectangular area dA in Fig. A-4 is known, the second moment I_x for the complete area A is simply $I_x = \int_A dI_x$.

3. If both the location of the centroid of the element and the second moment of the element about its centroidal axis parallel to the axis of interest for the complete area are known, the parallel axis theorem can often be used to simplify the solution of a problem. For example, consider the area shown in Fig. A-5. If both the distance d_x and the second moment dI_{xC} for the rectangular element dA are known, then by the parallel axis theorem $dI_x = dI_{xC} + d_x^2 dA$. The second moment for the complete area A is then simply $I_x = \int_A dI_x$.

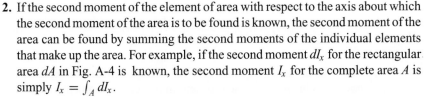

Figure A-4

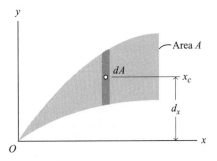

Figure A-5

From the previous discussion it is evident that either single or double integration may be required for the determination of second moments of area, depending on the element of area dA selected. When double integration is used, all parts of the element will be the same distance from the moment axis, and the second moment of the element can be written directly. Special care must be taken in establishing the limits for the two integrations to see that the correct area is included. If a strip element is selected, the second moment can usually be obtained by a single integration, but the element must be properly selected in order for its second moment about the reference axis to be either known or readily calculated by using the parallel axis theorem. The following examples illustrate the procedure for determining the second moments of areas by integration.

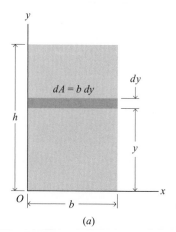

(a)

Figure A-6(a)

▶ The second moment I_x can be computed with a single integration if a thin strip element parallel to the x-axis is used.

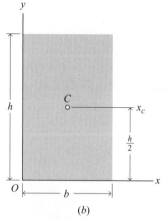

(b)

Figure A-6(b)

▶ If an area has an axis of symmetry, the centroid is located on that axis; if the area has two axes of symmetry, the centroid is located at the point of intersection of the two axes.

Example Problem A-1   Determine the second moment for the rectangle shown in Fig. A-6a with respect to

(a) The base of the rectangle.
(b) An axis through the centroid parallel to the base.
(c) An axis through the centroid normal to the area.

SOLUTION

(a) An element of area $dA = b\, dy$, as shown in Fig. A-6a, will be used. Since all parts of the element are located a distance y from the x-axis, Eq. A-1 can be used directly for the determination of the second moment I_x about the base of the rectangle. Thus,

$$I_x = \int_A y^2 dA = \int_0^h y^2\, b\, dy = \left[\frac{by^3}{3}\right]_0^h = \frac{bh^3}{3} \qquad \textbf{Ans.}$$

This result will be used frequently in later examples, when elements of the type shown in Fig. A-4 are used to determine second moments about the x-axis.

(b) The parallel axis theorem (Eq. A-6) will be used to determine the second moment I_{xC} about an axis that passes through the centroid of the rectangle (see Fig. A-6b) and is parallel to the base. Thus,

$$I_{xC} = I_x - d_x^2 A = \frac{bh^3}{3} - \left(\frac{h}{2}\right)^2 (bh) = \frac{bh^3}{12} \qquad \textbf{Ans.}$$

This result will be used frequently in later examples when elements of the type shown in Fig. A-5 are used to determine second moments about the x-axis.

(c) The second moment I_{yC} for the rectangle can be determined in an identical manner. It can also be obtained from the preceding solution by interchanging b and h; that is,

$$I_{yC} = \frac{hb^3}{12}$$

The polar second moment J_{zC} about the z-axis through the centroid of the rectangle is given by Eq. A-3 as

$$J_{zC} = I_{xC} + I_{yC} = \frac{bh^3}{12} + \frac{hb^3}{12} = \frac{bh}{12}(h^2 + b^2) \qquad \text{Ans.}$$

Example Problem A-2 Determine the second moment of area for the circle shown in Fig. A-7 with respect to a diameter of the circle.

SOLUTION
Polar coordinates are convenient for this problem. An element of area $dA = \rho\, d\theta\, d\rho$, as shown in Fig. A-7, will be used. If the x-axis is selected as the diameter about which the second moment of area is to be determined, then $y = \rho \sin \theta$. Application of Eq. A-1 yields

$$I_x = \int_A y^2\, dA = \int_0^{2\pi} \int_0^R (\rho \sin \theta)^2 (\rho\, d\theta\, d\rho)$$

$$= \int_0^{2\pi} \int_0^R \rho^3 \sin^2 \theta\, d\rho\, d\theta = \frac{R^4}{4}\left[\frac{\theta}{2} - \frac{\sin 2\theta}{4}\right]_0^{2\pi} = \frac{\pi R^4}{4} \qquad \text{Ans.}$$

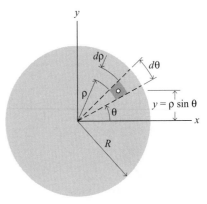

Figure A-7

▶ Polar coordinates are usually more efficient when circular boundaries are involved.

Example Problem A-3 Determine the polar second moment of area for the circle shown in Fig. A-8 with respect to

(a) An axis through the center of the circle and normal to the plane of the area.
(b) An axis through the edge of the circle and normal to the plane of the area.

SOLUTION

(a) Polar coordinates are convenient for this problem. An element of area $dA = 2\pi\rho\, d\rho$, as shown in Fig. A-8, will be used. Since all parts of the element are located a constant distance ρ from the center of the circle, Eq. A-2 can be used directly for the determination of the polar second moment J_z about an axis through the center of the circle and normal to the plane of the area. Thus,

$$J_z = \int_A r^2\, dA = \int_0^R \rho^2 (2\pi\rho\, d\rho) = \int_0^R 2\pi\rho^3\, d\rho = \frac{\pi R^4}{2} \qquad \text{Ans.}$$

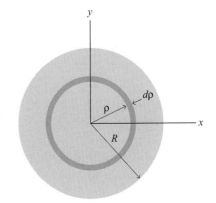

Figure A-8

▶ The second moment J_z can be computed with a single integration if a thin annular element at a constant distance ρ from the z-axis is used.

This result could have been obtained from the solution of Example Problem A-2 and use of Eq. A-3. Thus,

$$J_z = I_x + I_y = \frac{\pi R^4}{4} + \frac{\pi R^4}{4} = \frac{\pi R^4}{2}$$

(b) The parallel axis theorem (Eq. A-5) will be used to determine the polar second moment $J_{z'}$ about an axis that passes through the edge of the circle and is normal to the plane of the area. Thus,

$$J_{z'} = J_{zC} + d_z^2 A = \frac{\pi R^4}{2} + R^2(\pi R^2) = \frac{3\pi R^4}{2}$$ **Ans.**

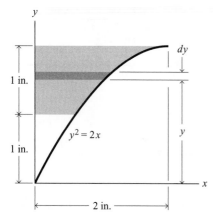

Figure A-9

▶ The second moment I_x can be computed with a single integration if a thin strip element parallel to the x-axis is used.

Example Problem A-4   Determine the second moment of area for the shaded region of Fig. A-9 with respect to

(a) The x-axis.

(b) An axis through the origin of the xy-coordinate system and normal to the plane of the area.

SOLUTION

(a) An element of area $dA = x\, dy = (y^2/2)\, dy$, as shown in Fig. A-3$b$, will be used. Since all parts of the element are located a distance y from the x-axis, Eq. A-1 can be used directly for the determination of the second moment I_x about the x-axis. Thus,

$$I_x = \int_A y^2\, dA = \int_A y^2\left(\frac{y^2}{2}\right) dy = \int_1^2 \frac{y^4}{2}\, dy = \left[\frac{y^5}{10}\right]_1^2 = 3.10\ \text{in.}^4$$ **Ans.**

(b) The same element of area can be used to obtain the second moment I_y if the result of Example Problem A-1 is used as the known value for dI_y. Thus,

$$dI_y = \frac{bh^3}{3} = \frac{dy(x)^3}{3} = \frac{dy(y^2/2)^3}{3} = \frac{y^6}{24}\, dy$$

Summing all such elements yields

$$I_y = \int_A dI_y = \int_1^2 \frac{y^6}{24}\, dy = \left[\frac{y^7}{168}\right]_1^2 = \frac{127}{168} = 0.756\ \text{in.}^4$$

Once I_x and I_y are known for the area, the polar second moment for an axis through the origin of the xy-coordinate system and normal to the plane of the area is obtained by using Eq. A-3. Thus,

$$J_z = I_x + I_y = 3.10 + 0.756 = 3.856 \cong 3.86\ \text{in.}^4$$ **Ans.**

A-2-3   Radius of Gyration of Areas   Since the second moment of an

area has the dimensions of length to the fourth power, it can be expressed as the area A multiplied by a length k squared. Thus, from Eqs. A-1 and A-2,

$$I_x = \int_A y^2 \, dA = Ak_x^2 \qquad k_x = \sqrt{\frac{I_x}{A}}$$

$$I_y = \int_A x^2 \, dA = Ak_y^2 \qquad k_y = \sqrt{\frac{I_y}{A}} \qquad \text{(A-7)}$$

$$J_z = \int_A r^2 \, dA = Ak_z^2 \qquad k_z = \sqrt{\frac{J_z}{A}}$$

and from Eq. A-3,

$$k_z^2 = k_x^2 + k_y^2 \qquad \text{(A-8)}$$

The distance k is called the *radius of gyration*. The subscript denotes the axis about which the second moment of area is taken. The radius of gyration of an area with respect to an axis can be visualized as the distance from the axis to the point where the area could be concentrated in order to produce the same second moment of area with respect to the axis, as does the actual area.

The parallel axis theorem for second moments of area was discussed early in Section A-2. A corresponding relation exists between the radii of gyration of the area with respect to two parallel axes, one of which passes through the centroid of the area. Thus, from Eqs. A-4 and A-7,

$$k_{x'}^2 = k_{xC}^2 + d_x^2$$

and

$$k_{y'}^2 = k_{yC}^2 + d_y^2 \qquad \text{(A-9)}$$

Similarly for polar second moments of area and radii of gyration,

$$k_{z'}^2 = k_{zC}^2 + (d_x^2 + d_y^2) = k_{zC}^2 + d_z^2 \qquad \text{(A-10)}$$

Solving Eqs. A-9 and A-10 for rectangular and polar radii of gyration for arbitrary and centroidal x-, y-, and z-axes yields

$$k_{x'} = \sqrt{k_{xC}^2 + d_x^2} \qquad k_{xC} = \sqrt{k_{x'}^2 - d_x^2}$$

$$k_{y'} = \sqrt{k_{yC}^2 + d_y^2} \qquad k_{yC} = \sqrt{k_{y'}^2 - d_y^2} \qquad \text{(A-11)}$$

$$k_{z'} = \sqrt{k_{zC}^2 + d_z^2} \qquad k_{zC} = \sqrt{k_{z'}^2 - d_z^2}$$

The following examples illustrate the concepts discussed in this section.

Example Problem A-5   For the shaded area shown in Fig. A-10a, determine

(a) The radii of gyration k_x, k_y, and k_z.
(b) The radius of gyration for an axis passing through the centroid and parallel to the y-axis.

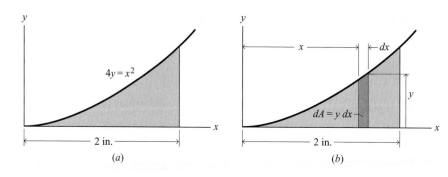

Figure A-10

SOLUTION

(a) The quantities required for the determination of k_x, k_y, and k_z are the area A and the second moments I_x and I_y. Since none of these quantities are readily available from known solutions, they will be determined by integration using the element of area shown in Fig. A-10b. For area A,

$$A = \int_A y \, dx = \int_0^2 \frac{x^2}{4} dx = \left[\frac{x^3}{12} \right]_0^2 = \frac{2}{3} \text{ in.}^2$$

▶ The second moment I_x can be computed with a single integration if a thin vertical strip element and the known results for a rectangle are used.

For the second moment I_x, the results of Example Problem A-1 can be used. In Example Problem A-1 it was shown that the second moment I about the base of a rectangle having a base b and a height h was $I = bh^3/3$. Thus, for the shaded element of area dA of Fig. A-10b, which has a base dx and a height y,

$$dI_x = \frac{1}{3}(dx)(y)^3 = \frac{1}{3}(dx)(x^2/4)^3 = \frac{x^6}{192} dx$$

and

$$I_x = \int_A dI_x = \int_0^2 \frac{x^6}{192} dx = \left[\frac{x^7}{1344} \right]_0^2 = \frac{2}{21} \text{ in.}^4$$

For the second moment I_y,

$$I_y = \int_A x^2 \, dA = \int_A x^2 y \, dx = \int_0^2 x^2 \left(\frac{x^2}{4} \right) dx = \int_0^2 \frac{x^4}{4} dx = \left[\frac{x^5}{20} \right]_0^2 = \frac{8}{5} \text{ in.}^4$$

Once A, I_x, and I_y are known, the radii of gyration k_x and k_y are obtained from Eqs. A-7. Thus,

$$k_x = \left[\frac{I_x}{A} \right]^{1/2} = \left[\frac{2/21}{2/3} \right]^{1/2} = 0.3780 = 0.378 \text{ in.} \qquad \textbf{Ans.}$$

$$k_y = \left[\frac{I_y}{A} \right]^{1/2} = \left[\frac{8/5}{2/3} \right]^{1/2} = 1.5492 \cong 1.549 \text{ in.} \qquad \textbf{Ans.}$$

The polar radius of gyration k_z obtained from k_x and k_y by using Eq. A-8 is

$$k_z = \sqrt{k_x^2 + k_y^2} = \sqrt{(0.3780)^2 + (1.5492)^2} = 1.595 \text{ in.} \qquad \textbf{Ans.}$$

(b) In order to determine the radius of gyration k_{yC}, the distance between the y-axis and the centroid of the area must be determined. Thus,

$$Ax_C = \int_A x \, dA = \int_0^2 x\left(\frac{x^2}{4}\right) dx = \int_0^2 \frac{x^3}{4} dx = \left[\frac{x^4}{16}\right]_0^2 = 1.000 \text{ in.}^3$$

from which

$$d_y = x_C = \frac{\int_A x \, dA}{A} = \frac{1.000}{2/3} = 1.500 \text{ in.}$$

The radius of gyration k_{yC} is then obtained by using Eqs. A-11. Thus,

$$k_{yC} = \sqrt{k_y^2 - d_y^2} = \sqrt{(1.5492)^2 - (1.500)^2} = 0.387 \text{ in.} \qquad \textbf{Ans.}$$

Example Problem A-6   Determine the radii of gyration of the area of the isosceles triangle shown in Fig. A-11a with respect to

(a) Horizontal and vertical centroidal axes.
(b) The x'- and y'-axes shown on the figure.

SOLUTION

(a) The quantities required for the determination k_{xC} and k_{yC} are the area A and the second moments I_{xC} and I_{yC}. The area A for the triangle is

$$A = \frac{1}{2}bh = \frac{1}{2}(50)(45) = 1125 \text{ mm}^2$$

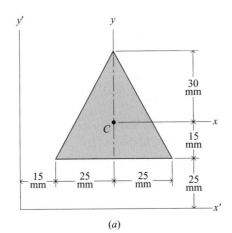

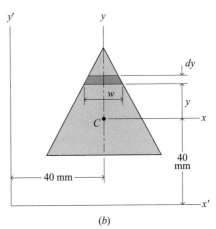

(a)   (b)

Figure A-11

▶ The second moment I_x can be computed with a single integration if a thin strip element parallel to the x-axis is used.

▶ The width of the element can be expressed as a function of y by observing that the entire triangle is similar to the triangle that lies above the shaded element.

The second moments I_{xC} and I_{yC} will be determined, by integration, using the element of area $dA = w\,dy$ shown in Fig. A-11b. From similar triangles,

$$\frac{w}{30-y} = \frac{50}{45} \qquad \text{from which} \qquad w = \frac{50}{45}(30-y)$$

$$dA = w\,dy = \frac{50}{45}(30-y)\,dy$$

Since all parts of the element $dA = w\,dy$ are located a distance y from the x_C-axis, the definition of a second moment can be used to determine I_{xC}. Thus,

$$I_{xC} = \int_A y^2\,dA = \frac{50}{45}\int_{-15}^{+30}(30-y)y^2\,dy$$

$$= \frac{50}{45}\left[\left(\frac{30y^3}{3} - \frac{y^4}{4}\right)\right]_{-15}^{+30} = 126{,}563 \text{ mm}^4$$

For the second moment I_{yC}, the results of Example Problem A-1 can be used. In Example Problem A-1 it was shown that the second moment I about a centroidal axis parallel to the base of a rectangle having a base b and a height h was $I = bh^3/12$. Thus, for the shaded element of Fig. A-11b, which has a base dy and a height w,

$$dI_{yC} = \frac{1}{12}(dy)(w)^3 = \frac{1}{12}(dy)\left[\frac{50}{45}(30-y)\right]^3 = \frac{1}{12}\left[\frac{50}{45}(30-y)\right]^3 dy$$

from which

$$I_{yC} = \int_A dI_{yC} = \int_{-15}^{+30}\frac{1}{12}\left[\frac{50}{45}(30-y)\right]^3 dy$$

$$= -\frac{1}{48}\left(\frac{45}{50}\right)\left\{\left[\frac{50}{45}(30-y)\right]^4\right\}_{-15}^{+30} = 117{,}188 \text{ mm}^4$$

From the definition of a radius of gyration,

$$k_{xC} = \sqrt{\frac{I_{xC}}{A}} = \sqrt{\frac{126{,}563}{1125}} = 10.607 \text{ mm} \cong 10.61 \text{ mm} \qquad \textbf{Ans.}$$

$$k_{yC} = \sqrt{\frac{I_{yC}}{A}} = \sqrt{\frac{117{,}188}{1125}} = 10.206 \text{ mm} \cong 10.21 \text{ mm} \qquad \textbf{Ans.}$$

(b) The radii of gyration $k_{x'}$ and $k_{y'}$ can be determined by using the parallel axis theorem for radii of gyration. Thus,

$$k_{x'} = \sqrt{k_{xC}^2 + d_x^2} = \sqrt{(10.607)^2 + (40)^2} = 41.4 \text{ mm} \qquad \textbf{Ans.}$$

$$k_{y'} = \sqrt{k_{yC}^2 + d_y^2} = \sqrt{(10.206)^2 + (40)^2} = 41.3 \text{ mm} \qquad \textbf{Ans.}$$

A-2-4 Second Moments of Composite Areas

The second moments I_x, I_y, and J_z of an area A with respect to any set of x-, y-, and z-coordinate axes were defined as

$$I_x = \int_A y^2 \, dA \qquad I_y = \int_A x^2 \, dA \qquad J_z = \int_A r^2 \, dA$$

Frequently in engineering practice, an irregular area A will be encountered which can be broken up into a series of simple areas $A_1, A_2, A_3, \ldots, A_n$ for which the integrals have been evaluated and tabulated. The second moment of the irregular area, the composite area, with respect to any axis is equal to the sum of the second moments of the separate parts of the area with respect to the specified axis. For example,

$$
\begin{aligned}
I_x &= \int_A y^2 \, dA \\
&= \int_{A_1} y^2 \, dA_1 + \int_{A_2} y^2 \, dA_2 + \int_{A_3} y^2 \, dA_3 + \cdots + \int_{A_n} y^2 \, dA_n \\
&= I_{x1} + I_{x2} + I_{x3} + \cdots + I_{xn}
\end{aligned}
$$

When an area such as a hole is removed from a larger area, its second moment must be subtracted from the second moment of the larger area to obtain the resulting second moment. Thus, for the case of a square plate with a hole,

$$I_{\blacksquare} = I_{\boxdot} + I_{\square}$$

Therefore,

$$I_{\boxdot} = I_{\blacksquare} - I_{\square}$$

Table A-1 contains a listing of the values of the integrals for frequently encountered shapes such as rectangles, triangles, circles, and semicircles. Tables listing second

Table A-1 Second Moments of Plane Areas

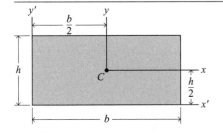

	$I_x = \dfrac{bh^3}{12}$	$I_{x'} = \dfrac{bh^3}{3}$ $A = bh$
	$I_x = \dfrac{bh^3}{36}$	$I_{x'} = \dfrac{bh^3}{12}$ $A = \dfrac{1}{2}bh$

Table A-1   Second Moments of Plane Areas (*continued*)

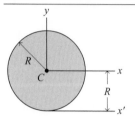

$$I_x = \frac{\pi R^4}{4}$$

$$I_{x'} = \frac{5\pi R^4}{4}$$

$$A = \pi R^2$$

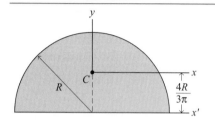

$$I_x = \frac{\pi R^4}{8} - \frac{8R^4}{9\pi}$$

$$I_y = \frac{\pi R^4}{8}$$

$$I_{x'} = \frac{\pi R^4}{8}$$

$$A = \frac{1}{2}\pi R^2$$

$$I_x = \frac{\pi R^4}{16} - \frac{4R^4}{9\pi}$$

$$I_{x'} = \frac{\pi R^4}{16}$$

$$A = \frac{1}{4}\pi R^2$$

$$I_x = \frac{R^4}{4}\left(\theta - \frac{1}{2}\sin 2\theta\right)$$

$$I_y = \frac{R^4}{4}\left(\theta + \frac{1}{2}\sin 2\theta\right)$$

$$x_C = \frac{2}{3}\frac{R\sin\theta}{\theta}$$

$$A = \theta R^2$$

---

moments of area and other properties for the cross sections of common structural shapes are found in engineering handbooks and in data books prepared by industrial organizations such as the American Institute of Steel Construction. An abbreviated listing is also included in Appendix B.

In some instances the second moments I_{xC}, I_{yC}, and I_{zC} of a composite shape with respect to centroidal x-, y-, and z-axes of the composite may be required. These quantities can be determined by first evaluating the second moments $I_{x'}$, $I_{y'}$, and $I_{z'}$ of the composite with respect to any convenient set of parallel x'-, y'-, and z'-axes and then transferring these second moments to the centroidal axes by using the parallel axis theorem.

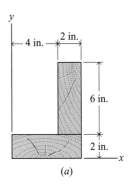

Example Problem A-7   A beam having the cross section shown in Fig. A-12a is constructed by gluing a 2 × 6-in. wooden plank 10 ft long to a second 2 × 6-in. wooden plank also 10 ft long. Determine the second moment of the cross-sectional area with respect to

(a) The x-axis.

(b) The y-axis.

(c) The x_C-axis, which passes through the centroid of the area and is parallel to the x-axis.

SOLUTION

(a) As shown in Fig. A-12b, the cross-sectional area can be divided into two simple rectangles. Since both rectangles have an edge along the x-axis, their second moments of area are just $bh^3/3$. Therefore, the second moment of area of the entire area with respect to the x-axis is

$$I_x = I_{x1} + I_{x2} = \frac{1}{3}(4)(2)^3 + \frac{1}{3}(2)(8)^3 = 352 \text{ in.}^4 \qquad \textbf{Ans.}$$

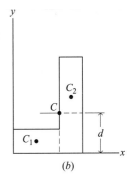

Figure A-12

(b) Using the same division of areas as in part (a), the first area has an edge along the y-axis and its second moment of area is given by $bh^3/3$. However, the parallel axis theorem is needed for the second rectangle, since neither the centroid of the rectangle nor either edge is along the y-axis. Therefore, the second moment of area of the entire area with respect to the y-axis is

$$I_y = I_{y1} + I_{y2}$$

$$= \frac{1}{3}(2)(4)^3 + \left[\frac{1}{12}(8)(2)^3 + 16(5)^2\right] = 448 \text{ in.}^4 \qquad \textbf{Ans.}$$

▶ The second moment of a composite area with respect to a given axis equals the sum of the second moments of the separate parts of the area with respect to the same axis.

(c) The centroid will be located using the same division of areas as above:

$$d(8 + 16) = 1(8) + 4(16) \qquad d = 3 \text{ in.}$$

Then, using the parallel axis theorem for both rectangles, the second moment of area of the entire area with respect to the x_C-axis is

$$I_{xC} = I_{xC1} + I_{xC2}$$

$$= \left[\frac{1}{12}(4)(2)^3 + 8(3 - 1)^2\right] + \left[\frac{1}{12}(2)(8)^3 + 16(4 - 3)^2\right] = 136.0 \text{ in.}^4$$

$$\textbf{Ans.}$$

Note that, since the x_C-axis passes through the centroid of the entire area, the second moments of area I_x and I_{xC} are related by the parallel axis theorem

$$I_x = I_{xC} + Ad^2$$
$$= 136.0 + 24(3)^2 = 352 \text{ in.}^4$$

Example Problem A-8   Determine the second moment of the shaded area shown in Fig. A-13a with respect to

(a) The x-axis.

(b) The y-axis.

(c) An axis through the origin O of the xy-coordinate system and normal to the plane of the area.

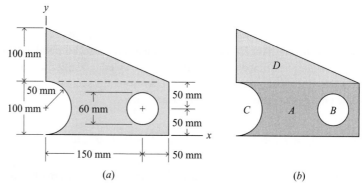

(a)   (b)

Figure A-13

SOLUTION

As shown in Fig. A-13b, the shaded area can be divided into a 100×200-mm rectangle (A), with a 60-mm-diameter circle (B) and a 100-mm-diameter half-circle (C) removed, and a 100×200-mm triangle (D). The second moments for these areas, with respect to the x- and y-axes, can be obtained by using information from Table A-1, as follows:

(a) For the rectangle (shape A),

$$I_{x1} = \frac{bh^3}{3} = \frac{200(100^3)}{3} = 66.667(10^6)\,\text{mm}^4$$

For the circle (shape B),

$$I_{x2} = I_{xC} + d_x^2 A = \frac{\pi R^4}{4} + d_x^2(\pi R^2)$$

$$= \frac{\pi(30^4)}{4} + (50^2)(\pi)(30^2) = 7.705(10^6)\,\text{mm}^4$$

For the half-circle (shape C),

$$I_{x3} = I_{xC} + d_x^2 A = \frac{\pi R^4}{8} + d_x^2\left(\frac{\pi R^2}{2}\right)$$

$$= \frac{\pi(50^4)}{8} + (50)^2\left[\frac{\pi(50)^2}{2}\right] = 12.272(10^6)\,\text{mm}^4$$

For the triangle (shape D),

$$I_{x4} = I_{xC} + d_x^2 A$$

$$= \frac{bh^3}{36} + d_x^2 \left(\frac{bh}{2}\right)$$

$$= \frac{200(100^3)}{36} + \left(100 + \frac{100}{3}\right)^2 \left[\frac{200(100)}{2}\right] = 183.333(10^6)\,\text{mm}^4$$

For the composite area,

$$I_x = I_{x1} - I_{x2} - I_{x3} + I_{x4}$$
$$= 66.667(10^6) - 7.705(10^6) - 12.272(10^6) + 183.333(10^6)$$
$$= 230.023(10^6) \cong 230(10^6)\,\text{mm}^4 \qquad \textbf{Ans.}$$

▶ When a hole is present in a larger area, its second moment must be subtracted from the second moment of the larger area in order to obtain the required second moment.

(b) For the rectangle (shape A),

$$I_{y1} = \frac{b^3 h}{3} = \frac{200^3(100)}{3} = 266.667(10^6)\,\text{mm}^4$$

For the circle (shape B),

$$I_{y2} = I_{yC} + d_y^2 A$$

$$= \frac{\pi R^4}{4} + d_y^2 (\pi R^2)$$

$$= \frac{\pi(30^4)}{4} + (150^2)(\pi)(30^2) = 64.253(10^6)\,\text{mm}^4$$

For the half-circle (shape C),

$$I_{y3} = \frac{\pi R^4}{8} = \frac{\pi(50^4)}{8} = 2.454(10^6)\,\text{mm}^4$$

For the triangle (shape D),

$$I_{y4} = \frac{bh^3}{12} = \frac{100(200^3)}{12} = 66.667(10^6)\,\text{mm}^4$$

For the composite area,

$$I_y = I_{y1} - I_{y2} - I_{y3} + I_{y4}$$
$$= 266.667(10^6) - 64.253(10^6) - 2.454(10^6) + 66.667(10^6)$$
$$= 266.627(10^6) \cong 267(10^6)\,\text{mm}^4 \qquad \textbf{Ans.}$$

(c) For the composite area,

$$J_z = I_x + I_y$$
$$= 230.023(10^6) + 266.627(10^6)$$
$$= 496.650(10^6) \cong 497(10^6)\,\text{mm}^4 \qquad \textbf{Ans.}$$

Example Problem A-9   A column with the cross section shown in Fig. A-14a is constructed from a W24 × 84 wide-flange section and a C12 × 30 channel section. Determine the second moments and radii of gyration of the cross-sectional area with respect to horizontal and vertical axes through the centroid of the cross section.

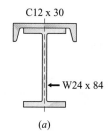

C12 x 30

W24 x 84

(a)

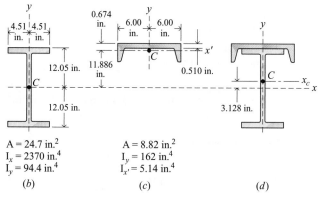

A = 24.7 in.^{2}
I$_x$ = 2370 in.^{4}
I$_y$ = 94.4 in.^{4}

(b)

A = 8.82 in.^{2}
I$_y$ = 162 in.^{4}
I$_{x'}$ = 5.14 in.^{4}

(c)

(d)

Figure A-14

SOLUTION
Properties and dimensions for the structural shapes can be obtained from Appendix B. The properties and dimensions for the wide-flange section are shown in Fig. A-14b. The properties and dimensions for the channel section are shown in Fig. A-14c. In Figs. A-14b and c, the x-axis passes through the centroid of the wide-flange section and a parallel x'-axis passes through the centroid of the channel. A centroidal x_C-axis for the composite section (see Fig. A-14d) can be located by using the principle of moments as applied to areas. The total area A_T for the composite section is

$$A_T = A_{WF} + A_{CH} = 24.7 + 8.82 = 33.52\,\text{in.}^2$$

The moment of the composite area about the x-axis is

$$A_T(y_C)_T = A_{WF}(y_C)_{WF} + A_{CH}(y_C)_{CH} = 24.7(0) + 8.82(11.886) = 104.835 \text{ in.}^3$$

▶ The principle of moments can be used to locate the centroid of any composite body if first moments of the individual parts are known or can be determined.

The distance $(y_C)_T$ from the x-axis to the centroid of the composite section is

$$(y_C)_T = \frac{104.835}{33.52} = 3.128 \text{ in.}$$

The second moment $(I_{xC})_{WF}$ for the wide-flange section about the centroidal x_C-axis of the composite section is determined by using the parallel axis theorem. Thus,

▶ The parallel axis theorem for second moments is valid only for transfers to or from a parallel axis through the centroid of the area.

$$(I_{xC})_{WF} = (I_x)_{WF} + (y_C)_{WF}^2 A_{WF} = 2370 + (3.128)^2(24.7) = 2611.7 \text{ in.}^4$$

Similarly, the second moment $(I_{xC})_{CH}$ for the channel about the centroidal x_C-axis of the composite section is

$$\begin{aligned}(I_{xC})_{CH} &= (I_{x'})_{CH} + [(y_C)_{CH} - (y_C)_T]^2 A_{CH} \\ &= 5.14 + (11.886 - 3.128)^2(8.82) = 681.7 \text{ in.}^4\end{aligned}$$

For the composite area,

$$\begin{aligned}(I_{xC})_T &= (I_{xC})_{WF} + (I_{xC})_{CH} \\ &= 2611.7 + 681.7 = 3293.4 \cong 3290 \text{ in.}^4\end{aligned} \qquad \textbf{Ans.}$$

The y-axis passes through the centroid of both areas; therefore, the second moment $(I_{yC})_T$ for the composite section is

$$\begin{aligned}(I_{yC})_T &= (I_{yC})_{WF} + (I_{yC})_{CH} \\ &= 94.4 + 162 = 256.4 \cong 256 \text{ in.}^4\end{aligned} \qquad \textbf{Ans.}$$

The radius of gyration about the x_C-axis for the composite section is

$$(k_{xC})_T = \left[\frac{(I_{xC})_T}{A_T}\right]^{1/2} = \left(\frac{3293.4}{33.52}\right)^{1/2} = 9.912 \cong 9.91 \text{ in.} \qquad \textbf{Ans.}$$

The radius of gyration about the y_C-axis for the composite section is

$$(k_{yC})_T = \left[\frac{(I_{yC})_T}{A_T}\right]^{1/2} = \left(\frac{256.4}{33.52}\right)^{1/2} = 2.766 \cong 2.77 \text{ in.} \qquad \textbf{Ans.}$$

A-2-5 Mixed Second Moments of Areas

The mixed second moment (commonly called the area product of inertia) dI_{xy} of the element of area dA shown in Fig. A-15 with respect to the x- and y-axes is defined as the product of the two coordinates of the element multiplied by the area of the element; thus

$$dI_{xy} = xy \, dA$$

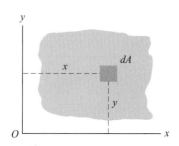

Figure A-15

The mixed second moment (area product of inertia) of the total area A about the x- and y-axes is the sum of the mixed second moments of the elements of the area; thus,

$$I_{xy} = \int_A dI_{xy} = \int_A xy\, dA \qquad \text{(A-12)}$$

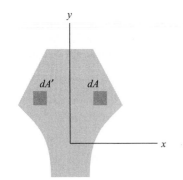

The dimensions of the mixed second moment are the same as the dimensions for the rectangular or polar second moments, but since the product xy can be either positive or negative, the mixed second moment can be positive, negative, or zero. Recall that rectangular or polar second moments are always positive.

The mixed second moment of an area with respect to any two orthogonal axes is zero when either of the axes is an axis of symmetry. This statement can be demonstrated by means of Fig. A-16, which is symmetrical with respect to the y-axis. For every element of area dA on one side of the axis of symmetry, there is a corresponding element of area dA' on the opposite side of the axis such that the mixed second moments of dA and dA' will be equal in magnitude but opposite in sign. Thus, they will cancel each other in the summation and the resulting mixed second moment for the total area will be zero.

Figure A-16

The parallel axis theorem for mixed second moments can be derived from Fig. A-17 in which the x- and y-axes pass through the centroid C of the area and are parallel to the x'- and y'-axes. The mixed second moment with respect to the x'- and y'-axes is

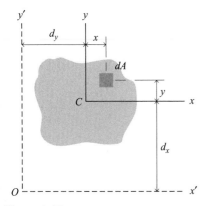

$$I_{x'y'} = \int_A x'y'\, dA = \int_A (d_y + x)(d_x + y)\, dA$$

$$= d_x d_y \int_A dA + d_y \int_A y\, dA + d_x \int_A x\, dA + \int_A xy\, dA$$

since d_x and d_y are the same for every element of area dA. The second and third integrals in the preceding equation are zero since x and y are centroidal axes. The last integral is the mixed second moment with respect to the centroidal axes. Consequently, the mixed second moment about a pair of axes parallel to a pair of centroidal axes is

Figure A-17

$$I_{x'y'} = I_{xyC} + d_x d_y A \qquad \text{(A-13)}$$

where the subscript C indicates that the x- and y-axes are centroidal axes. The parallel axis theorem for mixed second moments (area products of inertia) can be stated as follows; The mixed second moment of an area with respect to any two perpendicular axes x and y in the plane of the area is equal to the mixed second moment of the area with respect to a pair of centroidal axes parallel to the x- and y-axes added to the product of the area and the two centroidal distances from the x- and y-axes. The parallel axis theorem for mixed second moments is used most frequently in determining mixed second moments for composite areas. Values from Eq. A-12 for some of the shapes commonly used in these calculations are listed in Table A-2.

The determination of the mixed second moment (area product of inertia) is illustrated in the next two examples.

Table A-2   Mixed Second Moments of Plane Areas

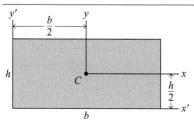

$$I_{xy} = 0 \qquad\qquad I_{x'y'} = \frac{b^2 h^2}{4}$$

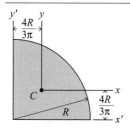

$$I_{xy} = -\frac{b^2 h^2}{72} \qquad\qquad I_{x'y'} = \frac{b^2 h^2}{24}$$

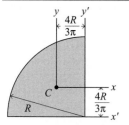

$$I_{xy} = \frac{b^2 h^2}{72} \qquad\qquad I_{x'y'} = -\frac{b^2 h^2}{24}$$

$$I_{xy} = \frac{(9\pi - 32)R^4}{72\pi} \qquad\qquad I_{x'y'} = \frac{R^4}{8}$$

$$I_{xy} = -\frac{(9\pi - 32)R^4}{72\pi} \qquad\qquad I_{x'y'} = -\frac{R^4}{8}$$

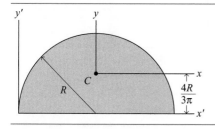

$$I_{xy} = 0 \qquad\qquad I_{x'y'} = \frac{2R^4}{3}$$

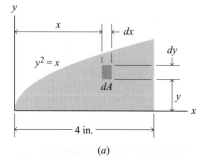

(*a*)

Figure A-18(a)

▶ The mixed second moment of an area with respect to any two orthogonal axes is zero when either of the axes is an axis of symmetry.

Example Problem A-10 Determine the mixed second moment (area product of inertia) of the shaded area shown in Fig. A-18*a* with respect to the *x*- and *y*-axes by using

(a) Double integration.

(b) The parallel axis theorem and single integration.

SOLUTION

(a) The mixed second moment (area product of inertia) dI_{xy} of the element of area $dA = dy\,dx$ shown in Fig. A-18*a* with respect to the *x*- and *y*-axes is defined as $dI_{xy} = xy\,dA$. Therefore,

$$I_{xy} = \int_A xy\,dA = \int_0^4 \int_0^{\sqrt{x}} xy\,dy\,dx$$

$$= \int_0^4 \left[\frac{y^2}{2}\right]_0^{\sqrt{x}} x\,dx$$

$$= \int_0^4 \frac{x^2}{2}dx = \left[\frac{x^3}{6}\right]_0^4 = 10.67\text{ in.}^4 \qquad \textbf{Ans.}$$

(b) The mixed second moment (area product of inertia) dI_{xy} of the element of area $dA = y\,dx$ shown in Fig. A-18*b* with respect to axes through the centroid of the element parallel to the *x*- and *y*-axes is zero. Thus, the mixed second moment of the element with respect to the *x*- and *y*-axes (see Fig. A-18*c*) is

$$dI_{xy} = dI_{xyC} + d_xd_y\,dA = 0 + \frac{y}{2}x(y\,dx) = \frac{x^2}{2}dx$$

Therefore,

$$I_{xy} = \int_A dI_{xy} = \int_0^4 \frac{x^2}{2}dx = \left[\frac{x^3}{6}\right]_0^4 = 10.67\text{ in.}^4 \qquad \textbf{Ans.}$$

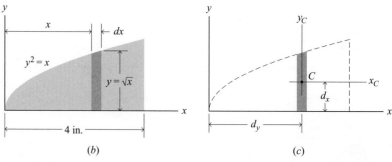

(*b*)   (*c*)

Figure A-18(b–c)

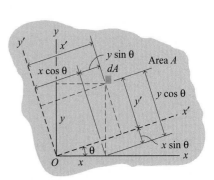

Example Problem A-11   Determine the mixed second moment (area product of inertia) of the quadrant of a circle shown in Fig. A-19 with respect to

(a) The x- and y-axes.
(b) A pair of axes through the centroid of the area and parallel to the x- and y-axes.

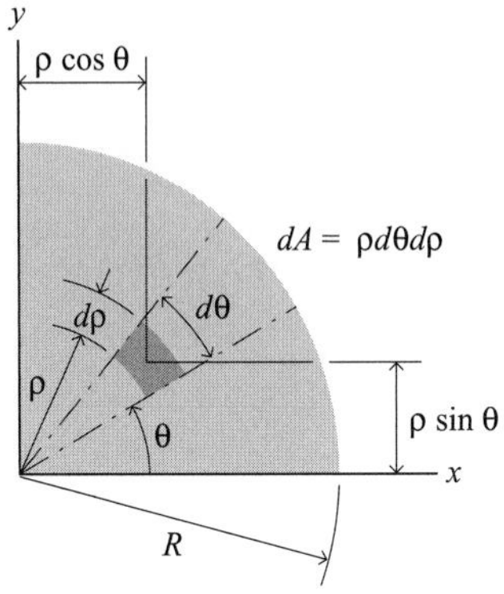

SOLUTION

(a) The mixed second moment (area product of inertia) dI_{xy} of the element of area $dA = \rho\, d\theta\, d\rho$ shown in Fig. A-19 with respect to the x- and y-axes is defined as $dI_{xy} = xy\, dA$. Therefore,

$$dI_{xy} = (\rho \cos \theta)(\rho \sin \theta)(\rho\, d\theta\, d\rho)$$

and

$$
I_{xy} = \int_A dI_{xy} = \int_0^R \int_0^{\pi/2} (\rho \cos \theta)(\rho \sin \theta)\rho\, d\theta\, d\rho
$$

$$
= \int_0^R \left[\frac{\sin^2 \theta}{2} \right]_0^{\pi/2} \rho^3\, d\rho
$$

$$
= \int_0^R \frac{\rho^3}{2}\, d\rho = \left[\frac{\rho^4}{8} \right]_0^R = \frac{R^4}{8}
$$ **Ans.**

(b) Once the mixed second moment is known with respect to a pair of axes, the mixed second moment with respect to a parallel set of axes through the centroid of the area can be found by using the parallel axis theorem. For the quarter circle, $d_x = d_y = 4R/3\pi$; therefore,

$$
I_{xyC} = I_{xy} - d_x d_y A
$$

$$
= \frac{R^4}{8} - \left(\frac{4R}{3\pi} \right)\left(\frac{4R}{3\pi} \right)\left(\frac{\pi R^2}{4} \right) = \frac{(9\pi - 32)R^4}{72\pi}
$$ **Ans.**

Figure A-19

▶ Polar coordinates are usually more efficient when circular boundaries are involved.

▶ The parallel axis theorem for second moments is valid only for transfers to or from a parallel axis through the centroid of the area.

A-3 PRINCIPAL SECOND MOMENTS

The second moment of the area A in Fig. A-20 with respect to the x'-axis through O will, in general, vary with the angle θ. The x- and y-axes used to obtain the polar second moment J_z about a z-axis through O (Eq. A-2) were any pair of orthogonal axes in the plane of the area passing through O; therefore,

$$
J_z = I_x + I_y = I_{x'} + I_{y'}
$$

where x' and y' are any pair of orthogonal axes through O. Since the sum of $I_{x'}$ and $I_{y'}$ is a constant, $I_{x'}$ will be maximum and the corresponding $I_{y'}$ will be minimum for one particular value of θ.

Figure A-20

The set of axes for which the second moments are maximum and minimum are called the principal axes of the area through point O and are designated as the u- and v-axes. The second moments of the area with respect to these axes are called the principal second moments of the area (principal area moments of inertia) and are designated I_u and I_v. There is only one set of principal axes for any point in an area unless all axes have the same second moment, such as the diameters of a circle. Principal axes are important in problems dealing with stresses and deformations in beams and columns.

The principal second moments of an area can be determined by expressing $I_{x'}$ as a function of I_x, I_y, I_{xy}, and angle θ and setting the derivative of $I_{x'}$ with respect to θ equal to zero to obtain the values of θ which give the maximum and minimum second moments. From Fig. A-20,

$$dI_{x'} = y'^2 dA = (y \cos \theta - x \sin \theta)^2 dA$$

Therefore,

$$I_{x'} = \int_A dI_{x'} = \cos^2 \theta \int_A y^2 dA + \sin^2 \theta \int_A x^2 dA - 2 \sin \theta \cos \theta \int_A xy \, dA$$

$$= I_x \cos^2 \theta + I_y \sin^2 \theta - 2 I_{xy} \sin \theta \cos \theta \tag{A-14}$$

Equation A-14 can be expressed in terms of the double angle 2θ by using the trigonometric identities

$$\sin 2\theta = 2 \sin \theta \cos \theta$$
$$\cos 2\theta = \cos^2 \theta - \sin^2 \theta$$

Thus,

$$I_{x'} = \frac{1}{2}(I_x + I_y) + \frac{1}{2}(I_x - I_y) \cos 2\theta - I_{xy} \sin 2\theta \tag{A-15}$$

The angle 2θ for which $I_{x'}$ is a maximum (or a minimum) can be obtained by setting the derivative of $I_{x'}$ with respect to θ equal to zero; thus,

$$\frac{dI_{x'}}{d\theta} = -(I_x - I_y) \sin 2\theta - 2 I_{xy} \cos 2\theta = 0$$

from which

$$\tan 2\theta_p = -\frac{2 I_{xy}}{(I_x - I_y)} \tag{A-16}$$

where θ_p represents the two values of θ that locate the principal axes u and v. Equation A-16 gives two values of $2\theta_p$ that are $180°$ apart, and thus two values of θ_p that are $90°$ apart. The principal second moments can be obtained by substituting these values of θ_p in Eq. A-14. Thus,

$$I_{u,v} = \frac{I_x + I_y}{2} \pm \left[\left(\frac{I_x - I_y}{2} \right)^2 + I_{xy}^2 \right]^{1/2} \tag{A-17}$$

The mixed second moment (area product of inertia) of the element of area in Fig. A-20 with respect to the x'- and y'-axes is

$$dI_{x'y'} = x'y' \, dA = (x \cos \theta + y \sin \theta)(y \cos \theta - x \sin \theta) \, dA$$

Therefore,

$$
\begin{aligned}
I_{x'y'} &= \int_A dI_{x'y'} \\
&= \sin \theta \cos \theta \int_A (y^2 - x^2) \, dA + (\cos^2 \theta - \sin^2 \theta) \int_A xy \, dA \\
&= (I_x - I_y) \sin \theta \cos \theta + I_{xy}(\cos^2 \theta - \sin^2 \theta) \quad\quad \text{(A-18)}
\end{aligned}
$$

Or in terms of the double angle 2θ,

$$I_{x'y'} = I_{xy} \cos 2\theta + \frac{1}{2}(I_x - I_y) \sin 2\theta \quad\quad \text{(A-19)}$$

The mixed second moment $I_{x'y'}$ will be zero when

$$\tan 2\theta = -\frac{2I_{xy}}{(I_x - I_y)} \quad\quad (a)$$

Equation (a) is the same as Eq. A-16. This fact indicates that mixed second moments with respect to principal axes are zero. Since mixed second moments are zero with respect to any axis of symmetry, it follows that any axis of symmetry must be a principal axis for any point on the axis.

The following example illustrates the procedure for determining second moments of area with respect to the principal axes.

Example Problem A-12   Determine the maximum and minimum second moments for the triangular area shown in Fig. A-21a with respect to axes through the centroid of the area.

SOLUTION
The second moments I_x, I_y, and I_{xy} can be determined by using the properties listed in Tables A-1 and A-2. Thus,

$$I_x = \frac{bh^3}{36} = \frac{24(48)^3}{36} = 73{,}728 \text{ mm}^4$$

$$I_y = \frac{hb^3}{36} = \frac{48(24)^3}{36} = 18{,}432 \text{ mm}^4$$

$$I_{xy} = \frac{b^2h^2}{72} = \frac{(24)^2(48)^2}{72} = 18{,}432 \text{ mm}^4$$

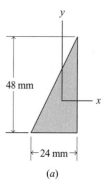

(a)

Figure A-21(a)

The principal angles θ_p are

$$\theta_p = \frac{1}{2} \tan^{-1}\left[-\frac{2I_{xy}}{I_x - I_y}\right]$$

$$= \frac{1}{2} \tan^{-1}\left[-\frac{2(18{,}432)}{73{,}728 - 18{,}432}\right] = -16.85° \text{ or } 73.15°$$

With $\theta_p = -16.85°$, Eq. A-14 yields

$$I_{x'} = I_x \cos^2 \theta + I_y \sin^2 \theta - 2I_{xy} \sin \theta \cos \theta$$
$$= 73{,}728 \cos^2(-16.85°) + 18{,}432 \sin^2(-16.85°)$$
$$-2(18{,}432) \sin (-16.85°) \cos (-16.85°) = 79{,}309 \text{ mm}^4$$

With $\theta_p = 73.15°$, Eq. A-14 yields

$$I_{x'} = I_x \cos^2 \theta + I_y \sin^2 \theta - 2I_{xy} \sin \theta \cos \theta$$
$$= 73{,}728 \cos^2 (73.15°) + 18{,}432 \sin^2 (73.15°)$$
$$-2(18{,}432) \sin (73.15°) \cos (73.15°) = 12{,}851 \text{ mm}^4$$

▶ Second moments associated with principal axes are the maximum and minimum values for all axes in the plane of the area that pass through the point.

Therefore, with respect to the x-axis,

$$I_u = I_{max} = 79{,}300 \text{ mm}^4 \text{ at } \theta_p = -16.85° \qquad \textbf{Ans.}$$

$$I_v = I_{min} = 12{,}850 \text{ mm}^4 \text{ at } \theta_p = 73.15° \qquad \textbf{Ans.}$$

The principal second moments can also be determined by using Eq. A-17:

$$I_{u,v} = \frac{I_x + I_y}{2} \pm \left[\left(\frac{I_x - I_y}{2}\right)^2 + I_{xy}^2\right]^{1/2}$$

$$= \frac{73{,}728 + 18{,}432}{2} \pm \left[\left(\frac{73{,}728 - 18{,}432}{2}\right)^2 + (18{,}432)^2\right]^{1/2}$$

$$= 46{,}080 \pm 33{,}229 \cong 79{,}300 \text{ mm}^4 \quad \text{and} \quad 12{,}850 \text{ mm}^4$$

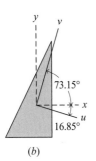

(b)

Figure A-21(b)

The orientations of the principal axes are shown in Fig. A-21b.

Appendix B
Tables of Properties

Table B-1 Wide-Flange Beams (U.S. Customary Units)

Designation*	Area (in.^{2})	Depth (in.)	Flange Width (in.)	Flange Thickness (in.)	Web Thickness (in.)	Axis X–X I (in.^{4})	Axis X–X S (in.^{3})	Axis X–X r (in.)	Axis Y–Y I (in.^{4})	Axis Y–Y S (in.^{3})	Axis Y–Y r (in.)
W36 × 230	67.6	35.90	16.470	1.260	0.760	15000	837	14.9	940	114	3.73
× 160	47.0	36.01	12.000	1.020	0.650	9750	542	14.4	295	49.1	2.50
W33 × 201	59.1	33.68	15.745	1.150	0.715	11500	684	14.0	749	95.2	3.56
× 152	44.7	33.49	11.565	1.055	0.635	8160	487	13.5	273	47.2	2.47
× 130	38.3	33.09	11.510	0.855	0.580	6710	406	13.2	218	37.9	2.39
W30 × 132	38.9	30.31	10.545	1.000	0.615	5770	380	12.2	196	37.2	2.25
× 108	31.7	29.83	10.475	0.760	0.545	4470	299	11.9	146	27.9	2.15
W27 × 146	42.9	27.38	13.965	0.975	0.605	5630	411	11.4	443	63.5	3.21
× 94	27.7	26.92	9.990	0.745	0.490	3270	243	10.9	124	24.8	2.12
W24 × 104	30.6	24.06	12.750	0.750	0.500	3100	258	10.1	259	40.7	2.91
× 84	24.7	24.10	9.020	0.770	0.470	2370	196	9.79	94.4	20.9	1.95
× 62	18.2	23.74	7.040	0.590	0.430	1550	131	9.23	34.5	9.80	1.38
W21 × 101	29.8	21.36	12.290	0.800	0.500	2420	227	9.02	248	40.3	2.89
× 83	24.3	21.43	8.355	0.835	0.515	1830	171	8.67	81.4	19.5	1.83
× 62	18.3	20.99	8.240	0.615	0.400	1330	127	8.54	57.5	13.9	1.77
W18 × 97	28.5	18.59	11.145	0.870	0.535	1750	188	7.82	201	36.1	2.65
× 76	22.3	18.21	11.035	0.680	0.425	1330	146	7.73	152	27.6	2.61
× 60	17.6	18.24	7.555	0.695	0.415	984	108	7.47	50.1	13.3	1.69
W16 × 100	29.4	16.97	10.425	0.985	0.585	1490	175	7.10	186	35.7	2.52
× 67	19.7	16.33	10.235	0.665	0.395	954	117	6.96	119	23.2	2.46
× 40	11.8	16.01	6.995	0.505	0.305	518	64.7	6.63	28.9	8.25	1.57
× 26	7.68	15.69	5.500	0.345	0.250	301	38.4	6.26	9.59	3.49	1.12
W14 × 120	35.3	14.48	14.670	0.940	0.590	1380	190	6.24	495	67.5	3.74
× 82	24.1	14.31	10.130	0.855	0.510	882	123	6.05	148	29.3	2.48
× 43	12.6	13.66	7.995	0.530	0.305	428	62.7	5.82	45.2	11.3	1.89
× 30	8.85	13.84	6.730	0.385	0.270	291	42.0	5.73	19.6	5.82	1.49
W12 × 96	28.2	12.71	12.160	0.900	0.550	833	131	5.44	270	44.4	3.09
× 65	19.1	12.12	12.000	0.605	0.390	533	87.9	5.28	174	29.1	3.02
× 50	14.7	12.19	8.080	0.640	0.370	394	64.7	5.18	56.3	13.9	1.96
× 30	8.79	12.34	6.520	0.440	0.260	238	38.6	5.21	20.3	6.24	1.52
W10 × 60	17.6	10.22	10.080	0.680	0.420	341	66.7	4.39	116	23.0	2.57
× 45	13.3	10.10	8.020	0.620	0.350	248	49.1	4.33	53.4	13.3	2.01
× 30	8.84	10.47	5.810	0.510	0.300	170	32.4	4.38	16.7	5.75	1.37
× 22	6.49	10.17	5.750	0.360	0.240	118	23.2	4.27	11.4	3.97	1.33
W8 × 40	11.7	8.25	8.070	0.560	0.360	146	35.5	3.53	49.1	12.2	2.04
× 31	9.13	8.00	7.995	0.435	0.285	110	27.5	3.47	37.1	9.27	2.02
× 24	7.08	7.93	6.495	0.400	0.245	82.8	20.9	3.42	18.3	5.63	1.61
× 15	4.44	8.11	4.015	0.315	0.245	48.0	11.8	3.29	3.41	1.70	0.876
W6 × 25	7.34	6.38	6.080	0.455	0.320	53.4	16.7	2.70	17.1	5.61	1.52
× 16	4.74	6.28	4.030	0.405	0.260	32.1	10.2	2.60	4.43	2.20	0.967
W5 × 16	4.68	5.01	5.000	0.360	0.240	21.3	8.51	2.13	7.51	3.00	1.27
W4 × 13	3.83	4.16	4.060	0.345	0.280	11.3	5.46	1.72	3.86	1.90	1.00

Courtesy of the American Institute of Steel Construction.

*W means wide-flange beam, followed by the nominal depth in inches, then the weight in pounds per foot of length.

Table B-2   Wide-Flange Beams (SI Units)

Designation*	Area (mm²)	Depth (mm)	Flange Width (mm)	Flange Thickness (mm)	Web Thickness (mm)	Axis X–X I (10⁶ mm⁴)	Axis X–X S (10³ mm³)	Axis X–X r (mm)	Axis Y–Y I (10⁶ mm⁴)	Axis Y–Y S (10³ mm³)	Axis Y–Y r (mm)
W914 × 342	43610	912	418	32.0	19.3	6245	13715	378	391	1870	94.7
× 238	30325	915	305	25.9	16.5	4060	8880	366	123	805	63.5
W838 × 299	38130	855	400	29.2	18.2	4785	11210	356	312	1560	90.4
× 226	28850	851	294	26.8	16.1	3395	7980	343	114	775	62.7
× 193	24710	840	292	21.7	14.7	2795	6655	335	90.7	620	60.7
W762 × 196	25100	770	268	25.4	15.6	2400	6225	310	81.6	610	57.2
× 161	20450	758	266	19.3	13.8	1860	4900	302	60.8	457	54.6
W686 × 217	27675	695	355	24.8	15.4	2345	6735	290	184	1040	81.5
× 140	17870	684	254	18.9	12.4	1360	3980	277	51.6	406	53.8
W610 × 155	19740	611	324	19.1	12.7	1290	4230	257	108	667	73.9
× 125	15935	612	229	19.6	11.9	985	3210	249	39.3	342	49.5
× 92	11750	603	179	15.0	10.9	645	2145	234	14.4	161	35.1
W533 × 150	19225	543	312	20.3	12.7	1005	3720	229	103	660	73.4
× 124	15675	544	212	21.2	13.1	762	2800	220	33.9	320	46.5
× 92	11805	533	209	15.6	10.2	554	2080	217	23.9	228	45.0
W457 × 144	18365	472	283	22.1	13.6	728	3080	199	83.7	592	67.3
× 113	14385	463	280	17.3	10.8	554	2395	196	63.3	452	66.3
× 89	11355	463	192	17.7	10.5	410	1770	190	20.9	218	42.9
W406 × 149	18970	431	265	25.0	14.9	620	2870	180	77.4	585	64.0
× 100	12710	415	260	16.9	10.0	397	1915	177	49.5	380	62.5
× 60	7615	407	178	12.8	7.7	216	1060	168	12.0	135	39.9
× 39	4950	399	140	8.8	6.4	125	629	159	3.99	57.2	28.4
W356 × 179	22775	368	373	23.9	15.0	574	3115	158	206	1105	95.0
× 122	15550	363	257	21.7	13.0	367	2015	154	61.6	480	63.0
× 64	8130	347	203	13.5	7.7	178	1025	148	18.8	185	48.0
× 45	5710	352	171	9.8	6.9	121	688	146	8.16	95.4	37.8
W305 × 143	18195	323	309	22.9	14.0	347	2145	138	112	728	78.5
× 97	12325	308	305	15.4	9.9	222	1440	134	72.4	477	76.7
× 74	9485	310	205	16.3	9.4	164	1060	132	23.4	228	49.8
× 45	5670	313	166	11.2	6.6	99.1	633	132	8.45	102	38.6
W254 × 89	11355	260	256	17.3	10.7	142	1095	112	48.3	377	65.3
× 67	8580	257	204	15.7	8.9	103	805	110	22.2	218	51.1
× 45	5705	266	148	13.0	7.6	70.8	531	111	6.95	94.2	34.8
× 33	4185	258	146	9.1	6.1	49.1	380	108	4.75	65.1	33.8
W203 × 60	7550	210	205	14.2	9.1	60.8	582	89.7	20.4	200	51.8
× 46	5890	203	203	11.0	7.2	45.8	451	88.1	15.4	152	51.3
× 36	4570	201	165	10.2	6.2	34.5	342	86.7	7.61	92.3	40.9
× 22	2865	206	102	8.0	6.2	20.0	193	83.6	1.42	27.9	22.3
W152 × 37	4735	162	154	11.6	8.1	22.2	274	68.6	7.12	91.9	38.6
× 24	3060	160	102	10.3	6.6	13.4	167	66.0	1.84	36.1	24.6
W127 × 24	3020	127	127	9.1	6.1	8.87	139	54.1	3.13	49.2	32.3
W102 × 19	2470	106	103	8.8	7.1	4.70	89.5	43.7	1.61	31.1	25.4

*W means wide-flange beam, followed by the nominal depth in mm, then the mass in kg per meter of length.

Table B-3   American Standard Beams (U.S. Customary Units)

Designation*	Area (in.²)	Depth (in.)	Flange Width (in.)	Flange Thickness (in.)	Web Thickness (in.)	Axis X–X I (in.⁴)	Axis X–X S (in.³)	Axis X–X r (in.)	Axis Y–Y I (in.⁴)	Axis Y–Y S (in.³)	Axis Y–Y r (in.)
S24 × 121	35.6	24.50	8.050	1.090	0.800	3160	258	9.43	83.3	20.7	1.53
× 106	31.2	24.50	7.870	1.090	0.620	2940	240	9.71	77.1	19.6	1.57
× 100	29.3	24.00	7.245	0.870	0.745	2390	199	9.02	47.7	13.2	1.27
× 90	26.5	24.00	7.125	0.870	0.625	2250	187	9.21	44.9	12.6	1.30
× 80	23.5	24.00	7.000	0.870	0.500	2100	175	9.47	42.2	12.1	1.34
S20 × 96	28.2	20.30	7.200	0.920	0.800	1670	165	7.71	50.2	13.9	1.33
× 86	25.3	20.30	7.060	0.920	0.660	1580	155	7.89	46.8	13.3	1.36
× 75	22.0	20.00	6.385	0.795	0.635	1280	128	7.62	29.8	9.32	1.16
× 66	19.4	20.00	6.255	0.795	0.505	1190	119	7.83	27.7	8.85	1.19
S18 × 70	20.6	18.00	6.251	0.691	0.711	926	103	6.71	24.1	7.72	1.08
× 54.7	16.1	18.00	6.001	0.691	0.461	804	89.4	7.07	20.8	6.94	1.14
S15 × 50	14.7	15.00	5.640	0.622	0.550	486	64.8	5.75	15.7	5.57	1.03
× 42.9	12.6	15.00	5.501	0.622	0.411	447	59.6	5.95	14.4	5.23	1.07
S12 × 50	14.7	12.00	5.477	0.659	0.687	305	50.8	4.55	15.7	5.74	1.03
× 40.8	12.0	12.00	5.252	0.659	0.462	272	45.4	4.77	13.6	5.16	1.06
× 35	10.3	12.00	5.078	0.544	0.428	229	38.2	4.72	9.87	3.89	0.980
× 31.8	9.35	12.00	5.000	0.544	0.350	218	36.4	4.83	9.36	3.74	1.00
S10 × 35	10.3	10.00	4.944	0.491	0.594	147	29.4	3.78	8.36	3.38	0.901
× 25.4	7.46	10.00	4.661	0.491	0.311	124	24.7	4.07	6.79	2.91	0.954
S8 × 23	6.77	8.00	4.171	0.426	0.441	64.9	16.2	3.10	4.31	2.07	0.798
× 18.4	5.41	8.00	4.001	0.426	0.271	57.6	14.4	3.26	3.73	1.86	0.831
S7 × 20	5.88	7.00	3.860	0.392	0.450	42.4	12.1	2.69	3.17	1.64	0.734
× 15.3	4.50	7.00	3.662	0.392	0.252	36.7	10.5	2.86	2.64	1.44	0.766
S6 × 17.25	5.07	6.00	3.565	0.359	0.465	26.3	8.77	2.28	2.31	1.30	0.675
× 12.5	3.67	6.00	3.332	0.359	0.232	22.1	7.37	2.45	1.82	1.09	0.705
S5 × 14.75	4.34	5.00	3.284	0.326	0.494	15.2	6.09	1.87	1.67	1.01	0.620
× 10	2.94	5.00	3.004	0.326	0.214	12.3	4.92	2.05	1.22	0.809	0.643
S4 × 9.5	2.79	4.00	2.796	0.293	0.326	6.79	3.39	1.56	0.903	0.646	0.569
× 7.7	2.26	4.00	2.663	0.293	0.193	6.08	3.04	1.64	0.764	0.574	0.581
S3 × 7.5	2.21	3.00	2.509	0.260	0.349	2.93	1.95	1.15	0.586	0.468	0.516
× 5.7	1.67	3.00	2.330	0.260	0.170	2.52	1.68	1.23	0.455	0.390	0.522

Courtesy of The American Institute of Steel Construction.

*S means standard beam, followed by the nominal depth in inches, then the weight in pounds per foot of length.

Table B-4 American Standard Beams (SI Units)

Designation*	Area (mm²)	Depth (mm)	Flange Width (mm)	Flange Thickness (mm)	Web Thickness (mm)	Axis X–X I (10⁶ mm⁴)	Axis X–X S (10³ mm³)	Axis X–X r (mm)	Axis Y–Y I (10⁶ mm⁴)	Axis Y–Y S (10³ mm³)	Axis Y–Y r (mm)
S610 × 180	22970	622.3	204.5	27.7	20.3	1315	4225	240	34.7	339	38.9
× 158	20130	622.3	199.9	27.7	15.7	1225	3935	247	32.1	321	39.9
× 149	18900	609.6	184.0	22.1	18.9	995	3260	229	19.9	216	32.3
× 134	17100	609.6	181.0	22.1	15.9	937	3065	234	18.7	206	33.0
× 119	15160	609.6	177.8	22.1	12.7	874	2870	241	17.6	198	34.0
S508 × 143	18190	515.6	182.9	23.4	20.3	695	2705	196	20.9	228	33.8
× 128	16320	515.6	179.3	23.4	16.8	658	2540	200	19.5	218	34.5
× 112	14190	508.0	162.2	20.2	16.1	533	2100	194	12.4	153	29.5
× 98	12520	508.0	158.9	20.2	12.8	495	1950	199	11.5	145	30.2
S457 × 104	13290	457.2	158.8	17.6	18.1	358	1690	170	10.0	127	27.4
× 81	10390	457.2	152.4	17.6	11.7	335	1465	180	8.66	114	29.0
S381 × 74	9485	381.0	143.3	15.8	14.0	202	1060	146	6.53	91.3	26.2
× 64	8130	381.0	139.7	15.8	10.4	186	977	151	5.99	85.7	27.2
S305 × 74	9485	304.8	139.1	16.7	17.4	127	832	116	6.53	94.1	26.2
× 61	7740	304.8	133.4	16.7	11.7	113	744	121	5.66	84.6	26.9
× 52	6645	304.8	129.0	13.8	10.9	95.3	626	120	4.11	63.7	24.1
× 47	6030	304.8	127.0	13.8	8.9	90.7	596	123	3.90	61.3	25.4
S254 × 52	6645	254.0	125.6	12.5	15.1	61.2	482	96.0	3.48	55.4	22.9
× 38	4815	254.0	118.4	12.5	7.9	51.6	408	103	2.83	47.7	24.2
S203 × 34	4370	203.2	105.9	10.8	11.2	27.0	265	78.7	1.79	33.9	20.3
× 27	3490	203.2	101.6	10.8	6.9	24.0	236	82.8	1.55	30.5	21.1
S178 × 30	3795	177.8	98.0	10.0	11.4	17.6	198	68.3	1.32	26.9	18.6
× 23	2905	177.8	93.0	10.0	6.4	15.3	172	72.6	1.10	23.6	19.5
S152 × 26	3270	152.4	90.6	9.1	11.8	10.9	144	57.9	0.961	21.3	17.1
× 19	2370	152.4	84.6	9.1	5.9	9.20	121	62.2	0.758	17.9	17.9
S127 × 22	2800	127.0	83.4	8.3	12.5	6.33	99.8	47.5	0.695	16.6	15.7
× 15	1895	127.0	76.3	8.3	5.4	5.12	80.6	52.1	0.508	13.3	16.3
S102 × 14	1800	101.6	71.0	7.4	8.3	2.83	55.6	39.6	0.376	10.6	14.5
× 11	1460	101.6	67.6	7.4	4.9	2.53	49.8	41.7	0.318	9.41	14.8
S76 × 11	1425	76.2	63.7	6.6	8.9	1.22	32.0	29.2	0.244	7.67	13.1
× 8.5	1075	76.2	59.2	6.6	4.3	1.05	27.5	31.2	0.189	6.39	13.3

*S means standard beam, followed by the nominal depth in mm, then the mass in kg per meter of length.

Table B-5 Standard Channels (U.S. Customary Units)

Designation*	Area (in.²)	Depth (in.)	Flange Width (in.)	Flange Thickness (in.)	Web Thickness (in.)	Axis X–X I (in.⁴)	Axis X–X S (in.³)	Axis X–X r (in.)	Axis Y–Y I (in.⁴)	Axis Y–Y S (in.³)	Axis Y–Y r (in.)	x_C (in.)
†C18 × 58	17.1	18.00	4.200	0.625	0.700	676	75.1	6.29	17.8	5.32	1.02	0.862
× 51.9	15.3	18.00	4.100	0.625	0.600	627	69.7	6.41	16.4	5.07	1.04	0.858
× 45.8	13.5	18.00	4.000	0.625	0.500	578	64.3	6.56	15.1	4.82	1.06	0.866
× 42.7	12.6	18.00	3.950	0.625	0.450	554	61.6	6.64	14.4	4.69	1.07	0.877
C15 × 50	14.7	15.00	3.716	0.650	0.716	404	53.8	5.24	11.0	3.78	0.867	0.798
× 40	11.8	15.00	3.520	0.650	0.520	349	46.5	5.44	9.23	3.37	0.886	0.777
× 33.9	9.96	15.00	3.400	0.650	0.400	315	42.0	5.62	8.13	3.11	0.904	0.787
C12 × 30	8.82	12.00	3.170	0.501	0.510	162	27.0	4.29	5.14	2.06	0.763	0.674
× 25	7.35	12.00	3.047	0.501	0.387	144	24.1	4.43	4.47	1.88	0.780	0.674
× 20.7	6.09	12.00	2.942	0.501	0.282	129	21.5	4.61	3.88	1.73	0.799	0.698
C10 × 30	8.82	10.00	3.033	0.436	0.673	103	20.7	3.42	3.94	1.65	0.669	0.649
× 25	7.35	10.00	2.886	0.436	0.526	91.2	18.2	3.52	3.36	1.48	0.676	0.617
× 20	5.88	10.00	2.739	0.436	0.379	78.9	15.8	3.66	2.81	1.32	0.692	0.606
× 15.3	4.49	10.00	2.600	0.436	0.240	67.4	13.5	3.87	2.28	1.16	0.713	0.634
C9 × 20	5.88	9.00	2.648	0.413	0.448	60.9	13.5	3.22	2.42	1.17	0.642	0.583
× 15	4.41	9.00	2.485	0.413	0.285	51.0	11.3	3.40	1.93	1.01	0.661	0.586
× 13.4	3.94	9.00	2.433	0.413	0.233	47.9	10.6	3.48	1.76	0.962	0.669	0.601
C8 × 18.75	5.51	8.00	2.527	0.390	0.487	44.0	11.0	2.82	1.98	1.01	0.599	0.565
× 13.75	4.04	8.00	2.343	0.390	0.303	36.1	9.03	2.99	1.53	0.854	0.615	0.553
× 11.5	3.38	8.00	2.260	0.390	0.220	32.6	8.14	3.11	1.32	0.781	0.625	0.571
C7 × 14.75	4.33	7.00	2.299	0.366	0.419	27.2	7.78	2.51	1.38	0.779	0.564	0.532
× 12.25	3.60	7.00	2.194	0.366	0.314	24.2	6.93	2.60	1.17	0.703	0.571	0.525
× 9.8	2.87	7.00	2.090	0.366	0.210	21.3	6.08	2.72	0.968	0.625	0.581	0.540
C6 × 13	3.83	6.00	2.157	0.343	0.437	17.4	5.80	2.13	1.05	0.642	0.525	0.514
× 10.5	3.09	6.00	2.034	0.343	0.314	15.2	5.06	2.22	0.866	0.564	0.529	0.499
× 8.2	2.40	6.00	1.920	0.343	0.200	13.1	4.38	2.34	0.693	0.492	0.537	0.511
C5 × 9	2.64	5.00	1.885	0.320	0.325	8.90	3.56	1.83	0.632	0.450	0.489	0.478
× 6.7	1.97	5.00	1.750	0.320	0.190	7.49	3.00	1.95	0.479	0.378	0.493	0.484
C4 × 7.25	2.13	4.00	1.721	0.296	0.321	4.59	2.29	1.47	0.433	0.343	0.450	0.459
× 5.4	1.59	4.00	1.584	0.296	0.184	3.85	1.93	1.56	0.319	0.283	0.449	0.457
C3 × 6	1.76	3.00	1.596	0.273	0.356	2.07	1.38	1.08	0.305	0.268	0.416	0.455
× 5	1.47	3.00	1.498	0.273	0.258	1.85	1.24	1.12	0.247	0.233	0.410	0.438
× 4.1	1.21	3.00	1.410	0.273	0.170	1.66	1.10	1.17	0.197	0.202	0.404	0.436

Courtesy of The American Institute of Steel Construction.

*C means channel, followed by the nominal depth in inches, then the weight in pounds per foot of length.

†Not part of the American Standard Series.

Table B-6   Standard Channels (SI Units)

			Flange		Web	Axis X–X			Axis Y–Y			
Designation*	Area (mm²)	Depth (mm)	Width (mm)	Thickness (mm)	Thickness (mm)	I (10^6 mm⁴)	S (10^3 mm³)	r (mm)	I (10^6 mm⁴)	S (10^3 mm³)	r (mm)	x_c (mm)
C457 × 86	11030	457.2	106.7	15.9	17.8	281	1230	160	7.41	87.2	25.9	21.9
× 77	9870	457.2	104.1	15.9	15.2	261	1140	163	6.83	83.1	26.4	21.8
× 68	8710	457.2	101.6	15.9	12.7	241	1055	167	6.29	79.0	26.9	22.0
× 64	8130	457.2	100.3	15.9	11.4	231	1010	169	5.99	76.9	27.2	22.3
C381 × 74	9485	381.0	94.4	16.5	18.2	168	882	133	4.58	61.9	22.0	20.3
× 60	7615	381.0	89.4	16.5	13.2	145	762	138	3.84	55.2	22.5	19.7
× 50	6425	381.0	86.4	16.5	10.2	131	688	143	3.38	51.0	23.0	20.0
C305 × 45	5690	304.8	80.5	12.7	13.0	67.4	442	109	2.14	33.8	19.4	17.1
× 37	4740	304.8	77.4	12.7	9.8	59.9	395	113	1.86	30.8	19.8	17.1
× 31	3930	304.8	74.7	12.7	7.2	53.7	352	117	1.61	28.3	20.3	17.7
C254 × 45	5690	254.0	77.0	11.1	17.1	42.9	339	86.9	1.64	27.0	17.0	16.5
× 37	4740	254.0	73.3	11.1	13.4	38.0	298	89.4	1.40	24.3	17.2	15.7
× 30	3795	254.0	69.6	11.1	9.6	32.8	259	93.0	1.17	21.6	17.6	15.4
× 23	2895	254.0	66.0	11.1	6.1	28.1	221	98.3	0.949	19.0	18.1	16.1
C229 × 30	3795	228.6	67.3	10.5	11.4	25.3	221	81.8	1.01	19.2	16.3	14.8
× 22	2845	228.6	63.1	10.5	7.2	21.2	185	86.4	0.803	16.6	16.8	14.9
× 20	2540	228.6	61.8	10.5	5.9	19.9	174	88.4	0.733	15.7	17.0	15.3
C203 × 28	3555	203.2	64.2	9.9	12.4	18.3	180	71.6	0.824	16.6	15.2	14.4
× 20	2605	203.2	59.5	9.9	7.7	15.0	148	75.9	0.637	14.0	15.6	14.0
× 17	2180	203.2	57.4	9.9	5.6	13.6	133	79.0	0.549	12.8	15.9	14.5
C178 × 22	2795	177.8	58.4	9.3	10.6	11.3	127	63.8	0.574	12.8	14.3	13.5
× 18	2320	177.8	55.7	9.3	8.0	10.1	114	66.0	0.487	11.5	14.5	13.3
× 15	1850	177.8	53.1	9.3	5.3	8.87	99.6	69.1	0.403	10.2	14.8	13.7
C152 × 19	2470	152.4	54.8	8.7	11.1	7.24	95.0	54.1	0.437	10.5	13.3	13.1
× 16	1995	152.4	51.7	8.7	8.0	6.33	82.9	56.4	0.360	9.24	13.4	12.7
× 12	1550	152.4	48.8	8.7	5.1	5.45	71.8	59.4	0.288	8.06	13.6	13.0
C127 × 13	1705	127.0	47.9	8.1	8.3	3.70	58.3	46.5	0.263	7.37	12.4	12.1
× 10	1270	127.0	44.5	8.1	4.8	3.12	49.2	49.5	0.199	6.19	12.5	12.3
C102 × 11	1375	101.6	43.7	7.5	8.2	1.91	37.5	37.3	0.180	5.62	11.4	11.7
× 8	1025	101.6	40.2	7.5	4.7	1.60	31.6	39.6	0.133	4.64	11.4	11.6
C76 × 9	1135	76.2	40.5	6.9	9.0	0.862	22.6	27.4	0.127	4.39	10.6	11.6
× 7	948	76.2	38.0	6.9	6.6	0.770	20.3	28.4	0.103	3.82	10.4	11.1
× 6	781	76.2	35.8	6.9	4.6	0.691	18.0	29.7	0.082	3.31	10.3	11.1

*C means channel, followed by the nominal depth in mm, then the mass in kg per meter of length.

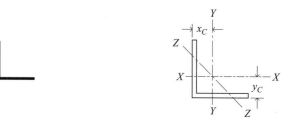

Table B-7   Equal Leg Angles (U.S. Customary Units)

Size and Thickness (in.)	Weight (lb/ft)	Area (in.²)	Axis X–X or Y–Y				Axis Z–Z
			I (in.⁴)	S (in.³)	r (in.)	x_C or y_C (in.)	r (in.)
L8 × 8 × 1	51.0	15.0	89.0	15.8	2.44	2.37	1.56
× 7/8	45.0	13.2	79.6	14.0	2.45	2.32	1.57
× 3/4	38.9	11.4	69.7	12.2	2.47	2.28	1.58
× 5/8	32.7	9.61	59.4	10.3	2.49	2.23	1.58
× 1/2	26.4	7.75	48.6	8.36	2.50	2.19	1.59
L6 × 6 × 1	37.4	11.0	35.5	8.57	1.80	1.86	1.17
× 7/8	33.1	9.73	31.9	7.63	1.81	1.82	1.17
× 3/4	28.7	8.44	28.2	6.66	1.83	1.78	1.17
× 5/8	24.2	7.11	24.2	5.66	1.84	1.73	1.18
× 1/2	19.6	5.75	19.9	4.61	1.86	1.68	1.18
× 3/8	14.9	4.36	15.4	3.53	1.88	1.64	1.19
L5 × 5 × 7/8	27.2	7.98	17.8	5.17	1.49	1.57	0.973
× 3/4	23.6	6.94	15.7	4.53	1.51	1.52	0.975
× 5/8	20.0	5.86	13.6	3.86	1.52	1.48	0.978
× 1/2	16.2	4.75	11.3	3.16	1.54	1.43	0.983
× 3/8	12.3	3.61	8.74	2.42	1.56	1.39	0.990
L4 × 4 × 3/4	18.5	5.44	7.67	2.81	1.19	1.27	0.778
× 5/8	15.7	4.61	6.66	2.40	1.20	1.23	0.779
× 1/2	12.8	3.75	5.56	1.97	1.22	1.18	0.782
× 3/8	9.8	2.86	4.36	1.52	1.23	1.14	0.788
× 1/4	6.6	1.94	3.04	1.05	1.25	1.09	0.795
L3½ × 3½ × 1/2	11.1	3.25	3.64	1.49	1.06	1.06	0.683
× 3/8	8.5	2.48	2.87	1.15	1.07	1.01	0.687
× 1/4	5.8	1.69	2.01	0.794	1.09	0.968	0.694
L3 × 3 × 1/2	9.4	2.75	2.22	1.07	0.898	0.932	0.584
× 3/8	7.2	2.11	1.76	0.833	0.913	0.888	0.58
× 1/4	4.9	1.44	1.24	0.577	0.930	0.84	.592
L2½ × 2½ × 1/2	7.7	2.25	1.23	0.724	0.739	0.806	0.487
× 3/8	5.9	1.73	0.984	0.566	0.753	0.762	0.487
× 1/4	4.1	1.19	0.703	0.394	0.769	0.717	0.491
L2 × 2 × 3/8	4.7	1.36	0.479	0.351	0.594	0.636	0.389
× 1/4	3.19	0.938	0.348	0.247	0.609	0.592	0.391
× 1/8	1.65	0.484	0.190	0.131	0.626	0.546	0.398

Courtesy of The American Institute of Steel Construction.

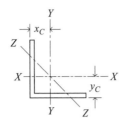

Table B-8 Equal Leg Angles (SI Units)

Size and Thickness (mm)	Mass (kg/m)	Area (mm²)	Axis X–X or Y–Y				Axis Z–Z
			I (10^6 mm⁴)	S (10^3 mm³)	r (mm)	x_C or y_C (mm)	r (mm)
L203 × 203 × 25.4	75.9	9675	37.0	259	62.0	60.2	39.6
× 22.2	67.0	8515	33.1	229	62.2	58.9	39.9
× 19.1	57.9	7355	29.0	200	62.7	57.9	40.1
× 15.9	48.7	6200	24.7	169	63.2	56.6	40.1
× 12.7	39.3	5000	20.2	137	63.5	55.6	40.4
L152 × 152 × 25.4	55.7	7095	14.8	140	45.7	47.2	29.7
× 22.2	49.3	6275	13.3	125	46.0	46.2	29.7
× 19.1	42.7	5445	11.7	109	46.5	45.2	29.7
× 15.9	36.0	4585	10.1	92.8	46.7	43.9	30.0
× 12.7	29.2	3710	8.28	75.5	47.2	42.7	30.0
× 9.5	22.2	2815	6.61	57.8	47.8	41.7	30.2
L127 × 127 × 22.2	40.5	5150	7.41	84.7	37.8	39.9	24.7
× 19.1	35.1	4475	6.53	74.2	38.4	38.6	24.8
× 15.9	29.8	3780	5.66	63.3	38.6	37.6	24.8
× 12.7	24.1	3065	4.70	51.8	39.1	36.3	25.0
× 9.5	18.3	2330	3.64	39.7	39.6	35.3	25.1
L102 × 102 × 19.1	27.5	3510	3.19	46.0	30.2	32.3	19.8
× 15.9	23.4	2975	2.77	39.3	30.5	31.2	19.8
× 12.7	19.0	2420	2.31	32.3	31.0	30.0	19.9
× 9.5	14.6	1845	1.81	24.9	31.2	29.0	20.0
× 6.4	9.8	1250	1.27	17.2	31.8	27.7	20.2
L89 × 89 × 12.7	16.5	2095	1.52	24.4	26.9	26.9	17.3
× 9.5	12.6	1600	1.19	18.8	27.2	25.7	17.4
× 6.4	8.6	1090	0.837	13.0	27.7	24.6	17.6
L76 × 76 × 12.7	14.0	1775	0.924	17.5	22.8	23.7	14.8
× 9.5	10.7	1360	0.732	13.7	23.2	22.6	14.9
× 6.4	7.3	929	0.516	9.46	23.6	21.4	15.0
L64 × 64 × 12.7	11.5	1450	0.512	11.9	18.8	20.5	12.4
× 9.5	8.8	1115	0.410	9.28	19.1	19.4	12.4
× 6.4	6.1	768	0.293	6.46	19.5	18.2	12.5
L51 × 51 × 9.5	7.0	877	0.199	5.75	15.1	16.2	9.88
× 6.4	4.75	605	0.145	4.05	15.5	15.0	9.93
× 3.2	2.46	312	0.079	2.15	15.9	13.9	10.1

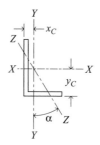

Table B-9   Unequal Leg Angles (U.S. Customary Units)

Size and Thickness (in.)	Weight (lb/ft)	Area (in.²)	Axis X–X				Axis Y–Y				Axis Z–Z	
			I (in.⁴)	S (in.³)	r (in.)	y_C (in.)	I (in.⁴)	S (in.³)	r (in.)	x_C (in.)	r (in.)	Tan α
L9 × 4 × 5/8	26.3	7.73	64.9	11.5	2.90	3.36	8.32	2.65	1.04	0.858	0.847	0.216
× 1/2	21.3	6.25	53.2	9.34	2.92	3.31	6.92	2.17	1.05	0.810	0.854	0.220
L8 × 6 × 1	44.2	13.0	80.8	15.1	2.49	2.65	38.8	8.92	1.73	1.65	1.28	0.543
× 3/4	33.8	9.94	63.4	11.7	2.53	2.56	30.7	6.92	1.76	1.56	1.29	0.551
× 1/2	23.0	6.75	44.3	8.02	2.56	2.47	21.7	4.79	1.79	1.47	1.30	0.558
L8 × 4 × 1	37.4	11.0	69.6	14.1	2.52	3.05	11.6	3.94	1.03	1.05	0.846	0.247
× 3/4	28.7	8.44	54.9	10.9	2.55	2.95	9.36	3.07	1.05	0.953	0.852	0.258
× 1/2	19.6	5.75	38.5	7.49	2.59	2.86	6.74	2.15	1.08	0.859	0.865	0.267
L7 × 4 × 3/4	26.2	7.69	37.8	8.42	2.22	2.51	9.05	3.03	1.09	1.01	0.860	0.324
× 1/2	17.9	5.25	26.7	5.81	2.25	2.42	6.53	2.12	1.11	0.917	0.872	0.335
× 3/8	13.6	3.98	20.6	4.44	2.27	2.37	5.10	1.63	1.13	0.870	0.880	0.340
L6 × 4 × 3/4	23.6	6.94	24.5	6.25	1.88	2.08	8.68	2.97	1.12	1.08	0.860	0.428
× 1/2	16.2	4.75	17.4	4.33	1.91	1.99	6.27	2.08	1.15	0.987	0.870	0.440
× 3/8	12.3	3.61	13.5	3.32	1.93	1.94	4.90	1.60	1.17	0.941	0.877	0.446
L6 × 3½ × 1/2	15.3	4.50	16.6	4.24	1.92	2.08	4.25	1.59	0.972	0.833	0.759	0.344
× 3/8	11.7	3.42	12.9	3.24	1.94	2.04	3.34	1.23	0.988	0.787	0.767	0.350
L5 × 3½ × 3/4	19.8	5.81	13.9	4.28	1.55	1.75	5.55	2.22	0.977	0.996	0.748	0.464
× 1/2	13.6	4.00	9.99	2.99	1.58	1.66	4.05	1.56	1.01	0.906	0.755	0.479
× 3/8	10.4	3.05	7.78	2.29	1.60	1.61	3.18	1.21	1.02	0.861	0.762	0.486
× 1/4	7.0	2.06	5.39	1.57	1.62	1.56	2.23	0.830	1.04	0.814	0.770	0.492
L5 × 3 × 1/2	12.8	3.75	9.45	2.91	1.59	1.75	2.58	1.15	0.829	0.750	0.648	0.357
× 3/8	9.8	2.86	7.37	2.24	1.61	1.70	2.04	0.888	0.845	0.704	0.654	0.364
× 1/4	6.6	1.94	5.11	1.53	1.62	1.66	1.44	0.614	0.861	0.657	0.663	0.371
L4 × 3½ × 1/2	11.9	3.50	5.32	1.94	1.23	1.25	3.79	1.52	1.04	1.00	0.722	0.750
× 3/8	9.1	2.67	4.18	1.49	1.25	1.21	2.95	1.17	1.06	0.955	0.727	0.755
× 1/4	6.2	1.81	2.91	1.03	1.27	1.16	2.09	0.808	1.07	0.909	0.734	0.759
L4 × 3 × 1/2	11.1	3.25	5.05	1.89	1.25	1.33	2.42	1.12	0.864	0.827	0.639	0.543
× 3/8	8.5	2.48	3.96	1.46	1.26	1.28	1.92	0.866	0.879	0.782	0.644	0.551
× 1/4	5.8	1.69	2.77	1.00	1.28	1.24	1.36	0.599	0.896	0.736	0.651	0.558
L3½ × 3 × 1/2	10.2	3.00	3.45	1.45	1.07	1.13	2.33	1.10	0.881	0.875	0.621	0.714
× 3/8	7.9	2.30	2.72	1.13	1.09	1.08	1.85	0.851	0.897	0.830	0.625	0.721
× 1/4	5.4	1.56	1.91	0.776	1.11	1.04	1.30	0.589	0.914	0.785	0.631	0.727
L3½ × 2½ × 1/2	9.4	2.75	3.24	1.41	1.09	1.20	1.36	0.760	0.704	0.705	0.534	0.486
× 3/8	7.2	2.11	2.56	1.09	1.10	1.16	1.09	0.592	0.719	0.660	0.537	0.496
× 1/4	4.9	1.44	1.80	0.755	1.12	1.11	0.777	0.412	0.735	0.614	0.544	0.506
L3 × 2½ × 1/2	8.5	2.50	2.08	1.04	0.913	1.00	1.30	0.744	0.722	0.750	0.520	0.667
× 3/8	6.6	1.92	1.66	0.810	0.928	0.956	1.04	0.581	0.736	0.706	0.522	0.676
× 1/4	4.5	1.31	1.17	0.561	0.945	0.911	0.743	0.404	0.753	0.661	0.528	0.684
L3 × 2 × 1/2	7.7	2.25	1.92	1.00	0.924	1.08	0.672	0.474	0.546	0.583	0.428	0.414
× 3/8	5.9	1.73	1.53	0.781	0.940	1.04	0.543	0.371	0.559	0.539	0.430	0.428
× 1/4	4.1	1.19	1.09	0.542	0.957	0.993	0.392	0.260	0.574	0.493	0.435	0.440
L2½ × 2 × 3/8	5.3	1.55	0.912	0.547	0.768	0.813	0.514	0.363	0.577	0.581	0.420	0.614
× 1/4	3.62	1.06	0.654	0.381	0.784	0.787	0.372	0.254	0.592	0.537	0.424	0.626

Table B-10   Unequal Leg Angles (SI Units)

Size and Thickness (mm)	Mass (kg/m)	Area (mm²)	Axis X–X				Axis Y–Y				Axis Z–Z	
			I (10⁶ mm⁴)	S (10³ mm³)	r (mm)	y_C (mm)	I (10⁶ mm⁴)	S (10³ mm³)	r (mm)	x_C (mm)	r (mm)	Tan α
L229 × 102 × 15.9	39.1	4985	27.0	188	73.7	85.3	3.46	43.4	26.4	21.8	21.5	0.216
× 12.7	31.7	4030	22.1	153	74.2	84.1	2.88	35.6	26.7	20.6	21.7	0.220
L203 × 152 × 25.4	65.8	8385	33.6	247	63.2	67.3	16.1	146	43.9	41.9	32.5	0.543
× 19.1	50.3	6415	26.4	192	64.3	65.0	12.8	113	44.7	39.6	32.8	0.551
× 12.7	34.2	4355	18.4	131	65.0	62.7	9.03	78.5	45.5	37.3	33.0	0.558
L203 × 102 × 25.4	55.7	7095	29.0	231	64.0	77.5	4.83	64.6	26.2	26.7	21.5	0.247
× 19.1	42.7	5445	22.9	179	64.8	74.9	3.90	50.3	26.7	24.2	21.6	0.258
× 12.7	29.2	3710	16.0	123	65.8	72.6	2.81	35.2	27.4	21.8	22.0	0.267
L178 × 102 × 19.1	39.0	4960	15.7	138	56.4	63.8	3.77	49.7	27.7	25.7	21.8	0.324
× 12.7	26.6	3385	11.1	95.2	57.2	61.5	2.72	34.7	28.2	23.3	22.1	0.335
× 9.5	20.2	2570	8.57	72.8	57.7	60.2	2.12	26.7	28.7	22.1	22.4	0.340
L152 × 102 × 19.1	35.1	4475	10.2	102	47.8	52.8	3.61	48.7	28.4	27.4	21.8	0.428
× 12.7	24.1	3065	7.24	71.0	48.5	50.5	2.61	34.1	29.2	25.1	22.1	0.440
× 9.5	18.3	3230	5.62	54.4	49.0	49.3	2.04	26.2	29.7	23.9	22.3	0.446
L152 × 89 × 12.7	22.8	2905	6.91	69.5	48.8	52.8	1.77	26.1	24.7	21.2	19.3	0.344
× 9.5	17.4	2205	5.37	53.1	49.3	51.8	1.39	20.2	25.1	20.0	19.5	0.350
L127 × 89 × 19.1	29.5	3750	5.79	70.1	39.4	44.5	2.31	36.4	24.8	25.3	19.0	0.464
× 12.7	20.2	2580	4.16	49.0	40.1	42.2	1.69	25.6	25.7	23.0	19.2	0.479
× 9.5	15.5	1970	3.24	37.5	40.6	40.9	1.32	19.8	25.9	21.9	19.4	0.486
× 6.4	10.4	1330	2.24	25.7	41.1	39.6	0.928	13.6	26.4	20.7	19.6	0.492
L127 × 76 × 12.7	19.0	2420	3.93	47.7	40.4	44.5	1.07	18.8	21.1	19.1	16.5	0.357
× 9.5	14.6	1845	3.07	36.7	40.9	43.2	0.849	14.6	21.5	17.9	16.6	0.364
× 6.4	9.82	1250	2.13	25.1	41.1	42.2	0.599	10.1	21.9	16.7	16.8	0.371
L102 × 89 × 12.7	17.7	2260	2.21	31.8	31.2	31.8	1.58	24.9	26.4	25.4	18.3	0.750
× 9.5	13.5	1725	1.74	24.4	31.8	30.7	1.23	19.2	26.9	24.3	18.5	0.755
× 6.4	9.22	1170	1.21	16.9	32.3	29.5	0.870	13.2	27.2	23.1	18.6	0.759
L102 × 76 × 12.7	16.5	2095	2.10	31.0	31.8	33.8	1.01	18.4	21.9	21.0	16.2	0.543
× 9.5	12.6	1600	1.65	23.9	32.0	32.5	0.799	14.2	22.3	19.9	16.4	0.551
× 6.4	8.63	1090	1.15	16.4	32.5	31.5	0.566	9.82	22.8	18.7	16.5	0.558
L89 × 76 × 12.7	15.2	1935	1.44	23.8	27.2	28.7	0.970	18.0	22.4	22.2	15.8	0.714
× 9.5	11.8	1485	1.13	18.5	27.7	27.4	0.770	13.9	22.8	21.1	15.9	0.721
× 6.4	8.04	1005	0.795	12.7	28.2	26.4	0.541	9.65	23.2	19.9	16.0	0.727
L89 × 64 × 12.7	14.0	1775	1.35	23.1	27.7	30.5	0.566	12.5	17.9	17.9	13.6	0.486
× 9.5	10.7	1360	1.07	17.9	27.9	29.5	0.454	8.70	18.3	16.8	13.6	0.496
× 6.4	7.29	929	0.749	12.4	28.4	28.2	0.323	6.75	18.7	15.6	13.8	0.506
L76 × 64 × 12.7	12.6	1615	0.866	17.0	23.2	25.4	0.541	12.2	18.3	19.1	13.2	0.667
× 9.5	9.82	1240	0.691	13.3	23.6	24.3	0.433	9.52	18.7	17.9	13.3	0.676
× 6.4	6.70	845	0.487	9.19	24.0	23.1	0.309	6.62	19.1	16.8	13.4	0.684
L76 × 51 × 12.7	11.5	1450	0.799	16.4	23.5	27.4	0.280	7.77	13.9	14.8	10.9	0.414
× 9.5	8.78	1115	0.637	12.8	23.9	26.4	0.226	6.08	14.2	13.7	10.9	0.428
× 6.4	6.10	768	0.454	8.88	24.3	25.2	0.163	4.26	14.6	12.5	11.0	0.440
L64 × 51 × 9.5	7.89	1000	0.380	8.96	19.5	20.7	0.214	5.95	14.7	14.8	10.7	0.614
× 6.4	5.39	684	0.272	6.24	19.9	20.0	0.155	4.16	15.0	13.6	10.8	0.626

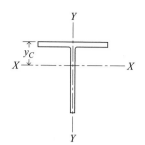

Table B-11 Structural Tees (U.S. Customary Units)

Designation*	Area (in.²)	Depth of Tee (in.)	Flange Width (in.)	Flange Thickness (in.)	Stem Thickness (in.)	Axis X–X I (in.⁴)	Axis X–X S (in.³)	Axis X–X r (in.)	Axis X–X y_C (in.)	Axis Y–Y I (in.⁴)	Axis Y–Y S (in.³)	Axis Y–Y r (in.)
WT18 × 115	33.8	17.950	16.470	1.260	0.760	934	67.0	5.25	4.01	470	57.1	3.73
× 80	23.5	18.005	12.000	1.020	0.650	740	55.8	5.61	4.74	147	24.6	2.50
WT15 × 66	19.4	15.155	10.545	1.000	0.615	421	37.4	4.66	3.90	98.0	18.6	2.25
× 54	15.9	14.915	10.475	0.760	0.545	349	32.0	4.69	4.01	73.0	13.9	2.15
WT12 × 52	15.3	12.030	12.750	0.750	0.500	189	20.0	3.51	2.59	130	20.3	2.91
× 47	13.8	12.155	9.065	0.875	0.515	186	20.3	3.67	2.99	54.5	12.0	1.98
× 42	12.4	12.050	9.020	0.770	0.470	166	18.3	3.67	2.97	47.2	10.5	1.95
× 31	9.11	11.870	7.040	0.590	0.430	131	15.6	3.79	3.46	17.2	4.90	1.38
WT9 × 38	11.2	9.105	11.035	0.680	0.425	71.8	9.83	2.54	1.80	76.2	13.8	2.61
× 30	8.82	9.120	7.555	0.695	0.415	64.7	9.29	2.71	2.16	25.0	6.63	1.69
× 25	7.33	8.995	7.495	0.570	0.355	53.5	7.79	2.70	2.12	20.0	5.35	1.65
× 20	5.88	8.950	6.015	0.525	0.315	44.8	6.73	2.76	2.29	9.55	3.17	1.27
WT8 × 50	14.7	8.485	10.425	0.985	0.585	76.8	11.4	2.28	1.76	93.1	17.9	2.51
× 25	7.37	8.130	7.070	0.630	0.380	42.3	6.78	2.40	1.89	18.6	5.26	1.59
× 20	5.89	8.005	6.995	0.505	0.305	33.1	5.35	2.37	1.81	14.4	4.12	1.57
× 13	3.84	7.845	5.500	0.345	0.250	23.5	4.09	2.47	2.09	4.80	1.74	1.12
WT7 × 60	17.7	7.240	14.670	0.940	0.590	51.7	8.61	1.71	1.24	247	33.7	3.74
× 41	12.0	7.155	10.130	0.855	0.510	41.2	7.14	1.85	1.39	74.2	14.6	2.48
× 34	9.99	7.020	10.035	0.720	0.415	32.6	5.69	1.81	1.29	60.7	12.1	2.46
× 24	7.07	6.985	8.030	0.595	0.340	24.9	4.48	1.87	1.35	25.7	6.40	1.91
× 15	4.42	6.920	6.730	0.385	0.270	19.0	3.55	2.07	1.58	9.79	2.91	1.49
× 11	3.25	6.870	5.000	0.335	0.230	14.8	2.91	2.14	1.76	3.50	1.40	1.04
WT6 × 60	17.6	6.560	12.320	1.105	0.710	43.4	8.22	1.57	1.28	172	28.0	3.13
× 48	14.1	6.355	12.160	0.900	0.550	32.0	6.12	1.51	1.13	135	22.2	3.09
× 36	10.6	6.125	12.040	0.670	0.430	23.2	4.54	1.48	1.02	97.5	16.2	3.04
× 25	7.34	6.095	8.080	0.640	0.370	18.7	3.79	1.60	1.17	28.2	6.97	1.96
× 15	4.40	6.170	6.520	0.440	0.260	13.5	2.75	1.75	1.27	10.2	3.12	1.52
× 8	2.36	5.995	3.990	0.265	0.220	8.70	2.04	1.92	1.74	1.41	0.706	0.773
WT5 × 56	16.5	5.680	10.415	1.250	0.755	28.6	6.40	1.32	1.21	118	22.6	2.68
× 44	12.9	5.420	10.265	0.990	0.605	20.8	4.77	1.27	1.06	89.3	17.4	2.63
× 30	8.82	5.110	10.080	0.680	0.420	12.9	3.04	1.21	0.884	58.1	11.5	2.57
× 15	4.42	5.235	5.810	0.510	0.300	9.28	2.24	1.45	1.10	8.35	2.87	1.37
× 6	1.77	4.935	3.960	0.210	0.190	4.35	1.22	1.57	1.36	1.09	0.551	0.785
WT4 × 29	8.55	4.375	8.220	0.810	0.510	9.12	2.61	1.03	0.874	37.5	9.13	2.10
× 20	5.87	4.125	8.070	0.560	0.360	5.73	1.69	0.988	0.735	24.5	6.08	2.04
× 12	3.54	3.965	6.495	0.400	0.245	3.53	1.08	0.999	0.695	9.14	2.81	1.61
× 9	2.63	4.070	5.250	0.330	0.230	3.41	1.05	1.14	0.834	3.98	1.52	1.23
× 5	1.48	3.945	3.940	0.205	0.170	2.15	0.717	1.20	0.953	1.05	0.532	0.841
WT3 × 10	2.94	3.100	6.020	0.365	0.260	1.76	0.693	0.774	0.560	6.64	2.21	1.50
× 6	1.78	3.015	4.000	0.280	0.230	1.32	0.564	0.861	0.677	1.50	0.748	0.918
WT2 × 6.5	1.91	2.080	4.060	0.345	0.280	0.526	0.321	0.524	0.440	1.93	0.950	1.00

Courtesy of The American Institute of Steel Construction.

*WT means structural T-section (cut from a W-section), followed by the nominal depth in inches, then the weight in pounds per foot of length.

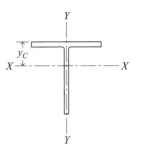

Table B-12   Structural Tees (SI Units)

Designation*	Area (mm²)	Depth of Tee (mm)	Flange Width (mm)	Flange Thickness (mm)	Stem Thickness (mm)	Axis X–X I (10⁶ mm⁴)	S (10³ mm³)	r (mm)	y_C (mm)	Axis Y–Y I (10⁶ mm⁴)	S (10³ mm³)	r (mm)
WT457 × 171	21805	455.9	418.3	32.0	19.3	389	1098	133	102	196	936	94.7
× 119	15160	457.3	304.8	25.9	16.5	308	914	142	120	61.2	403	63.5
WT381 × 98	12515	384.9	267.8	25.4	15.6	175	613	118	99.1	40.8	305	57.2
× 80	10260	378.8	266.1	19.3	13.8	145	524	119	102	30.4	228	54.6
WT305 × 77	9870	305.6	323.9	19.1	12.7	78.7	328	89.2	65.8	54.1	333	73.9
× 70	8905	308.7	230.3	22.2	13.1	77.4	333	93.2	75.9	22.7	197	50.3
× 63	8000	306.1	229.1	19.6	11.9	69.1	300	93.2	75.4	19.6	172	49.5
× 46	5875	301.5	178.8	15.0	10.9	54.5	256	96.3	87.9	7.16	80.3	35.1
WT229 × 57	7225	231.3	280.3	17.3	10.8	29.9	161	64.5	45.7	31.7	226	66.3
× 45	5690	231.6	191.9	17.7	10.5	26.9	152	68.8	54.9	10.4	109	42.9
× 37	4730	228.5	190.4	14.5	9.0	22.3	128	68.6	53.8	8.32	87.7	41.9
× 30	3795	227.3	152.8	13.3	8.0	18.6	110	70.1	58.2	3.98	51.9	32.3
WT203 × 74	9485	215.5	264.8	25.0	14.9	32.0	187	57.9	44.7	38.8	293	63.8
× 37	4755	206.5	179.6	16.0	9.7	17.6	111	61.0	48.0	7.74	86.2	40.4
× 30	3800	203.3	177.7	12.8	7.7	13.8	87.7	60.2	46.0	5.99	67.5	39.9
× 19	2475	199.3	139.7	8.8	6.4	9.78	67.0	62.7	53.1	2.00	28.5	28.4
WT178 × 89	11420	183.9	372.6	23.9	15.0	21.5	141	43.4	31.5	103	552	95.0
× 61	7740	181.7	257.3	21.7	13.0	17.1	117	47.0	35.3	30.9	239	63.0
× 51	6445	178.3	254.9	18.3	10.5	13.6	93.2	46.0	32.8	25.3	198	62.5
× 36	4560	177.4	204.0	15.1	8.6	10.4	73.4	47.5	34.3	10.7	105	48.5
× 22	2850	175.8	170.9	9.8	6.9	7.91	58.2	52.6	40.1	4.07	47.7	37.8
× 16	2095	174.5	127.0	8.5	5.8	6.16	47.7	54.4	44.7	1.46	22.9	26.4
WT152 × 89	11355	166.6	312.9	28.1	18.0	18.1	135	39.9	32.5	71.6	459	79.5
× 71	9095	161.4	308.9	22.9	14.0	13.3	100	38.4	28.7	56.2	364	78.5
× 54	6840	155.6	305.8	17.0	10.9	9.66	74.4	37.6	25.9	40.6	265	77.2
× 37	4735	154.8	205.2	16.2	9.4	7.78	62.1	40.6	29.7	11.7	114	49.8
× 22	2840	156.7	165.6	11.2	6.6	5.62	45.1	44.5	32.3	4.25	51.1	38.6
× 12	1525	152.3	101.3	6.7	5.6	3.62	33.4	48.8	44.2	0.587	11.6	19.6
WT127 × 83	10645	144.3	264.5	31.8	19.2	11.9	105	33.5	30.7	49.1	370	68.1
× 65	8325	137.7	260.7	25.1	15.4	8.66	78.2	32.3	26.9	37.2	285	66.8
× 45	5690	129.8	256.0	17.3	10.7	5.37	49.8	30.7	22.5	24.2	188	65.3
× 22	2850	133.0	147.6	13.0	7.6	3.86	36.7	36.8	27.9	3.48	47.0	34.8
× 9	1140	125.3	100.6	5.3	4.8	1.81	20.0	39.9	34.5	0.454	9.03	19.9
WT102 × 43	5515	111.1	208.8	20.6	13.0	3.80	42.8	26.2	22.2	15.6	150	53.3
× 30	3785	104.8	205.0	14.2	9.1	2.39	27.7	25.1	18.7	10.2	99.6	51.8
× 18	2285	100.7	165.0	10.2	6.2	1.47	17.7	25.4	17.7	3.80	46.0	40.9
× 13	1695	103.4	133.4	8.4	5.8	1.42	17.2	29.0	21.2	1.66	24.9	31.2
× 7	955	100.2	100.1	5.2	4.3	0.895	11.7	30.5	24.2	0.437	8.72	21.4
WT76 × 15	1895	78.7	152.9	9.3	6.6	0.733	11.4	19.7	14.2	2.76	36.2	38.1
× 9	1150	76.6	101.6	7.1	5.8	0.549	9.24	21.9	17.2	0.624	12.3	23.3
WT51 × 10	1230	52.8	103.1	8.8	7.1	0.219	5.26	13.3	11.2	0.803	15.6	25.4

*WT means structural T-section (cut from a W-section), followed by the nominal depth in mm, then the mass in kg per meter of length.

Table B-13   Properties of Standard Steel Pipe (U.S. Customary Units)

Dimensions				Properties				
Nominal Diam.	Outside Diam.	Inside Diam.	Wall Thickness	Weight				
d	do	di	t	w	A	I	S	r
(in.)	(in.)	(in.)	(in.)	(lb/ft)	(in.^{2})	(in.^{4})	(in.^{3})	(in.)
Standard Weight								
$\frac{1}{2}$	0.840	0.622	0.109	0.85	0.250	0.017	0.041	0.26
$\frac{3}{4}$	1.050	0.824	0.113	0.13	0.333	0.037	0.071	0.33
1	1.315	1.049	0.133	1.68	0.494	0.087	0.133	0.42
$1\frac{1}{4}$	1.660	1.380	0.140	2.27	0.669	0.195	0.235	0.54
$1\frac{1}{2}$	1.900	1.610	0.145	2.72	0.799	0.310	0.326	0.62
2	2.375	2.067	0.154	3.65	1.075	0.666	0.561	0.79
$2\frac{1}{2}$	2.875	2.469	0.203	5.79	1.704	1.530	1.064	0.95
3	3.500	3.068	0.216	7.58	2.228	3.017	1.724	1.16
$3\frac{1}{2}$	4.000	3.548	0.226	9.11	2.680	4.787	2.39	1.34
4	4.500	4.026	0.237	10.79	3.174	7.233	3.21	1.51
5	5.563	5.047	0.258	14.62	4.300	15.16	5.45	1.88
6	6.625	6.065	0.280	18.97	5.581	28.14	8.50	2.25
8	8.625	7.981	0.322	28.55	8.399	72.49	16.81	2.94
10	10.750	10.020	0.365	40.48	11.91	160.7	29.9	3.67
12	12.750	12.000	0.375	49.56	14.58	279.3	43.8	4.38
Extra Strong								
$1\frac{1}{2}$	1.900	1.500	0.200	3.63	1.068	0.391	0.412	0.61
2	2.375	1.939	0.218	5.02	1.477	0.868	0.731	0.77
$2\frac{1}{2}$	2.875	2.323	0.276	7.66	2.254	1.924	1.338	0.92
3	3.500	2.900	0.300	10.25	3.016	3.894	2.23	1.14
4	4.500	3.826	0.337	14.98	4.407	9.610	4.27	1.48
6	6.625	5.761	0.432	28.57	8.405	40.49	12.22	2.20
Double Extra Strong								
$1\frac{1}{2}$	1.900	1.100	0.400	6.41	1.885	0.568	0.564	0.55
2	2.375	1.503	0.436	9.03	2.656	1.311	1.104	0.70
$2\frac{1}{2}$	2.875	1.771	0.552	13.69	4.028	2.871	1.997	0.84
3	3.500	2.300	0.600	18.58	5.466	5.993	3.42	1.05
4	4.500	3.152	0.674	27.54	8.101	15.28	6.79	1.37
6	6.625	4.897	0.864	53.16	15.64	66.33	20.0	2.06

Table B-14   **Properties of Standard Steel Pipe (SI Units)**

	Dimensions			Properties				
Nominal Diam. d (mm)	Outside Diam. do (mm)	Inside Diam. di (mm)	Wall Thickness t (mm)	Mass m (kg/m)	A (mm^2)	I (10^6) (mm^4)	S (10^3) (mm^3)	r (mm)
Standard Weight								
13	21.3	15.8	2.77	1.264	161.3	0.007	0.672	6.6
19	26.7	20.9	2.87	1.681	214.8	0.015	1.163	8.5
25	33.4	26.6	3.38	2.499	318.7	0.036	2.179	10.7
32	42.2	35.1	3.56	3.376	431.6	0.081	3.851	13.7
38	48.1	40.9	3.68	4.045	515.5	0.129	5.342	15.8
51	60.3	52.5	3.91	5.428	693.5	0.277	9.193	20.0
64	73.0	62.7	5.16	8.611	1099	0.637	17.44	24.1
76	88.9	77.9	5.49	11.27	1437	1.256	28.25	29.5
89	101.6	90.1	5.74	13.55	1729	1.992	39.17	34.0
102	114.3	102.3	6.02	16.05	2048	3.011	52.60	38.4
127	141.3	128.2	6.55	21.74	2774	6.310	89.31	47.8
152	168.3	154.1	7.11	28.21	3600	11.71	139.3	57.2
203	219.1	202.7	8.18	42.46	5419	30.2	275.5	74.7
254	273.1	254.5	9.27	60.20	7684	66.9	490	93.2
305	323.9	304.8	9.53	73.71	9406	116.3	718	111.3
Extra Strong								
38	48.3	38.1	5.08	5.399	689	0.163	6.75	15.4
51	60.3	49.3	5.54	7.466	953	0.361	11.98	19.5
64	70.0	59.0	7.01	11.39	1454	0.801	21.93	23.5
76	88.9	73.7	7.62	15.24	1946	1.621	36.54	29.0
102	114.3	97.2	8.56	22.28	2843	4.000	69.67	37.6
152	168.3	146.3	10.97	42.49	5423	16.85	200	55.9
Double Extra Strong								
38	48.3	27.9	10.16	9.53	1216	0.236	0.564	13.9
51	60.3	38.2	11.07	13.43	1714	0.546	1.104	17.9
64	70.0	45.0	14.02	20.36	2600	1.195	1.997	21.4
76	88.9	58.4	15.24	27.63	3526	2.494	3.42	26.7
102	114.3	80.1	17.12	40.96	5226	6.360	6.79	34.8
152	168.3	124.4	21.95	79.06	10090	27.61	20.0	52.3

Table B-15 Properties of Standard Structural Timber (U.S. Customary Units)

| Dimensions* | | Properties | | | |
Nominal Size $b \times h$ (in.)	Dressed Size (in.)	Weight w (lb/ft)	Area A (in.²)	Second Moment I (in.⁴)	Section Modulus S (in.³)
2 × 4	$1\frac{5}{8} \times 3\frac{5}{8}$	1.64	5.89	6.45	3.56
6	$5\frac{5}{8}$	2.54	9.14	24.1	8.57
8	$7\frac{1}{2}$	3.39	12.2	57.1	15.3
10	$9\frac{1}{2}$	4.29	15.4	116	24.4
12	$11\frac{1}{2}$	5.19	18.7	206	35.8
4 × 4	$3\frac{5}{8} \times 3\frac{5}{8}$	3.65	13.1	14.4	7.94
6	$5\frac{5}{8}$	5.66	20.4	53.8	19.1
8	$7\frac{1}{2}$	7.55	27.2	127	34.0
10	$9\frac{1}{2}$	9.57	34.4	259	54.5
12	$11\frac{1}{2}$	11.6	41.7	459	79.9
6 × 6	$5\frac{1}{2} \times 5\frac{1}{2}$	8.40	30.3	76.3	27.7
8	$7\frac{1}{2}$	11.4	41.3	193	51.6
10	$9\frac{1}{2}$	14.5	52.3	393	82.7
12	$11\frac{1}{2}$	17.5	63.3	697	121
14	$13\frac{1}{2}$	20.6	74.3	1128	167
8 × 8	$7\frac{1}{2} \times 7\frac{1}{2}$	15.6	56.3	264	70.3
10	$9\frac{1}{2}$	19.8	71.3	536	113
12	$11\frac{1}{2}$	23.9	86.3	951	165
14	$13\frac{1}{2}$	28.0	101	1538	228
16	$15\frac{1}{2}$	32.0	116	2327	300
10 × 10	$9\frac{1}{2} \times 9\frac{1}{2}$	25.0	90.3	679	143
12	$11\frac{1}{2}$	30.3	109	1204	209
14	$13\frac{1}{2}$	35.6	128	1948	289
16	$15\frac{1}{2}$	40.9	147	2948	380
18	$17\frac{1}{2}$	46.1	166	4243	485
12 × 12	$11\frac{1}{2} \times 11\frac{1}{2}$	36.7	132	1458	253
14	$13\frac{1}{2}$	43.1	155	2358	349
16	$15\frac{1}{2}$	49.5	178	3569	460
18	$17\frac{1}{2}$	55.9	201	5136	587
20	$19\frac{1}{2}$	62.3	224	7106	729

*Properties and weights are for dressed sizes.

Table B-16   **Properties of Standard Structural Timber (SI Units)**

Dimensions*		Properties			
Nominal Size $b \times h$ (mm)	Dressed Size (mm)	Mass m (kg/m)	Area A (10^3 mm^2)	Second Moment I (10^6 mm^4)	Section Modulus S (10^3 mm^3)
51×102	41×92	2.44	3.77	2.68	58.3
152	140	3.78	5.86	10.03	140
203	191	5.04	7.83	23.8	251
254	241	6.38	9.88	48.3	400
305	292	7.72	12.0	85.7	587
102×102	92×92	5.43	8.46	5.99	130
152	140	8.42	13.2	22.4	313
203	191	11.2	17.6	52.9	557
254	241	14.2	22.2	107.8	893
305	292	17.3	26.9	191.1	1310
152×152	140×140	12.5	19.6	31.8	454
203	191	17.0	26.7	80.3	846
254	241	21.6	33.7	163.6	1350
305	292	26.0	40.9	290	1980
356	343	30.6	48.0	470	2740
203×203	191×191	23.2	36.5	110	1150
254	241	29.4	46.0	223	1850
305	292	35.5	55.8	396	2700
356	343	41.6	65.5	640	3740
406	394	47.6	75.3	969	4920
254×254	241×241	37.2	58.1	283	2340
305	292	45.1	70.4	501	3420
356	343	52.9	82.7	810	4740
406	394	60.8	95.0	1227	6230
457	445	68.6	107	1766	7950
305×305	292×292	54.6	85.3	607	4150
356	343	64.1	100	981	5720
406	394	73.6	115	1486	7540
457	445	83.1	130	2138	5620
508	495	92.7	145	2958	11950

*Properties and masses are for dressed sizes.

Table B-17 Properties of Selected Materials (U.S. Customary Units)

Exact values may vary widely with changes in composition, heat treatment, and mechanical working. More precise information can be obtained from manufacturers.

Materials	Specific Weight (lb/in.^{3})	Elastic Strength[a] Tension (ksi)	Comp. (ksi)	Shear (ksi)	Ultimate Strength Tension (ksi)	Comp. (ksi)	Shear (ksi)	Endurance Limit[c] (ksi)	Modulus of Elasticity (1000 ksi)	Modulus of Rigidity (1000 ksi)	Percent Elongation in 2 in.	Coefficient of Thermal Expansion (10^{-6}/°F)
Ferrous metals												
Wrought iron	0.278	30	b		48	b	25	23	28		30[d]	6.7
Structural steel	0.284	36	b		66	b		28	29	11.0	28[d]	6.6
Steel, 0.2% C hardened	0.284	62	b		90	b			30	11.6	22	6.6
Steel, 0.4% C hot-rolled	0.284	53	b		84	b		38	30	11.6	29	
Steel, 0.8% C hot-rolled	0.284	76	b		122	b			30	11.6	8	
Cast iron—gray	0.260				25	100		12	15		0.5	6.7
Cast iron—malleable	0.266	32	b		50	b			25		20	6.6
Cast iron—nodular	0.266	70			100				25		4	6.6
Stainless steel (18-8) annealed	0.286	36	b		85	b		40	28	12.5	55	9.6
Stainless steel (18-8) cold-rolled	0.286	165	b		190	b		90	28	12.5	8	9.6
Steel, SAE 4340, heat-treated	0.283	132	145		150	b	95	76	29	11.0	19	
Nonferrous metal alloys												
Aluminum, cast, 195-T6	0.100	24	25		36		30	7	10.3	3.8	5	
Aluminum, wrought, 2014-T4	0.101	41	41	24	62	b	38	18	10.6	4.0	20	12.5
Aluminum, wrought, 2024-T4	0.100	48	48	28	68	b	41	18	10.6	4.0	19	12.5
Aluminum, wrought, 6061-T6	0.098	40	40	26	45	b	30	13.5	10.0	3.8	17	12.5
Magnesium, extrusion, AZ80X	0.066	35	26		49	b	21	19	6.5	2.4	12	14.4
Magnesium, sand cast, AZ63-HT	0.066	14	14		40	b	19	14	6.5	2.4	12	14.4
Monel, wrought, hot-rolled	0.319	50	b		90	b		40	26	9.5	35	7.8
Red brass, cold-rolled	0.316	60			75				15	5.6	4	9.8
Red brass, annealed	0.316	15	b		40	b			15	5.6	50	9.8
Bronze, cold-rolled	0.320	75			100				15	6.5	3	9.4
Bronze, annealed	0.320	20	b		50	b			15	6.5	50	9.4
Titanium alloy, annealed	0.167	135	b		155	b			14	5.3	13	
Invar, annealed	0.292	42	b		70	b			21	8.1	41	0.6
Nonmetallic materials												
Douglas fir, green[e]	0.022	4.8	3.4		3.9	0.9			1.6			
Douglas fir, air dry[e]	0.020	8.1	6.4		7.4	1.1			1.9			
Red oak, green[e]	0.037	4.4	2.6		3.5	1.2			1.4			1.9
Red oak, air dry[e]	0.025	8.4	4.6		6.9	1.8			1.8			
Concrete, medium strength	0.087		1.2		3.0				3.0			6.0
Concrete, fairly high strength	0.087		2.0		5.0				4.5			6.0

[a]Elastic strength may be represented by proportional limit, yield point, or yield strength at a specified offset (usually 0.2 percent for ductile metals).

[b]For ductile metals (those with an appreciable ultimate elongation), it is customary to assume the properties in compression have the same values as those in tension.

[c]Rotating beam.

[d]Elongation in 8 in.

[e]All timber properties are parallel to the grain.

Table B-18   Properties of Selected Materials (SI Units)
Exact values may vary widely with changes in composition, heat treatment, and mechanical working. More precise information can be obtained from manufacturers.

Materials	Density (mg/m³)	Elastic Strength[a] Tension (MPa)	Comp. (MPa)	Shear (MPa)	Ultimate Strength Tension (MPa)	Comp. (MPa)	Shear (MPa)	Endurance Limit[c] (MPa)	Modulus of Elasticity (GPa)	Modulus of Rigidity (GPa)	Percent Elongation in 50 mm	Coefficient of Thermal Expansion (10⁻⁶/°C)
Ferrous metals												
Wrought iron	7.70	210	b		330	b	170	160	190		30[d]	12.1
Structural steel	7.87	250	b		450	b		190	200	76	28[d]	11.9
Steel, 0.2% C hardened	7.87	430	b		620	b			210	80	22	11.9
Steel, 0.4% C hot-rolled	7.87	360	b		580	b		260	210	80	29	
Steel, 0.8% C hot-rolled	7.87	520	b		840	b			210	80	8	
Cast iron—gray	7.20				170	690		80	100		0.5	12.1
Cast iron—malleable	7.37	220	b		340	b			170		20	11.9
Cast iron—nodular	7.37	480			690				170		4	11.9
Stainless steel (18-8) annealed	7.92	250	b		590	b		270	190	86	55	17.3
Stainless steel (18-8) cold-rolled	7.92	1140	b		1310	b		620	190	86	8	17.3
Steel, SAE 4340, heat-treated	7.84	910	1000		1030	b	650	520	200	76	19	
Nonferrous metal alloys												
Aluminum, cast, 195-T6	2.77	160	170		250		210	50	71	26	5	
Aluminum, wrought, 2014-T4	2.80	280	280	160	430	b	260	120	73	28	20	22.5
Aluminum, wrought, 2024-T4	2.77	330	330	190	470	b	280	120	73	28	19	22.5
Aluminum, wrought, 6061-T6	2.71	270	270	180	310	b	210	93	70	26	17	22.5
Magnesium, extrusion, AZ80X	1.83	240	180		340	b	140	130	45	16	12	25.9
Magnesium, sand cast, AZ63-HT	1.83	100	96		270	b	130	100	45	16	12	25.9
Monel, wrought, hot-rolled	8.84	340	b		620	b		270	180	65	35	14.0
Red brass, cold-rolled	8.75	410			520				100	39	4	17.6
Red brass, annealed	8.75	100	b		270	b			100	39	50	17.6
Bronze, cold-rolled	8.86	520			690				100	45	3	16.9
Bronze, annealed	8.86	140	b		340	b			100	45	50	16.9
Titanium alloy, annealed	4.63	930	b		1070	b			96	36	13	
Invar, annealed	8.09	290	b		480	b			140	56	41	1.1
Nonmetallic materials												
Douglas fir, green[e]	0.61	33	23			27	6.2		11			
Douglas fir, air dry[e]	0.55	56	44			51	7.6		13			
Red oak, green[e]	1.02	30	18			24	8.3		10			3.4
Red oak, air dry[e]	0.69	58	32			48	12.4		12			
Concrete, medium strength	2.41		8			21			21			10.8
Concrete, fairly high strength	2.41		14			34			31			10.8

[a]Elastic strength may be represented by proportional limit, yield point, or yield strength at a specified offset (usually 0.2 percent for ductile metals).

[b]For ductile metals (those with an appreciable ultimate elongation), it is customary to assume the properties in compression have the same values as those in tension.

[c]Rotating beam.

[d]Elongation in 200 mm.

[e]All timber properties are parallel to the grain.

Table B-19   Beam Deflections and Slopes

Case	Load and Support (Length L)	Slope at End ($+ \measuredangle$)	Maximum Deflection ($+$ upward)	Equation of Elastic Curve ($+$ upward)
1		$\theta = -\dfrac{PL^2}{2EI}$ at $x = L$	$v_{max} = -\dfrac{PL^3}{3EI}$ at $x = L$	$v = -\dfrac{Px^2}{6EI}(3L - x)$
2		$\theta = -\dfrac{wL^3}{6EI}$ at $x = L$	$v_{max} = -\dfrac{wL^4}{8EI}$ at $x = L$	$v = -\dfrac{wx^2}{24EI}(x^2 - 4Lx + 6L^2)$
3		$\theta = -\dfrac{wL^3}{24EI}$ at $x = L$	$v_{max} = -\dfrac{wL^4}{30EI}$ at $x = L$	$v = -\dfrac{wx^2}{120EIL}(10L^3 - 10L^2x + 5Lx^2 - x^3)$
4		$\theta = +\dfrac{ML}{EI}$ at $x = L$	$v_{max} = +\dfrac{ML^2}{2EI}$ at $x = L$	$v = \dfrac{Mx^2}{2EI}$
5		$\theta_1 = -\dfrac{Pb(L^2 - b^2)}{6LEI}$ at $x = 0$ $\theta_2 = +\dfrac{Pa(L^2 - a^2)}{6LEI}$ at $x = L$	$v_{max} = -\dfrac{Pb(L^2 - b^2)^{3/2}}{9\sqrt{3}LEI}$ at $x = \sqrt{(L^2 - b^2)/3}$ $v_{\substack{center \\ not\ max}} = -\dfrac{Pb(3L^2 - 4b^2)}{48EI}$ $v = -\dfrac{Pa^2b^2}{3EIL}$ at $x = a$	$v = -\dfrac{Pbx}{6EIL}(L^2 - b^2 - x^2)$ $0 \le x \le a$
6		$\theta_1 = -\dfrac{PL^2}{16EI}$ at $x = 0$ $\theta_2 = +\dfrac{PL^2}{16EI}$ at $x = L$	$v_{max} = -\dfrac{PL^3}{48EI}$ at $x = L/2$	$v = -\dfrac{Px}{48EI}(3L^2 - 4x^2)$ $0 \le x \le \dfrac{L}{2}$
7		$\theta_1 = -\dfrac{wL^3}{24EI}$ at $x = 0$ $\theta_2 = +\dfrac{wL^3}{24EI}$ at $x = L$	$v_{max} = -\dfrac{5wL^4}{384EI}$ at $x = L/2$	$v = -\dfrac{wx}{24EI}(x^3 - 2Lx^2 + L^3)$
8		$\theta_1 = -\dfrac{ML}{6EI}$ at $x = 0$ $\theta_2 = +\dfrac{ML}{3EI}$ at $x = L$	$v_{max} = -\dfrac{ML^2}{9\sqrt{3}EI}$ at $x = L/\sqrt{3}$ $v_{\substack{center \\ not\ max}} = -\dfrac{ML^2}{16EI}$	$v = -\dfrac{Mx}{6EIL}(L^2 - x^2)$

Table B-20   Properties of Areas for Curved Beams

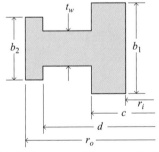	$A = b(r_o - r_i)$ $\displaystyle\int_A \frac{dA}{\rho} = b \ln\frac{r_o}{r_i}$	$r_C = \dfrac{r_i + r_o}{2}$

$$A = b(r_o - r_i) \qquad r_C = \frac{r_i + r_o}{2}$$

$$\int_A \frac{dA}{\rho} = b \ln\frac{r_o}{r_i}$$

$$A = \frac{b}{2}(r_o - r_i) \qquad r_C = \frac{2r_i + r_o}{3}$$

$$\int_A \frac{dA}{\rho} = \frac{br_o}{r_o - r_i} \ln\frac{r_o}{r_i} - b$$

$$A = \frac{b_1 + b_2}{2}(r_o - r_i)$$

$$r_C = \frac{r_i(2b_1 + b_2) + r_o(b_1 + 2b_2)}{3(b_1 + b_2)}$$

$$\int_A \frac{dA}{\rho} = \frac{b_1 r_o - b_2 r_i}{r_o - r_i} \ln\frac{r_o}{r_i} - b_1 + b_2$$

$$A = \pi b^2$$

$$\int_A \frac{dA}{\rho} = 2\pi\left(r_C - \sqrt{r_C^2 - b^2}\right)$$

$$A = \pi bh$$

$$\int_A \frac{dA}{\rho} = \frac{2\pi b}{h}\left(r_C - \sqrt{r_C^2 - h^2}\right)$$

$$A = (c - r_i)b_1 + (d - c)t_w + (r_o - d)b_2$$

$$r_C = \frac{1}{2A}\left[b_1\left(c^2 - r_i^2\right) + t_w(d^2 - c^2) + b_2\left(r_o^2 - d^2\right)\right]$$

$$\int_A \frac{dA}{\rho} = b_1 \ln\frac{c}{r_i} + t_w \ln\frac{d}{c} + b_2 \ln\frac{r_o}{d}$$

Index

AVERAGE PROPERTIES OF SELECTED ENGINEERING MATERIALS (INTERNATIONAL SYSTEM OF UNITS)

Exact values may vary widely with changes in composition, heat treatment, and mechanical working. More precise information can be obtained from manufacturers.

Materials	Density (Mg/m³)	Elastic Strength[a] Tension (MPa)	Comp. (MPa)	Shear (MPa)	Ultimate Strength Tension (MPa)	Comp. (MPa)	Shear (MPa)	Endurance Limit[c] (MPa)	Modulus of Elasticity (GPa)	Modulus of Rigidity (GPa)	Percent Elongation in 50 mm	Coefficient of Thermal Expansion (10⁻⁶/°C)
Ferrous metals												
Wrought iron	7.70	210	b		330	b	170	160	190		30[d]	12.1
Structural steel	7.87	250	b		450	b		190	200	76	28[d]	11.9
Steel, 0.2% C hardened	7.87	430	b		620	b			210	80	22	11.9
Steel, 0.4% C hot-rolled	7.87	360	b		580	b		260	210	80	29	
Steel, 0.8% C hot-rolled	7.87	520	b		840	b			210	80	8	
Cast iron—gray	7.20				170	690		80	100		0.5	12.1
Cast iron—malleable	7.37	220	b		340	b			170		20	11.9
Cast iron—nodular	7.37	480			690				170		4	11.9
Stainless steel (18-8) annealed	7.92	250	b		590	b		270	190	86	55	17.3
Stainless steel (18-8) cold-rolled	7.92	1140	b		1310	b		620	190	86	8	17.3
Steel, SAE 4340, heat-treated	7.84	910	1000		1030	b	650	520	200	76	19	
Nonferrous metal alloys												
Aluminum, cast 195-T6	2.77	160	170		250		210	50	71	26	5	
Aluminum, wrought, 2014-T4	2.80	280	280	160	430	b	260	120	73	28	20	22.5
Aluminum, wrought, 2024-T4	2.77	330	330	190	470	b	280	120	73	28	19	22.5
Aluminum, wrought, 6061-T6	2.71	270	270	180	310	b	210	93	70	26	17	22.5